THIRD CONFERENCE ON VITAMIN C

AF808686

ANNALS OF THE NEW YORK ACADEMY OF SCIENCES
Volume 498

THIRD CONFERENCE ON VITAMIN C

Edited by John J. Burns, Jerry M. Rivers, and Lawrence J. Machlin

The New York Academy of Sciences
New York, New York
1987

Copyright © 1987 by the New York Academy of Sciences. All rights reserved. Under the provisions of the United States Copyright Act of 1976, individual readers of the Annals *are permitted to make fair use of the material in them for teaching or research. Permission is granted to quote from the* Annals *provided that the customary acknowledgment is made of the source. Material in the* Annals *may be republished only by permission of the Academy. Address inquiries to the Executive Editor at the New York Academy of Sciences.*

Copying fees: *For each copy of an article made beyond the free copying permitted under Section 107 or 108 of the 1976 Copyright Act, a fee should be paid through the Copyright Clearance Center, Inc., 21 Congress Street, Salem, MA 01970. For articles of more than 3 pages the copying fee is $1.75.*

Library of Congress Cataloging-in-Publication Data

Conference on Vitamin C (3rd : 1986 : New York, N.Y.)
Third Conference on Vitamin C.

(Annals of the New York Academy of Sciences, ISSN 0077-8923 ; v. 498)
Conference held by the New York Academy of Sciences on Oct. 8-10, 1986, in New York, N.Y.
Includes bibliographies and index.
1. Vitamin C—Physiological effect—Congresses.
2. Vitamin C—Therapeutic use—Congresses. 3. Vitamin C—Metabolism—Congresses. I. Burns, John H., 1920- II. Rivers, Jerry M. III. Machlin, Lawrence J., 1927- . IV. New York Academy of Sciences. V. Title. VI. Series. [DNLM: 1. Ascorbic Acid—congresses. W1 AN626YL v.498/QU 120 C748 1986t]
Q11.N5 vol. 498 500 s [612'.399] 87-14050
[QP772.A8]
ISBN 0-89766-391-8
ISBN 0-89766-392-6 (pbk.)

PCP
Printed in the United States of America
ISBN 0-89766-391-8 (cloth)
ISBN 0-89766-392-6 (paper)
ISSN 0077-8923

ANNALS OF THE NEW YORK ACADEMY OF SCIENCES

Volume 498
July 7, 1987

THIRD CONFERENCE ON VITAMIN C[a]

Editors and Conference Chairs
JOHN J. BURNS, JERRY M. RIVERS, AND LAWRENCE J. MACHLIN

CONTENTS

[a] This volume is the result of the Conference on Vitamin C, which was held by the New York Academy of Sciences on October 8-10, 1986, in New York, New York.

Part II. Health Findings Based on Epidemiology

Part III. Biochemistry and Immunology

Part IV. Health and Disease

Part V. Xenobiotics, Free Radicals, and Methodology

Part VI. Metabolism, Requirements, and Safety

Poster Papers

Financial assistance was received from:

- AMERICAN INSTITUTE FOR CANCER RESEARCH
- CIBA CONSUMER PHARMACEUTICALS
- GLAXO INC.
- HOFFMANN-LA ROCHE INC.
- NATIONAL INSTITUTE OF DIABETES AND DIGESTIVE AND KIDNEY DISEASES—NATIONAL INSTITUTES OF HEALTH
- PFIZER INC.
- PROCTER AND GAMBLE COMPANY
- SANDOZ RESEARCH INSTITUTE

The New York Academy of Sciences believes it has a responsibility to provide an open forum for discussion of scientific questions. The positions taken by the participants in the reported conferences are their own and not necessarily those of the Academy. The Academy has no intent to influence legislation by providing such forums.

Preface

JOHN J. BURNS

The Rockefeller University
New York, New York 10021

JERRY M. RIVERS[a]

Cornell University
Ithaca, New York 14853

LAWRENCE J. MACHLIN

Hoffmann-La Roche Inc.
Nutley, New Jersey 07110

This volume contains the proceedings of the Third New York Academy of Sciences Conference on Vitamin C held on October 8-10, 1986, in New York City. The first conference was held in October 1960 and the second in October 1974. In view of the recent advancements in our knowledge of vitamin C, especially in the neurochemical, nutritional, epidemiological, and clinical areas, this conference was most timely.

The conference brought together scientists specializing in a variety of disciplines from a number of research institutions around the world. They presented new information concerning the functions of the vitamin at the molecular and cellular levels as well as findings relevant to human health. In particular the conference focused on new developments since 1974: the role of vitamin C in neuromodulation, neuropeptide synthesis, collagen synthesis, immune response, and aspects of metabolism relevant to nutritional requirements and the health of the elderly. The papers presented are herein published as part of the *Annals* series, as was done for the previous two conferences, and the information will thus be available to scientists throughout the world.

We felt that the conference went extremely well and on this point the participants and audience would undoubtedly concur. For that outcome we extend our gratitude to the organizational and preparatory work of Ms. Ellen Marks and her staff at the New York Academy of Sciences.

[a]Present address: Graduate Division of Nutrition, University of Texas, Austin, Texas 78712.

PART I. NEUROCHEMISTRY

Ascorbic Acid Distribution Patterns in Human Brain

A Comparison with Nonhuman Mammalian Species[a]

ARVIN F. OKE,[b] LESLIE MAY,[c] AND RALPH N. ADAMS[b]

[b] *Department of Chemistry*
University of Kansas
Lawrence, Kansas 66044

[c] *Department of Chemistry*
Indiana University
Bloomington, Indiana 47901

INTRODUCTION

Neural tissue has been recognized to contain ascorbic acid (AA) concentrations that rank among the highest of mammalian biological tissue.[1,2] Autoradiographic studies of labeled AA[1] and endogenous assays of regional specific areas[3–6] suggest that varying neuroanatomical locations concentrate AA in differing degrees. Certain brain locations, including the hippocampus and hypothalamus, consistently show high AA values in both man and animals when compared with other structures within the central nervous system (CNS).

In a series of experiments, Spector and coworkers[7–9] have traced the entrance of AA from blood to brain. Ascorbate molecules leave the blood and enter ventricular fluids through the choroid plexus. These molecules presumably then diffuse passively across ventricular walls into neighboring CNS tissue. Beyond this, it is not known if brain tissue in close proximity to ventricular walls concentrates high AA levels possibly at the expense of those structures more distant to it. Is it only incidental that these structures lay juxtaposed to the ventricular system containing cerebral spinal fluid? Does the brain in fact resemble a diffusing gradient away from ventricular walls? To address these questions a topographical mapping was undertaken of human brain and small animal brains.

[a] The support of this work by the NIH through grant no. NS08740 is gratefully acknowledged. Human tissue for this research was in part obtained from the National Neurological Research Bank, VA Wadsworth Medical Center, Los Angeles, Calif., which is sponsored by the NINCDS/NIMH, the National Multiple Sclerosis Society, the Hereditary Disease Foundation, and the Veterans Administration.

TISSUE PREPARATION AND ASCORBIC ACID ANALYSIS

Small animal brains were quickly removed and stored at −70°C until just prior to tissue slicing, which followed procedures described earlier.[10] Coronal slices ca. 1 mm in thickness were gridded into a matrix ca. 1.5 × 1.5 mm. Grid lines were recorded on reproduced graphics of the slice.

For human brain analysis, a coronal slice containing the hippocampus was selected. This frozen slice was photographed before and after a 4 × 4 mm to 5 × 5 mm grid matrix was etched onto the surface. The latter size matrix replaced the former when it became apparent the greater detail provided little added pattern information. Matrix units within the area of the hippocampus were further subdivided.

Analysis of tissue AA utilized electrochemical detection before and after the addition of ascorbic acid oxidase (AAO). Two techniques, both incorporating flow injection analysis, have been used. An initial method switched a sample between two sepharose beds, one of which had AAO bound to the sepharose moiety.[11] A later method bypassed the sepharose and utilized an automatic sampling unit that permitted sampled before and after the addition of solution AAO. In both cases, the difference between peak heights was the AA determination.

Tissue sonicated in pH 5.5 acetate buffer was followed by 5-min centrifugation at 10,000 rpm. Samples were kept on ice and quickly analyzed, usually within 20 min.

TOPOGRAPHIC DISTRIBUTION OF ASCORBIC ACID IN SMALL ANIMAL BRAINS

The pattern of distribution of AA throughout the full rostral to caudal extent of rat brain is represented in FIGURE 1. A consistent pattern was observed. Telencephalic structures that make up the cortical mantle (*i.e.*, cerebral cortex, hippocampus, and amygdala) contained the highest concentrations: levels up to 600 μg/g of tissue wet weight. Just underneath the cortical mantle (and/or the boundary imposed by the corpus callosum) AA concentrations generally shifted suddenly to levels reduced by approximately 50%. Neural tissue making up the core portion of the forebrain appears therefore to be circumscribed by a wall of dense AA. Within these core diencephalic structures, only the hypothalamus and especially the anterior medial regions of the hypothalamus contain values approximating those seen in the cortical mantle.

As we have reported in an earlier study,[5] CNS tissues with rostro-caudal extensions show decreasing AA concentrations toward the posterior limits. This is clearly seen when the percentage of the AA values in the most anterior position of a region is plotted against varying distances posterior to it (FIG. 2). Anterior-posterior (AP) decrements of ca. 30% are maintained within such structures as the caudate-putamen, thalamus, and the hippocampus. The hypothalamus has a more drastic decreasing gradient of ca. 60%. Neocortical values indicate a progressive 20% AP decrement. The cortex of the cerebellum, although anatomically quite separate and distinct from the cerebral neocortex, has an ascorbate value which suggests it is an extension of the decreasing values evident in the neocortex.

The portrait of ascorbate distribution in rat brain thus consists of diencephalic and mesencephalic core tissue with low concentration surrounded by a high telencephalic border. Nestled within this pattern are dense pockets whose boundaries closely

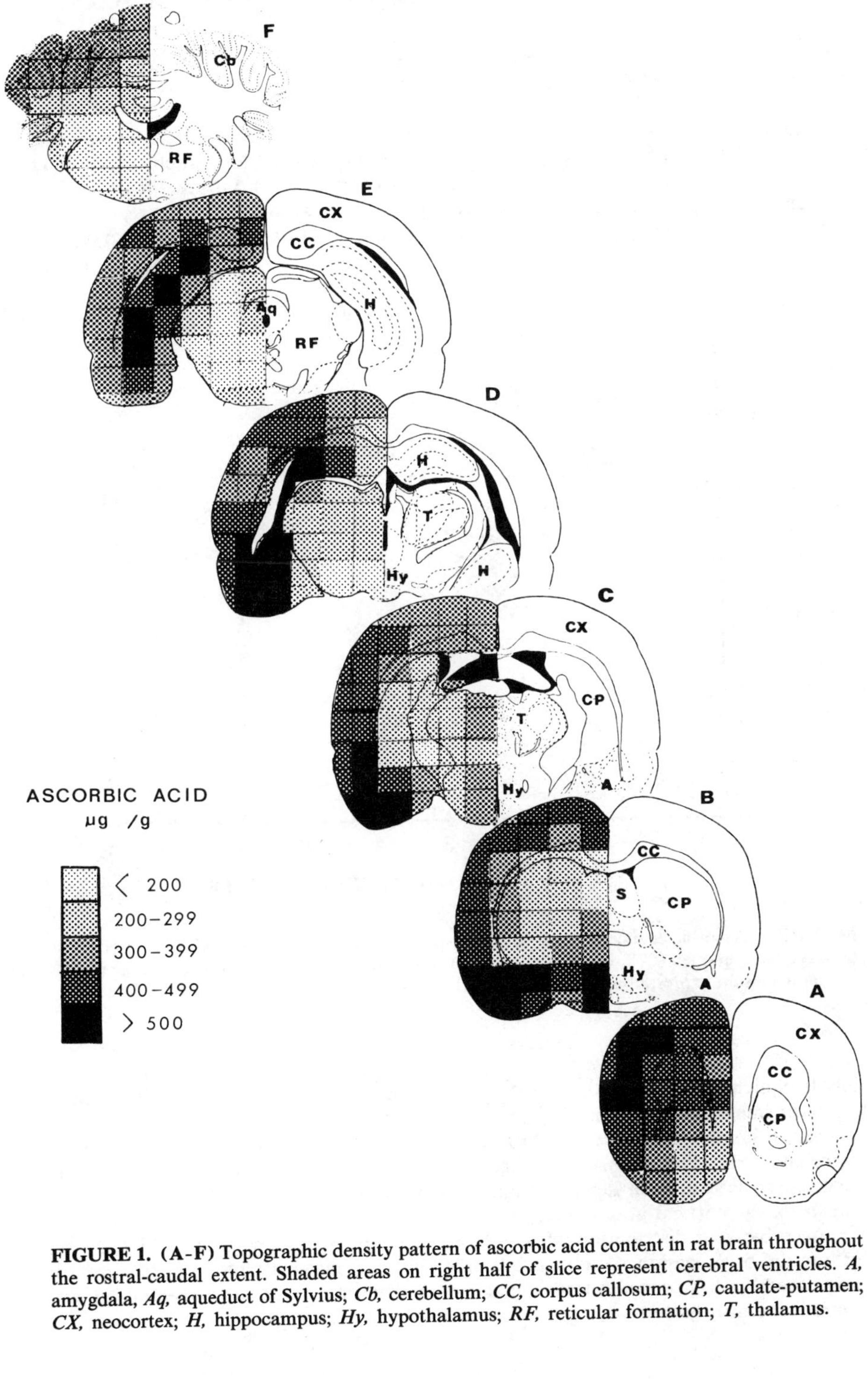

FIGURE 1. (A-F) Topographic density pattern of ascorbic acid content in rat brain throughout the rostral-caudal extent. Shaded areas on right half of slice represent cerebral ventricles. *A*, amygdala, *Aq*, aqueduct of Sylvius; *Cb*, cerebellum; *CC*, corpus callosum; *CP*, caudate-putamen; *CX*, neocortex; *H*, hippocampus; *Hy*, hypothalamus; *RF*, reticular formation; *T*, thalamus.

delineate the topography of an underlying neuroanatomical structure, as is the case in the amygdala, hippocampus, and hypothalamus. Most regions have an AP decreasing gradient.

Preliminary data on guinea pig AA distribution provides a pattern closely approximating that of the rat. Although a full three-dimensional topographical mapping has not been done, selected sections concur with the results from rat. FIGURE 3 is a chimeric composite with one hemisphere representing rat and the other representing guinea pig. The guinea pig hemisphere has been sufficiently reduced so as to appear

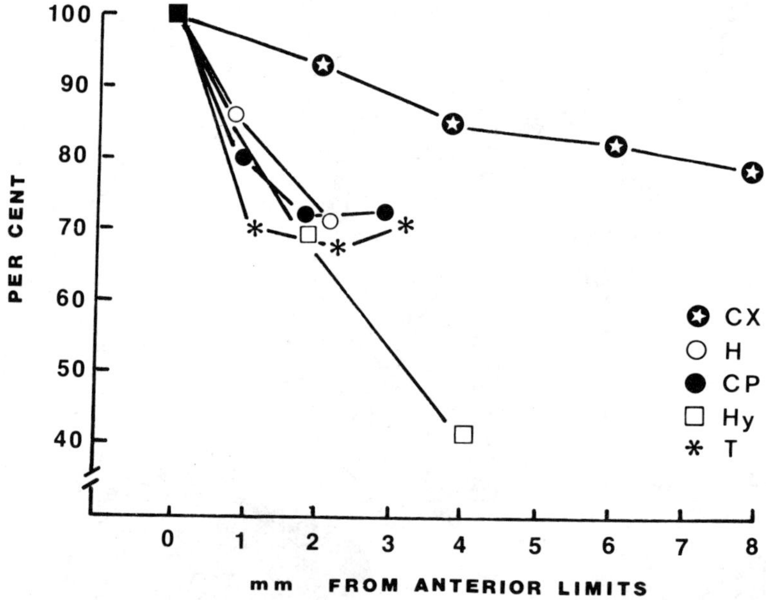

FIGURE 2. Ascorbic acid content of varying distances within the neocortex (*CX*), hippocampus (*H*), caudate-putamen (*CP*), hypothalamus (*Hy*), and thalamus (*T*) of rat brain along the rostral-caudal extension expressed as a percentage of the value found in the most anterior limits.

falsely equivalent in size with the rat. The ascorbate distribution on one side is almost a mirror image of the other, suggesting equivalent pattern layouts.

Similarly, the bovine brain (FIG. 4) demonstrates concentric zones of differing ascorbate concentrations. An enriched cortical mantle is separated from core diencephalic structures by a depleted zone (*i.e.*, internal capsule) of fibers projecting to and from the cortical gray matter.

Ascorbic acid distribution within these vertebrate species suggests a high degree of internal organization. Gradients are maintained. Neuroanatomical regions are distinguished by AA concentration boundaries. Furthermore, the three species sampled share a uniform pattern.

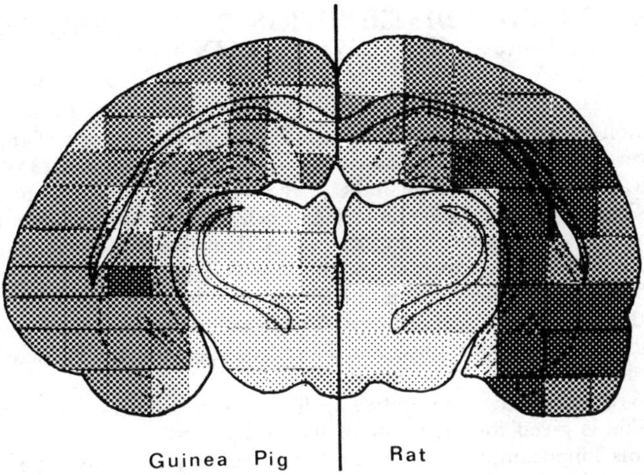

FIGURE 3. Chimeric representation of topographic density pattern of ascorbic acid content in closely matched sections of rat and guinea pig brains. For shading see FIGURE 1.

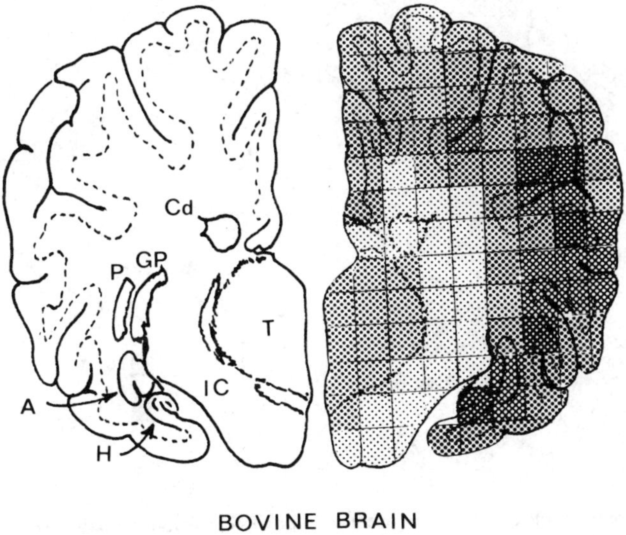

FIGURE 4. Topographic density pattern of ascorbic acid for bovine brain. For shading see FIGURE 1.

TOPOGRAPHIC DISTRIBUTION OF ASCORBIC ACID IN HUMAN BRAIN

Distribution patterns of AA in 10 human brains fell within one of three patterns. The most prevalent one seen (7 of 10 cases) was a dorsal-to-ventral (DV) increasing gradient (FIG. 5A). Concentration shifts for the most part disregard neuroanatomical landmarks. Cortical gray matter, consisting of cell bodies and projection terminals to them, do not necessarily contain any more AA than the white myelinated fibers of passage to them. With the exception of the hippocampus, no subcortical neuroanatomical grouping showed delineation due to ascorbate levels, a condition much unlike that seen with classical neurotransmitters. The ascorbate tissue value for any neuronal region, outside of the hippocampus, was dependent solely on the DV axis and not on neuroanatomical substrate.

FIGURE 6 represents the quantitative profile for seven brains in this group. A mean (+SEM) value is given for each row of the grid matrix. Any unit within the grid which contains hippocampal tissue is excluded from calculations. Typically, dorsal row AA content for the DV gradient group was less than 100 μg/g. Comparatively,

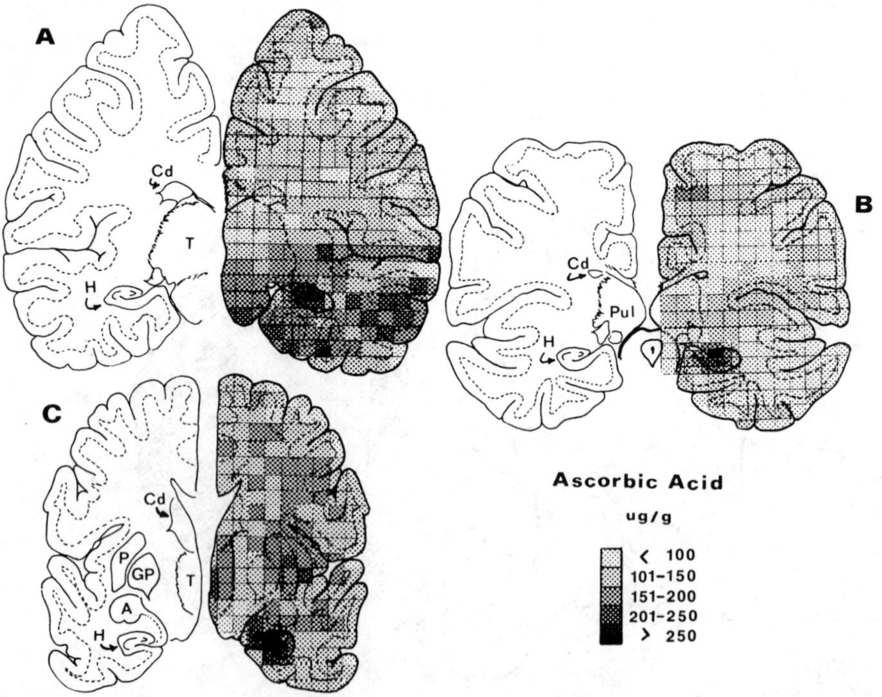

FIGURE 5. Topographic density of ascorbic acid for three distribution patterns in human brain. (**A**) Dorso-ventral increasing gradient pattern. (**B**) Uniform low density pattern. (**C**) Uniform high density pattern. *A,* amygdala; *Cd,* caudate nucleus; *GP,* globus pallidus; *P,* putamen; *Pul,* pulvinar, *T,* thalamus.

ventral rows show mean values between 150-200 μg/g. Any single brain within this group will quite commonly contain a two- to threefold DV increase.

A second pattern noted in 2 of 10 cases was (hippocampus exempted) a uniformly low content of AA. FIGURE 5B represents one such case. Row means, ranging from 82 μg/g to 122 μg/g are within 25% of the total slice mean of 94 μg/g. The other instance showed a slice mean of 106 μg/g, with row means ranging from 87 μg/g to 120 μg/g (see FIG. 6).

The third pattern is seen in FIGURE 5C. High row means extend almost the full DV axis. Row means noted in FIGURE 6 ranging from 102 to 164 μg/g are within 25% of the slice mean (140 μg/g).

As previously stated, the one neuroanatomical locus maintaining consistent AA values from brain to brain is the hippocampus, specifically the area within it called the dentate gyrus. FIGURE 7 shows three reconstructed drawings from the hippocampus, the area represented inside the box within the insert. FIGURES 7A and 7B are hippocampi from brains with DV gradients. FIGURE 7C comes from a brain uniformly low in AA. The high values in the dentate gyrus fall between 275-325 μg/g in all brains. Concentrations there did not in any way indicate the pattern distribution of the slice. The dentate gyrus, therefore, appears to maintain a high level of AA irrespective of surrounding tissue concentration.

Pattern variability is seemingly not correlated with known variables. TABLE 1 lists age, sex, and time from death to autopsy for each case. Age or sex do not appear to be related to any particular grouping. Death-to-autopsy does suggest a distinct group difference. Although it represents only one case, the unusually short 3-h death-to-autopsy interval is substantially lower than those seen in either of the other groups. It is conceivable that pattern variations result simply from postmortem oxidative decay inequalities, *i.e.*, are due to regionally different rates of AA decay. Vulnerability to oxidation would be greatest in the dorsal areas, followed by ventral regions, then the hippocampus. Accordingly, the group showing the greatest decay would logically also have the longest death-to-autopsy intervals. This is clearly not the case.

Assuming AA decay rates are equivalent in both human and small animal brains, a rat brain still within the skull was allowed to remain at room temperature for 18 h before removal. This condition is presumably more severe than an equivalent death-to-autopsy period since in the latter case the environment is not room temperature but 4°C. FIGURE 8 is a comparison of AA pattern distribution of a selected section in a rat brain immediately removed and frozen with the 18-h delayed brain. No indication of AA decay is seen as a consequence of the time delay. Nor is there a noticeable loss in pattern distinctiveness. We conclude that pattern variability is not a function of regional decay rates.

The full nature of these varying patterns of AA distribution within human brain requires a more extensive sampling than is available at the present time. Further, it is not known whether any particular pattern is maintained throughout the full rostrocaudal extent of the brain, although additional slices for a limited number of brains suggest the pattern seen in the midbrain region is also observed in frontal and caudal slices. Within any particular brain, these patterns may remain static or may be reflective of a point in time of continuous changing flux due to metabolic demands and dietary supply.

The extensive grid mapping carried out in this study is, we believe, of considerable significance in quantitative chemical neuroanatomy studies. It provides information that is not obtained by sampling methods that utilize whole anatomical nuclei or pooled homogenates of brain regions. Within subnuclei, or throughout the brain extent, it provides gradients and patterns that may be of considerable significance. Secondly, as an a priori sampling mode, it is unbiased and can be used to establish if chemical

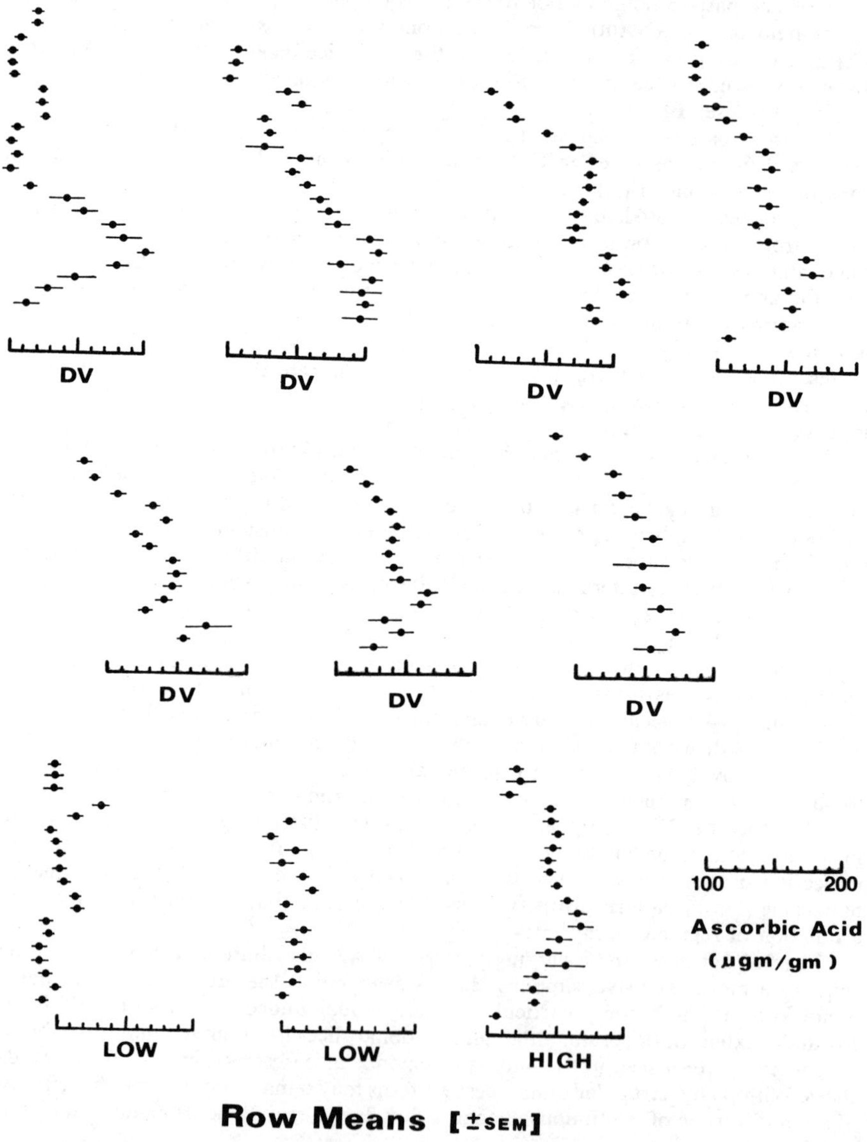

FIGURE 6. Row means (± SEM) of ascorbic acid from human coronal slice. Group identification is: *DV,* dorso-ventral increasing gradient; *LOW,* unifrom low density; *HIGH,* uniform high density.

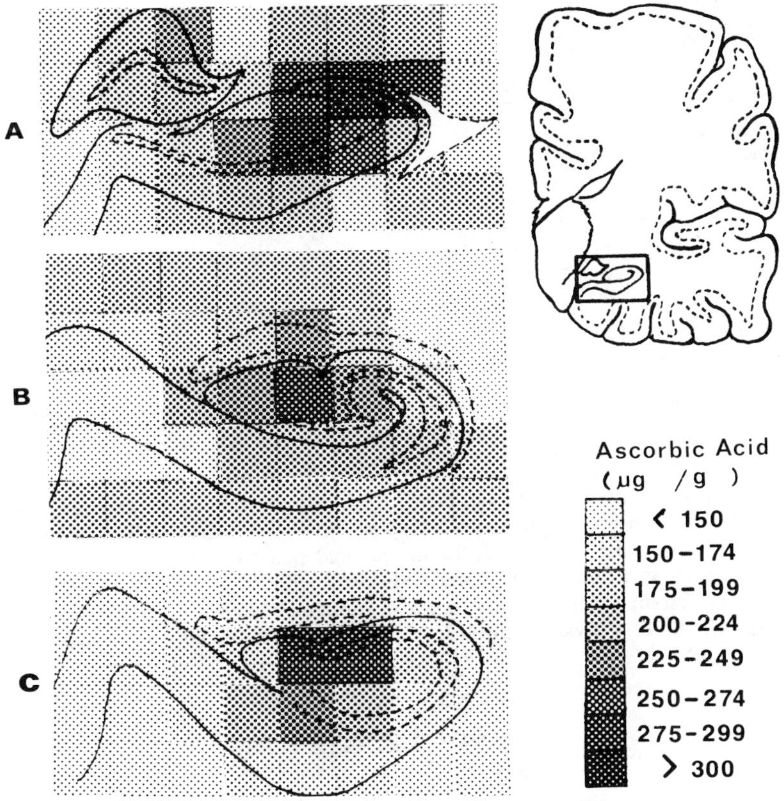

FIGURE 7. Topographic density pattern of ascorbic acid in human hippocampus (*boxed area*) for (**A,B**) dorso-ventral gradient group and (**C**) uniform low group.

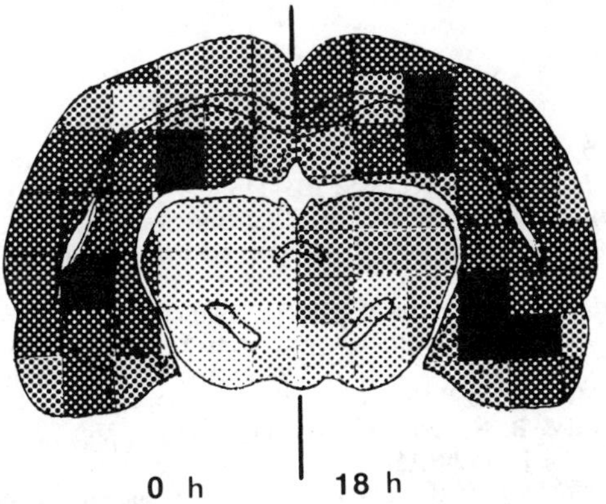

FIGURE 8. Topographic density pattern of ascorbic acid in a coronal section of rat brain. Left side contains values from a brain frozen immediately following removal. Right side represents values following 18 h at room temperature. For shading see FIGURE 1.

TABLE 1. Group, Age, Sex and Death-to-Autopsy Time for Human Brain Ascorbic Acid Analysis

Group	Age	Sex	Death-to-Autopsy (h)
Dorsal-ventral gradient	22	F	29
	34	F	18
	37	M	29
	45	M	22
	46	M	24
	50	F	63
	74	M	76
Uniform			
High density	56	M	23
	74	M	37
Low density	58	M	3

and classical neuroanatomy are similar or disparate. A somewhat surprising example of the latter is seen in the present work in which the AA distribution in rat, guinea pig, and bovine brains does conform to classical neuroanatomical delineations, but human brain AA patterns appear markedly independent of brain structure with the exception of the hippocampus. The latter in humans has an AA distribution anatomically specific and its invariant high level may be of functional significance.

ACKNOWLEDGMENTS

M. Leeson and J. Comack carried out parts of the ascorbate assays included in this study.

REFERENCES

1. HORNIG, D. 1975. Distribution of ascorbic acid, metabolites and analogues in man and animals. Ann. N. Y. Acad. Sci. **258:** 103-115.
2. RAJALAKSHMI, R., J. MALATHY & C. V. RAMAKRISHNAN. 1967. Effect of dietary protein content on regional distribution of ascorbic acid in the rat brain. J. Neurochem. **14:** 161-167.
3. SCHAUS, R. 1957. The ascorbic acid content of human pituitary, cerebral cortex, heart, and skeletal muscle and its relation to age. Am. J. Clin. Nutr. **5:** 39-41.
4. KRATZING, C. C., J. D. KELLEY & B. A. OELRICHS. 1982. Ascorbic acid in neural tissues. J. Neurochem. **39:** 625-627.
5. MILBY, K., A. OKE & R. N. ADAMS. 1982. Detailed mapping of ascorbate distribution in rat brain. Neurosci. Lett. **28:** 15-20.
6. MEFFORD, I. N., A. OKE & R. N. ADAMS. 1981. Regional distribution of ascorbate in human brain. Brain Res. **212:** 223-226.
7. SPECTOR, R. & A. V. LORENZO. 1973. Ascorbic acid homeostasis in the central nervous system. Am. J. Physiol. **225:** 757-763.

8. SPECTOR, R. & A. V. LORENZO. 1974. Specificity of ascorbic acid transport system of the central nervous system. Am. J. Physiol. **226:** 1468-1473.
9. SPECTOR, R. 1981. Penetration of ascorbic acid from cerebrospinal fluid into brain. Exp. Neurol. **72:** 645-653.
10. OKE, A., R. KELLER & R. N. ADAMS. 1978. Dopamine and norepinephrine enhancement in discrete rat brain regions following neonatal 6-hydroxydopamine treatment. Brain Res. **148:** 245-250.
11. BRADBERRY, C. W., R. T. BORCHARDT & C. J. DECEDUE. 1982. Immobilization of ascorbic acid oxidase. FEBS Lett. **146:** 348-352.

DISCUSSION OF THE PAPER

M. LEVINE (*National Institutes of Health, Bethesda, Md.*): Do you have any additional data on the distribution of ascorbate within the nervous tissue itself?

A. F. OKE (*University of Kansas, Lawrence, Kans.*): No, I don't.

E. J. DILIBERTO (*Wellcome Research Laboratories, Research Triangle Park, N.C.*): Has anybody looked at different brain areas to see if that pattern is really reflecting a difference in the mechanism of uptake into the different neurons?

A. F. OKE: No, we haven't.

S. D. VARMA (*University of Maryland at Baltimore, Baltimore, Md.*): Do you think that the high concentration of ascorbate in the brain is due to an active transport across the blood-brain barrier?

A. F. OKE: Spector's work indicates that ascorbic acid enters the brain not through the blood-brain barrier but through the choroid plexus of the ventricles and then passively diffuses. In fact that was one of the initial directions we had in looking at this, we wanted to find out if there is a concentration gradient away from the ventricular fluids of the brain. We are looking at a homeostatic mechanism here rather than an instantaneous input of the species.

S. D. VARMA: Well if the same kind of transport is not involved and ascorbic acid is really permeable, then why should you have a high concentration in one compartment, much higher than the blood or the cerebral spinal fluid?

A. F. OKE: That's an intriguing question: Why does the brain concentrate in some areas? Why does the hippocampus tend to find its allotment and maintain its allotment, while the rest of the tissue may suffer, or the rest of the tissues is trying to gain its allotment? We can't answer those questions well.

E. J. DILIBERTO: Did you say that you felt that the ascorbic acid was freely permeable?

S. D. VARMA: There are many reports that dehydroascorbic acid is actively transported and then is reduced by glutathione to ascorbic acid.

E. J. DILIBERTO: But I'm not sure if that's really been shown for the brain. In fact the reduction system that I think is most important there is the semidehydroascorbic reductase. It's not glutathione dependent.

M. CHEN (*University of Washington, Seattle, Wash.*): I think your finding is very interesting. I just wonder about those patterns in the ten individuals. Do you have any clinical data about how they died and what the status of their death was? How much stress had they been under, and were you able to separate them into different groups? Do you know the cause of death?

A. E. OKE: Yes, we do have the cause of death and although that was not listed, the majority of them were from pulmonary distress and heart attack. Certainly before any of these questions can be answered a lot more work will have to be done.

B. LADU (*University of Michigan Medical School, Ann Arbor, Mich.*): I wonder about the dietary ascorbic acid intake on your subjects, and certainly that could be tested with the guinea pigs on different levels of ascorbic acid in order to see how that affects the uniformity or nonuniformity of the distribution.

A. F. OKE: Yes, that's something that we certainly plan on doing.

A. B. KITABCHI (*University of Tennessee, Memphis, Tenn.*): It has always been stated that of the three organs, adrenal, ovaries and pituitary, adrenal and pituitary concentrate ascorbic acid to a larger degree. Did you evaluate posterior versus anterior pituitary in these subjects?

A. F. OKE: No, I did not.

M. CHEN: Did you have any diabetics in the ten subjects?

A. F. OKE: No.

Ascorbate-Like Factor from Embryonic Brain

Role in Collagen Formation, Basement Membrane Deposition, and Acetylcholine Receptor Aggregation by Muscle Cells[a]

ZVI VOGEL,[b] MATHEW P. DANIELS,[d]
THERESA CHEN,[d] ZHAO-YONG XI,[d] ESTHER BACHAR,[b]
LIAT BEN-DAVID,[b] NOA ROSENBERG,[b]
MICHAEL KRAUSE,[b] DAN DUKSIN,[c,e] AND
CHAYA KALCHEIM[b]

[b] *Department of Neurobiology*
[c] *Department of Biophysics*
Weizmann Institute of Science
Rehovot 76100, Israel

[d] *Laboratory of Biochemical Genetics*
National Heart, Lung, and Blood Institute
National Institutes of Health
Bethesda, Maryland 20892

INTRODUCTION

Innervation has a number of inductive effects on the differentiation of the skeletal muscle cell surface. Acetylcholine receptors (AcChR) are found over the entire surface of muscle prior to innervation or following denervation. AcChRs become localized into a high-density array on the region of the myotube plasma membrane that comes in contact with the nerve.[1-3] Receptor clusters on muscle cells in culture can also be produced by various factors from neural tissues or cells. Such receptor aggregation factors have been found in culture media conditioned by various neuronal cells[4,5] as well as in extracts of a variety of neuronal tissues.[3,6-11]

[a] This work was supported by research grants from the Muscular Dystrophy Association of America (to ZV), Minerva Foundation, Munich, FRG (to ZV), and the U.S.-Israel Binational Science Foundation (to ZV and MPD).

[e] Present address: Makor Chemicals Ltd., P.O. Box 6570, Jerusalem, Israel.

A relatively small amount of basement membrane is present on young myotubes in culture[12,13] or in muscle fibers before innervation.[14–16] When the nerve makes contact with muscle, basement membrane components appear at virtually all junctional areas as soon as they form.[14–17] The amount of basement membrane on the muscle continues to increase in the days following innervation and the entire surface of the muscle fiber is eventually covered with basement membrane.[14–16] Innervation also influences the qualitative composition of the muscle basement membrane. Antibodies have been obtained against several specific components that are selectively associated with the synaptic basement membrane rather than the extrajunctional basement membrane.[18]

Neuronal tissues contain factors that affect basement membrane deposition on muscle cell surfaces. Sanes *et al.* have shown that adult rat brain extract increases the accumulation of various basement membrane antigens on the myotube surface; in particular, the number of patches which contain "synaptic" basement membrane antigens is increased.[19] In this article we will review previous published results and present original data showing that extracts of various embryonic neuronal tissues contain an ascorbate-like factor that increases the production of collagen as well as the deposition of laminin and basement membrane on the muscle cells. In addition, this factor as well as ascorbic acid increase the aggregation of AcChR on the muscle cell surface.

RESULTS AND DISCUSSION

Accumulation of Collagen on Muscle Cells: The Collagen-stimulating Factor

We have shown that incubation of rat muscle cultures with extracts of embryonic rat brain results in a dose-dependent increase (3- to 15-fold) in the amount of collagenous proteins secreted into the medium, as well as an increase of two- to threefold in the amount of extracellular insoluble collagen accumulated on the cells.[8,20] The collagen-stimulating activity is found in rat, sheep, and fetal calf brains as well as in fetal rat spinal cord and chick ciliary ganglia. Extracts of fetal rat muscle and liver as well as brain extract and sera of adult rat are much less effective.[8,21] The collagen-stimulating activity is of low molecular weight and has been partially purified from brain homogenates of 18- to 20-day-old rat embryos by membrane (Diaflo PM 10) and gel (Biogel P-10) filtrations. The resulting preparation has been designated collagen-stimulating factor.[22,23]

Biochemical analysis of the newly synthesized proteins in cultures treated with collagen-stimulating factor demonstrated the increased secretion of various collagen precursors, including procollagens and intermediate chains in the conversion of procollagens, as well as increased secretion and deposition of the final processed collagens of types I, III, and V.[22] Immunofluorescent staining of muscle cultures treated with the factor revealed a large increase in staining intensity of collagen types I, III, IV, and V compared to control cultures and showed a distinct pattern of pericellular staining on the surface of the treated myotubes.[22] The increase in pericellular staining was especially evident (see FIG. 1) for collagens of types IV and V, known to be part of the basement membrane.[24,25]

We have been interested in the mechanism by which the factor stimulates the production of collagen by cultured muscle cells. Using the reticulocyte cell-free translation system, we found that the level of procollagen mRNA was not affected by treatment with the factor. On the other hand, using muscle cultures prelabeled with [^{3}H]proline, we found[23] that the factor enhances the secretion of prelabeled procollagen macromolecules, thereby demonstrating that the factor acts on collagen production at a posttranslational step. In addition, the factor stimulates the hydroxylation of proline residues in the secreted proteins (TABLE 1). These two observations agree well with previous data, which showed that hydroxylation of proline residues is required for procollagen secretion.[26] Indeed, incubation of muscle cells with the factor markedly increased (by 5- to 50-fold) the apparent activity of the enzyme prolyl hydroxylase (EC 1.14.11.2), compared to the activity observed in control muscle cultures.[23]

Regarding the nature of the collagen-stimulating factor, we found[21,23] that the factor resembles ascorbic acid in various biochemical and functional properties: (1) The factor elutes very close to ascorbate in two gel filtration systems (Sephadex G-10 and Biogel P-2 columns). (2) The factor acts as a reducing agent in the Fe-dipyridyl reaction system, which is quite specific to ascorbate[27]; the reducing activity and the collagen-stimulating factor coeluted from the columns. (3) Both the factor and ascorbic acid stimulate the hydroxylation of proline in procollagen and the secretion of procollagen into the culture medium (see FIG. 2 and TABLE 2). (4) Both the biological and reducing activities of the factor preparation (see TABLE 2) are abolished by the enzyme ascorbate oxidase (EC 1.10.3.3), which converts ascorbic acid to dehydroascorbic acid.

Interestingly, the factor was approximately fourfold more active than ascorbate when compared on the basis of equal absorbance units in the dipyridyl reaction. This might suggest that the two compounds are similar but not identical. Alternatively, the difference may be due to the presence of other compounds in the factor preparation, which either act synergistically with the ascorbate-like activity, or stabilize it against inactivation.

In agreement with this possibility, we have found by using high-pressure liquid chromatography that the major absorbing compound in the purified factor preparation comigrated with ascorbic acid. The material purified by the HPLC, however, had a low stability and low biological activity compared to the factor loaded on the column.

Ciliary Ganglia and Spinal Cord Explants Stimulate Collagen Production in Muscle Cultures

Increased collagen production was also observed when muscle cells were cocultured with neuronal explants such as ciliary ganglia (CG) taken from 8-day-old chick embryos or sections of spinal cord (SC) obtained from 15- to 18-day-old rat embryos. In muscle-CG cocultures there is an increase, compared to control, in the deposition of type IV collagen on the muscle cell surface. Near-complete inhibition of the collagen deposition, as seen by immunohistochemical staining, was obtained by including the enzyme ascorbate oxidase in the culture medium (FIG. 3).

Proline hydroxylation is a good indicator for the formation of mature collagen, since hydroxyproline is found exclusively in collagens and in a few other proteins that contain collagen-like amino acid sequences.[28] Coculturing muscle cells together with SC explants resulted in an 8- to 25-fold increase in proline hydroxylation. A 5- to 10-

fold stimulation was obtained in muscle cocultured with CG taken from 8-day-old chick embryos[21] (see TABLE 3). The capacity of the CG to stimulate proline hydroxylation decreased with time in coculture; the major drop occurred at coculture periods longer than 24 h, suggesting that the active factor might be unstable or depleted with time in culture. Culture media conditioned by sections of SC or CG were also

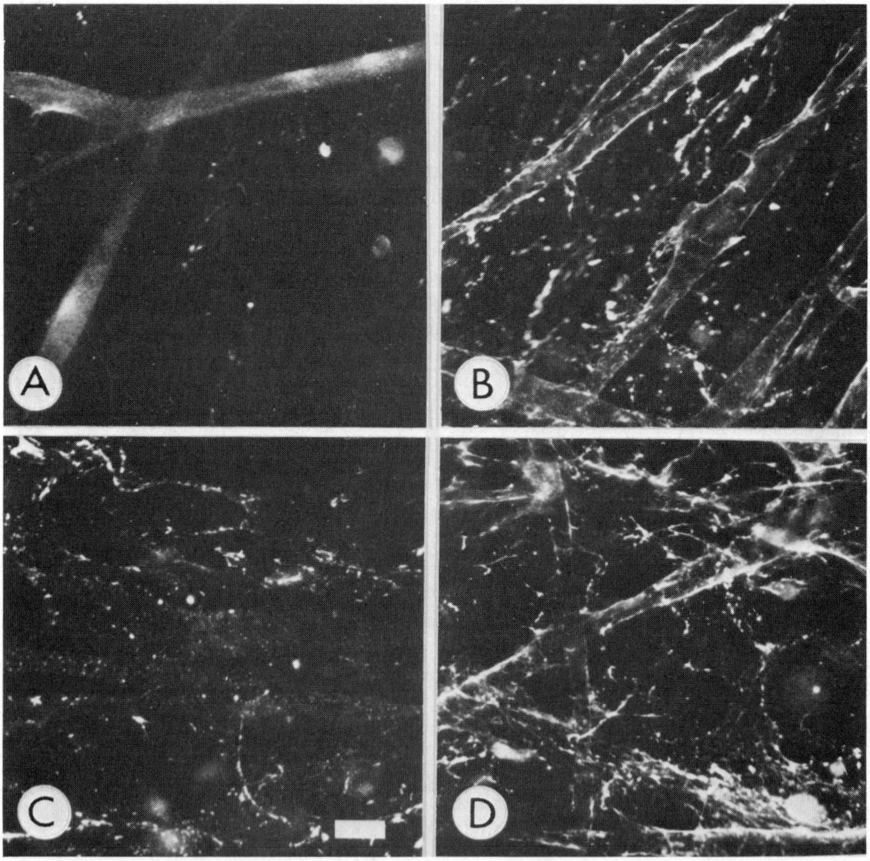

FIGURE 1. Immunofluorescence staining of collagen types IV and V in control and factor-treated muscle cells. Rat muscle cultures, 6 days in culture (**A** and **C**) and muscle cultures treated for the last 16 h with 50 μl/ml collagen-stimulating factor (**B** and **D**) were fixed with 2% paraformaldehyde and stained with rabbit antisera to collagen type IV (**A** and **B**) and type V (**C** and **D**). *Bar* represents 10 μm. (For additional details see Duksin *et al.*[22])

effective in stimulating proline hydroxylation in muscle cultures. As described above, however, the released factor was unstable in the culture medium and lost its activity with time. Incubation of the muscle-nerve cocultures with ascorbate oxidase markedly reduced the observed increase in proline hydroxylation. Thus, the results show that the cultured explants release ascorbic acid (or an ascorbic-acid-like factor compound),

TABLE 1. Stimulation of Proline Hydroxylation by Neural Extracts Added to Muscle Cultures

Treatment of Muscle Cultures[a]	Hyp/(Pro+Hyp) (%)
Control muscle cultures	1.5
+ 300 μg embryonic brain extract	30.0
+ 25 μg embryonic spinal cord extract	31.0
+ 40 μl/ml collagen-stimulating factor[b]	23.4

[a] Muscle cells, 5 days in culture, were incubated with the extracts specified for 16 h, the last 4 h in the presence of 10 μCi/ml [^{3}H]proline. Proteins secreted into the medium were submitted to amino acid analysis for determination of labeled proline and hydroxyproline.[21]

[b] Factor was prepared and brought to a volume of 1 ml per g original brain weight.[23]

which promotes proline hydroxylation and collagen production by muscle. In agreement with these results, we have found that both tissues contain high concentrations of a material which reacts in the dipyridyl reaction and coelutes with ascorbic acid in HPLC. The amounts (in ascorbate equivalents per mg protein) found in CG and SC were 19 and 13 μg, respectively. Brains of 18- to 20-day-old rat embryos contain 10.5 μg ascorbate equivalents per mg protein. Medium conditioned by SC sections also contains ascorbic acid. The amount found by HPLC after 3 h of conditioning was equivalent to 2.5 μg of ascorbate per mg of SC protein.

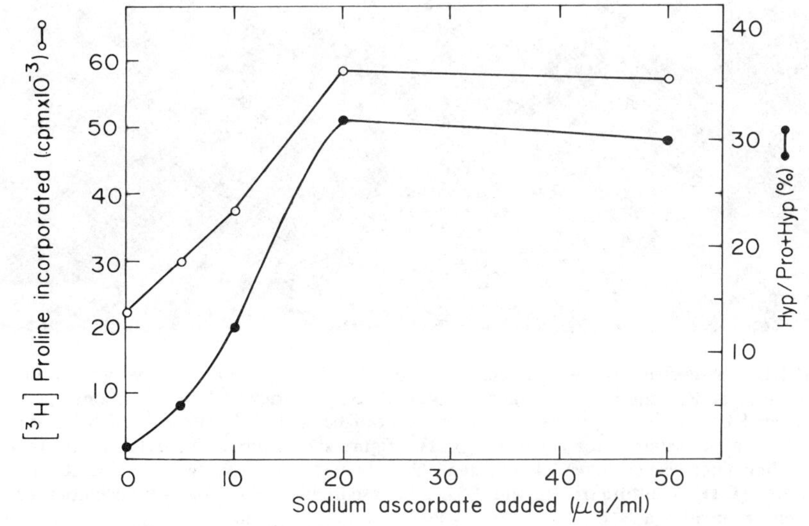

FIGURE 2. Stimulation of proline incorporation and proline hydroxylation by sodium ascorbate. Incubation was for 16 h; [^{3}H]proline was then added for additional 4 h. For additional details see Kalcheim *et al.*[21] and TABLE 1.

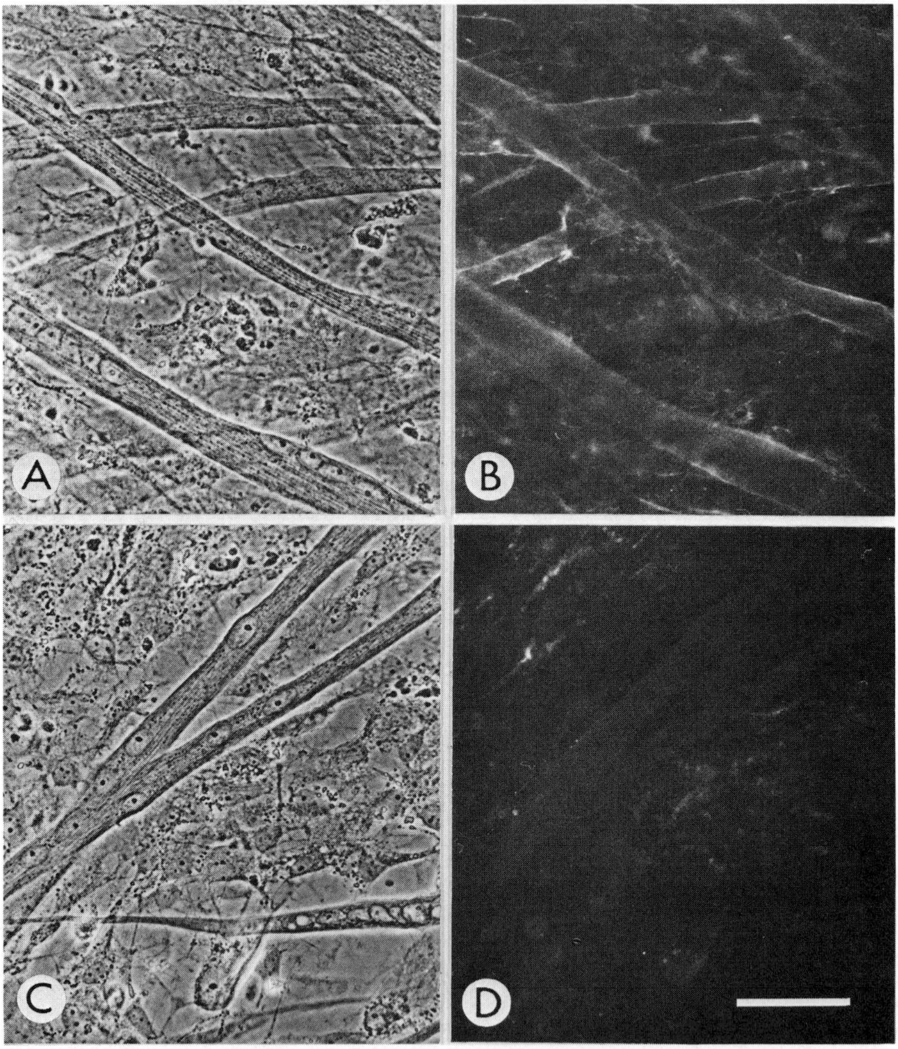

FIGURE 3. Ascorbate oxidase inhibits deposition of collagen type IV on myotubes in muscle-CG cocultures. Rat muscle cells, 6 days in culture, were cocultured for two additional days with CG from 8-day-old chick embryos. Cultures were fixed with 2% paraformaldehyde and immunofluorescence stained for collagen type IV. Figure shows phase micrographs and the corresponding fluorescence taken at a distance less than 1 mm from ganglion. (**A,B**) control coculture: (**C,D**) coculture treated with 5 U/ml of ascorbate oxidase for entire coculture period. *Bar* represents 50 μm.

TABLE 2. Pretreatment of Collagen-stimulating Factor and Sodium Ascorbate with Ascorbate Oxidase: Effect on Dipyridyl Reaction and on the [^{3}H]Proline Incorporated into Secreted Proteins

Treatment of Muscle Cultures[a]	Proline Incorporated (cpm $\times$ 10^{-3}/Dish)	Dipyridyl Reaction ($A_{525\ nm}$)
Collagen-stimulating factor (30 μl)		
− ascorbate oxidase	209	0.21
+ ascorbate oxidase	6	0.02
Sodium ascorbate (15 μg)		
− ascorbate oxidase	130	0.56
+ ascorbate oxidase	10	0.06

[a] Ascorbate and factor were incubated for 1 h at 22°C with 18 U/ml of ascorbate oxidase. Reaction mixtures were boiled and aliquots taken to determine effect on proline incorporation by muscle cultures and to determine the reducing activity by the dipyridyl reaction. Background values of radioactivity and absorbance were subtracted.[23]

Accumulation of Laminin and Basement Membrane on Muscle Cells

Laminin is known to be an important component of the basement membrane.[24,25] Utilizing specific antibodies against laminin, we have shown that ascorbic acid (and the ascorbic-acid-like factor present in brain extract) increases the amount of laminin found on the surface of the myotubes. Control muscle cultures stained with antibodies against laminin reveal a weak staining of the extracellular matrix and a diffuse staining of the myotubes with occasional patches containing high concentrations of laminin (FIG. 4B). Very frequently these patches correlate with AcChR aggregates (see Daniels *et al.*[12]). In cultures treated with crude embryonic brain extract or with the more purified collagen-stimulating factor, however, the myotubes appear heavily stained in a pericellular pattern (see FIG. 4C). In contrast, fibroblasts present in the same culture were not significantly stained for laminin. Pretreatment of the factor (or crude em-

TABLE 3. Proline Hydroxylation in Nerve-Muscle Cocultures and in Muscle Cultures Treated with Medium Conditioned by Rat Spinal Cord Explants

Cultures Tested[a]	Hyp/(Pro+Hyp) (%) Without Ascorbate Oxidase	Hyp/(Pro+Hyp) (%) With Ascorbate Oxidase[b]
Control muscle cultures	2.6	2.5
Muscle-CG cocultures	23.8	5.0
Muscle-SC cocultures	20.0	5.2
Muscle treated with 50 μl of medium conditioned by SC explants[c]	15.2	

[a] Culture and assay conditions as described.[21]

[b] Ascorbate oxidase, 5 U/ml, added together with explants and present throughout incubation.

[c] Medium conditioned by incubating SC sections (equivalent to 1 mg protein) in 1 ml culture medium for 3 h at 37°C.

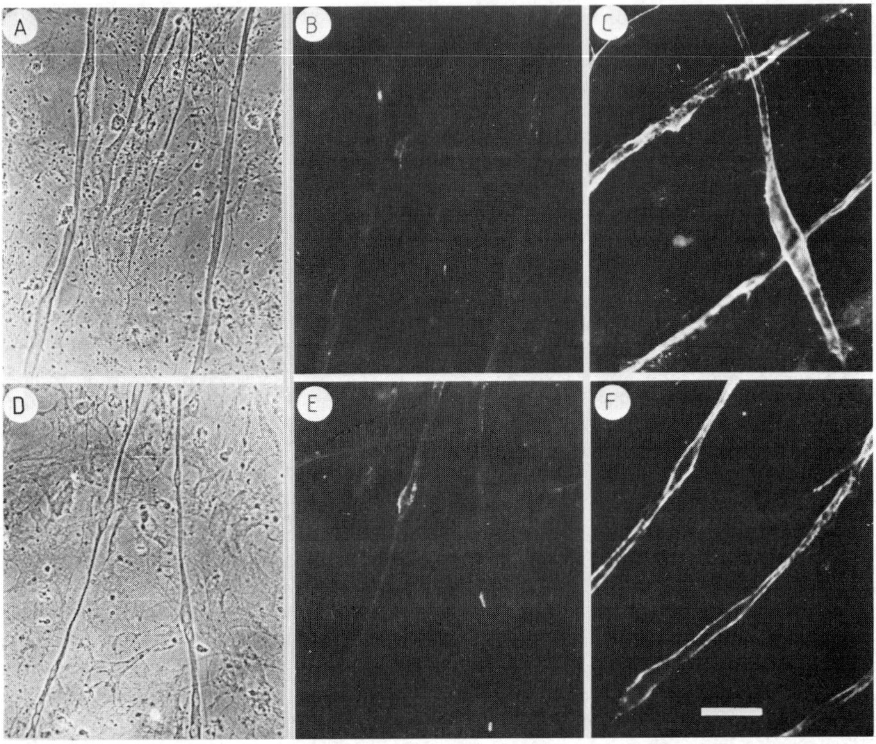

FIGURE 4. Effect of embryonic rat brain extract, ascorbic acid, and ascorbate oxidase on laminin deposition on muscle. Photomicrographs show laminin immunostaining on 5-day-old rat muscle cultures. The last 24 h included incubation with: (**B**) control conditions; (**C**) embryonic rat brain extract, 50 μl/ml; (**E**) extract pretreated with 5 U/ml ascorbate oxidase; (**F**) ascorbic acid, 50 μg/ml. **A** and **D** are phase-contrast views of **B** and **E**, respectively. *Bar* represents 50 μm.

bryonic brain extract) with ascorbate oxidase sharply reduced the amount of laminin found on the cells (FIG. 4E), thus suggesting that ascorbic acid is involved in laminin deposition. In agreement with this observation, we have shown (see FIG. 4F) that the addition of ascorbic acid to the cultures markedly increased the amount of laminin on the cells. Similarly, in cocultures of CG with muscle (data not shown), there was a large increase in laminin staining over the myotube surface (especially in the vicinity of explant). This deposition of laminin was again reduced to a large extent by ascorbate oxidase.

The mechanism by which ascorbic acid or the ascorbic-acid-like factor increases laminin deposition on the muscle cells is not yet clear. Laminin is known to interact with collagen type IV[29,30]; it remains to be tested whether the enhanced deposition of laminin is due to the primary effect of the factor on the accumulation of collagenous protein, or alternatively, whether this deposition results from an effect of the factor on the synthesis, secretion, or deposition of laminin.

Electron microscopy of treated and untreated muscle cultures and cocultures revealed that conditions which increase deposition of collagen and laminin also increase the amount of basal lamina on the muscle cell surface and the number of collagen fibrils in the surrounding spaces. These increases were reflected in a threefold increase in the mean percentage of the myotube surface covered with basement membrane (as seen by electron microscopy) and by a striking increase in the abundance of collagen fibrils. Ascorbate oxidase sharply inhibited these effects. FIGURE 5 shows, for example, the effect of one-day treatment of muscle cultures with ascorbate compared to control. TABLE 4 summarizes some of the results regarding the effect of ascorbate oxidase on the extent of the myotube surface covered with basement membrane under the various experimental conditions. These results are consistent with a role for ascorbate in the regulation of extracellular matrix formation in muscle.

Effect of Ascorbate on Aggregation of AcChR

Nerve cells and neuronal tissues contain and secrete factors that increase the number of AcChRs as well as the number of receptor aggregates on the surface of cultured myotube.[3–11] Such a factor was also found in association with the extracellular

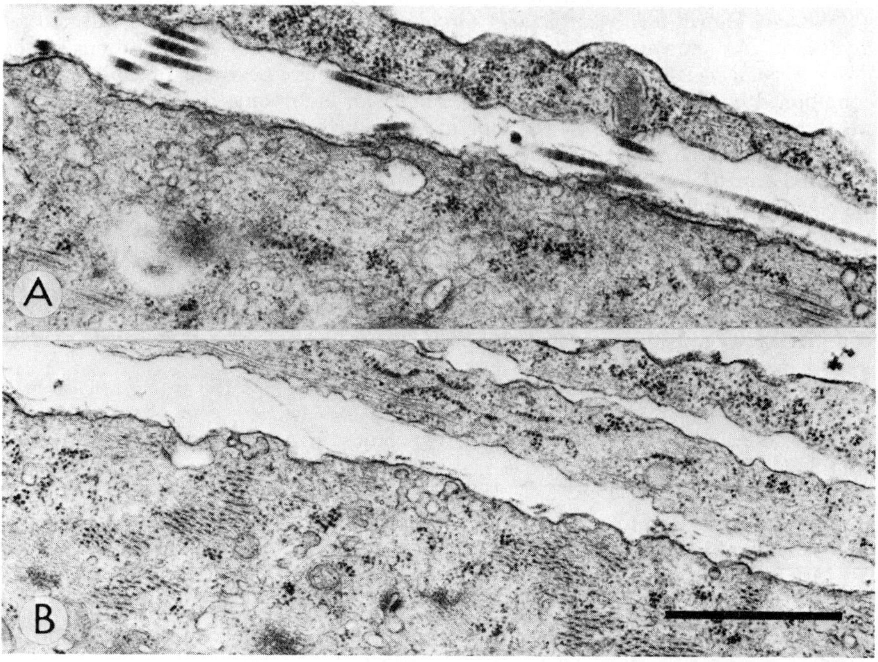

FIGURE 5. Electron micrographs of control and ascorbate-treated muscle cells. Rat muscle cells, 7 days in culture, were incubated for an additional 24 h under the following conditions: (**A**) with 50 μg/ml ascorbic acid; (**B**) control conditions. Cultures were fixed with 2.5% glutaraldehyde containing 0.15% tannic acid and were processed for electron microscopy. *Bar* represents 1 μm.

TABLE 4. Effect of Ascorbate Oxidase on the Induction of Myotube Basement Membrane

Culture Conditions[b]	Percent of Myotube with Basement Membrane[a] — Ascorbate Oxidase	+ Ascorbate Oxidase[c]
Control	15.9 ± 3.3	4.1 ± 1.3
+ Ascorbic acid (50 μg)	47.7 ± 9.1	10.9 ± 1.9
+ Embryonic rat brain extract (50 μl)	56.5 ± 12.3	13.5 ± 2.9
+ Low MW fraction of brain extract (50 μl)	55.2 ± 5.6	19.9 ± 4.6
Muscle cocultured with CG	48.2 ± 11.4	21.8 ± 4.0

[a] Mean ± SEM from transverse section of 6-11 myotubes.
[b] Preparation of extract and cultures (refs. 8, 21). Muscle-CG cocultured for 48 h; other treatments were for 24 h.
[c] 5 U/ml.

matrix of *Torpedo californica* electric organ.[31,32] The molecular weights reported for the different factor preparations vary considerably, but they can be divided into two groups, one of high molecular weight (> 40,000 daltons) and the other of low molecular weight (< 5,000 daltons).

We have shown that ascorbic acid stimulates receptor aggregation on cultured rat myotubes.[8] As shown above, ascorbic acid is present in high concentrations in the low molecular weight fraction of brain extract as well as in extracts of (and in media conditioned by) embryonic chick ciliary ganglia or embryonic rat spinal cord. Here we show that the increase in receptor aggregation obtained by the low molecular weight fractions of embryonic rat or chick brain extracts is due largely to ascorbic acid and can be abolished by ascorbate oxidase. On the other hand, the receptor aggregation activity present in the high molecular weight fraction of these extracts is not sensitive to this enzyme (see TABLE 5).

In addition, using CG-muscle cocultures we could show that, as expected, the CG explants cause a striking increase in the number of AcChR aggregates. This increase (see FIG. 6) is markedly reduced, but not eliminated, by ascorbate oxidase. The considerable aggregation response still occurring in the presence of ascorbate oxidase

TABLE 5. Effects of Various Factors on Aggregation of AcChR on Chick Muscle Cultures

Factors[a]	Aggregates/Field (Control = 1.0) — Ascorbate Oxidase	+ Ascorbate Oxidase[b]
Low MW factor (25 μl)	2.2	1.1
High MW factor (25 μl)	2.55	2.5
High MW factor (25 μl) + Low MW factor (50 μl)	3.95	2.5

[a] Extract of 10-day-old embryonic chick brain (0.3 g tissue per ml) was filtered through Diaflo PM-10[8]; volume of retentate (high MW factor) was adjusted to that of the ultrafiltrate (low MW factor). Factors were incubated with muscle cultures for 16 h.
[b] Preincubation of factors with 5 U/ml ascorbate oxidase for 1 h at room temperature.

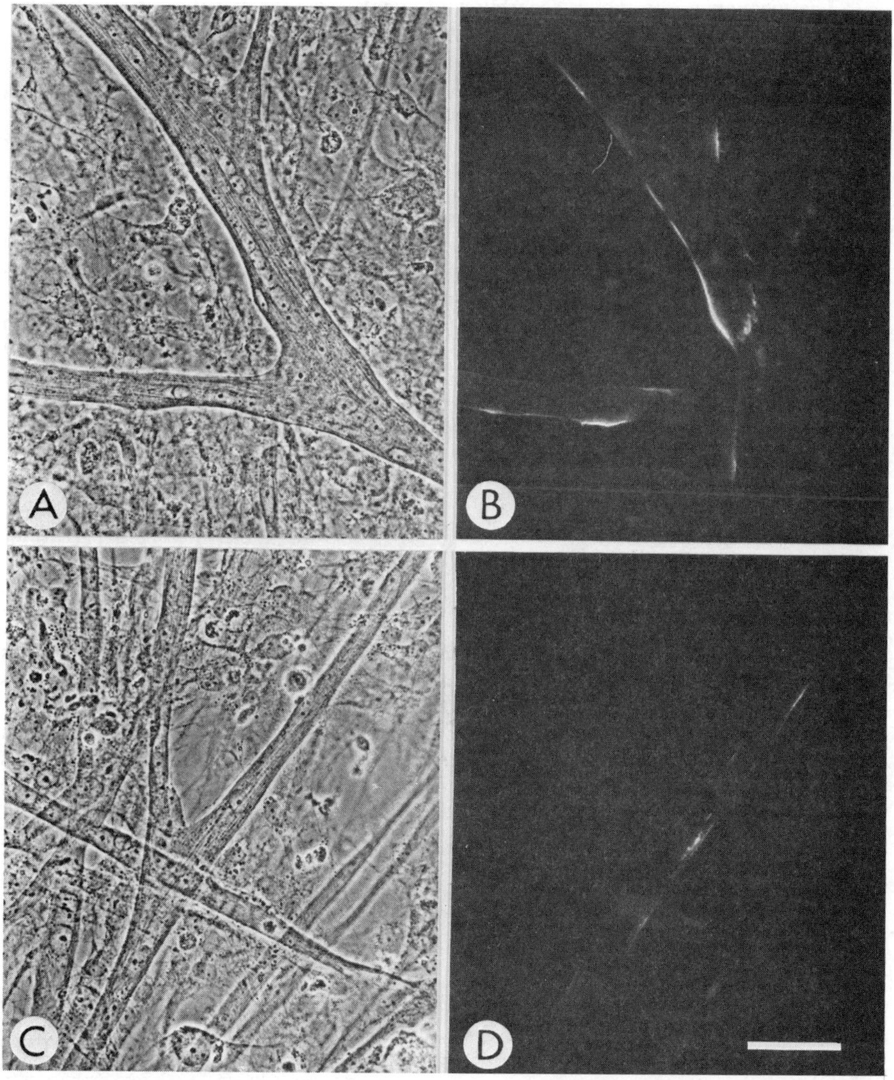

FIGURE 6. Effect of ascorbate oxidase on AcChR aggregation in muscle-CG cocultures. Muscle-CG cocultures (see FIG. 1) were incubated with 5×10^{-8} M α-bungarotoxin tetramethylrhodamine, fixed with paraformaldehyde, and observed by phase (**A,C**) or fluorescence (**B,D**) microscopy at a distance less than 1 mm from the CG explant. (**A,B**) control cocultures; (**C,D**) ascorbate-oxidase-treated (5 U/ml) cocultures. *Bar* represents 50 μm.

could be accounted for by high molecular weight aggregation factors released from the CG explants.

The mechanism by which ascorbic acid affects receptor distribution is not clear. As described above, ascorbic acid (as well as the ascorbic-acid-like factor of brain extract and of neuronal explants) increase collagen, laminin, and basement membrane deposition on muscle. We have previously shown that conditions which stimulate or decrease collagen accumulation on the surface of the cells have corresponding effects on AcChR aggregation.[8,20,23,34] Moreover, we have shown that addition of exogenous laminin increases receptor aggregation and stimulates the receptor aggregation obtained with the high molecular weight aggregation factor secreted by a neuroblastoma x glioma cell line.[35] Molecules associated with the basement membrane of injured frog muscle,[36] as well as those associated with the extracellular matrix of *Torpedo* electric organ, are involved in receptor aggregation.[31,32] It is, thus, an appealing hypothesis that ascorbic acid through its contribution to the assembly of basement membrane changes the number or the characteristics of the sites at which AcChR aggregation could take place. Knaack *et al.*,[37] using the rat muscle cell line L5, have similarly shown the effect of ascorbic acid on receptor aggregation (as well as on receptor site density) but suggested that ascorbic acid affects the AcChR biosynthetic pathway. It remains to be examined whether these two effects of ascorbic acid are related or are independently induced.

CONCLUSIONS

Embryonic brain extracts contain a low molecular weight factor that increases aggregation of acetylcholine receptors on cultured myotubes. The factor also stimulates release of collagenous proteins (by enhancing proline hydroxylation) and their deposition on the muscle surface. The factor increases the amount of laminin found on muscle and the surface area covered with basement membrane. A similar activity is released from explants of ciliary ganglia cocultured with muscle. Ascorbate oxidase blocks these effects in the treated cultures and reduces them in the ciliary ganglia-myotube cocultures. HPLC indicates that the factor contains a high concentration of ascorbate. Ascorbate increases receptor aggregation and mimics the effects of the factor on collagen and laminin deposition and basement membrane assembly on muscle. These results are consistent with a role for ascorbate in neural regulation of extracellular matrix formation and receptor aggregation in muscle.

REFERENCES

1. DENNIS, M. J. 1981. Development of the neuromuscular junction: Inductive interactions between cells. Ann. Rev. Neurosci. **4:** 48-68.
2. RUBIN, L. L. & K. F. BARALD. 1983. Neuromuscular development in tissue culture. *In* Somatic and Autonomic Nerve-Muscle Interactions. G. Burnstock, R. O'Brien & G. Vrbova, Eds.: 109-151. Elsevier. Amsterdam.
3. SALPETER, M. M. & R. H. LORING. 1985. Nicotinic acetylcholine receptors in vertebrate muscle: Properties, distribution and neural control. Prog. Neurobiol. **25:** 297-325.
4. CHRISTIAN, C. N., M. P. DANIELS, H. SUGIYAMA, Z. VOGEL, L. JAQUES & P. G. NELSON.

1978. A factor from neurons increases the number of acetylcholine receptor aggregates on cultured muscle cells. Proc. Natl. Acad. Sci. U.S.A. **75:** 4011-4015.
5. SCHAFFNER, A. E. & M. P. DANIELS. 1982. Conditioned medium from cultures of embryonic neurons contains a high molecular weight factor which induces acetylcholine receptor aggregation on cultured myotubes. J. Neurosci. **2:** 623-632.
6. PODLESKI, T. R., D. AXELROD, P. RAVDIN, I. GREENBERG, M. M. JOHNSON & M. M. SALPETER. 1978. Nerve extract induces increase and redistribution of acetylcholine receptors on cloned muscle cells. Proc. Natl. Acad. Sci. U.S.A. **75:** 2035-2039.
7. JESSELL, T. M., R. E. SIEGEL & G. D. FISCHBACH. 1979. Induction of acetylcholine receptors on cultured skeletal muscle by a factor extracted from brain and spinal cord. Proc. Natl. Acad. Sci. U.S.A. **76:** 5397-5401.
8. KALCHEIM, C., Z. VOGEL & D. DUKSIN. 1982. Embryonic brain extract induces collagen biosynthesis in cultured muscle cells: Involvement in acetylcholine receptor aggregation. Proc. Natl. Acad. Sci. U.S.A. **79:** 3077-3081.
9. BUC-CARON, M. H., P. NYSTROM & G. D. FISCHBACH. 1983. Induction of acetylcholine receptor synthesis and aggregation: Partial purification of low molecular weight activity. Dev. Biol. **95:** 378-386.
10. NEUGEBAUER, K., M. M. SALPETER & T. R. PODLESKI. 1985. Differential responses of L5 and rat primary muscle cells to factors in rat brain extract. Brain Res. **346:** 58-69.
11. KNAACK, D. & T. PODLESKI. 1985. Ascorbic acid mediates acetylcholine receptor increase induced by brain extract on myogenic cells. Proc. Natl. Acad. Sci. U.S.A. **82:** 575-579.
12. DANIELS, M. P., M. VIGNY, P. SONDEREGGER, H. C. BAUER & Z. VOGEL. 1984. Association of laminin and other basement membrane components with regions of high acetylcholine receptor density on cultured myotubes. Int. J. Dev. Neurosci. **2:** 87-99.
13. SASSE, J., H. VON DER MARK, U. KUHL, W. DESSAU & K. VON DER MARK. 1981. Origin of collagen types I, III, and V in cultures of avian skeletal muscle. Dev. Biol. **83:** 79-89.
14. CHIU, A. Y. & J. R. SANES. 1984. Development of basal lamina in synaptic and extrasynaptic portions of embryonic rat muscle. Dev. Biol. **103:** 456-467.
15. KELLY, A. M. & S. I. ZACKS. 1969. The fine structure of motor endplate morphogenesis. J. Cell Biol. **12:** 154-169.
16. JACOB, M. & T. L. LENTZ. 1979. Localization of acetylcholine receptors by means of horseradish peroxidase-α-bungarotoxin during the formation and development of the neuromuscular junction in the chick embryo. J. Cell Biol. **82:** 195-211.
17. ANDERSON, M. J., F. G. KLIER & K. E. TANGUAY. 1984. Acetylcholine receptor aggregation parallels the deposition of a basal lamina proteoglycan during development of the neuromuscular junction. J. Cell Biol. **99:** 1769-1784.
18. SANES, J. R. & Z. W. HALL. 1979. Antibodies that bind specifically to synaptic sites on muscle fiber basal lamina. J. Cell Biol. **83:** 357-370.
19. SANES, J. R., D. H. FELDMAN, J. M. CHENEY & J. C. LAWRENCE, JR. 1984. Brain extract induces synaptic characteristics in the basal lamina of cultured myotubes. J. Neurosci. **4:** 464-473.
20. KALCHEIM, C., D. DUKSIN & Z. VOGEL. 1982. Involvement of collagen in the aggregation of acetylcholine receptors on cultured muscle cells. J. Biol. Chem. **257:** 12722-12727.
21. KALCHEIM, C., E. BACHAR, D. DUKSIN & Z. VOGEL. 1985. Ciliary ganglia and spinal cord explants release an ascorbate-like compound which stimulates proline hydroxylation and collagen formation in muscle cultures. Neurosci. Lett. **58:** 219-224.
22. DUKSIN, D., C. KALCHEIM & Z. VOGEL. 1983. Characterization and localization of collagens synthesized by cultured muscle cells stimulated with collagen-inducing factor from embryonic brain extracts. J. Biol. Chem. **258:** 14585-14591.
23. KALCHEIM, C., D. DUKSIN, E. BACHAR & Z. VOGEL. 1985. Collagen-stimulating factor from embryonic brain has ascorbate-like activity and stimulates prolyl hydroxylation in cultured muscle cells. Eur. J. Biochem. **146:** 227-232.
24. HAY, E. D., Ed. 1981. Cell Biology of Extracellular Matrix. Plenum Press. New York and London.
25. KLEINMAN, H. K., R. J. KLEBE & G. R. MARTIN. 1981. Role of collagenous matrices in the adhesion and growth of cells. J. Cell Biol. **88:** 473-485.

26. PROCKOP, D. J., K. I. KIVIRIKKO, L. TUDERMAN & N. A. GUZMAN. 1979. The biosynthesis of collagen and its disorders. N. Engl. J. Med. **301:** 13-23.
27. OMAYE, S. T., J. D. TURNBULL & H. E. SAUBERLICH. 1979. Selected methods for the determination of ascorbic acid in animal cells, tissue and fluids. *In* Methods in Enzymology, Vol. 62. D. B. McCormick & L. D. Wright, Eds.: 3-14. Academic Press. New York.
28. KIVIRIKKO, K. I. & R. MYLLYLÄ. 1980. The hydroxylation of prolyl and lysyl residues. *In* The Enzymology of Post-translational Modifications of Proteins. R. B. Freedman & H. C. Hawkins, Eds.: 53-104. Academic Press. New York.
29. TERRANOVA, V. P., D. H. ROHRBACH & G. R. MARTIN. 1980. Role of laminin in the attachment of PAM 212 (epithelial) cells to basement membrane collagen. Cell **22:** 719-726.
30. KLEINMAN, H. K., M. L. MCGARVEY, J. R. HASSELL, V. L. STAR, F. B. CANNON, G. W. LAURIE & G. R. MARTIN. 1986. Basement membrane complexes with biological activity. Biochemistry **25:** 312-318.
31. GODFREY, E. W., R. M. NITKIN, B. G. WALLACE, L. L. RUBIN & U. J. MCMAHAN. 1984. Components of *Torpedo* electric organ and muscle that cause aggregation of acetylcholine receptors on cultured muscle cells. J. Cell Biol. **99:** 615-627.
32. WALLACE, B. G. 1986. Aggregating factor from *Torpedo* electric organ induces patches containing acetylcholine receptors, acetylcholinesterase, and butyrylcholinesterase on cultured myotubes. J. Cell Biol. **102:** 783-794.
33. USDIN, T. B. & G. D. FISCHBACH. 1986. Purification and characterization of a polypeptide from chick brain that promotes the accumulation of acetylcholine receptors in chick myotubes. J. Cell Biol. **103:** 493-507.
34. KALCHEIM, C., D. DUKSIN & Z. VOGEL. 1982. Aggregation of acetylcholine receptors in nerve-muscle cocultures is decreased by inhibitors of collagen production. Neurosci. Lett. **31:** 265-270.
35. VOGEL, Z., C. N. CHRISTIAN, M. VIGNY, H. C. BAUER, P. SONDEREGGER & M. P. DANIELS. 1983. Laminin induces acetylcholine receptor aggregation on cultured myotubes and enhances the receptor aggregation activity of a neuronal factor. J. Neurosci. **3:** 1058-1068.
36. BURDEN, S. J., P. B. SARGENT & U. J. MCMAHAN. 1979. Acetylcholine receptors in regenerating muscle accumulate at original synaptic sites in the absence of the nerve. J. Cell Biol. **82:** 412-425.
37. KNAACK, D., I. SHEN, M. M. SALPETER & T. R. PODLESKI. 1986. Selective effects of ascorbic acid on acetylcholine receptor number and distribution. J. Cell Biol. **102:** 795-802.

DISCUSSION OF THE PAPER

J. STUNART (*Walter Reed Institute of Research, Washington, D.C.*): There is a substantial body of literature on the clustering of acetylcholine receptors which people believe to be dependent on the cytoskeleton. Have you done any experiments under conditions where you have disrupted the microfilament system or the microtubule system in order to dissect out whether this is a purely collagen basement-type membrane side effect of something that's acting in concert with the cytoskeleton, and whether ascorbic acid has any effect on the cytoskeletal structures also?

Z. VOGEL (*Weizmann Institute of Science, Rehovot, Israel*): This is a very good question; various laboratories looked into this and there is evidence that the receptor is also bound from the inside of the cell.

In addition there is evidence that the extracellular matrix can determine the way the cytoskeletal elements are being connected into the plasma membrane.

J. STUNART: Yes, as far as I remember, John Connolly's work demonstrated a number of years ago that the microfilaments are necessary for the clustering effect of the receptors, but the microtubules are required for the disappearance of the receptors eventually.

Z. VOGEL: It's very possible that actin either directly or through microfilament binding proteins (such as vinculin and talin) interacts with the cytoplasmic part of specific receptors to basement membrane components (*e.g.*, laminin receptor or the fibronectin receptor) which penetrate the plasma membrane and interact with the basement membrane on top of the cell.

M. LEVINE (*National Institutes of Health, Bethesda, Md.*): I have two questions. First, where is the ascorbate to get your effect? In other words, you've shown very nicely that at least one of your neuronal factors seems to be ascorbate. Does it work simply extracellularly or must it be incorporated into your cells, and if so where does it go? The second question is, Is there a time lag or do you get the effect immediately?

Z. VOGEL: Embryonic brain as shown here contains a high concentration of ascorbic acid, so when we make the extract we actually purify the ascorbic acid from the brain. There is relatively little ascorbic acid present in one-week-old cultured muscle cells prior to addition of factor or ascorbic acid. Experiments using ^{14}C-ascorbate demonstrated that the ascorbate is incorporated into the cells. Moreover, ascorbate activates the enzyme prolyl hydroxylase, which is located inside the cell.

The aggregation of the receptor takes about 4 to 8 hours to be seen. The effect observed in collagen formation depends on the incubation period. The prominent effect during the first 2 hours is a large increase in procollagen released from the cells. If one waits longer the procollagen released is processed into collagen which sediments on the muscle cell surface.

S. L. ROMNEY (*Albert Einstein College of Medicine, New York, N.Y.*): Have you looked at basement membrane formation in embryonic tissue other than muscle?

Z. VOGEL: No, we didn't do that but other people have already shown that ascorbic acid will induce formation of basement membrane on other cells as well. So I don't think muscle is anything specific. We are interested in synaptogenesis and muscle is the target of the formation of the synapse with the nerve.

E. J. DILIBERTO: To put things in perspective, this turns out to be a very exciting finding because it's really extracellular action, although it has an intracellular function. It's an extracellular action of ascorbic acid, so the neuron is essentially priming the muscle cells for synaptogenesis. So, in other words, the neuron is actually secreting the ascorbic acid. I hope someone will look at synaptogenesis in the brain, which is turning out to be a very exciting area, and also whether or not that's one of the reasons why ascorbic acid might be released from neurons in the brain in such a diverse pattern, as Dr. Oke has already pointed out to us.

Adrenomedullary Chromaffin Cells as a Model to Study the Neurobiology of Ascorbic Acid: From Monooxygenation to Neuromodulation

EMANUEL J. DILIBERTO, JR.,[a] FRANK S. MENNITI,[a] JANE KNOTH,[a] ALEJANDRO J. DANIELS,[a] J. STEPHEN KIZER,[b] AND O. HUMBERTO VIVEROS[a]

[a] *Section of Neuroscience*
Department of Medicinal Biochemistry
The Wellcome Research Laboratories
Research Triangle Park, North Carolina 27709

[b] *Departments of Medicine and Pharmacology*
and
Biological Sciences Research Center
School of Medicine
University of North Carolina
Chapel Hill, North Carolina 27514

Perhaps the greatest recent growth in our knowledge of the biological roles of ascorbic acid has occurred in the area of neuroendocrine function. Within these systems, ascorbic acid has many sites of action that may be conveniently separated into well-defined intracellular and less clearly defined extracellular functions. Twenty-five years ago ascorbic acid was proposed to be the intracellular cofactor for norepinephrine synthesis in the adrenal medulla.[1,2] More recently, ascorbic acid has been considered as an essential cofactor in peptide neurotransmitter-hormone biosynthesis. Indeed, the recent discovery of the ascorbic-acid-dependent peptidylglycine α-amidating monooxygenase (PAM) has established the importance of this vitamin in peptide messenger synthesis (see Mains *et al.*[3] and Glembotski, this volume[69]). That ascorbic acid may play critical intracellular roles is further suggested by the high concentration in brain and endocrine glands[14] and by the presence of a complex system regulating its intracellular and intraventricular availability.[18,22] The extracellular functions of ascorbic acid are less clear, although there is extensive literature suggesting that ascorbic acid can effect cellular functions by acting on plasma membrane proteins such as receptors and enzymes.[4–6] This volume brings these diverse areas of investigation into focus.

In the past ten years, we have investigated both the intra- and extracellular functions of ascorbic acid using adrenomedullary chromaffin cells, an excellent model for the study of presynaptic events.[50] Our investigations of the intracellular roles of ascorbic acid have not only concentrated on its function as cofactor for two monooxygenases (dopamine β-hydroxylase and peptidylglycine α-amidating monooxygenase) in the adrenal medulla, but also on the various cellular processes that govern ascorbic acid availability for neurotransmitter-hormone synthesis. Because of the recent proposal that ascorbic acid functions as a chemical messenger, we have also examined the mechanisms of neuroendocrine secretion of this vitamin. To satisfy one of several criteria for consideration as a chemical messenger, ascorbic acid must be secreted from cells when challenged with an appropriate stimulus. We will review in this report some of our studies and more recent findings regarding (1) the roles of ascorbic acid in neurotransmitter synthesis and (2) the mechanism(s) of its secretion from adrenomedullary chromaffin cells.

Studies on the Characterization of the Enzymic Oxidation Product of Ascorbic Acid by Dopamine β-Hydroxylase

Dopamine β-hydroxylase (3,4-dihydroxyphenylethylamine, ascorbate: oxygen oxidoreductase, hydroxylating; EC 1.14.17.1; DBH) catalyzes the final step in the biosynthesis of norepinephrine. The identification of molecular oxygen as the source of the hydroxyl oxygen and the demonstration of a stoichiometric oxidation of the electron donor established dopamine β-hydroxylase as a monooxygenase.[1,2,7] These studies also showed that ascorbic acid was the most effective electron donor for the enzyme *in vitro,* providing evidence to consider ascorbic acid as the putative cofactor for this system. Originally, the reaction mechanism was thought to involve the initial transfer of two electrons from one molecule of ascorbic acid to the enzyme-bound copper with the release of dehydroascorbate.[2] Hydroxylation then occurs in a complex set of reactions involving the reduced form of the enzyme, molecular oxygen, and substrate, resulting in a reoxidation of the enzyme and simultaneous formation of the β-hydroxylated product and water.[2,7]

Although dehydroascorbate was postulated to be a product of dopamine β-hydroxylation, no evidence had been gathered to support the existence of a system for the intracellular reduction of dehydroascorbate within the adrenal medulla or noradrenergic neurons. In view of studies on the mechanism of catechol oxidation by DBH,[9] we investigated the possibility that formation of fully reduced enzyme during dopamine β-hydroxylation required a series of one-electron transfers. Such a mechanism for the reduction of DBH with ascorbate would generate the ascorbate free radical, semidehydroascorbate (FIG. 1). By using two different partially purified preparations of semidehydroascorbate reductase (NADH: semidehydroascorbate acid oxidoreductase, EC 1.6.5.4; SDR), the generation of semidehydroascorbate has been demonstrated during dopamine β-hydroxylation.[10,11] This initial observation was confirmed spectrophotometrically by Skotland and Ljones.[12] Using the coupled reaction of DBH with SDR, we also proved that ascorbic acid free radical formation was proportional to the rate of β-hydroxylation irrespective of the concentration of tyramine or DBH, or the presence or absence of inhibitors and activators of DBH and SDR. Furthermore, at low rates of hydroxylation (concentration of tyramine or DBH reduced), two molecules of semidehydroascorbate were formed from two ascorbic acid molecules for every molecule of octopamine that was formed from tyramine. Finally, incubations containing limited amounts of ascorbic acid did not stop forming β-hydroxylated product. In these experiments, by 15 min, 4.3-fold more octopamine was formed than could be accounted for by the initial concentration of ascorbic acid,

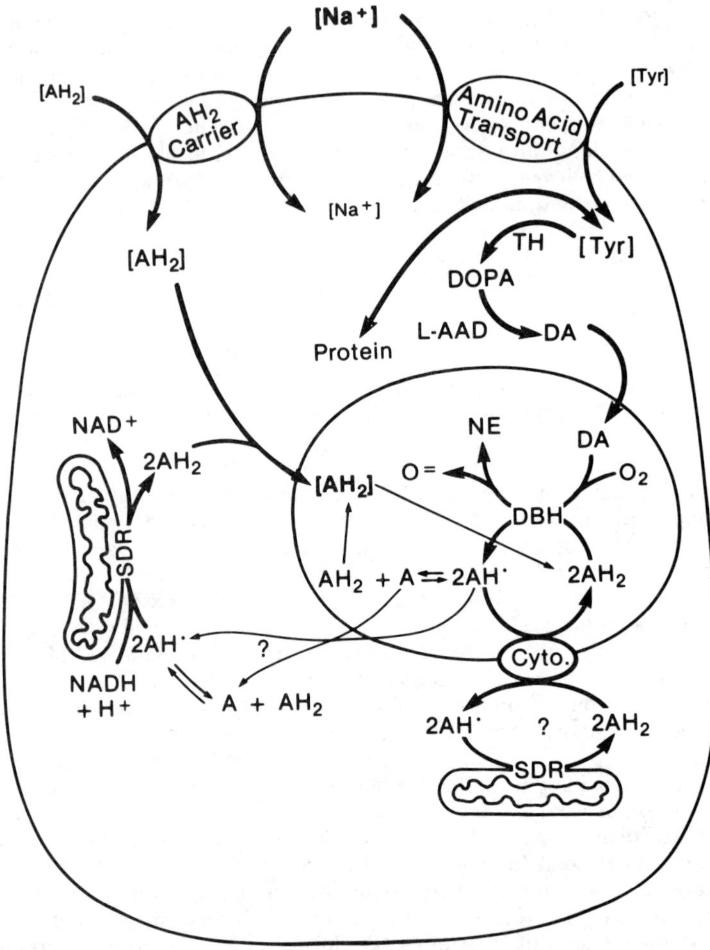

FIGURE 1. The interrelationship of ascorbic acid metabolism and norepinephrine biosynthesis in the chromaffin cell. *DBH,* dopamine β-hydroxylase; *SDR,* semidehyroascorbate reductase; *Cyto.,* cytochrome b_{561}; AH_2, ascorbate; $AH^{\cdot}$, semidehydroascorbate; *A,* dehydroascorbate; *TH,* tyrosine hydroxylase; and *L-AAD,* L-amino acid decarboxylase.

indicating that ascorbate must be recycled (through the SDR reaction).[11] These results establish that semidehydroascorbate is the exclusive product of ascorbic acid oxidation by DBH (FIG. 1).

The mechanism for dopamine β-hydroxylation and the evidence for ascorbic acid recycling imply that an enzyme system different from dehydroascorbate reductase may be involved in the regeneration of ascorbic acid for hydroxylation. Thus, the localization and regulation of semidehydroascorbate reductase become of added significance. It should be mentioned parenthetically, however, that the well-recognized human requirement for ascorbic acid in the prevention of scurvy[13] is evidence that cellular processes other than recycling of semidehydroascorbate are also involved in maintaining levels of ascorbic acid for dopamine β-hydroxylation.

Ascorbic Acid Transport into Adrenomedullary Chromaffin Cells and Its Subcellular Distribution

The capacity to produce ascorbic acid is limited to the liver of most mammals. At some point in evolution certain mammals, such as man, other primates, and guinea pigs have lost the ability to synthesize ascorbic acid.[14–16] Thus, to satisfy the intracellular requirement(s) for ascorbic acid, tissues other than liver (and in animals unable to synthesize ascorbic acid, all cells) must possess an uptake and/or regeneration system for this vitamin. Indeed, ascorbic acid transport has been demonstrated in a variety of tissues and is an energy-dependent, Na^+-sensitive process.[16,17] To investigate the mechanism of ascorbic acid transport in the adrenal medulla, ascorbic acid uptake by primary cultures of bovine adrenomedullary chromaffin cells was measured using either high-performance liquid chromatography with electrochemical detection or radiometric techniques (L-[1-^{14}C]ascorbic acid). Ascorbic acid transport is highly temperature- and energy-dependent and exhibits Michaelis-Menten kinetics with an apparent K_m of 29 μM (FIG. 2). Further characterization of this system

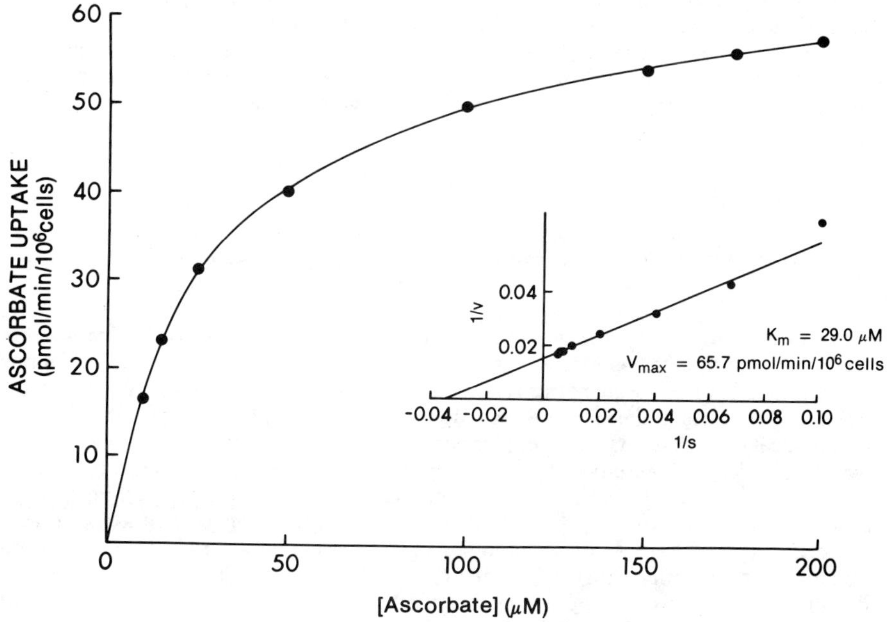

FIGURE 2. Uptake of L-[1-^{14}C]ascorbic acid into bovine chromaffin cells as a function of extracellular concentration of ascorbate. Bovine adrenomedullary chromaffin cells were isolated, plated in 35-mm plates at a density of 5.2×10^5 cells/cm^2 and maintained in primary culture until use with culture medium containing 10% fetal calf serum.[66] Cells after 3 days in culture were incubated in normal Na^+ (150 mM) medium for 15 min at 37°C with L-[1-^{14}C]ascorbic acid, 8.4 mCi/mmol, at varying concentrations. Incubations were carried out and terminated as previously described.[18] The *inset* is a Lineweaver-Burk plot of the reciprocal of uptake as a function of the reciprocal of ascorbate concentration: *v* in pmol/min/10^6 cells; *s*, ascorbic acid concentration (μM). The above experiment was repeated on six different cell preparations with identical results. (From Diliberto *et al.*[18])

showed that ascorbic acid transport by these cells is an active process driven by the Na^+ electrochemical gradient. Kinetic analysis of this cotransport system fits an "affinity type" model where binding of Na^+ to the transporter increases the affinity of the transporter for ascorbic acid and vice versa.[18]

Can such a cell membrane transport system satisfy the requirements for ascorbic acid in catecholamine biosynthesis? In the synthesis of norepinephrine, the only ascorbic-acid-dependent step is catalyzed by DBH. This enzyme is exclusively localized within the catecholamine-containing chromaffin vesicles. Thus, for ascorbic acid to function as cofactor for norepinephrine synthesis, ascorbic acid must first enter the chromaffin vesicle. Because of the evidence suggesting that isolated chromaffin vesicles do not take up ascorbic acid,[19] we decided to reinvestigate this apparent paradox by examining ascorbic acid uptake in primary cultures of chromaffin cells. In a series of pulse chase experiments using labeled L-[1-^{14}C]ascorbic acid followed by subcellular fractionation, we demonstrated a relatively slow accumulation of ascorbic acid into chromaffin vesicles.[18,20] When cells labeled with a 4-h pulse of L-[^{14}C]ascorbic acid are chased for 20 h in the absence of ascorbic acid, complete equilibration of newly acquired ascorbic acid and resident ascorbic acid occurs (FIG. 3).[18,20] In fact, in these experiments, the subcellular distribution of the labeled and cold ascorbic acid in the cultures was identical to the distribution of the vitamin in freshly isolated tissue, with 40% of the intracellular ascorbic acid contained in chromaffin vesicles.[20,21] Thus, even though ascorbic acid is rapidly taken up by chromaffin cells, the rate of transfer into chromaffin vesicles is very slow. Would this system be adequate to meet the ascorbic acid requirements for catecholamine biosynthesis in the intact cell?

Rate of Dopamine β-Hydroxylation as a Function of the Ascorbic Acid Concentration in Adrenomedullary Chromaffin Cells in Culture

Catecholamine biosynthesis was examined in primary cultures of bovine adrenomedullary chromaffin cells after six to seven days. To characterize the dopamine β-hydroxylation step in the pathway, the rate of conversion of tyramine to octopamine was monitored using high-performance liquid chromatography with coulometric electrochemical detection. The intracellular concentration of ascorbic acid in these cultures was modified by taking advantage of two properties of these cells in culture: (1) the decline in ascorbic acid levels with time in culture when cells are maintained in the absence of the cofactor ($t_{1/2}$ of 31 h)[18] and (2) the active transport of ascorbic acid (described above) capable of maintaining high intracellular ascorbic acid concentrations when the vitamin is added exogenously.[18] Thus, as shown in TABLE 1, after 7 days in culture, cells maintained in media supplemented twice daily with 200 μM ascorbic acid had ascorbic acid levels several thousand times higher than those maintained in unsupplemented media. This difference in intracellular ascorbic acid was expressed as a rate of octopamine synthesis fivefold higher in supplemented as compared to unsupplemented cultures. Perhaps of greater importance is the fact that in both ascorbic-acid-supplemented and ascorbic-acid-depleted cells, the amount of octopamine synthesized was greater than the amount of ascorbate present in the cells at all times. The ratio of octopamine to ascorbic acid reached approximately 3:1 in supplemented cells and 3500:1 in depleted cells. Moreover, in both supplemented and unsupplemented cells, the levels of ascorbic acid did not change for the duration of the incubations (TABLE 1). These observations support the postulate that at least

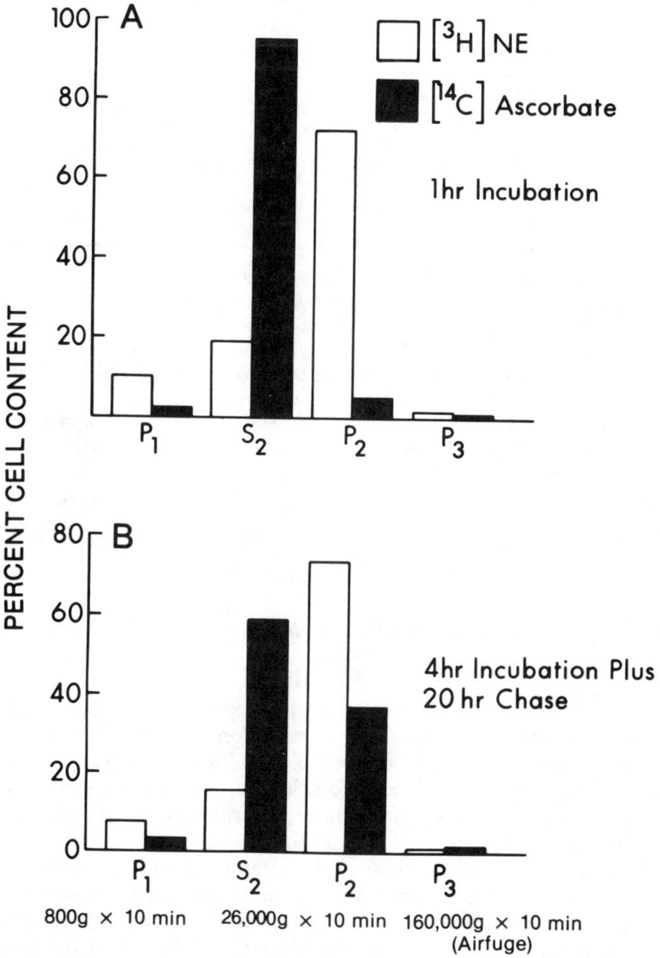

FIGURE 3. Subcellular distributions of newly acquired [^{14}C]ascorbic acid and [^{3}H]norepinephrine (*NE*) in isolated chromaffin cells. Cells were incubated in the presence of 200 μM [^{14}C]ascorbic acid and 0.37 μM [^{3}H]NE at 37°C. (**A**) Cells were incubated for 1 h (day 3), washed, homogenized, and fractions prepared as previously described.[20] (**B**) Cells were incubated for 4 h (day 2), washed, and incubated for a further 20 h at 37°C in a balanced salt solution (BSS: 150 mM NaCl, 4.2 mM KCl, 1.0 mM NaH_2PO_4, 11.2 mM glucose, 10 mM Hepes, 0.7 mM $MgCl_2$, 2 mM $CaCl_2$, pH 7.4); at this time, cells were processed as above. The amount of the radioactive material present in the homogenates, as calculated from their specific activities, was 20.6 and 14.6 nmol of [^{14}C]ascorbic acid, and 0.096 and 0.467 nmol of [^{3}H]NE for **A** and **B**, respectively. The distribution of endogenous catecholamines was identical to that of [^{3}H]NE. The distribution of resident nonradioactive ascorbic acid was the same for both **A** and **B** and the same as the [^{14}C]ascorbic acid pattern for **B**. (From Daniels *et al.*[20])

TABLE 1. Effect of Ascorbic Acid Supplementation on Intracellular Ascorbic Acid Content and Octopamine Formation in Adrenomedullary Chromaffin Cells in Culture[a]

Medium Supplement	Tyramine Incubation Time (h)	Catecholamine	Ascorbate	Octopamine	Molar Ratio Octopamine/ Ascorbate
None	2	79.2	0.0006	1.27	2120.0
	4	75.9	0.0006	2.10	3500.0
Ascorbate	2	91.3	3.82	7.53	1.9
	4	86.1	4.05	11.65	2.9

[a] Results are expressed in nmol/10^6 cells. Primary cultures of adrenomedullary chromaffin cells received twice daily additions of 200 μM ascorbic acid (final concentration) or no addition beginning three days after plating. On day 7 (24 h after the last addition) cultures were incubated in balanced salt solution containing 5 mM tyramine for 2 or 4 h at 37°C. Monoamine and ascorbic acid concentrations were determined as previously described.[22] Values are the mean of duplicate plates. The intracellular tyramine content was approximately 35 nmol/10^6 cells and was not differentially affected by prior ascorbate additions or duration of incubation.

under certain conditions a system for regenerating ascorbic acid is necessary during dopamine β-hydroxylation.[22]

Regulation of Dopamine β-Hydroxylation by Intravesicular Ascorbic Acid

These experiments also allowed us to examine the rate of dopamine β-hydroxylation as a function of the intracellular concentration of ascorbic acid. First, based on experiments where the number of ascorbic acid additions was varied as well as the time interval between the last ascorbic acid addition and the start of incubation with tyramine, we concluded that under those experimental conditions ascorbic acid was the only intravesicular electron donor for DBH and that the DBH activity was proportional to the concentration of the vitamin inside the vesicle.[22] Moreover, in experiments where ascorbic acid and tyramine were added simultaneously to the cultures, the rapid increase in the extravesicular (cytosolic) concentration of ascorbic acid did not effect the rate of dopamine β-hydroxylation. Our interpretation of these results is in contrast to a report by Levine *et al.*,[23] who suggested that extravesicular ascorbic acid had a direct effect on dopamine β-hydroxylase activity. This direct correlation between (1) the slow rate of transfer of ascorbic acid into chromaffin vesicles as seen in subcellular distribution studies and (2) the increase in the rate of dopamine β-hydroxylation implies that the intravesicular availability of the cofactor may be a regulatory mechanism for β-hydroxylation.

In a more extensive study, where several concentrations of ascorbic acid (22-200 μM) were added to the culture medium either once or twice daily (the last addition was made 18 h prior to starting tyramine incubation), a wide range of intravesicular concentrations of ascorbic acid was obtained ranging from 1 to 31 mM. Double reciprocal plots of the rate of octopamine synthesis vs. the intravesicular concentration of ascorbic acid revealed an apparent K_m of 15.0 ± 2.0 mM and a V_{max} of 5.8 ± 0.4 nmol/h/10^6 cells (FIG. 4 and inset). These results not only support the contention that the rate of dopamine β-hydroxylation varies with the intravesicular concentration of ascorbic acid, but also argues that the *in situ* affinity of DBH for its cofactor is markedly less (approximately 30-fold) than the value determined with purified or partially purified DBH (Menniti, Knoth and Diliberto, unpublished observations). Since this *in situ* K_m value is close to the estimated endogenous intravesicular concentration of ascorbic acid, DBH should be susceptible to regulation through changes in the levels of ascorbic acid in the organelle. To investigate this possibility further, we studied octopamine synthesis in isolated chromaffin vesicles and chromaffin-vesicle ghosts. Again, similar apparent K_m values for DBH with respect to ascorbic acid were obtained in both the vesicles and the ghosts that are preferentially depleted of small molecular weight vesicular components. In conclusion, although the affinity of DBH for its cofactor is influenced by factors residing within the chromaffin vesicle, it must be independent of the presence of high concentrations of small molecules (see FIGURE 6 below; Menniti, Knoth and Diliberto, unpublished observations).

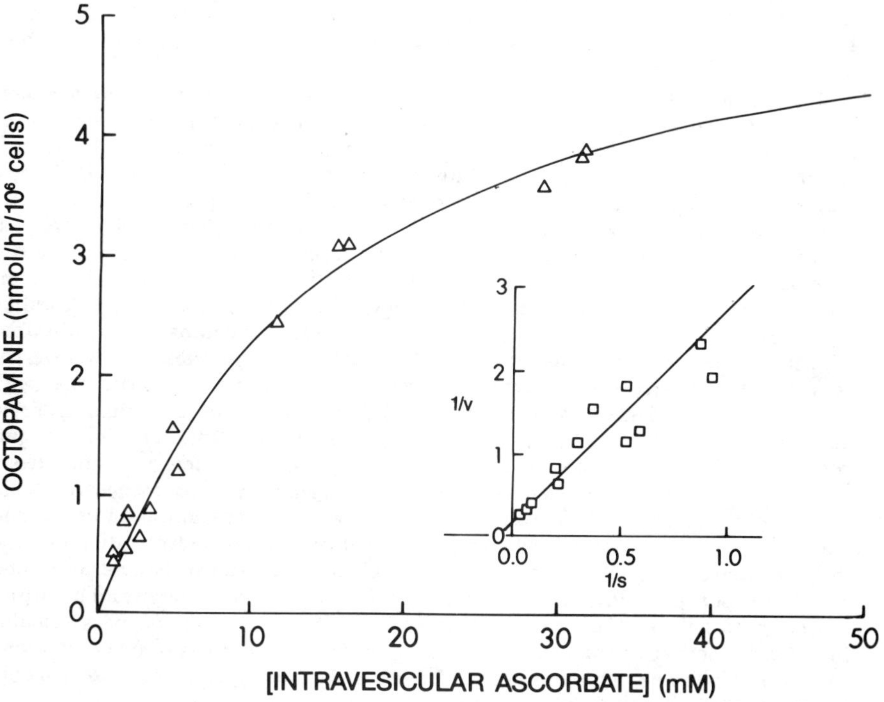

FIGURE 4. The rate of octopamine synthesis as a function of intravesicular ascorbate concentration in bovine adrenomedullary chromaffin cells in culture. The intravesicular ascorbate concentration of chromaffin cells was modified by varying the number and concentration of ascorbate additions to the culture media between days 3 and 6 in culture. On day 7 in culture, the rate of octopamine synthesis was measured as previously described.[22] Each point represents the mean of three similarly treated plates. The curve is fit to the equation $v = (V_{max} * S)/(K_m + S)$, where K_m and V_{max} were determined from a double-reciprocal plot of the rate of octopamine synthesis versus intravesicular ascorbate concentration (*inset*) according to the method of Cleland.[67]

Regeneration of Intravesicular Ascorbic Acid in Isolated Chromaffin Vesicles during Dopamine β-Hydroxylation

A regeneration system involving the transfer of ascorbic acid across the vesicle membrane could be ruled out because of its slow rate of vesicular accumulation as described above. Thus, to postulate a recycling system for ascorbic acid during catecholamine synthesis in adrenal chromaffin cells poses another question: Is there a link between our earlier observation that semidehydroascorbate is the product of dopamine β-hydroxylation and chromaffin-vesicle membrane electron transfer for regeneration of intravesicular ascorbic acid? A mechanism for regeneration of ascorbic acid from intravesicular semidehydroascorbate has been offered by Njus *et al.*,[24] who demonstrated that membranes of chromaffin vesicle contain an electron transport system. When chromaffin-vesicle ghosts are loaded with ascorbic acid and incubated with different external electron acceptors, a membrane potential will develop due to transmembrane electron transfer,[24–26] indicating the thermodynamic feasibility of electron transfer through the vesicle membrane. Recently, the putative vesicle membrane electron transport protein, cytochrome b_{561}, has been isolated,[26–29] and the spectral changes associated with oxidation and reduction of this protein by semidehydroascorbate and ascorbic acid have been measured.[30]

To test the hypothesis that vesicular transmembrane electron transport mediates the intravesicular reduction of semidehydroascorbate generated during dopamine β-hydroxylation, we isolated intact chromaffin vesicles by differential centrifugation.[22] Chromaffin vesicles were preincubated for 10 min at 37°C in the presence or absence of an external reductant, and then tyramine was added. Octopamine synthesis was observed in the absence of reductant, but the rate of synthesis progressively decreased. Simultaneously, the intravesicular ascorbic acid content fell (FIG. 5).[22] That the decrease of intravesicular ascorbic acid is caused by β-hydroxylation can be demonstrated by a blockade of depletion by inhibitors of dopamine β-hydroxylase like KCN and by the omission of tyramine. The molar ratio of octopamine synthesized to intravesicular ascorbate depleted was 1.35 ± 0.25 in the absence of external reductant (FIG. 5, inset). Inclusion of the ascorbic acid analogue, glucoascorbic acid, in the incubation mixture protected against the depletion of intravesicular ascorbic acid during octopamine synthesis (FIG. 5A). At the highest concentrations of glucoascorbic acid, the decrease in octopamine formation with time was completely prevented (FIG. 5B). Moreover, the ratio of octopamine formed to ascorbic acid catabolized was increased by extravesicular glucoascorbic acid addition (FIG. 5B and inset). Analysis of vesicular contents by high-performance liquid chromatography, which separates ascorbic acid from glucoascorbic acid, indicated that the protection in the depletion of intravesicular ascorbic acid was not due to permeation of glucoascorbic acid through the vesicle membrane.[22] Thus, the effects of extravesicular glucoascorbic acid are most likely mediated by an electron transport mechanism across the vesicle membrane. Similar studies by Beers *et al.*[31] using extravesicular ascorbic acid rather than glucoascorbic acid support this postulate.

This model system incorporating external glucoascorbic acid to protect the depletion of intravesicular ascorbic acid permitted reexamination of the catalytic properties of DBH in intact chromaffin vesicles and the demonstration that electron transfer across the vesicle membrane is not a rate-limiting step for catecholamine synthesis within the vesicle. When glucoascorbic acid was added to incubations of isolated chromaffin vesicles at various times after the addition of 5 mM tyramine, chromaffin vesicles with progressively less intravesicular ascorbic acid were obtained (FIG. 6). Samples were taken immediately after glucoascorbic acid addition and after a further 10 min of incubation; it was possible to demonstrate that glucoascorbic acid arrested

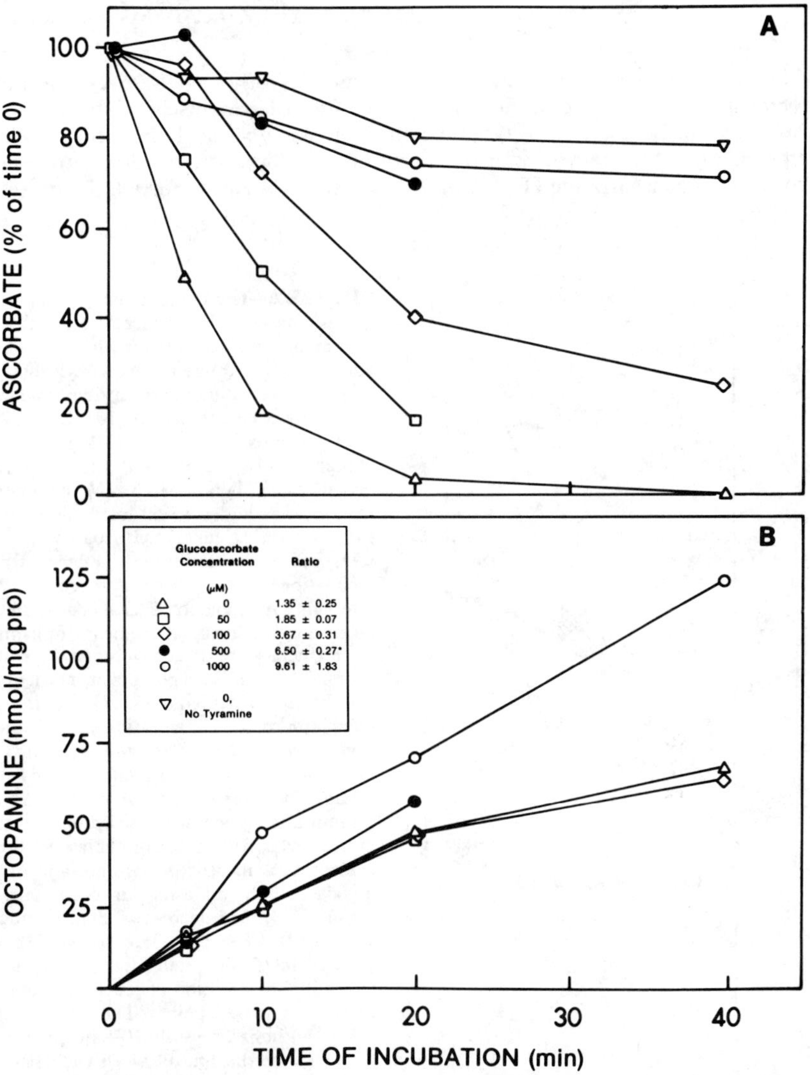

FIGURE 5. Effect of glucoascorbate on octopamine synthesis and intravesicular ascorbate concentration in isolated chromaffin vesicles. Chromaffin vesicles (2.0 mg protein/ml) were preincubated for 10 min at 37°C in 2.5 ml of 0.3 M sucrose, 10 mM Hepes buffer, pH 7.2, containing 5 mM MgATP and 1-1000 μM glucoascorbate. Incubations were initiated by the addition of tyramine (5 mM, final concentration) or vehicle and at the times indicated samples were taken for determination of (**A**) ascorbate and (**B**) octopamine content as previously described.[22] Ascorbate content at each time point is expressed as the percentage of the intravesicular ascorbate content immediately after tyramine addition (the ascorbate contents at time 0 were 47-65 nmol/mg protein). Each point represents the mean of duplicate incubations; ▽, no tyramine; glucoascorbate concentrations: △, 0 μM; □, 50 μM; ◇, 100 μM; ●, 500 μM; ○, 1 mM. *Inset,* The molar ratio of octopamine formed to ascorbate depleted was calculated for each condition. The amount of intravesicular ascorbate depleted was calculated as the difference in ascorbate content immediately following tyramine addition and at 5, 10, 20, or 40 min afterwards. The ratio of octopamine formed to ascorbate depleted at each time point for each condition was averaged and is expressed as the mean (± SEM). The ratio marked * was calculated without the five min point for 500 μM glucoascorbate. (From Menniti *et al.*[22])

further depletion of ascorbic acid during octopamine synthesis (FIG. 6A). The amount of octopamine synthesized in the latter 10 min interval is inversely related to the time delay between the addition of tyramine and glucoascorbic acid (FIG. 6B). A double reciprocal plot of the rate of octopamine synthesis and its corresponding intravesicular ascorbic acid concentration (FIG. 6C) revealed an apparent K_m for DBH with respect

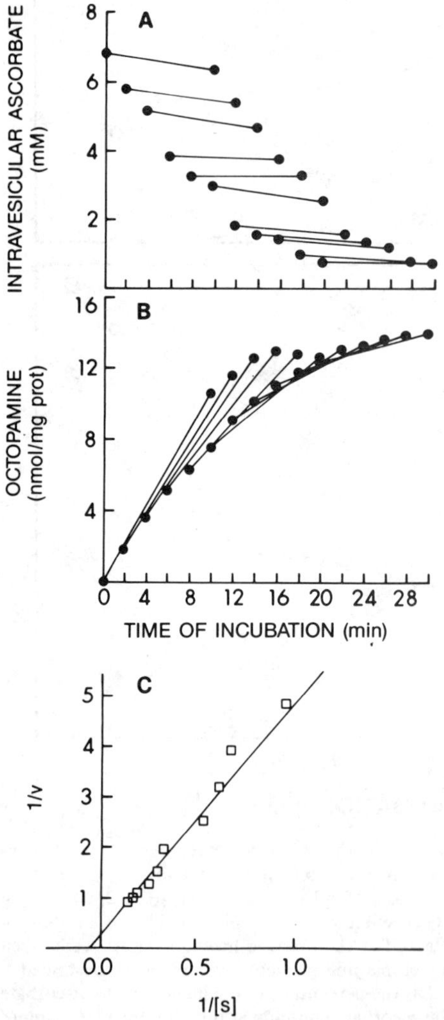

FIGURE 6. The octopamine content and intravesicular ascorbate concentration of isolated chromaffin vesicles as a function of time after addition of tyramine and glucoascorbate. Isolated bovine adrenomedullary chromaffin vesicles were preincubated for 10 min at 37°C in 0.3 M sucrose and 10 mM Hepes (pH 7.2) containing 5 mM MgATP (2.32 mg vesicle protein/ml). Tyramine (5 mM, final concentration) was added to all incubations to start the reaction (time 0). Glucoascorbate (500 μM, final concentration) was added to different incubations at time 0 or at successive 2-min intervals. Aliquots of each incubation were taken as previously described[22] immediately after glucoascorbate addition and after an additional 10 min. Each point is the mean of three simultaneous incubations. (**A**) The intravesicular ascorbate concentration of each incubation immediately after and 10 min after glucoascorbate addition is indicated by paired points. The decrease in ascorbate concentration 10 min after glucoascorbate addition was calculated as a percentage of the ascorbate concentration measured immediately after glucoascorbate addition for each pair. The mean ($\pm$ SEM) decrease for all incubations was 10.4 $\pm$ 1.8%. (**B**) The octopamine content of each incubation immediately after and 10 min after the addition of glucoascorbate is indicated by paired points. The rate of octopamine synthesis for each 10-min interval was the slope of the line between each pair. (**C**) The reciprocal of the rate of octopamine synthesis (v) obtained from **B** is plotted as a function of the reciprocal of its corresponding intravesicular ascorbate concentration(s) obtained from **A** immediately after glucoascorbate addition. K_m and V_{max} were estimated from this plot according to the method of Cleland.[67]

to ascorbic acid of 17.0 $\pm$ 5.1 mM, a value that closely agrees with that obtained for DBH in the intact chromaffin cells (FIG. 4). These results further demonstrate that electron transfer from an extravesicular reductant to intravesicular semidehydroascorbate is probably part of the mechanism for cofactor regeneration during dopamine β-hydroxylation.

Interestingly, the normal electrochemical gradient between chromaffin vesicle and chromaffin cell cytosol favors electron transport from extravesicular ascorbic acid to intravesicular semidehydroascorbate. The vesicle-membrane ATPase generates a proton electrochemical gradient such that the vesicle interior is acidic (approximately pH 5.5) and positively charged with respect to the cytosol.[32,33] Under these conditions, the redox potential of the vesicular ascorbate-semidehydroascorbate couple inside the vesicle is shifted to more positive values with respect to this couple in the cytosol.[34] Furthermore, the positive potential inside the vesicle would favor electron movement toward its interior.

Semidehydroascorbate Reductase as a Component of the Ascorbic Acid Regeneration System in Chromaffin Cells

When intact chromaffin cells in culture are incubated with tyramine in the absence of external ascorbic acid, there are no changes in intracellular levels of the cofactor during the synthesis of octopamine (TABLE 1), implying that a system for regeneration through reduction of oxidized species of ascorbic acid must exist in these cells (FIG. 1). Indeed, high levels of semidehydroascorbate reductase are found within the adrenal medulla. The enzyme activity is enriched in the mitochondrial fraction.[10] Submitochondrial fractionation studies indicate that the enzyme is an outer membrane protein capable of reducing cytosolic semidehydroascorbate.[35] Thus, it appears that the final components of the system for regeneration of oxidized intravesicular ascorbic acid are cytosolic ascorbic acid and mitochondrial semidehydroascorbate reductase. In fact, this enzyme, by decreasing the cytosolic semidehydroascorbate concentration, will shift the redox potential of the ascorbate-semidehydroascorbate couple to more negative values and thus promote electron flow to the vesicle interior.

Peptidyl Glycine α-Amidating Monooxygenase: A Newly Discovered Ascorbic-Acid-Dependent Enzyme Involved in Neurotransmitter and Hormone Synthesis

As the mechanism for ascorbic acid regeneration in chromaffin vesicles was unfolding, a similar, but less complete, picture was emerging from studies on secretory vesicles from other tissues. The secretory vesicles of the anterior, posterior, and intermediate lobes of the pituitary, like the chromaffin vesicles, contain a proton-translocating MgATPase which creates an electrochemical gradient (acid and positive inside).[36] More recently, Russel *et al.*[37] demonstrated that the neurosecretory vesicles of the posterior pituitary have a transmembrane electron carrier similar to that described above for chromaffin vesicles. They also confirmed the presence of a high concentration of ascorbic acid in these vesicles. Furthermore, the putative electron transport protein, cytochrome b_{561}, has been immunologically identified as a component of the membranes of these vesicles.[38,39] Finally, the pituitary is very rich in semidehydroascorbate reductase.[35]

It is not surprising, therefore, that a new ascorbic-acid-dependent monooxygenase has been discovered in the pituitary.[40,41] This enzyme, peptidylglycine α-amidating monooxygenase, catalyzes the conversion of peptides ending in AA_n-Gly-COOH to the corresponding α-amidated peptide AA_n-$CONH_2$ by a mechanism similar to DBH.[42,43] The wide distribution of α-amidated peptides throughout the nervous and endocrine systems suggests that PAM is involved in the processing of a variety of peptide neurohormones; indeed, its presence in brain has also been described.[44] For a more extensive review of this subject, see the paper by Glembotski in this volume.[69]

The Presence and Regulation of Peptidylglycine α-Amidating Monooxygenase for Opioid Peptide Processing in the Adrenal Medulla

Two amidated opioid peptides (amidorphin and metorphamide), derived from the posttranslational processing of proenkephalin A, have been identified in brain and in adrenal medulla.[45–48] In view of these reports, we examined the subcellular distribution of PAM in the adrenal medulla (TABLE 2). The enzyme is predominately localized in the chromaffin vesicles where most of the activity is associated with the chromaffin vesicle soluble components (CVS), although at least 10% of the activity is on the vesicle membrane. A substantial amount of activity is also associated with the microsomal fraction (P_3). A similar subcellular distribution has been obtained from bovine adrenomedullary chromaffin cells in culture. In the rat, however, the enzyme activity found in the chromaffin vesicle fraction distributes approximately in equal amounts between the membrane and soluble fractions (Diliberto, Kizer and Viveros, unpublished observations).

These findings have important implications with respect to opioid peptide processing in the adrenal medulla. Processing of opioid peptides from proenkephalin A requires several enzymes including endopeptidases, aminopeptidases, and carboxypeptidases. In the adrenal medulla, where this processing is incomplete, a variety of opioid peptides are produced. The α-amidated peptides, metorphamide and amidorphin, are probably formed from the proenkephalin-A-derived peptides E and F, respectively. Thus, PAM is apparently also involved in opioid peptide processing. Studies of the regulation of α-amidation suggest that treatments which induce opioid peptide synthesis and processing also increase α-amidating enzyme activity. Activation of the rat adrenal medulla *in vivo* by insulin-induced hypoglycemia for 2 h progressively increases PAM activity in the large granular fraction of the adrenal medulla which peaks at 24 h and remains elevated until 120 h after insulin (FIG. 7). This rise in PAM activity slightly precedes the increase in opioid peptides, but occurs after the increase in proenkephalin A mRNA.[49] It is probable that neurogenic stimulation, in addition to increasing proenkephalin A biosynthesis and processing, activates and/or induces PAM in the chromaffin vesicle.

In summary, two monooxygenases are present within the same chromaffin vesicle that require the regeneration of ascorbic acid for peptide and hormone synthesis. Thus, while man and certain mammals have a dietary requirement for this vitamin, neurons and endocrine cells from all species require an active cellular transport and vesicular regeneration systems for ascorbic acid to keep pace with the demand for chemical messenger synthesis. The existence in other tissues of subcellular elements similar to those found in the chromaffin cell suggest that other ascorbic-acid-dependent monooxygenases are yet to be found.

Secretion of Newly Acquired Ascorbic Acid by Adrenomedullary Chromaffin Cells

During our studies on the entry of ascorbic acid into the chromaffin vesicle compartment, we examined the cosecretion of newly acquired ascorbic acid with endogenous catecholamines from adrenomedullary chromaffin cells in primary culture. Secretion of catecholamines occurs by exocytosis, with the all-or-none release of the entire vesicle content into the extracellular space. Cells were labeled by incubating at 37°C for 30 min with L-[1-^{14}C]ascorbic acid followed by an extensive wash. The cells were then stimulated with various secretagogues for 10 min at room temperature. Secretion from bovine adrenomedullary cells is predominately nicotine-receptor mediated,[50] and nicotine (10 μM) induced the release of newly acquired ascorbic acid concomitantly with endogenous catecholamines in the presence but not in the absence of external Ca^{2+}.[51] Similar results have been reported by Levine *et al.*[52] using other secretagogues. Like the catecholamines, the secretion of newly acquired ascorbic acid exhibits a nicotine- and Ca^{2+}-concentration dependency and responds only to nicotinic and not to muscarinic receptor agonists or antagonists.[20] Although the catecholamines and newly acquired ascorbic acid cosecretion suggested that release originated from the chromaffin vesicle, on closer examination it was found that the fraction of labeled

TABLE 2. Subcellular Distribution of Peptidylglycine α-Amidating Monooxygenase (PAM) in the Bovine Adrenal Medulla

	PAM Activity		Catecholamines	
Fractions	Total[a] (nmol/h)	Specific Activity (pmol/h/mg protein)	Total (μmol)	Specific Activity (μmol/mg protein)
H	9.0	79	170.6	0.83
P_1	8.0	269	24.2	0.32
S_2	7.6	148	28.0	0.33
S_3	11.0	215	25.9	0.37
P_2	26.9	965	88.9	1.78
CV	19.6	1,860	—	—
CVM	4.8	610	3.6	0.19
CVS	28.8	6,787	86.5	8.21
P_3	8.9	1,075	5.6	0.30

[a] Based on 2.7 g medulla (wet weight).

Bovine adrenal medullae were homogenized 1:10 (wt/vol) in 0.3 M sucrose containing 1 mM EDTA and 10 mM Hepes, pH 7.2. Subcellular fractions of the homogenate were prepared by differential and discontinuous sucrose gradient centrifugations as previously described.[65] Membranes of the chromaffin vesicle fraction were prepared by osmotic lysis of the particles 1:20 (wt/vol) in 50 mM Hepes, pH 7.2; after centrifugation at 30,000 for 30 min the membranes were resuspended 1:10 (wt/vol) in the Hepes buffer. Aliquots of each fraction were taken for the determination of catecholamines,[22] proteins,[18] and peptidylglycine α-amidating monooxygenase activity.[44] Abbreviations: H, homogenate; P_1, nuclear fraction; S_2, post large granular supernatant; S_3, postmicrosomal supernatant; P_2, large granular fraction; CV, chromaffin vesicle; CVM, chromaffin vesicle membrane; CVS, chromaffin vesicle soluble fraction; P_3, microsomal fraction.

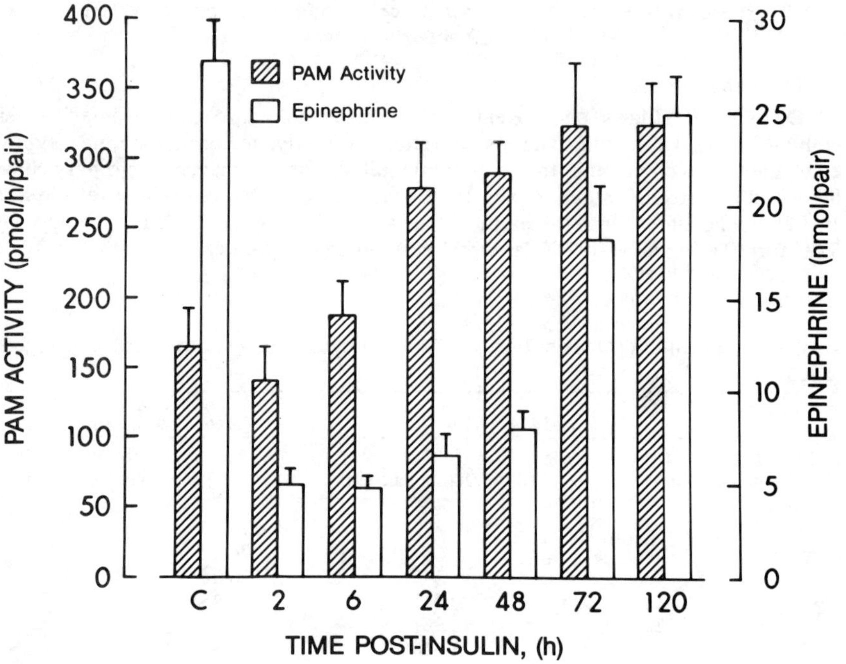

FIGURE 7. Peptidylglycine α-amidating monooxygenase activity in the rat adrenal medulla following 2 h of insulin-induced hypoglycemia. After 2 h of insulin (10 U/kg, i.p.) hypoglycemia, rats were recovered by administration of sucrose and food ad libitum and sacrificed at different times thereafter. Adrenal glands were rapidly removed from the animals following decapitation and the medulla freed of cortex under a dissecting microscope on a cold plate. Adrenal medullae from each animal were homogenized in 75 μl of 0.3 M sucrose containing 10 mM Hepes and 1 mM EDTA, pH 7.2, and centrifuged at 800 ×g for 10 min. The pellet was rehomogenized in 50 μl buffered sucrose and centrifuged again at 800 ×g for 10 min. The combined supernatants were centrifuged at 26,000 ×g for 20 min to obtain a crude chromaffin vesicle (P_2) fraction and supernatant (S_2). Aliquots of each fraction were taken for determination of catecholamine,[22] proteins,[18] and peptidylglycine α-amidating monooxygenase (PAM) activity.[44] Results are presented as the epinephrine content and PAM activity of the P_2 fraction from six animals per group (mean ± SEM).

ascorbic acid released was always less than the fractional secretion of catecholamines. The possibility that newly acquired ascorbic acid was released from compartments other than the chromaffin vesicle had to be considered. Nonvesicular release of newly acquired ascorbic acid was verified by comparing the changes in the rates of secretion of catecholamines and L-[1-^{14}C]ascorbic acid in the presence of nicotine, high K^+, and veratridine (FIG. 8). With these three secretagogues, the catecholamine secretion rate peaked within 2 min of stimulation. The secretion of newly acquired ascorbic acid, however, peaked considerably later,[20] arguing that catecholamines and newly acquired ascorbic acid are released from different compartments. In addition, the release of newly acquired ascorbic acid and catecholamines exhibited differential sensitivities to various Ca^{2+} channel blockers and other secretagogues.[20] More compelling evidence for this proposal was obtained by studying the subcellular distribution of newly acquired and endogenous ascorbic acid and catecholamines.[18,20,52] After a brief pulse with L-[1-^{14}C]ascorbic acid, most of the label was in the 26,000 × g supernatant, with only minimal amounts in chromaffin vesicles (FIG. 3).

In further studies of the differential release of newly acquired ascorbic acid and endogenous catecholamines, we emphasized the following: (1) the location of newly acquired ascorbic acid within the chromaffin cell, and (2) its mechanism of secretion. Cells were loaded with α-[methyl-^{3}H]aminoisobutyric acid (AIB), a nonmetabolizable amino acid (cytoplasmic marker), along with L-[1-^{14}C]ascorbic acid. Catecholamine secretion from cultured chromaffin cells was modified by the osmolality of the incubation medium.[53] Under hyperosmotic conditions, the release of catecholamines, newly acquired ascorbic acid, and AIB by 1,1-dimethyl-4-phenylpiperazinium (DMPP), a nicotinic agonist, was progressively inhibited. The degree of inhibition, however, was much greater for newly acquired ascorbic acid and AIB. Also, release of newly acquired ascorbic acid and AIB was maximal under hypoosmotic conditions, whereas catecholamine release was greatest in an isotonic medium.[54]

Chromaffin cells permeabilized by the addition of detergents such as digitonin and saponin[55–57] or by exposure to an intense electric field[58] are a useful model for the study of the postreceptor molecular steps leading to exocytosis. Release of catechol-

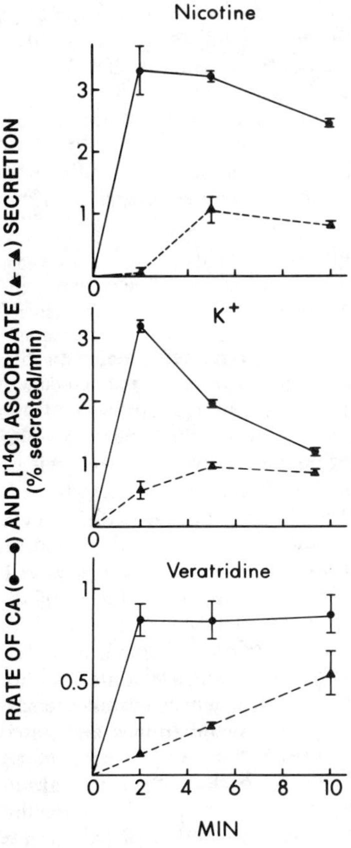

FIGURE 8. Changes in the secretion rate of catecholamines and newly acquired [^{14}C]ascorbate in response to various secretagogues. Chromaffin cells (plated in 16-mm multiwell plates at a density of 1 × 10^{6} cells/well) were labeled with L-[1-^{14}C]ascorbic acid for 30 min at 37°C; the cells were then washed and stimulated with nicotine (10 μM), K^+ (56 mM), or veratridine (100 μM) for different periods of time, and rates were calculated by dividing the amount secreted during each successive period and the computed time. Basal release rate at each time point has been subtracted. Results are expressed as the mean ± standard deviation (five wells per treatment). Initial cell content was 22,401 ± 2071 dpm of [^{14}C]ascorbic acid and 78.2 ± 4.3 nmol of endogenous CA (mean ± standard deviation). (From Daniels *et al.*[20])

amines in these cells occurs only in the presence of extracellular MgATP and Ca^{2+}. In digitonin-permeabilized cells, release of newly acquired ascorbic acid and AIB occurs in both the presence and absence of Ca^{2+} and is detergent-concentration- and time-dependent (FIG. 9). Thus, under several conditions (*i.e.*, changes in the osmolality of the external medium and in cells permeabilized by digitonin), the release profiles for newly acquired ascorbic acid and catecholamines are dissimilar. On the other hand, release of newly acquired ascorbic acid and AIB are nearly identical under these same conditions, suggesting corelease from the same or very similar compartment(s).

Although studies of permeabilized cells indicate that newly acquired ascorbic acid is released from a "free" cytosolic pool, it was unclear whether the cisternae of the endoplasmic reticulum formed part of this pool. AIB is considered to be a cytosolic marker based on its distribution into the postmicrosomal supernatant fraction.[59,60] Therefore, because the techniques employed for subcellular fractionation cannot separate fragile subcellular organelles that are disrupted during processing, both newly acquired ascorbic acid and AIB could be localized to the cisternae of the endoplasmic reticulum. A soluble isoenzyme of acetylcholinesterase (AchE) is localized exclusively to the cisternae of the endoplasmic reticulum, and can be shown to be released with catecholamines following stimulation of the adrenal medulla.[61,62] Furthermore, Mizobe and Livett[63] and Mizobe *et al.*[64] have demonstrated nicotinic-receptor-mediated exocytotic release of AchE from chromaffin cells in culture that is Ca^{2+}-dependent and proportional to, but independent of, catecholamine release, indicating that AchE is released from an extravesicular compartment. To explore the possibility that AIB and newly acquired ascorbic acid are sequestered in the endoplasmic reticulum, release of soluble AchE, AIB, and catecholamines from chromaffin cells was studied under various osmotic conditions. The release profile for AchE closely resembles that for catecholamines, where maximal release was obtained under isosmotic conditions but was different from AIB release.[54] AchE release in saponin-permeabilized cells was also very similar to Ca^{2+}-dependent catecholamine release (Knoth, Viveros and Diliberto, unpublished observations). These data support the proposal that newly acquired ascorbic acid and AIB are sequestered within a cytosolic compartment and subsequently coreleased following stimulation of the cells. The precise location of this compartment within the cytosol, however, is unknown.

We also examined the effects of metabolic inhibitors and ionic dependence on AIB and ascorbic acid release to test the energy dependency and possible mechanism of the secretory process. We found that in general, newly acquired ascorbic acid and AIB release is much more susceptible to metabolic inhibitors than catecholamine release.[54] Furthermore, both Cl^- and Na^+ substitution markedly attenuated release of newly acquired ascorbic acid and AIB, whereas catecholamine release was only partially diminished (FIG. 10). Ascorbic acid and AIB release were drastically reduced in the presence of sucrose, whereas catecholamine release was minimally affected. Interestingly, the anion channel blocker, 4,4-diisothiocyano-2,2′-disulfonic acid stilbene, had no effect on catecholamine release, but produced partial inhibition of the stimulated release of newly acquired ascorbic acid (approximately 40%).[54]

The above as well as other evidence suggest that the release of newly acquired ascorbic acid and AIB involves a transporter. Experiments with ouabain, an inhibitor of the $(Na^+ + K^+)$ATPase, showed that AIB release was particularly sensitive to this inhibitor (for both basal and stimulated release), as was the release of newly acquired ascorbic acid.[54] AIB is presumably actively taken up into the cell via a Na^+-dependent amino acid transporter, thus the sensitivity of AIB release to changing Na^+ concentrations (and/or gradient) implies that release may occur via a reversal of this transporter. Because release of newly acquired ascorbic acid and AIB release are

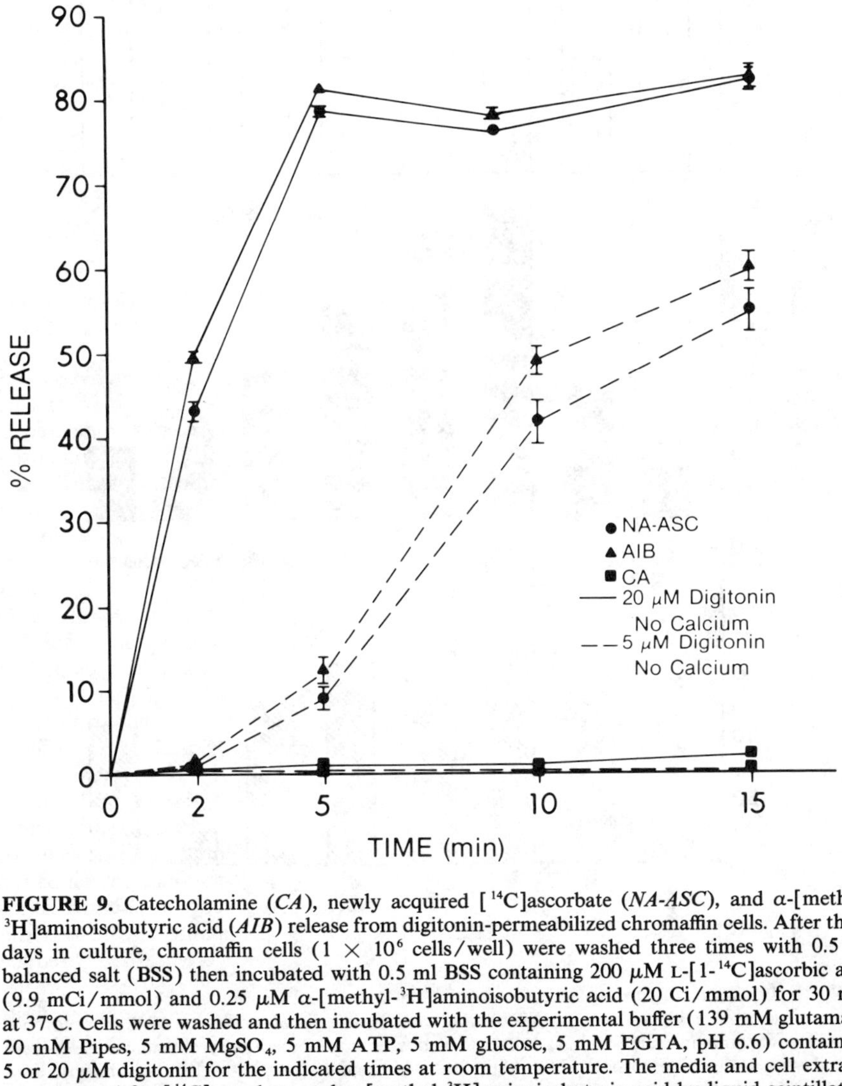

FIGURE 9. Catecholamine (*CA*), newly acquired [^{14}C]ascorbate (*NA-ASC*), and α-[methyl-^{3}H]aminoisobutyric acid (*AIB*) release from digitonin-permeabilized chromaffin cells. After three days in culture, chromaffin cells (1 × 10^6 cells/well) were washed three times with 0.5 ml balanced salt (BSS) then incubated with 0.5 ml BSS containing 200 μM L-[1-^{14}C]ascorbic acid (9.9 mCi/mmol) and 0.25 μM α-[methyl-^{3}H]aminoisobutyric acid (20 Ci/mmol) for 30 min at 37°C. Cells were washed and then incubated with the experimental buffer (139 mM glutamate, 20 mM Pipes, 5 mM $MgSO_4$, 5 mM ATP, 5 mM glucose, 5 mM EGTA, pH 6.6) containing 5 or 20 μM digitonin for the indicated times at room temperature. The media and cell extracts were assayed for [^{14}C]ascorbate and α-[methyl-^{3}H]aminoisobutyric acid by liquid scintillation counting and endogenous catecholamine content by the fluorometric trihydroxyindole method.[68] Release is expressed as percentage of total cell content minus basal release (n = 3; mean ± SEM).

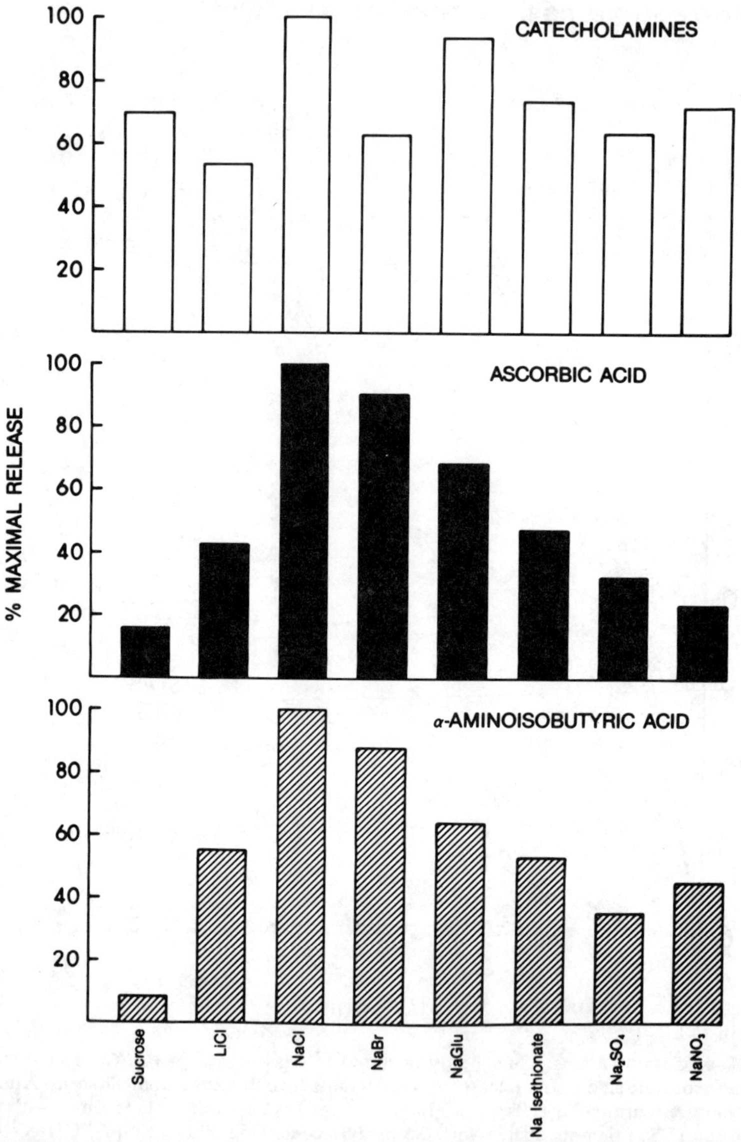

FIGURE 10. Effect of ion replacement on the release of endogenous catecholamines, newly acquired ascorbic acid, and α-aminoisobutyric acid from primary cultures of bovine adrenomedullary chromaffin cells. Primary cultures of bovine adrenomedullary chromaffin cells were labeled with L-[1-^{14}C]ascorbate (NA-ASC; 200 μM) and α-[methyl-^{3}H]aminoisobutyric acid (AIB; 0.25 μM) for 30 min at 37°C, washed and then stimulated with 100 μM veratridine for 20 min (Cl^- replacement) or 10 μM DMPP for 15 min (Na^+ replacement). The various replacements were made by complete substitution of Cl^- or Na^+ and normalized according to 100% release in NaCl medium adjusted to 270 mOsM (Cl^- substitution) or to 320 mOsM (Na^+ substitution). Actual values for percentage release in 270 mOsM NaCl medium were: catecholamines (CA), 28.24 ± 2.6; NA-ASC, 20.82 ± 3.4; and AIB, 14.50 ± 2.0. Actual values for percentage release in 320 mOsM NaCl medium were: CA, 30.79 ± 1.90; NA-ASC, 5.32 ± 0.43; and AIB, 3.27 ± 0.67. Percentage release is equal to percentage release of total cell content stimulated minus unstimulated percentage release; mean ± SEM, n = 3. Cells were labeled, washed, and assayed as described in FIGURE 9.

similar, it is tempting to speculate that newly acquired ascorbic acid was released by reversal of its uptake transporter. For proof of this hypothesis, newly acquired ascorbic acid release should be: (1) highly temperature sensitive (as has been shown for its uptake system), (2) enhanced by the presence of external ascorbate, and (3) blocked by the transport inhibitor phloridzin. Release of newly acquired ascorbic acid is markedly enhanced when the temperature is raised from 25°C to 37°C (Knoth, Viveros and Diliberto, unpublished observations). Furthermore, as shown in FIGURE 11, at this elevated temperature there is a concentration-dependent increase in newly acquired ascorbic acid release with respect to external ascorbic acid. Moreover, this external ascorbic-acid-induced release was found to be inhibited by phloridzin (Knoth, Viveros and Diliberto, unpublished observations). These data provide support for the interesting possibility that release of ascorbic acid is accomplished by reversal of its transporter. The nature of the coupling of this release process with activation of adrenomedullary secretion and its modulation by second messengers is an exciting area for future study.

Depletion of Ascorbic Acid from the Adrenal Medulla in Vivo *in the Rat in Response to Hypoglycemic Shock*

Insulin-induced hypoglycemia in the rat causes a rapid increase in splanchnic nerve discharge and activation of the adrenal medulla. This specific neurogenic stimulation of epinephrine-containing chromaffin cells has long been used as a model to study the mechanism of adrenomedullary secretion *in vivo.* Adrenomedullary secretion of catecholamines, in response to insulin-induced (10 U/kg, i.p.) hypoglycemia, is accompanied by a decrease in the content of ascorbic acid (50% of control). By 24 h, however, the medullary ascorbic acid content is completely restored while the catecholamine content remains depressed (Daniels, Menniti, Viveros, and Diliberto, unpublished observations). Because ascorbic acid is released from multiple compartments within cultured chromaffin cells,[20] the subcellular distribution of ascorbic acid during recovery from hypoglycemic shock in comparison with that of catecholamines is of considerable interest. As shown in FIGURE 12, adrenomedullary ascorbic acid content is rapidly replenished, with a 50% recovery as early as 6 h and a full recovery by 12 h. Recovery of ascorbic acid content in the soluble (S_2) and vesicular (P_2) fractions is similar. Most (95%) of the ascorbic acid is contained in an extravesicular compartment, whereas the catecholamines are found wholly in the vesicular compartment.

Ascorbic acid secretion from cultured chromaffin cells is mediated by nicotinic receptor activation.[20,51,52] Therefore, it was important to verify that this hypoglycemia-induced secretion of ascorbic acid was mediated by the splanchnic nerves and nicotinic receptors. Whereas administration of the ganglionic blocking agent, chlorisondamine, either alone or in combination with atropine, causes a nearly complete or complete, respectively, blockade of catecholamine secretion, neither treatment affects the insulin-induced depletion of ascorbic acid (Daniels, Menniti, Viveros and Diliberto, unpublished observations). To further test the hypothesis that ascorbic acid depletion from the adrenal medulla is nonneurogenic, ascorbic acid depletion following hypoglycemia was studied in unilaterally denervated adrenals. FIGURE 13 shows that hypoglycemia does not induce secretion of catecholamines from the denervated gland, whereas depletion of ascorbic acid is unchanged, confirming the presence of a nonneurogenic component for adrenomedullary depletion of ascorbic acid.

To further study the mechanism for the depletion of ascorbic acid by hypoglycemia, insulin was administered to hypophysectomized and sham-operated rats. After hypophysectomy, insulin fails to deplete ascorbic acid from the adrenal medulla. Furthermore, porcine ACTH at high doses, but not synthetic ACTH (1-23), slightly depletes medullary ascorbic acid, an effect that is potentiated by hypophysectomy, suggesting that some component of the porcine ACTH preparation may be responsible for depletion of ascorbic acid (Daniels, Menniti, Viveros, and Diliberto, unpublished observations).

In summary, hypoglycemia, in addition to reflexly inducing adrenomedullary secretion of catecholamines, also results in ascorbic acid depletion from the rat adrenal medulla. Whereas catecholamine secretion is neurogenic, the depletion of ascorbic acid appears to be under hormonal control *in vivo* in the rat. This is in striking contrast to cultured adrenomedullary chromaffin cells, where secretion is mediated through the nicotinic receptor. It should be emphasized that a small quantity of ascorbic acid released from the chromaffin vesicles would not have been detected in our later experiments, since most intracellular ascorbic acid in the rat adrenal medulla is nonvesicular (FIG. 12). To resolve this issue, the subcellular distribution of ascorbic acid after pharmacological blockade or splanchnic denervation is necessary. Nonetheless, results from both models (the *in vivo* rat and the bovine chromaffin cell) suggest that ascorbic acid is released from multiple compartments. The difference in stimulus secretion coupling may be a result of species differences or changes in response to culture conditions. Each of these possibilities is currently under investigation.

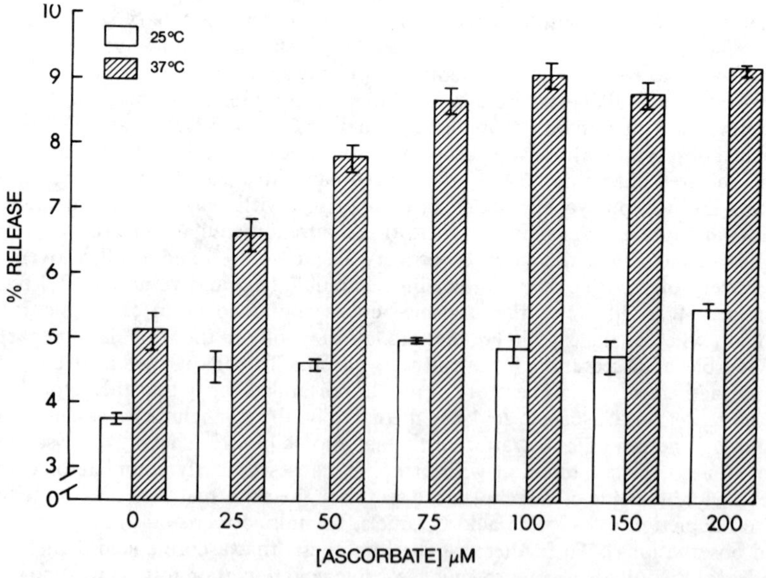

FIGURE 11. Effect of external ascorbate on newly acquired [^{14}C]ascorbate (NA-ASC) release at 37°C and 25°C. Chromaffin cells were labeled with [^{14}C]ascorbate (100 μM) for 30 min at 37°C, washed as described in legend to FIGURE 8, and basal release measured 20 min following the addition of various concentrations of ascorbic acid at 37°C and 25°C. The media and cell extracts were assayed for [^{14}C]ascorbate and endogenous catecholamine content as described in FIGURE 9. Release is expressed as the percentage of total cell content (n = 3; mean ± SEM).

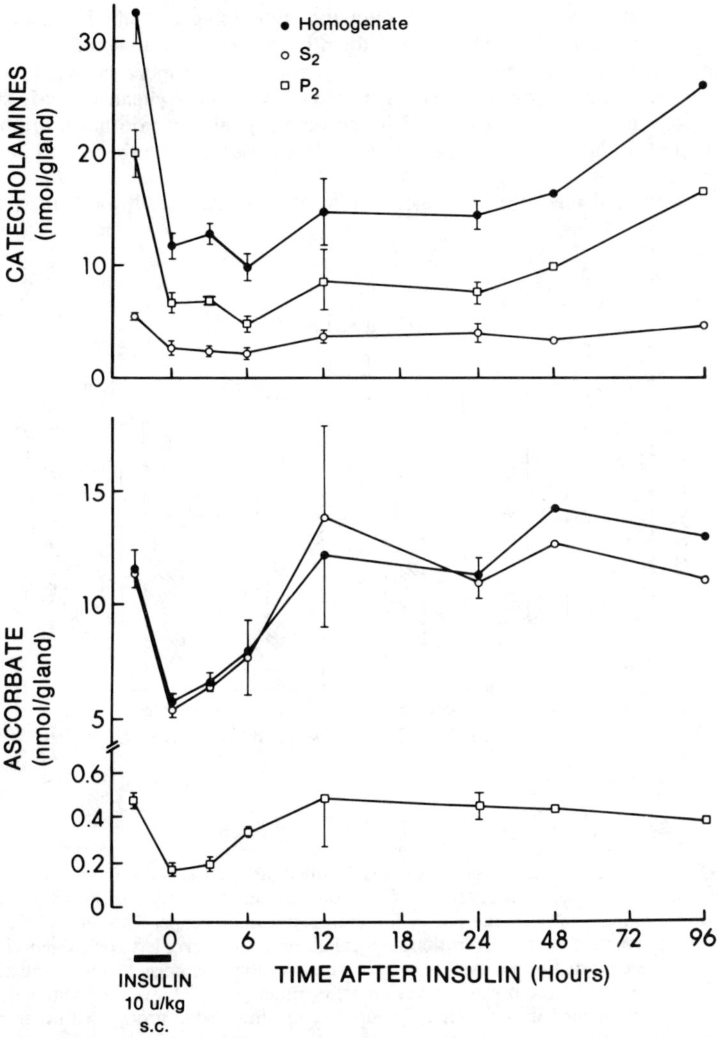

FIGURE 12. Depletion and replenishment of catecholamines and ascorbic acid in different subcellular fractions of the adrenal medulla after hypoglycemic shock in the rat. Administration of insulin to rats, removal and dissection of the adrenal gland, and subcellular fractionation of the adrenal medullae were performed as described in legend to FIGURE 7. The results are expressed as nmol/gland to minimize variations when pooling data from different experiments. The data is shown as the mean ± SD of at least six animals per time point from two different experiments.

Concluding Remarks

Ascorbic acid participates in neurotransmitter and hormone synthesis at several cellular loci. Ascorbic acid is not only a cofactor for monooxygenases involved in messenger synthesis, but is also a component of a system to reduce the intravesicular oxidation product of ascorbic acid following monooxygenation. Thus, cytosolic ascorbic acid, maintained in the reduced form by mitochondrial semidehydroascorbate reductase, donates electrons to the vesicle-membrane transport system to reduce the intravesicular semidehydroascorbate generated by monooxygenation. Moreover, the presence of various components of this regenerating system in other secretory tissues suggests that other monooxygenases involved in messenger synthesis will be identified in the near future.

In the adrenal medulla, ascorbic acid is released from multiple compartments by

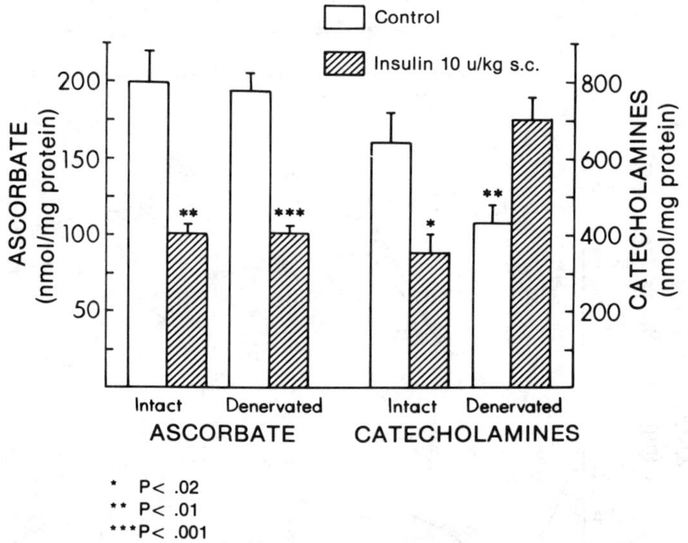

FIGURE 13. Effect of denervation on the insulin-mediated release of adrenomedullary ascorbic acid and catecholamines. Denervation of the rat left adrenal was performed under sodium pentobarbital anesthesia with the aid of a stereoscopic microscope. The left splanchnic nerve was localized at its emergence from the diaphragm, and the nerve, left coeliac ganglia, and the branches going toward the adrenal were removed, avoiding damage to the circulation of the gland. The contralateral gland was used as control (intact gland). Eleven days after denervation, the animals were divided into a control group and insulin-treated group. All other procedures are as indicated in legend to FIGURE 7, except that medullae were homogenized in 50 mM perchloric acid containing 0.01% EDTA. Results are expressed as the mean ± SD of at least ten denervated animals.

two apparently different mechanisms, implying a new order of complexity for adrenomedullary secretion and function. Equally significant inferences come from the use of the adrenal medulla as a model to study presynaptic mechanisms. In-depth studies of adrenal ascorbic acid secretion suggest a novel mechanism for the cosecretion of bioactive molecules from multiple compartments of a single cell. Such a complex arrangement of secreting compartments for the export of messengers may be a common feature of other neurons and endocrine cells, suggesting a much greater degree of flexibility and plasticity of these systems.

REFERENCES

1. LEVIN, E. Y., B. LEVENBERG & S. KAUFMAN. 1960. J. Biol. Chem. **235:** 2080-2086.
2. FRIEDMAN, S. & S. KAUFMAN. 1965. J. Biol. Chem. **240:** 4763-4773.
3. MAINS, R. E., A. C. MYERS & B. A. EIPPER. 1985. Endocrinology **116:** 2505-2515.
4. HADJICONSTANTINOU, M. & N. H. NEFF. 1983. Neuropharmacology **22:** 939-943.
5. HEIKKILA, R. E., F. S. CABBAT & L. MANZINO. 1982. Neurochem. **38:** 1000-1006.
6. GARDINER, T. W., M. ARMSTRONG-JAMES, A. W. CAAN, R. M. WIGHTMAN & G. V. REBEC. 1985. Brain. Res. **344:** 181-185.
7. KAUFMAN, S., W. F. BRIDGERS, F. EISENBERG & S. FRIEDMAN. 1962. Biochem. Biophys. Res. Commun. **9:** 497-502.
8. KAUFMAN, S. & S. FRIEDMAN. 1965. Pharmacol. Reviews **17:** 71-100.
9. DILIBERTO, E. J., JR. & S. KAUFMAN. 1978. Fed. Proc. **37:** 1421.
10. DILIBERTO, E. J., JR. & P. L. ALLEN. 1980. Mol. Pharmacol. **17:** 421-426.
11. DILIBERTO, E. J., JR. & P. L. ALLEN. 1981. J. Biol. Chem. **256:** 3385-3393.
12. SKOTLAND, T. & T. LJONES. 1980. Biochim. Biophys. Acta **630:** 30-35.
13. HODGES, R. E., J. HOOD, J. E. CANHAM, H. E. SAUBERLICH & E. M. BAKER. 1971. Am. J. Clin. Nutr. **24:** 432-443.
14. HORNIG, D. 1975. Ann. N.Y. Acad. Sci. **258:** 103-118.
15. CHATTERJEE, I. B. 1973. Science **182:** 1271-1272.
16. CHATTERJEE, I. B. 1978. World Rev. Nutr. Diet. **30:** 69-87.
17. HORNIG, D. 1975. World Rev. Nutr. Diet. **23:** 225-258.
18. DILIBERTO, E. J., JR., G. D. HECKMAN & A. J. DANIELS. 1983. J. Biol. Chem. **258:** 12886-12894.
19. TIRRELL, J. G. & E. W. WESTHEAD. 1979. Neuroscience **4:** 181-186.
20. DANIELS, A. J., G. DEAN, O. H. VIVEROS & E. J. DILIBERTO, JR. 1983. Mol. Pharmacol. **23:** 437-444.
21. INGEBRETSEN, O. C., O. TERLAND & T. FLATMARK. 1980. Biochim. Biophys. Acta **628:** 182-189.
22. MENNITI, F. S., J. KNOTH & E. J. DILIBERTO, JR. 1986. J. Biol. Chem. **261:** 16901-16908.
23. LEVINE, M., K. MORITA & H. POLLARD. 1985. J. Biol. Chem. **260:** 12942-12947.
24. NJUS, D., J. KNOTH, C. COOK & P. M. KELLEY. 1983. J. Biol. Chem. **258:** 27-30.
25. HARNADEK, G. J., R. E. CALLAHAN, A. R. BARONE & D. NJUS. 1985. Biochemistry **24:** 384-389.
26. HARNADEK, G. J., E. A. RIES & D. NJUS. 1985. Biochemistry **24:** 2640-2644.
27. SRIVASTAVA, M., L. T. DUONG & P. J. FLEMING. 1984. J. Biol. Chem. **259:** 8072-8075.
28. FLATMARK, T. & M. GRØNBERG. 1981. Biochem. Biophys. Res. Commun. **99:** 292-301.
29. DUONG, L. T. & P. J. FLEMING. 1984. Arch. Biochem. Biophys. **228:** 332-341.
30. KELLY, P. M. & D. NJUS. 1986. J. Biol. Chem. **261:** 6429-6432.
31. BEERS, M. F., R. G. JOHNSON & A. SCARPA. 1986. J. Biol. Chem. **261:** 2529-2535.
32. HOLZ, R. W. 1978. Proc. Natl. Acad. Sci. U.S.A. **75:** 5190-5194.
33. JOHNSON, R. G. & A. SCARPA. 1979. J. Biol. Chem. **254:** 3750-3760.
34. IYANAGI, T., I. YAMAZAKI & K. F. ANAN. 1985. Biochim. Biophys. Acta **806:** 255-261.
35. DILIBERTO, E. J., JR., G. DEAN, C. CARTER & P. L. ALLEN. 1982. J. Neurochem. **39:** 563-568.
36. RUSSELL, J. T. & R. W. HOLZ. 1981. J. Biol. Chem. **256:** 5950-5953.
37. RUSSELL, J. T., M. LEVINE & D. NJUS. 1985. J. Biol. Chem. **260:** 226-231.
38. DUONG, L. T., P. J. FLEMING & J. T. RUSSELL. 1984. J. Biol. Chem. **259:** 4885-4889.
39. TROTTA, K., R. M. PRUSS, R. A. DELELLIS, H. J. WOLFE & A. S. TISCHLER. 1986. Lab. Invest. **54:** 64A.
40. BRADBURY, A. F., M. D. A. FINNIE & D. G. SMYTH. 1982. Nature **298:** 686-688.
41. EIPPER, B. A., R. E. MAINS & C. C. GLEMBOTSKI. 1983. Proc. Natl. Acad. Sci. U.S.A. **80:** 5144-5148.
42. BATEMAN, R. C., JR., W. W. YOUNGBLOOD, W. H. BUSBY, JR. & J. S. KIZER. 1985. J. Biol. Chem. **260:** 9088-9091.
43. KIZER, J. S., R. C. BATEMAN, JR., C. R. MILLER, J. HUMM, W. H. BUSBY, JR. & W. W. YOUNGBLOOD. 1986. Endocrinology **118:** 2262-2267.

44. KIZER, J. S., W. H. BUSBY, JR., C. COTTLE & W. W. YOUNGBLOOD. 1984. Proc. Natl. Acad. Sci. U.S.A. **81:** 3228-3232.
45. MATSUO, H., A. MIYATA & K. MIZUNO. 1983. Nature **305:** 721-723.
46. MIYATA, A., K. MIZUNO, N. MINAMINO & H. MATSUO. 1984. Biochem. Biophys. Res. Commun. **120:** 1030-1036.
47. WEBER, E., F. S. ESCH, P. BÖHLEN, S. PATERSON, A. D. CORBETT, A. T. MCKNIGHT, H. W. KOSTERLITZ, J. D. BARCHAS & C. J. EVANS. 1983. Proc. Natl. Acad. Sci. U.S.A. **80:** 7362-7366.
48. SEIZINGER, B. R., D. C. LIEBISCH, C. GRAMSCH, A. HERZ, E. WEBER, C. J. EVANS, F. S. ESCH & P. BÖHLEN. 1985. Nature **313:** 57-59.
49. KANAMATSU, T., C. D. UNSWORTH, E. J. DILIBERTO, JR., O. H. VIVEROS & J.-H. HONG. 1986. Proc. Natl. Acad. Sci. U.S.A. In press.
50. VIVEROS, O. H. 1975. *In* Handbook of Physiology, Sect. 7: Endocrinology, Vol. 6. Adrenal Gland. H. Blaschko, G. Sayers & A. D. Smith, Eds.: 389-426. American Physiological Society. Washington, D.C.
51. DANIELS, A. J., G. DEAN, O. H. VIVEROS & E. J. DILIBERTO, JR. 1982. Science **216:** 737-739.
52. LEVINE, M., A. ASHER, H. POLLARD & O. ZINDER. 1983. J. Biol. Chem. **258:** 13111-13115.
53. HAMPTON, R. Y. & R. W. HOLZ. 1983. J. Cell Biol. **96:** 1082-1088.
54. KNOTH, J., O. H. VIVEROS & E. J. DILIBERTO, JR. 1987. Ann. N. Y. Acad. Sci. **493:** 480-488.
55. DUNN, L. A. & R. W. HOLZ. 1983. J. Biol. Chem. **258:** 4989-4993.
56. WILSON, S. P. & N. KIRSHNER. 1983. J. Biol. Chem. **258:** 4994-5000.
57. BROOKS, J. C. & S. TREML. 1983. J. Neurochem. **40:** 468-473.
58. KNIGHT, D. E. & P. F. BAKER. 1982. Membrane Biol. **68:** 107-140.
59. ORREGO, F., J. JANKELEVICH, L. CERUTI & E. FERRERA. 1974. Nature **251:** 55-56.
60. VARGAS, O. & F. ORREGO. 1976. J. Neurochem. **26:** 31-34.
61. CHUBB, I. W. & A. D. SMITH. 1975. Proc. R. Soc. London, Ser. B. **191:** 245-261.
62. SOMOGYI, P., I. W. CHUBB & A. D. SMITH. 1975. Proc. R. Soc. London, Ser. B. **191:** 271-283.
63. MIZOBE, F. & B. G. LIVETT. 1983. J. Neurosci. **3:** 871-876.
64. MIZOBE, F., M. IWAMOTO & B. G. LIVETT. 1984. J. Neurochem. **42:** 1433-1438.
65. DILIBERTO, E. J., JR., O. H. VIVEROS & J. AXELROD. 1976. Proc. Natl. Acad. Sci. U.S.A. **73:** 4050-4054.
66. WILSON, S. P. & O. H. VIVEROS. 1981. Exp. Cell. Res. **133:** 159-169.
67. CLELAND, W. W. 1963. Biochim. Biophys. Acta **67:** 104-137.
68. ANTON, A. H. & D. F. SAYRE. 1962. J. Pharmacol. Exp. Ther. **138:** 360-375.
69. GLEMBOTSKI, C. C. 1987. The role of ascorbic acid in the biosynthesis of the neuroendocrine peptides α-MSH and TRH. This volume.

DISCUSSION OF THE PAPER

UNIDENTIFIED SPEAKER: Were you able to adjust the ascorbate concentration? You said that this was a calculated concentration. Could you explain what you mean by that?

E. J. DILIBERTO: The experiment you are talking about is in the isolated vesicles. What we have demonstrated is that upon the addition of the external reductant there is a very rapid reduction of intravesicular semidehydroascorbic acid. Thus, we add tyramine, to deplete intravesicular ascorbic acid; as soon as we add a reductant it arrests further depletion of the ascorbic acid. So by delaying the addition of glucoascorbic acid at various times after the addition of tyramine we can end up with chromaffin vesicles in suspension that have varying concentrations of ascorbic acid.

D. J. MORRÉ (*Purdue University, West Lafayette, Ind.*): The ascorbate molecule seems to uniquely provide an opportunity for a reutilizable electron acceptor inside these vesicles. My question is a philosophical one. How would you regard this as a candidate for an essential function of the vitamin?

E. J. DILIBERTO: Well I think that is simple, because in man what is missing is the synthetic pathway to form ascorbic acid, so you have to start with a baseline level. There seem to be two main systems for accumulation of the vitamin and it varies from cell to cell. One is an uptake system and the predominant one in most cells is a sodium-dependent uptake system and a system for the reduction of the free radical which is the semidehydroascorbate reductase.

Now in blood elements there's still another system as well as in liver, which is glutathionine dependent, that regenerates dehydroascorbate acid.

D. J. MORRÉ: I guess we have a slightly different definition of essentiality. What I am looking for is an essential role for the vitamin. In other words, something where this vitamin carries out a biochemical function in the cell where it cannot be replaced by some other molecule.

E. J. DILIBERTO: I think that ascorbic acid can do it so much better than anything else.

D. J. MORRÉ: Ascorbic acid is uniquely designed as something which can get inside the vesicles and then can stay there and can be regenerated. And I don't know of any other molecule that can duplicate this.

E. J. DILIBERTO: I didn't want to throw this in because it gets complicated, but catechols themselves can donate electrons but you still need another reductant on the outside.

M. RICE (*New York University Medical Center, New York, N.Y.*): When you talk about a depletion of ascorbate from the vesicles, was the ascorbate you were measuring within the vesicles carbon-14 labeled?

E. J. DILIBERTO: No. The ascorbic acid in the isolated vesicles is resident in fresh tissue. They are prepared by differential centrifugation to isolate a large granular fraction. The isolated vesicles contain the ascorbic acid that was originally in the gland from the animal. We incubate those vesicles with tyramine, the substrate for dopamine β-hydroxylase that is ascorbate dependent, and we obtain a disappearance of the ascorbic acid in the vesicles unless we add an external reductant.

M. RICE: Are you familiar with the studies of Aschenk *et al.* in brain tissue? They have shown that ascorbate in the extracellular fluid governs the amount of ascorbate taken up by the neuronal elements.

E. J. DILIBERTO: That is cellular transport.

M. RICE: It picks it up at greater than a certain concentration in the extracellular fluid, so this extracellular concentration is important also. I wonder if rather than just being a redox reaction, the glucoascorbate is telling the membrane that a certain extracellular or extravesicular concentration of ascorbate is present; therefore, ascorbate shouldn't be lost to the extracellular medium.

E. J. DILIBERTO: No, this disappearance is due to oxidation.

D. HORNIG (*Hoffmann-La Roche & Company, Basel, Switzerland*): Can you comment on the question of whether isoascorbic acid is a substrate for either the carrier system or for the semidehydroascorbate reductase system?

E. J. DILIBERTO: As far as I know it is for the reductase. It's very important that we are able to differentiate the ascorbate analogue with ascorbic acid, otherwise we don't have concrete evidence that the reductant is not moving through.

The Role of Ascorbic Acid in the Biosynthesis of the Neuroendocrine Peptides α-MSH and TRH[a]

CHRISTOPHER C. GLEMBOTSKI

Molecular Biology Institute
and
Department of Biology
San Diego State University
San Diego, California 92182

The role of ascorbic acid as a cofactor in many biological processes such as collagen biosynthesis is well established.[1,2] However, there are studies that have indicated the importance of this vitamin in other cellular functions such as catecholamine biosynthesis and the regulation of cholinergic and monoamine receptor levels.[3] Our recent findings have indicated that in addition to these functions, ascorbic acid is also an essential cofactor in the biosynthesis of many neuropeptides and endocrine (neuroendocrine) peptides.[4] Thus, it is apparent that ascorbate plays a central role in the proper functioning of both the nervous and endocrine systems.

In this paper the evidence supporting the participation of ascorbic acid in neuroendocrine peptide biosynthesis is presented. Two model systems have been used in our studies: the biosynthesis of α-melanocyte-stimulating hormone (α-MSH) by the intermediate pituitary, and the biosynthesis of thyrotropin-releasing hormone (TRH) by the hypothalamus. By using these model systems, data have been gathered that point to the key role played by ascorbic acid in the COOH-terminal α-amidation of neuroendocrine peptides.

MECHANISMS OF THE BIOSYNTHESIS OF NEUROENDOCRINE PEPTIDES

Neuroendocrine peptides are synthesized initially on the rough endoplasmic reticulum (RER) as parts of larger precursor polypeptides. Amino-terminal signal sequences direct the cotranslational translocation of the nascent polypeptide into the lumen of the RER.[5,6] As the signal sequence enters the RER lumen, it is cleaved before the remainder of the precursor has been made and inserted. In addition to the proteolytic cleavage of the signal sequence, several other cotranslational modifications

[a] This work was supported by NIH grant no. NS 20396 and by NIMH grant no. MH 41252. The author is an Established Investigator of the American Heart Association (no. 860173); funds were contributed in part by the California Affiliate of the American Heart Association.

frequently occur. These include core glycosylation, phosphorylation, and sulfation.[7,8] Upon the completion of translation, the precursor polypeptide is transported from the RER to the Golgi apparatus. It is within the Golgi apparatus that the glycosyl moieties mature and the initial events of the endoproteolytic removal of the mature secreted peptide from the precursor begin. Various processing events continue as the precursor-derived molecules and processing enzymes are packaged into secretory granules. These processing events include further endoproteolysis (often at pairs of basic amino acids), COOH-terminal exoproteolytic trimming, NH_2-terminal α-*N*- and *O*-acetylation, NH_2-terminal pyroglutamate formation, and COOH-terminal α-amidation.[8-10] Only a few of the many posttranslational modifications are found to occur frequently during the biosynthesis of neuroendocrine peptides. Among the frequent posttranslational modifications is COOH-terminal α-amidation.

The presence of COOH-terminal α-amide moieties on peptides seems in many cases to be required for the optimal interaction of the mature peptide with its receptors. Additionally, it is thought that blocking the COOH terminus of a peptide with an α-amide imparts resistance to carboxypeptidase degradation after secretion, thus enhancing the bioactivity of the peptide by increasing its half-life in the synapse and/or the circulation. Until recently, very little was known of the mechanism of peptidyl α-amidation. When our work on peptidyl α-amidation began, the only information available regarding the mechanism of this posttranslational modification was that peptide sequencing and molecular cloning studies had demonstrated that precursors to α-amidated peptides possess a glycine residue COOH-terminal to the amino acid that will be the α-amidated residue in the mature molecule.[8] Because of the well-characterized nature of the model system, we selected the biosynthesis of α-MSH by the rat intermediate pituitary for studies of the mechanism of peptidyl α-amidation.

ASCORBIC ACID AND THE α-AMIDATION OF α-MSH

α-MSH is one of two COOH terminally α-amidated peptides derived from pro-ACTH/endorphin.[11] It is composed of the NH_2-terminal 13 amino acids of ACTH and possesses a COOH-terminal valineamide. Studies utilizing primary cultures of rat intermediate pituitary demonstrated that as a function of time *in vitro,* the cells lost the ability to form α-MSH, and instead accumulated the COOH terminally glycine-extended form of the peptide.[12,13] There was no alteration of any other pro-ACTH/endorphin posttranslational processing event as a function of time *in vitro.* This rather selective loss of α-amidation ability occurred with a half-time of approximately 24 h. Further studies were undertaken to characterize the enzyme(s) responsible for the α-amidation of α-MSH in hopes of determining why the primary cultures lost α-amidation ability.

The search for the α-amidation enzyme activity utilized the previously characterized synthetic substrate, D-Tyr-Val-Gly.[14,15] By utilizing ^{125}I-D-Tyr-Val-Gly, it was shown that secretory granules isolated from rat intermediate pituitary contained an enzymatic activity capable of forming ^{125}I-D-Tyr-Val-NH_2.[4] Interestingly, this activity was found to be associated with secretory granules from several neuroendocrine tissues including the anterior, intermediate, and posterior lobes of the pituitary as well as hypthalamic synaptic vesicles.[4,16-19] Since these initial studies, an enzyme with comparable activities has been localized to nearly all neuroendocrine tissues.

In characterizing the pituitary α-amidation enzyme, it was shown that various peptides possessing COOH-terminal glycine residues were effective competitive inhibitors of the conversion of ^{125}I-D-Tyr-Val-Gly to ^{125}I-D-Tyr-Val-NH_2. Those peptides ending COOH terminally with an α-amide or an amino acid other than glycine were ineffective as inhibitors. Further studies demonstrated that the partially purified peptidyl α-amidation enzyme from rat anterior pituitary was capable of converting the COOH terminally glycine-extended form of αMSH to the amidated form of the peptide.[20] Of the many metals tested, only copper [Cu(II)] significantly enhanced the activity of the pituitary α-amidation enzyme. It was also shown that among several redox cofactors, ascorbic acid was capable of stimulating the α-amidation enzyme activity. Furthermore, it was shown that the α-amidation reaction occurred only in the presence of molecular oxygen. Thus, it was apparent that like another secretory granule-associated enzyme, dopamine β-hydroxylase (DBH), the α-amidation enzyme required Cu(II), ascorbic acid, and O_2 for optimal activity. The α-amidation enzyme has been named peptidylglycine α-amidating monooxygenase, or PAM.[19]

Since the medium for maintenance of the primary pituitary cells did not contain ascorbic acid, it was hypothesized that the lack of the vitamin was the cause of the loss of cellular α-amidation ability. When cultured intermediate pituitary cells were maintained in the presence of ascorbic acid, they retained the ability to form α-amidated αMSH.[21–23] Pulse-chase labeling experiments were performed in order to verify the precursor-product relationship between glycine-extended and α-amidated αMSH. Intermediate pituitary cells that had been maintained for 10 days without ascorbic acid were incubated with [^{3}H]trytophan also in the absence of the vitamin. The only form of labeled αMSH produced by the cells during the label period was glycine-extended. Following the label period (pulse) the cells were incubated with unlabeled medium containing ascorbic acid (chase). It was shown that as a function of chase time, the labeled glycine-extended αMSH was converted to the COOH terminally α-amidated form of the peptide. The formation of labeled α-amidated αMSH was dependent on the quantity of ascorbic acid present in the chase medium such that an EC_{50} of approximately 25 μM ascorbic acid was observed.[22] Oxidized ascorbic acid was incapable of supporting the α-amidation of αMSH by the cultured cells.

In order to establish further the direct role of ascorbic acid in peptidyl α-amidation, uptake studies were carried out. By using ^{14}C-labeled ascorbic acid, it was shown that corticotropic pituitary tumor cells (AtT-20) possess a high affinity transport system for the vitamin.[24,25] The AtT-20 cells displayed the ability to assimilate ascorbic acid with a K_m of about 35 μM, similar to circulating levels of the vitamin. The cells did not possess a high affinity uptake system for dehydro- or fully oxidized ascorbic acid. The assimilation of ascorbic acid was also shown to be sodium dependent.

Returning to the primary intermediate pituitary cells, it was shown that the ability of ascorbic acid to reestablish α-amidation ability was also dependent on sodium.[26] Similar sodium-dependent ascorbic acid uptake systems have been observed for primary adrenal chromaffin cells as well as pheochromocytoma cells. Thus, it was apparent that the assimilation of ascorbic acid, via a sodium-dependent high-affinity transport system, was required in order for cultured intermediate pituitary cells to continue α-amidation.

ASCORBIC ACID AND THE α-AMIDATION OF TRH

Because PAM is present in the hypothalamus as well as the pituitary, studies were undertaken to determine whether peptides produced in neurons were also α-amidated in an ascorbic-acid-dependent manner. The synthesis of TRH (pyro-Glu-His-Pro-NH_2) by primary cultures of hypothalamic cells was used as a model system. When this study was undertaken, the structure of the precursor of TRH was not known. Thus, it was not clear whether the direct precursor to TRH would possess a COOH-terminal Gly. Utilizing an antiserum that cross-reacts with TRH but not TRH-Gly, it was shown that the inclusion of ascorbic acid in the culture medium resulted in a marked increase in cellular levels of immunoactive TRH compared to those hypothalamic cultures maintained without ascorbic acid.[27] This increase in cellular TRH content displayed a dependence on the concentration of ascorbic acid in the medium with an EC_{50} of approximately 20 μM (FIG. 1). Oxidized ascorbic acid was not

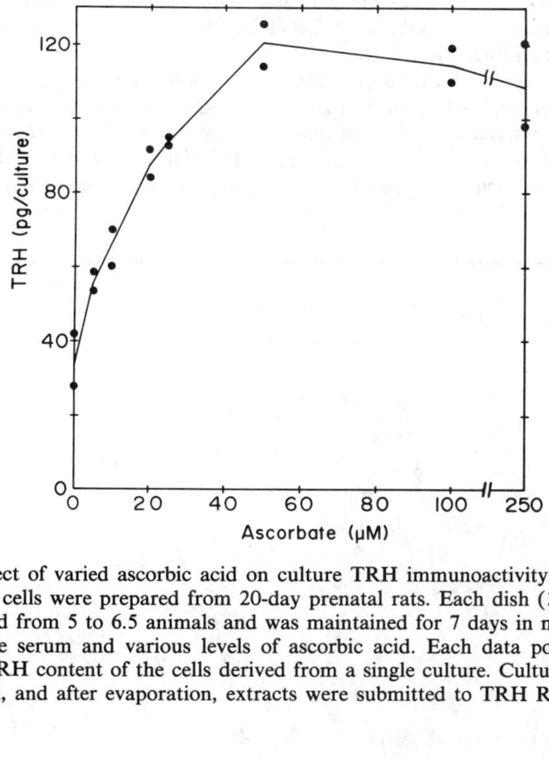

FIGURE 1. Effect of varied ascorbic acid on culture TRH immunoactivity. Primary cultures of hypothalamic cells were prepared from 20-day prenatal rats. Each dish (35 mm) contained the cells prepared from 5 to 6.5 animals and was maintained for 7 days in medium containing 10% fetal bovine serum and various levels of ascorbic acid. Each data point represents the immunoactive TRH content of the cells derived from a single culture. Cultures were extracted in PBS-methanol, and after evaporation, extracts were submitted to TRH RIA. (From Glembotski *et al.*[27]).

capable of producing an increase in cellular TRH immunoactivity. However, D-isoascorbic acid, an isomer of ascorbic acid recently shown to serve as a cofactor for PAM *in vitro,* was capable of increasing the cellular content of immunoactive TRH.

In order to investigate whether a glycine-extended form of TRH was produced as an alternate end product by hypothalamic cells maintained under ascorbic-acid-free conditions, an antiserum was raised that was capable of detecting synthetic TRH-Gly but not TRH. When cultures were maintained in the absence of ascorbic acid, the ratio of TRH to TRH-Gly was eightfold lower than that in cultures maintained in the presence of ascorbic acid (TABLE 1).[27] Moreover, when the TRH-sized material from rat hypothalamic extracts was analyzed by reversed phase high-performance

TABLE 1. Effect of Ascorbic Acid on Levels of TRH and TRH-Gly Immunoactivity in Cultured Hypothalamic Cells

Treatment	TRH (pg/hypo)	TRH-Gly (pg/hypo)	TRH/TRH-Gly
+ Ascorbate	100	106	0.94
− Ascorbate	35	310	0.11

Cells were prepared and maintained with or without ascorbic acid as described in FIGURE 1. Levels of immunoactive TRH and TRH-Gly were determined by RIA using antisera directed against each peptide. (From Glembotski *et al.*[27])

liquid chromatography, it was shown that immunoactive material eluting with a retention time identical to synthetic TRH-Gly was detectable with the TRH-Gly-specific antiserum (FIG. 2).

During the course of this study the cDNA sequence for rat prepro-TRH was reported.[28] It was established that in rats, pro-TRH possesses five copies of the TRH sequence that are flanked NH_2-terminally by paired basic amino acids, a common signal for endoproteolytic maturation of peptide hormones, and COOH-terminally by glycine residues followed by paired basic amino acids. Combined with the molecular

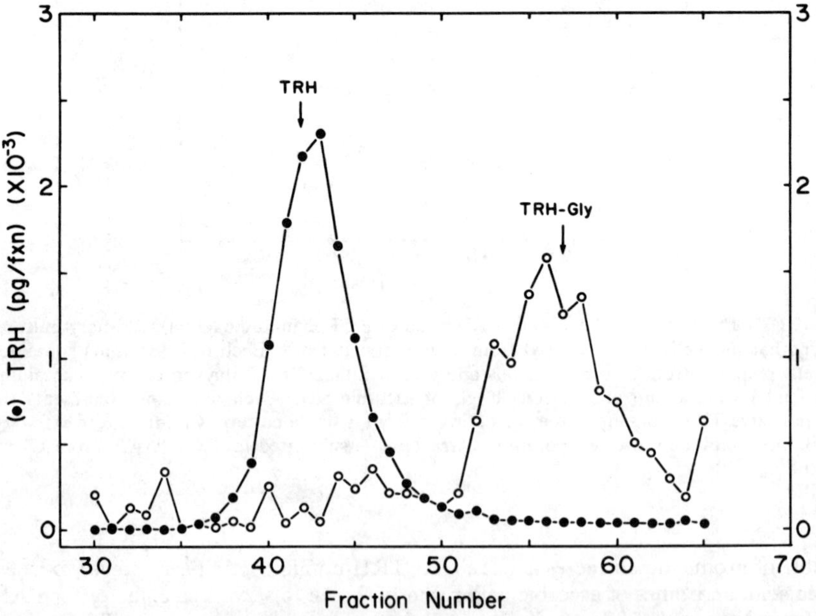

FIGURE 2. Reversed phase high-performance liquid chromatographic characterization of TRH and TRH-Gly. Rat hypothalami were extracted and analyzed first by G-75 Sephadex gel filtration in 10% formic acid containing 0.1 mg/ml BSA. The TRH-related material was vacuum evaporated and chromatographed by RP-HPLC in a previously described TFA/acetonitrile buffer system. The fraction size was 500 μl. The *arrows* mark the elution positions of standard synthetic TRH and TRH-Gly as localized by UV absorbance in a parallel analysis. (From Glembotski *et al.*[27]).

cloning data, our studies indicate that Gln-His-Pro-Gly is excised from pro-TRH by a trypsin-like enzyme, and following COOH-terminal trimming by a carboxypeptidase E-like enzyme, and the NH_2-terminal clyclization of Gln to form pyro-Glu, PAM converts TRH-Gly to TRH in an ascorbic-acid-dependent reaction (FIG. 3).

CONCLUSION

PAM is among the best characterized neuroendocrine peptide biosynthetic enzymes. It has been purified to apparent homogeneity from several sources and displays

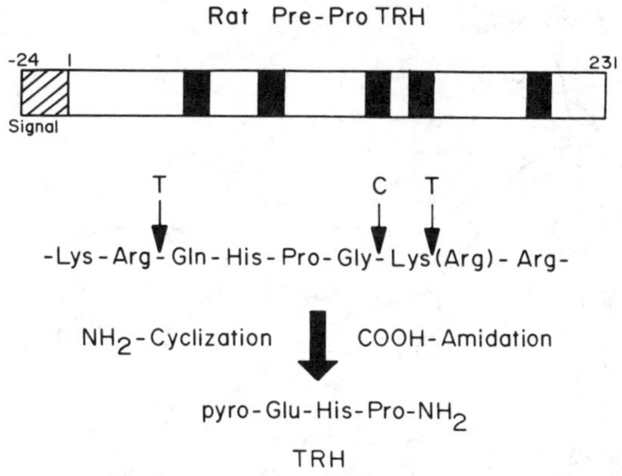

FIGURE 3. Diagram of rat pre-pro TRH and hypothesized processing of the precursor. The complete cDNA sequence indicates that pro-TRH is composed of 231 amino acids.[28] The black boxes mark the approximate locations of TRH sequences within the precursor. The probable posttranslational processing of a region of the TRH precursor is also shown. The TRH sequences in the precursor are all flanked NH_2 terminally with Lys-Arg pairs, and COOH terminally with Gly-Lys-Arg or Gly-Arg-Arg. *T* and *C* mark the proposed sites of cleavage by trypsin-like and carboxypeptidase-E-like enzymes, respectively, in the posttranslational processing of pro-TRH. The mechanism of pyro-Glu formation is not precisely known.

an approximate M_r of 35 to 50 kDa.[29] It has been shown to utilize many different peptide substrates, providing they possess COOH-terminal glycine residues; however, the rate of catalysis and optimal pH range are quite variable depending on other aspects of the peptide structure.[25] Furthermore, it is apparent that PAM from the pituitary and the brain utilizes ascorbic acid as a cofactor such that one mole of the vitamin is consumed for every mole of peptide that becomes α-amidated.[25]

Two secretory-granule-associated enzymes, PAM and dopamine β-hydroxylase (DBH), possess relatively rare but identical cofactor requirements. However, unlike DBH, PAM appears to be localized to many neuroendocrine tissues including the brain, pituitary, thyroid, and submaxillary gland. Interestingly, PAM has also been found in the circulation and the cerebral spinal fluid.[25] The PAM derived from each of these locations possesses similar substrate specificities, cofactor requirements, and

M_r characteristics. Thus, the widespread occurrence of α-amidated peptides correlates with the relatively ubiquitous tissue distribution of PAM.

The discovery of this new ascorbic-acid-requiring peptide processing enzyme adds further to the list of biological reactions in which ascorbic acid participates. From our studies, it is apparent that all neurosecretory cells likely possess a high-affinity, sodium-dependent ascorbic acid transport system that provides reducing equivalents for the eventual α-amidation of secreted peptides. Our present hypothesis is that ascorbic acid is assimilated from interstitial spaces surrounding the neuroendocrine cell, accumulates in the cytosol, and then gains access to the lumen of the secretory granule in order to participate in peptidyl α-amidation (FIG. 4). To date, few studies have addressed directly whether ascorbic acid enters the peptide processing compartment as the Golgi apparatus is maturing to form secretory granules, or whether

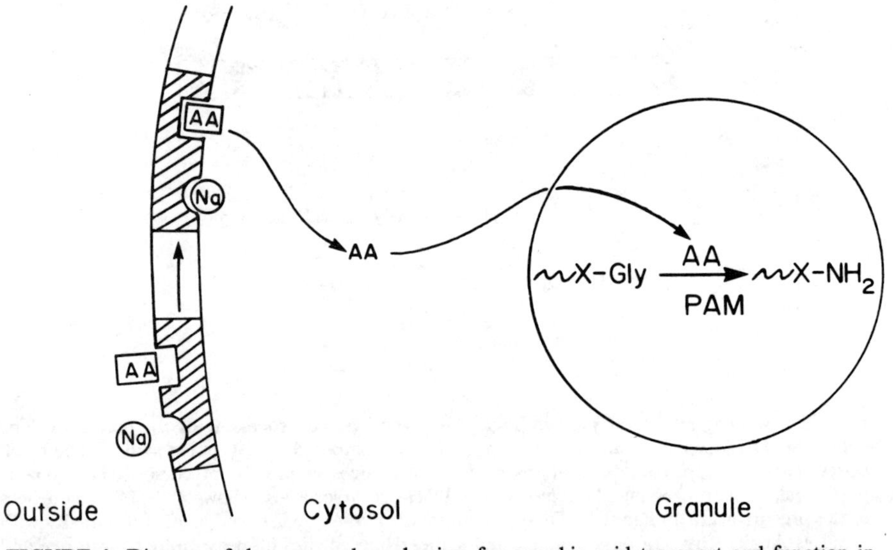

FIGURE 4. Diagram of the proposed mechanism for ascorbic acid transport and function in neuroendocrine peptidyl α-amidation.

the cofactor enters the granules after maturation. Pulse-chase labeling experiments have indicated that ascorbic acid added to cultured cells can fully support the α-amidation of labeled peptides that are contained within mature secretory granules.[22,26] Thus, ascorbic acid can either be assimilated by mature neurosecretory granules, or it can shuttle reducing equivalents into the granule lumen.

In the future, it will be important to investigate animal models of ascorbic acid depletion in light of the recent discovery of PAM. It will be interesting to determine whether the symptoms resulting from ascorbic acid deficiency are in part a result of the inability to form α-amidated peptides. Since so many neuropeptides rely on α-amidation for optimal biological activity (*e.g.,* hypothalamic releasing factors), there undoubtedly are CNS function deficits during ascorbic acid depletion that can be ascribed to the lack of PAM activity. It is also apparent that since various endocrine hormones (*e.g.*, cholecystokinin) require α-amidation for biological activity, endocrine disorders will also be determined to be due to the lack of PAM activity.

It is therefore apparent that ascorbic acid serves many functions in the biosynthesis and posttranslational processing of proteins, thus emphasizing further its importance as a nutritional component for animals, like man, that are incapable of biosynthesizing this compound.

REFERENCES

1. Basu, T. K. & C. J. Schorah. 1982. Vitamin C in Health and Disease. Avi Publishing Co. Westport, Conn.
2. Nobile, S. & J. M. Woodhill. 1981. Vitamin C. MTP Press. Boston.
3. Levin, E. Y., B. Levenberg & S. Kaufman. 1960. J. Biol. Chem. **235:** 2080-2086.
4. Eipper, B. A., R. E. Mains & C. C. Glembotski. 1983. Proc. Natl. Acad. Sci. U.S.A. **80:** 5144-5148.
5. Perara, E., R. E. Rothman & V. R. Lingappa. 1986. Science **232:** 348-352.
6. Wickner, W. T. & H. F. Lodish. 1985. Science **230:** 400-407.
7. Kornfeld, R. & S. Kornfeld. 1985. Ann. Rev. Biochem. **54:** 631-644.
8. Mains, R. E., B. A. Eipper, C. C. Glembotski & R. M. Dores. 1983. Trends Neruosci. **6:** 229-235.
9. Gainer, H. T., J. R. Russell & Y. P. Loh. 1985. Neuroendocrinology **40:** 171-184.
10. Lynch, D. R. & S. H. Snyder. 1986. Ann. Rev. Biochem. **55:** 773-799.
11. Eipper, B. A. & R. E. Mains. 1980. Endocrine Rev. **1:** 1-27.
12. Eipper, B. A., C. C. Glembotski & R. E. Mains. 1983. J. Biol. Chem. **258:** 7292-7298.
13. Glembotski, C. C., B. A. Eipper & R. E. Mains. 1983. J. Biol. Chem. **258:** 7299-7304.
14. Bradbury, A. F. & D. G. Smyth. 1983. Biochem. Biophys. Res. Commun. **112:** 372-377.
15. Landymore-Lim, A. E. N., A. F. Bradbury & D. G. Smyth. 1983. Biochem. Biophys. Res. Commun. **117:** 289-293.
16. Glembotski, C. C., B. A. Eipper & R. E. Mains. 1984. J. Biol. Chem. **259:** 6385-6392.
17. Mains, R. E., C. C. Glembotski & B. A. Eipper. 1984. Endocrinology **114:** 1522-1530.
18. Eipper, B. A., C. C. Glembotski & R. E. Mains. 1983. Peptides **4:** 921-928.
19. Emeson, R. B. 1984. J. Neurosci. **4:** 2604-2613.
20. Glembotski, C. C. 1985. Arch. Biochem. Biophys. **241:** 673-683.
21. Eipper, B. A., R. E. Mains & C. C. Glembotski. 1985. Biogenetics of Neurohormonal Peptides. R. Hakanson & J. Thorell, Eds.: 187-209. Academic Press. London.
22. Glembotski, C. C. 1984. J. Biol. Chem. **259:** 13041-13048.
23. May, V. & B. A. Eipper. 1985. J. Biol. Chem. **260:** 16224-16231.
24. Shields, P. P., T. R. Gibson & C. C. Glembotski. 1986. Endocrinology **118:** 1452-1460.
25. Mains, R. E., E. I. Cullen, V. May & B. A. Eipper. 1987. Ann. N.Y. Acad. Sci. **293:** 278-291.
26. Glembotski, C. C. 1986. Endocrinology **118:** 1461-1468.
27. Glembotski, C. C., S. Manaker, A. Winokur & T. R. Gibson. 1986. J. Neurosci. **6:** 1796-1802.
28. Lechan, R. M., P. Wu, I. M. D. Jackson, H. Wolf, S. Cooperman, G. Mandel & R. H. Goodman. 1986. Science **231:** 159-161.
29. Murthy, A. S. N., R. E. Mains & B. A. Eipper. 1986. J. Biol. Chem. **261:** 1815-1822.

DISCUSSION OF THE PAPER

H. SPRINCE (*Jefferson Medical College, Philadelphia, Pa.*): Do you know of any work on the effect of aldehydes on the ascorbate amidation reaction?

C. C. GLEMBOTSKI: To the best of my knowledge, none has been published.

H. SPRINCE: The reason I bring this up is we published years ago in quite an entirely different context that ascorbic acid could protect against aldehyde toxicity and I'm just extrapolating these findings to your system.

M. RICE (*New York University Medical Center, New York, N.Y.*): You mentioned the EC_{50} for your hypothalamic cells; the concentration was similar to the extracellular concentration in that region. *In vivo* electro-chemical studies have demonstrated that throughout the brain the extracellular concentration of ascorbate is 300-400 micromolar. If the ascorbate is sequestered vesicularly, as it is in adrenal chromaffin vesicles, then the intracellular free ascorbate concentrations may be very important.

C. C. GLEMBOTSKI: I want to detract from the notion that the cellular EC_{50} is relevant to what occurs inside the granules. Others have determined the concentration of ascorbic acid in pituitary granules to be in the mM range. So whatever is going on in the granules could be altering the apparent EC_{50} as the enzyme exists in its normal milieu, and what we're looking at here with cells in culture is a composite of several ascorbate transport systems.

E. J. DILIBERTO (*Wellcome Research Laboratories, Research Triangle Park, N.C.*): What might be the kinetics of the reaction is involving more the cellular transport mechanism. The kinetics of transport into chromaffin cells which is very similar to what you had found in the cultured pituitary is in the order of 25-30 micromolar, so in fact the transport *in vivo* is usually saturated.

M. RICE: This would be a maximum effect certainly.

C. C. GLEMBOTSKI: If we observe the enzyme *in vitro* and determine the K_m for ascorbate, it appears to be on the millimolar level.

E. J. DILIBERTO: Mark Levine has found very high concentrations of ascorbic acid in the vesicles. So it does concentrate similarly to what we find in the bovine tissue. There's a ten-to-one gradient from the vesicle to the cytosol of the bovine chromaffin cells and I think similar findings are coming out for the pituitary.

J. BLONDIN (*Tufts University, Boston, Mass.*): In your talk you mentioned that there were other posttranslational modifications that you'd studied, such as acetylation and endoproteolysis. Did you find any effects of ascorbic acid in those two systems?

C. C. GLEMBOTSKI: No, those reactions are virtually unchanged whether or not there is ascorbate in the medium.

Ascorbic Acid, Redox Cycling, Lipid Peroxidation, and the Binding of Dopamine Receptor Antagonists[a]

RICHARD E. HEIKKILA AND
LAWRENCE MANZINO

Department of Neurology
University of Medicine and Dentistry of New Jersey
Robert Wood Johnson Medical School
Piscataway, New Jersey 08854

The mammalian brain contains relatively high concentrations of ascorbic acid.[1-3] In the rat, the ascorbic acid concentration appears to be at a maximum at the time of birth and then decreases somewhat as the animal grows older.[4] A detailed mapping of the ascorbic acid concentration has demonstrated a rather homogeneous distribution throughout both human and rat brain.[1,2] There appears to be no obvious correlation between the pattern of distribution of ascorbic acid and any known neurotransmitter. The ascorbic acid level in rat substantia nigra, the brain area examined that had the lowest level, was found to be 190 ± 11 μg/g of tissue. The nucleus accumbens, the brain area studied with the highest ascorbic acid concentration, had a level of 464 ± 30 μg/g of tissue.[2] This approximate twofold difference in concentration of ascorbic acid differs markedly from the regional distribution of most neurotransmitters, the concentration of which can vary several hundredfold from one brain region to another. As an example, it has been reported that the ascorbate-to-catecholamine ratio varies from about 20 in the rat caudate putamen to about 905 in the hippocampus.[2] These different ratios obviously occur as a result of differences in the catecholamine content and not in the ascorbic acid content. Although there has been considerable speculation, the precise function of ascorbic acid in the central nervous system is at present unexplained. Most theories focus in some way on its reducing properties.

Ascorbic acid is a relatively good reducing agent, and because of this property, it is often utilized to prevent unwanted oxidation processes. As an example, ascorbic acid is often added to buffers containing catecholamines, such as dopamine or norepinephrine, which are relatively unstable molecules, "to prevent their oxidative decomposition." As such, this approach is valid and in general quite sound, however, there are certain circumstances in which the routine, indiscriminate use of ascorbic acid as a reducing agent may not be appropriate. As a case in point, 6-hydroxydopamine (6-OHDA) is a widely used neurotoxin that is extremely unstable in aqueous media.[5] It spontaneously reacts with dissolved oxygen at a very rapid rate and almost instantaneously, the addition of 6-OHDA to a buffer at pH 7.4 and 37°C can change a colorless solution into a bright red solution. This bright red color, which has an

[a] This work was supported by grants from the United States Public Health Service, Hoffmann-La Roche, Inc., and from the Smokeless Tobacco Research Council.

absorbancy maximum at 490 nm, is due to a quinoidal oxidation product of 6-OHDA. Perhaps not surprisingly, the addition of ascorbic acid together with 6-OHDA results in a colorless solution. At face value, ascorbic acid has retarded the autoxidation of 6-OHDA.[1] It has functioned as a reducing agent, as it should. If one studies this chemical reaction described above with the use of an oxygen electrode, however, one will draw very different conclusions. The addition of ascorbic acid to a solution of 6-OHDA results in a greatly accelerated rate of oxygen consumption. By this criterion, ascorbic acid, a reducing agent, has promoted an oxidation reaction. This seemingly confusing situation, which appears to be a paradox, is in fact not. A description of these phenomena are presented in FIGURE 1. It may be seen that 6-OHDA reacts with oxygen to form a quinone and hydrogen peroxide (H_2O_2). The addition of ascorbic acid (or of several other reducing agents) results in the reduction of the quinone of 6-OHDA back to 6-OHDA. As long as excess ascorbate and oxygen are present, this reaction, which has been termed redox cycling, can repeat itself again and again. And strikingly, the addition of a reducing agent like ascorbic acid to a solution can result in the formation of large amounts of an oxidizing agent like H_2O_2. One can readily see that with a small, almost catalytic, amount of 6-OHDA in the presence of a large excess of O_2 and ascorbate, a very large amount of H_2O_2 can be formed (FIG. 2, bottom). Many compounds, and particularly several known cytotoxins, can similarly engage in redox cycling. And it is widely felt that the reducing agent-promoted formation of oxidizing species accounts for the observed cytotoxicity.[6]

It has recently been speculated that another dopaminergic neurotoxin, namely 1-methyl-4-phenyl-1,2,3,6-tetrahydropyridine (MPTP) and its two-electron oxidation product 1-methyl-4-phenyl-2,3-dihydropyridine ($MPDP^+$) or four-electron oxidation product 1-methyl-4-phenylpyridine (MPP^+) may engage in redox cycling (see FIG. 2, top). However, it has also been demonstrated that this is unlikely, in that MPP^+ is extremely difficult to reduce.[7] Moreover, a cycle between MPTP and $MPDP^+$ is unlikely, in that MPTP is quite stable under normal laboratory conditions and does not autoxidize to form $MPDP^+$. Ideally, a good redox cycle consists of a compound that readily oxidizes to a product, which in turn can readily be reduced back to the starting compound by an appropriate reducing agent. 6-OHDA clearly meets the criteria; MPTP does not.

Because it is widely accepted that ascorbic acid is a relatively good reducing agent and because several commonly used ligands are unstable, ascorbic acid is often routinely added as a component of the buffer in binding studies. The rationale is that ascorbic acid will, by its prevention of oxidative decomposition of the ligand being

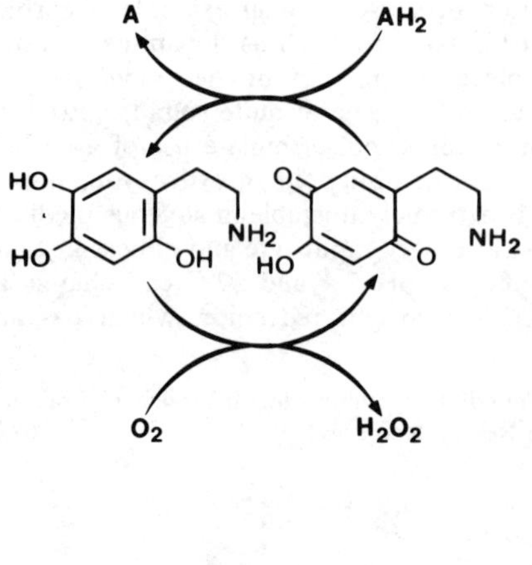

FIGURE 1. A redox cycling scheme for 6-hydroxydopamine (6-OHDA) and its corresponding quinone. AH_2 represents a reducing agent such as ascorbic acid.

$$\begin{array}{lcl} MPP^{+} + AH_2 & \longrightarrow & MPDP^{+} + A \\ MPDP^{+} + O_2 & \longrightarrow & MPP^{+} + H_2O_2 \\ \hline AH_2 + O_2 & \longrightarrow & A + H_2O_2 \end{array}$$

$$\begin{array}{lcl} \text{6-OHDA} + O_2 & \longrightarrow & \text{Quinone} + H_2O_2 \\ \text{Quinone} + AH_2 & \longrightarrow & \text{6-OHDA} + A \\ \hline AH_2 + O_2 & \longrightarrow & A + H_2O_2 \end{array}$$

FIGURE 2. Theoretical redox cycling reaction schemes for l-methyl-4-phenyl-pyridinium (MPP^+, *top*) and 6-hydroxydopamine (6-OHDA, *bottom*). AH_2 represents a reducing agent such as ascorbic acid. Note that the cycle can begin with either an oxidation or a reduction reaction.

studied, maintain the integrity of the ligand and thus be useful. All of this is based on the assumption that ascorbic acid will otherwise be innocuous. As will be discussed below, this assumption is often correct, but unfortunately, it is also often erroneous.

MATERIAL AND METHODS

Binding of Tritiated Dopamine Antagonists in Neostriatal Membrane Preparations

The binding procedure is a modification of that used by Burt *et al.*[8] which we have previously utilized.[9,10] Male Sprague-Dawley rats were killed and the left and right neostriata were dissected from the brains. The striata were homogenized with a Teflon-glass homogenizer in 0.3 M cold sucrose (2.5 ml per striatum) and centrifuged at 700 × g for 10 min in a Sorvall RC-5B centrifuge (used for all centrifugation). The supernatant was centrifuged at 32,000 ×g for 15 min. The resulting pellet was homogenized in 50 mM Tris buffer (pH 7.7, 2.5 ml per striata) and recentrifuged at 32,000 ×g for 10 min. The final pellet was rehomogenized in appropriate volumes of 50 mM Tris cation buffer (pH 7.7, containing 120 mM NaCl, 5 mM KCl, 2 mM $CaCl_2$, 0.2 mM $MgCl_2$, and 0.01 mM pargyline), so that the final concentration of tissue was 5-6 mg/ml, based on original weight. Aliquots of 1 ml of tissue preparation were added to test tubes. Some tubes then received the nonradioactive agent that was used to define specific binding (10^{-6} M D-butaclamol in ethanol) or 10 μl of the appropriate vehicle. Ascorbic acid (0-6 mM) was then added and all tubes were preincubated for 10 min at 37°C in a water bath. In some experiments, various agents were added immediately prior to ascorbic acid. The ^{3}H-labeled ligand (New England Nuclear, Boston, Mass.) was added to a concentration of 2 nM; the tubes were then incubated with shaking for 15 min. The solutions from the tubes were filtered through

Gooch crucibles containing GFC filter papers, and the tubes were rinsed twice with 5 ml cold Tris buffer. The filter paper containing the tissue with the bound radioactivity was placed in 3 ml absolute ethanol for the extraction of radioactivity. After 15 min, 10 ml Scintiverse (Fisher, Springfield, N.J.) was added and the radioactivity was measured.

The specific binding of [^{3}H]spiroperidol or [^{3}H]domperidone was defined as total binding minus binding in the presence of 10^{-6} M nonradioactive D-butaclamol. The specific binding of the ligands in the presence of ascorbic acid was compared with control samples containing no ascorbic acid. The data were calculated as a percentage of control.

Lipid Peroxidation

The thiobarbituric acid (TBA) test was used to determine the extent of lipid peroxidation.[11] Tissue samples were prepared as above. Ascorbic acid (0-6 mM) was added to 1 ml of the tissue preparations. In some experiments, various agents were added just prior to the ascorbic acid. Samples were then incubated at 37°C for 10 min in a water bath. Then, 1 ml of cold 10% trichloroacetic acid (wt/vol) was added, and the samples were allowed to stand for 5 min, followed by centrifugation at 8000 ×g for 10 min in the Sorvall centrifuge. To 1 ml of the supernatant, 1.3 ml of 0.5% thiobarbituric acid (wt/vol: Sigma, St. Louis, Mo.) was added. The samples were then placed in a boiling water bath for 20 min, removed, and allowed to cool to room temperature. The absorbancy was read at 532 nm on a Gilford model 300 flow-through spectrophotometer. The amount of lipid peroxide was calculated by using a molar extinction coefficient of 1.5×10^5 as described previously.[11]

RESULTS AND DISCUSSION

The addition of L-ascorbic acid, sodium ascorbate, or D-isoascorbic acid (the epimer of ascorbic acid) all cause an extensive lipid peroxidation in neostriatal membrane preparations (FIG. 3). The similar results with ascorbic acid and sodium ascorbate are not surprising in that the two will exist in the same form in this experimental system, which is adequately buffered at pH 7.7, well above the pK_a for ascorbic acid. Interestingly, the dose-response relationship is not simple, *i.e.,* higher concentrations do not necessarily cause greater effects. For example, there is a more extensive lipid peroxidation at the two intermediate ascorbate concentrations (0.06 and 0.6 mM) than at the lowest and highest concentrations tested (0.006 and 6 mM respectively). This complex U-shaped dose-response relationship for an ascorbate-induced lipid peroxidation has often been demonstrated in many experimental systems. Ascorbic acid, sodium ascorbate, and isoascorbic acid also caused an extensive loss of the stereoselective binding of the dopamine receptor antagonist [^{3}H]spiroperidol in neostriatal membrane preparations (FIG. 4). As was the case with lipid peroxidation, there was a U-shaped dose-response relationship. The effects at the two intermediate ascorbate concentrations (0.06 and 0.6 mM) were greater than the effects at the lowest and highest concentrations (0.006 and 6 mM respectively). In other experiments (not shown) ascorbate caused a similar loss of binding of another dopamine receptor antagonist, [^{3}H]domperidone.[9] This decrement of binding caused by ascorbate was not due to the physical presence of ascorbate in that there was no reversal on washing the tissue preparation free of ascorbate and resuspension of the tissue prior to measurement of the binding.

The addition of the well-known inhibitor of lipid peroxidation, propyl gallate, at 10^{-5} M prevented both the ascorbic-acid-induced lipid peroxidation (FIG. 5) and loss of [^{3}H]spiroperidol binding to the membrane preparation (FIG. 6). Again, note the similarities in the dose-response relationship for ascorbate in FIGURES 5 and 6. The ascorbate-induced increase in lipid peroxidation is also presented in TABLE 1. Mean control values and values in the presence of ascorbate are 23 ± 29 and 923 ± 150

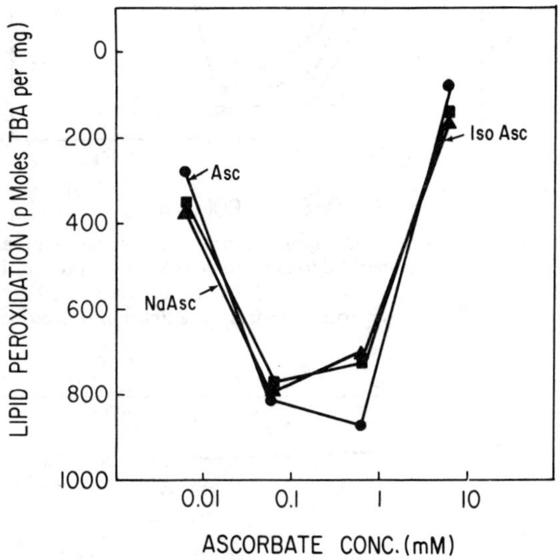

FIGURE 3. The effects of ascorbic acid (*Asc*), sodium ascorbate (*NaAsc*), and isoascorbic acid (*IsoAsc*) on the formation of lipid peroxides in neostriatal membrane preparations. Data are given in picomoles of lipid peroxide formed per milligram of tissue (based on original weight). Data represent the mean results for at least four experiments. (From Heikkila *et al.*[9] Reprinted by permission from the *Journal of Neurochemistry.*)

pmol/mg of tissue respectively. This effect of ascorbate was partially prevented by the iron-chelating agent EDTA, and was completely prevented by the iron-chelating agent DETAPAC and by the widely used inhibitors of lipid peroxidation, α-naphthol, butylated hydroxyanisole (BHA) and cobalt chloride (TABLE 1). Similarly, these same agents either partially protected (EDTA) or completely protected (DETAPAC, α-naphthol, BHA, and cobalt chloride) against the ascorbate-induced decrement in [^{3}H]spiroperidol binding (TABLE 2). Higher concentrations of EDTA (*e.g.*, 5×10^{-4} M) completely prevented both the ascorbate-induced increase in lipid peroxidation and ascorbate-induced decrement in [^{3}H]spiroperidol binding.

In other experiments, several other reducing agents or ascorbic acid analogues were tested at 0.1 mM for their effects on lipid peroxidation or on [^{3}H]spiroperidol

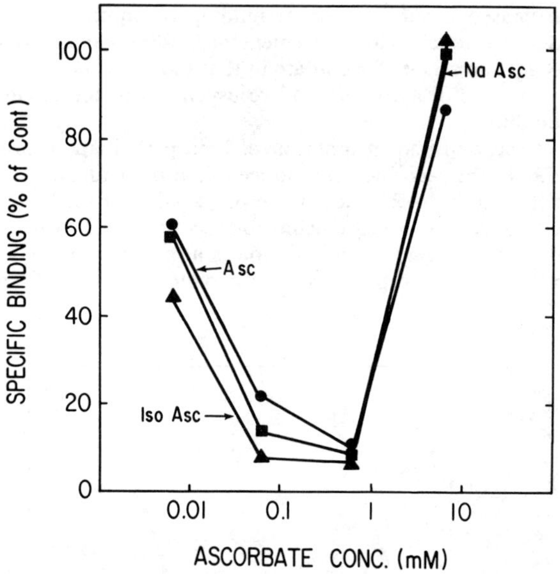

FIGURE 4. The effects of ascorbic acid (*Asc*), sodium ascorbate (*NaAsc*), and isoascorbic acid (*IsoAsc*) on the binding of [^{3}H]spiroperidol to neostriatal membrane preparations. Data represent the mean specific binding as percentage of control for at least four experiments with each agent. (From Heikkila *et al.*[9] Reprinted by permission from the *Journal of Neurochemistry.*)

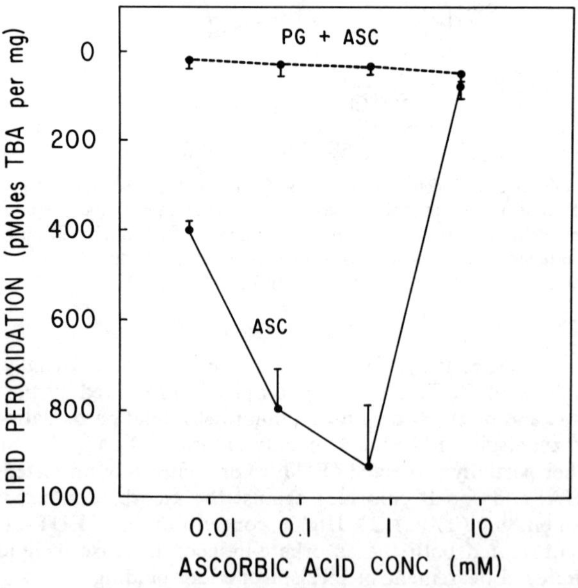

FIGURE 5. The prevention by propyl gallate (10^{-5} M) of the effects of ascorbic acid on lipid peroxidation in neostriatal membrane preparations.

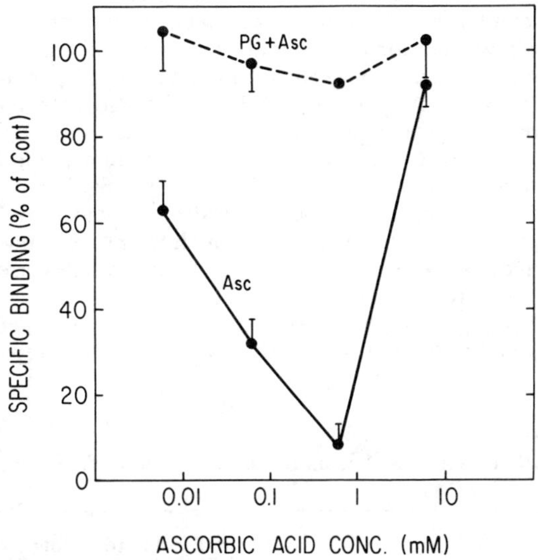

FIGURE 6. The prevention by propyl gallate (10^{-5} M) of the effects of ascorbic acid on specific [^{3}H]spiroperidol binding in neostriatal membrane preparations. Data represent the mean percentage of control binding ± SD for four experiments. (From Heikkila *et al.*[9] Reprinted by permission from the *Journal of Neurochemistry.*)

TABLE 1. The Effects of Various Agents on Lipid Peroxidation Induced in Neostriatal Membrane Preparations by Ascorbic Acid (0.6 mM)[a]

	Lipid Peroxide Formed (pmol/mg Tissue)	
Addition (10^{-4} M)	No Ascorbic Acid	Plus Ascorbic Acid
None	23 ± 29	923 ± 150
EDTA	10 ± 14	447 ± 11
DETAPAC	18 ± 12	75 ± 34
α-Napthol	51 ± 28	33 ± 33
BHA[b]	8 ± 10	55 ± 43
Cobalt chloride	20 ± 15	36 ± 42

[a] Data represent the amount of lipid peroxide formed (mean pmol ± SD per mg of tissue, based on original tissue weight). There were three to five experiments for each group; experiments were run as described in the text. EDTA, DETAPAC, and cobalt chloride were dissolved in water, the others were dissolved in ethanol, which added in the same volume (10 μl) had no effect on binding or on lipid peroxidation.

[b] Butylated hydroxyanisole.

binding.[10] Some of the agents (*e.g.*, ascorbic acid hemicalcium) were found to behave similarly to ascorbate. Others (dithiothreitol, sodium metabisulfate, and glutathione) were qualitatively similar, but quantitatively weaker, than ascorbate. In general, these weaker agents caused an enhanced lipid peroxidation and a greater loss of [^{3}H]spiroperidol binding when exogenous iron (either $FeSO_4$ or $FeCl_3$) was added to the incubation medium. Addition of the iron salts also potentiated the effects of ascorbate. In contrast, other compounds (*i.e.*, dehydroascorbic acid or ascorbic acid 2-sulfate), even in the presence of added iron, caused no increase in lipid peroxidation or loss of [^{3}H]spiroperidol binding.[10] It should be emphasized that although the type of effect described above can always be demonstrated (sometimes only with added iron), the effects are quantitatively different in different experimental systems. Of particular importance is the manner in which tissue samples are prepared.[10] For example, certain investigators have found no significant effect of ascorbate on [^{3}H]spiroperidol binding.[12] We have repeated some of these experiments using the published methods and found that the addition of iron to the preparation containing

TABLE 2. The Effects of Various Agents on the Inhibition of [^{3}H]Spiroperidol Binding by Ascorbic Acid (0.6 mM)[a]

	Specific Binding (dpm/mg Tissue)	
Addition (10^{-4} M)	No Ascorbic Acid	Plus Ascorbic Acid
None	1082 ± 127	65 ± 114
EDTA	1060 ± 170	205 ± 99
DETAPAC	1057 ± 156	1104 ± 228
α-Naphthol	919 ± 115	989 ± 119
BHA	856 ± 80	852 ± 88
Cobalt chloride	1059 ± 195	894 ± 99

[a] Data represent the specific [^{3}H]spiroperidol binding (mean dpm ± SD per mg of tissue, based on original tissue weight). There were four to six experiments for each group; experiments were run in triplicate, as described in the text.

ascorbate in the absence of an iron-chelating agent can bring about the loss of binding and the increased lipid peroxidation.[10]

The present study clearly demonstrates that, at certain concentrations, the reducing agent ascorbic acid can promote lipid peroxidation. Interestingly, the capacity of ascorbate to promote lipid peroxidation is greater at intermediate than at higher or lower concentrations (FIGS. 3 and 5). The inhibitory effects of the iron-chelating agents EDTA and DETAPAC, combined with the fact that exogenous iron can enhance the capacity of ascorbate to promote lipid peroxidation,[10] suggests that this effect of ascorbate is iron-dependent. In all likelihood, the capacity of ascorbate to reduce iron from the ferric (Fe^{3+}) state to the ferrous (Fe^{2+}) state is an important feature of this experimental system. For a description of some of the reactions involved, see Liebler *et al.*[13] In a sense, an ascorbate-dependent cycling of iron between oxidized and reduced states is similar to the redox cycling reaction schemes depicted in FIGURES 1 and 2. It would appear that at the highest concentration tested (6 mM) ascorbate acts only as a reducing agent and not as a prooxidant in that it does not promote lipid peroxidation. Interestingly, the concentration of ascorbate found in the brain is also high, around 2 mM.

A comparison of the data presented above demonstrates a very close relationship between lipid peroxidation and loss of [^{3}H]spiroperidol binding. In general, whenever ascorbate promoted an increase in lipid peroxidation there was a concomitant decrement in [^{3}H]spiroperidol binding. A simple conclusion for all of the data is that the ascorbate-induced lipid peroxidation is responsible for the loss of [^{3}H]spiroperidol binding. It would appear as if the binding of the [^{3}H]spiroperidol is dependent upon the lipid environment of the receptor and its structural integrity. This receptor has been classified as the D-2 dopamine receptor by several investigators.[14,15] Alternatively, the ascorbate-induced lipid peroxidation may lead to some damage in the protein component of the receptors, which accounts for the decrement in binding.

This destructive effect of ascorbate is by no means unique to the dopamine receptor. It has been shown, for example, that an ascorbate-induced lipid peroxidation can lead to a loss in binding of various ligands to opiate receptors,[16–18] serotonin receptors,[19-23] α-adrenergic receptors,[24,25] β-adrenergic receptors,[26,27] muscarinic cholinergic receptors,[17] and phorbol ester receptors,[28] as well as to dopamine receptors.[9,10,29,30] It should be emphasized, however, that because certain ligands that are used in receptor binding experiments are unstable, it still may be necessary to use reducing agents like ascorbic acid as a component of the buffers used in binding studies. It has, for example, been shown recently that binding of serotonin (5-hydroxytryptamine) to its receptors may require agents like ascorbic acid.[22,23] It has been described that the binding of [^{3}H]serotonin carried out in the absence of reducing agents was artifactually high, due to the binding of an oxidation product of the [^{3}H]serotonin to the receptor preparation.[22,23] It would appear as if a number of options are available so that both lipid peroxidation and ligand decomposition might be avoided. These would include: use of an ascorbate concentration that prevents ligand decomposition but does not promote lipid peroxidation, combined use of ascorbate with either a chelating agent (*e.g,* EDTA, DETAPAC) or an inhibitor of lipid peroxidation (*e.g.,* propyl gallate, BHA), or use of another reducing agent that prevents ligand decomposition but does not promote lipid peroxidation. In any case, it would appear as if a wise course of action would be to use the lowest concentrations of all of the agents so that the desired action is obtained without the undesired action. It seems obvious that one cannot just add a chemical to an experimental system without extremely careful consideration of all of the complex possibilities. As a further emphasis, the U-shaped dose-response curve for ascorbate indicates that careful consideration must be given to the appropriateness of the concentration being used. In some cases more of a compound (*i.e.,* ascorbate) might be desirable and in certain cases it clearly would not.

REFERENCES

1. MEDFORD, I. N., A. F. OKE & R. N. ADAMS. 1981. Regional distribution of ascorbate in human brain. Brain Res. **212:** 223-226.
2. MILBY, K., A. OKE & R. N. ADAMS. 1982. Detailed mapping of ascorbate distribution in rat brain. Neurosci. Lett. **28:** 15-20.
3. KRATZING, C. C., J. D. KELLY & B. A. OELRICHS. 1982. Ascorbic acid in neural tissues. J. Neurochem. **39:** 625-627.
4. KRATZING, C. C., J. D. KELLY & J. E. KRATZING. 1985. Ascorbic acid in fetal rat brain. J. Neurochem. **44:** 1623-1624.

5. HEIKKILA, R. E. & G. COHEN. 1972. Further studies on the generation of hydrogen peroxide by 6-hydroxydopamine: Potentiation by ascorbic acid. Mol. Pharmacol. **8:** 241-248.
6. H. KAPPUS. 1986. Overview of enzyme systems involved in bio-reduction of drugs and in redox cycling. Biochem. Pharmacol. **35:** 1-6.
7. SAYRE, L. M., P. K. ARORA, S. C. FEKE & F. L. URBACH. 1986. Mechanism of induction of Parkinson's disease by 1-methyl-4-phenyl-1,2,3,6-tetrahydropyridine (MPTP). Chemical and electrochemical characterization of a geminal-dimethyl-blocked analogue of a postulated toxic metabolite. J. Am. Chem. Soc. **108:** 2464-2466.
8. BURT, D. R., I. CREESE & S. H. SNYDER. 1976. Properties of [^{3}H]haloperidol and [^{3}H]dopamine binding associated with dopamine receptors in calf brain membranes. Mol. Pharmacol. **12:** 800-812.
9. HEIKKILA, R. E., F. S. CABBAT & L. MANZINO. 1982. Inhibitory effects of ascorbic acid on the binding of [^{3}H]dopamine antagonists to neostriatal membrane preparations: Relationship to lipid peroxidation. J. Neurochem. **38:** 1000-1006.
10. HEIKKILA, R. E. & F. S. CABBAT. 1983. Ascorbate-induced lipid peroxidation and inhibition of [^{3}H]spiroperidol binding in neostriatal membrane preparations. J. Neurochem. **41:** 1384-1392.
11. KOVACHICH, G. B. & O. P. MISHRA. 1980. Lipid peroxidation in rat brain cortical slices as measured by the thiobarbituric acid test. J. Neurochem. **35:** 1449-1452.
12. BACOPOULOS, N. G. 1982. Dopamine receptors in rat brain regions. Optimal conditions for ^{3}H-agonist binding, pH dependency and lack of inhibition by ascorbic acid. Biochem. Pharmacol. **31:** 3085-3091.
13. LIEBLER, D. C., D. S. KLING & D. J. REED. 1986. Antioxidant protection of phospholipid bilayers by α-tocopherol. J. Biol. Chem. **261:** 12114-12119.
14. LEFF, S. E. & I. CREESE. 1983. Dopamine receptors re-explained. Trends Pharmacol. Sci. **4:** 463-467.
15. GRIGORIADIS, D. & P. SEEMAN. 1984. The dopamine/neuroleptic receptor. Can. J. Neurol. Sci. **11:** 108-113.
16. DUNLAP, C. E. III, F. M. LESLIE, M. RADO & B. M. COX. 1979. Ascorbate destruction of opiate stereospecific binding in guinea pig brain homogenate. Mol. Pharmacol. **16:** 105-119.
17. LESLIE, F. M., C. E. DUNLAP III & B. M. COX. 1980. Ascorbate decreases ligand binding to neurotransmitter receptors. J. Neurochem. **34:** 219-221.
18. COX, B. M., F. M. LESLIE & C. E. DUNLAP III. 1980. The use of ascorbate as a probe of opioid receptor structure: Evidence for two independent mechanisms of receptor destruction by ascorbate. J. Rec. Res. **1:** 329-354.
19. MUAKKASSAH-KELLY, S. F., J. W. ANDRESEN, J. C. SHIH & P. HOCHSTEIN. 1982. Decreased [^{3}H]serotonin and [^{3}H]spiperone binding consequent to lipid peroxidation in rat cortical membranes. Biochem. Biophys. Res. Commun. **104:** 1003-1010.
20. WEINER, N., N. AROLD & W. WESEMANN. 1982. Possible hazards in serotonin binding assays in human and rat brain: Effects of ascorbate, pargyline, and buffer concentration. J. Neurosci. Meth. **5:** 41-45.
21. MUAKKASSAH-KELLY, S. F., J. W. ANDRESEN, J. C. SHIH & P. HOCHSTEIN. 1983. Dual effects of ascorbate on serotonin and spiperone binding in rat cortical membranes. J. Neurochem. **41:** 1429-1439.
22. ANDRESEN, J. W. & J. C. SHIH. 1986. Necessity of ascorbic acid in the radioligand binding assay for [^{3}H]5-hydroxytryptamine. Neuropharmacol. **25:** 869-875.
23. PEROUTKA, S. J., P. I. ISON, D. U. LIU & R. W. BARRETT. 1986. Artifactual high-affinity and saturable binding of [^{3}H]5-hydroxytryptamine induced by radioligand oxidation. J. Neurochem. **47:** 38-45.
24. WOLFMAN, C., M. LEVI DE STEIN & E. DE ROBERTIS. 1983. Time course of changes in [^{3}H]prazosin binding after catecholaminergic denervation by 6-hydroxydopamine in rat cerebral cortex. Neuropharmacology **22:** 1061-1064.
25. JONES, S. B. & D. B. BYLUND. 1986. Ascorbic acid inhibition of alpha-adrenergic receptor binding. Biochem. Pharmacol. **35:** 595-599.
26. HEIKKILA, R. E. 1983. Ascorbate-induced lipid peroxidation and the binding of [^{3}H]dihydroalprenolol. Eur. J. Pharmacol. **93:** 79-85.

27. BRUININK, A. & W. LICHTENSTEIGER. 1984. Effects of pH and ascorbic acid on ^{3}H-spiperone and ^{3}H-dihydroalprenolol binding in rat forebrain homogenates. J. Rec. Res. **4:** 127-139.
28. DELCLOS, K. B. & P. M. BLUMBERG. 1982. Identification of ascorbic acid as the heat-stable factor from brain which inactivates the phorbol ester receptor. Cancer Res. **42:** 1227-1232.
29. LAZARENO, S. & S. R. NAHORSKI. 1982. Selective labelling of dopamine (D_2) receptors in rat striatum by [^{3}H]domperidone but not by [^{3}H]spiperone. Eur. J. Pharmacol. **81:** 273-285.
30. MADRAS, B. K. & B. CHAN. 1983. Modulation of dopamine receptor binding by ascorbic acid. *In* CNS Receptors: From Molecular Pharmacology to Behavior. P. Mandel & F. V. DeFeudis, Eds.: 275-287. Raven Press. New York.

DISCUSSION OF THE PAPER

E. J. DILIBERTO (*Wellcome Research Laboratories, Research Triangle Park, N.C.*): In addition to what you said in terms of your ligand binding effects, I think it's important to realize that this might not be an aberrant reaction. It might be something that we hopefully will understand a little bit more about later on in the conference, and this might be the major link between the discussion today and what we're going to know in terms of the biochemistry and in terms of the lipid peroxidation reaction.

How is that really an aberrant reaction or is that something that's expected in the number of systems? Whether or not it may be some of the extracellular functions of ascorbic acid may be by way of lipid peroxidation, and perhaps maybe we can even speculate a little further and say that that might in fact be a receptor in its own right.

R. E. HEIKKILA (*UMDNJ, Piscataway, N.J.*): I would fully agree that there is something very fascinating about this kind of reaction in which you initially have the starting material and you go through a whole reaction sequence and you end up back with your starting material. If you study the reaction in one way you can draw one conclusion, and if you study the reaction in another way you can draw another conclusion. There are really many different chemical reactions going on.

E. J. DILIBERTO: Actually I'm looking for criticism for this being a model for the neuromodulation for ascorbic acid.

UNIDENTIFIED SPEAKER: Well instead of talking about the model I want to talk about the very simple chemistry or biochemistry involved. It is true that ascorbic acid can act as an oxidant or prooxidant as well as reductant, but I think the manner in which it acts depends upon the redox potentials of the component coupling reactions involved.

First I would like to know whether the 6-hydroxydopamine oxidation can be inhibited by SOD. Second, of the various reactions that you wrote on the oxidation of ascorbic acid by dopamine, did you do a thermodynamic calculation to determine whether the catalytic protection of dopamine oxidation by ascorbic acid is in accordance with the redox potential or is it because of the reversal of the redox reactions because the concentrations of the various ingredients that are used are far apart form the concentrations that are used for the normal measurements of the potentials?

R. E. HEIKKILA: With respect to your first question, 6-hydroxydopamine autoxidation does go through the superoxide radical. If you take a solution of 6-hydroxydopamine and you add superoxide dismutase which inactivates the superoxide radical, you slow down greatly the autoxidation of 6-hydroxydopamine. So the net stoichiometry in that reaction cycle did involve a superoxide radical and the one-electron oxidation product of 6-hydroxydopamine. These are indeed formed in solution. So 6-hydroxydopamine itself undergoes a catalytic cycle dependent upon the superoxide radical but the net result is the two-electron oxidation to form peroxide. You form a superoxide radical as a one-electron product, then you form peroxide as a two-electron product.

In terms of your comments about oxidation and reduction you clearly can convert the quinone of 6-hydroxydopamine back to 6-hydroxydopamine by adding a reducing agent like ascorbic acid. In terms of whether we've made the physical chemical measurements, we have not. But Dr. Adams and Dr. Oke in their laboratories have made these measurements and studied the oxidation-reduction potential of all of these different molecules.

6-Hydroxydopamine is an extremely reactive molecule. Within seconds it reacts with oxygen to form these products. It's extremely unstable. In terms of a comparison, dopamine is much harder to oxidize. The neurotransmitter dopamine, found in the brain at a high concentration, if put in solution at pH 7.4 at 37°C will decompose completely in about an hour; 6-hydroxydopamine will decompose in around 30 seconds. So there is a tremendous difference between the ease of oxidation of 6-hydroxydopamine, for example, and the catecholeamines which are found in the brain.

Z. VOGEL (*Weizmann Institute of Science, Rehovot, Israel*): I understand you did the reactions with the receptors, different receptors in membrane preparations. Have you repeated them on complete cells because there the redox potentials might be a little different?

R. E. HEIKKILA: These experiments I've described are done in the way that people do receptor binding experiments. The common way that people do this is to take the brain and add a hypotonic medium and homogenize and go through multiple centrifugation steps. We've not done these experiments in whole cells. It's anybody's guess what will happen under those conditions.

UNIDENTIFIED SPEAKER: This reaction is important for various reasons, breaking down lipid membranes producing arachidonic acid and the products of prostaglandins. So I'd like to pursue that question a little bit, not from that viewpoint but to get back to your use of the word "cycle" here. My understanding of the cycle is that you have an enzyme that takes the reaction in one direction and then something converts it back in the other direction so you end up with your original substrate and it seems as if the ascorbic acid is acting more as a substrate. And this is sort of going on inside or outside the cell, but then to keep this going without keeping on feeding ascorbic acid you'd have to bring it back to its original state. Could you give us a clue as to the rest of the cycle, besides the part that you showed us?

R. E. HEIKKILA: I think Dr. Diliberto alluded to that earlier this morning in terms of how once you form the oxidation products of ascorbic acid, there are different mechanisms and there are enzymes which cause the reduction of these things back towards ascorbic acid. I was really focusing on ascorbic acid reducing these kinds of molecules and forming dehydroascorbate, but obviously as you pointed out it's very important, it doesn't stop with the ascorbic acid, or the dehydroascorbic acid. Something else happens to these and there's another kind of a cycle. And in terms of the ascorbate, you can look upon it as a catalyst in that it promotes the autoxidation of the 6-hydroxydopamine, for example, or you can look at it the other way around.

The U-shaped dose-response curve for ascorbic acid shows that at low concentrations you get a minimal effect; at intermediate concentrations, you get a greater effect, and at higher concentrations the effect is gone. So in fact at high concentrations, on the order of one to five or six millimolar, ascorbic acid only acts as reducing agent.

The concentration of ascorbate Dr. Oke mentioned in the first talk this morning that is present in the brain is around two millimolar. The concentrations that prevent autooxidation or prevent the lipid peroxidation in our system are on the same order of magnitude. So for example, a high ascorbate would be simply a reducing agent under these circumstances.

UNIDENTIFIED SPEAKER: Does this mean that people who are taking high doses of ascorbic acid are really taking doses which have a net overall effect on reducing systems in the body?

R. E. HEIKKILA: I can only speculate. Ascorbic acid at high dose is a very good reducing agent so if one attained those high doses after *in vivo* ingestion you'd be left with net reduction.

E. J. DILIBERTO: The answer to that question is probably starting to emerge in some other laboratories as well as Dr. Heikkila's where they are actually finding that administration of ascorbic acid does effect certain receptor systems in the brain. Mark Wightman, for one, has shown that you can show changes in the neuronal activity in the striatal tissue for instance.

So there probably will be effects to megadoses of vitamin C that may in fact override some systems that we consider a function of ascorbic acid.

UNIDENTIFIED SPEAKER: This just brings one thing to mind, basically Linus Pauling just may have a point about the role of ascorbic acid in mental illness.

UNIDENTIFIED SPEAKER: You have very nicely redemonstrated that rather low amounts of ascorbic acid in the presence of iron is very useful in producing lipid peroxidation and you have shown that this may have consequences *in vivo* with regard to receptors. My question is, Are we ever able to obtain *in vivo* such low concentrations in the brain? As far as I know the brain will retain the ascorbic acid even after prolonged ascorbic acid deficiency. The brain is different from all the other tissues that will be depleted in deficiency. So do you think that in marginal chronic ascorbic acid deficiency there is any chance of increased peroxidation or whether even in prolonged ascorbic acid marginal deficiency there is still enough ascorbic acid which remains an antioxidant rather than a prooxidant?

R. E. HEIKKILA: Yes, in terms of a general statement I'm not here to say that all these reactions are going to take place and ascorbic acid is bad. What I'm really doing is pointing out some of the interesting chemistry of ascorbic acid and how one can use this compound as a reducing agent. The net concentration of ascorbate under normal circumstances is probably about 1-3 millimolar in the brain. The intracellular localization of the concentration, the concentration of the synaptic area is anybody's guess. What the regional localized concentration of ascorbate is would be difficult to say. I don't think our brain is peroxidizing away under normal circumstances. I think there are lots of defense mechanisms. I'm merely pointing out that these are theoretical reactions which can take place. I'm not saying that ascorbic acid is only a prooxidant, it's a very effective reducing agent. At high concentrations ascorbate may be maintaining tocopherol levels. Tocopherol interferes with and can break this chain reaction of lipid peroxidation in these systems. When lipid peroxidation is going on we must be losing our tocopherol, so the ascorbic acid (by maintaining the tocopherol concentration) is protective. Thus under certain conditions it's a prooxidant, under other conditions it's an antioxidant. And some of these complicated reactions which are taking place at these intermediate concentrations may, as Dr. Diliberto pointed out, be important reactions.

The cycling reaction obviously has a general aspect about it. Paraquat is an herbicide that was used to spray marijuana crops. It has caused lung damage when people have ingested it; it engages in redox cycling. So for these general reactions that I'm talking about, whether or not it's ascorbic acid or another reducing agent or an enzyme is not important. And I was not necessarily making the statement that ascorbic acid is going to do this or do that under these various conditions.

The brain is extremely hard to deplete of ascorbate. So your points are well taken.

UNIDENTIFIED SPEAKER: Some of the studies show that as you increase the dose the actual amount that's getting into the body decreases and also because of saturation of tubular reabsorption mechanisms the exposure to the body is considerably less than the actual dose that's ingested.

Ascorbic Acid and Acetylcholine Receptor Expression[a]

DAVID KNAACK,[b] THOMAS R. PODLESKI, AND MIRIAM M. SALPETER

Section of Neurobiology and Behavior
Cornell University
Ithaca, New York 14853

Adult skeletal muscle contains comparatively low levels of ascorbic acid.[1] Embryonic levels are more than twice those observed in adult, with adult levels being reached a few days prior to hatching in the chick,[2] but not until two weeks postpartum in the rat.[3] Denervation of adult skeletal muscle results in a return to embryonic ascorbic acid levels.[4,5] The timing of these changes in cellular ascorbic acid is similar to that seen for changes in the synthesis of the receptor for the neurotransmitter acetylcholine. Acetylcholine receptor (AchR) synthesis is high in embryonic muscle, declining to adult levels prior to hatching in the chick[6] and after two weeks postnatally in the rat.[7] AchR synthesis also increases dramatically following skeletal muscle denervation in all species studied.[8–10]

Experimental evidence from our laboratories has also indicated a relationship between exogenous ascorbic acid and AchR synthesis in the skeletal-muscle-derived L5 and L6 cell lines. Some years ago Podleski and Salpeter found that addition of rat brain and spinal cord extracts to the growth medium of L6 cells led to an increase in the site density of surface membrane AchR on differentiated myotubes.[11] A primary active component was subsequently purified and found to be identical with commercially available ascorbic acid by a number of criteria, including ultraviolet spectrum, mass spectrum, migration on thin-layer chromatography, and nuclear magnetic resonance spectroscopy (NMR).[12] NMR homonuclear decoupling techniques were employed to study the spatial relationships among the carbons in the 4-, 5-, and 6-positions. The homonuclear decoupling data (FIG. 1) confirm that the purified factor was not an ascorbic acid isomer with an alternative side chain configuration (see figure legend for details and interpretation). Commercially available ascorbic acid also increased AchR expression on L5 cells with the same dose-response characteristics as the purified factor.[12,13]

The response by L5 cells to ascorbic acid was further characterized and compared to the response to unfractionated brain extracts.[13] The increase in AchR expression

[a]This work was supported by grant no. GM 10422 from the National Institutes of Health (to MMS) and by a grant to TRP, MMS, and R. Oswald from the Cornell University Biotechnology Program, sponsored by the New York State Science and Technology Foundation and a consortium of industries.

[b]Present address: Laboratory of Molecular and Cellular Cardiology, Howard Hughes Medical Institute, Department of Cardiology, The Children's Hospital, Harvard Medical School, Boston, Massachusetts 02115.

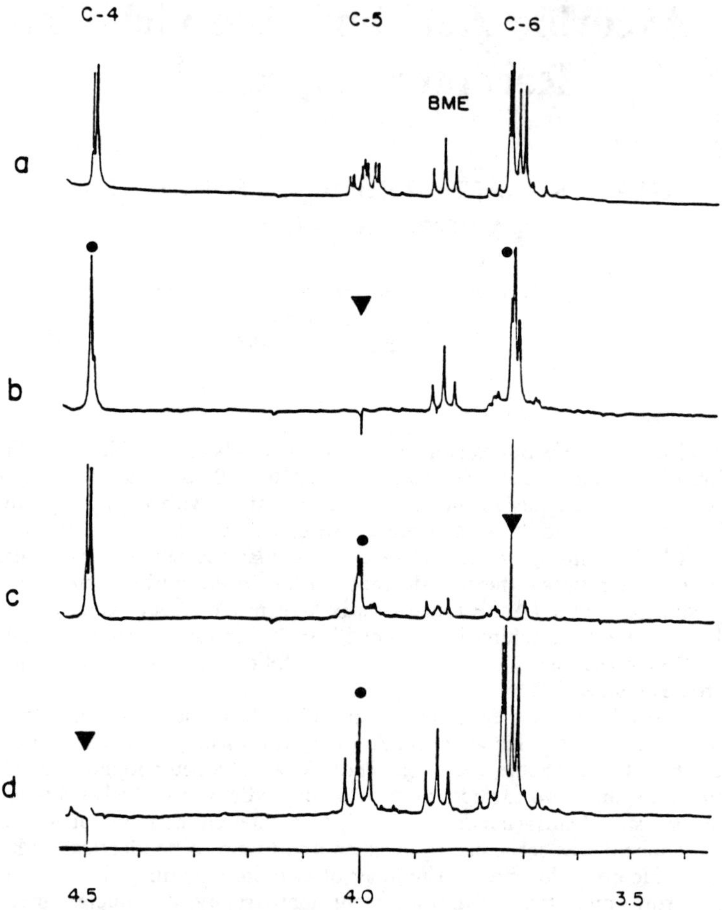

FIGURE 1. ^{1}H-NMR homonuclear decoupling experiments on AchR-inducing activity purified from fetal calf brain. Establishment of the same side chain configuration as ascorbic acid. NMR of purified AchR-inducing activity, previously identified as ascorbic acid or an ascorbic acid isomer. (**a**) The C-4-H, C-5-H, and C-6-H resonances of the purified fraction are as indicated and are the same as published values for authentic ascorbic acid.[21] Resonances of the methylene protons from residual 2-mercaptoethanol used in the purification are labeled in the figure. (**b,c,d**) Homonuclear decoupling (HD) experiments. The ▼ indicates the irradiated resonance, and ● indicates the associated decoupled resonances. (**b**) HD of the H-5 resonance demonstrates its coupling to both the H-6 and H-4 resonances. Irradiation of the H-5 resonance collapses the H-4 doublet to a singlet, and the H-6 multiplet to a triplet. Thus C-5 is adjacent to both C-6 and C-4. (**c**) Irradiation of the H-6 resonance collapses only the H-5 multiplet to a doublet. Thus C-6 is adjacent only to C-5. The vertical line at 3.75 ppm is an artifact resulting from irradiation at that frequency. The decreased amplitude of the 2-mercaptoethanol triplet in this experiment is due to the increased irradiation power needed to properly decouple the H-6 nuclei. Note there is no change in the coupling constant of the 2-mercaptoethanol resonance, indicating, as expected, no coupling to the irradiated C-6 proton. (**d**) Irradiation of the H-4 resonance collapses only the H-5 multiplet. Thus C-4 is adjacent to C-5, but not C-6, confirming the linear arrangement of the side chain carbons with C-4.

in ascorbic-acid-treated cells is due to a 2.8-fold increase in the rate of AchR insertion into the surface membrane and is not the result of a change in surface membrane AchR degradation. The AchR induction is not accompanied by a corresponding increase in creatine kinase activity, total protein levels, or myoblast fusion, and occurs in the presence of the mitotic inhibitor cytosine arabinoside (AraC). When compared to the biological activity of unfractionated brain extract, the activity of commercially available ascorbic acid can account for the increase in both AchR surface membrane site density and total culture AchR. Ascorbic acid and unfractionated brain extracts also had similar effects with regard to a number of other cellular parameters and neither had an appreciable effect on creatine kinase levels. Although ascorbic acid also caused some AchR clustering, it was only about one-fifth the level of that caused by brain extract. Preliminary characterization of the clustering activity of brain extracts suggests that another factor or factors of apparent high molecular weight is (are) the primary source(s) of clustering activity.[14]

The finding that ascorbic acid elevates AchR expression in L5 cells suggests an involvement of ascorbic acid in AchR biosynthesis or in a process that directly affects AchR biosynthesis. Our current investigation into the relationship between ascorbic acid and AchR synthesis has been directed along two lines. The first is to study further the specificity of the L5 cells' response to exogenously applied ascorbic acid in terms of effects on cellular RNA levels. The second involves the characterization of endogenous levels of cellular ascorbic acid in L5 cells and primary muscle cell cultures. These two issues will be considered separately.

EFFECTS OF ASCORBIC ACID ON CELLULAR TRANSCRIPT LEVELS

We have obtained cDNA probes for a number of muscle-differentiation-specific RNAs. They are: α-actin, creatine kinase, myosin light chain 2 (MLC2) and the α-subunit of the AchR. These probes were used to compare transcript levels in total RNA preparations from L5 cells either treated with ascorbic acid or receiving no treatment. FIGURE 2 shows autoradiograms of a Northern blot probed for the α-subunit of the AchR alone (lanes 1 and 2) and the same blot reprobed for MLC2 (lanes 3 and 4). Densitometric analysis of autoradiograms from Northern blots of four separate extractions of total RNA (see legend to FIG. 2 for details) indicates that the α-subunit transcript increases 3.8-fold by 72 hours after ascorbate addition. This increase is of approximately the same magnitude as the increase in AchR observed in the surface membrane.[12,13] At the same time, transcript levels for MLC2 are decreased by approximately 50%. We have used slot blots as our primary means of quantification of these changes in RNA levels. The results of the slot blot experiments are given in TABLE 1. The increase in AchR mRNA determined by slot blots was approximately 30% lower than that found by Northern blots. MLC2 changes, on the other hand, were approximately the same by either measure. To summarize this information, AchR transcript levels increase about the same magnitude as the increase in the AchR protein. The transcript levels of both MLC2 and α-actin are decreased and those of creatine kinase are unchanged.

We conclude from these data that the increased expression of surface membrane AchR caused by ascorbic acid is due to an increase in ribosomal synthesis of the protein. This induction is specific in that not all differentiation-specific proteins are

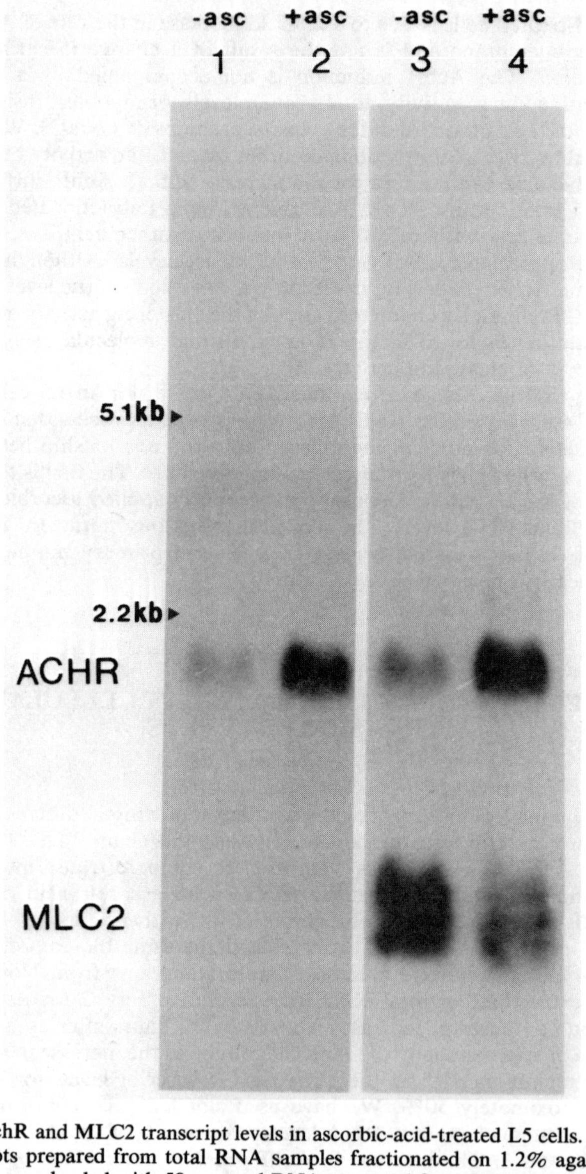

FIGURE 2. AchR and MLC2 transcript levels in ascorbic-acid-treated L5 cells. Autoradiogram of Northern blots prepared from total RNA samples fractionated on 1.2% agarose denaturing gels. Lanes 1-4 were loaded with 50 μg total RNA extracted from L5 cells 7 days after plating. Lanes 1 and 3 are RNA from untreated control cells. Lanes 2 and 4 were from ascorbic-acid-treated cells. Probes: Lanes 1 and 2 were probed for α-subunit AchR mRNA using a pA59-derived SP6 RNA probe (see below). Lanes 3 and 4 are the identical blots as lanes 1 and 2, reprobed for MLC2 mRNA with the cDNA probe p103. Preparation of the SP6 plasmid[22] containing an insert from an *Ava*I-*Pst*I digest of pA59 cDNA, as well as preparation of ^{32}P-labeled probe by reverse transcription, was done using riboprobe kits (Promega Biotec) according to manufacturer's directions. cDNA probes were nick-translated with ^{32}P according to the method described by Maniatis *et al.*[23] Hybridization of RNA probes was for 6 hours at 60°C in 50%

affected, and among those that are, the transcript levels of at least two are decreased. Further characterization of the response with regard to other differentiation-specific and integral membrane proteins will be important to establish the scope of the cellular response to ascorbic acid.

COMPARISON OF ENDOGENOUS ASCORBATE LEVELS IN L5 CELLS AND PRIMARY MUSCLE CELL CULTURES

Evidence presented above on L5 cells suggests that ascorbic acid may play a role in the synthesis of AchR by skeletal muscle. In contrast to L5 cells, it has been reported for both rat and chick primary cultures that AchR levels on differentiated myotubes do not increase in response to exogenously applied ascorbic acid[15,16] (although preliminary data from our laboratory suggest a chick primary response to ascorbic acid.) Furthermore, rat primary myotubes express almost five times as much surface membrane AchR as L5 and chick primary myotubes (ref. 18 and unpublished observations). This high level of AchR expression occurs in the absence of exogenously applied ascorbic acid. Any hypothesis concerning an involvement of ascorbic acid in AchR biosynthesis must ultimately be reconciled with these observations. One approach to understanding the differences between the cloned L5 cells and primary cultures is the determination of endogenous cellular ascorbate levels in the culture systems.

Our L5 cultures are grown in Dulbecco's modified growth medium plus 10% heat-inactivated fetal calf serum. Although there is no ascorbic acid in Dulbecco's modified Eagle's medium, we determined that ascorbic acid is present in fetal calf serum in a concentration range of 1-2 μg/ml. Thus medium concentrations of ascorbic acid are about 150 ng/ml. This concentration is about an order of magnitude less than that needed for a minimal AchR response by L5 cells,[12,13] but the presence of fetal-calf-serum-derived ascorbate would be expected to contribute to baseline cellular ascorbate acid levels. Measurable levels of ascorbic acid were indeed present in L5 cells. These levels fell exponentially over the culture period with a half-time of less than 2 days (FIG. 3). Similar measurements of ascorbic acid in rat primary cultures, however, led to a surprising finding. Data from two experiments demonstrated that starting levels of cellular ascorbic acid were variable, even when normalized to total protein, but were in the range of initial cellular ascorbate levels found in L5 cells.

---

formamide, 6× SSPE, 4× Denhardts, 0.5% SDS, 100 μg/ml salmon sperm DNA. DNA probes were hybridized overnight at 65°C in 1% SDS, 1 M NaCl, 10% dextran sulfate, 100 μg/ml denatured salmon sperm DNA. Blots were washed for 2 hours in 0.1× SSC, 0.1% SDS at 70°C. RNA: Total RNA was prepared by LiCl-urea extraction according to the method of Auffray *et al.*[24] L5 cells were extracted directly from the tissue culture dishes and extracts vortexed vigorously. All samples were extracted for a minimum of 24 h at 4°C and frozen at −70°C prior to centrifugation and phenol-chloroform extraction. Cell and tissue treatments: Differentiating L5 cells were dosed with ascorbic acid for two consecutive days (concentration in growth medium = 30 μg/ml) as reported previously.[12] Total RNA was prepared 72 hours after the first addition of ascorbic acid. Agarose gels and blotting: 1.2% denaturing agarose gels with formaldehyde were run overnight and blotted onto Gene Screen Plus (du Pont NEF-976) following the manufacturer's directions. Scanning densitometry: Preflashed X-ray film (Kodak XAR-2) was exposed overnight and developed according to manufacturer's directions. Autoradiograms were scanned on a Zenith soft laser scanning densitometer with integrator.

Unlike L5 cells, however, ascorbic acid levels in the rat primary cultures increase over the culture period (FIG. 4). Without normalization to total protein, the rate of increase in total culture ascorbic acid (FIG. 5) determined from two separate experiments was 93.7 ng/day ($r = 0.81$). A single experiment was also performed on chick primary cultures (FIG. 6) with similar results. In this case, the rate of increase was 52.3 ng/day ($r = 0.93$). By comparison, total ascorbate levels in L5 cells decrease with time in culture (FIG. 7). It is important to point out that rat primary cultures receive fresh media the day after plating and are not refed over the experimental period. Thus after day 2, no additional ascorbic acid is introduced to the cultures. (The chick cell feeding schedule was that used by Jessel *et al.*[26] and had two feedings.)

Although not yet conclusively demonstrated, we believe the cellular ascorbic acid detected in primary cultures is within differentiated myotubes. This is likely to be the case for the following reason: the mitotic inhibitor AraC is added as 10 μl of a 10^{-3} M stock solution on day 3 or 4 and the experiments are carried out in its presence. Fibroblasts do persist in culture even in the presence of AraC, however in much lower numbers than in the absence of AraC. Yet cultures grown in the presence of AraC

TABLE 1. Ascorbic Acid Effects on Transcript Levels of Some Differentiation-Specific Proteins in L5 Cells

Differentiation-Specific Transcript	Transcript Level Relative to Control (± SEM)	Number of Experiments
Acetylcholine receptor	2.4 ± 0.22	3
Myosin light chain 2	0.53 ± 0.08	3
α-Actin	0.80 ± 0.18	2
Creatine kinase	1.0 ± 0.28	2

All data is derived from slot blot analysis of total RNA preparations. See legend to FIGURE 2 for RNA workup. cDNA probes used were as follows: AchR, pA59[27]; MLC2, p103[28]; α-actin, pα[29]; creatine kinase, pckm-1.[30]

had the same ascorbate levels (higher when normalized to protein levels), as sister cultures in the absence of AraC. If fibroblasts were the source of the cellular ascorbate, AraC free cultures should have contained more ascorbate.

In spite of some variability in starting levels, the pattern of change in cellular ascorbate is reproducible for both cell types, even without normalization (see FIGS. 5 and 7). Thus, levels of endogenous ascorbate decline in L5 cells but increase in primary cultures derived from either chick or rat skeletal muscle. While the increase in cellular ascorbate may be due to *de novo* synthesis of ascorbic acid, other potential explanations exist and must be explored. They are: (1) Fetal-calf-serum-derived ascorbic acid in the growth medium of primary cultures might be more stable than in the medium of L5 cultures, resulting in prolonged ascorbic acid uptake and accumulation by primary cells. (2) Cellular ascorbic acid is known to exist in both bound and soluble forms. It might be possible that a repartitioning of cellular ascorbic acid occurs over time in primary cultures such that the existing ascorbate is more readily solubilized by TCA at later times. Although we cannot yet make conclusions regarding the mechanism of the increase in cellular ascorbic acid that occurs in the primary cultures, we have identified a clear difference between primary and L5 cells in their

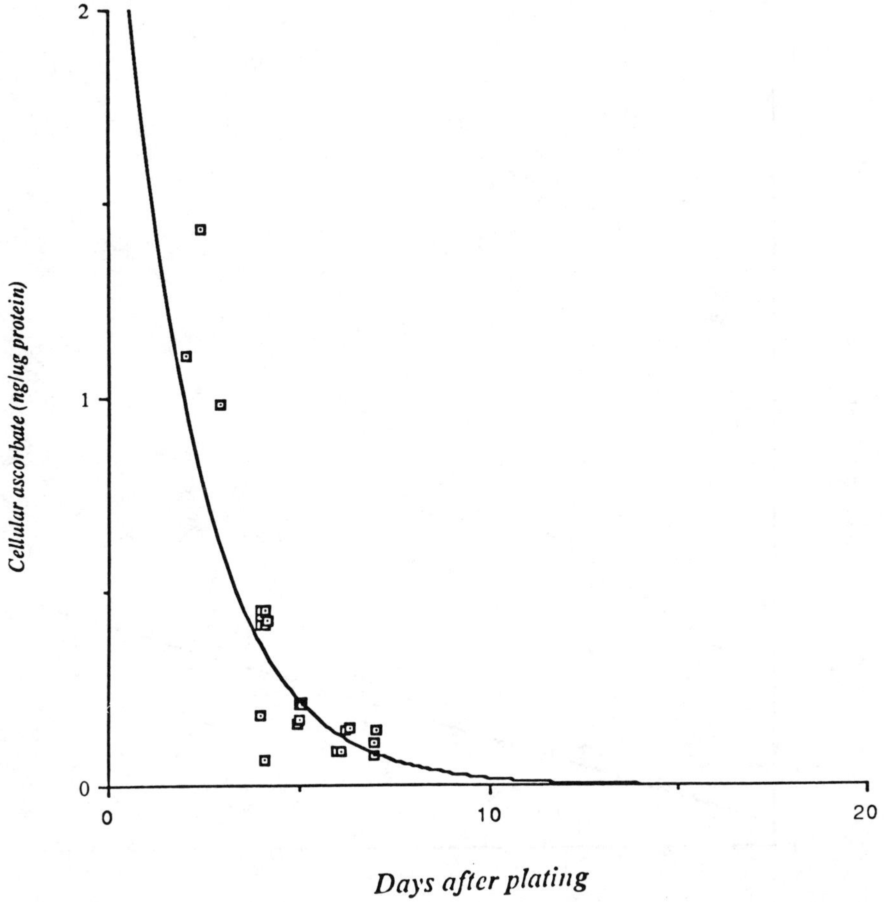

FIGURE 3. L5 cells: endogenous cellular ascorbic acid levels normalized to total protein. Ascorbic acid levels (ng/μg protein) are given as a function of days in culture. The half-life of the curve (generated from Cricket Graph, Cricket Software Co.) is 1.4 days ($r = 0.84$). Each data point represents the quotient of the means of duplicate determinations of ascorbate and total protein levels. Data are pooled from five experiments. Either 60 or 100 mm dishes were harvested at the indicated time points in 0.4 ml of 0.01% trichloroacetic acid (TCA). Homogenates were vigorously vortexed and an equal volume of 10% TCA was then added. Samples were centrifuged in an Eppendorf table-top centrifuge, supernates decanted, and stored at −70°C until assayed for ascorbic acid content by the method of Omaye *et al.*[20] TCA precipitates were dissolved in 50 μl of 6 N NaOH, heated at 38°C for approximately 2 h, then diluted to 0.5 ml with H_2O. A portion of this diluted sample was assayed for total protein by the method of Lowry.[25] Bovine serum albumin was used as a standard. Tissue culture was as reported previously for L5 cells.[12]

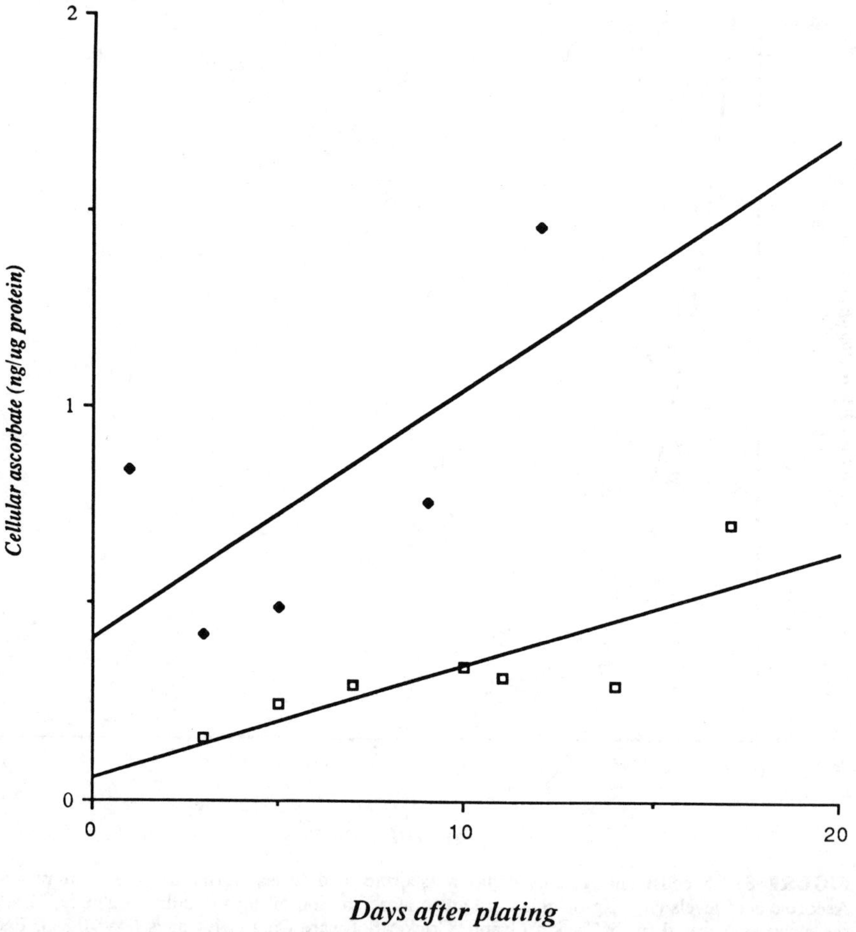

FIGURE 4. Primary rat myogenic cultures: endogenous cellular ascorbic acid levels normalized to total protein. Ascorbic acid levels (ng/μg protein) are given as a function of days in culture for two separate platings of rat primary cultures. The slopes and correlation coefficients for the two experiments are as follows. ◆ : slope = 0.06 ng/μg × day, $r = 0.69$; □: slope = 0.03 ng/μg × day, $r = 0.2$. Each data point is the quotient of the means for ascorbic acid and total protein from duplicate plates. Cell harvesting and ascorbic acid and total protein assays were as given in the legend to FIGURE 3. Rat primary cultures were prepared as reported previously,[18] except that the dishes were not coated with collagen.

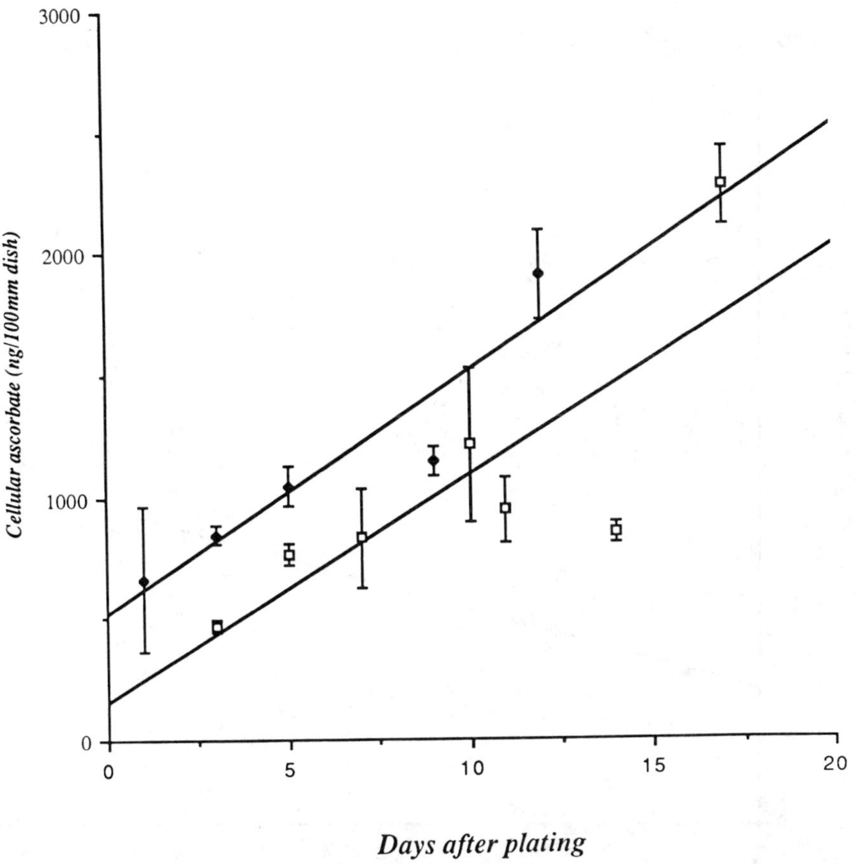

FIGURE 5. Primary rat myogenic cultures: total cellular ascorbate with time in culture. Ascorbic acid levels are given as total cellular ascorbate per 100 mm dish. Data are from two separate experiments. Each data point represents the mean of duplicate determinations (*error bars* = range). Linear regressions were determined for each experiment (Cricket graph). Slopes and correlation coefficients are as follows: □: slope = 93.2 ng/day, r = 0.80; ◆: slope = 99.8 ng/day, r = 0.94. Data were subsequently pooled by normalizing the data to the y intercept for each data set (y intercept = 0). The new regression equation was then calculated from the pooled normalized data giving the slope and correlation coefficient as indicated in the text.

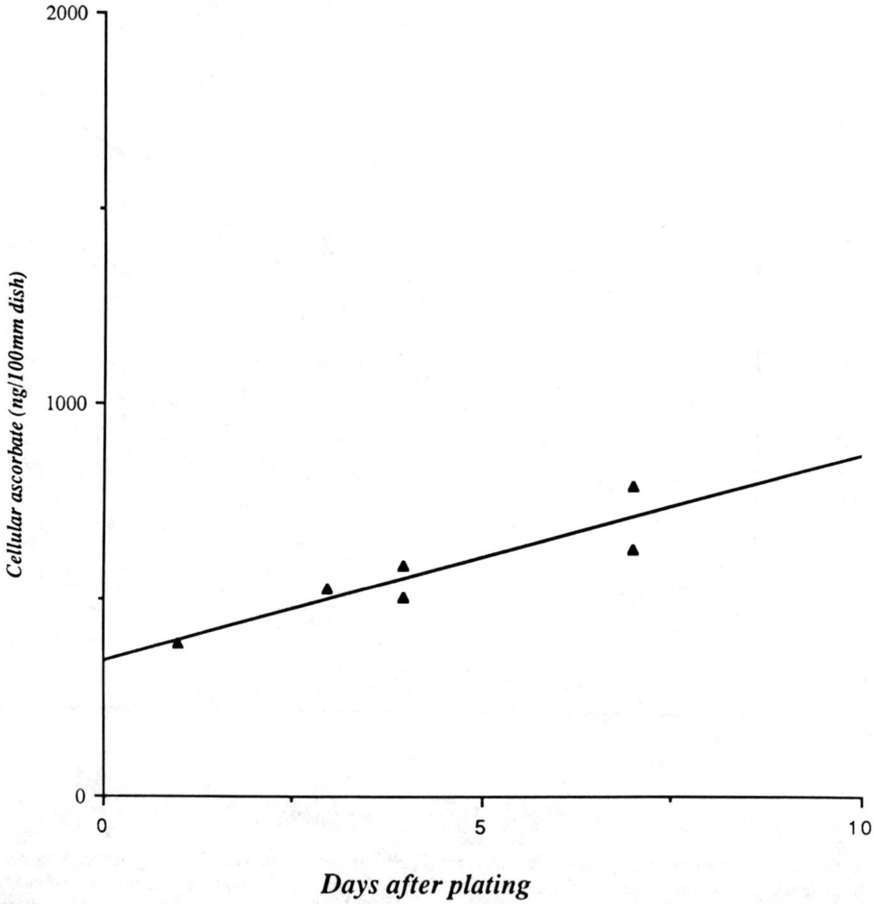

FIGURE 6. Primary chick myogenic cultures: total cellular ascorbate with time in culture. Time course data are from a single experiment. Methods are as given in the legends to FIGURES 3 through 5, except tissue culture conditions were according to the method of Jessel *et al.*,[26] and cultures were grown on collagen.

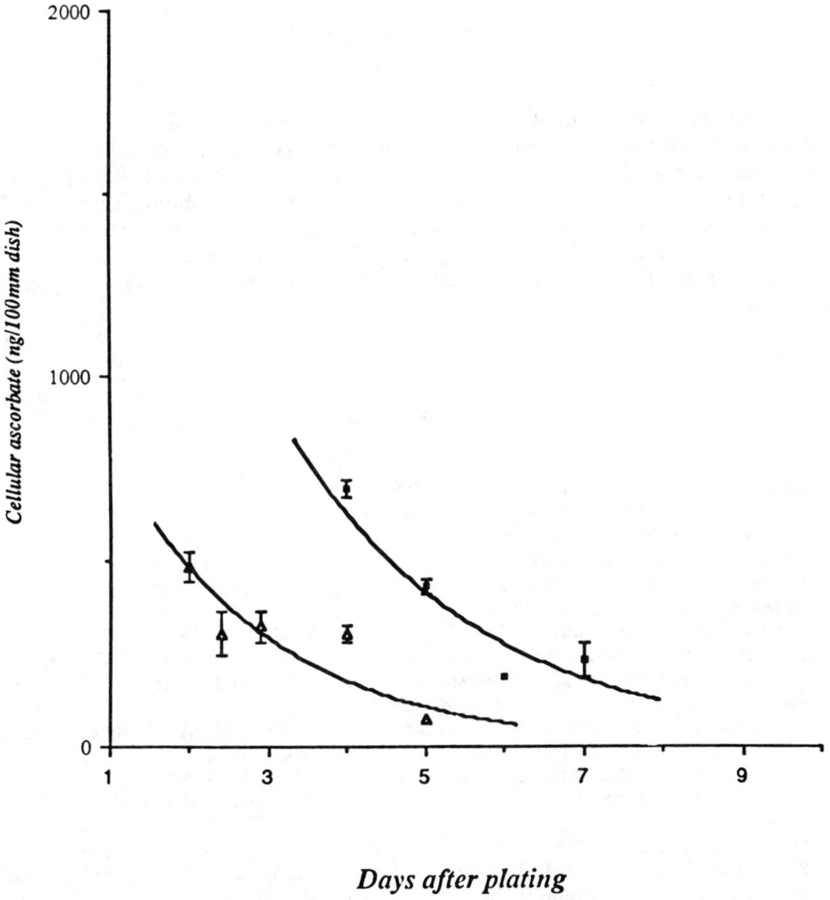

FIGURE 7. L5 cells: total cellular ascorbate with time in culture. Ascorbic acid levels are given as total cellular ascorbate per 100 mm dish. Data are from two separate experiments. Each data point represents the mean of duplicate determinations (*error bars* = range). Data are fitted to exponentials (Cricket graph); however, correlation coefficients were not significantly different when fitted by linear regression. ■: slope = −163 ng/day, r = 0.91; △: slope = −104 ng/day, r = 0.88.

ability to maintain cellular ascorbic acid levels. How this difference may be related to AchR expression is not clear but merits further investigation.

ACKNOWLEDGMENTS

The authors wish to thank Smadar Admon for her invaluable assistance and guidance in all phases of this work. We gratefully acknowledge the gifts of cDNA probes from the following individuals: David Yaffe for the myosin light chain 2 (p103), Joe Calvo for the α-actin (pα) probes, John Merlie for the α-subunit AchR (pA59) probe, and Pamela Benfield and the Du Pont Corporation for the creatine kinase probe (pckm-1). Gaetano Monteleone generously performed the NMR analysis. In addition we thank Dorothy Bell and Iris Greenberg for help in primary myogenic culture experiments.

REFERENCES

1. HORNIG, D. 1975. Ann. N.Y. Acad. Sci. **258:** 103-118.
2. FABBRO, S. & L. M. RINALDINI. 1965. Dev. Biol. **11:** 468-488.
3. PATNAIK, B. & M. KANUNGO. 1966. Biochem. J. **100:** 59-62.
4. GRAFF, G. L. A., A. J. HUDSON & K. P. STRICKLAND. 1965. Can. J. Biochem. **43:** 705-710.
5. KRISHNAMOORTHY, R. V. & P. SATYAM. 1969. Experientia **25:** 267-268.
6. BURDEN, S. 1979. Dev. Biol. **57:** 317-329.
7. BEVAN, S. & J. STEINBACH. 1977. J. Physiol. (Lond.) **267:** 195-213.
8. MERLIE, J., K. ISENBERG, S. RUSSELL & J. SANES. 1984. J. Cell Biol. **99:** 332-335.
9. GOLDMAN, D., J. BOULTER, S. HEINEMANN & J. PATRICK. 1985. J. Neurosci. **5:** 2553-2558.
10. FAMBROUGH, D. 1979. Physiol. Rev. **59:** 165-227.
11. PODLESKI, T. R., D. AXELROD, P. RAVDIN, I. GREENBERG, M. M. JOHNSON & M. M. SALPETER. 1978. Proc. Natl. Acad. Sci. U.S.A. **75:** 2035-2039.
12. KNAACK, D. & T. PODLESKI. 1985. Proc. Natl. Acad. Sci. U.S.A. **82:** 575-579.
13. KNAACK, D., I. SHEN, M. M. SALPETER & T. R. PODLESKI. 1986. J. Cell Biol. **102:** 795-802.
14. NEUGEBAUER, K., M. M. SALPETER & T. R. PODLESKI. 1985. Brain Res. **346:** 58-69.
15. KALCHEIM, C., Z. VOGEL & D. DUSKIN. 1982. Proc. Natl. Acad. Sci. U.S.A. **79:** 3077-3081.
16. USDIN, T. & G. D. FISCHBACH. 1986. J. Cell Biol. **103:** 493-507.
17. LAND, B., T. R. PODLESKI, E. E. SALPETER & M. M. SALPETER. 1977. J. Physiol. (Lond.) **269:** 155-176.
18. SALPETER, M. M., S. SPANTON, K. HOLLEY & T. R. PODLESKI. 1982. J. Cell Biol. **93:** 417-425.
19. ROE, J. H. & C. A. KUETHNER. 1943. J. Biol. Chem. **147:** 399.
20. OMAYE, S. T., J. D. TURNBULL & H. E. SAUBERLICH. 1979. Methods Enzymol. **62:** 7-8.
21. PAUKSTELIS, J. V., D. D. MUELLER, P. A. SEIB & D. W. LILLARD, JR.. 1982. *In* Ascorbic Acid: Chemistry, Metabolism, and Uses. P. A. Seib & B. M. Tolbert, Eds.: 125-152. American Chemical Society. Washington, D.C.
22. MELTON, D. A., P. A. KRIEG, M. R. REBAGLIATI, T. MANIATIS, K. ZINN & M. R. GREEN. 1984. Nucleic Acids Res. **12:** 7035-7056.
23. MANIATIS, T., A. JEFFREY & D. KLEID. 1975. Proc. Natl. Acad. Sci. U.S.A. **72:** 1184-1188.
24. AUFFRAY, C., R. NAGEOTTE, B. CHAMBROUD & F. ROUGEON. 1980. Nucleic Acids Res. **8:** 1231.

25. Lowry, O. H., N. J. Roseborough, A. L. Farr & R. J. Randall. 1951. J. Biol. Chem. **193:** 265-275.
26. Jessel, T. M., R. E. Siegel & G. D. Fischbach. 1979. Proc. Natl. Acad. Sci. U.S.A. **76:** 5397-5401.
27. Merlie, J. P., R. Sebbane, S. Gardner & J. Lindstrom. 1983. Proc. Natl. Acad. Sci. U.S.A. **80:** 3845-3849.
28. Katcoff, D., U. Nudel, D. Zevin-Sonkin, Y. Carmon, M. Shani, H. Lehrach, A. M. Frischauf & D. Yaffe. 1980. Proc. Natl. Acad. Sci. U.S.A. **77:** 960-964.
29. Melloul, D., B. Aloni, J. Calvo, D. Yaffe & U. Nudel. 1984. EMBO J. **3:** 983-990.
30. Benfield, P. A., R. A. Zivin, L. S. Miller, R. Sowder, G. W. Smythers, L. Henderson, S. Oroszlan & M. L. Pearson. 1984. J. Biol. Chem. **259:** 14979-14984.

DISCUSSION OF THE PAPER

M. Levine (*National Institutes of Health, Bethesda, Md.*): I have two questions again. The first is a simple one: Is the newly synthesized receptor the same as the receptor that was previously there? Do you have any way of knowing that?

D. Knaack (*Cornell University, Ithaca, N.Y.*): We haven't explored that issue. However, current information regarding developmentally regulated alternative subunit expression by the receptor *in vivo* suggests that would be an important avenue of investigation.

M. Levine: The second question relates to the Northern blots. If you inhibit RNA synthesis and then do your analysis, what happens?

D. Knaack: We haven't done that. I think the more direct phrasing of that question is whether the increase in acetylcholine receptor mRNA is due to increased transcription or is it simply transcript stabilization. We don't have an answer. The copy number of acetylcholine receptor is very low, and it's very hard to do runoff experiments which would then tell us about the transcriptional status of these messages in these systems. All we can say right now is cellular levels of RNA increase and it could be due to stabilization or a true induction.

E. J. Diliberto: (*Wellcome Research Laboratories, Research Triangle Park, N. C.*) Were the primary culture data that you showed with ascorbic acid added to the medium or not?

D. Knaack: No, with the caveat that there's fetal calf serum 10% in there, and that makes available something like 1500 nanograms of ascorbate in a 100 mm dish.

E. J. Diliberto: What you get with the L5 cells is identical to what we get with the chromaffin cells, which is a half-life of about 31 hours.

D. Knaack: Have you checked if that's the same as the medium half-life, which is what we assume it is?

E. J. Diliberto: Well I'm talking about in cultured cells. When we first put them in culture it disappears in the same profile that you've seen, almost identical. Even in the presence of fetal calf serum we do get a decline. It's been checked by a number of people, and there is no synthesis of ascorbic acid in that tissue. But you raised a very interesting question in terms of the ganglion—whether or not that should be really checked.

Ascorbic Acid Intakes and Plasma Levels in Healthy Elderly[a]

PHILIP J. GARRY, DOROTHY J. VANDERJAGT, AND WILLIAM C. HUNT

Clinical Nutrition Laboratory
Department of Pathology
University of New Mexico School of Medicine
Albuquerque, New Mexico 87131

INTRODUCTION

The human need for ascorbic acid to prevent scurvy is approximately 10 mg/day. The current recommended daily dietary allowance (RDA) for ascorbic acid is 60 mg, and in the judgement of the Committee on Dietary Allowances of the Food and Nutrition Board, this amount will meet the known needs of practically all healthy Americans and assure adequate body reserves of ascorbic acid.[1]

The RDA for some vitamins, including ascorbic acid, has been reexamined in recent years. For example, Levine[2] has questioned whether a daily intake of 60 mg is the optimal amount. Levine suggests that the optimal dietary level of ascorbic acid may be different from the RDA, and that it is important to determine this level. Recently, Schneider *et al.*[3] suggested that age-related alterations in physiology and metabolism can change human nutritional requirements for certain vitamins and minerals. Their concern is that RDAs for adults are established without sufficient scientific information about the interaction between nutrition and aging. While Schneider *et al.*[3] did not specifically discuss whether the current RDA for ascorbic acid is adequate to meet the needs of our older citizens, they stress that research related to the formation of RDAs for older populations is important and that this research may also yield new insights into the mechanisms of aging and age-dependent diseases.

Unfortunately, there is little information on the pharmacokinetics of ascorbic acid in the elderly that is similar to that which has been obtained in younger males by Kallner *et al.*[4] Most studies of elderly individuals have employed plasma and/or leukocyte levels as an indirect assessment of the ascorbic acid needs of the elderly. Assessments of the risk for developing clinical symptoms associated with hypovitaminosis C have been based primarily on serum or plasma ascorbic acid concentrations. Plasma ascorbate determinations have been the favored measurement of the ascorbic acid status in individuals and populations because it is technically more difficult to measure leukocyte ascorbic acid, and also because plasma ascorbate concentrations have shown a better correlation with intake of vitamin C than leukocytes.[5]

[a]This work was supported by grants from the United States Public Health Service (nos. AG02049 and RR-00997-05,06) and a grant-in-aid from Hoffmann-La Roche.

Previous studies of the ascorbic acid needs of the elderly, using plasma ascorbic acid levels, have reported conflicting results and this has been summarized in the recent review by Cheng *et al.*[6] For example, a few reports have shown that adult women have higher plasma levels than men at given intakes of ascorbic acid. It has also been reported that plasma ascorbic acid levels decrease with advancing age. However, these effects have not always been seen when elderly subjects have been studied.[6]

In this presentation we will examine the relationship between ascorbic acid intake and plasma ascorbic acid concentrations in a large group of healthy elderly men and women. Understanding this relationship in the healthy elderly may help determine whether age should be considered in future recommendations of ascorbic acid intake. We have summarized results from two separate studies. The first is a longitudinal study in which intakes and plasma levels of ascorbic acid were obtained from a group of 270 elderly men and women in 1980, 1981, 1984, and 1985. Part of this data has been published previously.[7] The second study is a clinical trial involving eight men and nine women in which daily intake of ascorbic acid was strictly controlled at six different levels ranging from 30 to 280 mg, and these results have also been reported.[8]

From the information obtained in these two studies, we hypothesize that the sex difference we find in plasma ascorbic acid concentration is related to a difference in the renal absorptive process for ascorbic acid. Also, we propose a desirable dietary intake of ascorbic acid to assure adequacy in the elderly.

SUBJECTS AND METHODS

In 1979, 304 healthy elderly men and women were recruited for a longitudinal study of nutrition in the elderly. One of the aims of this study was to examine the relationship between intake (diet and supplements) and plasma concentration of various essential nutrients. Only volunteers 60 years of age and older with no diagnosed major medical illnesses and taking no prescription medication (other than occasional hypnotic, laxative, or analgesic) were selected. The median age was 72 years in 1980. Although entrance was not restricted to any ethnic group, all volunteers were Caucasian; only 3% were Hispanic. A more complete description of the population was given previously.[9]

Volunteers were seen as outpatients in the Clinical Research Center at the University of New Mexico Hospital. The volunteers came to the clinic after an overnight fast and approximately 20 ml of heparinized blood was obtained for various biochemical measurements. Samples for ascorbic acid analysis were prepared by mixing equal volumes of plasma and metaphosphoric acid (10% wt/vol). After thorough mixing, the samples were frozen at −20°C until assayed (within 2 weeks). Plasma ascorbic acid concentrations were measured by an automated 2,6-dichlorophenolindophenol procedure.[10]

Dietary intake was assessed from 3-day food records collected during three successive weekdays that coincided with obtaining blood samples from the volunteers. At the end of the 3-day food recording period, a dietitian visited the volunteers' homes to check completeness of the food records. If vitamin and/or mineral supplements were used by the subject, the brand name, contents, and amounts of each nutrient were recorded for inclusion in determining total intakes. Dietary records that were incomplete or otherwise unreliable were not used. A more complete description of the dietary methodology used in this study is given in our previous report.[9] Dietary records were collected from the same volunteers at four different time periods, 1980, 1981, 1984, and 1985. Blood samples were obtained yearly beginning in 1980.

In 1985 a clinical trial on the effect of graded intakes of ascorbic acid on plasma and leukocyte levels was conducted in a subgroup of the elderly volunteers. These volunteers, eight men and nine women, were required to adhere to a dietary intake of ascorbic acid of 30 ± 10 mg/day for a period of 21 weeks. After the initial 3-week period, blood samples were taken for ascorbic acid analysis. The volunteers were then given a supply of capsules containing 15 mg ascorbic acid and instructed to take one capsule per day for a period of three weeks. At the end of this period blood samples were again obtained for ascorbic acid analysis. This same procedure was repeated five more times and at the end of each 3-week period the volunteers were given a new supply of capsules containing, in sequential order, 30, 60, 120, and 250 mg ascorbic acid. During the last 3-week period the capsule contained no ascorbic acid (placebo). Compliance to the dietary restriction of ascorbic acid was determined by obtaining a 24-hour dietary recall from the subjects on two occasions during each 3-week period. Blood samples were processed as described above, and all the plasma samples were analyzed for ascorbic acid and for total vitamin C (ascorbic acid plus dehydroascorbic acid) by two different high-performance liquid chromatographic methods described in the published results of this study.[8] Both studies described above were approved by the Human Research Review Committee of the University of New Mexico School of Medicine. Informed consent was obtained from each participant.

RESULTS AND DISCUSSION

The distribution of ascorbic acid intake for each year that dietary information was collected are presented in TABLES 1 and 2. Differences in mean ascorbic acid intake for men and women are slight for both dietary and supplemental sources. Although there is some variability in dietary intake over time, the mean intake of ascorbic acid remained approximately twice the current RDA of 60 mg/day for adults.[1] Analysis showed that the mean dietary intakes of ascorbic acid were the same for nonsupplemented individuals in comparison to those taking an ascorbic acid supplement. The

TABLE 1. Reported Daily Ascorbic Acid Intake (mg) from Diet

	Year	*n*	Mean	SD	Median	Q_1[a]	Q_3[b]
Men	1980	125	140	69	135	91	174
	1981	113	133	68	122	89	162
	1984	94	131	62	123	84	171
	1985	90	127	57	129	86	160
Women	1980	143	139	73	132	87	172
	1981	137	124	63	113	80	157
	1984	121	125	61	116	83	160
	1985	101	162	63	117	84	160

[a] Q_1 = 25th percentile.
[b] Q_3 = 75th percentile.

TABLE 2. Reported Daily Ascorbic Intake (mg) from Supplements

	Year	*n*	Percent[a]	Median	Q_1[b]	Q_3[c]
Men	1980	70	56	500	184	1000
	1981	68	60	500	200	937
	1984	56	60	500	105	688
	1985	59	66	500	90	1000
Women	1980	85	59	450	195	700
	1981	80	58	500	200	994
	1984	73	60	400	90	960
	1985	60	59	500	150	950

[a] Percentage of subjects taking a supplement containing ascorbic acid.
[b] Q_1 = 25th percentile.
[c] Q_3 = 75th percentile.

decrease in number of subjects from 1980 to 1985 is due to several factors, such as deaths, relocation, and an unwillingness by some volunteers to complete the 3-day diet record. TABLE 2 shows that the median intake of supplemental ascorbic acid was approximately 500 mg/day, although the range of intake is quite large as demonstrated by the 25th and 75th percentile values. Approximately 60% of elderly men and women consumed an ascorbic acid supplement. These particular elderly men and women are very health conscious and the high use of ascorbic acid supplements in this group of elderly cannot be generalized to the elderly population of the United States as a whole.

FIGURE 1 shows the relationship of estimated total daily intake of ascorbic acid (diet plus supplement) and plasma ascorbic acid concentration. The data in this figure represents 924 pairs of values collected over 4 years. To construct FIGURE 1, the intake axis was cut at every fifth percentile of intake and each point represents the median plasma ascorbic acid concentration for males and females within these cuts. A smooth curve was used to interpolate the points. Although some individuals had intakes above 700 mg, the graph has been truncated so that the relationship at lower levels of intake can be seen more clearly. We noted a sharp rise in plasma ascorbic acid concentration over a range of daily ascorbic acid intake of 20 to 100 mg, followed by a plateauing at higher intakes. A sex difference is clearly indicated in this figure and the difference is persistent throughout the range of intake.

In order to quantify more precisely the relationship between plasma level and intake of ascorbic acid, we regressed plasma ascorbic acid concentration on the square root of the reciprocal of ascorbic acid intake. This transformation was chosen to make the relationship approximately linear. The regression model is shown in FIGURE 2 for the data collected in 1984. The regression lines for men and women are nearly parallel, suggesting that women have higher plasma levels than men at all levels of intake. This regression model was applied separately to each of the four years and the results were very similar. For this reason we pooled the data from all four years. The estimated plasma levels at a daily intake of 1500 mg were 1.56 and 1.38 for women and men, respectively. The slopes are -5.58 for women and -5.96 for men, a statistically insignificant difference. R^2 for the model was 0.32.

In the previously described regressions we excluded individuals with daily ascorbic acid intakes greater than 1500 mg in order to eliminate the effects of plasma concentrations greater than 2.0 mg/dl found at higher levels of intake. These elevated plasma levels may not truly reflect steady-state plasma values at the time the blood samples

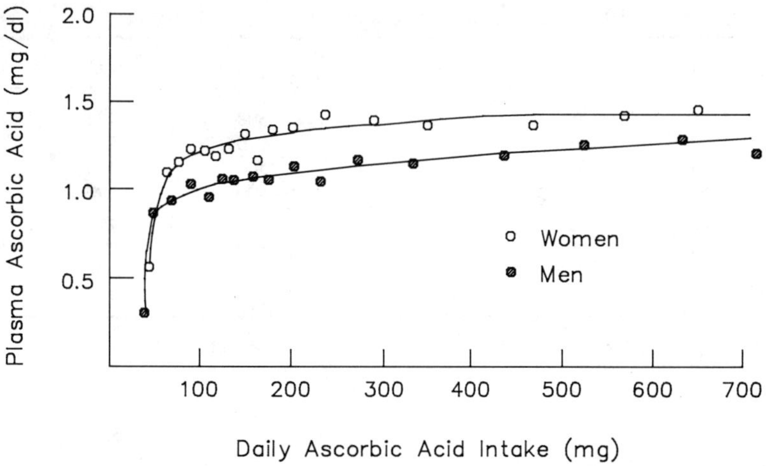

FIGURE 1. Median plasma ascorbic acid concentration by sex for each 5th percentile of intake. Median plasma ascorbic acid concentrations are plotted at the median ascorbic acid intake for each percentile. The data represents 914 pairs of values collected over 4 years. Daily intakes of ascorbic acid above 700 mg are not shown in this figure.

were collected, even though blood was obtained after an overnight fast. Melethil *et al.*[11] have recently shown that younger adults taking 500, 1000, or 2000 mg of ascorbic acid per day for 7 days required more than 12 hours for plasma concentrations to return to predose steady-state levels. The reason for this is that renal clearance of ascorbic acid is dependent on plasma concentration as well as time. This observation by Melethil *et al.*[11] may be exaggerated in the elderly who have been shown to have reduced kidney function in comparison to younger adults.[12]

We hypothesize that a difference in the renal tubular reabsorption maximum (T_m) between elderly men and women is responsible for the sex difference in plasma ascorbic acid levels at similar intakes. Evidence for a difference in T_m between elderly men and women in this population has recently been demonstrated for phosphorous.[13]

TABLE 3 shows that the median plasma ascorbic acid level in men who were not taking supplemental ascorbic acid was approximately 1.0 mg/dl. These men had a median daily dietary intake in the range of 121 to 135 mg (TABLE 1). TABLE 3 also shows that over a period of 6 years, when the average age of both the elderly male and female population increased from 72 to 77 years, there was no trend, either decreasing or increasing, in plasma ascorbic acid levels. Dietary intakes of ascorbic acid also showed no change over that period of time. This suggests that the age-related change in plasma ascorbic acid levels reported by others[14,15] may be related to a decreased intake of ascorbic acid with increasing age, and not to aging per se.

As can be noted in FIGURE 1, the relationship between estimated intake of ascorbic acid and plasma concentration is poorly defined at lower intakes because very few individuals consumed less than 75 mg per day (see TABLE 1). In order to define this relationship more precisely, we conducted a separate clinical trial in which ascorbic acid intake was controlled at levels ranging from 30 to 280 mg/day. FIGURE 3 shows that the median plasma ascorbic acid level increased rapidly for intakes above 45 mg/day in both elderly males and females. Steady-state plasma ascorbic acid concentration in males approached a maximum level of approximately 1.0 mg/dl at a daily intake in the range of 135 to 150 mg. Female subjects reached a plasma level of 1.0 mg/dl at a daily intake of approximately 75 to 80 mg, and as the intake increased to 280 mg, plasma concentrations gradually increased to a maximum of approximately 1.3 mg/dl. Analysis of variance showed that the average plasma ascorbic acid concentration in men was significantly lower than that in women and that the sex difference did not vary significantly by amount of supplementation. This controlled clinical trial confirmed the results in our previous study in which 3-day diet records were used to estimate dietary ascorbic acid intake.

It should be noted that plasma ascorbic acid concentrations in the controlled clinical study were determined by an HPLC procedure using an electrochemical detector, whereas in the previously described study we used an automated dichlorophenolindophenol (DCIP) method. When the HPLC and DCIP procedures were

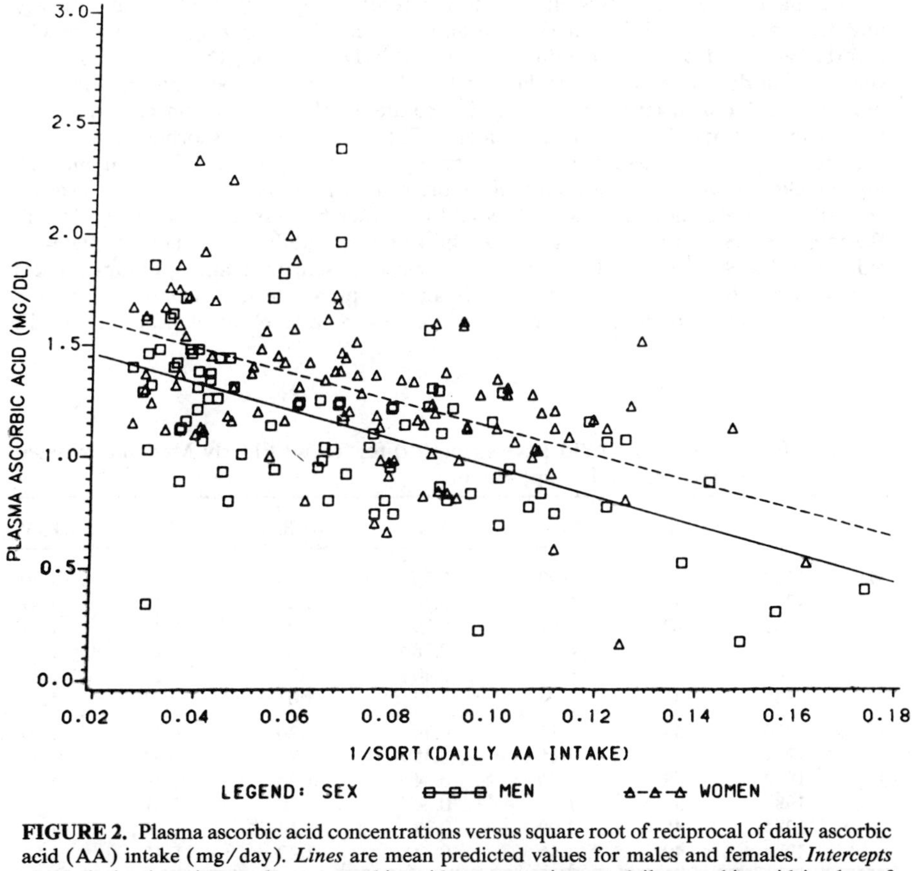

FIGURE 2. Plasma ascorbic acid concentrations versus square root of reciprocal of daily ascorbic acid (AA) intake (mg/day). *Lines* are mean predicted values for males and females. *Intercepts* on vertical axis estimate plasma ascorbic acid concentrations at daily ascorbic acid intakes of 1500 mg. Figure is from data collected in 1984 and is representative of similar findings found in 1980, 1981, and 1985.

compared, we found that they gave comparable results (r=0.97) over a wide range of plasma values.[16] However, the DCIP procedure gave higher values for plasma ascorbic acid concentrations below 1.0 mg/dl, with an average difference of 0.07 mg/dl. This difference becomes important when assessing plasma ascorbic acid concentration at low intakes of ascorbic acid. Because of the greater specificity of the HPLC procedure compared to colorimetric methods, the HPLC procedure is recommended in controlled clinical trials examining plasma concentrations at low intakes of ascorbic acid.

Recently, the Committee on Dietary Allowances of the Food and Nutrition Board, National Academy of Sciences, prepared a draft report that called for lowering the RDA for ascorbic acid from the present value of 60 mg/day for adults to 40 mg for men and 30 mg for women.[17] FIGURE 3 shows that with a daily intake of 40 mg, the median plasma ascorbic concentration in elderly men would be below 0.4 mg/dl, a level that is often used to identify individuals at risk for hypovitaminosis.[18] Even at a daily ascorbic acid intake of 60 mg, the majority of elderly men in this study had plasma levels below 0.4 mg/dl.

The data presented in this chapter do not resolve the question of optimal dietary intakes of ascorbic acid for elderly men and women. If maximizing body reserves of ascorbic acid is a desirable condition for the elderly, however, then it is possible to speculate on desirable dietary intakes for the elderly. This can be achieved when the plasma ascorbic acid level approaches the maximum steady-state concentration (1.0 mg/dl in men and 1.2 mg/dl in women). The daily dietary ascorbic acid intakes required to maintain these plasma levels are approximately 135 mg for men and 115 mg for elderly women. The intake of ascorbic acid required to maintain a plasma level of 1.0 mg/dl in elderly women would be somewhat less, approximately 75 mg. Based on our results the suggested RDA of 40 mg for adult men is not adequate for elderly males even though this amount will prevent scurvy. Until optimum intakes can be clearly defined, preferably using functional parameters, it would be desirable to assure adequacy by maintaining plasma ascorbic acid levels at or near 1.0 mg/dl.

TABLE 3. Plasma Ascorbic Acid Levels (mg/dl) for Those Elderly Men and Women Not Taking an Ascorbic Acid Supplement

	Year	n	Mean	SD	Median	P5[a]	P95[b]
Men	1980	55	0.90	0.36	0.94	0.22	1.50
	1981	45	0.92	0.38	1.06	0.18	1.39
	1982	42	1.01	0.35	1.02	0.31	1.55
	1983	36	0.98	0.36	1.03	0.24	1.46
	1984	38	0.95	0.40	0.94	0.17	1.83
	1985	31	1.04	0.32	1.03	0.32	1.58
Women	1980	58	1.14	0.36	1.20	0.19	1.60
	1981	57	1.15	0.35	1.23	0.37	1.58
	1982	63	1.20	0.34	1.26	0.36	1.60
	1983	57	1.20	0.64	1.20	0.21	1.89
	1984	48	1.09	0.37	1.12	0.20	1.61
	1985	40	1.16	0.34	1.29	0.45	1.59

[a] P5 = 5th percentile.
[b] P95 = 95th percentile.

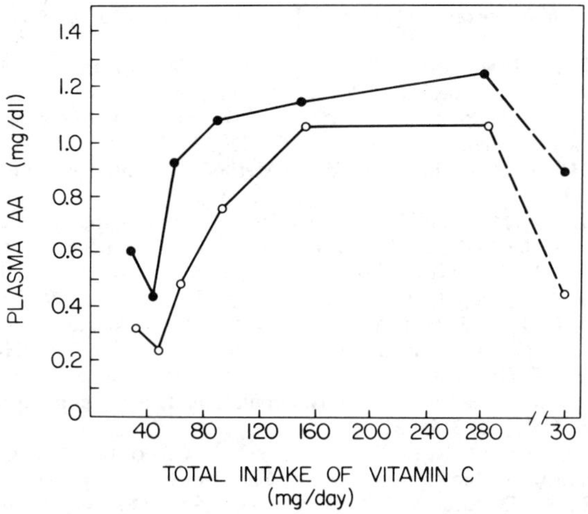

FIGURE 3. Median plasma ascorbic acid (AA) concentration versus total daily intake of vitamin C; ○: men, n = 8; ●: women, n = 9. (From VanderJagt *et al.*[8])

SUMMARY

The data presented show that elderly men have significantly lower plasma ascorbic acid concentrations than elderly women at similar intakes of ascorbic acid. We hypothesize that this difference is due to lower renal tubular reabsorption of ascorbic acid in elderly men.

The finding that more than half of the healthy elderly men had plasma ascorbic acid levels less than 0.3 mg/dl with daily intakes of ascorbic acid in the range of 40 to 50 mg raises concern about the rationale of lowering the RDA for vitamin C from the current value of 60 mg to 40 mg for men and 30 mg for women. This recommendation is inconsistent with efforts to improve the health status of elderly Americans by assuring adequate intakes of all essential nutrients.

Our data do not allow us to determine an optimal intake of ascorbic acid for the elderly. The possibility of increased health benefits associated with intakes that exceed the current RDA of 60 mg needs to be examined. We believe, however, that a prudent approach to both of these questions would be to recommend that elderly receive a daily ascorbic acid intake that would allow plasma concentration to be maintained at or near 1.0 mg/dl. The daily level of intake required to maintain this level would be approximately 125 mg and 75 mg in healthy elderly men and women respectively. These intakes would assure adequate body reserves of ascorbic acid.

REFERENCES

1. FOOD AND NUTRITION BOARD. 1980. Recommended Dietary Allowances. 9th ed. National Academy Press. Washington, DC. pp. 72-82.

2. LEVINE, M. 1986. New concepts in the biology and biochemistry of ascorbic acid. N. Engl. J. Med. **314:** 892-902.
3. SCHNEIDER, E. L., E. M. VINING, E. C. HADLEY & S. A. FARNHAM. 1986. Recommended dietary allowances and health of the elderly. N. Engl. J. Med. **314:** 157-160.
4. KALLNER, A., D. HARTMAN & D. HORNIG. 1979. Steady-state turnover and body pool of ascorbic acid in man. Am. J. Clin. Nutr. **32:** 530-539.
5. SAUBERLICH, H. E. 1975. Vitamin C status: Methods and findings. Ann. N. Y. Acad. Sci. **258:** 438-450.
6. CHENG, L., M. COHEN & H. N. BHAGAVAN. 1985. Vitamin C and the elderly. *In* Handbook of Nutrition in the Aged. R. R. Watson, Ed. CRC Press. Boca Raton, Fla.
7. GARRY, P. J., J. S. GOODWIN, W. C. HUNT & B. A. GILBERT. 1982. Nutritional status in a healthy elderly population: Vitamin C. Am. J. Clin. Nutr. **36:** 332-339.
8. VANDERJAGT, D. J., P. J. GARRY & H. N. BHAGAVAN. Relationship between ascorbic acid intake and plasma levels in healthy elderly. Am. J. Clin. Nutr. In press.
9. GARRY, P. J., J. S. GOODWIN, W. C. HUNT, E. M. HOOPER & A. G. LEONARD. 1982. Nutritional status in a healthy elderly population: Dietary and supplemental intakes. Am. J. Clin. Nutr. **36:** 319-331.
10. GARRY, P. J., G. M. OWEN, D. W. LASHLEY & P. C. FORD. 1974. Automated analysis of plasma and whole blood ascorbic acid. Clin. Biochem. **7:** 131-145.
11. MELETHIL, S., W. D. MASON & C. J. CHANG. 1986. Dose-dependent absorption and excretion of vitamin C in humans. Int. J. Pharmaceut. **31:** 83-89.
12. ROWE, J. W., R. ANDRES, J. D. TOBIN, A. H. NORRIS & N. W. SHOCK. 1976. The effect of age on creatinine clearance in men: A cross-sectional and longitudinal study. J. Gerontol. **31:** 155-163.
13. ENDRES, D. B., C. H. MORGAN, P. J. GARRY & J. L. OMDAHL. Age related changes in serum immunoreactive parathyroid hormone and its biological action in healthy men and women. J. Clin. Endocrinol. Metab. In press.
14. SASAKI, R., T. KUROKAWA, T. KOBAYASI & S. TERO-KUBOTA. Influence of sex and age on serum ascorbic acid. 1983. Tohoku J. Exp. Med. **140:** 97-104.
15. BURR, M. L., P. C. ELWOOD, D. J. HALE, R. J. HURLEY & R. E. HUGHES. 1974. Plasma and leukocyte ascorbic acid levels in the elderly. Am. J. Clin. Nutr. **27:** 144-151.
16. VANDERJAGT, D. J., P. J. GARRY & W. C. HUNT. 1986. Ascorbate plasma as measured by liquid chromatography and by dichlorophenolindophenol colorimetry. Clin. Chem. **32:** 1004-1006.
17. PEAR, R. 1985. Lower nutrient levels proposed in draft report on American diet. Nutr. Today **20:** 5-9.
18. NUTRITION CANADA INTERPRETIVE STANDARDS. 1973. National survey. Information Canada. Ottawa. p. 46.

DISCUSSION OF THE PAPER

B. LANE (*Columbia University School of Public Health, New York, N. Y.*): I just want to reinforce your last comments. In 1980 I reported in Copenhagen on a study in which we looked at ascorbic acid in adults of all ages, and what I found was that looking at intraocular pressure those subjects that were taking in a total of less than 85 mg of ascorbic acid a day as a group, with a group mean of 74.6 mg ascorbic acid per day, had intraocular pressures of 22.5 mm of mercury pressure, which is an undesirably elevated intraocular pressure.

On the other hand when we looked at subjects in the study who had a total intake of vitamin C greater than 150 mg per day, their intraocular pressures averaged 15.15

and this was significant at the 0.001 level. Your comment was that for women about 125 mg should be the minimum, and about 175-180 mg for men. It's interesting that when we looked at this latter group who have the more normal intraocular pressures their mean ascorbic acid intake was 205 mg with or without supplementation. But still it does reinforce your comment that at least for a normal average, typical American population in the New York, New Jersey, and Connecticut metropolitan area, less than 85 mg seems to be too little on a standard diet.

J. L. SMITH (*Strategic Marketing Associates, Mt. Laurel, N. J.*): What is the smoking history of your subjects, particularly the differences in ratios for males and females?

P. J. GARRY (*University of New Mexico School of Medicine, Albuquerque, N. M.*): Since we had so few of these elderly men and women who smoked, we couldn't even look at that question, probably less than five men and only about a couple of women routinely did any heavy smoking.

UNIDENTIFIED SPEAKER: In view of the fact that the absorption of the vitamin is also saturable, how can you hypothesize that it's the renal transport that's in question and absorption is not relevant?

P. J. GARRY: I think most of the low levels at about 125 mg would be absorbed. It could be absorption but I have a stronger feeling for the renal handling of ascorbic acid.

M. LEVINE (*National Institutes of Health, Bethesda, Md.*): Do you think that plasma ascorbic acid is a good marker for total body stores?

D. J. GARRY: Yes, I think from the pharmacokinetic studies that Dr. Kallner has done, one can make a pretty good estimate of total body stores from plasma levels, at least at saturation levels.

M. LEVINE: With respect to the mechanism of the decrease in plasma levels in the elderly men, and you were postulating that the mechanism would be similar to that found with decreased tubular reabsorption of phosphorus, do you think the actual mechanism occurs at the same site?

D. J. GARRY: No, I wouldn't think so. I only gave the phosphorus data to show that there can be a difference in tubular maximum for another analyte, and between elderly men and elderly women.

B. M. TOLBERT (*University of Colorado, Boulder, Colo.*): What analytical method did you use for ascorbic acid?

D. J. GARRY: We used two different analytical methods, one was a dichlorophenolindophenol automated assay and in the control clinical trials they were all done by HPLC with an electrical chemical detector.

B. M. TOLBERT: I found your figures very interesting on the miligram intake, and I wondered what would be the figure if you wanted to take care of 95% of the population reaching down further on the standard deviations on your curves? Does that make sense?

P. J. GARRY: I guess you need about 135 for men and about 75 or 80 for women.

Y. BERNER (*Memorial Sloan-Kettering Cancer Center, New York, N.Y.*): I think that using the plasma level of vitamin C as you used for calculating the requirement for consumption by old people is not the best way because we know about the interrelation of the ascorbic acid in the intestinal tract with other nutrients like iron, so when looking at requirements we have to take this into account. Therefore, the increased amount that you recommend is quite understandable.

P. J. GARRY: We're not really talking about requirements, I'm only talking about a desirable level of intake.

Vitamin C and Blood Lipoproteins in an Elderly Population

PAUL F. JACQUES, STUART C. HARTZ, ROBERT B. McGANDY, ROBERT A. JACOB, AND ROBERT M. RUSSELL

USDA Human Nutrition Research Center on Aging
Tufts University
Boston, Massachusetts 02111

Since Willis[1] first reported that ascorbic acid (AA) deficiency produced atherosclerotic lesions in guinea pigs in 1953, numerous investigations have examined the possible relationship between ascorbic acid and blood cholesterol levels in both animals[2–6] and humans.[2,7–21] Many of these reports suggest that increasing ascorbic acid intake may raise HDL cholesterol (HDL-C) and lower total cholesterol (T-C). However, the nature of the association between AA and cholesterol metabolism in humans remains unclear.

This uncertainty is due, in part, to problems in study design in many of these investigations. Few of the published clinical studies to date were double-blinded, randomized, placebo-controlled trials. Many of the trials did not have controls. Of greater importance, however, is that the clinical trials that failed to detect an association between ascorbic acid and cholesterol had very low statistical power because of the small number of subjects studied.[13–16] Even some of the larger observational studies lacked the statistical power to detect modest, but potentially important, correlations.[17,21]

As part of a large nutritional status survey of elderly in the Boston area, information was obtained on nutrient intake including vitamin supplement use, plasma vitamin levels, and plasma lipid levels. Thus, we were able to explore the relationship of AA to HDL-C and T-C.

SUBJECTS AND METHODS

Study Population

Between November 1981 and March 1984, 761 noninstitutionalized individuals aged 60 through 100 years residing in the greater Boston area were interviewed and examined. Subjects were volunteers who were recruited through direct mail, media, and discussions with groups of elderly individuals. Of 761 subjects who completed the study, 15 were excluded for medical reasons, and 66 were excluded because of incomplete information on one or more of the primary study variables (55 had missing information on AA intake—dietary and/or supplement use—and 11 had missing HDL-C, T-C, or plasma ascorbic acid determinations).

Interview and Examination

Each subject was seen at two visits scheduled within a 1-month period. At the subject's first visit, a fasting (overnight) blood sample was obtained for nutrient biochemical, hematological, and routine clinical analyses. Subjects then were instructed by a dietitian to keep a complete 3-day record of all foods, beverages, and nutritional supplements consumed. At the second visit, the food diary was reviewed for accuracy, completeness, and clarity. Subjects were interviewed by a nurse practitioner to assess personal health habits (including tobacco use, exercise, and alcohol consumption), medical history, current and past nutrient supplement use (including brand name, dosage, frequency of use, and duration of use), and current and past medication use (both physician prescribed and over-the-counter). A brief physical examination was also performed. This study is described in greater detail elsewhere.[22]

Covariates were included in the statistical analyses because of their potential confounding or modifying effects on the associations of AA with HDL-C AND T-C. These covariates included age, sex, race, triceps skinfold thickness, fasting blood glucose, low-density and very-low-density lipoprotein cholesterol, alcohol intake, cholesterol intake, fat intake, carbohydrate intake, current smoking status, regular exercise, education, history of myocardial infarction, pulse rate, and systolic and diastolic blood pressure.

Usual alcohol intake was determined using the following estimates of beverage alcohol content from the U.S. Department of Agriculture (USDA) Handbook 456:[26] 13.1 grams of alcohol per 12 ounces of beer, 12.2 grams of alcohol per 3.5-ounce glass of wine, and 15.5 grams of alcohol per 1.5 ounces of liquor (86) proof. All cocktails were assumed to contain 1.5 ounces of liquor.

Biochemical Determinations

Plasma AA was determined by the dinitrophenylhydrazine method, modified by Roe and Kuether.[23] The procedure for isolation of high-density lipoprotein cholesterol (HDL-C) is based on the procedure described by Lopes-Virella *et al.*[24] Plasma total cholesterol (T-C) and HDL-C were determined by an enzymatic procedure described by Allain *et al.*[25]

Statistical Analysis

The SPSS-X PEARSON CORR procedure[27] was used to examine the unadjusted (bivariate) associations between cholesterol and AA. Multiple regression procedures were employed to examine the influence of plasma AA on plasma HDL-C and T-C adjusted for other potential confounding or modifying factors. The SPSS-X REGRESSION procedure[23] was employed for these analyses. The multivariate analyses included 672 subjects, as 8 subjects were missing information on one or more of the covariates. HDL-C and T-C were transformed for the multivariate analyses using the natural logarithm because both factors were positively skewed.

RESULTS

The age-sex breakdown of the 680 subjects is presented in TABLE 1. Thirty-five percent of the sample was male. Mean ages were 72 and 73 years for males and females, respectively. Plasma AA levels ranged from 0.18 to 2.63 mg/dl for this sample. Mean plasma AA levels were 1.3 mg/dl for females and 1.1 mg/dl for males.

FIGURE 1 displays the unadjusted relationship between plasma AA and HDL-C for females and males. This figure indicates a positive association between HDL-C and plasma AA. The correlation coefficients are 0.13 ($p<0.01$) for females and 0.12 ($p=0.07$) for males. The regression lines indicate that for each plasma AA increment of 1 mg/dl the mean HDL-C level increased by 4.7 mg/dl in females and 3.6 mg/dl in males. FIGURE 2 displays the unadjusted association between plasma AA and T-C. No association is apparent for either females ($r=0.02, p>0.10$) or males ($r=0.04$, $p>0.10$).

The association between plasma AA and HDL-C does not appear to be the consequence of confounding, as the relationship remained after adjusting for a number of known or suspected determinants of HDL-C using regression models. The partial correlation coefficient for the adjusted analysis was 0.13 ($p<0.01$). Likewise, the lack of association between plasma AA and T-C does not appear to be the consequence of confounding as no relationship was observed after adjustment ($r=0.05$, $p>0.10$).

Age appears to be a strong modifier of the observed relationship between plasma AA and HDL-C. partial correlation coefficients (TABLE 2) indicate a statistically significant ($p<0.05$) correlation for those aged 60-69 years ($r=0.14$), but not for those aged 70-79 years ($r=0.08$) or those aged 80 years and older ($r=0.01$). Because the natural logarithm of HDL-C was used as the dependent variable, the regression coefficient is interpreted, not as the absolute change in HDL-C per unit change in plasma AA, but rather as the relative change in HDL-C per unit change in plasma AA. The magnitude of the AA effect on HDL-C, as measured by the antilogarithm of the regression coefficient, represents the percentage increase (or decrease) in HDL-C per unit increase in plasma AA. The regression models suggest that each 0.1 mg/

TABLE 1. Age and Sex Distribution of the 680 Subjects

	Age (Years)				
Sex	60-69	70-79	80-89	90-99	All Ages
Males	100	94	36	5	235
Females	156	175	96	18	445
Combined	256	269	132	23	680

TABLE 2. Multiple Regression (*B*) and Partial Correlation (*r*) Coefficients[a] for HDL Cholesterol (HDL-C) and Plasma Ascorbic Acid (AA) and Change in HDL-C per 0.1 mg/dl Increase in Plasma AA by Age Group

Age Group (Years)	*B*	*r*	Change in HDL-C
60-69	0.092[b]	0.14[b]	1%[b]
70-79	0.052	0.08	0.5%
80-99	0.010	0.01	0.001%

[a] Covariates considered for this analysis include age, sex, race, triceps skinfold thickness, serum glucose, LDL and VLDL-C, alcohol intake, cholesterol intake, fat intake, carbohydrate intake, current smoking status, regular exercise, education, history of MI, pulse rate, and blood pressure.

[b] $p < 0.05$.

dl increase in plasma AA will increase HDL-C by 1% in those aged 60-69, 0.5% in those aged 70-79, and 0.01% in those aged 80 and older (TABLE 2).

DISCUSSION

Our results strongly suggest that plasma AA is related to HDL-C, but not to T-C. Although the effect of AA is modest, it appears to be of practical significance in the age group 60-69 years, but not in the very old (80 years and older). For example, an 0.8 mg/dl increase in plasma AA could increase HDL-C by approximately 8% in persons 60-69 years old. Such an increase would translate into a 4% increase in HDL-C levels in persons 70-79 years old and a negligible change in those over 80 years old.

Even if this model is appropriate, is it possible to effect a large enough change in plasma AA to make the change in HDL-C clinically important in persons under age 70? Data from our sample suggest that by modifying AA intake, plasma AA may increase by at least 0.8 mg/dl in a substantial proportion of the population. Although a 0.8 mg/dl change in plasma AA may seem large, it is only about one third of the range of plasma AA seen in our sample (0.18 to 2.63 mg/dl). In addition, the difference in plasma AA between individuals with AA intakes of 1000 mg/day or greater and those consuming less than 120 mg/day is 0.8 mg/dl in this sample. (About 30% of our study sample consumed less than 120 mg per day.) Those with AA intakes under 60 mg/day (about 10% of our sample) had plasma AA levels about 1 mg/dl below that of those consuming 1000 mg or more of AA daily.

If we can increase HDL-C levels by 8-10%, is this change sufficient to influence the risk of coronary heart disease (CHD)? An 8-10% increase in HDL-C would translate approximately into a 4-5 mg/dl increase in our sample. Castelli *et al.*[28] reported that a difference of 3-4 mg/dl separated persons with CHD and those without CHD in the Cooperative Lipoprotein Phenotyping Study. In addition, logistic regression models from the Framingham Heart Study[29] suggest that a 4-5 mg/dl increase

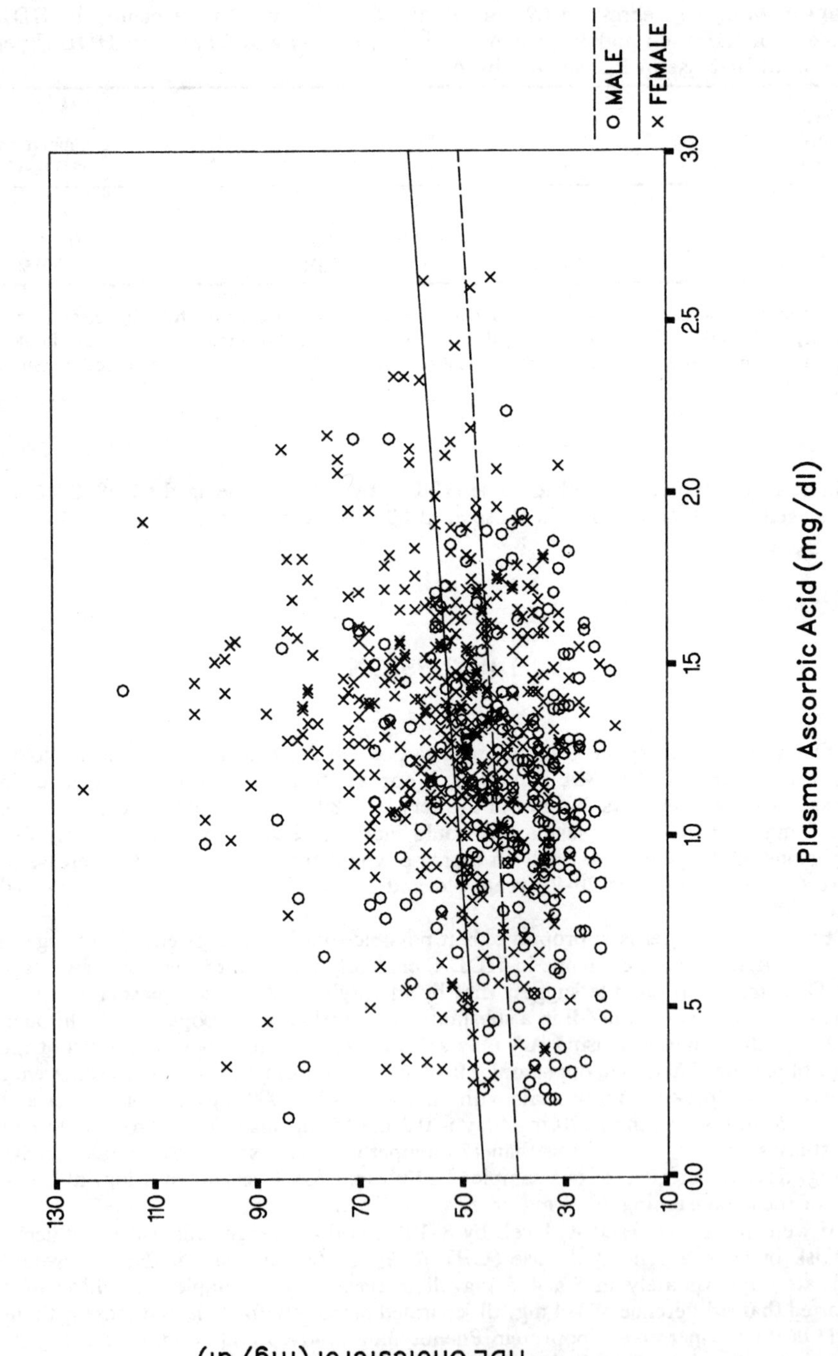

FIGURE 1. Relationship between HDL cholesterol and plasma ascorbic acid.

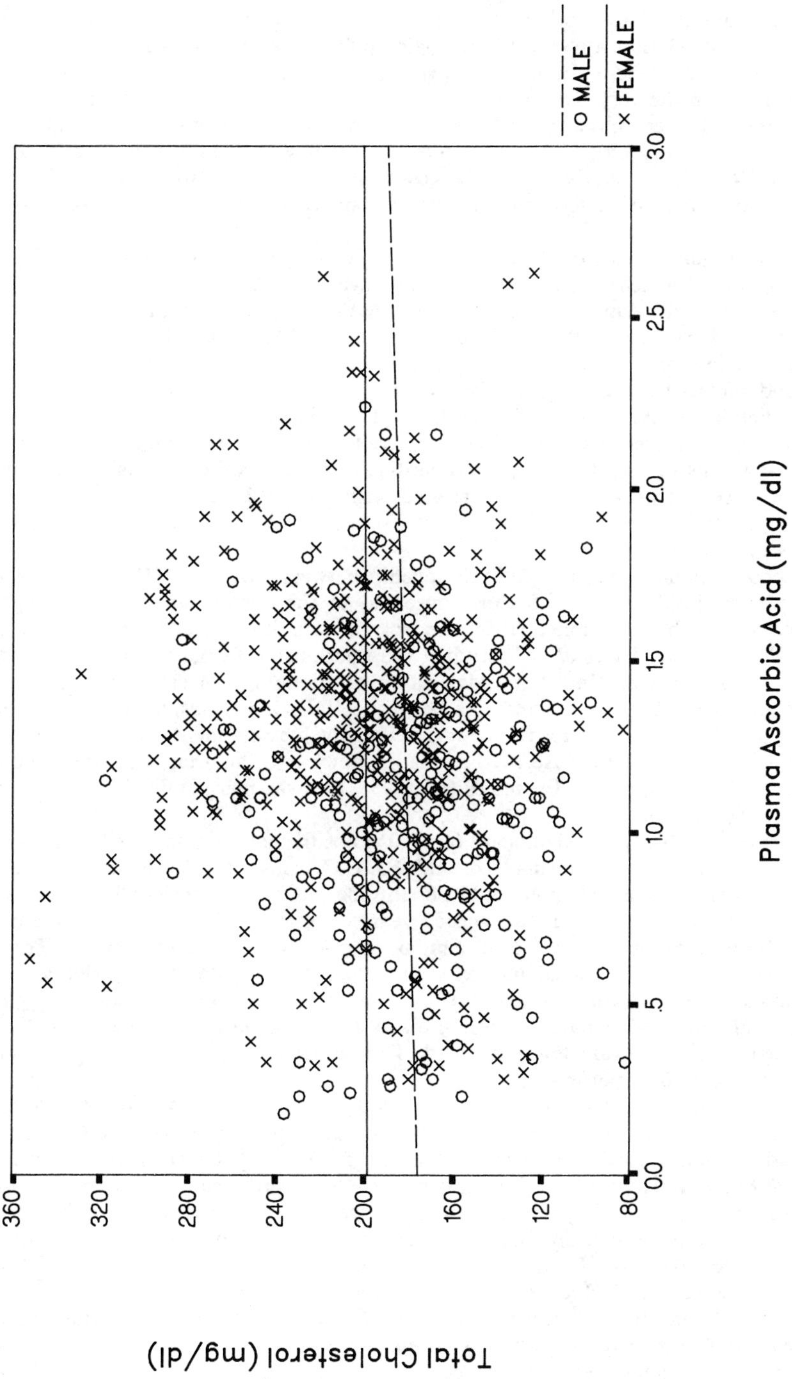

FIGURE 2. Relationship between total plasma cholesterol and plasma ascorbic acid.

in HDL-C could translate into a 10-20% decrease in CHD risk. Thus, it would appear that this modest relationship could have major public health implications.

We have no data from the current investigation to suggest a reason for the diminished effect in the very old. However, there may be a simple explanation for this phenomenon. The very old are survivors who may be a biologically different group. Therefore, in the very old, HDL cholesterol levels may be determined by different etiologic factors. These survivors may be genetically predisposed to elevated HDL-C levels. In such individuals, AA may not play a significant role in determining HDL-C levels.

There is a substantial literature on the relationship of AA to HDL-C and T-C. Guinea pigs fed AA-deficient diets showed an elevation of cholesterol in plasma and tissue compared with animals with sufficient AA even when the diet was free of cholesterol.[2-4] Treatment of these deficient animals with AA resulted in a decrease of total body cholesterol.[4] AA administration also decreased blood and tissue cholesterol in rats and rabbits fed diets high in cholesterol.[5] Conversely, atherogenic (high cholesterol) diets have shown to reduce the level of AA in tissues of guinea pigs.[6]

Prior human studies of this relationship have taken two forms: intervention trials involving administration of AA by the investigator,[7-16] and observational studies in which the observed plasma AA is correlated with plasma lipid levels.[2,17-21] The intervention trials have shed little light on the nature of this relationship because of problems in study design.

The observational studies of the relationship between AA and HDL cholesterol have produced results similar to those observed in the current analyses, although interpretation has been complicated by the lack of statistical power in some of these studies. For example, Bates *et al.*[17] studied the relationship of plasma ascorbic acid and HDL-C in 109 male and 228 female noninstitutionalized Welsh residents aged 70 years and older. Positive correlations between HDL-C and plasma AA for males and females were 0.12 and 0.13, respectively. These values are identical to those seen in the current analyses. Because of the smaller number of subjects, the correlation coefficient was not significant for males ($p > 0.05$), but was marginally significant for the females ($p = 0.05$).

Burr *et al.*[18,19] observed a significant ($p < 0.05$) positive correlation of 0.21 for both men and women between plasma AA and HDL-C using only 97 males and 121 females. This stronger association could possibly be due to the considerably younger ages of these subjects. The mean age was 51 years for males and 46 years for females.

The study of Hooper *et al.*[21] was similar to our study in many aspects. They studied healthy volunteers aged 60 years and older, but no statistically significant correlation between plasma AA and HDL-C was noted. Given the smaller sample size (125 males and 145 females), it was unlikely that this study would have detected the correlation if it is truly less than 0.19. The actual values of the nonsignificant correlations were not reported.

Observational studies have also reported correlations between AA and T-C. Greco and LaRocca[20] reported an inverse correlation between plasma AA and T-C in 100 males and females aged 70 to 84 years. Hanck and Weiser[2] described results from the Third Basel study in which T-C and plasma AA were negatively correlated in 159 males who used vitamin preparations. This association was not seen in 487 males who did not use any vitamin preparation. Burr *et al.*[18,19] observed a significant negative correlation with T-C in males only. No relationship between T-C and plasma AA was seen in the report of Hooper *et al.*[21] or in the current study.

It is possible that the different findings among these studies are due to differences in plasma AA status of the populations studied. For example, the mean plasma AA levels for males and females in the Greco and LaRocca report were 0.46 and 0.62

mg/dl, respectively. Burr *et al.* reported a plasma AA of 0.78 mg/dl for males. However, the current study, and that of Hooper *et al.* both report plasma AA of well over 1 mg/dl for both males and females. In contrast, Hanck and Weiser[2] only observed a relationship among supplement users, whom one would expect to have higher plasma AA levels than nonusers.

Our findings present evidence of an effect of AA on HDL-C. As most of the prior clinical trials suffer from some deficiency in study design, a well-designed trial of long duration and with appropriate sample size is now needed. Given the magnitude of effect observed in the current report, the potential impact on a large population in terms of CHD is quite significant. An added benefit is the relative ease and safety with which plasma AA can be increased.

ACKNOWLEDGMENTS

The authors would like to express their appreciation to all the staff of the Nutritional Status Survey of Boston Elderly, and particularly to Conni Otradovec, the Project Manager. Special thanks go to Dan Abrams, Saul Tannenbaum, Lorry Scura, Dorothy Beaton, and all the staff of the HNRC Division of Scientific Computing for this assistance in data management, programming, and analysis. We would also like to recognize the effort of the staff of the HNRC Nutrition Evaluation Laboratory on this project.

REFERENCES

1. Willis, G. C. 1953. An experimental study of the intimal ground substance in atherosclerosis. Can. Med. Assoc. J. **69:** 17-22.
2. Hanck, A. & H. Weiser. 1979. The influence of vitamin C on lipid metabolism in man and animals. Int. J. Vit. Nutr. Res. (Suppl.) **19:** 83-93.
3. Hanck, A. & H. Weiser. 1977. Vitamin C and lipid metabolism. Int. J. Vit. Nutr. Res. (Suppl.) **16:** 67-81.
4. Zloch, Z. 1982. Changes in the cholesterol content of the high-density plasma lipoproteins in guinea pigs in the presence of a raised exogenous cholesterol intake and different degrees of vitamin C saturation. Physiol. Bohemoslov. **31:** 557-562.
5. Sokoloff, B., M. Hori, C. Saelhof, B. McConnell & T. Imai. 1967. Effect of ascorbic acid on certain blood fat metabolism factors in animals and man. J. Nutr. **91:** 107-118.
6. Nambisan, B. & P. A. Kurup. 1975. Ascorbic acid and glycosaminoglycan and lipid metabolism in guinea pigs fed normal and atherogenic diets. Atherosclerosis **22:** 447-461.
7. Spittle, C. R. 1971. Atherosclerosis and vitamin C. Lancet **2:** 1280-1281.
8. Ginter, E., I. Kajaba & O. Nizner. 1970. The effect of ascorbic acid on cholesterolemia in healthy subjects with seasonal deficit of vitamin C. Nutr. Metabol. **12:** 76-86.
9. Ginter, E., O. Cerna, J. Budlovshy, V. Balaz, F. Hruba, V. Roch & E. Sasko. 1977. Effect of ascorbic acid on plasma cholesterol in humans in a long-term experiment. Int. J. Vit. Nutr. Res. **47:** 123-134.
10. Fidanza, A., M. Audisio & P. Mastroiacovo. 1982. Vitamin C and cholesterol. Int. J. Vit. Nutr. Res. (Suppl.) **23:** 153-171.
11. Heine, H. & C. Norden. 1979. Vitamin C therapy in hyperlipoproteinemia. Int. J. Vit. Nutr. Res. (Suppl.) **19:** 45-54.

12. HORSEY, J., B. LIVESLEY & J. W. T. DICKERSON. 1981. Ischaemic heart disease and aged patients: Effects of ascorbic acid on lipoproteins. J. Hum. Nutr. **35:** 53-58.
13. PETERSON, V. E., P. A. CRAPO, J. WEINIGER, H. GINSBERG & J. OLEFSKY. 1975. Quantification of plasma cholesterol and triglyceride levels in hypercholesterolemic subjects receiving ascorbic acid supplements. Am. J. Clin. Nutr. **28:** 584-587.
14. KHAN, A. R. & F. A. SEEDARNEE. 1981. Effect of ascorbic acid on plasma lipids and lipoproteins in healthy young women. Atherosclerosis **39:** 89-95.
15. JOHNSON, G. E. & S. S. OBENSHAIN. 1981. Nonresponsiveness of serum high-density lipoprotein-cholesterol to high dose ascorbic acid administration in normal men. Am. J. Clin Nutr. **32:** 2088-2091.
16. WAHLBERG, G. & G. WALLDIUS. 1982. Lack of effect of ascorbic acid on serum lipoprotein concentrations in patients with hypertriglyceridemia. Atherosclerosis **43:** 283-288.
17. BATES, C. J., M. K. BURR & A. S. ST. LEGER. 1979. Vitamin C, high density lipoproteins and heart disease in elderly subjects. Age Ageing **8:** 177-182.
18. BURR, M. L., C. J. BATES, J. E. MILBANK & J. W. G. YARNELL. 1982 The relationship between plasma ascorbate and lipid concentrations in fasting men. Hum. Nutr.: Clin. Nutr. **36C:** 135-139.
19. BURR, M. L., C. J. BATES, P. M. SWEETNAM & M. E. BARASI. 1982. Plasma ascorbate and HDL-cholesterol in women. Hum. Nutr.: Clin. Nutr. **36:** 399-400.
20. GRECO, A. M. & L. LAROCCA. 1982. Correlation between chronic hypovitaminosis C in old age and plasma levels of cholesterol and triglycerides. Int. J. Vit. Nutr. Res. (Suppl.) **23:** 129-136.
21. HOOPER, P. L., E. M. HOOPER, W. C. HUNT, P. J. GARRY & J. S. GOODWIN. 1983. Vitamins, lipids and lipoproteins in a healthy elderly population. Int. J. Vit. Nutr. Res. **53:** 412-419.
22. MCGANDY, R. B., R. M. RUSSELL, S. C. HARTZ, R. A. JACOB, S. TANNENBAUM, H. PETERS, N. SAHYOUN & O. L. OTRADOVEC. 1986. Nutritional status survey of healthy noninstitutionalized elderly: Energy and nutrient intakes from three-day diet records and nutrient supplements. Nutr. Res. **6:** 785-798.
23. ROE, J. H. & C. A. KEUTHER. 1943. The determination of ascorbic acid in whole blood and urine through the 2,4-dinitrophenylhydrazine derivative of dehydroascorbic acid. J. Biol. Chem. **147:** 399-407.
24. LOPES-VIRELLA, M. F., P. STONE, S. ELLIS & J. A. COLWELL. 1977. Cholesterol determination in high-density lipoproteins separated by three different methods. Clin. Chem. **23:** 882-884.
25. ALLAIN, C.C., L. S. POON, C. S. G. CHAN, W. RICHMOND & P. C. FU. 1974. Enzymatic determination of total serum cholesterol. Clin. Chem. **20:** 470-75.
26. ADAMS, C. 1975. Agricultural Handbook No. 456: Nutritive Value of American Foods in Common Units. Agricultural Research Services, U.S. Department of Agriculture. Washington, D.C.
27. SPSS, INC. 1983. SPSS-X User's Guide. McGraw-Hill. New York.
28. CASTELLI, W. P., J. T. DOYLE, T. GORDON, C. G. HAMES, M. C. HJORTLAND, S. B. HULLEY, A. KAGAN & W. J. ZUKEL. 1977. HDL cholesterol and other lipids in coronary heart disease. The Cooperative Lipoprotein Phenotyping Study. Circulation **55:** 767-772.
29. GORDON, T., W. P. CASTELLI, M. C. HJORTLAND, W. B. KANNEL & T. R. DAWBER. 1977. High density lipoprotein as a protective factor against coronary heart disease. The Framingham Study. Am. J. Med. **62:** 707-714.

DISCUSSION OF THE PAPER

P. J. GARRY (*University of New Mexico School of Medicine, Albuquerque, N. M.*): I assume that many of your elderly were taking supplements. Did you ever look at the difference between men and women, or men who were taking supplements versus nonsupplemented women?

P. F. JACQUES (*Tufts University, Boston, Mass.*): I think 35% of our males and 43% of our females were taking vitamin C supplements. We had looked at this relationship in men and women not taking supplements and we did see this relationship even at lower levels.

E. GINTER (*Institute of Human Nutrition Research, Bratislava, Czechoslovakia*): There was a very small percentage of people with lower levels of vitamin C, maybe less than 5% were in a state of marginal deficiency.

P. F. JACQUES: Yes, we did have some people who were subnormal in vitamin C.

E. GINTER: The reason is that in your country the intake of vitamin C is relatively high and maybe this is also the reason that you haven't found a correlation between total cholesterol and vitamin C level. In my country the intake of vitamin C is much lower and therefore there's no problem in finding in a sample maybe 30% of people with marginal vitamin C deficiency, and also in finding direct correlation between vitamin C and cholesterol. For instance, in a group of 600 healthy blood donors, there was a very clear correlation between vitamin C and total cholesterol.

P. F. JACQUES: Yes, most of the studies that I have seen on relationships of cholesterol and vitamin C were not done in the United States. In the Greco and LaRocca study in Italy, they had a large percentage of people with very low vitamin C intake. The Burr *et al.* studies were done in South Wales, and again they had a large percentage of people with low vitamin C intake. In both the studies in this country, the Hooper *et al.* studies and our studies, the plasma vitamin C levels were very high. The average in our study was 1.3 mg for males and 1.1 mg for females, and I believe that was very similar to the Hooper *et al.* data.

Relationship of Plasma Level of Vitamin C to Mortality from Ischemic Heart Disease

K. FRED GEY [a–c]

[a]*Vitamin Research Department*
F. Hoffmann-La Roche & Co., Ltd.
Basel, Switzerland

HANNES B. STÄHELIN [a]

Geriatric Clinic
Kantonsspital
University of Basel
Basel, Switzerland

PEKKA PUSKA [b,c]

Public Health Institute
Helsinki, Finland

ALUN EVANS[c]

World Health Organization
MONICA Center
Belfast, Northern Ireland

[a]On behalf of the prospective Basel Study, with Georg B. Brubacher, Hans Riggenbach, Florence Bernasconi, and Jean Paul Vuilleumier of F. Hoffmann-La Roche & Co., Ltd., Basel, Switzerland, as co-workers.

[b]In experimental collaboration with Flaminio Fidanza, Perugia, Italy; Horst Georgi, Basel, Switzerland; Mario Mancini, Naples, Italy; Tapio Nikkari, Tampere, Finland; Michael F. Oliver, Edinburgh, Scotland; Rudolph A. Riemersma, Edinburgh, Scotland; Günther Ritzel, Basel, Switzerland; Paolo Rubba, Naples, Italy; Matti Salo, Tampere, Finland; Erkki Vartiainen, Helsinki, Finland; Jean Paul Vuilleumier, Basel, Switzerland; and David A. Wood, Edinburgh, Scotland, as principal investigators of the International Collaborative Study on the Fatty Acid-Antioxidant Hypothesis of Arteriosclerosis.

[c]Optional Study on Antioxidant Vitamins, WHO-MONICA Project.

INTRODUCTION

Ischemic heart disease (IHD) is predicted by hitherto established risk factors such as hypercholesterolemia, hypertension, and smoking only to about 50-60%. Among candidates as complementary direct risk factors are aggressive oxygen species (superoxide anion, hydroxyl radical, etc.). Thus their scavengers, the antioxidants, are inverse risk factors.[1]

The direct measurement of the aggressive short-lived oxygen radicals *in vivo* is, however, rather difficult or almost impossible. Thus evidence for the involvement of short-lived oxygen free radicals can mainly be obtained indirectly. One possibility is to study the acute excess of exogenous radicals, or the burst of endogenous radicals under special *in situ* situations, *e.g.*, in the reperfusion experiment. Indeed, hypoxia with subsequent reoxygenation results in a sudden burst of oxygen radicals that can overload protective enzymes such as superoxide dismutase. Then the oxygen radicals will seriously damage blood vessels as soon as the physiological radical scavengers such as the vitamins E, C, and glutathione are exhausted. The reperfusion situation proves that the acute attack by oxygen radicals can cause arterial damage.[1]

Since IHD develops rather slowly, however, *i.e.*, in years or decades, longer-lived intermediates of lipid peroxidation (such as hydroperoxides) or of relatively stable end products (*e.g.*, hydroxy derivatives of polyunsaturated fatty acids, PUFAs, or hydroxyalkenals) are presumably of greater pathophysiological importance, particularly since their lower reactivity permits them to travel over relatively long distances, *i.e.*, to more distant intracellular sites or even to neighboring cells.[1,2] Exogenous peroxidized PUFAs damage the endothelium and heart muscle cells and provoke proliferation of smooth muscle cells, and even cause sudden death in animals. This cytotoxicity can be prevented by antioxidants such as vitamins C or E. Endothelial cells or platelets are able to modify LDL (low-density lipoproteins) *in vitro*, presumably by peroxides and/or alkenals such as malondialdehyde. The modified LDL is, in contrast to normal LDL, taken up by macrophages that are thereby transformed into arteriosclerotic lipid-laden foam cells. Antioxidants such as vitamin E can prevent this LDL transformation. Peroxides of LDL and/or PUFA-hydroperoxides, *e.g.*, 15-hydroperoxyeicosatetraenoic acid, also inhibit the arterial formation of prostacyclin. The quantitative importance of these phenomena in the presence of antioxidants *in vivo* is, however, unclear. The direct measurement of metastable metabolites in tissues is hampered by technical problems as well. For instance, lipid peroxides have been demonstrated in arteriosclerotic plaques, but artifactual autoxidation postmortem has not been excluded. Nevertheless, every demonstration of lipid peroxides in excised tissues suggests at least an increased susceptibility towards peroxidation. Patients with ischemic heart disease contain increased plasma levels of thiobarbituric-acid-reactive material, but it is rather doubtful whether this is true malondialdehyde, and if so, whether is stems from lipid peroxidation or the prostaglandin metabolism.[1]

When the body's defense system against oxygen radical injury is deteriorated by chronic marginal deficiency of vitamin C or E, arteriosclerosis-like lesions have been found in rodents and piglets.[1]

The demonstration in men of a corresponding effect of a comparably poor status of these essential antioxidants has been missing. Interestingly, it has been found that acute experimental scurvy in man can suddenly cause cardiomegaly, electrographic abnormalities, and acute cardiac emergency. Arteriographically established IHD was also reported to be inversely related to the status of vitamin C, but this could be a consequence of the disease rather than a cause.[1] The previous epidemiological evidence suggests that the vegetarian type of diet is associated with a lower mortality from IHD. Specifically, the standardized mortality from IHD in England, Wales, Scotland, Norway and Israel seems to be inversely correlated to the calculated ascorbic acid intake from fresh fruits and green vegetables. But a calculated commodity consumption is not conclusive for the *in vivo* status of vitamins.[1]

In summary, most of the above arguments lead to the question: Does a poor plasma level of antioxidant vitamins really occur in westernized countries, and if so, is it associated with an increased risk of IHD? There is a possible three-step route to an answer: (1) cross-cultural epidemiology with measurement of plasma levels, (2) corresponding prospective studies, and (3) intervention trials as potential proof. The presently available evidence for an inverse correlation of the plasma level of vitamin C and E with incidence of IHD is based mainly on cross-sectional data.

SUBJECTS AND METHODS

In the cross-sectional surveys, groups of apparently healthy males, 40-49 years of age, from regions with different mortality from IHD were compared during the period January through April with regard to the plasma status of the principal essential antioxidants, *i.e.,* the vitamins A, C, and E, β-carotene, and selenium. The study populations were, in accordance with the well-known North-South gradient of IHD in the northern hemisphere, representative for regions with different incidence of death levels:

1. Highest incidence (yearly >350 deaths from IHD [ICD 410-414] per 100,000 males, 40-59 years of age): North Karelia, Finland ($n = 92$), Southwest Finland ($n = 82$) and Edinburgh, Scotland ($n = 108$).
2. Medium incidence (around 250/100,000): Belfast, Northern Ireland ($n = 124$).
3. Low incidence (<130/100,000): Thun, prealpine Switzerland ($n = 220$) and Sapri, Southern Italy ($n = 77$).

The subjects were selected at random (within the International Collaborative Study on the Fatty Acid-Antioxidant Hypothesis of Arteriosclerosis or the subsequent Optional Study on Antioxidant Vitamins, WHO/MONICA Project), except the large Swiss cohort of Federal industrial employees in Thun who were in every respect representative for this age group of the working population of German-speaking Switzerland. Details and analytical methods have been described elsewhere.[1,3,4] The PEARSON coefficient did not show any significant correlations ($p > 0.05$) of plasma parameters except for total cholesterol and the lipid-soluble vitamin E ($r^2 = 0.2$ to 0.5). In the present study plasma vitamin E was tentatively standardized by linear adjustment to 5.7 mM cholesterol.[5]

In the prospective Basel Study, about 3000 apparently healthy male employees of pharmaceutical companies volunteered for a baseline measurement in 1971 through 1973 (disregarding the seasonal variations of plasma ascorbic acid). Plasma was almost immediately analyzed for all principal antioxidants, the values of which were evaluated with regard to age- and smoking-standardized mortality within the subsequent 7 years.[5,6]

RESULTS AND DISCUSSION

Cross-Sectional Evidence

As mentioned initially, total plasma cholesterol (and its major lipoprotein carriers respectively) cannot fully account for the substantial epidemiological differences in the incidence of IHD. This is also true for the six European study populations (FIG. 1). Thus, in confirmation of previous data, Italy (with a low incidence of IHD) had a low cholesterol median and North Karelia (with the greatest incidence of IHD) revealed the highest cholesterol median. The remaining study populations, however, had very similar cholesterol levels. Medians were around 5.8 mM, roughly corresponding to the European average. In consequence, sites with similar plasma cholesterol may be particularly suited to the search for risk factors other than cholesterol.

The medians of plasma vitamin C (FIG. 2) from both areas with low mortality from IHD showed even in the most critical season, *i.e.,* in the first months of the year, a fairly good vitamin C status, and their lower quartiles were practically out of the critical range of accepted marginal vitamin C deficiency, *i.e.,* less than 0.4 mg/dl or less than 22.7 μM.[7] In contrast, regions with medium-to-high coronary mortality had a significantly poorer vitamin C status. Their medians were early in the year on average 33% lower ($p<0.01$), and with the exception of Southwest Finland very near to the accepted borderline of marginal vitamin C deficiency. Interestingly, Southwest Finland had a higher vitamin C and a lower IHD mortality in comparison to North Karelia. In the British Isles the lowest quartile of subjects was even below the level where the risk of overt scurvy begins, *i.e.,* at 11 μM.[7] The present data confirm previous assumptions of a notoriously or at least seasonally critical ascorbic acid status in Finland and the British Isles.[1] Although the vitamin C status in Britain and North Karelia[3] is likely to become less critical during summer and autumn, a corresponding rise may be expected for Italy and Switzerland, and thus the relative differences will probably persist all over the year. In conclusion the present findings strongly support the idea that a poor plasma status of vitamin C (near marginal vitamin C deficiency) may be correlated with an increased risk of IHD.

Since the absolute plasma level of the lipid-soluble α-tocopherol has been known for long to vary to some extent with the plasma lipids, only lipid-standardized vitamin E is a reliable index of the plasma status.[1,8] The cholesterol-standardized vitamin E (FIG. 3), as used in the present study, indicates actually the vitamin E concentration within cholesterol-carrying plasma lipoproteins and may be an acceptable indicator of the vitamin E level in membranes since vitamin E can penetrate into all lipids by thermodynamic partitioning. In regions with low and medium mortality from IHD, *i.e.,* Italy, Switzerland, and Northern Ireland, the plasma value of cholesterol-standardized vitamin E was fairly high, *i.e.,* between 26 and 28 μM. In contrast, in regions with most frequent IHD, *i.e.,* Finland and Scotland, the medians of the cholesterol-standardized vitamin E were significantly ($p<0.01$) on average about 25% lower than in regions with medium-to-rare IHD, and there was also only a small overlap of distributions. Thus, the lower quartile of the areas with low-to-medium risk was above the upper quartile of sites with highest risk. The present finding of low vitamin E levels in Finland and Scotland actually confirms previous measurements,[1] but a correlation between the vitamin E status and IHD might have been overlooked since in previous papers the vitamin E value was not lipid-standardized.

If the cholesterol-standardized plasma vitamin E were of any importance for IHD, the threshold of risk might be below 25 μM, and above or equal to the average of

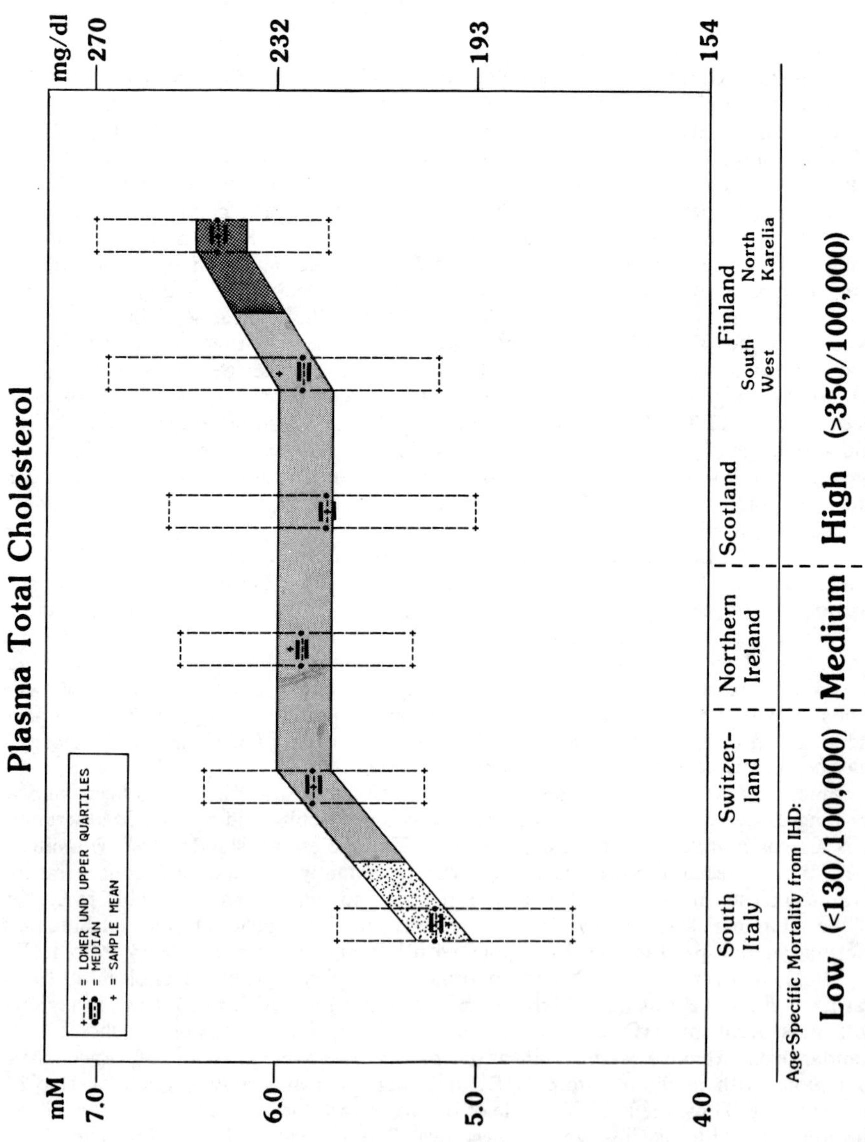

FIGURE 1. Cross-sectional comparison of plasma cholesterol levels in European populations with different mortality from IHD.

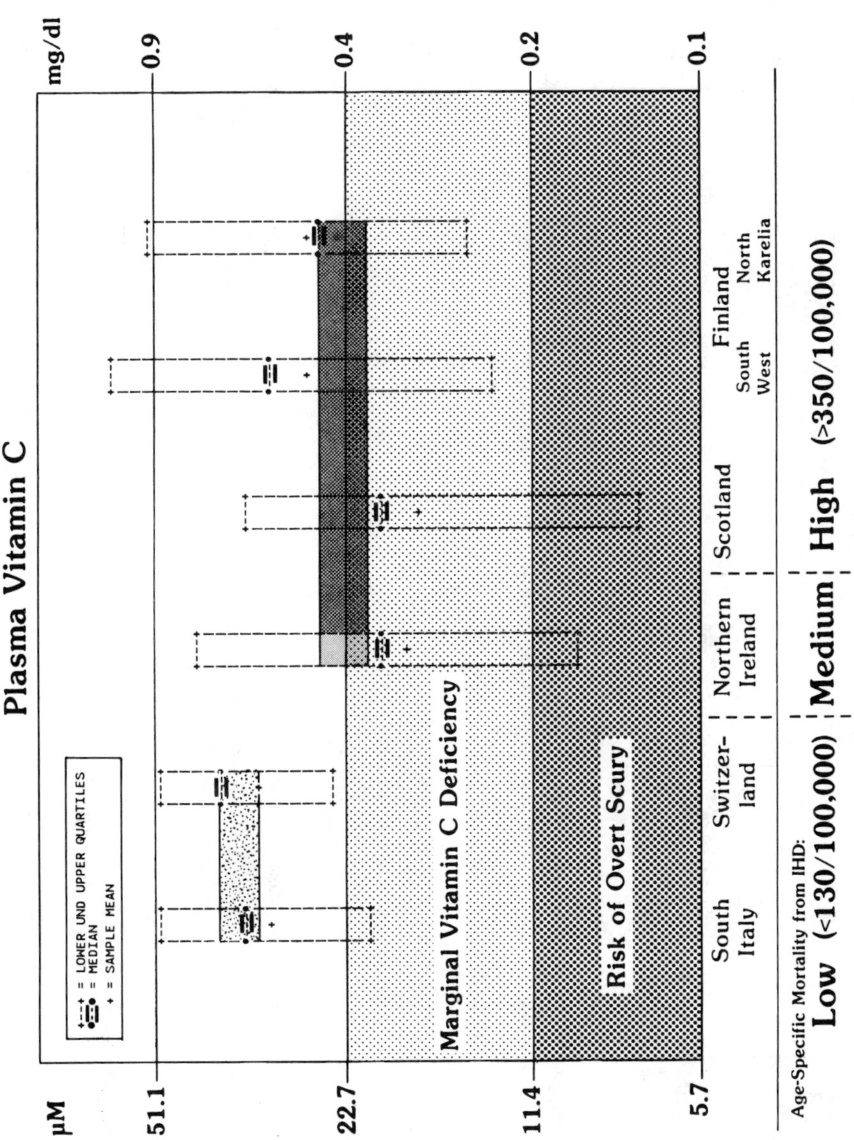

FIGURE 2. Cross-sectional comparison of plasma vitamin C in European populations with different mortality from IHD.

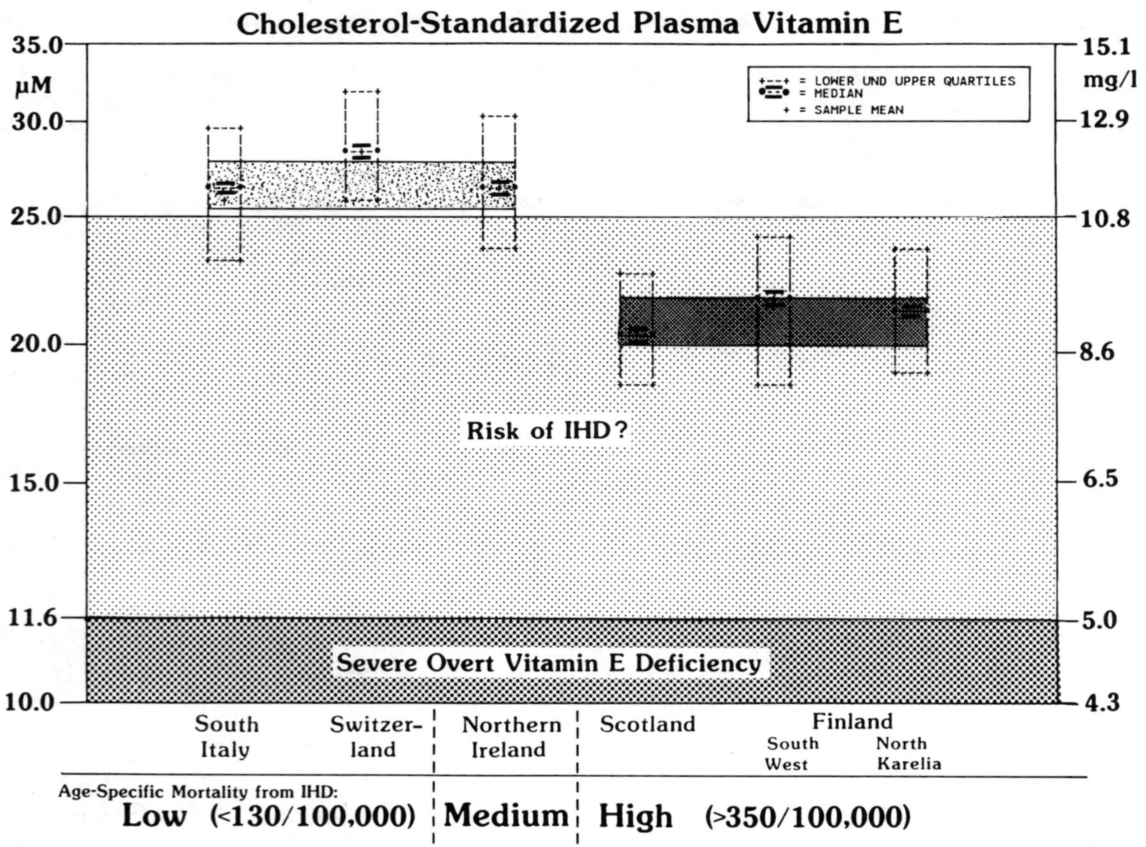

FIGURE 3. Cross-sectional comparison of cholesterol-standardized plasma vitamin E (α-tocopherol) in European populations with different mortality from IHD.

the three high-risk regions, *i.e.,* 21 μM. This threshold would be about twice as high as the level at which the classical vitamin E deficiency in men with severe neuromuscular disorders will develop.[8] In any case it may be concluded that a relatively poor plasma status of cholesterol-standardized vitamin E shows a significant association with a very high risk of IHD, at least in countries such as Scotland and Finland.

Of course, this association of vitamin E with IHD remains to be differentiated from that of polyunsaturated fatty acids[4] from vegetable oils, which are a variable albeit major dietary source of vitamin E.[1]

The plasma values of the other antioxidants, *i.e.,* vitamin A, β-carotene, and selenium did not reveal marked regular differences. This does not, however, exclude their potential involvement as minor additional risk factors.[1] Thus, among the evaluated plasma parameters the vitamins C and E as well as cholesterol may be the most important.

If we assume that the variables of the present study are not only statistically but also biologically independent, and if their correlations with IHD are linear, the mortality by IHD can (by negating other risk factors, *e.g.,* hypertension, smoking, etc.) very tentatively be estimated by the formula IHD mortality $= 759 + 978 \times$ median cholesterol $- 19 \times$ medium vitamin C $- 679 \times$ median standardized vitamin E.

Such an approach, however, may be premature with only six study populations, and the synergistic interrelations between vitamins C and E remain to be elucidated in more details. In any case, the fact remains that within these European study populations both a medium risk of IHD (as in Northern Ireland) and a high risk (as in Scotland and Finland) were associated with a poorer vitamin C status (20-31 μM, on average 25 μM), whereas the high-risk regions showed a poor status in standardized vitamin E as well (20-21.5 μM, on average 21.2 μM).

Prospective Studies

The presently available evidence from prospective studies is scarce since vitamin C has only been measured in the prospective Basel Study, where the baseline level of all principal antioxidant vitamins in plasma was measured in almost 3000 male employees of three pharmaceutical companies in the period 1971-73.[5,6] The Basel Study population showed fairly high mean levels of vitamin C and abundance of cholesterol-standardized vitamin E, in comparison to other countries (the plasma samples of which were measured in the same laboratory), being even higher than in the semialpine Swiss population of the cross-sectional study discussed above. In consequence, if the antioxidant hypothesis for IHD were correct, low levels of vitamin C and E should not be quantitatively important risk factors of IHD in the Basel population, since the mortality from IHD is of a similarly low order in both Swiss regions. Indeed, the subjects who died in Basel from IHD in the subsequent 7-year period ($n = 67$) did not have significantly ($p > 0.05$) lower mean values of vitamin C and of cholesterol-standardized vitamin E than the survivors ($n = 2707$).[5]

In a prospective Dutch study in which the cholesterol-standardized vitamin E was at baseline probably as high as in the Basel population, vitamin E was not associated with an increased risk for IHD either.[9] Nevertheless, classification of vitamin C values in the Basel Study suggested for the range of marginal vitamin C deficiency a 1.6-fold risk of death from IHD (TABLE 1), but statistical significance could hardly be expected for the few cases after measuring plasma vitamin C all year round, *i.e.,* with

TABLE 1. Seven-Year Follow-up of Prospective Basel Study: Plasma Vitamin C in Cases of IHD

Plasma Vitamin C	Mortality		Relative Risk	p
	Observed	Expected[a]		
Deficient 11-23 μM (n = 12)	0.0490	0.0306	1.60	> 0.05
Sufficient > 23 μM (n = 53)	0.0198	0.0211	0.94	> 0.05

[a] After adjustment for age and smoking.

the well-known seasonal variations (which were excluded in the cross-sectional study by a standardized sampling period).

These data on vitamin C in the Basel population fit at least the assumption that a sufficient vitamin C status avoids an increased risk of IHD. Since in the IHD cases of the Basel Study hypercholesterolemia still remained a significant risk factor ($p<0.01$ in trend analysis), it is likely that the inverse association of plasma vitamin C and/or cholesterol-standardized vitamin E is, at least in part, independent of the risk due to hypercholesterolemia.

Conceivable Mechanisms of Action

The antioxidant hypotheses[1] offers a plausible explanation for a synergistic link of both antioxidants (FIG. 4). Vitamin E is the major chain-breaking antioxidant in lipids, *i.e.*, it primarily protects the PUFAs in membrane lipid bilayers and in plasma lipoproteins. When an endogenous oxygen free radical has escaped the enzymatic defense mechanisms it will be able to hit a PUFA molecule and produce a PUFA-peroxyl radical. This can be scavenged by vitamin E, which thereby becomes transiently a radical itself, the relatively less reactive vitamin E free radical (chromanoxyl free radical). It has been assumed that one molecule of vitamin E can protect roughly 3000 PUFA molecules against injury by free oxygen radicals. Clearly, if vitamin E falls below a critical level the well-known chain reaction of lipid peroxidation will progress. Thus, we may speculate as to whether in subjects with poor vitamin E status relatively more lipids become peroxidized and will thus contribute to the very high mortality from IHD, as can be expected from animal experiments. There is growing and substantial evidence[10] that the chromanoxyl radical can be reconverted into vitamin E by interaction with the water-soluble vitamin C at the interphase. Since the vitamin C free radical, which is rather stable and almost unreactive, can correspondingly be reconverted to vitamin C, we deal with a cascade of free radicals, postulated by TAPPEL and complemented by WILLSON.[11] If this scheme were operating *in vivo*, a critically poor vitamin C status should diminish the regeneration of vitamin E and consequently unblock the chain reaction of lipid peroxidation as well. Therefore, it may be speculated whether in subjects with a poor vitamin C status the reconversion of the chromanoxyl radical into vitamin E is diminished, or whether finally the vitamin E pools will even be materially reduced. Chronic marginal deficiency in vitamin C decreases the tissue level of vitamin E, at least in animals.[1] A combined deficiency of vitamins E and C is very likely to amplify the susceptibility towards lipid peroxidation with its subsequent health hazards.

Apart from the antioxidant hypothesis, various alternative (for IHD potentially important) mechanisms of action are, however, conceivable as well, particularly for vitamin C. Thus, the metabolism of procollagen and collagen[12,13] and mucopolysaccharides depends on vitamin C, and correspondingly scurvy is characterized by severe defects of perivascular connective tissue. Furthermore, vitamin C is required for many other hydroxylation reactions as well, *e.g.,* in cholesterol transformation into bile acids[14] (which could potentially alter cholesterol levels[15]), in carnitine biosynthesis, in cytochrome P-450 enzymes, but also in the cyclooxygenase and lipoxygenase[16] pathways. Vitamin C is able to affect some immune responses as well as the reactivity of monocytes and macrophages. A strong inverse association was even found in the National Health and Nutrition Examination Survey I for systolic blood pressure and serum vitamin C.[1] Clearly, the mechanisms of action remain to be differentiated in more detail.

Need of Proper Intervention Trials

In any case, final proof of any causal relationship of vitamins C and/or E and IHD awaits properly designed intervention studies. Almost all previous short-term, rather controversial clinical supplementation studies on symptoms or risk factors of IHD[1] were unfortunately inconclusive since they failed to characterize the respective vitamin status prior to and during supplementation.

SUMMARY

The present epidemiological data support and extend previous evidence in men and animals. Thus, a poor plasma status of vitamin C ($<23\ \mu M = 0.4$ mg/dl) and/

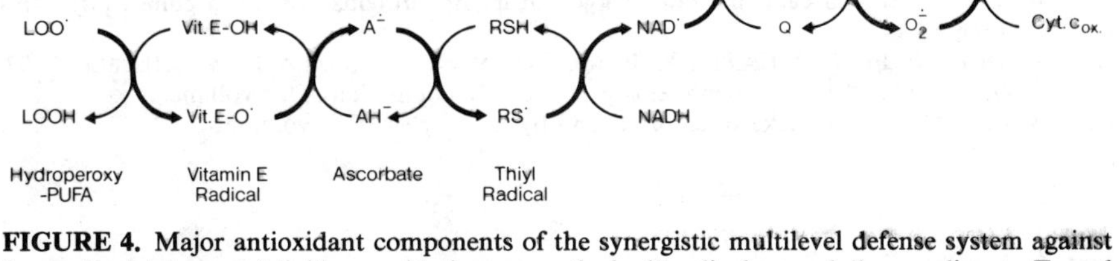

FIGURE 4. Major antioxidant components of the synergistic multilevel defense system against free radical injury and lipid peroxidation respectively ("radical cascade" according to Tappel and Willson[11]).

or of cholesterol-standardized vitamin E (<20-21.5 μM = 9 mg/l) occurs in westernized countries with an increased risk of IHD. A poor status in the major essential antioxidants may be a hitherto underrated,' at least permissive, risk factor of IHD that could, at least in some European countries, substantially complement the previously established risk factors such as hypercholesterolemia.

REFERENCES

1. Gey, K. F. 1986. On the antioxidant hypothesis with regard to arteriosclerosis. Bibl. Nutr. Dieta **37:** 53-91.
2. Slater, T. F. 1976. Panminerva Medica **18:** 381-390.
3. Vartiainen, E., P. Puska, J. Huttunen, H. Karppanen & K. F. Gey. 1983. Vitamiinit ja sepelvattimotauti. Finn. Med. J. **35:** 3146-3150.
4. Riemersma, R. A., D. A. Wood, S. Butler, R. A. Elton, M. Oliver, M. Salo, T. Nikkari, E. Vartiainen, P. Puska, K. F. Gey, P. Rubba, M. Mancini & F. Fidanza. 1986. Linoleic acid content in adipose tissue and coronary heart disease. Brit. Med. J. **292:** 1423-1427.
5. Gey, K. F., H. B. Stähelin, G. B. Brubacher, H. Riggenbach, F. Bernasconi & L. K. Widmer. 1987. Inverse association of antioxidant vitamins in plasma with subsequent risk of cancer. Initial level of vitamins A, C, E and β-carotene in 2975 men followed-up for 7 years in the prospective Basel Study. Submitted.
6. Stähelin, H. B., K. F. Gey & G. B. Brubacher. 1987. Plasma vitamin C and cancer death. This volume.
7. Bureau of Nutritional Sciences, Department of National Health and Welfare, Canada. 1983. Recommended nutrient intake for Canadians. Canadian Government Publications Centre. pp. 64-76.
8. Machlin, L. J. Vitamin E. 1984. *In* Handbook of Vitamins. L. J. Machlin, Ed.: 99-145. Dekker. New York and Basel.
9. Kok, F. J., A. M. de Bruijn, R. Vermeeren, A. Hofman, A. van Laar, M. de Bruin, R. J. J. Hermus & H. A. Valkenburg. 1986. Serum selenium, vitamin antioxidants and cardiovascular mortality. A 9-year follow-up study in the Netherlands. Am. J. Epidemiol. In press.
10. Niki, E. 1987. Interaction of ascorbate and α-tocopherol. This volume.
11. Willson, R. L. 1983. Free radical protection: Why vitamin E, not vitamin C, β-carotene or glutathione? *In* Ciba Foundation Symposium 101. Pitman. London. pp. 19-37.
12. Buzina, R., J. Aurer-Kozelj, K. Srdak-Jorgic, E. Bühler & K. F. Gey. 1987. Increase of gingival hydroxyproline and proline by improvement of ascorbic acid status in man. Int. J. Vit. Nutr. Res. **56:** 367-372.
13. Schwarz, R. I., P. Kleinman & N. Owens. 1987. Ascorbate can act as an inducer of the collagen pathway because most steps are tightly coupled. This volume.
14. Ginter, E., M. Jurčovičová & J. Jurcovicova. 1987. Chronic vitamin C deficiency lowers fractional catabolic rate of low-density lipoproteins (LDL) in guinea pigs. This volume.
15. Jacques, P. F., S. C. Hartz, R. B. McGandy, R. A. Jacob & R. M. Russell. 1987. Vitamin C and blood lipoproteins in an elderly population. This volume.
16. Moser, U. 1987. Uptake of ascorbic acid by leukocytes. This volume.

DISCUSSION OF THE PAPER

E. J. DILIBERTO (*Wellcome Research Laboratories, Research Triangle Park, N. C.*): It's clear that the reperfusion and associated with it the oxygen toxicity and free radicals might be benefited by ascorbic acid to curtail or reduce the oxygen radicals. Is it clear that there's also a correlation in terms of the incidence? I can see what you're dealing with mainly is the mortality. What about the incidence or the initial insult of the heart disease?

K. F. GEY: The morbidity data are much weaker than the mortality data since the reliability of the morbidity register shows great differences from one country to another. Therefore the present cross-sectional studies within the WHO-MONICA project are concentrating on the standardized mortality rate.

H. SPRINCE (*Jefferson Medical College, Philadelphia, Pa.*): I would gather from this talk that the northernmost countries are more prone to ischemic heart disease.

K. F. GEY: Right.

H. SPRINCE: Have any studies been done on cardiovascular disease of this type in Eskimos?

K. F. GEY: Unfortunately the mortality data from Eskimos are very poor. The data I know showed that about 40% of the Eskimos die by accident, presumably before they get a heart attack if they ever get any. It is a statistical problem which considerably hampers the evaluation of the presumably beneficial effect of the omega-3 fatty acids as well. Nevertheless it's very likely that the vitamin C status and the vitamin E status of the Eskimos is very good since Eskimo means (in the language of their former Indian neighbors) "people-who-eat-raw-meat." Actually, if Eskimos catch fish or birds they will consume a lot of their catch (including offals) in the raw state, which is thus still very rich in ascorbic acid. They eat the adrenals, they eat the brain, and this is presumably their most important vitamin C source. Furthermore, the meat of the fishes, whale, or seal they consume is also relatively rich in vitamin E, richer anyway than is the meat from animals.

H. SPRINCE: I'm also wondering what happens with such studies made in people living in equatorial regions, say in Central America, or Brazil?

K. F. GEY: These countries have very low IHD mortality according to statistics, but they also have, as the Eskimos, conditions that are not common in westernized countries. Many die by infections or they have a vitamin A deficiency. That's the most common vitamin deficiency all over the world, and so studies in third-world countries may be hampered by other factors that will make the analysis very difficult.

A. B. KITABCHI (*University of Tennessee, Memphis, Tenn.*): This is a very provocative study. It's interesting that none of your subjects surveyed actually had scurvy but they had borderline vitamin C deficiency. Vitamin C has, as you are well aware, a prooxidant effect at low concentration and an antioxidant effect at a high level. Therefore, a low concentration in fact could act as a prooxidant and could catalyze lipid peroxidation.

Obviously these studies do not establish cause or relationship but simply show correlation. Of interest would be the level of unsaturated fatty acid that these subjects use. For example, Italians with olive oil and oleic acid versus a more polyunsaturated fat which is a substrate for peroxidation may be also equally important. And so these two effects may of course be in addition to catecholamines, which may be an important phenomenon of stress.

K. F. GEY: First, with regard to the possibility of a prooxidative effect at low levels of antioxidants in the extracellular space: when chronic marginal deficiency of vitamin C results in very low plasma levels, *e.g.*, 10% of normal, a metal-catalyzed prooxidative effect (corresponding to *in vitro* experiments requiring the addition of iron or copper ions) cannot be excluded because of the presence of metals in plasma. Such a prooxidant effect would, however, only be indirect (attributable to the reduction of the metal catalyst) and could not even be caused by the ascorbyl free radical.[1]

Second, inverse correlations found in cross-sectional comparisons can only suggest but not establish a causal relationship. Such correlations may, however, justify intervention trials since only the latter can prove preventive potentials of essential antioxidants.

Now with regard to the polyunsaturated fatty acids from vegetable oils which are simultaneously a major, although very variable, dietary source of vitamin E. I'm at present unable to review this interrelationship in more detail. The study was in the first part actually done with additional biopsies from the adipose tissue, but unfortunately we have only Italy, Scotland, and Finland in this part of the study. The data showed that in Italy, indeed, the level of linoleic acid or the P/S ratio was higher than in the northern countries. Unfortunately the fatty acids of adipose tissue are not very informative, since the higher fatty acids are not present in considerable quantities. So we changed the protocol and are now going to analyze the erythrocyte membrane which contains a lot of fatty acids with 20 and 22 carbon atoms, and hopefully this will show the relative importance for IHD of polyunsaturated fatty acids and the P/S ratio (or of n-6 and n-3 fatty acids respectively) in relation to antioxidant vitamins.

S. L. ROMNEY (*Albert Einstein College of Medicine, Bronx, N. Y.*): The geographic epidemiologic implications of your study are truly impressive, but I would ask: Do you have any impression as to the value or siginificance of an isolated plasma determination in terms of either the mechanistic involvement in the chronicity of heart disease or any other pathophysiologic state?

K. F. GEY: You touched on the problem of whether a single measurement is representative. In the cross-sectional studies we standardized the sampling period that was January through April since we wanted to get the most critical period for the ascorbic acid status. In some countries, for instance in both Finnish regions, we have repeated the measurements in fall, and we found (as in the Basel study) that the majority of subjects remained in the same vitamin category. In the Basel study we have repeated about a tenth of the measurements several years later. In spite of this long time interval the major part, around 50%, of the subjects remained in the same category (quartiles). We feel that taking one blood sample gives you a fair impression of the gross status of the vitamin. You will, however, not be able to get a very precise correlation, but the gross status is maintained. Of course it will be better to repeat the measurements, but this is technically very difficult as you can imagine.

L. MACHLIN (*Hoffmann-La Roche, Nutley, N. J.*): I would question your comments about the prooxidant effects of the low levels of vitamin C *in vivo.* In the review of the literature there's only one instance of a prooxidant effect, that is when vitamin C was infused intravenously at enormously high levels into a G-6-PD deficient individual and hemolysis occurred. Other than that there's ample evidence for an antioxidant effect in animals and to some extent in man *in vivo.* So I think it's a little premature to extrapolate some of the *in vitro* findings that we discussed this morning to *in vivo* findings. I'm sure at the biochemical level the prooxidant effect may be important in certain chemical reactions, but to consider it a prooxidant in the circumstance to the point of a hazard is misleading.

K. F. GEY: I guess that I wasn't clear enough. I do not consider at all any prooxidant effect at physiological levels of vitamin C *in vivo.* Nevertheless, from the scientific point of view we should at least consider the possibility that a metal-mediated prooxidant effect of very low levels of vitamin C could to some extent occur in certain compartments in addition to the lipid peroxidation, which may result from the pure lack of the antioxidant. I fully agree, however, that there is lack of evidence *in vivo* of this possibility.

E. J. DILIBERTO: I actually wanted to preface Dr. Machlin's remarks. There are two things I would want to clarify, one with regard to what is known about the etiology of the reperfusion toxicity. I think that it's because of the readministration of oxygen when there has been deprivation of oxygen in the tissue and the enzymic reactions that have been changed and it's actually like the xanthine oxidase system that starts generating superoxide ions. At least that's one of the possible mechanisms. So you don't need to go into any detail with regard to the prooxidant effects of ascorbic acid.

In addition, there's a different situation in terms of what we were discussing this morning, there the idea is that because of the capacity of neurons and certain cells to take up the vitamins, you actually do have very low concentrations in the extracellular fluid surrounding those cells like you would see in the synaptic cleft. So those kinds of reactions can become a reality in terms of affecting the lipid layer of the exterior of the cell and not the interior of the cell.

S. D. VARMA (*University of Maryland, Baltimore, Md.*): I think the prooxidant effect of ascorbic acid which happens at low levels is indeed metal dependent and under *in vivo* situations there are only trace amounts of free metals, almost all of them 99% chelated. There is a paper in the recent issue of the *Communications in Free Radicals,* which studies in detail the oxidation of ascorbic acid in the absence and presence of metals, and copper is the only element which really substantially accelerates the oxidation of ascorbic acid producing dangerous radicals.

UNIDENTIFIED SPEAKER: Did you look at the enzymes which remove oxygen radicals or lipid peroxides like SOD and glutathione peroxidase?

K. F. GEY: It would be very important to do that since the defense mechanisms against the radical injury includes of course a lot of enzymes and they should be measured too, but unfortunately it is technically impossible.

Plasma Vitamin C and Cancer Death: The Prospective Basel Study

HANNES B. STÄHELIN,[a] K. FREDERICK GEY,[b] AND GEORG BRUBACHER[b]

[a]*Geriatric Clinic*
Kantonsspital Basel
and
[b]*Department of Vitamin Research*
F. Hoffmann-La Roche Ltd.
Basel, Switzerland

INTRODUCTION

There are a number of good arguments in favor of a beneficial action of vitamin C in preventing cancer. Indeed, shortly after the elucidation of the chemical structure of ascorbic acid such an effect was suggested. Subsequently a number of epidemiological studies addressed this question by analyzing food composition and cancer incidence. These indirect studies based on the consumption of food rich in vitamin C indicated that the high intake of vitamin C may lower the incidence of certain cancers.[1] Serveral investigators noted a lower cancer incidence of the upper gastrointestinal tract in subjects with a high vitamin C intake compared to low vitamin C users.[2–4] On the other hand, no clear association of vitamin C intake with other cancer sites has been established thus far.[5] The report of a beneficial effect of pharmacological doses of ascorbic acid on advanced carcinoma of the large bowel[6] raised great hopes. A carefully controlled trial by Moertel *et al.*,[7] however, could not confirm the claims made by previous workers. Meanwhile experimental work in animals and cell systems elucidated many of the actions of vitamin C and led to several hypotheses on how ascorbic acid interacts with tumor induction and/or promotion.[5]

Pertinent epidemiological studies are not abundant and the majority of them are retrospective. The diagnostic accuracy of many studies is furthermore limited to younger age groups, since multimorbidity in elderly patients and a low autopsy rate may impair the correct identification of cases and associations may be missed by lumping all cancer cases together.

The Basel Study,[8] initially planned in 1960 as a survey of cardiovascular disease in a group of about 6300 healthy volunteers all employed by pharmaceutical companies at Basel, took account of plasma vitamins in its third examination in 1971-73. The mortality follow-up until 1980 allows an analysis of the association of vitamin C alone and in conjunction with vitamin A, β-carotene, and vitamin E, as well as other factors known to influence carcinogenesis and mortality from cancer at various sites.

POPULATION AND METHODS

The Basel Study began in 1960 with special emphasis on peripheral vascular diseases. Details have been given previously.[8,9] The cohort was reinvestigated in 1965 and 1971-73. In the last follow-up examination a survey was made of certain eating habits at the time the plasma vitamin concentration was measured. A complete mortality follow-up was obtained until 1980.

The causes of death were established according to the death certificates. The cancer registry and autopsy data allowed control of the accuracy of the death certificate (autopsy rate 56%). In instances where a tumor coexisted but death occurred due to other causes (*e.g.*, coronary heart disease) these cases were classified with the cancer group.

GROUPING

Cancer deaths were categorized for this analysis in three groups: cancer of the lung, stomach cancer, and remaining sites. In analyzing the risk ratio for different vitamin levels, gastrointestinal cancer (stomach and large bowel) were considered together. Clinical and laboratory data used for this evaluation were detailed previously.[9] Venipuncture occurred on a voluntary basis, therefore not all participants with clinical data also had laboratory data. Thus of the 3756 men and women of the third Basel Study (BS III), 2975 men had a complete data set and are included in the present analysis. The 688 women are not included because of the low mortality and the low incidence of cancer cases.

A replicate analysis of all vitamin parameters with similar laboratory methods was carried out in 1978-79 in 355 participants of the BS III during an adhering family study. These results indicate that individuals remain in the gross category of their vitamin status. Though the correlation between the first and second vitamin C measurement 6 years later is highly significant ($p < 0.001$), the correlation coefficient is low ($r = 0.29$). Of the subjects in the lowest quartile, 41% remain low and 18% are found in the highest quartile after 6 years. This indicates that a single measurement of plasma vitamin C does not necessarily identify all subjects at risk.

Statistical analysis involved the comparison of group means with survivors after age adjustment and for vitamin E after adjustment for plasma cholesterol concentration. The risk ratio comparing different concentrations of vitamins was calculated using a logistic model taking into consideration age and smoking. The procedure is detailed elsewhere.[10] The cumulative index of antioxidant vitamins (CIAVIT) was computed as a product of the molar concentrations of vitamin C $\times$ vitamin E $\times$ β-carotene.

RESULTS

Of the 2975 men at risk participating in the BS III for whom there was a complete data set, 268 died by 1980 (9%). The 102 total cancer cases comprised 37 lung cancer,

17 stomach cancer, 9 colorectal cancer and 39 cancer of other sites not analyzed in further details. The average age at death was rather similar: for carcinoma of the lung 62.7 ± 6.1, for stomach cancer 60.8 ± 7.5, and for colorectal cancer 64.6 ± 6.7. The mean survival period was 5.1 years (4.9 and 4.2 years for lung and stomach cancer respectively).

The mean plasma concentration of vitamin C was significantly ($p<0.05$) lower in all death cases compared to survivors, all cancer cases, and stomach cancer. However, after adjustment for smoking, the impact of vitamins on mortality is greatly diminished and for vitamin C a significantly lower value was only observed for stomach cancer (TABLE 1).

An acceptable vitamin C plasma concentration is assumed to be 0.4 mg/dl or higher. This cutoff value was therefore chosen to compute risk ratios for different plasma vitamin concentrations. Indeed there is a trend, albeit not significant to an inverse association with vitamin C adjusted for age and smoking and all death and all cancer, and a positive association with vitamin C status and survival ($p=0.08$). Again a clearly inverse association was observed for vitamin C and gastrointestinal cancer ($p<0.05$, FIG. 1). It will be noted from TABLE 1 and FIGURE 1 that similar inverse correlations are observed for vitamin A, E, and CIAVIT. The marked inverse association of β-carotene in gastrointestinal cancer is lost if smoking status is considered. For stomach cancers the p value becomes 0.056, whereas low vitamin A and stomach cancer correlate strongly ($p<0.0005$). Association of vitamin E and stomach cancer is $p=0.057$ and for CIAVIT $p<0.05$.

Here the question arises of whether vitamin C merely reflects the intake of other nutrients or vitamins important in cancer induction or promotion. No correlation of vitamin C and vitamin A was observed in our sample (Pearson's correlation coefficient 0.04). The relative risk of stomach cancers in subjects with low β-carotene and low vitamin C increases to 2.19 and with higher levels the risk diminishes to 0.52. This indicates that both antioxidants act independently but synergistically. Low vitamin C and E enhanced the risk ratio to 2.39, whereas with higher levels the risk drops to 0.68. Again a synergistic mechanism is suggested. When the logistic model accounted for either vitamin it appears that both, C and E, cannot fully substitute and may even require an adequate status of either vitamin C or E to achieve optimal risk lowering. Thus the association of vitamin C with gastrointestinal cancer is not explained by other antioxidant vitamins.

DISCUSSION

The first question to be raised addresses the following problem: To what degree does a single plasma vitamin C concentration reflect long-term vitamin C intake? There is in our population only a small number with marginal vitamin C intake. The fact that the participants were all steadily employed males participating in a prospective study since 1960 almost excludes the persons with inadequate nutrient intake. Nevertheless, participants with a low status tended to stay low ($p<0.0001$). Furthermore, vitamin C correlated best with citrus food intake[10] and in an initial case control analysis stomach cancer cases had a lower citrus fruit intake compared to controls ($p=0.08$).[11] It is conceivable that in our sample the impact of vitamin C and carcinogenesis is less obvious due to the well supplied population on one hand and the relative uncertainty of a single plasma vitamin C determination as a measurement of

TABLE 1. Mean Plasma Vitamin Concentration: Comparison of Different Mortality Groups with Survivors

Cause of Death	n	Vitamin C (μmol/l)	β-Carotene (μmol/l)	Vitamin A (μmol/l)	Vitamin E (μmol/l)	CIAVIT[a] U
All causes	268	43([b,e])	0.30[c]([b,e])	2.82	34.9[b]	450[c]([b,e])
All cancer	102	41([b,e])	0.28[b]([c,e])	2.77[b]	34.2	400[b]([c,e])
Lung	37	43	0.25[b]([c,e])	2.85	33.7	362([c,e])
Stomach	17	36[b]	0.23([d,e])	2.48[c]	33.1[b]	268([d,e])
Remaining	48	42	0.34	2.83	35.0	571
Survivors	2707	47	0.35	2.83	35.2	580

[a] CIAVIT = cumulative index of antioxidant vitamins.
[b] $p < 0.05$.
[c] $p < 0.01$.
[d] $p < 0.001$.
[e] Without adjustment to smoking.

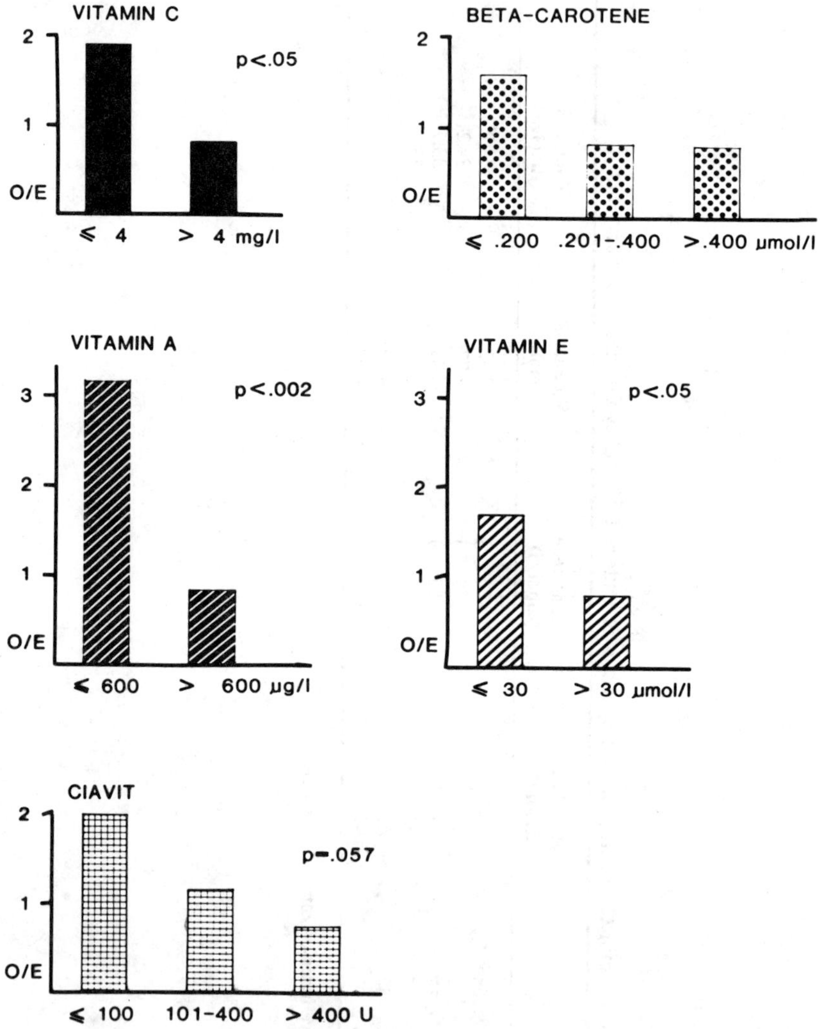

FIGURE 1. The risk ratio of observed over expected gastrointestinal cancers (17 stomach, 9 large bowel) for different vitamin concentrations is shown. All values are computed after adjustment for age and smoking (vitamin E is also adjusted for cholesterol).

long-term vitamin C intake on the other hand. Our well-documented observation of an inverse association of vitamin C with cancer of the gastrointestinal tract thus fully supports the previous findings made by others.[1]

Obviously low vitamin C acts not solely but synergistically with other factors, two of the most important possibly being smoking and alcohol. Although smoking and alcohol consumption did not correlate with vitamin C,[9] the inclusion of smoking in the logistic model lowers the impact of antioxidant vitamins on mortality and all cancer deaths, but not of vitamin C on gastrointestinal cancer. This may partly reflect the well-known finding that smokers have a lower vitamin C plasma concentration.[12] Interestingly, the effect of β-carotene seems strongly linked to smoking (TABLE 1).[10] Whether this association implies a pathogenetic link in the disease remains to be elucidated.

In our study we found no association of vitamin C with other cancer sites. Since the number of women is so small we had to exclude this group from analysis, so no information on cervical cancer and vitamin C is available.[13] Our results are essentially in agreement with Hinds *et al.*,[14] who found no association of dietary vitamin C intake with lung cancer in Hawaii.

Vitamin C exerts numerous metabolic effects, some of which are only partly known. Thus the question arises of which action accounts for tumor prevention. From animal experiments, tissue cultures, and studies in humans several hypotheses emerge. Probably the most popular is the model proposed by Tannenbaum,[15] based on the observation that ascorbic acid inhibits the formation of *N*-nitrosamine in the gastrointestinal tract.[16] Nitrosamines may come from exogenous or endogenous sources. In human the endogenous formation is a dominant source.[5] Ascorbic acid is a potent inhibitor of *N*-nitrosamine formation,[16–18] which may be a major problem in hypochlorhydric subjects.[19] In another study, supplementation of the diet with vitamin C and α-tocoferol reduced the mutagenic compounds in the human feces to about one fourth.[20] Our own data are in agreement with the proposed model by Tannenbaum.

The antioxidant properties of ascorbic acid are well established and a possible mode of action could be the protection from free radicals. Indeed, animal experiments are consistent with this interpretation.[5] The fact that other antioxidant vitamins (CIAVIT) and to some degree vitamin A act synergistically in our group lends support to this view. The lack of association of vitamin C alone with bronchus cancer or other cancer sites but the highly significant negative association of CIAVIT with total cancer mortality could indicate that vitamin C has a role as part of the antioxidant system. Ascorbic acid has various other metabolic effects, notably on connective tissue and immunity, which may interfere with tumor promotion.[21] From an epidemiological point of view this mode of action remains controversial.[22] In our sample, there were no differences in subjects with short and long survival after the vitamin determination. However, the number is rather small and the ongoing mortality analysis comparing all cases until 1985 is more apt to elucidate this question. In conclusion therefore, the results of the Basel Study suggest an independent role of vitamin C in the development of cancer of the gastrointestinal tract, particularly in stomach cancer. An effect of vitamin C in cancer protection as part of the body's antioxidant system seems possible but less firmly established.

Correa[23] proposed models of cancerigenesis involving the interaction of several risk factors. Our data fit to this interpretation. In the presence of a low vitamin C (and E) intake *N*-nitrosamine formation is insufficiantly prevented and a poor status in β-carotene, vitamin E, and/or vitamin A could be another prerequisite for the devlopment of dysplastic and metaplastic tissue changes leading ultimately to cancer. Further studies will be required in order to establish optimal vitamin C supply and the daily allowances recommended for the prevention of overt vitamin deficiencies may be inadequate for the maintenance of optimum health.

REFERENCES

1. NATIONAL ACADEMY OF SCIENCES, NATIONAL RESEARCH COUNCIL. 1982. Diet, Nutrition and Cancer. National Academy Press. Washington, D.C.
2. MEINSMA, L. 1964. Nutrition and cancer. Voeding **2:** 357-365.
3. HIGGINSON, J. 1966. Etiological factors in gastrointestinal cancer in man. J. Natl. Cancer Inst. **37:** 527-545.
4. HAENSZEL, W. & P. CORREA. 1975. Developments in the epidemiology of stomach cancer over the past decade. Cancer Res. **35:** 3452-3457.
5. GLATTHAAR, B. E., D. H. HORNIG & U. MOSER. The role of vitamin C in carcinogenesis. JNCI Monograph Series. In press.
6. CAMERON, E. & L. PAULING. 1976. Supplemental ascorbate in the supportive treatment of cancer: Prolongation of survival times in terminal human cancer. Proc. Natl. Acad. Sci. **73:** 3685-3689.
7. MOERTEL, C. G., T. R. FLEMING, E. T. CREAGAN, *et al.* 1985. High dose vitamin C versus placebo in the treatment of patients with advanced cancer who have had no prior chemotherapy. N. Engl. J. Med. **312:** 137-141.
8. WIDMER, L. K., H. B. STÄHELIN, C. NISSEN, *et al.* 1981. Basler Studie: Venen-Arterienkrankheiten, coronare Herzkrankheit bei Berufstätigen. Huber. Bern.
9. STÄHELIN, H. B., F. RÖSEL, E. BUESS & G. BRUBACHER. 1984. Cancer, vitamins and plasma lipids: Prospective Basel Study. JNCI **73:** 1463-1468.
10. GEY, F. K., H. B. STÄHELIN, G. BRUBACHER, *et al.* Inverse association of antioxidant vitamins in plasma with subsequent risk of cancer. Submitted.
11. STÄHELIN, H. B., F. RÖSEL, E. BUESS & G. BRUBACHER. 1986. Dietary risk factors for cancer in the Basel Study. Bibl. Nutr. Dieta **37:** 144-155.
12. RITZEL, G. 1975. Evaluation von Ernährungserhebungen im Rahmen der Basler Studie III. *In* Zur Ernährungssituation der schweizerischen Bevölkerung. G. Brubacher & G. Ritzel, Eds.: 57-82. Huber. Bern.
13. WASSERTHEIL-SMOLLER, S., S. L. ROMNEY, J. WYLIE-ROSETT, *et al.* 1981. Dietary vitamin C and uterine cervical dysplasia. Am. J. Epidemiol. **114:** 714-724.
14. HINDS, M. W., L. N. KOLONEL, J. H. HANKIN & J. LEE. 1984. Dietary vitamin A, carotene, vitamin C, and risk of lung cancer in Hawaii U.S.A. Am. J. Epidemiol. **119:** 227-237.
15. TANNENBAUM, S. R. 1965. N-nitroso compounds. A perspective on human exposure. Lancet **1:** 629-632.
16. IVANKOVIC, S., R. PREUSSMANN, D. SCHMÄHL & J. W. ZELLER. Prevention by ascorbic acid of in vivo formation of N-nitroso compounds. *In* N-Nitroso Compounds in the Environment. IARC Scientific Publication No. 9. P. Bogorski & E. A. Walker, Eds.: 101-102. International Agency for Research on Cancer. Lyon, France.
17. MIRVISH, S. S. 1975. Formation of N-nitroso compounds: Chemistry, kinetics and in vivo occurrence. Toxicol. Appl. Pharmacol. **31:** 325-351.
18. OSHIMA, H. & H. BARTSCH. 1981. The influence of vitamin C on the in vivo formation of nitrosamines. *In* Vitamin C (Ascorbic Acid). J. N. Counsell & D. H. Hornig, Eds.: 215-224. Applied Science Publishers. London.
19. REED, P. J., K. SUMMER, P. L. R. SMITH, *et al.* 1983. Effect of ascorbic acid treatment on gastric juice nitrite and N-nitroso compound concentrations in achlorhydric subjects. Gut **24:** 492-493.
20. DION, P. W., E. B. BRIGHT-SEE, C. C. SMITH & W. R. BRUCE. 1982. The effect of dietary ascorbic acid and alpha-tocopherol on fecal mutagenicity. Mutat. Res. **102:** 27-37.
21. SCHAMBERGER, R. J. 1984. Nutrition and Cancer. Plenum Press. New York.
22. WITTES, R. E. 1985. Vitamin C and cancer. N. Engl. J. Med. **312:** 178-179.

23. CORREA, P., W. HAENSZEL, C. CUELLO, *et al.* 1975. A model for gastric cancer epidemiology. Lancet **2:** 58-60.

DISCUSSION OF THE PAPER

UNIDENTIFIED SPEAKER: I wonder if there are a couple of things that sampling at one point or even at multiple points leaves out. Is there a difference in vitamin C metabolic demand? And the other question is, Is there a circadian rhythm for vitamin C because when you sample your population you are sampling at some limited time period?

H. B. STÄHELIN: (*Kantonsspital, Basel, Switzerland*): To answer the last question first, the vitamin C was sampled always on a Tuesday morning, I think between eight o'clock and ten o'clock every week. There are of course some seasonal variations, however, accounting for this seasonal aberration did not really change the picture. And we don't really know about the vitamin C requirements of our subjects. I don't think that we could answer this question with the data we have.

B. LANE: (*Columbia University School of Public Health, New York, N.Y.*): In your introduction you mentioned the carefully controlled trial by the Mayo Clinic could not confirm the claims of Cameron and Pauling and that's probably true, but there's an observation that needs to be made about that trial. They compared the control group of people on what they contended were excellent diets, so consequently they were looking at a group that many nutritionists would feel had an excellent vitamin C status, and they were comparing them against a high-dose ascorbic acid group. The question of whether low or deficient ascorbic acid is a factor was really not addressed. Would you comment?

H. B. STÄHELIN: Yes, I actually mentioned this study only just to indicate that vitamin C was also used as a pharmaceutical agent and that there is no real study proving that in the megadoses there is an effect on retardation of tumors compared to, as you say, naturally or to good eating habits. In our study we have by and large a very well-nourished population.

H. SPRINCE: (*Jefferson Medical College, Philadelphia, Pa.*): Would this effect of vitamin C on stomach cancer involve the nitrosamine store?

H. B. STÄHELIN: Well I think this is a very elegant explanation of something we observe. There are other possibilities too. It seems from our data that they also have inadequate vitamin E, β-carotene, and vitamin A status, and it is conceivable that β-carotene and vitamin A prevent the occurrence of dysplastic or metaplastic tissue changes and then if this is low the effect of vitamin C is much more dangerous.

Plasma Reduced and Total Ascorbic Acid in Human Uterine Cervix Dysplasias and Cancer

SEYMOUR L. ROMNEY,[a] JAYASRI BASU,[a] STEN VERMUND,[b] PRABHUDAS R. PALAN,[a] AND CHANDRALEKHA DUTTAGUPTA[c]

[a]*Department of Obstetrics and Gynecology*
Albert Einstein College of Medicine
Bronx, New York 10461

[b]*Department of Epidemiology and Social Medicine*
Albert Einstein College of Medicine
Bronx, New York 10461

[c]*Cytology Research Centre (ICMR)*
New Delhi, India 110002

INTRODUCTION

Deficiencies in essential nutrients are of intense interest in studies of carcinogenesis.[1] Of the 14 known vitamins, ascorbic acid is of special interest because it (1) is involved in collagen synthesis,[2] (2) detoxifies chemical carcinogens,[3] (3) interferes with the formation of chemical carcinogens *in vivo,*[4] and (4) modulates the immune system.[5] These biological processes may be altered or impaired when a tumor is initiated and followed through sequential development.[6–8] However, any role of vitamin C as a possible risk factor for the development of cancer remains speculative. Clinical studies of patients with malignant neoplasms suggest that many have minimal tissue stores of ascorbic acid, and some may even show physical signs suggestive of scurvy.[9,10] The critically important issue then is whether low vitamin C participates in the carcinogenic process or is merely a consequence of it. An advantage of studying asymptomatic cervical dysplasia is that a nutritional finding, particularly any negative association, is more likely to be related to the causative factor rather than to result from the anorexic effects of the neoplasm itself.

We have reported an unexpected finding in a case-control study designed to investigate the role of vitamin A as a putative factor for cervical dysplasias. It was demonstrated that ascorbic acid (AA) was a significant nutritional risk factor.[11,12] Intake was measured using dietary recall. Dietary questionnaires, diaries, and histories have been used to assess an individual's food intake, but have the limitations of recall bias inherent in subjective responses.[13] Our more recent approach has been to assess plasma AA levels biochemically to eliminate subject recall bias. We followed up the unexpected epidemiologic findings with biochemical measurements of reduced AA (RAA) in 34 normal women and 46 women with varying degrees of cervical abnormalities.[14] Significantly lower RAA levels were seen in women with cervical abnormalties, but no dose-response effect was seen. The present study is an effort to use optimal spectrophotometric techniques to assess both total AA (TAA) and RAA in over 200 women, some of whom were normal, others of whom had various levels of abnormal cervix findings. The hypothesis, developed from our earlier work, is that low plasma vitamin C levels occur in asymptomatic women with cervix dysplasias.

MATERIAL AND METHODS

Pap Smear and Colposcopy

Routine techniques were employed in obtaining Pap smears and conducting colposcopy examinations.[15] For the Pap smear, a cotton-tipped applicator was rotated in the cervical os and a wooden Ayre spatula rotated 360° about the exocervix. The yield of exfoliated epithelial cells was plated on glass slides, promptly fixed, coded, stained, and submitted to staff cytopathologists for interpretation. Colposcopy examinations were carried out with a Zeiss photocolposcope with a fiber optic light source. A bivalve vaginal speculum exposed the lower endocervical canal by everting the lips of the cervix. Tissue contrast was achieved by application of 4% acetic acid and the use of a green light filter. Special attention was directed to the colposcopically identified squamous metaplasia in the transformation zone, situated between the native squamous epithelium of the exocervix and the columnar epithelium of the endocervical canal. Foci of white epithelium and nuclear-cytoplasmic changes characterized as punctation, mosaicism, and abnormal vascular patterns are visually perceived when transilluminated. Such abnormalities were biopsied under colposcopic direction. Staff pathologists employed the criteria of the World Health Organization as the basis for their cytologic and histopathologic interpretations.[16]

Selection of Subjects

The recruitment of subjects, both women with cervical abnormalities or those considered normal, were all drawn from the same cachement area and receive their health care at the Bronx Municipal Hospital Center (BMHC). These women have similar socioeconomic backgrounds.

Normal Women

Fifty-six women attending the family planning clinics were recruited with informed consent as controls if they had two consecutive negative Pap smears and a cervix that was colposcopically assessed as normal. A peripheral venous blood sample was withdrawn at the time of the colposcopy examination.

Women with Cervical Dysplasias and Carcinoma-in-Situ (CIS)

One hundred twenty-two asymptomatic women with atypical, suspicious, or positive Pap smears who also had abnormalities noted on colposcopy were recruited into the study. Colposcopically directed biopsies were histopathologically interpreted as *dysplasias* of various degrees or *carcinoma-in-situ*. Peripheral venous blood samples were withdrawn at the time of biopsy.

Women with Cervix Cancer

Women with squamous cell carcinoma of the cervix were admitted for surgery to a hospital adjacent to the BMHC, the Hospital of the Albert Einstein College of Medicine. Peripheral venous blood samples with informed consent were obtained from 24 women with invasive cancer prior to the initiation of any therapy. Many of these women were drawn from the BMHC cachement area, but several were referred to the hospital from elsewhere.

Posttherapy Subgroups

Peripheral venous blood samples were obtained from 27 women who underwent surgical conization for cervical lesions diagnosed as either severe dysplasias or CIS. Fourteen women, hysterectomized for having cervix cancer, were also included in the posttreatment groups. The blood samples were withdrawn at the time of follow-up care, minimally 6 weeks after therapy.

Pathology

The histopathologic interpretations of the staff pathologists of the colposcopically directed cervical biopsies and the cervical tissues of the hysterectomized uteri were all reviewed by a senior cytopathologist who served as a referee. The referee's diagnoses were employed as the basis for the final stratification of the histopathologic findings, and it was the referee's determination that was employed in the data analysis.

Biochemical Procedures

Peripheral venous blood samples were obtained in heparinized tubes, coded, and promptly sent to the biochemistry laboratory for blinded analyses. The determinations were carried out on the same day within hours after the samples were obtained. Upon being brought to the laboratory, each blood specimen was centrifuged at 800 ×g for 15 minutes. Plasma obtained was pipetted out into separately labeled polypropylene tubes, and promptly processed.

Extraction of Ascorbic Acid from the Plasma

To a known volume of plasma in the test tube a known volume of 40% trichloroacetic acid was added. Following mixing, the mixture was left in ice for 10 minutes to ensure complete precipitation of protein, and then centrifuged at 800 ×g for 10 minutes. Supernatant obtained was used for both reduced and total ascorbic acid analyses.

Reduced Ascorbic Acid Analyses of the Plasma

Reduced ascorbic acid was determined by 2-2′dipyridyl method.[17] The following reagents were added in sequence to a known volume of deproteinized samples: 85% orthophosphoric acid, 1% aqueous solution of 2-2′dipyridyl (solubilized by heating at 60° C for 5 minutes) and 3% ferric chloride. The assay mixture was incubated at room temperature for 15 minutes and the absorbance at 525 nm was measured against a reagent blank.

Total Ascorbic Acid of the Plasma

Total ascorbic acid was determined by the 2,4-dinitrophenylhydrazine (DNPH) method.[18] Deproteinized plasma sample was first treated with bromine for 10 minutes at room temperature. Excess bromine was removed by passing water-saturated air through the samples. To a known volume of the resultant solution, a known volume of 2% 2,4-DNPH solution, containing 4% thiourea, was added. The mixture was incubated at 37°C for 3 hours, after which a known volume of 65% H_2SO_4 was added, keeping the tubes immersed in ice. The assay mixture was incubated for 30 minutes at room temperature and the absorbance at 540 nm was measured against a reagent blank. Acidified solutions of ascorbic acid were prepared fresh daily and standard curves were run simultaneously for each set of determinations. Each assay was done in duplicate and the results were the mean of the two determinations.

Data Management and Analysis

The results of the blood analyses were classified according to the histopathologic interpretations of the project referee: mild, moderate, and severe cervical dysplasias, carcinoma-in-situ, or cervix cancer. The normal women had two negative Pap smears, negative colposcopy examination, and no known gynecologic dysfunction. No woman with abnormal colposcopy had normal histology on biopsy. The group means and variances of the biochemically determined AA concentrations, categorized by cervical morphology, were compared using a one-way analysis of variance (ANOVA). Results for total AA were analyzed separately from those for reduced AA. Both arithmetic and geometric means were calculated.

Posttreatment groups were analyzed separately, and compared to normals in an ANOVA. This was done as pretreatment samples were not available and because these were women with multiple diagnoses, *i.e.,* they did not represent a single morphologic group.

RESULTS

Of the 202 women without prior cervical treatment, 56 were in the normal control group; 37 had mild dysplasia on biopsy; 39 had moderate, and 29 severe dysplasia. Carcinoma-in-situ (CIS) was seen in 17 women, and 24 subjects were cancer patients. Reduced ascorbic acid (RAA) concentrations measured in mg/dl were averaged with both arithmetic and geometric means, though for these nearly normally distributed values, only arithmetic means are presented (TABLE 1). Normal, mild dysplasia and moderate dysplasia patients had nearly identical mean RAA values, group means ranging from 0.62 to 0.69. Patients with severe dysplasia had higher than expected RAA, and CIS and cancer patients had lower than expected values (TABLE 1). ANOVA demonstrated these group means to be highly significantly different, $p<0.00001$. Results for TAA were quite parallel to those for RAA (TABLE 1), and group means were similarly different from one another by ANOVA, $p<0.00001$. Lower than expected TAA levels in CIS and cancer patients were observed, as was the case with RAA.

Posttreatment women after either cone biopsy or hysterectomy were compared to the normal control group for both TAA and RAA (TABLE 2). Posttreatment patients had lower values than the normal women, but the differences were only significant for the RAA values, $p = 0.02$ by ANOVA. TAA measurements were nearly the same in normal, postcone biopsy, and posthysterectomy patients.

DISCUSSION

In the present study, our interest has been to further investigate the plasma concentrations of both the reduced and total ascorbic acid in 202 women who were initially screened by a Pap test and subsequently determined to be either normal control subjects or were diagnosed to have histopathologic cervical dysplastic lesions of various grades, carcinoma-in-situ, or cervix cancer. Significant decreases in plasma RAA concentrations, previously reported by our group[14] in women with mild, moderate, and severe dysplasias, were not observed in the present study. In this larger sample of subjects with mild and moderate dysplasias, the RAA concentrations were in the normal range. A surprise finding in the more recent study was the significant increases in the RAA levels in women with severe dysplasias. The plasma TAA concentrations estimated in the same groups of women showed parallel results (TABLE 1).

In vitro studies carried out with reduced and oxidized ascorbic acid suggest that both forms favor cell multiplication at low doses, whereas inhibition occurs at high doses.[19] In an animal model system, induced stress was reported to result in depletion of ascorbic acid in the adrenal gland, liver, and plasma.[20,21] Hence, the paradoxical increases in both plasma RAA and TAA noted in women with severe dysplasia may reflect a decrease in urinary excretion, possibly to preserve the body's ascorbic acid pool and inhibit excess cellular proliferation. On the other hand, the significant decreases in plasma RAA and TAA seen in asymptomatic women with cervical CIS

TABLE 1. Plasma Reduced (RAA) and Total (TAA) Ascorbic Acid Concentrations in Women with Varying Cervical Status (Mean mg/dl ± SD)

Cervical Status	n	RAA[a]	TAA[b]
Normal	56	0.65 ± 0.28	0.69 ± 0.27
Dysplasias			
Mild	37	0.62 ± 0.27	0.69 ± 0.22
Moderate	39	0.69 ± 0.35	0.68 ± 0.30
Severe	29	0.83 ± 0.31	0.82 ± 0.28
CIS[c]	17	0.48 ± 0.32	0.52 ± 0.35
Carcinoma	24	0.33 ± 0.16	0.39 ± 0.20

[a] $p < 0.00001$, ANOVA.
[b] $p < 0.00001$, ANOVA.
[c] CIS, carcinoma-in situ.

may be attributed to metabolic stress or to the increase in utilization of ascorbic acid by the abnormal proliferating cells.

In women with squamous cell carcinoma of the uterine cervix, statistically significant reductions in both plasma RAA and TAA were observed (TABLE 1). Significant decreases in plasma ascorbic acid concentrations have frequently been reported in patients with malignant lesions of different organ sites.[22–24] The low plasma levels observed in women with cervix cancer may in part be secondary to anorexia associated with the malignant state. An alternative suggestion by other investigators is that preferential accumulation of this vitamin occurs in tumor tissues.[25,26]

We were also interested in measuring the changes in plasma RAA and TAA levels in women after surgical removal of the severely abnormal epithelial lesions, including *in situ* and cervix cancer. The plasma TAA values in surgically treated women were comparable to the levels determined in normal women (TABLE 2). It would be attractive to speculate that this elevation in plasma ascorbic acid level is a consequence of the removal of the abnormal cervical tissue. However, the plasma RAA levels in the same groups of women did not show similar results (TABLE 2), therefore, the

TABLE 2. Plasma Ascorbic Acid Concentrations in Women following Surgical Treatment Compared to Normals (Mean mg/dl ± SD)

Subjects	n	RAA[a]	TAA[b]
Normal	56	0.65 ± 0.28	0.69 ± 0.27
Postcone biopsy	27	0.51 ± 0.25	0.66 ± 0.28
Posthysterectomy	14	0.46 ± 0.22	0.64 ± 0.38

[a] $p = 0.02$, ANOVA.
[b] $p = 0.8$, ANOVA.

speculation remains unfounded. The women of these treated groups were not studied prior to their therapy; it is appropriate to investigate these observations further.

In an independent study Carr *et al.*[27] observed that RAA levels of tissues estimated by the 2-2′dipyridyl method were higher than the values estimated by either the DNPH method or the liquid chromatographic method using electrochemical detection. In 27% of our determinations, such higher RAA levels than that of TAA were observed. The distribution of these high values were determined in all groups of women, regardless of their cervical status. In spite of its limitation, the 2-2′dipyridyl method was pursued in the present study because we were equally interested in determining the amounts of reduced ascorbic acid in our study subjects. High correlation between RAA and TAA levels was observed ($R^2 = 0.73$). As ascorbic acid is known to be accumulated in most tissues in the reduced form,[28] the analogous RAA and TAA values observed in all subjects of our cross-sectional study groups may, therefore, suggest that in states of cervical dysplasias, carcinoma-in-situ and cancer, ascorbic acid remained predominantly in the reduced form.

The anatomic accessibility of the cervix and the existence of the Pap smear in conjunction with the ability to carry out colposcopy and colposcopically directed biopsies,[29] combined with the interest of asymptomatic women in participating in a cervical cancer screening program, provides an unusual opportunity to investigate early, presumably precancerous stages of a human neoplasm. This constitutes a unique system to conduct studies focused on evaluating the role of nutritional factors in the carcinogenesis of the uterine cervix. There are few comparable early opportunities to identify and recruit human subjects who are totally free of symptoms for extended investigations of the presumed natural history of a human cancer. Generally, symptoms bring a patient into the health care system when a lesion is already established. In addition, the organ site that may be harboring a precancerous lesion is not directly visible or anatomically easily accessible for biopsy.

To fully exploit this opportunity, studies are being initiated in which plasma vitamin C can be assessed in women with precancerous cervical lesions simultaneously with β-carotene levels, HPV detection, and sociodemographic-historical data collection of epidemiologic interests. In this way we hope to unravel the increasingly complex issue of ascorbic acid and cervical cancer and dysplasias. Furthermore, leukocyte ascorbic acid is currently being assessed in our laboratory to judge its comparability to RAA and TAA. Some investigators assert that leukocyte ascorbic acid is a better long-term measure of ascorbic acid intake.[30] Biochemical and dietary recall correlations are also of interest. Ultimately, should low vitamin C levels in asymptomatic CIS patients be confirmed as an independent observation, *i.e.*, not confounded by other nutrient intake, intervention studies are of interest.

We have conducted a 6-month pilot study in which 14 women given placebo and 14 women given 1 g in split doses of 500 mg vitamin C daily were followed to judge the impact on mild and moderate cervical dysplasias. Compliance, judged by biochemical monitoring, was good (FIGURES 1 and 2), and end point histopathology revealed a slight trend in favor of the vitamin C group. Side effects were nil.

In conclusion, low RAA and TAA in asymptomatic CIS patients were noted, while no decreases were seen in women with lower-grade dysplasias. Invasive cancer patients had the lowest values, but this is noted in largely symptomatic individuals. Posttreatment groups had intermediate values. An intervention trial with vitamin C for cervical dysplasia is of considerable interest. Subsequent work must account for vitamin C as confounded by or confounding for other factors,[31–34] particularly β-carotene.[35,36]

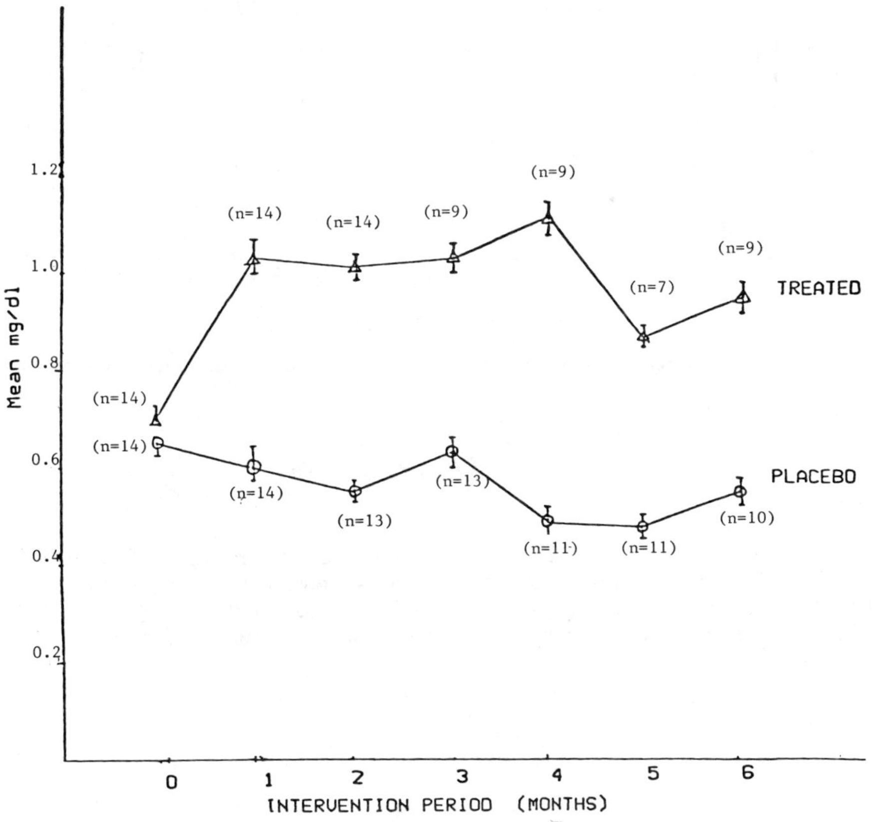

FIGURE 1. Plasma reduced ascorbic acid concentrations following 1 g/day vitamin C intervention.

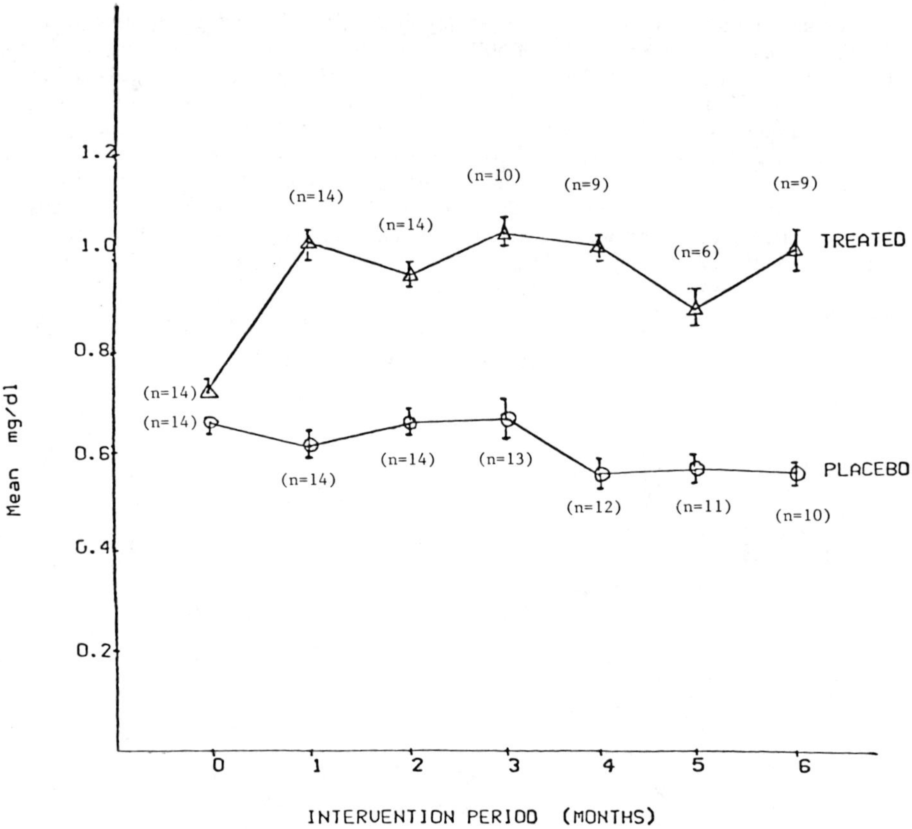

FIGURE 2. Plasma total ascorbic acid concentrations following 1 g/day vitamin C intervention.

ACKNOWLEDGMENTS

We thank several individuals who have supported this research effort: Annette Dwyer, R.N. has been an invaluable reserach nurse coordinator; Anna Kadish, M.D. has served as the pathologist referee for the project; Carla Romney gave technical assistance; and Virginia Fahey has provided invaluable help in the preparation of the manuscript.

REFERENCES

1. Committee on Diet, Nutrition and Cancer. 1982. Diet, Nutrition and Cancer. 1982. Assembly of Life Sciences and National Research Council. National Academy Press. Washington, D. C.

2. Kivirikko, K. I. & J. P. Darwin. 1967. Enzymatic hydroxylation of proline and lysine in protocollagen. Proc. Natl. Acad. Sci. U. S. A. **57:** 782-789.
3. Mirvish, S. S. 1975. Blocking the formation of N-nitroso compounds with ascorbic acid in vitro and in vivo. Ann. N. Y. Acad. Sci. **258:** 175-180.
4. Mirvish, S. S. 1981. Inhibition of the formation of carcinogenic N-nitroso compounds by ascorbic acid and other compounds. *In* Cancer. Achievements, Challenges and Prospects for the 1980's, Vol. 1. J. H. Burchenal & H. F. Oettgen, Eds.: 557-587. Grune & Stratton. New York.
5. Fraser, R. C., S. Pavlovic, C. G. Kurahara *et al.* 1978. The effect of variations in vitamin C intake on the cellular immune response of guinea pigs. Am. J. Clin. Nutr. **33:** 839-847.
6. Mirvish, S. S. 1975. Induction of mouse lung adenomas by amines or ureas plus nitrite and by N-nitroso compounds: Effect of ascorbate, sialic acid, thiocyanate, and caffeine. J. Natl. Cancer Inst. **55:** 633-636.
7. Reddy, B. S. & N. Hirota. 1979. Effect of dietary ascorbic acid on 1,2-dimethylhydrazine-induced colon cancer in rats. Fed. Proc. **38:** 2565.
8. Anthony, H. M. & C. J. Schorah. 1982. Severe hypovitaminosis C in lung-cancer patients: The utilization of vitamin C in surgical repair and lymphocyte-related host resistance. Br. J. Cancer **46:** 354-367.
9. Krasner, N. & I. W. Dymock. 1974. Ascorbic acid deficiency in malignant diseases: A clinical and biochemical study. Br. J. Cancer **30:** 142-145.
10. Crandon, J. H., C. C. Lund & D. B. Dill. 1940. Experimental human scurvy. N. Engl. J. Med. **223:** 353.
11. Wassertheil-Smoller, S., S. L. Romney, J. Wylie-Rosett, *et al.* 1981. Dietary vitamin C and uterine cervical dysplasia. Am. J. Epidemiol. **114:** 714-724.
12. Wylie-Rosett, J., S. L. Romney, N. S. Slagle, *et al.* 1984. Influence of vitamin A on cervical dysplasia and carcinoma in situ. Nutr. Cancer **6:** 49-57.
13. Morgan, R. W., M. Jain, A. B. Miller, *et al.* 1978. A comparison of dietary methods in epidemiologic studies. Am. J. Epidemiol. **107:** 488-498.
14. Romney, S. L., C. Duttagupta, J. Basu, *et al.* 1985. Plasma vitamin C and uterine cervical dysplasia. Am. J. Obst. Gynecol. **151:** 976-979.
15. Romney, S. L., M. J. Gray, A. Brian Little, J. A. Merrill, E. J. Quilligan & R. W. Stander, Eds. 1981. Gynecology and Obstetrics: The Health Care of Women. 2d Ed. McGraw-Hill. New York.
16. International Committee on Histological Definitions. 1961. Editorial. Acta Cytol. **6:** 235-236.
17. Zannoni, V., M. Lynch, S. Goldstein & P. Sato. 1974. A rapid micro method for the determination of ascorbic acid in plasma and tissues. Biochem. Med. **11:** 41-48.
18. Roe, J. H. & C. Kuether. 1943. The determination of ascorbic acid in whole blood and urine through the 2,4-dinitrophenylhydrazine derivative of dehydro ascorbic acid. J. Biol. Chem. **147:** 399-407.
19. Liotti, F. S., A. R. Meghini, P. Guerrieri, *et al.* 1984. Effects of ascorbic and dehydroascorbic acid on the multiplication of tumor ascites cells in vitro. J. Cancer Res. Clin. Oncol. **108:** 230-232.
20. Tariq, M., M. Ahmad & T. M. Salahuddin. 1975. Effect of combination of stresses on myocardial metabolism. Recent Adv. Stud. Card. Struct. Metab. **10:** 29-36.
21. Hamin Bin Rajikin, M. & T. T. Kechik. 1975. Effect of stress on the ascorbic acid content of various tissues in mice. Malays. Agric. Res. **4:** 69-76.
22. Basu, T. K., R. W. Raven, J. W. T. Dickerson & D. C. Williams. 1974. Leucocyte ascorbic acid and urinary hydroxyproline levels in patients bearing breast cancer with skeletal metastases. Eur. J. Cancer. **10:** 507-511.
23. Greco, A. M., M. Gentile, O. DiFilippo & A. Coppola. 1982. Study of blood vitamin C in lung and bladder cancer patients before and after treatment with ascorbic acid. A preliminary report. Acta Vitaminol. Enzymol. **4:** 155-162.
24. Kakar, S. & C. W. M. Wilson. 1974. Ascorbic acid metabolism in human cancer. Proc. Nutr. Soc. **33:** 110A.
25. Avtandilov, G. G., S. S. Kasabayan & V. Sgibneva. 1977. The microspectrophoto-

metric research of the content of ascorbic acid in carcinomas of the uterine cervix. Acta Histochem. **59:** 254-257.
26. MORIARTY, M. J., S. MULGREW, J. R. MALONE & M. K. O'CONNOR. 1977. Results and analysis of tumor levels of ascorbic acid. Ir. J. Med. Sci. **146:** 74-78.
27. CARR, R. S., M. BALLY, P. THOMAS & J. M. NEFF. 1983. Comparison of methods for determination of ascorbic acid in animal tissues. Anal. Chem. **55:** 1229-1232.
28. HAMMARSTROM, L. 1966. Autoradiographic studies on the disribution of ^{14}C labelled ascorbic acid and dehydroascorbic acid. Acta Physiol. Scand. **70** (Suppl. 289): 1-84.
29. YOUKELES, L., A. B. FORSYTHE & E. STERN. 1976. Evaluation of Papanicolaou smear and effect of sample biopsy in follow-up of cervical dysplasia. Cancer Res. **36:** 2080-2084.
30. LOH, H. S. & C. W. M. WILSON. 1971. Relationship between leucocyte and plasma ascorbic acid concentrations. Br. J. Med. **iii:** 733-735.
31. ROMNEY, S. L., P. R. PALAN, C. DUTTAGUPTA, *et al.* 1981. Retinoids and the prevention of cervical dysplasias. Am. J. Obst. Gynecol. **141:** 890-894.
32. BUTTERWORTH, C. E., K. D. HATCH, H. GORE, *et al.* 1982. Improvement in cervical dysplasia associated with folic acid therapy in users of oral contraceptives. Am. J. Clin. Nutr. **35:** 73-82.
33. KADISH, A., R. BURK, W. KRESS, *et al.* 1986. Human papillomavirus of different types in precancerous lesions of the uterine cervix. Hum. Pathol. **17:** 384-392.
34. DUTTAGUPTA, C., S. L. ROMNEY, P. R. PALAN & N. S. SLAGLE. 1982. Urinary cyclic nucleotides and cytopathology of human uterine cervical dysplasias. Cancer Res. **42:** 2938-2943.
35. PETO, R., R. DOLL, J. D. BUCKLEY & M. B. SPORN. 1981. Can dietary β-carotene materially reduce human cancer rates? Nature **290:** 201-207.
36. PALAN, P., S. L. ROMNEY, J. BASU, *et al.* Plasma β-carotene in cervical dysplasias and cancer. Unpublished.

DISCUSSION OF THE PAPER

W. LOHMANN (*University of Giessen, Giessen, FRG*): How about the ratio of the reduced and oxidized forms of ascorbic acid, did they change?

S. L. ROMNEY (*Albert Einstein College of Medicine, Bronx, N.Y.*): No real change, the predominant form appears to be reduced ascorbic acid.

M. LEVINE (*National Institutes of Health, Bethesda, Md.*): Do decreased plasma ascorbate levels that you measured in the patients correlate with decreased intake or is it something else, for example, utilization or change in pool size? Do you have any data on the intake of those patients?

S. L. ROMNEY: We would like very much to carry out a dietary survey in conjunction with those studies. We haven't done it in terms of the measurements that I've shown you here today.

M. LEVINE: For example in the patients with full-blown carcinomas, can some of that be accounted for by cachexia?

S. L. ROMNEY: Yes, I think that's the problem in terms of trying to clarify what may be a dietary contributing factor in the carcinogenic process and what the consequence of an established neoplasm is upon the food intake in diet. That hasn't been worked out.

S. MCPHERSON (*University of Texas School of Public Health, Houston, Tex.*): With the relationship of increased risks at low levels of β-carotene in cervical cancer, would an intervention trial just focusing on vitamin C indeed prove the true change, because so many of the women tend to have the lower levels of the other nutrients as well and other recent findings are also implicating that folic acid should be considered also?

S. L. ROMNEY: We've all thought about that in terms of trying to demonstrate a significant change in either regression rate or stability or progression in a placebo-controlled study if we venture into a multinutrient combination. That may be the first way to go but then the question would still be to try and filter on down and find out which nutrient or combination of nutrients is effective.

S. MCPHERSON: How were you going to monitor the diets of these women in vitamin C intervention while they are in the trial for that 6 months?

S. L. ROMNEY: Well we can monitor them by plasma total and reduced concentrations. We're quite prepared to accept the value we obtain as an indication of their dietary intake.

W. DEWYS (*Capital Area Permanente Medical Group, Springfield, Va.*): To return to your second question, the question of multiple nutrients, I think one consideration in terms of study design is the so-called factorial design where different groups will get either one nutrient or a combination and then of course the group receiving no intervention.

UNIDENTIFIED SPEAKER: At what time in relation to ingestion of your vitamin C tablet (the 1 g) did you take your blood levels?

S. L. ROMNEY: Very much like the Switzerland response. These women were seen on a fixed day, they returned for a morning clinic visit, they were not fasting. We drew the blood when they came for a repeat Pap smear colposcopy examination.

Serum Levels of Vitamin C in Relation to Dietary and Supplemental Intake of Vitamin C in Smokers and Nonsmokers

JACK L. SMITH

Strategic Marketing Associates, Inc.
Mt. Laurel, New Jersey 08054

ROBERT E. HODGES

University of California at Irvine
Irvine, California 92664

INTRODUCTION

Several studies of smokers and nonsmokers suggest that smokers may have increased needs for vitamin C. Calder *et al.*[1] found significantly lower levels of ascorbic acid in plasma or leukocytes of smokers than of nonsmokers. They contend that these are effects of long-term smoking: in a short-term test, plasma levels of vitamin C were not changed after 6 hours of smoking.

In a study by Brook and Grimshaw, lower plasma vitamin C values were found in male or female smokers than in their nonsmoking counterparts[2] up to 60 years of age. Brook and Grimshaw also found smokers had lower vitamin C levels in leukocytes, which was not significantly affected by age. Heavy smoking appeared to influence plasma vitamin C concentrations in the same way as increasing age by forty years.

Two experiments by Pelletier[3,4] showed lower blood and plasma vitamin C values in smokers prior to a vitamin C saturation test. Smokers required 630 mg more of the vitamin to reach saturation.[3] During loading of vitamin C, initially there was higher retention of C in smokers.[4]

Pelletier also examined serum vitamin C levels using data from the Nutrition Canada National Survey.[5] For smokers, serum C values were 30% lower than for nonsmokers. This difference increased to 40% for respondents who smoked over 20 cigarettes per day. More smokers than nonsmokers had less than 0.4 mg/dl serum C ("at risk" according to Nutrition Canada classifications). He suggested that there is less bioavailable vitamin C in smokers.

Among smokers and nonsmokers with similar plasma levels of ascorbic acid,

Kallner, Hartmann, and Hornig[6] found that smokers had a higher total turnover. To reach a total body pool of ascorbic acid comparable to nonsmokers, smokers would have to maintain total turnover about 30 mg/day higher than nonsmokers. Kallner, Hartmann, and Hornig suggest that smokers would need to consume 45-50 mg/day more of ascorbate than nonsmokers in order to sustain this higher turnover.

Hoefel found lower blood ascorbic acid levels in smokers, and found a sharper difference between female smokers and nonsmokers.[7] He suggested that the nicotine in cigarettes may be important as a possible root of this effect. Nicotine stimulates the release of catecholamines from the adrenal medulla. Ascorbic acid is a part of the synthesis pathway for catecholamines.

This study investigates differences in the serum vitamin C values of smokers and nonsmokers in relation to their intake of the vitamin through both diet and supplementation, based on data from the Second National Health and Nutrition Examination Survey (NHANES II). It extends previous reports by the authors.[8–10]

METHODS

NHANES II collected information from a probability sample of 20,322 persons ranging in age from 6 months to 74 years. This sample was representative of the U.S. noninstitutionalized civilian population of all 50 states and the District of Columbia. Through the application of sampling weights derived from Bureau of Census estimates, NHANES II data is projectable to a U.S. population of approximately 203.5 million people.

Serum vitamin C levels in NHANES II were determined by a modification of a 2,4-dinitrophenylhydrazine method.[11,12] Results were recorded to the nearest 0.1 mg/dl. This study used a sample consisting of respondents for whom serum vitamin C values were recorded.

A "smoker" was defined as a respondent who smoked one or more forms of tobacco (cigarettes, cigars, or pipe) at the time of the survey. This was determined by the question, "Do you smoke cigarettes now?" Similar questions were asked for cigar and pipe use. Smoking data was available only for subjects 18 years of age or older.

Carboxyhemoglobin (COHb) levels in whole blood were analyzed in half of the subjects between 3 and 74 years of age by a spectrophotometric method.[13] Frederiksen and Martin[14] found that nonsmokers had a COHb level less than 2% with an average of 1.5%, and that smokers had COHb levels ranging from 1.7% to over 18% with an average of 7.4%. Based on this report,[14] respondents 15 years of age or older were assigned to a low, moderate, or high COHb group with values 0-1.99%, 2-3.99%, and greater than 4.0% respectively. The low COHb group are likely to be nonsmokers and the high group likely to be smokers. Only the high and low COHb groups were used in comparisons of serum vitamin C by COHb level.

A 24-hour dietary recall was conducted by a dietitian in NHANES II.[15] The kind and quantity of each food and beverage the respondent reported consuming the previous day was recorded. Dietary intake of vitamin C was based on computer-calculated measurements of the recall data. Results were reported in terms of milligrams of dietary vitamin C consumed during the 24-hour period.

Vitamin C supplementation use was determined by responses to the question, "Are you taking vitamins or minerals"; if yes, "What kind?" "Vitamin C supplemented" respondents were defined as those who, at the time of the survey, were taking multiple vitamins, multiple vitamins with additional supplements, multiple vitamins and minerals, multiple vitamins and minerals with additional supplements, multiple vitamins with iron, vitamin C, or vitamin C with additional supplements. "Non-C-supplemented" respondents were those who were not taking one of the supplements listed above or were not taking any supplement.

For NHANES II, only the supplement first mentioned was coded into the data tapes released. Therefore, some people who did take vitamin C in addition to another supplement might have escaped classification as C-supplemented respondents. Their data would have been incorporated with data from non-C-supplemented subjects. It is unlikely that this had any major effect on the results since the numbers of such subjects are likely to be low.

RESULTS

Smoking Status by Carboxyhemoglobin

The weighted subsample (having COHb values, having serum C values, 15+ years of age) used to examine serum vitamin C by COHb was projectable to a population of 145.8 million people. When categorized by COHb level, 66.3% (96.6 million) had low COHb (presumably nonsmokers), 12.4% (18.1 million) had moderate COHb levels, and 21.3% (31.1 million) had high values (presumably smokers). Eighty and one half percent (25.0 million) of the high COHb group (smokers) were not vitamin C supplemented whereas 19.5% (6.1 million) took a vitamin-C-containing supplement. In the low COHb group (nonsmokers), 76.4% (73.8 million) took no C supplement. C supplement takers constituted 23.6% (22.8 million) of the low COHb group.

The carbon monoxide that binds with the hemoglobin to form carboxyhemoglobin can be derived from sources other than smoking, automotive exhaust being the most noteworthy. An effort was made to validate the smoking classification by comparing the respondent's response to the smoking questions (TABLE 1). Approximately 89% of the respondents in the low COHb group reported that they did not smoke. In the high COHb group, over 92% reported smoking. The moderate COHb group (which was not used in this analysis) was more heterogeneous: 78% reported smoking, 22% did not.

Earlier analysis had revealed lower average vitamin C intake among smokers[10] from diet. Due to the intake differences for vitamin C between smokers and nonsmokers, subsequent analyses compared groups of respondents having similar dietary and supplemental intakes of that vitamin.

When non-C-supplemented smokers (high COHb) and nonsmokers (low COHb) within each specified range of dietary C intake were compared, a larger percentage of smokers had serum vitamin C equal to or less than 0.3 mg/dl (FIG. 1). For each dietary range, the percentage of smokers consistently was about two or more times as high as nonsmokers having similar C intake. The percentage of C-supplemented smokers or nonsmokers with less than 0.3 mg/dl serum C was lower than the non-C-supplemented counterparts for every intake range (FIG. 2). In general, C-supplemented smokers and non-C-supplemented nonsmokers had similar percentages of subjects with serum C less than 0.3 mg/dl. C-supplemented nonsmokers had the lowest percentage of subjects with less than 0.3 mg/dl serum C in all intake ranges.

TABLE 1. Smoking Status by Carboxyhemoglobin

	COHb Level		
	Low (%)	Moderate (%)	High (%)
Smoker	10.8	77.5	92.2
Nonsmoker	89.2	22.5	7.8
Total	100.0	100.0	100.0

Smoking Status by Questionnaire

The subsample (having serum vitamin C values, 18+ years of age) based on smoking status in the NHANES II questionnaire projected to 133.9 million people. Smokers constituted 40.4% (54.1 million) of the population, and nonsmokers 59.6% (79.8 million). Among smokers, 21.9% (11.9 million) took some supplement containing vitamin C, whereas 78.1% (42.2 million) were non-C-supplemented. For nonsmokers, 24.0% (19.2 million) were vitamin C supplemented, 76.0% (60.6 million) were not.

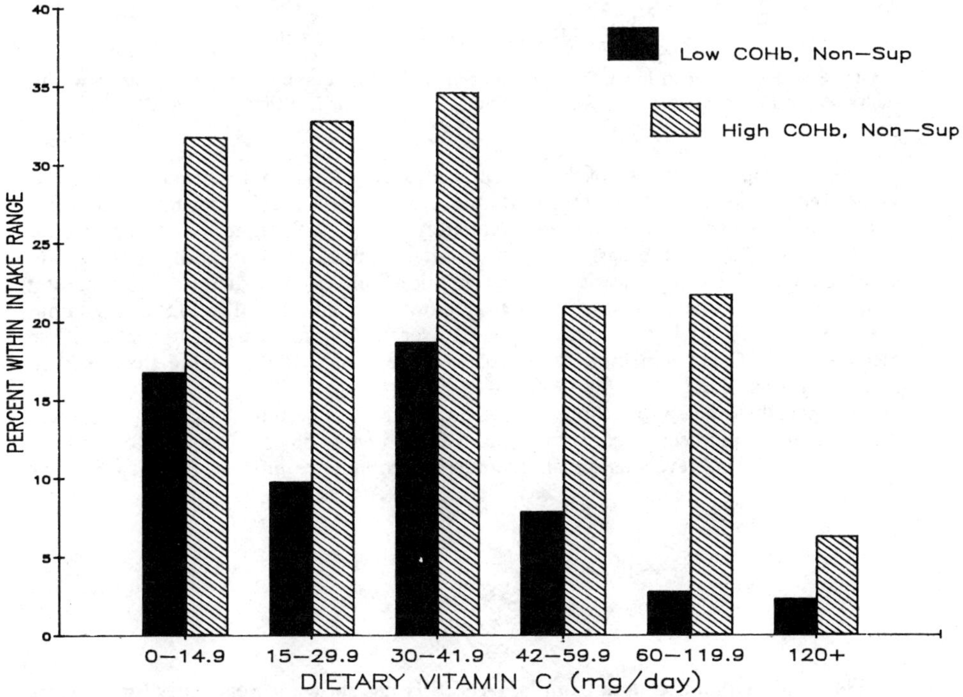

FIGURE 1. Percentage of low COHb group or high COHb group within each dietary intake range having serum vitamin C equal to or less than 0.3 mg/dl (non-C-supplemented only).

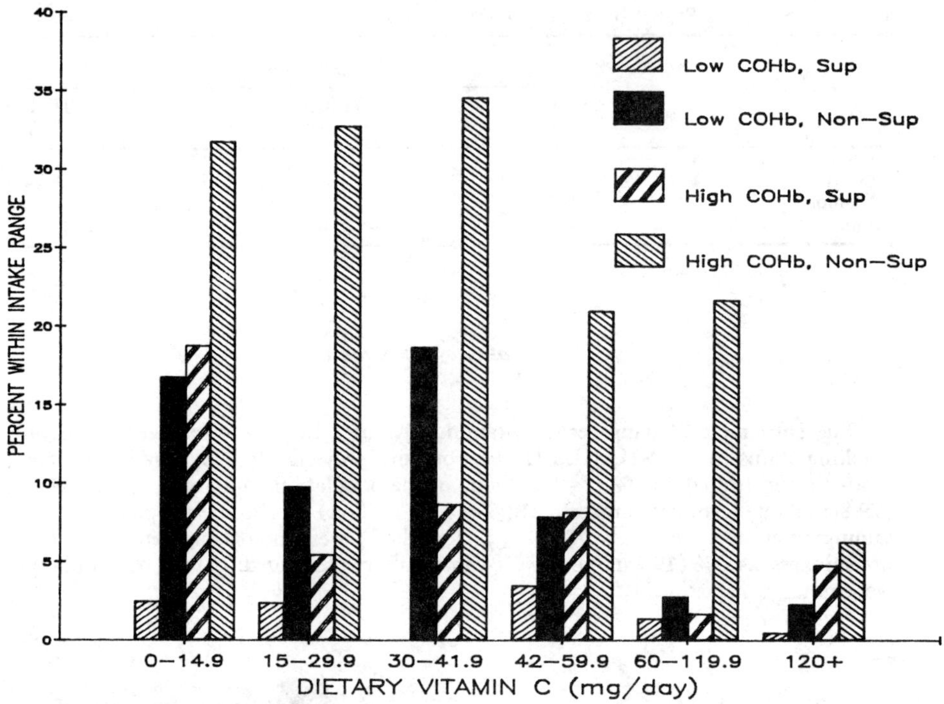

FIGURE 2. Percentage of low COHb group or high COHb group by C-supplementation within each dietary intake range having serum vitamin C equal to or less than 0.3 mg/dl.

Non-C-supplemented smokers and nonsmokers within each specified dietary C range were rank ordered by serum vitamin C level. The median serum C value for nonsmokers in each range was approximately 0.2 mg/dl higher than the smoker median serum C for the same dietary C intake (FIG. 3). For both smokers and nonsmokers, there was a positive linear relationship between dietary C intake and median serum C when dietary intake was below approximately 135 mg/day vitamin C (225% RDA). At higher dietary C levels, serum C values tended to plateau. The median serum C for nonsmokers tended to plateau at a higher value than median serum C for smokers.

The parallel bioassay method, applied to the median serum C analysis, estimated that smokers would need to consume an additional 59 mg/day of vitamin C to maintain equivalent serum C levels as nonsmokers (95% confidence interval of 52 to 68 mg/day).

DISCUSSION

The serum vitamin C data from NHANES II has shown lower values for smokers than for nonsmokers through comparisons based on either COHb levels or response to smoking status questions. Although the COHb level would be expected to reflect any source of carbon monoxide, low and high values appeared to be adequate estimates of smoking status.

In a previous report by Smith and Hodges[10] in which smoking status was assigned by response to the questionnaire, a larger portion of smokers had serum C less than 0.3 mg/dl than nonsmokers with similar dietary intake. Similar results were found whether or not they were taking a vitamin C supplement. These results agree with the results of the current comparison based on blood COHb levels. In fact, a lower percentage of C-supplemented smokers had serum C less than 0.3 mg/dl than non-C-supplemented nonsmokers.

Smith and Hodges[10] found a positive linear relation between vitamin C dietary intake and mean serum C level for each intake range when intake was below 140 mg/day for non-C-supplemented nonsmokers, below 170 mg/day for smokers. As in the current analysis for median serum C values, serum C tended to plateau with higher dietary intakes of vitamin C. Again, the difference between smoker and nonsmoker mean serum C within any intake range was approximately 0.2 mg/dl serum C. The parallel bioassay method applied to mean serum C values estimated that smokers need to consume an additional 65.4 mg/day of vitamin C on the average to maintain serum values equivalent to nonsmokers (95% confidence interval of 52.9 to 79.4 mg/day). This compares to a difference of 59 mg/day of C when analysis is based on median serum C. In the total adult population (18+ years of age), 40.9% of smokers were consuming less than 70% of the vitamin C RDA in their diet. Only 30.7% of nonsmokers received under 70% of the RDA for vitamin C.

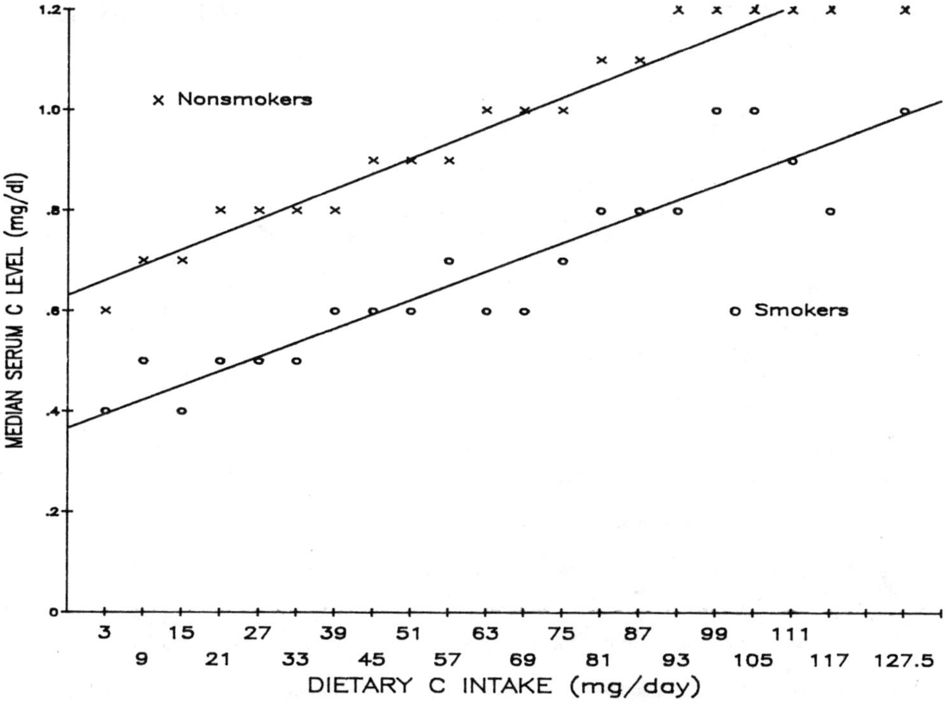

FIGURE 3. Median serum vitamin C value for each dietary intake range by smoking status.

Although the mechanisms through which a smoker's serum vitamin C is lowered are not known, it appears that smokers might maintain serum vitamin C levels comparable to nonsmokers by increasing vitamin C intake through diet or supplementation. This supports several other studies previously discussed that have suggested a greater potential need for vitamin C intake among smokers.

SUMMARY

Serum vitamin C levels were compared in smokers and nonsmokers in relation to dietary and supplemental intake of vitamin C, using data from the Second National Health and Nutrition Examination Survey. Smokers reported a lower mean daily intake (53 mg) of vitamin C than nonsmokers (65 mg) and a higher percentage (41% compared to 31%) consuming less than 70% RDA. Smoking status of respondents was judged by carboxyhemoglobin levels or by questionnaire. With both methods, percent of nonsupplemented smokers with serum vitamin C 0.3 mg/dl or less was two or more times as high as nonsupplemented nonsmokers at similar dietary intake levels. When smokers and nonsmokers with similar dietary vitamin C intake were ranked by serum C level, median and mean serum C for smokers was consistently lower than nonsmokers by approximately 0.2 mg/dl. By using the parallel bioassay method, it was estimated that smokers would need an additional 59 mg/day dietary vitamin C (95% confidence interval of 52-68 mg/day) based on median values or 65 mg/day (53-79 mg/day) based on mean values to attain serum C levels comparable to nonsmokers.

REFERENCES

1. Calder, J. H., R. C. Curtis & H. Fore. 1963. Comparison of vitamin C in plasma and leukocytes of smokers and nonsmokers. Lancet **1:** 556.
2. Brook, M. & J. J. Grimshaw. 1968. Vitamin C concentration of plasma and leukocytes as related to smoking habit, age, and sex of humans. Am. J. Clin. Nutr. **21**(11): 1254-1258.
3. Pelletier, O. 1968. Smoking and vitamin C levels in humans. Am. J. Clin. Nutr. **21**(11): 1259-1267.
4. Pelletier, O. 1970. Vitamin C status of cigarette smokers and nonsmokers. Am. J. Clin. Nutr. **23**(5): 520-524.
5. Pelletier, O. 1977. Vitamin C and tobacco. Int. J. Vit. Nutr. Res. Suppl. **16:** 147-170.
6. Kallner, A. B., D. Hartmann & D. H. Hornig. 1981. On the requirements of ascorbic acid in man: Steady-state turnover and body pool in smokers. Am. J. Clin. Nutr. **34:** 1347-1355.
7. Hoefel, O. S. 1977. Plasma vitamin C levels in smokers. Int. J. Vit. Nutr. Res. Suppl. **16:** 127-137.
8. Smith, J. L. 1984. The impact of smoking on serum vitamin C levels (abstract). Fed. Proc. **43:** 861.
9. Smith, J. L. & R. E. Hodges. 1985. Quantitative relationship between serum levels and dietary intake of vitamin C in smokers and nonsmokers (abstract). Fed. Proc. **44**(3): 778.
10. Smith, J. L. & R. E. Hodges. 1986. Serum levels of vitamin C in relation to dietary and supplemental intake of this vitamin by smokers and nonsmokers. Am. J. Clin. Nutr. In press.

11. ROE, J. H. 1967. Ascorbic acid. *In* The Vitamins: Physiology, Pathology, Methods, Vol. VII. 2d ed. P. Gyorgy & W. N. Pearson, Eds.: 27-49. Academic Press. New York.
12. ROE, J. H. & C. A. KUETHER. 1943. The determination of ascorbic acid in whole blood and urine through the 2,4-dinitrophenylhydrazine derivative of dehydro-ascorbic acid. J. Biol. Chem. **147:** 399-407.
13. SMALL, K. A., E. P. RADFORD & J. M. FRAZIER *et al.* 1971. A rapid method for simultaneous measurement of carboxy-and-methemoglobin in blood. J. Appl. Physiol. **31:** 154-160.
14. FREDERIKSEN, L. W. & J. E. MARTIN. 1979. Carbon monoxide and smoking behavior. Addictive Behaviors **4:** 21-30.
15. MCDOWELL, A., A. ENGEL, J. T. MASSEY & K. MAURER. 1981. Plan and operation of the Second National Health and Nutrition Examination Survey 1976-80. National Center for Health Statistics. Vital and Health Statistics, Series I, No. 15. Pub. No. (PHS)81-1317. Public Health Service. U.S. Government Printing Office. Washington.

DISCUSSION OF THE PAPER

C. K. CHOW (*University of Kentucky, Lexington, Ky.*): How did you measure carboxyhemoglobin and when was it measured? The carboxyhemoglobin has a very short half-life.

J. L. SMITH: We're reporting using NHANES data and I'm afraid I can't answer when it was measured on whole blood where the sample was drawn by the spectrophotometric method.

P. J. GARRY (*University of New Mexico School of Medicine, Albuquerque, N.M.*): With reference to the difference between smokers and nonsmokers, as the intakes went up in smokers why wouldn't you reach plasma levels eventually somewhere like those of the nonsmokers?

J. L. SMITH: Why do females or males not reach the same levels with females on your curves? I can't answer that other than to say this is what was there.

P. J. GARRY: I'll just hypothesize again, maybe smoking has something to do with the tubular maximum.

J. L. SMITH: Serum concentration is sort of a function of three factors, absorption, stores, and excretion, and I don't think the data that we have will allow you to eliminate any one of those, though I would agree with you that the TM is the most attractive to me but I don't have data to support that.

A. DUBOIS (*Yale University School of Medicine, New Haven, Conn.*): Could you explain why smoking would be associated with the difference between intake and plasma level? I'm not sure I understand the basic reason for asking that question in the first place, or formulating the hypothesis.

J. L. SMITH: When we say a difference in the serum levels between smokers and nonsmokers, that could be due to the fact that they smoke but it also could be due to the fact that they have a dietary intake that is different. If we're going to look at the effects related to smoking we then need to control for it and we've done this in a univariate way. NCHS has reported some multivariate effects which are consistent with what we've shown where they've controlled as far as dietary intake and serum levels are concerned.

A. KALLNER (*Karolinska Hospital, Stockholm, Sweden*): We have done some kinetic studies on vitamin C metabolism both in smokers and nonsmokers and the model we use allowed us to differentiate between the various phases. We found that smokers indeed required about 50% more vitamin C to reach the same level of vitamin C stores. But the model also allowed us to differentiate between tubular or renal handling and metabolism, and we found that what was happening in smokers was an increased metabolic turnover where the renal turnover was unchanged.

J. L. SMITH: Yes, that is interesting and certainly the theoretical suggestion would be that by the smoke creating some oxidizing-type metabolites through the lung that are reactive ascorbic acid would be used up faster.

UNIDENTIFIED SPEAKER: There is the study from Karolinska showing a definite increase in basal metabolic rate as you referred to, and it brings up the question, Is cigarette smoking a form of stress? This is possible because you see the same drop in plasma vitamin C levels, as you do in the various stressful situations.

The Role of Ascorbate in Biomembrane Energetics[a]

D. JAMES MORRÉ,[b] FREDERICK L. CRANE,[c]
IRIS L. SUN,[c] AND PLACIDO NAVAS[b,c]

[b] *Department of Medicinal Chemistry*
[c] *Department of Biological Sciences*
Purdue University
West Lafayette, Indiana 47907

The redox properties of ascorbic acid (vitamin C) and its dehydro and free radical (semidehydro or monodehydro) forms make the vitamin an important intracellular electron donor-acceptor. There are, however, few studies to relate this activity to an essential role of the vitamin.

Our interest in ascorbate-mediated electron transfer events has come from the discovery of a fixation-resistant NADH ferricyanide reductase[1,2] associated with the trans half of the Golgi apparatus stack,[3] transport vesicles,[4] and the plasma membrane.[5] This activity has subsequently been identified as an ascorbate free radical (mono- or semidehydroascorbate) oxidoreductase.[6] Additionally, ascorbate has been reported as a constituent present in clathrin-coated vesicles,[6,7] Golgi apparatus,[6] chromaffin granules,[8,9] and secretory vesicles,[10] and has been implicated as important to the secretory process.[6–10]

In this report, we summarize evidence for a role of ascorbate free radical as an electron acceptor in a redox system involving the NADH-free radical reductase that may reside uniquely in those cellular membranes (plasma membrane, coated vesicles, secretory vesicles, endocytic vacuoles) most often implicated in bulk translocations of cellular membranes and in membrane movements. The results are consistent with such a role as part of a mechanism to drive membrane translocations and for an involvement as an energy-producing enzyme (via generation of a proton gradient or of a membrane potential) in the generation of the force required for expression of cell motility. Additionally, transmembrane redox components may be important in the regulation of cell growth.[11] Thus, a role for the ascorbate free radical as a natural electron acceptor in the growth process also must be considered.

MATERIALS AND METHODS

Experimental Animals and Preparation of Isolated Membrane Fractions

Male Holtzman rats (200-250 g, Holtzman Co., Madison, Wis.) were provided with food and drinking water ad libitum. Animals were killed by decapitation following

[a]This work was supported in part by grants from Hoffmann-La Roche Inc.

light CO_2 anesthesia and drained of blood. Livers (ca. 40 g) were excised and homogenized using a Polytron 20 ST tissue homogenizer.

Golgi apparatus were prepared by the rapid procedure of Morré *et al.*[12] Coated vesicles were isolated as described[6] using 5-60% continuous sucrose gradients according to Pearse[13] and Mello *et al.*[14] The yield of coated vesicles was 1 to 2 mg protein per 40 g liver. Based on analyses of electron micrographs and measurements of various marker enzymes, the fractions were 80 ± 5% coated vesicles. Rapid and facile isolation of plasma membranes was carried out by the aqueous two-phase partitioning procedure of Albertsson *et al.*[15] Crude plasma membranes representing 10 g liver were combined with 6.4% (wt/wt) Dextran T500 (Pharmacia), 6.4% polyethylene glycol 3350 (Fisher), 0.25 M sucrose, and 5 mM potassium phosphate, pH 7.2. The contents of the two-phase system were then mixed by 40 inversions of the tubes in the cold (4°C). The two phases were next separated by centrifugation in a swinging bucket rotor at 150 g for 5 min. The upper phase, enriched in plasma membranes, was rewashed twice by centrifugation to separate the phases after each set of 40 inversions. A fresh upper phase was partitioned first against the original lower phase, followed by the two lower phases that washed the original upper phase. Finally, the two upper phases containing plasma membranes were combined, diluted with 1 mM sodium bicarbonate, and centrifuged at 16,000 g for 30 min. Cell fractionation procedures for other cell components are described in detail elsewhere together with criteria for determination of the purity of the fractions.[16]

Isolation of Golgi Apparatus Subfractions by Preparative Free-Flow Electrophoresis

By using the technique of preparative continuous free-flow electrophoresis, cisternae of isolated Golgi apparatus may be separated into those derived from the cis, intermediate, and trans Golgi apparatus regions of location within the polarized stack of dictyosome.[17] Purified, isolated Golgi apparatus are unstacked by a combination of enzymatic and mechanical disruption,[18] and the resulting single cisternae are used for the electrophoretic separation. In this technique, pioneered by Hannig and coworkers,[19] the mixture of cisternae to be separated is introduced as a fine jet into a separation buffer moving perpendicular to the flux lines of an electrical field. Membranes bearing different electrical charge densities will migrate different distances across the separation chamber and thus may be resolved. Fractions nearest the anode (farthest from the point of sample injection) are derived from the trans face while fractions farthest from the anode (nearest the point of sample injection) are enriched in Golgi apparatus elements derived from the cis face. Intermediate cisternae are located in intermediate fractions.

The electrophoresis medium (chamber buffer) contained 10 mM triethanolamine, 10 mM acetic acid, 5 mM glucose, 0.25 M sucrose, and 0.5 mM $MgCl_2$, final pH 6.5 (1 N NaOH), osmolarity 270 mOsM, conductivity 5.9×10^2 mho. The conditions for electrophoresis were (unless indicated otherwise): 167 mA (constant current), 131 ± 10% V/cm, buffer flow 2.75 ml per fraction per h, sample injection 3.5 ml/h, temperature 6°C, using a Vap-V continuous free-flow electrophoresis unit (Bender and Hobein, Munich, FRG). The fractions were collected by centrifugation for 2 min in a microfuge (Eppendorf model 5412) and resuspended in 0.5 to 1 ml distilled water for analysis.

Assays of Enzymes, Ascorbate, Ascorbate Free Radical, Pyridine Nucleotides, and Protein

NADH-Ferricyanide Reductase (EC 1.6.99.3). Assay of NADH-ferricyanide reductase, as described,[20] used 0.05 M sodium phosphate, pH 7.0 and about 35 μg protein in a final volume of 2.8 ml. The concentration of NADH was 15 μM and the ferricyanide concentration was 100 μM. Absorbance was monitored at 420 nm with reference at 500 nm.

NADH-Semidehydroascorbate Reductase (EC 1.6.5.5). Assays were in 50 mM Tris-HCl, pH 7.4 and a final volume of 2.8 mL. The NADH concentration was 0.17 mM. To this was added 3.3 mM of equal mixture of ascorbate and semidehydroascorbate, adjusted to pH 6.5 with 10 mM imidazole. Absorbance was read at 340 nm with reference at 500 nm.

Ascorbate Oxidase (EC 1.10.3.3). Assays were in 50 mM Tris-HCl, pH 7.4 with a final volume of 1.0 ml. The ascorbate concentration was 3.0 mM. Oxygen uptake was monitored using an oxygen electrode and reaction rates were recorded with a Sargent-Welch SRG recorder.

Pyridine Nucleotide Concentrations. Pyridine nucleotide concentrations were determined by the cycling assay procedure of Matsumura and Miyachi,[21] following extraction of NAD or NADP with perchloric acid and sodium hydroxide as described.[22]

Ascorbate, Ascorbate Free Radical, and Protein Determinations. Ascorbate was determined after derivatization with 2,4-dinitrophenylhydrazine according to Roe and Kuether.[23] Proteins were estimated according to Lowry *et al.*[24] Ascorbate free radical in Golgi apparatus and its reduction upon addition of NADH was demonstrated by the disappearance of the characteristic electron spin resonance signal of the ascorbate free radical, measured using a Varian E-109-ESR spectrophotometer at 9.4 GHz microwave frequency and in a cell compartment controlled thermostatically at 30°C.

Cytochemistry. Enzyme cytochemistry was as described[1] following the basic procedure for localization of ferricyanide reductases as described by Karnovsky and Roots.[25]

RESULTS

Enzyme Localization by Cytochemistry

NADH-ascorbate free radical oxidoreductase may be an important contributor to unspecific NADH-ferricyanide oxidoreductase activity as measured in the usual assays. The latter activity may be demonstrated cytochemically by coupling NADH oxidation to ferricyanide reduction in the presence of copper sulfate and sodium potassium tartrate to yield a reduced ferrocyanide-copper complex known as Hatchett's brown (FIGS. 1 and 2). This material is electron dense and may be visualized at sites of deposition in the electron microscope (FIGS. 1 and 2). With internal membranes of rodent liver, strongest deposits of Hatchett's brown are seen associated with the trans half of the Golgi apparatus (FIG. 2). Cis (immature) Golgi apparatus elements lack the reactivity entirely. Intermediate cisternae may show the reaction product in small patches only, whereas trans cisternae are completely reactive. Strong reactivity is also seen on membranes of mature secretory vesicles and on the plasma membrane. With

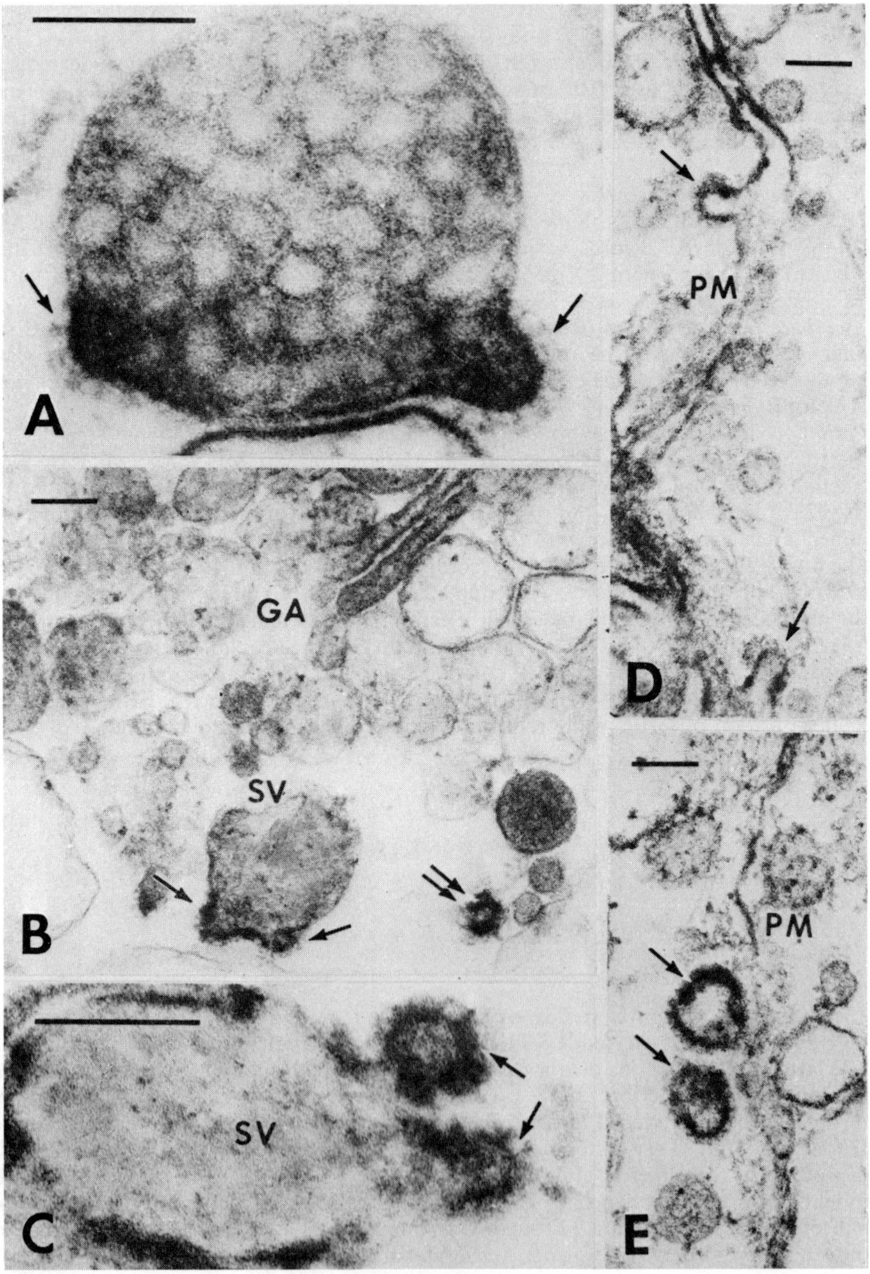
A
GA
SV
B
SV
C
PM
D
PM
E

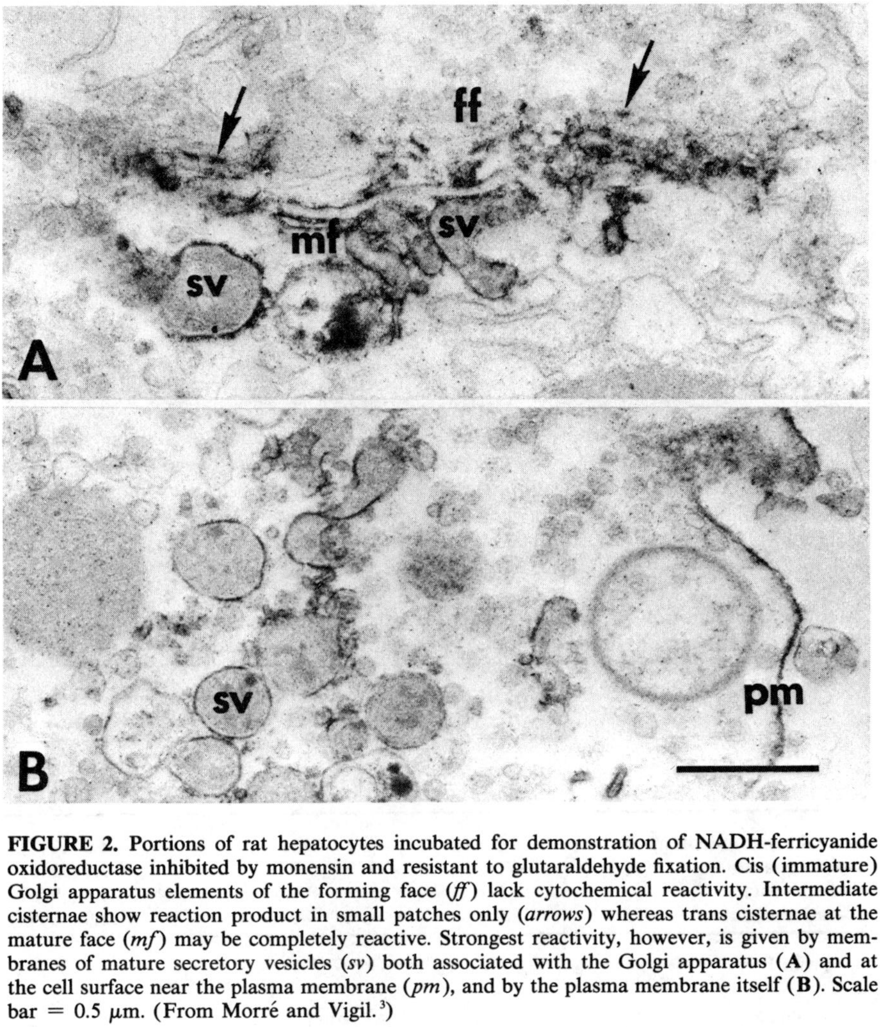

FIGURE 2. Portions of rat hepatocytes incubated for demonstration of NADH-ferricyanide oxidoreductase inhibited by monensin and resistant to glutaraldehyde fixation. Cis (immature) Golgi apparatus elements of the forming face (*ff*) lack cytochemical reactivity. Intermediate cisternae show reaction product in small patches only (*arrows*) whereas trans cisternae at the mature face (*mf*) may be completely reactive. Strongest reactivity, however, is given by membranes of mature secretory vesicles (*sv*) both associated with the Golgi apparatus (**A**) and at the cell surface near the plasma membrane (*pm*), and by the plasma membrane itself (**B**). Scale bar = 0.5 μm. (From Morré and Vigil.[3])

FIGURE 1. (**A-C**) Electron micrographs of portions of mouse liver parenchymal cells illustrating clathrin-type coats (*arrows*) that occur on portions of nearly all mature secretory vesicles attached to or near the Golgi apparatus stack. In **B** through **E**, preparations of broken cells were reacted cytochemically for a glutaraldehyde-resistant NADH-ferricyanide oxidoreductase (NADH-ascorbate free radical reductase) characteristic of the plasma membrane.[1] The coated portions of the vesicle membranes (*arrows*) as well as coated vesicles in the cytoplasm (*double arrows*) were reactive. Also reactive were coated membrane surfaces of the plasma membrane. Scale marker = 0.2 μm. (From a study with J. Kartenbeck, W. W. Franke and D. M. Morré.[4])

regard to the latter, however, strongest reactivity is always associated with membrane regions that underlie clathrin coats (FIG. 1). Clathrin-coated vesicles and pits are commonly associated with the cell surface where various roles in receptor-mediated endocytosis and the internalization of ligand-receptor complexes have been indicated.[26,27] A second population of coated membranes and vesicles is associated with the Golgi apparatus. The coat material of these Golgi apparatus and cell surface-associated vesicles has been identified as clathrin, a single 180,000 kDa peptide, by immunochemical procedures applied to electron microscope preparations.[48]

Enzyme Localization by Cell Fractionation Procedures

The NADH-ferricyanide reductase activity that resists cytochemical preparation and is also resistant to low concentrations of glutaraldehyde tissue fixatives[2] can now be equated with NADH-ascorbate free radical oxidoreductase based on studies of subcellular distributions. NADH-ascorbate free radical oxidoreductase activity is most highly concentrated among cellular membranes in those cell components (coated vesicles, Golgi apparatus, and plasma membranes) showing the strong cytochemical reactivity (TABLE 1). For the latter experiments, coated vesicles were isolated from rat liver. The fractions were approximately 85% coated vesicles as determined by electron microscope morphometry and a detailed analysis of marker enzymes. Other fractions analyzed also were at least 85% pure; plasma membrane fraction purity from mouse liver exceeded 95% without a reduction in NADH-ascorbate free radical oxidoreductase, indicating that the activity is not associated with a contaminating membrane.

TABLE 1. Distribution of NADH-Ascorbate Free Radical Oxidoreductase among Rat Liver Membranes and Inhibition by 20 μM Monensin[a]

	NADH-Ascorbate Free Radical Oxidoreductase (Specific Activity, nmol NADH Oxidized/min/mg Protein)		
Fraction	No Monensin	20 μM Monensin	Change Due to Monensin
Endoplasmic reticulum	24.0	21.0	3.0 (12%)
Golgi apparatus	4.3	0.8	3.5 (81%)
Coated vesicles	15.5	9.5	6.0 (39%)
Plasma membrane	12.0	0.0	12.0 (100%)

[a] From Sun *et al.*[6]

Further evidence for a relationship between NADH-ascorbate free radical oxidoreductase and the cytochemical reactivity comes from experiments where Golgi apparatus cisternae have been separated cis to trans across the stacked cisternae of the Golgi apparatus. The cis half of the stack (fractions 36-41, TABLE 2) has high NADH cytochrome *c* reductase activity, an endoplasmic-reticulum-associated activity, and low Mg^{2+}-ATPase, a plasma-membrane-associated activity. In contrast, the trans half of the stack (fractions 30-35 TABLE 2) showed the inverse relationship of low

TABLE 2. Distribution of Membrane Redox and Marker Enzymatic Activities among Golgi Apparatus Subfractions Prepared by Free-Flow Electrophoresis

Electrophoretic Fraction		NADH-Cyt *c* Reductase (nmol/min/mg Protein)	Mg^{2+} ATPase (nmol/h/mg Protein)	NADH-Ferricyanide Oxidoreductase (μmol/min/mg Protein)	NADH-Ascorbate Free Radical Oxidoreductase (μmol/min/mg Protein)	Ascorbate (μg/mg Protein)
Unfractionated		645	(6)	3.2	11.0	26
Cis face	(39-41)	915	6	7.9	8.8	14
	(36-38)	878	18	3.4	6.5	10
	(33-35)	220	39	4.3	9.0	14
Trans face	(30-32)	136	60	7.3	17.0	58

Fractions were pooled to correspond approximately to each of the four cisternae present in the intact Golgi apparatus of the starting material. Intact Golgi apparatus were unstacked enzymatically and mechanically prior to the electrophoretic separation.[18] The preparations were from rat liver.

NADH cytochrome *c* oxidoreductase, high ATPase, and a high NADH ascorbate free-radical oxidoreductase.

Characteristics of the NADH-Ascorbate Free Radical Oxidoreductase

Characteristics of the NADH-ascorbate free radical oxidoreductase of Golgi apparatus and coated vesicles have been determined. The pH optimum is over the range 7 to 7.5. A K_m for ascorbate free radical of 1.67 mM (ascorbate + dehydroascorbate) was found. The activity was not stimulated appreciably by treatment of the membranes with detergent (0.0015 to 0.03% Triton X-100), confirming the cytochemical observations that the activity is associated primarily with the cytoplasmic surface of Golgi apparatus and coated vesicles and not within the membrane interiors. The NADH-free radical oxido-reductase of the Golgi apparatus and coated vesicles was more strongly inhibited by the sodium-selective ionophore monensin at physiological concentrations of 10 to 20 μM than was that of endoplasmic reticulum membranes (TABLE 1). In plasma membranes, the activity was inhibited completely by monensin (TABLE 1). The activity in Golgi apparatus and coated vesicles but not in endoplasmic reticulum was stimulated by calmodulin and inhibited by blockers of calmodulin function.[28]

To provide evidence that intraluminal ascorbate free radical was present and serving as an acceptor for electrons from NADH, electron spin resonance measurements were carried out with isolated Golgi apparatus (FIG. 3). The upper trace (A) is the electron spin resonance signal at about $g = 2$ for the ascorbate radical present endogenously in isolated Golgi apparatus fractions. The lower trace (B) is the same preparation following the addition of NADH. Upon addition of NADH, the signal is very much attenuated as would be expected if the intraluminal ascorbate free radical was serving as an acceptor for electrons from NADH. The strong electron spin resonance signal shows the presence of ascorbate free radical inside the membrane vesicles and its attenuation upon the addition of NADH is indicative of monodehydroascorbate participation in electron transfer.

Subcellular Distribution of Ascorbate

With homogenates prepared in sucrose-containing media that preserve organelle structure, approximately 50% of the total ascorbate is particulate. With the particulate fraction, ascorbic acid was present in Golgi apparatus, coated vesicles, and endoplasmic reticulum. In coated vesicles and in Golgi apparatus it is enriched somewhat on a protein basis (TABLE 3).

Within the Golgi apparatus, a close relationship between ascorbate content and the activity of NADH-ascorbate free radical oxidoreductase of the different electrophoretic subfractions of TABLE 2 is seen (FIG. 4). These activities are highest with the most trans Golgi apparatus fraction (fractions 30-32), that portion of the Golgi apparatus most closely associated with the formation of coated vesicles. The activity, however, of fractions 30-32 is not due to contamination with coated vesicles since coated vesicles migrate to a much less electronegative portion of the electrophoretic separation, nearer to fractions 39-41.

Functional Relationship between NADH-Ascorbate Free Radical Oxidoreductase and Its Subcellular Location

One possibility to regulate NADH-ascorbate free radical oxidoreductase activity and to account for the cytochemical findings indicative of enhanced ferricyanide

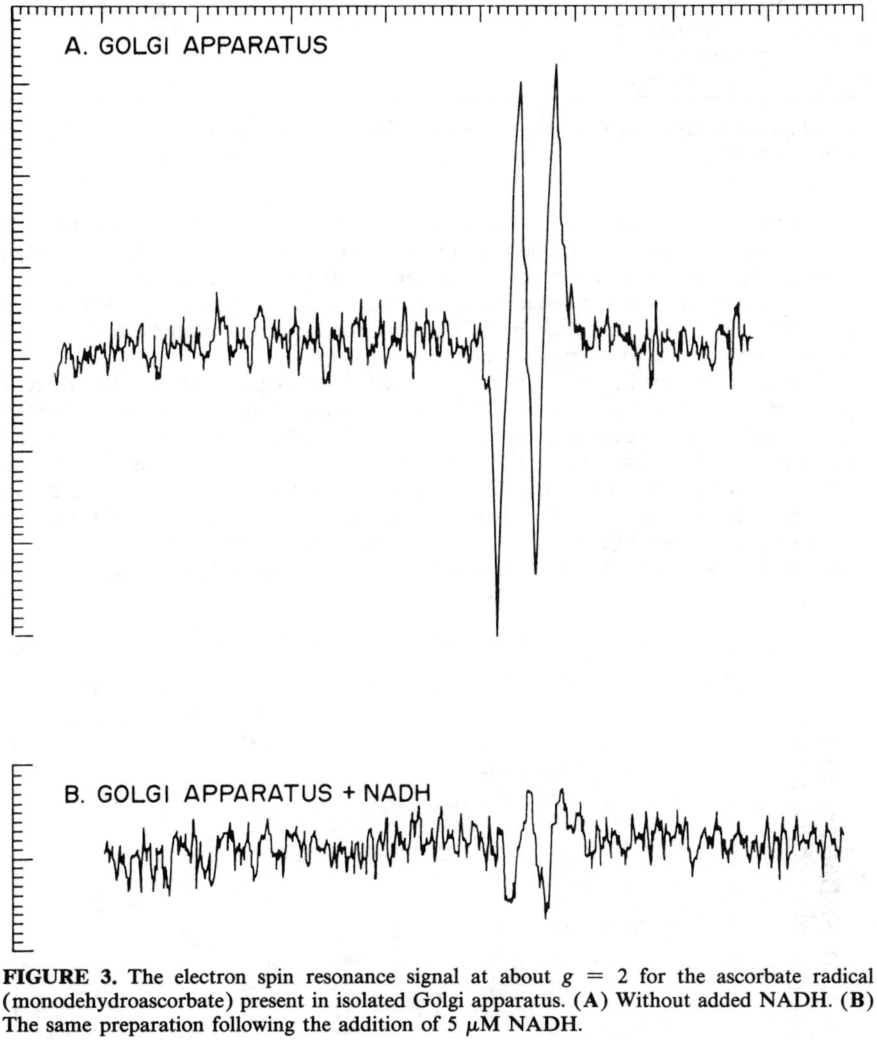

FIGURE 3. The electron spin resonance signal at about $g = 2$ for the ascorbate radical (monodehydroascorbate) present in isolated Golgi apparatus. (**A**) Without added NADH. (**B**) The same preparation following the addition of 5 μM NADH.

reduction associated with coated membrane surfaces[4] would be that some combination of the membrane-bound reductase with the coat protein clathrin resulted in the activation of the reductase. This, in turn, would provide for the enhanced electron transport to drive the movement of the coated vesicles and coated regions during both exo- and endocytosis as well as with intracellular membrane translocations.

TABLE 3. Subcellular Distribution of Ascorbic Acid

Fraction	Ascorbic Acid (μg/mg Protein)
Total homogenate	5.85 ± 1.1
100,000 × g pellet (total particulate fraction)	5.7 ± 0.9
100,000 × g supernatant	6.3 ± 0.8
Golgi apparatus	7.4 ± 3.2
Coated vesicles	9.7 ± 2.7
Endoplasmic reticulum	6.2 ± 2.9

Values are to be regarded as minimal since strict precautions to prevent losses of ascorbic acid during the preparation of the fractions were not taken. From Sun *et al.*[6]

To test this hypothesis, coated vesicles from rat liver were treated with 0.5 M tris (hydroxymethyl) methyl ammonium chloride, pH 7.2, to dissociate clathrin baskets. Stripped membranes were removed by centrifugation for 1 h at 100,000 g and both fractions, clathrin-depleted membranes and the clathrin-enriched supernatant, were analyzed for content of clathrin and membrane proteins by polyacrylamide gel electrophoresis[6] and for enzymatic activities (TABLE 4).

Following removal of the clathrin coats, ascorbate oxidase was enriched threefold in the membrane pellets as expected for a membrane-associated enzyme. However, neither NADH-ascorbate free radical nor NADH-Fe$(CN)_6$ oxidoreductase specific activity showed enrichment, nor did the supernatant containing the clathrin have activity. However, when the coated vesicles were reconstituted by combining the two fractions, the final specific activity based on total protein was nearer that of the starting coated vesicles prior to extraction and approximately threefold stimulated based on membrane protein for NADH-ascorbate free radical oxidoreductase. Ascorbate oxidase

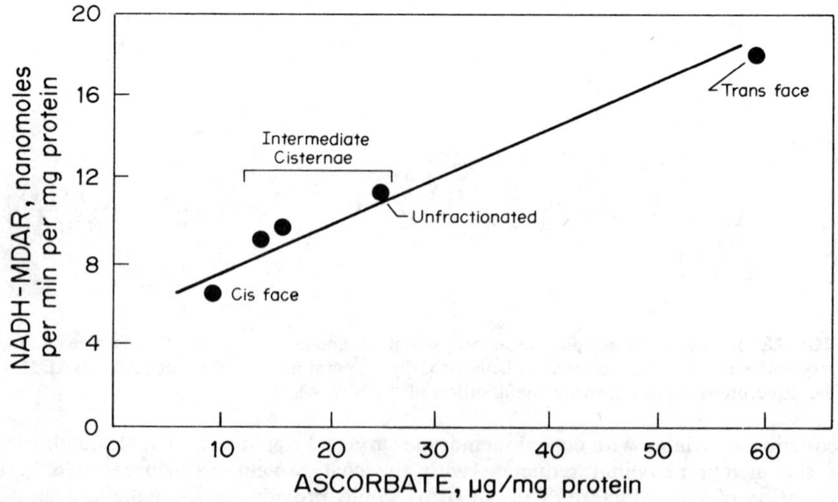

FIGURE 4. NADH-ascorbate free radical oxidoreductase as a function of ascorbate content comparing Golgi apparatus subfractions from TABLE 2.

was not so affected. NADH-$Fe(CN)_6$ reductase showed a trend similar to NADH-ascorbate free radical oxidoreductase, although not as striking. Addition of the clathrin-enriched fractions to Golgi apparatus membranes resulted in a nearly twofold stimulation of the NADH-ascorbate free radical oxidoreductase activity as well (TABLE 4). Additionally, the activity of NADH-ascorbate free radical oxidoreductase activity of coated vesicles and of Golgi apparatus but not that of the endoplasmic reticulum was stimulated by calmodulin and was inhibited by drugs (trifluroperazine, pimozide, and R24571) that are known inhibitors of calmodulin function.[28]

To test for changes in membrane potential in the coated vesicles from rat liver in response to the addition of monodehydroascorbate and NADH as well as ascorbate, experiments were conducted using fluorescence of carbocyanine dyes to estimate relative changes in membrane potential.[29] A response was noted (TABLE 5) and the combination of NADH plus monodehydroascorbate gave a greater response than either one alone. Ascorbate (5 mM) was similar to NADH alone.

TABLE 4. Subfractionation of Coated Vesicles and the "Activation" of NADH-Free Radical Oxidoreductase of Isolated and Clathrin-depleted Membranes by Addition of Fractions Enriched in Clathrin

Fraction	NADH-MDA Reductase	Ascorbate Oxidase	NADH-$Fe(CN)_6$ Reductase
Coated vesicles	17.0 ± 3.0	41 ± 9	1031
Clathrin-depleted membranes	14.0 ± 1.1	143 ± 24	672 ± 24
Clathrin-enriched supernatant	0.7 ± 0.2	10 ± 2	23 ± 1
Membranes + supernatant			
Based on total protein	15.2 ± 3.7	57 ± 9	22 ± 22
Based on membrane protein	30.4 ± 7.4	114 ± 18	750 ± 40
Golgi apparatus (GA)	1.6 ± 0.4	50 ± 9	1771 ± 5
GA + clathrin-enriched supernatant			
Based on total protein	1.34 ± 0.05		906 ± 78
Based on membrane protein	2.7 ± 0.1		1892 ± 156

Values are specific activities (nmol/min/mg protein) ± standard deviations from three different membrane preparations from Sun *et al.*[6]

The slow decline of fluorescence of the control vesicles was stopped by addition of KCl. Furthermore if NADH was added with the KCl, the rate of fluorescence decline was increased. Monodehydroascorbate also increased the fluorescence decline in the presence of KCl.

Proton pumping to maintain an electrochemical gradient is well known in the major cell compartments including coated vesicles.[30–32] Zang and Schneider[33] have found high ATPase activity in Golgi apparatus from rat liver. This ATPase could be inhibited by dicyclohexylcarbodiimide, which also resulted in a threefold accumulation of newly synthesized proteins in Golgi apparatus. Thus, the implication is that energy for Golgi apparatus, and perhaps for coated vesicle function, is supported in part by an electrogenic proton gradient. Additionally, Glickman *et al.*[34] found a proton pump in Golgi apparatus that functions in parallel to chloride conductance.

In our own studies[35] we have found that Golgi apparatus membranes isolated from mouse liver pump protons inwards when supplied either with ATP or NADH. Acidification of Golgi apparatus cisternae was detected with neutral red, a permeant dye at neutral pH absorbing at 600 nm that becomes protonated to a nonpermeant form absorbing at 550 nm within the interiors of acidified vesicles.

In other studies, the response of coated vesicles of rat hepatocytes and hepatomas to ascorbate supplied in the medium was studied. The results showed a concentration-dependent increase in the numbers of coated vesicles visible in the Golgi apparatus zone following addition of ascorbate to a minimal essential tissue culture medium.[36] The response was maximal between 30 and 60 min of treatment at an optimum concentration of about 30 mM ascorbate.

Relationship to Growth

While plasma membrane redox systems have been long implicated as important to regulation of growth processes,[11] a role for ascorbate free radical as the electron acceptor has been less clear. Recent findings from plants, however, strongly support such a role.

TABLE 5. Changes in Charge across Coated Vesicle Membranes Measured by Carbocyanine Dye Fluorescence[7]

Addition	Fluorescence (Arbitrary Units/min/mg Protein)
None	−2.2
+ 4 nM KCl	−0.7
45 μM NADH	−2.8
+ 4 mM KCl	−1.7
5 mM ascorbate radical[a]	−2.8
+ 4 mM KCl	−1.4
45 μM NADH + 5 mM ascorbate radical[a]	−3.3
+ 4 mM KCl	−1.1

[a] Equal mixture of ascorbate and dehydroascorbate.

With plasma membrane vesicles isolated by aqueous two-phase partition in the presence of ascorbate, some of the ascorbate trapped inside the vesicles is converted to ascorbate free radical (FIG. 5). NADH oxidation by the plasma membrane vesicles is enhanced by isolation in the presence of ascorbate and a slight further stimulation is obtained from addition of either ascorbate or monodehydroascorbate (equal mixture of ascorbate and dehydroascorbate) to the reaction mixture (TABLE 6). The activity is stimulated by addition of the plant growth hormone indole-3-acetic acid (IAA) or its synthetic analogue 2,4-dichlorophenoxyacetic acid (2,4-D) (TABLE 6) in a concentration-dependent manner (not shown) with an optimum near 1 μM. The near-optimum concentration of the hormone for promotion of elongation growth in excised stem segments also is near 1 μm.

DISCUSSION

Ascorbic acid is found naturally in the interiors of coated vesicles,[6,7] Golgi apparatus,[6] chromaffin granules,[8,9] and secretory granules.[10] It is a bivalent oxidation-

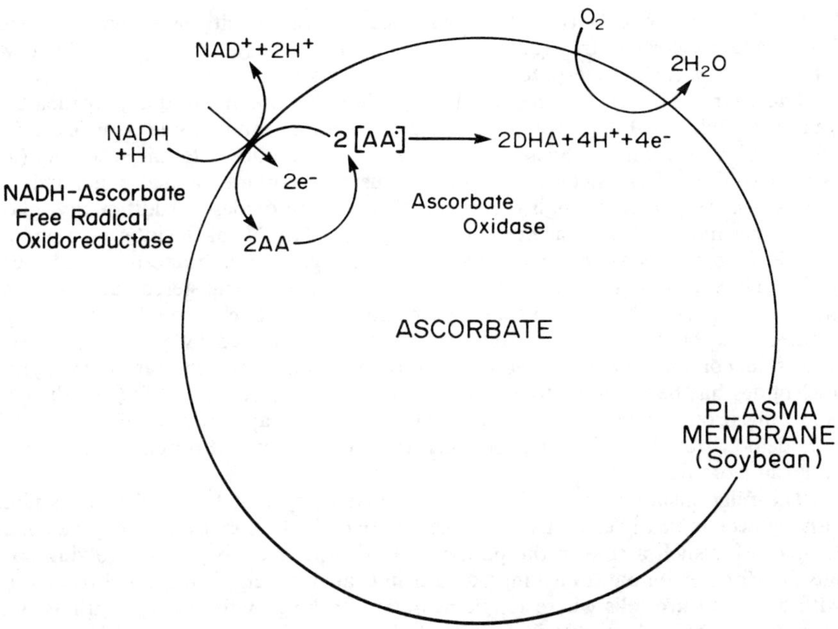

FIGURE 5. Diagram illustrating the design of the incubation system with plasma membrane vesicles from plant stems (hypocotyls) of etiolated soybeans, *Glycine max,* isolated by two-phase partition in the presence of ascorbate. Oxidation of external NADH stimulated by auxin is assumed to use as the electron acceptor the ascorbate radical produced from ascorbate trapped inside the vesicle. (From Morré *et al.*[47])

reduction molecule that acts via a free radical intermediate, monodehydroascorbate. The latter serves as a one-electron oxidant and as a one-electron reductant.[37] Monodehydroascorbate has been described as an electron acceptor for NADH oxidation in plasma membranes,[38–40] microsomal membranes,[41–43] Golgi apparatus,[6] and coated vesicles.[6,7] Additionally, ascorbic acid acts as a transmembrane electron donor to reduce cytochrome *c* or ferricyanide in pituitary neurosecretory granule vesicles.[44]

TABLE 6. Stimulation of NADH Oxidation by Soybean Plasma Membranes Isolated by Aqueous Two-Phase Partitioning From Homogenates Prepared in the Presence of 10 mM Ascorbate[47]

Additions	Concentration	NADH-Oxidase (nmol/min/mg Protein)
None		15.7
2,4-D	10 μM[a]	23.9
Ascorbate	1.7 mM	20.4
Ascorbate + 2,4-D	1.7 mM + 10 μM	30.6
Ascorbate radical[b]	1 mM	24.4
Ascorbate radical[b] + 2,4-D	1 mM + 10 μM	30.6

[a] Results with 1 μM 2,4-D were similar.
[b] A mixture of equal parts of ascorbate and dehydroascorbate.

During this latter process, the ascorbate induces an inside positive membrane potential. Also NADH-ascorbate free radical oxidoreductase activity has been associated with generation of a membrane potential in chromaffin granules.[8]

Thus, various internal endomembranes, including those involved in bulk membrane transport such as Golgi apparatus, secretory vesicles, and coated vesicles, have a transmembrane electron transport since they all can reduce external acceptors (such as ferricyanide, NADH, or cytochrome *c*) using ascorbate radical generated from internal ascorbic acid through the action of ascorbate oxidase. Additionally, plasma membranes may use a similar mechanism to oxidize NADH at the internal membrane surface. Electron transfer is indicated by the change in membrane potential created during the NADH oxidation. Thus electrons appear to be transferred across a sealed membrane vesicle (FIG. 6) and simple permeation by either electron donor or acceptor is ruled out. That vesicles are well sealed has been verified by various permeation studies in conjunction with electron microscopy. Lack of penetration by pyridine nucleotides has been shown by measurements with radioactive NAD (less than 10^{-3} pmol/mg protein taken up over 20 min) and lack of rapid release of ascorbate has been demonstrated by retention of intravesicular ascorbate through several washing cycles in isotonic media.

The relationship of such an electron transport system to membrane dynamics, although theoretically attractive, has been more difficult to investigate experimentally. Evidence for such a role in the pathway of adsorptive endocytosis involving coated pits and/or membrane trafficking of clathrin-coated vesicles is indicated from studies with tissue culture cells where supplementation of the growth medium with ascorbate altered the number and distribution of clathrin-coated membranes.[36] Also indicative of such a role is an apparent activation of the NADH-ascorbate free radical oxido-reductase of membranes when clathrin is present. The latter is evidenced both from cytochemical observations[1] (FIG. 1) and direct measurements of enzyme activity[6] (TABLE 4).

An intriguing possibility presently under investigation is a role for ascorbate free radical as an electron acceptor in elongation growth in both animals and in plants. In animals, there is considerable evidence for a role of transplasma membrane redox in growth control.[11] The intracellular donor is NADH[45] and an impermeant electron

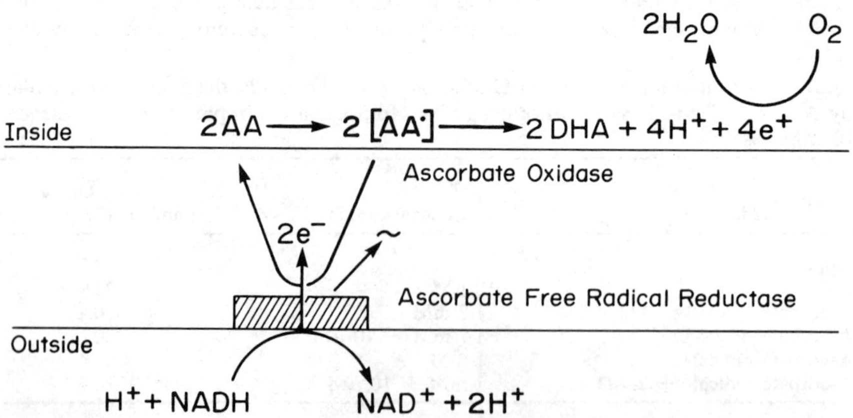

FIGURE 6. Schematic representation of the NADH-ascorbate free radical complex of Golgi apparatus and coated vesicles linked to oxygen via ascorbate. (Modified from Sun *et al.*[6])

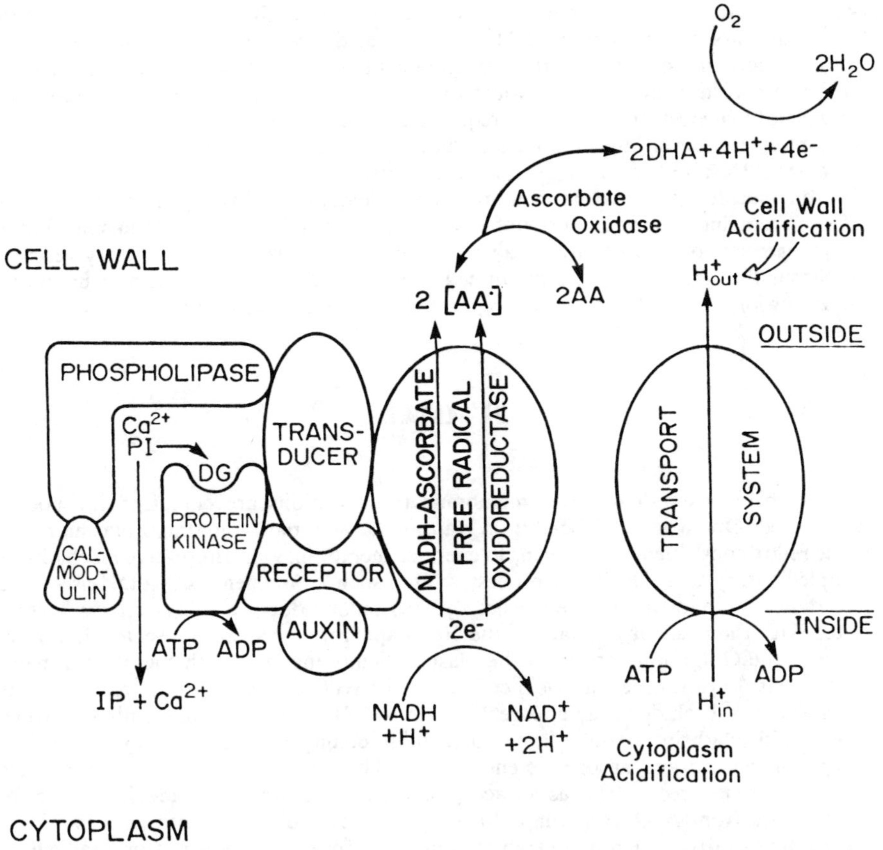

FIGURE 7. A model for control of hormone-stimulated elongation growth of plant cells in which activation of plasma-membrane-associated NADH-ascorbate free radical oxidoreductase is coupled to the oxidation of NADH with transfer of electrons to cytoplasmic ascorbate radical generated by ascorbate oxidase.

acceptor of the correct potential will stimulate both the electron flow and growth rate. One excellent candidate for a physiological acceptor is diferric transferrin.[46] In plants as well, NADH appears to be the electron donor and the ascorbate radical may provide a suitable external acceptor. Ascorbate is secreted by plants into the cell wall where abundant oxidative activities exist to generate ascorbate free radical. The activity appears to be directly responsive to regulation by plant growth hormones of the auxin type.[47] The model that is being tested currently is depicted in FIGURE 7.

In animals also, the plasma membrane redox system responds to hormones and other growth factors[5,11,39] and could, in principle, use externally generated ascorbate free radical as an electron donor to oxidize cytoplasmic NADH via a mechanism identical to that postulated for secretory granules, Golgi apparatus, and coated vesicles. With internal membranes, ascorbate free radical is generated within the membrane lumens to serve as an electron acceptor for cytoplasmic NADH. Assuming that the enzyme orientation present at the cell surface is the same as for internal endomembranes, any ascorbate free radical present external to the cell would act as an acceptor for oxidation of cytoplasmic NADH. However, at the cell surface the enzyme system would respond not only to ascorbate free radical but to a variety of electron acceptors. At the cell surface the situation would differ from that of internal membranes where a regeneratable electron acceptor is required and where ascorbate free radical derived from intervesicular or interluminal ascorbate emerges as the acceptor of choice utilized by the cell. Despite considerable work to establish the ascorbate radical as a potential electron acceptor for a plasma membrane transmembrane dehydrogenase, the obvious question remains as to whether or not ascorbate is a general cofactor for transmembrane dehydrogenases or related only to a special function in coated pits, secretory vesicles, or chromaffin granules where continued redox function depends strictly upon the utilization of a restricted pool of a regeneratable electron acceptor.

SUMMARY

The mechanism(s) whereby membrane translocations are energized are poorly understood. Our work has focused on transmembrane microsomal and plasma membrane redox constituents as a means to energize membranes via alternative mechanisms complementary to ATP-driven processes. One such component is NADH-ascorbate free radical (mono- or semidehydroascorbate) oxidoreductase. This activity is associated with the trans or exit face of the Golgi apparatus, transport vesicles that move between the Golgi apparatus and the plasma membrane, and with the plasma membrane itself. Various lines of evidence, mostly indirect, link this activity to membrane translocations. Included is an apparent activation of the reductase in membranes when coated with clathrin, a single large polypeptide chain involved in exocytosis and in receptor-mediated and absorptive endocytosis. The results are consistent with a role of the ascorbate free radical as an acceptor for electron transport-mediated transfer of electrons from NADH perhaps to oxygen by coated membranes as a part of a mechanism to drive membrane translocations via generation of a proton gradient or of a membrane potential. Additionally, plasma membrane redox may be important in the regulation of cell growth, but a strict dependence on ascorbate free radical for the latter seems less likely than with internal endomembranes, where redox function may strictly depend upon the restricted pool of regeneratable acceptor that the ascorbate free radical provides.

REFERENCES

1. MORRÉ, D. J., E. L. VIGIL, C. FRANTZ, H. GOLDENBERG & F. L. CRANE. 1978. Cytobiologie **18:** 213-230.
2. HUANG, C. M., H. GOLDENBERG, C. FRANTZ, D. J. MORRÉ, T. W. KEENAN & F. L. CRANE. 1979. Int. J. Biochem. **10:** 723-731.
3. MORRÉ, D. J. & E. L. VIGIL. 1979. J. Ultrastruct. Res. **68:** 317-324.
4. MORRÉ, D. J. 1981. *In* International Cell Biology 1980-1981. H. G. Schweiger, Ed.: 800-804. Springer-Verlag. Berlin/Heidelberg/New York.
5. GOLDENBERG, H., F. L. CRANE & D. J. MORRÉ. 1979. J. Biol. Chem. **254:** 2491-2498.

6. SUN, I., D. J. MORRÉ, F. L. CRANE, K. SAFRANSKI & E. M. CROZE. 1984. Biochim. Biophys. Acta **797:** 266-275.
7. MORRÉ, D. J., I. SUN & F. L. CRANE. 1985. *In* Vitamins and Cancer: Human Cancer Prevention by Vitamins and Micronutrients. F. L. Myeskins & K. N. Prasad, Eds.: 83-92. Humana Press. Clifton, NJ.
8. NJUS, D., J. KNOTH, C. COOK & P. M. KELLEY. 1983. J. Biol. Chem. **258:** 27-30.
9. SRIVASTAVA, M., L. T. DUONG & P. J. FLEMING. 1984. J. Biol. Chem. **259:** 8072-8075.
10. VAN ZASTROW, M., T. R. TRITTON & J. D. CASTLE. 1984. J. Biol. Chem. **259:** 11746-11750.
11. CRANE, F. L., I. L. SUN, M. G. CLARK & H. LÖW. 1985. Biochim. Biophys. Acta **811:** 233-264.
12. MORRÉ, D. J., R. D. CHEETHAM, S. E. NYQUIST & L. OVTRACHT. 1972. Prep. Biochem. **2:** 61-69.
13. PEARSE, B. M. F. 1976. Proc. Natl. Acad. Sci. U.S.A. **73:** 1255-1259.
14. MELLO, R. J., M. S. BROWN, J. L. GOLDSTEIN & R. G. W. ANDERSON. 1980. Cell **20:** 829-837.
15. ALBERTSSON, P. A., B. ANDERSSON, C. LARSSON & H. E. AKERLUND. 1982. Methods Biochem. Anal. **28:** 115-150.
16. MORRÉ, D. J. 1973. *In* Molecular Techniques and Approaches in Developmental Biology. M. J. Chrispeels, Ed.: 1-27. John Wiley & Sons. New York.
17. MORRÉ, D. J., K. E. CREEK, G. R. MATYAS, N. MINNIFIELD, I. SUN, P. BAUDOIN, D. M. MORRÉ & F. L. CRANE. 1984. BioTechniques **2:** 224-233.
18. MORRÉ, D. J., D. M. MORRÉ & H.-G. HEIDRICH. 1983. Eur. J. Cell Biol. **31:** 263-274.
19. HANNIG, K. & H.-G. HEIDRICH. 1977. *In* Cell Separation Methods, Part IV. Electrophoretic Methods. H. Bloemendahl, Ed.: 96-116. Elsevier/North Holland Biomedical Press. New York.
20. CRANE, F. L., J. L. GLENN & D. E. GREEN. 1956. Biochim. Biophys. Acta **22:** 457-487.
21. MATSUMURA, M. & S. MIYACHI. 1980. Methods Enzymol. **69:** 465-470.
22. JACOBSEN, E. L. & M. K. JACOBSEN. 1976. Arch. Biochem. Biophys. **175:** 627-634.
23. ROE, J. H. & C. A. KEUTHER. 1943. J. Biol. Chem. **147:** 399-407.
24. LOWRY, O. H., N. J. ROSEBROUGH, A. L. FARR & R. J. RANDALL. 1951. J. Biol. Chem. **193:** 265-275.
25. KARNOVSKY, M. J. & L. ROOTS. 1964. J. Histochem. Cytochem. **12:** 219-221.
26. ANDERSON, R. G. W., M. S. BROWN & J. L. GOLDSTEIN. 1977. Cell **10:** 351-354.
27. WILLINGHAM, M. C. & I. PASTAN. 1980. Cell **21:** 67-77.
28. SUN, I. L., F. L. CRANE & D. J. MORRÉ. 1983. Biochem. Biophys. Res. Commun. **115:** 952-957.
29. SIMS, P. J., A. S. WAGGONER, C. H. WANG & J. F. HOFFMAN. 1974. Biochemistry **13:** 3315-3330.
30. STONE, D. K., X.-S. XIE & E. RACKER. 1983. J. Biol. Chem. **258:** 4059-4062.
31. XIE, X.-S., D. K. STONE & E. RACKER. 1983. J. Biol. Chem. **258:** 14834-14838.
32. FORGAC, M., L. CANTLEY, B. WIEDEMANN, L. ALTSTIEL & D. BRANTON. 1983. Proc. Natl. Acad. Sci. U.S.A. **80:** 1300-1303.
33. ZANG, G. & D. L. SCHNEIDER. 1983. Biochem. Biophys. Res. Commun. **114:** 620-625.
34. GLICKMAN, J., K. CROEN, S. KELLY & Q. ALAWQUATI. J. Cell Biol. **97:** 1303-1308.
35. BARR, R., K. SAFRANSKI, I. L. SUN, F. L. CRANE & D. J. MORRÉ. 1984. J. Biol. Chem. **259:** 14604-14607.
36. MINNIFIELD, N. & D. J. MORRÉ. 1984. Cell Biol. Int. Rpts. **8:** 215-219.
37. IYANAGI, I., J. YAMAZAKI & K. F. ANAN. 1985. Biochim. Biophys. Acta **806:** 255-261.
38. ORRINGER, E. P. & M. E. S. ROER. 1979. J. Clin. Invest. **63:** 53-58.
39. GOLDENBERG, H. 1980. Biochem. Biophys. Res. Commun. **94:** 721-726.
40. GOLDENBERG, H., C. GREBING & H. LÖW. 1983. Biochem. Int. **6:** 1-10.
41. STAUDINGER, H., K. KIRSCH & S. LEONHAUSER. 1961. Ann. N.Y. Acad. Sci. **92:** 195-
42. LUMPLER, VON L., W. SCHNEIDER & H. STAUDINGER. 1967. Hoppe-Seyler's Z. P Chem. **348:** 323-328.
43. GEIP, D. & H.-U. SCHULZE. 1975. FEBS Lett. **60:** 374-379.
44. RUSSEL, J. T., M. LEVINE & D. NJUS. 1985. J. Biol. Chem. **260:** 226-231.
45. NAVAS, P., I. L. SUN, D. J. MORRÉ & F. L. CRANE. 1986. Biochem. F Commun. **135:** 110-115.

46. CRANE, F. L., I. L. SUN, D. J. MORRÉ & H. LÖW. 1985. J. Cell Biol. **101:** 483a.
47. MORRÉ, D. J., P. NAVAS, C. PENEL & F. J. CASTILLO. 1986. Protoplasma **133:** 195-197.
48. CROZE, E. M., D. J. MORRÉ, D. M. MORRÉ, J. KARTENBECK & W. W. FRANKE. 1982. Eur. J. Cell Biol. **28:** 130-138.

DISCUSSION OF THE PAPER

M. LEVINE (*National Institutes of Health, Bethesda, Md.*): You said that ascorbate enters the vesicles. How does ascorbate get into these vesicles?

D. J. MORRÉ (*Purdue University, West Lafayette, Ind.*): We have not been able to demonstrate an uptake of ascorbate by these vesicles in the cell free system. I suspect the uptake process might be fairly specific and it might require some sort of cofactor or something that we don't yet understand. That's one possibility. The second possibility is that ascorbate may enter much earlier into the transport process and that we've been looking at the wrong membrane. And of course when we return to that problem one of the first experiments that we'll do is to address the question of whether endoplasmic reticulum can take up ascorbate, and if it can, that may be part of the explanation.

M. LEVINE: So when you deal with your cell free system these vesicles already have prepackaged ascorbate?

D. J. MORRÉ: They are prepackaged. The only place where we've loaded has been in the plant cells, where we can make a plasma membrane with the right orientation so that we can actually trap ascorbate on the inside as the membrane breaks.

E. J. DILIBERTO (*Wellcome Research Labs, Research Triangle Park, N.C.*): Do you know what the mechanism of inhibition of monensin is on the enzyme?

D. J. MORRÉ: Monensin is a classic Golgi poison, and I'm wondering if the reason it's a classic Golgi poison is because it's inhibiting this enzyme and not for some other reason.

E. J. DILIBERTO: If you were to lyse the vesicles, would you still get the inhibition with monensin? Might you have affects secondary to ion changes associated?

D. J. MORRÉ: They are intact vesicles. The only thing I can tell you is that it is not due to the depolarization of the membrane because you can take other depolarizing agents and they have no effect.

E. J. DILIBERTO: So in effect you don't know whether or not the enzyme in the mitochondria is different from what you find in the coated vesicles?

D. J. MORRÉ: It is different but we haven't characterized the monensin response.

A. B. KITABCHI (*University of Tennessee, Memphis, Tenn.*): You mentioned that you have not demonstrated selective uptake in the cell free system for the ascorbate . . .

D. J. MORRÉ: No, we have not demonstrated uptake.

A. B. KITABCHI: You mentioned growth. Is it hyperplasia or hypertrophy?

D. J. MORRÉ: In the plant system it is pure hypertrophy.

A. B. KITABCHI: In the regenerating system do you expect ascorbate to have a affinity for a particular subcellular fraction versus stable hepatocyte?

D. J. MORRÉ: We haven't really talked about regenerating with tissue, but in tissue culture cells when you stimulate plasma membrane redox you ultimately do increase the cell number. This of course is something which can be driven by a whole variety of acceptors, ascorbate free radical only being one. The ascorbate free radical does work.

W. LOHMANN (*University of Giessen, Giessen, FRG*): I would like to follow up the first question. We have determined the permeation rate of the ascorbate free radical across the membranes. We started out with artificial membranes, the vesicles all the way up to natural membranes, and indeed it will pass the membrane but the permeation rate is one thousandth of the oxidized dehydroascorbic acid. The other question would be then, How do you produce a free radical?

D. J. MORRÉ: Ascorbate oxidase is in the membrane. All of these membranes that contain a reductase also contain ascorbate oxidase so they can generate the free radicals and it's kept reduced simply because of the NADH. Ascorbate is constantly being regenerated through electron transport from NADH.

W. LOHMANN: Have you studied the influence of other redox systems such as glutathione redox system in your system?

D. J. MORRÉ: Not enough to make any kind of definitive statement.

B. L. HORECKER (*Cornell University Medical College, New York, N.Y.*): What about NADPH?

D. J. MORRÉ: NADPH doesn't drive. There's a little bit of activity there but it doesn't drive any of these interesting processes. You can oxidize NADPH but it doesn't do anything physiological that we know of in the Golgi apparatus or coated vesicles.

W. LOHMANN: There's another well-known fact. If ascorbic acid is oxidized H_2O_2 is produced.

D. J. MORRÉ: That's right.

W. LOHMANN: What is H_2O_2 doing in your model?

D. J. MORRÉ: I hope nothing, because we've done those controls. We scavenged the H_2O_2; it works fine.

R. C. ROSE (*Pennsylvania State University College of Medicine, Hershey, Pa.*): Are you convinced that ascorbate oxidase actually oxidizes ascorbic acid through the free radical as opposed to through two-electron transfer?

D. J. MORRÉ: We haven't done that type of mechanistic work, although we've done it empirically; you throw in ascorbate oxidase and you get free radicals.

Ascorbate Can Act as an Inducer of the Collagen Pathway Because Most Steps Are Tightly Coupled[a]

RICHARD I. SCHWARZ, PAUL KLEINMAN, AND NANCY OWENS

Laboratory of Cell Biology
Division of Biology and Medicine
Lawrence Berkeley Laboratory
Berkeley, California 94720

INTRODUCTION

Ascorbate is best known for its ability to cure scurvy. Similarly, scurvy is well known as a disease that affects collagen production. Yet, the only direct impact of ascorbate on the collagen pathway is to alter the level of hydroxylation of proline residues. This raises the important question of whether the presence or absence of a posttranslational modification is sufficient to act as a regulatory element, or must one invoke new roles for the ascorbate molecule.

Previous studies both *in vivo* and in culture are difficult to interpret in light of the above question. Studies *in vivo* with the guinea pig support the idea that the drop in collagen production on ascorbate deprivation is a consequence of fairly rapid loss of appetite.[1,2] These studies point out the complexity of *in vivo* research and how the indirect effects of ascorbate can dominate. In cell culture, where theoretically one has more control over the environment of the cell, ascorbate with different cell types has been reported to stimulate, to inhibit, and to have no effect on collagen production.[3–5] The net result of such diverse conclusions was to place doubt on whether ascorbate could act as a specific inducer of the collagen pathway.

Part of the explanation for the inconsistency of ascorbate action in culture is that the cell culture environment is not a passive element but can itself suppress the inductive effect of ascorbate. Using primary avian tendon (PAT) cells, we have shown that high serum levels completely eliminate ascorbate's ability to induce procollagen production.[3] In low serum (0.2%), ascorbate stimulates PAT cells to raise collagen production from 12% to 50% of total protein synthesis.[3,6,7] Furthermore, repeated growth of PAT cells to confluent densities can irreversibly drop procollagen production to low levels, further negating any inductive effect of ascorbate.[8] Therefore, cell lines making very low levels of procollagen (around 1%) may best be viewed as being

[a]This work was supported in part by the National Institutes of Health (grant no. CA37958) and by the Office of Health and Environmental Research, Office of Energy Research, U.S. Department of Energy (under contract no. DE-AC03-76SF00098).

suppressed for procollagen production, and addition of ascorbate is not sufficient to restore high procollagen production.

Finally, while most steps are linked, there are exceptions. The level of the enzyme prolyl hydroxylase does not change with ascorbate addition. It does, however, increase dramatically with cell density.[22] This may account for the fact that ascorbate only induces cells at high density. Thus, a pathway that is highly coupled and producing at high levels becomes susceptible to subtle regulation at many locations. Perhaps this explains how collagen production can be so precisely regulated during development.

In any case, PAT cells grown in low serum medium can be used to finally answer the question of whether a change in a posttranslational modification is sufficient to induce procollagen production. In this report, pertinent old and new data have been combined to support the idea that ascorbate's only role is to stimulate hydroxylation of proline residues. This rapidly stimulates procollagen secretion, and because various steps in the pathway are tightly coupled a chain of events ensues leading to a fully induced cell 60 h after ascorbate addition. Furthermore, we present a model where several of the required feedback mechanisms between steps can be physically localized to the membrane of the endoplasmic reticulum.

METHODS

Cell Culture

PAT cells were isolated from 16-day chicken embryos by a modification[3] of the Dehm and Prockop[7] procedure. PAT cells were grown in F12 medium[10] with 0.2% fetal calf serum in 25-cm^2 flasks as described previously.[8]

Collagen Assay

PAT cells were labeled with [^{3}H]proline (5′; Amersham Corp.) as described in the figure legends. Collagen production was determined by a collagenase assay[11] modified from that described by Peterkofsky and Diegelmann.[12] Bacterial collagenase, used in this assay, recognizes a four-amino-acid sequence that is, to a first approximation, only found in the collagen part of the procollagen molecule. This highly repetitive sequence in the collagen part of the procollagen molecule is degraded by collagenase and thereby detected.[11] Our assay only measures collagen sequences, and we mathematically compensate for the fact that collagen is first made as a procollagen molecule. The collagenase was purified to the extent that 50- to 100-fold excess enzyme could be used in the assay without evidence of general proteolytic activity.[11]

Procollagen mRNA Assays

RNA extractions and dot blot assays were performed as previously described.[7]

α,α′-Dipyridyl Experiments

Cells growing with and without ascorbate were given 0.3 mM α,α′-dipyridyl[14] in fresh medium at time zero. At various times thereafter, cells were pulsed for 2 h with [^{3}H]proline (40 μCi per flask) to assay for procollagen synthesis.

Secretion Rates

PAT cells were labeled with 100 μCi of [^{3}H]proline for 2 h in 2 ml of growth medium. The cells were then washed with unlabeled medium (10 ml), and 2 ml of medium were added containing 100× normal proline concentration. At various time intervals, the medium was removed and new medium (with 100× proline) was added. The medium was assayed for counts incorporated into procollagen and noncollagen protein using a collagenase assay as previously described.[13]

Membrane-Bound Polysomes

The methods of Kao *et al.*[14] were used with the following differences. Detergent extraction was at room temperature in the presence of vanadyl ribonucleoside complexes.[15] This gave a higher yield of membrane-bound ribosomes. Second, a shorter (10 min) higher speed (25,000 rpm) first spin was used. This yielded a tighter pellet, making it easier to remove the supernatant. Finally, while many experiments quantitated the ribosome pellets by optical density as Kao *et al.* described, this required large numbers of cells. Alternatively, cells were labeled with [^{3}H]uridine (10 μCi/flask) with 10^{-5} M cold uridine for three days; then ribosome yield could be quantitated by dissolving the pellet in 0.04 M NaOH (1 h, 37°C), neutralizing with acetic acid and counting in a scintillation counter.

RESULTS

Kinetics of Ascorbate Induction

Ascorbate is a nearly ideal inducer of procollagen production in PAT cells. The induction is specific, causing a sixfold change in absolute procollagen production and only minor changes in the synthesis of other proteins.[6,7] The increase in procollagen synthesis occurs in addition to the protein synthesis of the uninduced cells. As a consequence, ascorbate has very little affect on the generation time of PAT cells, avoiding a potential complication. Another asset of the ascorbate induction process is that it acts slowly, requiring 48 h for the full increase in procollagen synthesis.[3,6,7] This has allowed us to resolve the various changes independently of each other.

The most rapid effect of ascorbate is the stimulation of the hydroxylation of proline residues in the procollagen molecule. The level of hydroxyproline rises from about 10% of all proline residues to almost 50%.[6] In less than 1 h after the addition of ascorbate, previously synthesized procollagen residing in the endoplasmic reticulum (ER) becomes fully hydroxylated and is secreted from the cell at a sixfold higher rate (FIG. 1).[13] The higher secretion rate is directly associated with the increase in proline hydroxylation that stabilizes the triple-helical conformation of the molecule, and this conformation is secreted at a more rapid rate.[7] A consequence of increasing secretion rates and not translation rates is a sixfold drop in the procollagen pool inside the endoplasmic reticulum.

Subsequently, translation rates increase with a biphasic curve: slowly increasing after 4 h, more rapidly after 12 h up to sixfold above the uninduced state (FIG. 2).[7] Procollagen mRNA levels are the last to increase. After a 12-h lag phase, mRNA levels increase linearly, requiring up to 72 h to reach a new steady-state level approximately sixfold above the uninduced state (FIG. 3).[7] The increase in mRNA level is due in part to increased transcription (threefold) and in part to a change in mRNA half-life from 10.5 to 20 h.[16] During the transition from the uninduced to the induced state, the relative increase in procollagen translation rates exceeds the relative increase in mRNA levels by almost twofold (FIG. 4).[7]

Effects of Inhibitors of Ascorbate Action

Withdrawal of ascorbate from the medium should theoretically cause the pathway to down-regulate by a similar sequence of events. Even though ascorbate has a short half-life in medium (about 1¼ h),[17] it is rapidly taken up by the cell in quantities sufficient to meet the needs of the cell for the next 36 hours.[7] Therefore, withdrawing ascorbate does not cause a synchronous drop in procollagen production. To observe the kinetics of down-regulation, one is forced to use inhibitors of ascorbate action. The most common is α, α'-dipyridyl, which chelates ferrous ion and thereby inhibits the proline hydroxylation reaction[18] and procollagen secretion.[18] Depriving a cell of ferrous ion has its negative consequences on other cellular functions; in particular, translation rates are inhibited 30-50%.[18] However, in a minus ascorbate control, the inhibition of protein synthesis is nonspecific, affecting procollagen and noncollagen proteins to the same degree. Adding α, α'-dipyridyl to an ascorbate induced cell causes a rapid and specific decline in procollagen synthesis (FIG. 5).[7] This correlates well with the decrease in procollagen secretion rates observed by Kao *et al.*[18] using this inhibitor.

Therefore, in down-regulation the coupling between secretion rates and translation is very tightly controlled. The effect of α,α'-dipyridyl is rapidly reversible by washing the cells and adding back ferrous ion (FIG. 6). Since α,α'-dipyridyl does not have a detrimental effect on procollagen mRNA levels (TABLE 1) in short-term experiments,[7] all the components are still present to synthesize large amounts of procollagen and the induced state is quickly reestablished.

There are two additional reasons for supporting the hypothesis that there is a tight linkage between secretion and translation rates. One point comes from the shape of the inhibition curve for α,α'-dipyridyl (FIG. 5). More precisely, the ratio of procollagen to noncollagen synthesis is not completely reduced to the uninduced level. However, on the basis of absolute procollagen synthesis, the inhibition approaches the uninduced level. In other words, α,α'-dipyridyl causes a specific reduction in procollagen synthesis

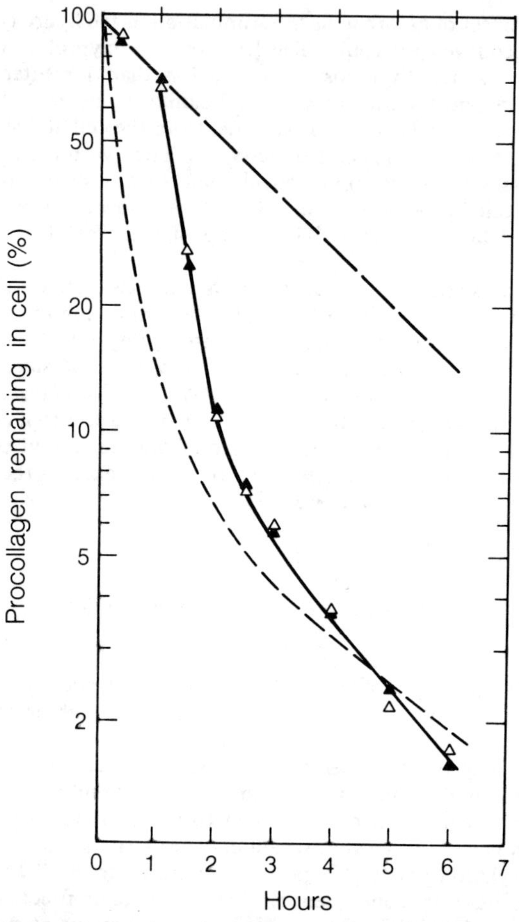

FIGURE 1. Secretion rates of procollagen into the medium for confluent cultures of PAT cells given ascorbate only at the time of the chase (*triangles*). The points are derived from independent analyses of two different flasks (*open and closed points*). Data from cells grown without ascorbate (*long dash*) and for cells given ascorbate at the time of the pulse (*short dash*) are also shown; in these cases, the actual data points have been omitted to simplify the figure. (From Schwarz.[13] Reprinted by permission from the *Journal of Biological Chemistry.*)

and a nonspecific reduction in the synthesis of all proteins. The sum of the two different types of inhibition on procollagen production reduces it to the rate at which the cell is able to secrete procollagen. A second reason supporting a tight linkage is that other inhibitors show similar effects. Colchicine addition to the medium is mildly inhibitory to procollagen secretion. The average half-life of procollagen secretion goes from 20 to 40 min. The ratio of procollagen to noncollagen synthesis is also inhibited twofold (TABLE 2). Thus, from several perspectives there is a coupling between secretion and translation rates.

Ascorbate Induction and Ribosome Binding to the Endoplasmic Reticulum

Very little is known about mechanisms for translational regulation, but one hypothesis would be that a sixfold induction of procollagen synthesis by ascorbate would require the formation of additional membrane-bound ribosomes.[14] The kinetics of this process could reveal important characteristics of the regulation of translation rates.

Experimentally, one can distinguish free- from membrane-bound ribosomes by differential centrifugation in the presence or absence of detergents.[14] Without detergents, membrane-bound polysomes remain attached to the ER and pellet in a low-speed spin. Free ribosomes (polysomes) require greater force, and in addition, a sucrose cushion is used to trap any lower density membrane-associated ribosomes carried over

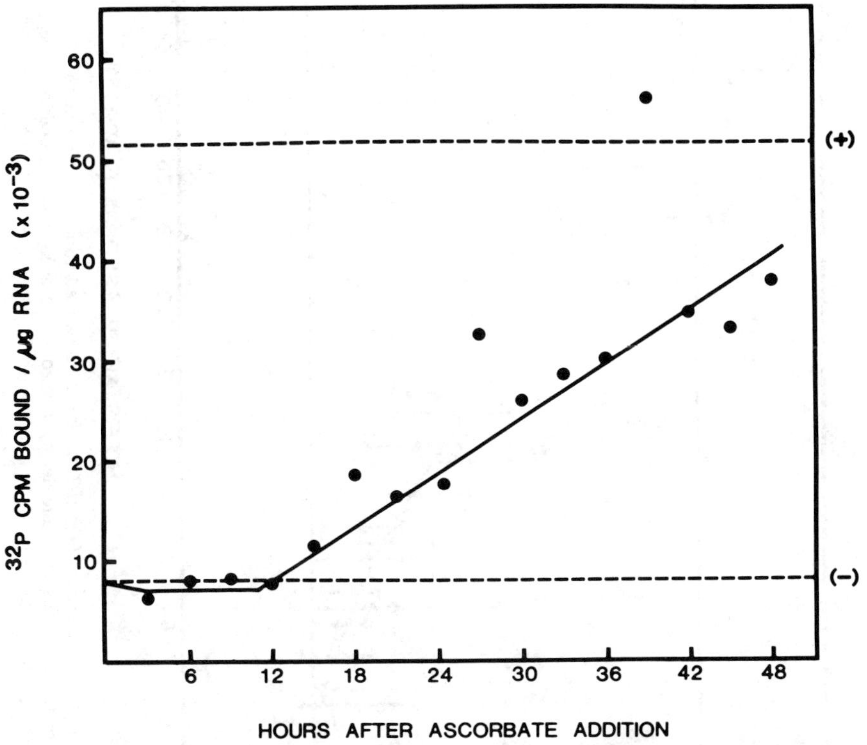

FIGURE 2. Time course for the ascorbate induction of the percentage of procollagen synthesis to total protein synthesis. The uninduced (−) and fully induced (+) levels are shown as *dashed lines.* PAT cells were given ascorbate at time zero and at various points up to 60 h the cells were pulsed with 40 μCi of [^{3}H]proline (1 ml of medium) for 2 h. The fraction of the incorporated counts in both the cells and the medium that were susceptible to collagenase was determined, corrected for the fact that collagen contains 5.2 times more proline than the average protein and then further corrected by 1.5 times for the fact that collagenase does not significantly degrade the propeptide portion of the procollagen molecule. (From Rowe and Schwarz.[7] Reprinted by permission from *Molecular and Cellular Biology.*)

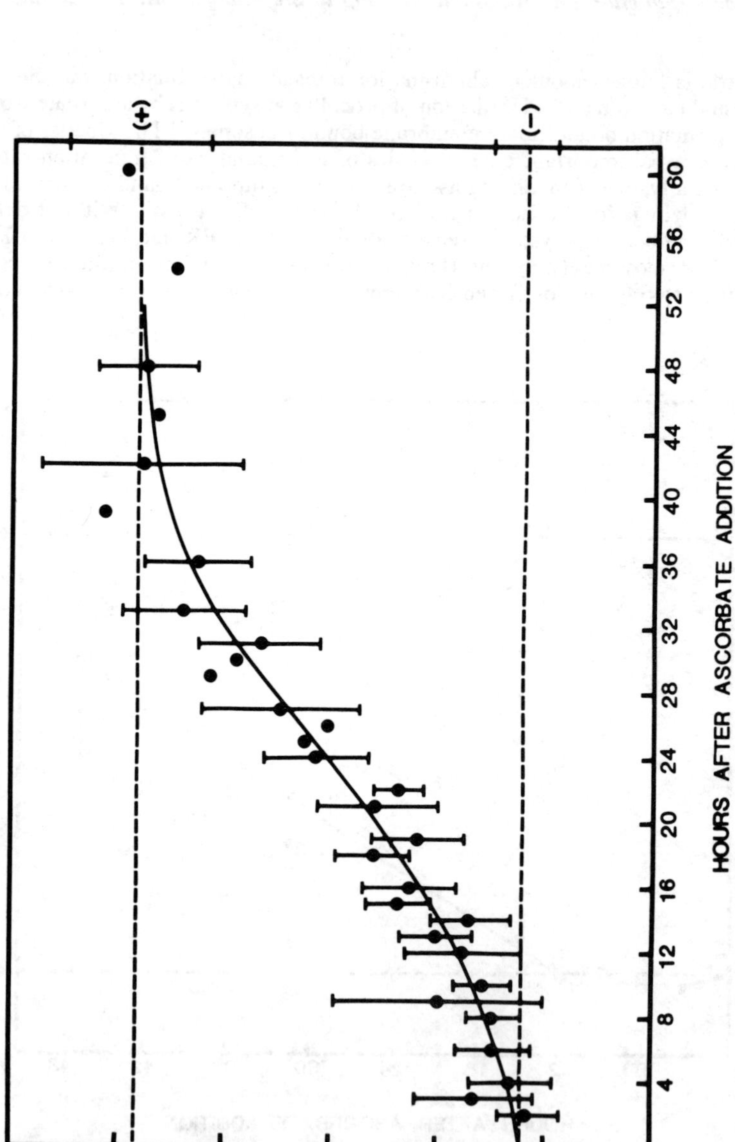

FIGURE 3. Accumulation of procollagen α_2 (I) mRNA after ascorbate addition to the medium. PAT cells were given ascorbate at time zero, and cytoplasmic RNA was extracted at various times and dot hybridized with a nick-translated cDNA probe, pCg45.[7,23] The uninduced level (−) and fully induced level (+) are shown as dashed lines.

from the first spin. In contrast to the expected result, we find that both induced and uninduced PAT cells have high percentages of ribosomes membrane-bound, approximately 70% (TABLE 3). We obtained similar results whether the ribosomes were detected and quantitated by long-term [^{3}H]uridine label or by optical density (TABLE 3). This value is also similar to that obtained from freshly isolated PAT cells; therefore, the high content of membrane-bound polysome is not a consequence of culturing the cells. Accordingly, at least over a 1-week period, the lack of ascorbate, while reducing procollagen production sixfold, does not affect the number of membrane-bound ribosomes.

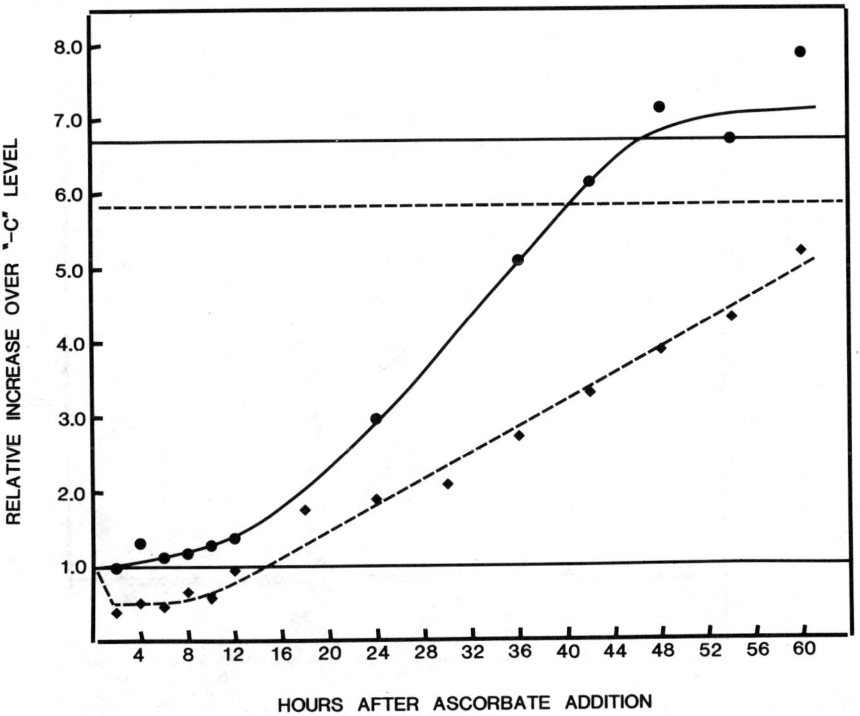

FIGURE 4. Comparison of the relative increase in the rate of procollagen synthesis (*solid line*) to the relative accumulation of procollagen mRNA (*dashed line*) after ascorbate addition. All values were normalized to uninduced cells. Average maximum values are indicated.

DISCUSSION

Several unique aspects of the collagen pathway allow a reducing agent, ascorbate, to act in a regulatory manner. Very few proteins require reducing potential in their production, but the large number of hydroxyproline residues in the procollagen molecule makes this essential. Formation of hydroxyproline residues stabilizes the helical structure,[19,20] and this conformation is secreted six times more rapidly. Most importantly, the pathway is tightly coupled so that an increase in secretion rates can trigger an increase in translation rates, and this can lead even more slowly to greater procollagen mRNA levels.

One of the difficulties in postulating a feedback mechanism is the underlying obligation to describe the linkage between the steps. On first thought, procollagen translation seems unrelated to the mechanism that secretes the molecule. However, in a physical sense both steps are quite close together. The rate-limiting step in secretion

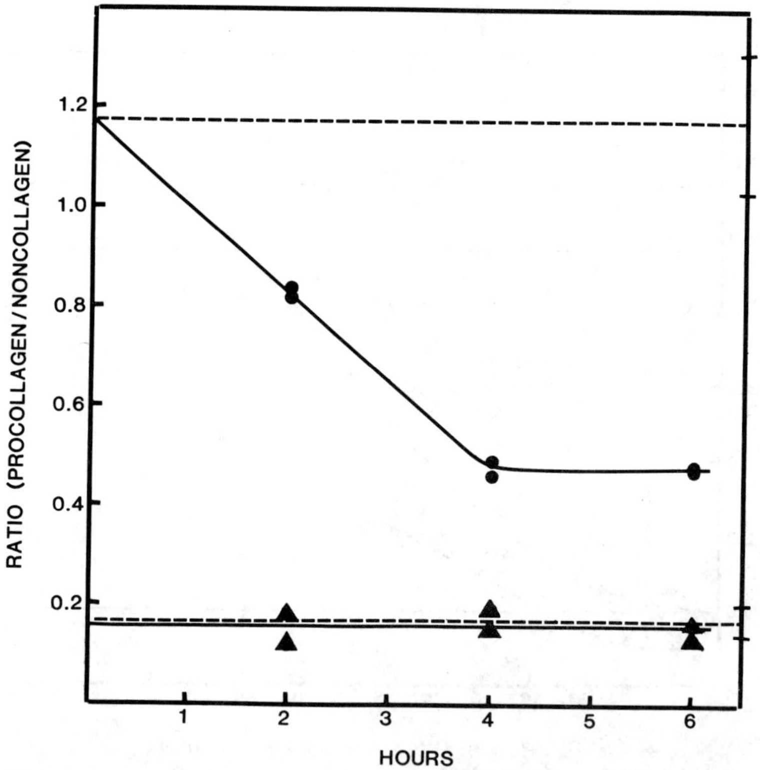

FIGURE 5. Effect of adding α,α'-dipyridyl (0.3 mM) to fully induced (*circles*) and uninduced (*triangles*) PAT cells. The rates of procollagen synthesis have been normalized to that of non-collagen protein synthesis to correct for nonspecific inhibition of protein synthesis (30-50% in 6 h). Each point in the graph reflects a 2-h pulse ending at the time indicated. The controls (without α,α'-dipyridyl) are shown as dashed lines ± SD.

is leaving the ER, and translation is occurring directly on the other side of the membrane. So these two steps need only signal across a membrane.

The form of that signal is unknown, but some indirect experimental data do reveal a few of its characteristics. Kao *et al.*[14] grew tendon cells in high serum and subcultured them several times without ascorbate. Procollagen production was low and membrane-bound polysomes had dropped to two thirds that of freshly isolated cells. On addition of ascorbate, the percentage of procollagen synthesis did not change, but the percentage

TABLE 1. Effect of α,α'-Dipyridyl on Procollagen (α_2) mRNA Levels[a]

	Cells ^{32}P (cpm) Bound per μg of RNA
−C	5,488 ± 1,121
−C + α,α'-dipyridyl	4,818 ± 1,021
+C	34,089 ± 3,848
+C + α,α'-dipyridyl	37,346 ± 12,589

[a] Cytoplasmic RNA was isolated from induced and uninduced cells that had been treated for 6 h with α,α'-dipyridyl (0.3 mM) and compared with untreated controls. The RNA was dot hybridized with nick-translated pCg45 probe.[23] The hybridized dots were cut out and counted on a scintillation counter and listed in the table as the mean ± SD; −C, uninduced cells; +C, cells fully induced with ascorbate.

of membrane-bound ribosomes was restored to the original value. We have found (Valmassoi and Schwarz, manuscript in preparation) that the high serum levels used in their experiments inhibits the ascorbate induction process at the translation step. The increase in the secretion rate constant by ascorbate still occurs. Therefore, in their tendon cells, the internal procollagen pool within the ER would be depleted. The implication of their results is that this drop in the procollagen pool may be sufficient for the stimulation of additional ribosome binding to the membrane of the ER. Furthermore, since ascorbate is such a specific inducer, this would suggest that there are specific sites on the ER for collagen production—a situation postulated by other groups[21]—that would be particularly valuable in coordinating the production

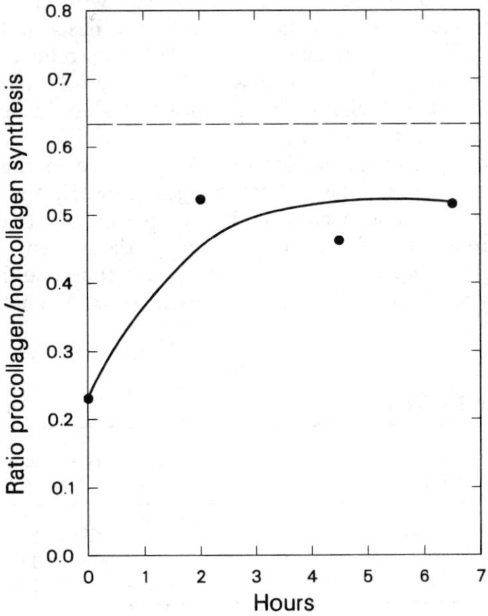

FIGURE 6. Fully induced PAT cells were given α,α'-dipyridyl (0.3 mM) for 3 h. At zero time the medium was changed and fresh medium containing 1 mM ferrous sulfate was added. Cells were assayed as described in legend to FIGURE 5. By 6.5 h, while the specific inhibition of procollagen synthesis was mostly reversed, overall protein synthesis had recovered to about 70% of normal.

TABLE 2. Effects of 10^{-7} M Colchicine on Procollagen Synthesis

	Synthesis Ratio (Procollagen/Noncollagen)
0	0.91
2 h	0.74
4 h	0.64
6 h	0.54

and association of the two α_1 and one α_2 chains that make up the procollagen molecule. There are components such as the β subunit of prolyl hydroxylase that are part of the ER membrane and are present in large quantities that could serve to orient ribosome binding, but no evidence exists to identify such a component.

Our data also show that the linkage between secretion and translation goes beyond ribosome binding because, in our case, maximal ribosome binding never changes. Instead, we see that the number of procollagen molecules that can be made from a procollagen mRNA increases about twofold during the transition from the uninduced to the induced state.[7] So the ability to deplete the pool on one side of the ER membrane can stimulate the rate of translation on the other side. In contrast, when we inhibit the induction process with α,α'-dipyridyl, our ability to extract procollagen mRNA with the nonionic detergent, TX-100, improved (data not shown). The implication is that inhibition of ascorbate action does directly affect the procollagen translation machinery. Again, the data is suggestive but not conclusive as to the nature of the coupling between secretion and translation rates.

Events on the ER membrane may affect mRNA levels as well. The approximately sixfold induction of mRNA levels with ascorbate is due partly to an increase in transcription rates (threefold) and partly to a decrease in degradation rates (twofold).[16] Changes in procollagen mRNA stability could relate to the common hypothesis that the more often an mRNA molecule participates in translation, the greater its relative stability: the ribosome offers protection to the stochastic process of RNA degradation. If this hypothesis turns out to be correct, then a change in secretion rates could be linked fairly directly to part of the increase in mRNA levels.

The greatest difficulty with the model presented is how to link a change in the procollagen secretion rate constant to a specific increase in procollagen transcription. Not only are they physically distinct, in that one occurs in the cytoplasm and the other in the nucleus, but temporally there is a 12-h lag in the coupling of these events. One can always invoke the possibility of a second messenger molecule that could stimulate procollagen transcription. An alternative hypothesis consistent with what

TABLE 3. Percentage of Membrane-Bound Ribosomes

	By Long-Term [^{3}H]Uridine	By Optical Density
Freshly isolated cells	—	69 ± 8
Plus ascorbate	70 ± 10	73 ± 13
Minus ascorbate	77 ± 13	81 ± 7

we have already observed is that the free pool of procollagen mRNA may serve to link these steps. When ascorbate stimulates translation rates, the free procollagen mRNA pool would be depleted as it becomes utilized in membrane-bound polysomes. This could feed back on the step, transcription, that controls the levels of the free mRNA pool. We are currently exploring whether there is a drop in the pool of free procollagen mRNA after ascorbate induction.

REFERENCES

1. CHOJKIER, M., R. SPANHEIMER & B. PETERKOFSKY. 1983. Specifically decreased collagen biosynthesis in scurvy dissociated from hydroxylation and correlated with body weight loss. In vitro studies in guinea pig calvarial bones. J. Clin. Invest. **72:** 826-835.
2. SPANHEIMER, R. G. & B. PETERKOFSKY. 1985. A specific decrease in collagen synthesis in acutely fasted vitamin C-supplemented guinea pigs. J. Biol. Chem. **260:** 3955-3962.
3. SCHWARZ, R. & M. J. BISSELL. 1977. Dependence of the differentiated state on the cellular environment: Modulation of collagen synthesis in tendon cells. Proc. Natl. Acad. Sci. U.S.A. **74:** 4453-4457.
4. BATES, C. J., A. J. BAILEY, C. J. PRYNNE & C. I. LEVENE. 1972. The effects of ascorbic acid on the synthesis of collagen precursor secreted by 3T6 mouse fibroblasts in culture. Biochim. Biophys. Acta **278:** 372-390.
5. PETERKOFSKY, B. 1972. The effect of ascorbic acid on collagen polypeptide synthesis and proline hydroxylation during the growth of cultured fibroblasts. Arch. Biochem. Biophys. **152:** 318-328.
6. SCHWARZ, R. I., R. B. MANDELL & M. J. BISSELL. 1981. Ascorbate induction of collagen synthesis as a means for elucidating a mechanism for quantitative control of tissue-specific function. Mol. Cell. Biol. **1:** 843-853.
7. ROWE, L. B. & R. I. SCHWARZ. 1983. Role of procollagen mRNA levels in controlling the rate of procollagen synthesis. Mol. Cell. Biol. **3:** 241-249.
8. SCHWARZ, R. I., D. A. FARSON & M. J. BISSELL. 1979. Requirements for maintaining the embryonic state of avian tendon cells in culture. In Vitro **15:** 941-946.
9. DEHM, P. & D. J. PROCKOP. 1971. Synthesis and extrusion of collagen by freshly isolated cells from chick embryo tendon. Biochim. Biophys. Acta **240:** 358-369.
10. HAM, R. G. 1965. Clonal growth of mammalian cells in chemically defined synthetic medium. Proc. Natl. Acad. Sci. U.S.A. **53:** 288-293.
11. SCHWARZ, R. I., L. COLARUSSO & P. DOTY. 1976. Maintenance of differentiation in primary cultures of avian tendon cells. Exp. Cell Res. **102:** 63-71.
12. PETERKOFSKY, B. & R. DIEGELMANN. 1971. Use of a mixture of protease-free collagenase for specific assay of radioactive collagen in the presence of other proteins. Biochemistry **10:** 988-994.
13. SCHWARZ, R. I. 1985. Procollagen secretion meets the minimum requirements for the rate-controlling step in the ascorbate induction of procollagen synthesis. J. Biol. Chem. **260:** 3045-3049.
14. KAO, W. W.-Y., J. G. FLAKS & D. J. PROCKOP. 1976. Primary and secondary effects of ascorbate on procollagen synthesis and protein synthesis by primary cultures of tendon fibroblasts. Arch. Biochem. Biophys. **173:** 638-648.
15. BERGER, S. L., D. M. WALLACE, G. P. SIEGAL, M. J. M. HITCHCOCK, C. S. BIRKENMEIER & S. B. REBER. 1981. Preparation of interferon messenger RNAs with the use of ribonucleoside-vanadyl complexes. Methods Enzymol. **79:** 59-68.
16. LYONS, B. L. & R. I. SCHWARZ. 1984. Ascorbate stimulation of PAT cells causes an increase in transcription rates and a decrease in degradation rates of procollagen mRNA. Nucleic Acids Res. **12:** 2569-2579.

17. BISSELL, M. J., C. HATIE, D. A. FARSON, R. I. SCHWARZ & W.-J. SOO. 1980. Ascorbic acid inhibits replication and infectivity of avian RNA tumor virus. Proc. Natl. Acad. Sci. U.S.A. **77:** 2711-2715.
18. KAO, W. W.-Y., D. J. PROCKOP & R. A. BERG. 1979. Kinetics for the secretion of nonhelical procollagen by freshly isolated tendon cells. J. Biol. Chem. **254:** 2234-2243.
19. JIMENEZ, S. A., M. HARSCH, L. MURPHY & J. ROSENBLOOM. 1974. Effects of temperature on conformation, hydroxylation, and secretion of chick tendon procollagen. J. Biol. Chem. **249:** 4480-4486.
20. JIMENEZ, S. A., M. HARSCH & J. ROSENBLOOM. 1973. Hydroxyproline stabilizes the triple helix of chick tendon collagen. Biochem. Biophys. Res. Commun. **52:** 106-114.
21. VEIS, A., S. J. LEIBOVICH, J. EVAN & T. Z. KIRK. 1985. Supramoleculear assemblies of mRNA direct the coordinated synthesis of type I procollagen chains. Proc. Natl. Acad. Sci. U.S.A. **82:** 3693-3697.
22. KAO, W.-Y. W., W.-C. C. KAO & R. I. SCHWARZ. 1985. Prolylhydroxylase production can be uncoupled from the regulation of procollagen synthesis. Exp. Cell Res. **157:** 265-270.
23. LEHRACH, H., A. M. FRISCHAUF, D. HANAHAN, J. WOZNEY, F. FULLER, R. CRKVENJAKOV, H. BOEDTKER & P. DOTY. 1978. Construction and characterization of a 2.5-kilobase collagen clone. Proc. Natl. Acad. Sci. U.S.A. **75:** 5417-5421.

DISCUSSION OF THE PAPER

M. LEVINE (*National Institutes of Health, Bethesda, Md.*): I would like you to clarify your results in light of Beverly Peterkofsky's at NIH. As I understand her results, she takes scorbutic animals and isolates collagen producing cells; there is decrease in proline hydroxylation but not a change in collagen biosynthesis. Could you explain why you clearly get an increase in collagen biosynthesis and yet at least from some of her data that I've seen she doesn't?

R. I. SCHWARZ (*Lawrence Berkeley Laboratory, Berkeley, Calif.*): I think you're referring to the cell line data that she has. I should just point out one thing, that if we grow a cell in high serum, for instance, we cannot get any ascorbate induction so that it's crucial under what conditions you are growing your cells, how they've been grown, and all of that. So part of that answer is how you've grown the cells and what cells we're using.

D. J. MORRÉ (*Purdue University, West Lafayette, Ind.*): With your transformed cells do you actually get an effect of ascorbate on secretion, can you stimulate that?

R. I. SCHWARZ: No, those cells were grown with ascorbate.

D. J. MORRÉ: So if you didn't give them ascorbate it would be even worse presumably?

R. I. SCHWARZ: You can't necessarily make something that isn't secreting well secrete worse. We don't see any effect of ascorbate on that.

D. J. MORRÉ: With reference to the α,α'-dipyridyl data, did you actually inhibit secretion?

R. I. SCHWARZ: By adding various levels of α,α'-dipyridyl you will change from the fast rate up to the slow rate in a proportional manner. If you inhibit the hydroxylation you inhibit secretion.

C. BOREK (*Columbia University, New York, N.Y.*): You mentioned something about high serum inhibiting translation. What is "high serum"?

R.I. SCHWARZ: We grow the primary avian tendon (PAT) cells in 0.2% serum. If we have it above that what in essence is happening is we change the kinds of proteins made by the cell and there's a tremendous stimulation of noncollagen proteins over collagen.

C. BOREK: The transformed cells usually require less serum than normal cells do. So if you're saying that there is an inhibition of translation in the transformed cells at the serum levels that you are using, is there the same inhibition of translation at lower serum levels which would be comparable to the amount required by the normal?

R. I. SCHWARZ: There's a similarity in essence between transformation and high serum; both of them stimulate noncollagen production. Noncollagen proteins in the transformed cells were increased about one and a half times, so there's a similarity in that transformed cells stimulate noncollagen production.

C. BOREK: But my question is a little bit different. The transformed cells that show this inhibition are grown at the same serum levels as normal, since transformed cells require lower levels of serum for comparable growth. Do you get the same inhibition if you lower the serum levels to lower levels than usual?

R. I. SCHWARZ: We haven't done that experiment.

J. NATH (*Walter Reed Army Institute of Research, Washington, D.C.*): I'm fascinated by the effect of ascorbate on your secretion. Ascorbate is known to have an effect on the cytoskeleton and the secretory process, and particularly in view of your transformed cell data the cytoskeletal changes are very significant in the transformed cells. So I think it's worth looking into whether this effect of ascorbate is really a general effect on the secretory process.

R. I. SCHWARZ: There are two things. The change with ascorbate is extremely fast. You've got to give ascorbate a little bit of time to get in and the hydroxylase a little bit of time to hydroxylate and if you take those into account then it's almost instantaneous. There is this hypothesis, the only one that makes any sense, that the triple-helical molecule gets out six times faster. When you hydroxylate it's clear that you do form the triple helix. Now you stabilize the conformation and it fits the data that there's a receptor for getting out of the endoplasmic reticulum which can recognize a triple-helical procollagen. How that occurs I don't know but the data does seem to fit extremely well.

On the effect of ascorbate on noncollagen secretion, there's no difference between plus and minus C, so noncollagen proteins come out with an average half-life of about 75 minutes and that is unchanged whether there is ascorbate there or not.

F. GEY (*Hoffmann-La Roche, Basel, Switzerland*): We have looked in collaboration with Buzina at gingival tissue in subjects that were at 0.4 milligram per deciliter ascorbic acid in plasma in comparison to those who had higher levels and found tht both the hydroxyproline and the proline increased, and this might fit your idea that ascorbic acid stimulates procollagen formation as well. The optimal level in these subjects was about 0.9 milligram per deciliter or higher.

Interaction of Ascorbate and α-Tocopherol

ETSUO NIKI

Department of Reaction Chemistry
Faculty of Engineering
University of Tokyo
Hongo, Tokyo 113, Japan

INTRODUCTION

Oxygen is a double-edged sword. We cannot live without oxygen, but at the same time we are continuously exposed to oxygen toxicity. Recently, the random, mostly nonenzymatic, free-radical-mediated peroxidations of biological molecules and tissues, especially lipids, have received much attention in connection with a variety of pathological events such as heart disease, rheumatoid arthritis, inflammatory disorders, cancer, and even the aging process. Many books and reviews on this subject have been published (see for example Pryor[1]). The aerobic organisms are protected from oxygen toxicity by an array of defense mechanisms. Two types of antioxidants are known: the preventive antioxidants such as catalase and peroxidase reduce the formation of free radicals by, for example, decomposing peroxide without generation of free radicals, while the chain-breaking antioxidants scavenge radicals to terminate free radical reactions and prevent chain propagation reactions. Water-soluble chain-breaking antioxidants such as ascorbic acid (vitamin C), uric acid, cysteine, and glutathione scavenge oxygen radicals in the aqueous phase, whereas lipophilic antioxidants such as tocopherols (vitamin E) scavenge oxygen radicals within the membrane. Since the water-soluble chainbreaking antioxidants cannot scavenge radicals in the membranes and since vitamin E may be the only protective agent present within the membranes, vitamin E may be the "last line" of defense against membrane peroxidation.[2] It has been pointed out that this precious antioxidant, vitamin E, is regenerated to maintain its level in biological systems. Golumic and Mattill[3] reported as early as in 1941 that vitamin C enhanced the antioxidant activity of vitamin E for lard and cottonseed oil. Tappel[4] has also observed the cooperative inhibition of oxidations by vitamins E and C and suggested that vitamin C could reduce oxidized vitamin E *in vivo.* Recently, the mechanism of cooperative and synergistic inhibition of oxidation by the combination of vitamins C and E has been studied. In this presentation, I would like to summarize the recent results.

INTERACTION BETWEEN VITAMIN C AND VITAMIN E RADICAL

Cooperative Inhibition of Oxidation by Vitamins E and C

Polyunsaturated fatty acids and their phospholipid esters are readily oxidized by a free radical chain mechanism to give peroxides (FIG. 1). FIGURE 2 shows the example of the oxidation of methyl linoleate in solution.[5] It shows that one initiating radical induced the oxidation of about 90 molecules of methyl linoleate to give four types of isomeric conjugated diene hydroperoxides. It also shows that substrate reacted, oxygen uptake, and hydroperoxides formed agree well within the experimental error.

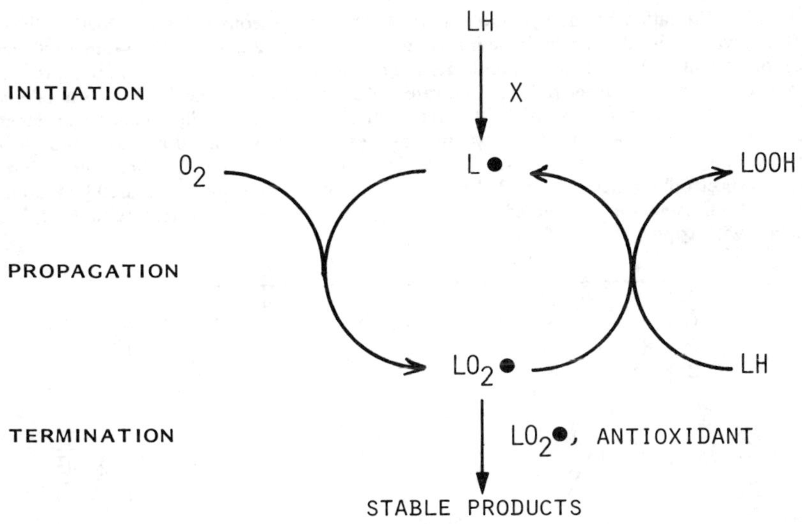

FIGURE 1. Free radical chain oxidation of lipids. *LH,* lipid; *L*·, lipid radical; $LO_2\cdot$, lipid peroxyl radical.

Both vitamins C and E function as chain-breaking antioxidants. For example, as shown in FIGURE 3, the oxidation of methyl linoleate was suppressed markedly and a clear inhibition period was observed when either vitamin C or vitamin E was added to the reaction mixture.[6] Vitamins C and E were consumed linearly with time during the inhibition period and fast oxidation took place when they were depleted. When both vitamins C and E were added together, the inhibition period was close to the sum of the inhibition period observed when either vitamin C or vitamin E was used alone.[7,8] FIGURE 4 shows the rates of consumption of vitamins E and C during the oxidation.[8] Initially, only vitamin C was consumed; vitamin E remained constant and was consumed after vitamin C disappeared. When both vitamins C and E were depleted, a fast oxidation proceeded. Similarly, the addition of vitamin C during the oxidation of methyl linoleate in the presence of vitamin E suppressed the consumption of vitamin E almost completely (FIG. 5).[8]

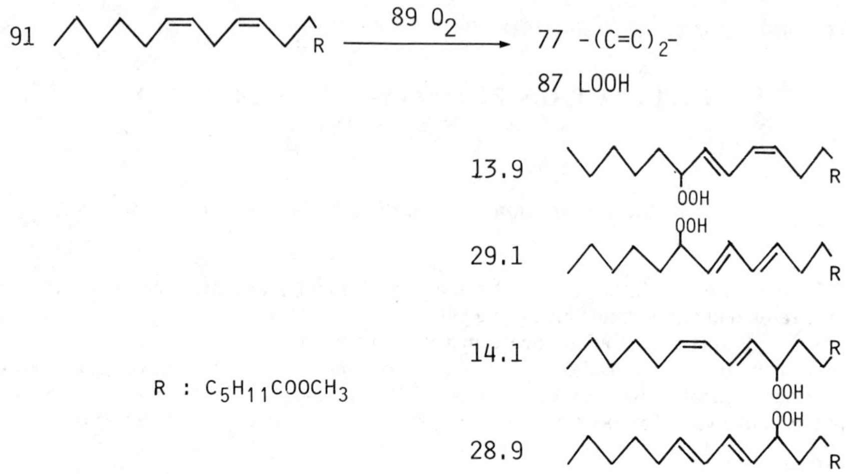

FIGURE 2. Oxidation of methyl linoleate (LH) in homogeneous solution.[5] Methyl linoleate (0.380 M) was oxidized in acetonitrile initiated with 22.6 mM 2,2-azobis (2, 4-dimethylvaleronitrile) for 60 min at 50°C. Numbers indicate the amounts in micromoles of compounds reacted and/or formed per initiating radical. The rates of chain initiation and of oxygen uptake were 5.03×10^{-8} M/s and 4.91×10^{-6} M/s, respectively, from which the kinetic chain length is calculated as 91. The amount of methyl linoleate reacted was measured by gas-liquid chromatography. Oxygen uptake was measured with Toepler pump. Conjugated diene formation was measured spectrophotometrically at 234 nm. Total hydroperoxides were measured by iodometric titration. Four isomeric conjugated diene hydroperoxides were measured with HPLC after reduction with triphenylphosphine.

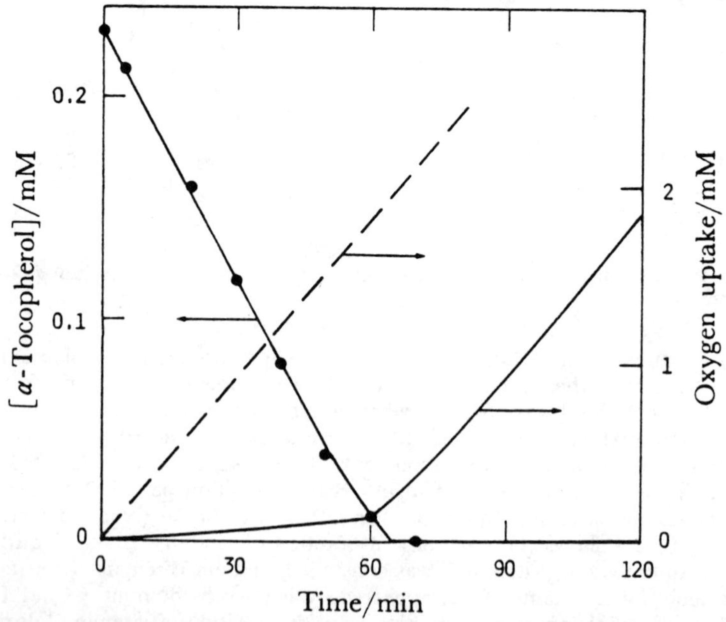

FIGURE 3. Rates of oxygen uptake and α-tocopherol decay in the oxidation of 144 mM methyl linoleate in benzene under air at 37°C initiated with 30.0 mM 2,2′-azobis (2,4-dimethylvaleronitrile). The *dashed line* shows the oxygen uptake in the absence of α-tocopherol.

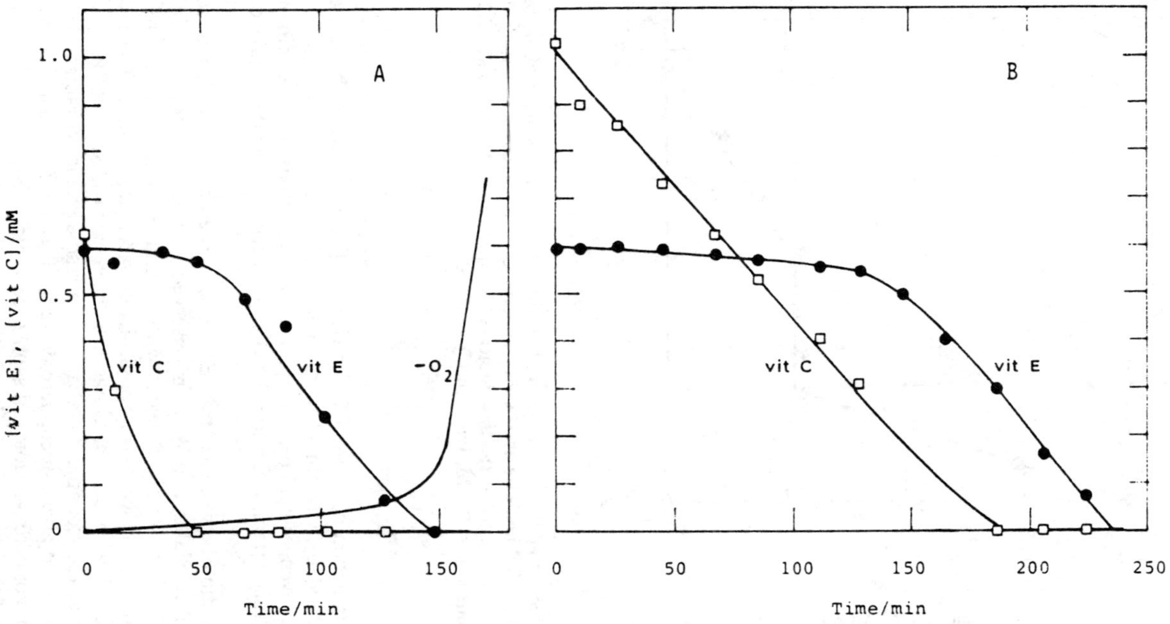

FIGURE 4. Disappearance of vitamin E (●) and vitamin C (□) in the oxidation of methyl linoleate at 37°C in *tert*-butyl alcohol/methanol (3:1 by volume). [LH] = 0.60 M, [AMVN] = 0.010 M; [vitamin E] = 0.595 mM, [vitamin C] = 0.620 mM **(A)** and 1.03 mM **(B)**.

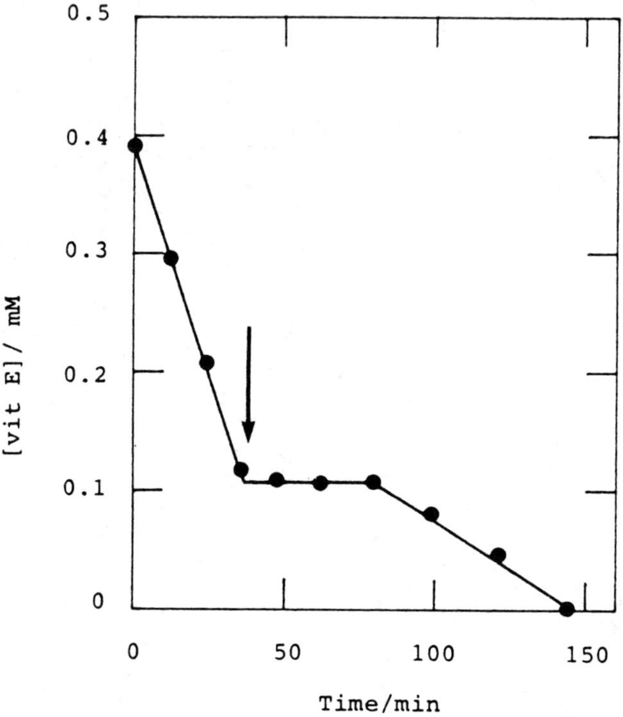

FIGURE 5. Effect of addition of vitamin C on the rate of disappearance of vitamin E during the oxidation of methyl linoleate at 37°C in *tert*-butyl alcohol/methanol (3:1 vol/vol). About 0.5 mM of vitamin C was added at the time indicated by an *arrow.* [LH] = 0.6 M, [2,2′-azobis(2,4-dimethylvaleronitrile)] = 0.1 M, [vitamin C] = 0.390 mM.

The sparing action of vitamin C for vitamin E is ascribed to the reduction of the vitamin E radical by vitamin C to regenerate vitamin E. As shown in FIGURE 6, vitamin E (**1**) scavenges lipid peroxyl radical ($LO_2\cdot$) to interrupt the chain propagation, and the resulting vitamin E radical (**2**) is reduced by ascorbate (**3**) to regenerate vitamin E.

This interaction between vitamin E radical and ascorbate has been proved experimentally. Packer, Slater, and Willson[9] observed spectrophotometrically in their pulse radiolysis study that vitamin C reacts with vitamin E radical with a rate constant of $1.55 \times 10^6\ M^{-1}s^{-1}$ in water-isopropyl alcohol-acetone solution. The interaction of vitamin E radical with ascorbate has also been confirmed by ESR spectroscopy.[10] FIGURE 7 shows the example.[10] Galvinoxyl, a stable phenoxyl radical, reacts rapidly with α-tocopherol to give α-tocopheroxyl radical. When 5,6-ascorbyl dipalmitate was added, the α-tocopheroxyl radical disappeared rapidly. Free ascorbic acid also reacts with α-tocopheroxyl radical rapidly. Other reducing agents such as glutathione also react with α-tocopheroxyl radical to give α-tocopherol and glutathione radical. The formation of glutathione radical is confirmed by its spin trapping with POBN.[10]

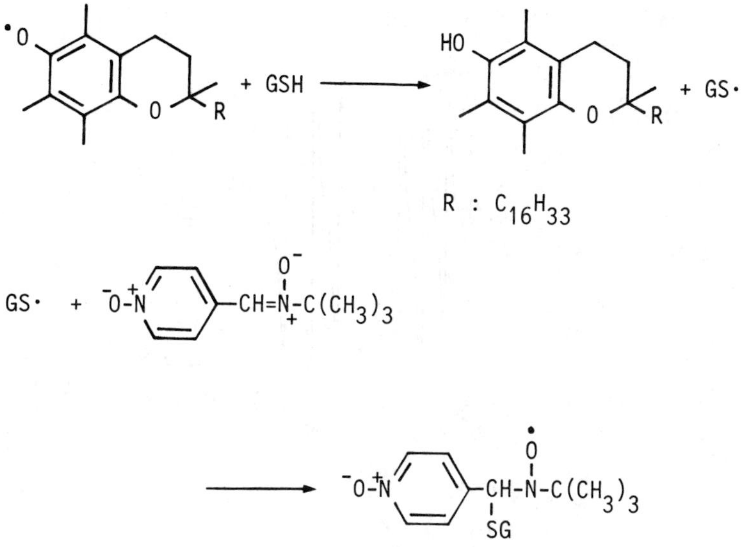

SYNERGISTIC INHIBITION OF OXIDATION BY VITAMINS C AND E

A question still remains whether such an interaction between vitamin C and vitamin E radical can take place in biological systems where lipophilic vitamin E resides within the membrane while hydrophilic vitamin C resides in the aqueous region. This question can be answered by controlling the site of initial radical production.

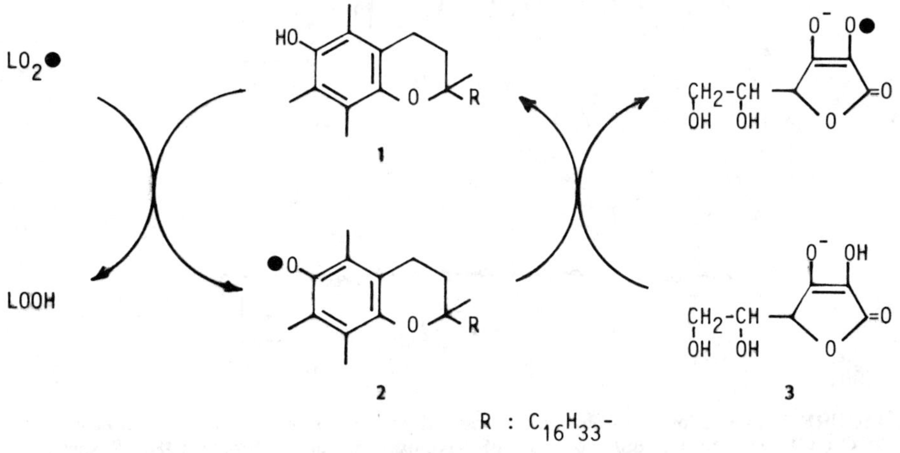

FIGURE 6. Regeneration of vitamin E from vitamin E radical by vitamin C.

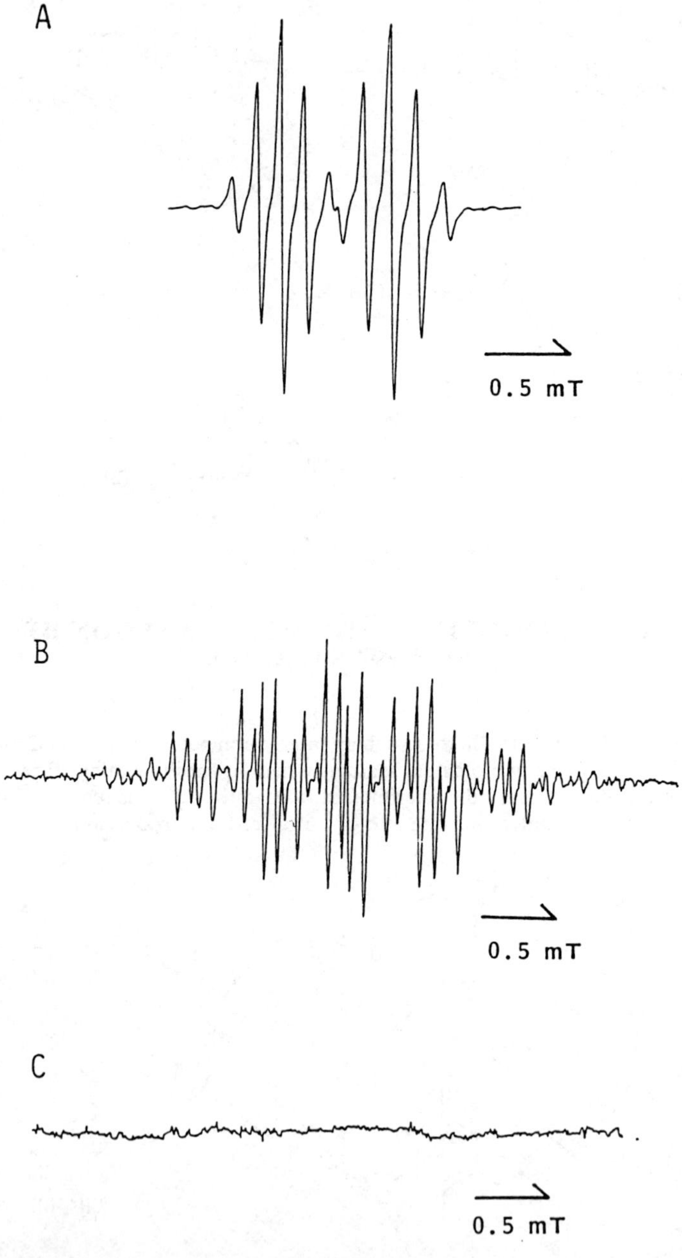

FIGURE 7. Interaction of α-tocopheroxyl radical with 5,6-ascorbyl dipalmitate in benzene at 25°C under vacuum. **(A)** ESR spectrum of galvinoxyl radical in benzene. **(B)** ESR spectrum observed when α-tocopherol was added to solution in **A.** **(C)** ESR spectrum observed when 5,6-ascorbyl dipalmitate was added to solution in **B.**

FIGURE 8 illustrates the profile of the oxidations of soybean PC liposomal membranes initiated with (A) a water-soluble radical initiator and (B) a lipid-soluble radical initiator, respectively.[11] In the absence of radical initiator, the rate of spontaneous oxidation of soybean PC liposomes was quite small, but the addition of radical initiator induced the free radical chain oxidation without any induction period. When a water-soluble initiator was used and the radicals were generated initially in the aqueous region, vitamin C suppressed the oxidation markedly and produced a clear induction period apparently by scavenging the radicals generated from the initiator before they attack the lipids in the membrane. Vitamin E incorporated into the liposomal mem-

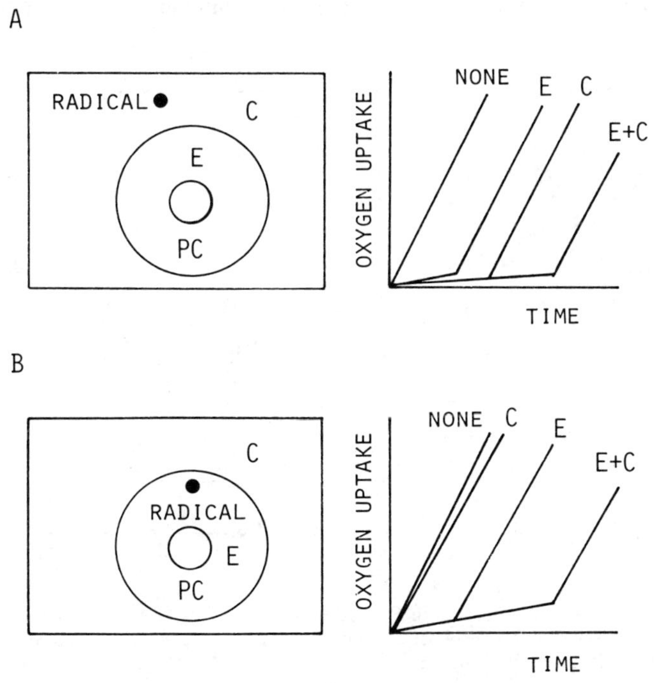

FIGURE 8. Profile of the oxygen uptake during the oxidations of soybean PC liposomes in aqueous dispersions initiated with (A) water-soluble and (B) lipid-soluble radical initiators, respectively. *None,* without antioxidant; *E,* with vitamin E; *C,* with vitamin C; *E + C,* with both vitamins E and C.

branes also suppressed the oxidation and produced an inhibition period. In the presence of both vitamin C and vitamin E, the inhibition period was close to the sum of their individual inhibition periods.

On the other hand, the lipid-soluble initiator generates free radicals within the membrane. Vitamin E could suppress the oxidation. However, as shown in FIGURE 8B, vitamin C could not inhibit the oxidation appreciably, suggesting that vitamin C located in the aqueous phase cannot scavenge radicals within the membrane. Nevertheless, the addition of vitamin C to the aqueous phase in the presence of vitamin E in the membrane lengthened the inhibition period, suggesting that, although vitamin

C could not scavenge radicals within the liposomal membranes, it could interact with vitamin E radical, probably at the surface of the membrane.

In accordance with this assumption, vitamin C spared the consumption of vitamin E. FIGURE 9 shows the rates of consumption of vitamins C and E during the oxidation of soybean PC liposomes initiated with lipid-soluble initiators.[11] In the absence of vitamin C, vitamin E was consumed linearly with time and when it was exhausted, the inhibition period was over and a rapid oxidation proceeded. On the other hand, when vitamin C was added to the aqueous phase, the rate of consumption of vitamin E was markedly reduced and vitamin C disappeared much faster than vitamin E. When vitamin C was depleted, vitamin E began to decrease at the similar rate as that in the absence of vitamin C.

The above results suggest that, although vitamin C located in the aqueous region

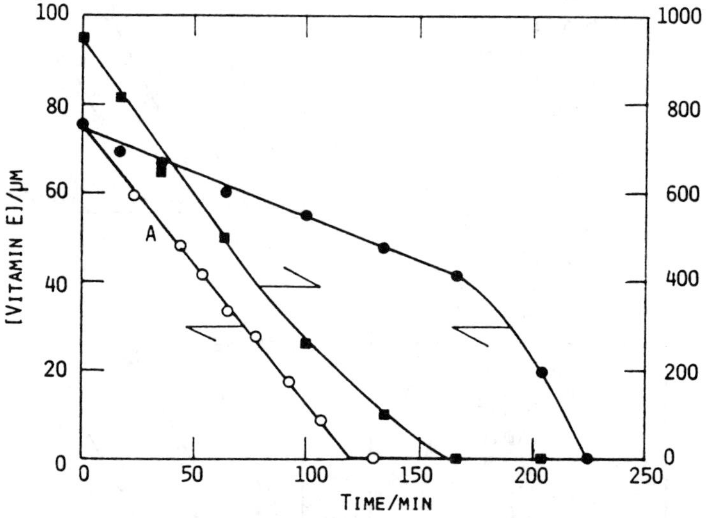

FIGURE 9. Consumption of vitamin E (●) and vitamin C (■) during the oxidation of 20 mM soybean PC liposome in 0.1 M NaCl aqueous dispersion initiated with 4.0 mM 2,2′-azobis(2,4-dimethylvaleronitrile) at 50°C under air. [vitamin E] = 77 μM, [vitamin C] = 950 μM. Line *A* (○) shows the rate of decay of vitamin E in the absence of vitamin C.

cannot scavenge peroxyl radicals within the liposomal membranes efficiently, it is more accessible to vitamin E radical and regenerates vitamin E. Probably, the phytyl side chain of vitamin E is buried in the bilayer of the membrane and the polar chromanoxyl head group must be located near the membrane-water boundary.

Similar results have been also observed by Doba, Burton, and Ingold in the oxidation of dilinoleoyl PC liposomes.[12] It has been also observed that vitamins E and C delay the formation of malondialdehyde, assumed to be the product of lipid peroxidation, during the oxidation of rat liver microsomes and phospholipid liposomes catalyzed by ferrous ion.[13] The ESR study by Scarpa *et al.*[14] also suggests the interaction between vitamins C and E in the oxidation of soybean PC liposomes.

The synergistic inhibition of oxidation by a combination of vitamin C and vitamin E has been also observed in micelle systems. Barclay *et al.*[15,16] observed the synergistic inhibition by vitamins C and E in the oxidation of linoleic acid in aqueous sodium dodecyl sulfate (SDS) dispersions.

EFFICIENCY OF INTERACTION BETWEEN VITAMIN C AND VITAMIN E RADICAL

As discussed above, vitamin C reduces vitamin E radical and regenerates vitamin E not only in homogeneous solution but also in liposomal membrane system. The efficiency or relative importance of vitamin C-vitamin E radical interaction is smaller in liposomal membrane system than in homogeneous solution, that is, vitamin C suppresses the consumption of vitamin E almost completely in homogeneous solution, whereas vitamin E is consumed appreciably even in the presence of vitamin C during the oxidation of liposomal membranes initiated with lipid-soluble radical initiator.

The lipid peroxyl radicals in the membrane must be scavenged exclusively by vitamin E. The vitamin E radical formed may undergo several competing reactions. It may scavenge another peroxyl radical to give stable product, react with another vitamin E radical to give a dimer, or interact with vitamin C to regenerate vitamin E. The lower the peroxyl radical concentration, the higher the efficiency of vitamin E regeneration. Apparently, this efficiency depends also on the accessibility of vitamin C to the vitamin E radical in the membrane.

$$LO_2^{\bullet} + \text{vit E} \longrightarrow LOOH + \text{vit E}^{\bullet}$$

$$\text{vit E}^{\bullet} + LO_2^{\bullet} \longrightarrow \text{stable products}$$

$$\text{vit E}^{\bullet} + \text{vit E}^{\bullet} \longrightarrow \text{dimer}$$

$$\text{vit E}^{\bullet} + \text{vit C} \longrightarrow \text{vit E} + \text{vit C}^{\bullet}$$

Other reducing compounds such as cysteine and glutathione can also reduce vitamin E radical and regenerate vitamin E, but with slower rate than vitamin C. The rate of reduction of galvinoxyl, a stable phenoxyl radical, incorporated into the phosphatidylcholine liposomal membranes decrease in the order ascorbic acid > cysteine > glutathione.[17] Therefore, the regeneration of vitamin E from vitamin E radical by cysteine and glutathione may not be practically significant.

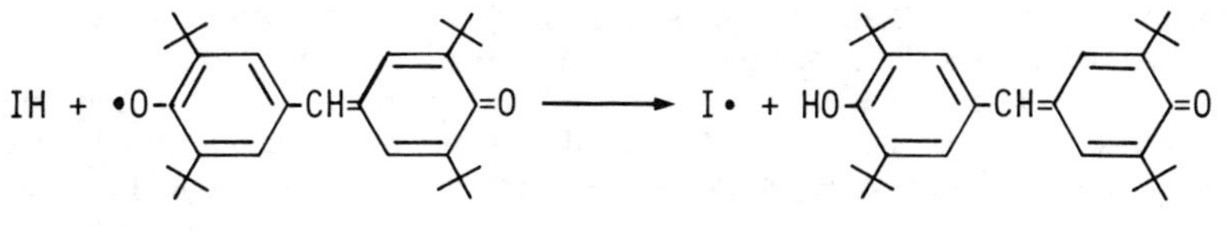

APPLICATION TO FOOD STABILIZATION

Foods such as vegetable oils that contain polyunsaturated fatty acids are oxidized readily and deteriorated. Various kinds of additives are used to prevent air oxidation of foods. Tocopherols are often used as natural and safe antioxidants. However, tocopherols, especially α-tocopherol, may act as prooxidants under certain conditions.[18–23] For example, FIGURE 10 shows the results of spontaneous oxidation of

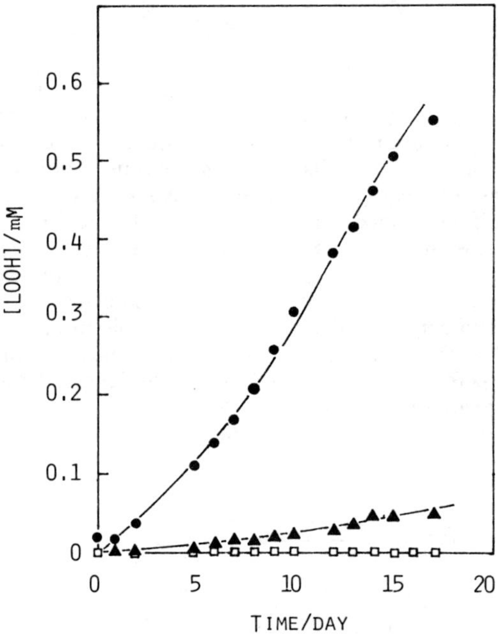

FIGURE 10. Effect of α-tocopherol and ascorbic acid on the spontaneous oxidation of methyl linoleate at room temperature in the dark. Little oxidation proceeded in the absence of α-tocopherol.

	●	▲	□
[α-tocopherol]/mM	9.86	0.196	0.190
[ascorbic acid]/mM	0	0	0.102

methyl linoleate in the presence and absence of α-tocopherol and ascorbic acid. The prooxidant effect may be ascribed to the hydrogen atom abstraction from the hydroperoxides by tocopheroxyl radical which is formed spontaneously from tocopherol. The resulting peroxyl radical can induce free radical chain oxidation of polyunsaturated fatty acids. When a small amount of ascorbic acid is present, it reduces tocopheroxyl radical and suppresses the prooxidant activity of α-tocopherol.

As described above, the ascorbate-α-tocopherol interactions may be important *in vivo* as well as *in vitro.* The vitamin C radical may be enzymatically reduced back to vitamin C by NADH-dependent systems. These interactions may account for the facts that vitamin E level is maintained *in vivo* and that vitamin E deficiency has seldom been observed clinically in man.

$LO_2^{\bullet} + LH \longrightarrow$ induce chain peroxidation

SUMMARY

Vitamins C and E function as water-soluble and lipid-soluble chain-breaking antioxidants, respectively, and protect lipids, proteins, and membranes from oxidative damage. Vitamin C scavenges oxygen radicals in the aqueous phase, whereas vitamin E scavenges oxygen radicals within the membranes. Vitamin C regenerates vitamin E by reducing vitamin E radicals formed when vitamin E scavenges the oxygen radicals. This interaction between vitamin C and vitamin E radicals can take place not only in homogeneous solutions but also in liposomal membrane systems where vitamins C and E reside separately outside and within the membranes respectively, and vitamin C can act as a synergist.

REFERENCES

1. PRYOR, W. A., Ed. 1976-1984. Free Radicals in Biology, Vols. I-VI. Academic Press. New York.
2. BURTON, G. W. & K. U. INGOLD. 1986. Vitamin E: Application of the Principles of Physical Organic Chemistry to the Exploration of Its Structure and Function. Acc. Chem. Res. **19:** 194-201.
3. GOLUMBIC, C. & H. A. MATTILL. 1941. Antioxidants and the autoxidation of fats. XIII. The antioxygenic action of ascorbic acid in association with tocopherols, hydroquinones and related compounds. J. Am. Chem. Soc. **63:** 1279-1280.
4. TAPPEL, A. L. 1968. Will antioxidant nutrients slow aging processes? Geriatrics **23:** 97-105.
5. YAMAMOTO, Y., E. NIKI & Y. KAMIYA. 1982. Oxidation of methyl linoleate in solution. Lipids **17:** 870-877.
6. NIKI, E., J. TSUCHIYA, Y. YOSHIKAWA, Y. YAMAMOTO & Y. KAMIYA. 1986. Antioxidant activities of α-, β-, γ-, and δ- tocopherols. Bull. Chem. Soc. Jpn. **59:** 497-501.
7. NIKI, E., T. SAITO & Y. KAMIYA. 1983. The role of vitamin C as an antioxidant. Chem. Lett. 631-632.
8. NIKI, E., T. SAITO, A. KAWAKAMI & Y. KAMIYA. 1984. Inhibition of oxidation of methyl linoleate in solution by vitamin E and vitamin C. J. Biol. Chem. **259:** 4177-4182.
9. PACKER, J. E., T. F. SLATER & R. L. WILLSON. 1979. Direct observation of a free radical interaction between vitamin E and vitamin C. Nature **278:** 737-738.

10. NIKI, E., J. TSUCHIYA, R. TANIMURA & Y. KAMIYA. 1982. Regeneration of vitamin E from α-chromanoxy radical by glutathione and vitamin C. Chem. Lett. 789-792.
11. NIKI, E., A. KAWAKAMI, Y. YAMAMOTO & Y. KAMIYA. 1985. Synergistic inhibition of oxidation of soybean phosphatidylcholine liposomes in aqueous dispersion by vitamin E and vitamin C. Bull. Chem. Soc. Jpn. **58:** 1971-1975.
12. DOBA, T., G. W. BURTON & K. U. INGOLD. 1985. Antioxidant and co-antioxidant activity of vitamin C. The effect of vitamin C, either alone or in the presence of vitamin E or a water-soluble vitamin E analogue, upon the peroxidation of aqueous multilamellar phospholipid liposomes. Biochim. Biophys. Acta **835:** 298-303.
13. LEUNG, H. W., M. J. VANG & R. D. MAVIS. 1981. The cooperative interaction between vitamin E and vitamin C in suppression of peroxidation of membrane phospholipids. Biochim. Biophys. Acta **664:** 266-272.
14. SCARPA, M., A. RIGO, M. MARIORINO, F. URSINI & C. GREGOLIN. 1984. Formation of α-tocopherol radical and recycling of α-tocopherol by ascorbate during peroxidation of phosphatidylcholine liposomes. An electron spin resonance study. Biochim. Biophys. Acta **801:** 215-219.
15. BARCLAY, L. R. C., S. J. LOCKE & J. M. MACNEIL. 1983. The autoxidation of unsaturated lipids in micelles. Synergism of inhibitors vitamins C and E. Can. J. Chem. **61:** 1288-1290.
16. BARCLAY, L. R. C., S. J. LOCKE & J. M. MACNEIL. 1985. Synergism of vitamin C with lipid soluble vitamin E and water soluble Trolox. Can. J. Chem. **63:** 366-374.
17. TSUCHIYA, J., T. YAMADA, E. NIKI & Y. KAMIYA. 1985. Interaction of galvinoxyl radical with ascorbic acid, cysteine, and glutathione in homogeneous solution and in aqueous dispersions. Bull. Chem. Soc. Jpn. **58:** 326-330.
18. KANNO, C., K. YAMAUCHI & T. TSUGO. 1970. Antioxidant effect of tocopherols on autoxidation of milk fat. II. Antioxidant activity of tocopherols in the churned and the solvent-extracted milk fat. Agric. Biol. Chem. **34:** 886-890.
19. CILLARD, J. & P. CILLARD. 1980. Behavior of alpha, gamma, and delta tocopherols with linoleic acid in aqueous media. J. Am. Oil Chem. Soc. **57:** 39-42.
20. CILLARD, J., P. CILLARD & M. CORMIER. 1980. α-Tocopherol prooxidant effect in aqueous media: Increased autoxidation rate of linoleic acid. J. Am. Oil Chem. Soc. **57:** 252-255.
21. CILLARD, J., P. CILLARD & M. CORMIER. 1980. Effect of experimental factors on the prooxidant behavior of α-tocopherol. J. Am. Chem. Soc. **57:** 255-261.
22. BAZIN, B., J. CILLARD, J.-P. KOSKAS & P. CILLARD. 1984. Arachidonic acid autoxidation in an aqueous media. Effect of α-tocopherol, cysteine, and nucleic acids. J. Am. Oil Chem. Soc. **61:** 1212-1215.
23. TERAO, J. & S. MATSUSHITA. 1986. The peroxidizing effect of α-tocopherol on autoxidation of methyl linoleate in bulk phase. Lipids **21:** 255-260.

DISCUSSION OF THE PAPER

C. K. CHOW (*University of Kentucky, Lexington, Ky.*): Congratulations for doing such nice work. Your work really helps to explain the interrelationship between vitamin E and vitamin C. I would like to ask about another factor called glutathione-dependent protein, which is also involved in the tocopherol regeneration.

E. NIKI (*University of Tokyo, Tokyo, Japan*): I don't think I have seen good data that strongly suggest the presence of something which regenerates vitamin E *in vivo* other than ascorbic acid.

E. GINTER (*Institute of Human Nutrition Research, Bratislava, Czechoslovakia*): I personally believe that such interactions between vitamin C and vitamin E exist *in vivo* also. When we have guinea pigs with marginal vitamin C deficiency they consume the food normally like a control, the same amount of food and the same food. That means they have the same consumption of vitamin E, and we have repeatedly shown that such marginally deficient guinea pigs have lower levels of α-tocopherol in their tissues, especially in oxygenated tissues, for instance, lungs, heart, and also liver.

On the other hand, there was no difference in vitamin E level in testes and in brain. Therefore, it is very probable that such interaction exists.

E. NIKI: Yes, I know the kind of nutritional data you are referring to. Thank you.

B. LANE (*Columbia University School of Public Health, New York, N.Y.*): You have noted that vitamin C is consumed first in the presence of tocopherol. I believe it was about 4 years ago (1982) at the New York Academy of Sciences symposium on vitamin E that Bhuyan *et al.*[a] showed significant enhanced levels of eye vitreous and aqueous ascorbate in rabbits with supplemented vitamin E (after both groups were challenged with 3-aminotriazole).

E. NIKI: Well, I have no comment.

A. BENDICH (*Hoffmann-La Roche, Nutley, N.J.*): Is there any evidence in your liposome models that ascorbate free radical could cause lipid peroxidation?

E. NIKI: In our system we did not use any iron and as far as I know ascorbic acid can act as a prooxidant only in the presence of metal.

A. B. KITABCHI (*University of Tennessee, Memphis, Tenn.*): Congratulations. One of the most exciting areas of your work is the application of your hypothesis to the ideal model system that's present in nature, namely, isolated adrenal cells which contain both high concentrations of vitamin E and high concentrations of vitamin C. Four years ago in a symposium on vitamin E it was reported that in a small dose ascorbate inhibits steroid synthesis in vitamin-E-deficient rat adrenal. However, at the higher concentration ascorbate has a bimodal effect and it works as an antioxidant.

So ascorbate in a low concentration does act as a prooxidant because there's always enough metal there but at the very high concentration it works exactly as you have suggested, synergizing or potentiating.

L. MACHLIN (*Hoffmann-La Roche, Nutley, N.J.*): I think this work is extremely important and it does explain the difficulty in the past of seeing *in vivo* a decrease in tocopherol in tissues that are exposed to high oxygen levels. Generally it has been hard to show either a decrease in vitamin E or the appearance of an oxidation product and the fact that the tocopherol radical can be regenerated by C goes a long way to explain that, particularly since vitamin C radical can be easily reduced back to ascorbic acid so you have quite a complete cycle that can keep the process going.

In some of our own work we have confirmed the work mentioned by Dr. Ginter, that in a guinea pig when you have low levels of C then you can start seeing a reduction in vitamin E levels in certain tissues.

[a]Bhuyan, K. C. *et al.* Ann. N. Y. Acad. Sci. **393:** 169-171 (1982), especially pp. 170 and 171 and Table 1 on p. 171, an excerpt of which was as follows:

Ascorbate (mM)	Preexposure	No Exposure	α-Tocopherol*
Vitreous	0.41 ± 0.07	0.17 ± 0.05	0.42 ± 0.12
Aqueous	1.17 ± 0.14	0.38 ± 0.03	1.01 ± 0.35

* In a dose of 50 mg/kg body weight administered intramuscularly daily for 2-6 weeks.

Uptake of Ascorbic Acid by Leukocytes

ULRICH MOSER

Department of Vitamin and Nutrition Research
F. Hoffmann-La Roche & Co., Ltd.
CH-4002 Basel, Switzerland

UPTAKE OF ASCORBIC ACID BY LEUKOCYTES

Ascorbic acid (AA) is accumulated by various tissues such as adrenal glands, pituitary gland, eye lens, as well as by leukocytes (TABLE 1).[1,2] In order to achieve a concentration of over 2 mM as it is found in these organs, AA has to be transported against a concentration gradient. If cellular functions depend on the homeostasis of AA, the characteristic of the transport could be used to derive the optimal daily intake from the plasma concentration that ensures a sufficient uptake in various cells.[3]

The transport of AA has been extensively studied in adrenocortical cells and in adrenal chromaffin cells.[4-8] Both cell types exhibit saturable transport kinetics with K_m values in the range of 17-30 μM. This is within physiological limits of plasma concentrations. In one case, a K_m of 103 μM was found, but this was probably due to experimental conditions.[6] Furthermore, the transport is found to be dependent on temperature, calcium, potassium, and sodium.

In human granulocytes and mononuclear cells dehydroascorbic acid (DHAA) and glucose diffusion are facilitated by a common transporter. 2-Deoxy-D-glucose and *3-O*-methyl-D-glucose competitively inhibit the transport of DHAA, which shows transport kinetics similar to the glucose analogues, the K_m being in the range of 2.0-3.2 mM.[9] Leukocytes isolated from diabetic patients exhibit a reduced transport rate of DHAA, *i.e., V_{max}* is lower than in normal cells.[10] In mononuclear leukocytes the AA level could be reduced by the administration of an intravenous glucose load.[11] By using ascorbate as substrate in human granulocytes, an inhibition of the transport by glucose could be found as well, however, the transport did not show a simple saturation kinetic.[12] Activation of human neutrophils with either opsonized zymosan or phorbol myristate acetate leads to an oxidation of 30-40% of AA to DHAA, but no AA is lost. Two hours after activation with *N*-formyl-L-methionyl-L-leucyl-L-phenylalanine, on the other hand, 20% of total AA is lost but not oxidized.[13] The aim of this paper will be to further elucidate the mechanism by which ascorbic acid is transported into leukocytes with respect to stereospecificity and to organ specificity.

METHODS

Preparation of Leukocytes

Human blood was obtained from male and female volunteers. Leukocytes have been isolated using dextran (2%) sedimentation followed by density centrifugation with percoll (d = 1.060 for mononuclear cells, d = 1.076 for polymorphonuclear cells [PMN]). Remaining erythrocytes have been removed by lysis in a hypotonic medium.[14,15]

TABLE 1. Concentration of Ascorbic Acid in Human Tissues

Tissue	Concentration (mM)	Ref.
Plasma	0.045 ± 0.022	2
Erythrocytes	0.043 ± 0.016	2
Platelets	1.9 ± 0.8	2
Granulocytes	1.2 ± 0.3	2
Mononuclear leukocytes	3.8 ± 1.0	2
Adrenals	1.9 – 2.5	1
Pituitary gland	2.5 – 3.2	1
Liver	0.6 – 1.0	1
Spleen	0.6 – 0.9	1
Lungs	0.4	1
Kidneys	0.3 – 0.9	1
Testes	0.2	1
Thyroid	0.1	1
Heart muscle	0.3 – 0.9	1
Skeletal muscle	0.2 – 0.3	1
Brain	0.8 – 0.9	1
Pancreas	0.6 – 0.9	1
Eye lens	1.6 – 2.0	1
Saliva	0.004 – 0.005	1

Preparation of Adrenocortical Cells

Porcine adrenals were obtained from the local slaughterhouse. Cortex cells were isolated according to Fredlund *et al.*[16] and Role and Perlman[17] by collagenase digestion and metrizamide (d = 1.079) centrifugation.

Measurement of Ascorbic Acid Uptake

Cells (1-5 $\times$ 10^6) were incubated in 1 ml Krebs-Ringer bicarbonate buffer containing 4% bovine serum albumin and 2-5 $\times$ 10^5 dpm [1-^{14}C]ascorbic acid in stoppered

Erlenmeyer flasks in a shaking water bath at 37°C for 30-45 minutes. The incubation was stopped on ice, the cells were washed, and the radioactivity was counted in INSTA-GEL after disruption of the cells in 1 N NaOH.

Measurement of Ascorbic Acid

Ascorbic acid has been measured spectroscopically using α,α'-bipyridyl[18] and ascorbate oxidase for the blank.[19]

Material

D-ascorbic acid has been prepared by G. Moine, F. Hoffmann-La Roche & Co., Ltd., Basel. All other chemicals were of purest available grade.

RESULTS

A net accumulation of ascorbic acid in cells could be the result of intracellular binding to membranes of proteins. In order to test this possibility, guinea pigs were fed either a normal diet containing 500 mg AA per kg diet (32 animals) or a marginal diet with 50 mg AA per kg diet (28 animals) for 4 weeks. The AA content was measured in the adrenal gland in the nuclear, microsomal, and mitochondrial fraction and in the cytosol. The distribution pattern of AA was expected to change in the marginally fed animals in favor of the binding compartment. The total AA content dropped from 1.28 mg AA/g tissue (control) to 0.19 mg AA/g (marginal). Ninety-two percent of AA (control) or 84% (marginal) could be found in the ultrafiltrate of the cytosol (FIG. 1). Furthermore, a tracer of [1-^{14}C]AA was given orally, which was supposed to label any binding sites, preferentially in the marginal group. Only insignificant amounts of labeled AA could be found in the different compartments of the adrenal glands (results not shown). Therefore it can be concluded that AA is dissolved in the cytosol and has to be transported against a concentration gradient to maintain an intracellular level in the millimolar range.

In subsequent experiments, the transport of AA was measured in isolated cells using L-[1-^{14}C]AA as marker. The results obtained from spectroscopic analysis of AA coincided well with the radiometric analysis.

In porcine adrenocortical cells, AA uptake exhibits a saturation kinetic according to the Michaelis-Menten relationship

$$v = \frac{V_{max} \times s}{K_m + s}, \tag{1}$$

with a K_m of 20.5 μM and a V_{max} of 7.3 nmol/10^8 cells $\times$ min (FIG. 2). These values are in good agreement with those found in bovine chromaffin cells where an apparent

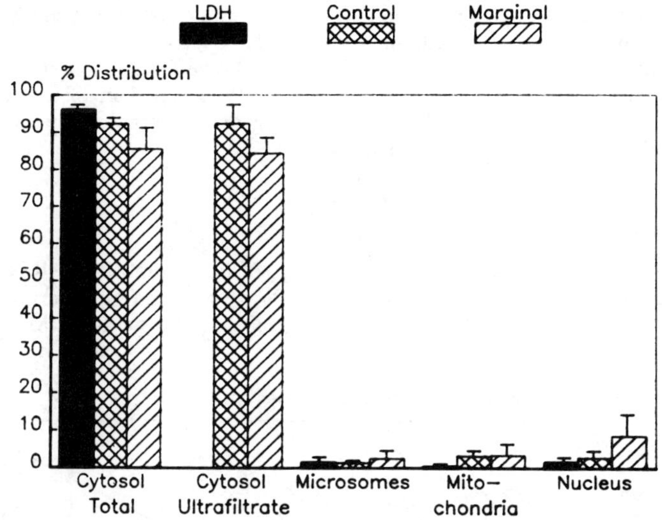

FIGURE 1. 32 guinea pigs (Himalayan spotted) were fed a normal diet containing 500 mg AA/kg diet and 28 guinea pigs received a marginal diet with 50 mg/kg for four weeks. Adrenal glands were removed, homogenized, and fractionated (600 g, 15,000 g, 100,000 g). The distribution of AA in each fraction was determined (mean ± SD) and compared with the distribution of a cytosolic marker, lactate dehydrogenase (LDH) (mean ± SD; $n = 5$). The cytosol was also passed through an ultrafilter to remove proteins.

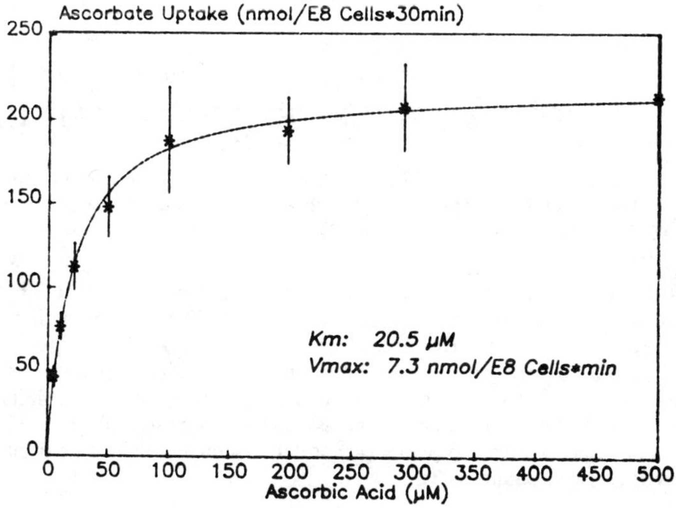

FIGURE 2. Adrenocortical cells (1–2 × 10^6) were prepared and incubated as described in METHODS in siliconized 10 ml Erlenmeyer flasks in 1 ml medium. After a preincubation of 15 minutes, [1-^{14}C]ascorbic acid (70,000–150,000 dpm, concentration as indicated) was added to the cells. The data (mean ± SEM, $n = 8$) were fitted according to equation **1** using the Marquardt iteration.

K_m of 29 μM and a V_{max} of 6.6 nmol/10^8 cells × min have been estimated,[7] and with the K_m of 16.6 μM found in adrenocortical cells.[4]

Human granulocytes and mononuclear cells, however, did not show a saturation even at extracellular concentrations up to 10 mM. A good fit of the data could be obtained assuming a complex kinetic consisting of a saturable active transport plus passive diffusion according to equation **2:**

$$v = \frac{V_{max} \times s}{K_m + s} + P \times s. \qquad (2)$$

For granulocytes a K_m of 38.7 μM, a V_{max} of 0.28 nmol/10^8 cells × min, and a permeability constant (P) of 1.75 μl/ 10^8 cells × min can be determined (FIG. 3). The corresponding values for mononuclear cells are K_m = 100 μM, V_{max} = 0.31

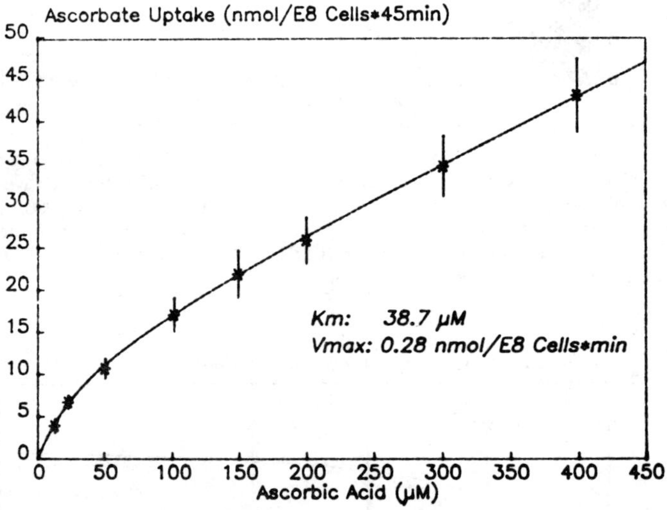

FIGURE 3. Granulocytes have been isolated and incubated as described in METHODS in the presence of 100,000–1,000,000 dpm [1-^{14}C]ascorbate for 45 minutes. The data (mean ± SEM, n = 6) were fitted according to equation **2** (see also FIG. 2).

nmol/10^8 cells × min, and P = 0.32 μl/10^8 cells × min respectively (FIG. 4). From these data it was calculated that in the range of 25-50 μM AA, which is a normal plasma concentration, the active transport contributes 64-72% in granulocytes and 87-89% in mononuclear cells of AA uptake respectively. The uptake is not dependent on sodium, potassium, or calcium (not shown). However, glucose inhibits the uptake (FIG. 5), the K_i being 3.7 mM as can be estimated from the Hill plot (FIG. 5, inset). The data can best be fitted to a kinetic model, which involves an inhibition of V_{max} as well as P (FIG. 6, equation **3**).

$$v = \frac{V_{max} \times s}{(K_m + s)(1 + i/K_i)} + \frac{P \times s}{1 + i/K_i} \qquad (3)$$

Similar results have been reported with dehydroascorbate as substrate in neutrophils[9] and red blood cells.[20] Phlorizin, a potent inhibitor of glucose transport, inhibits

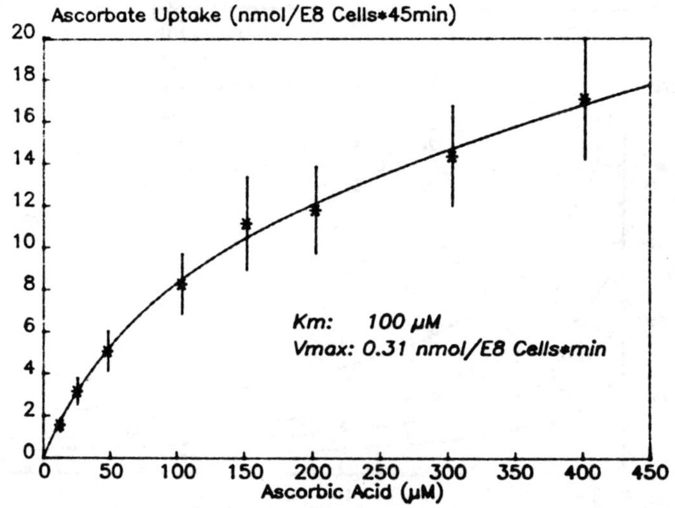

FIGURE 4. See legend to FIGURE 3. Data points are mean ± SEM (n = 8).

the ascorbate transport as well as glucose uptake in neutrophils. The inhibition by glucose is specific since fructose and galactose are without effect.[21]

The chemotactic activity of neutrophils can be stimulated by *N*-formyl peptides such as *N*-formyl-L-methionyl-L-leucyl-L-phenylalanine (FMLP).[22] Oxidizing agents like the myeloperoxidase/ H_2O_2/ halide system completely block the movement of

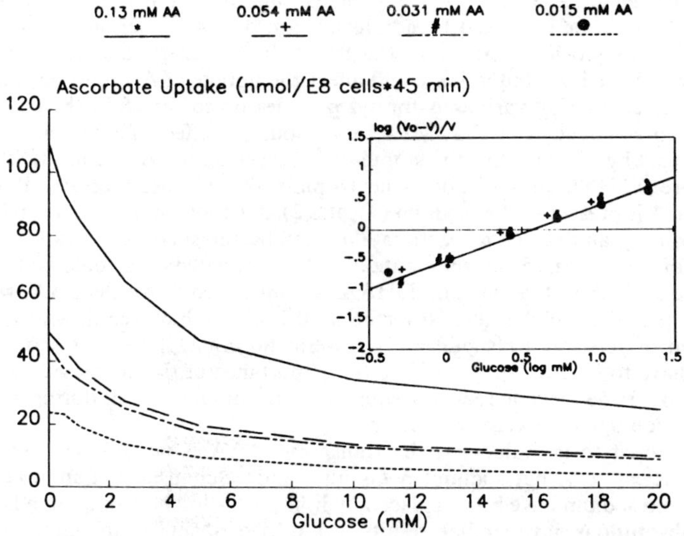

FIGURE 5. Granulocytes have been incubated in the presence of 0.015 mM, 0.031 mM, 0.054 mM, and 0.13 mM [1-^{14}C]AA plus varying amounts of glucose. From the Hillplot *(inset),* a K_i of 3.7 mM and a slope of 0.91 can be calculated.

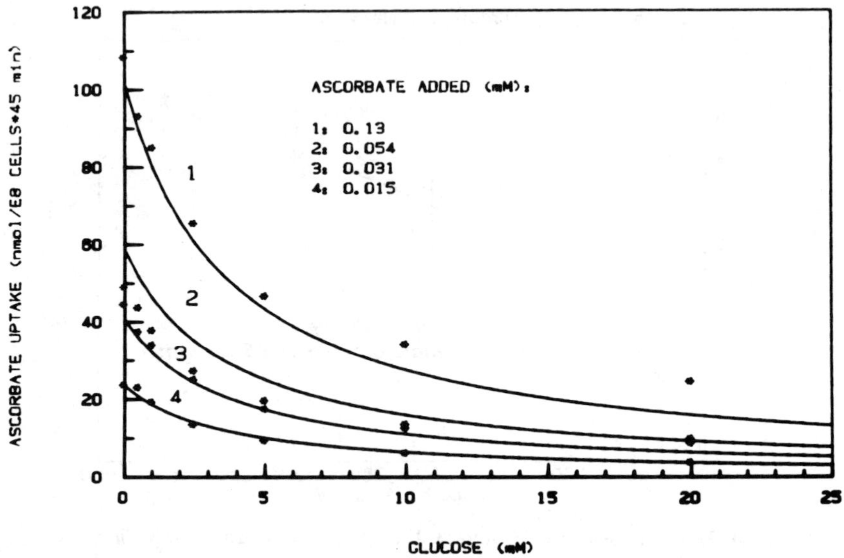

FIGURE 6. See legend to FIGURE 5. Kinetic data obtained in FIGURE 3 and the K_i from FIGURE 5 have been used in equation **3** to verify the type of inhibition of the AA uptake in granulocytes by glucose.

granulocytes, which can be reversed by reducing agents, *e.g.,* AA.[23] Impaired chemotaxis in premature infants[24] or in patients suffering from chronic granulomatous disease[25,26] could be improved by administration of AA, which was accompanied by increased staphylocidal activity in the latter. If AA exerts a function in activated granulocytes, such a stimulation could affect the uptake of AA. Indeed, the transport of AA is increased by various *N*-formyl peptides up to threefold (FIG. 7). The non-formylated analogue of FMLP, MLP, is without any effect. The uptake of glucose is not changed by FMLP, thus the stimulation is specific for AA (FIG. 8). The calcium ionophore A23187, an activator of neutrophils as well, also increases the uptake of AA if Ca^{2+} is present in the medium (TABLE 2). A similar stimulation can be obtained by exposing granulocytes to bacteria-like acetobacter suboxydans (not shown). One reason for the increased transport rate of AA in activated cells could be the oxidation of AA to DHAA in the medium. DHAA is transported faster than AA, which could lead to the higher uptake rate. Nevertheless 0.1 mM dithiothreitol, which inhibits the oxidation of AA, only insignificantly lowers the AA uptake (not shown). Further studies have to be done to investigate the importance of this increased uptake. There is also a higher demand on reducing equivalents in stimulated cells during inflammation as protection against oxygen radical damage.[27]

In contrast to granulocytes, the transport of AA is not inhibited by glucose in adrenocortical cells, but sodium, potassium, and calcium are required (FIG. 9). Potassium and sodium have been replaced by lithium or choline to maintain isosmolarity. Both substitutions gave similar results; 0.1 mM ouabain, an inhibitor of potassium transport, inhibits the AA uptake in adrenals, but not in granulocytes. Conversely, the transport is temperature dependent in granulocytes and adrenocortical cells with a maximum around 40°C and a negligible uptake below 20°C. From the Arrhenius

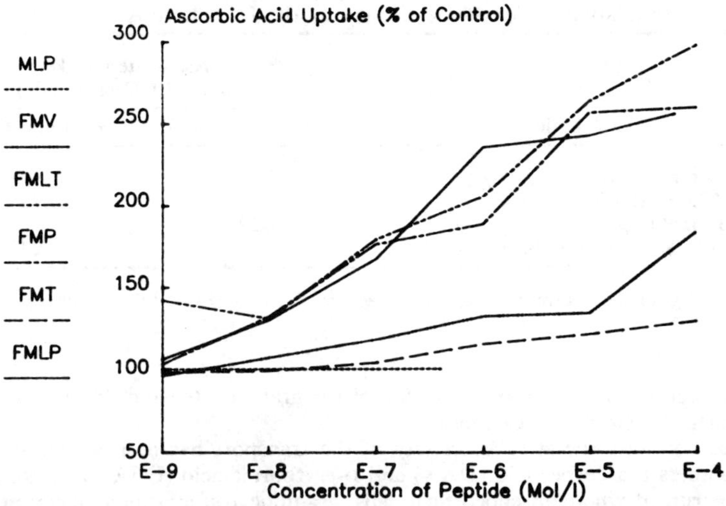

FIGURE 7. Granulocytes have been incubated in 1 ml medium containing 0.5% bovine serum albumin, 115 μM [1-^{14}C]AA, and the peptides as indicated dissolved in DMSO. *FMLP, N*-formyl-L-methionyl-L-leucyl-L-phenylalanine; *MLP,* L-methionyl-L-leucyl-L-phenylalanine; *FMT, N*-formyl-L-methionyl-L-tryptophan; *FMP, N*-formyl-L-methionyl-L-phenylalanine; *FMLT, N*-formyl-L-methionyl-L-leucyl-L-tyrosine, *FMV, N*-formyl-L-methionyl-L-valine.

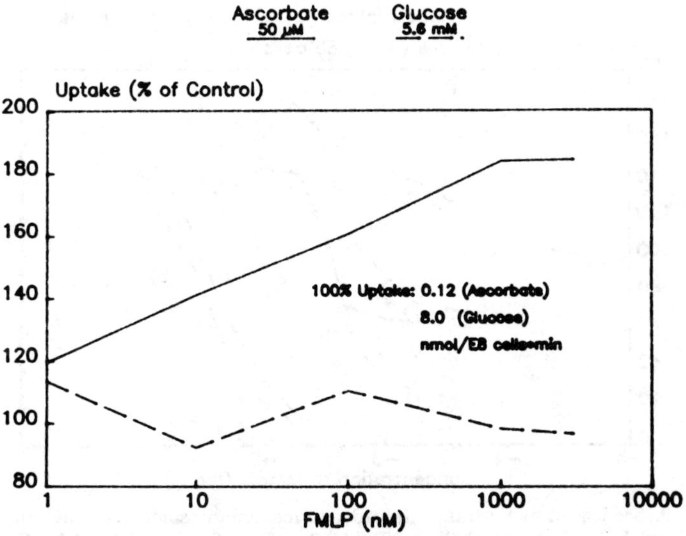

FIGURE 8. See legend to FIGURE 7. Glucose uptake was measured using 5.6 mM [U-^{14}C]glucose (700,000 dpm) and the AA transport with 50 μM [1^{14}C]AA (900,000 dpm).

TABLE 2. Stimulation of Ascorbate Uptake by the Ionophore A23187

Addition	Ascorbate Uptake (nmol/10^8 Cells × min) $-Ca^{2+}$	$+1.3$ mM Ca^{2+}
Control	0.27	0.40
A23187 (5 μM)	0.34	2.33
FMLP (1 μM)	0.58	1.16
A23187 (5 μM) + FMLP (1 μM)	0.54	2.27

5×10^6 granulocytes were incubated in presence of 0.28 mM L-[1-^{14}C]ascorbate plus additions as indicated.

plot an activation energy of 25 kcal/mol for granulocytes and 28 kcal/mol for adrenocortical cells can be calculated.

Furthermore, the stereospecificity of the transport has been analyzed using the two isomers D-ascorbic acid (DAA) and D-erythorbic acid (DEA). FIGURE 10 shows an experiment where adrenocortical cells, granulocytes, and mononuclear cells have been incubated in the presence of various concentrations of the isomers. The uptake rate of the highest concentration of AA has been set to 100% in each cell type. The isomers have been measured spectroscopically with the same method as AA, *i.e.*, it could not be distinguished between the isomers and AA. Therefore, the AA content was measured separately in a control sample and deducted from the value obtained in the experimental sample after the incubation. Only DAA and DEA could be found in significant amounts in adrenocortical cells when given at a concentration of 200

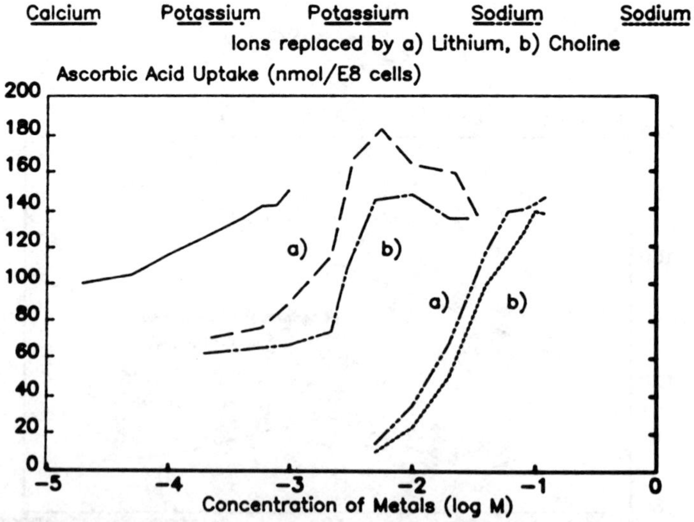

FIGURE 9. See legend to FIGURE 2. In order to maintain isosmolarity, NaCl resp. KCl have been replaced by LiCl *(a)* or choline chloride *(b)*. $NaHCO_3$ was replaced by $KHCO_3$ in the experiment with varying sodium concentrations. The uptake rate of AA in a medium lacking Ca^{2+}, resp. K^+, resp. Na^+ was 63.6%, 41.8%, and 1.7% of the rate measured in the standard medium.

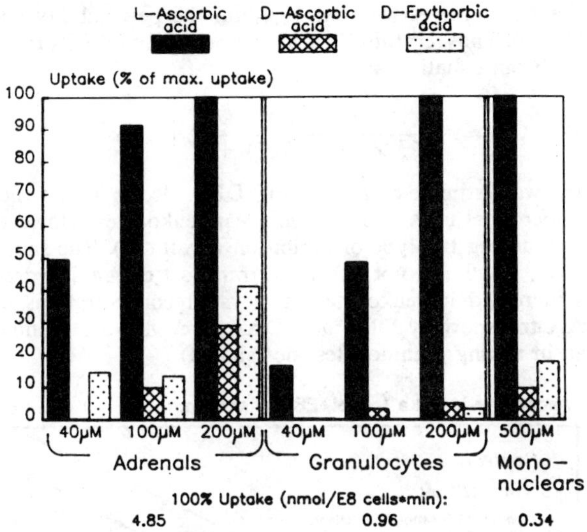

FIGURE 10. Adrenocortical cells and granulocytes have been incubated with 0 mM, 0.04 mM, 0.1 mM, and 0.2 mM of L-ascorbate, D-ascorbate, and D-erythorbate, mononuclear leukocytes with 0 mM and 0.5 mM of the isomers. Intracellular levels of the isomers have been measured by the spectroscopic method. The value obtained in the control sample (no additions) has been deducted from all determinations.

μM. The measurements of the isomers in granulocytes and mononuclears were within the limit of the analytical method. DAA and DEA inhibit the uptake of AA in both types of leukocytes (FIGS. 11 and 12). By calculating K_m, V_{max}, and P of the AA uptake from experiments with different concentrations of DEA, it is apparent that in

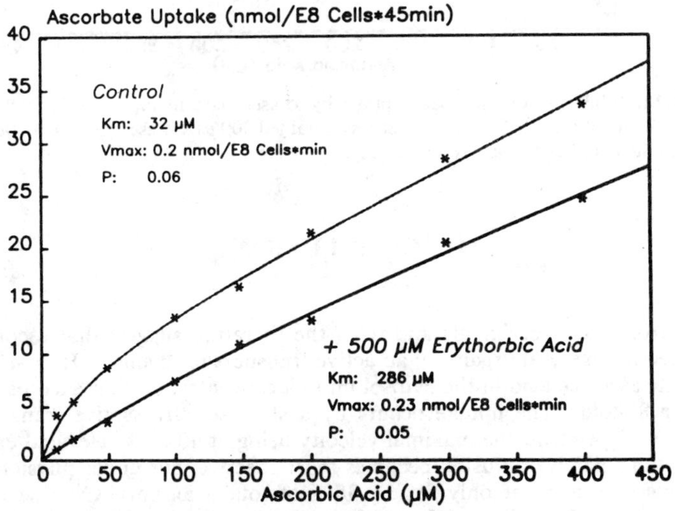

FIGURE 11. Inhibition of ascorbate uptake by erythorbate in granulocytes. See legend to FIGURE 3. In addition, 500 μM D-erythorbate was present in all samples of the lower curve. Kinetic parameters have been determined using equation 2.

granulocytes the K_m increases whereas V_{max} and the permeability coefficient remain unchanged (FIG. 13). Thus, erythorbic acid competitively inhibits the transport of AA in granulocytes. From equation **4**:

$$v = \frac{V_{max} \times s}{K_m \times (1 + i/K_i) + s} + P \times s, \qquad \textbf{(4)}$$

a K_i of 51.3 μM was estimated. In addition, DEA also increases the K_m of the AA uptake in adrenocortical cells and mononuclear leukocytes. However, the data are not conclusive regarding the type of inhibition (TABLE 3). The same is true for the influence of DAA, which did not affect the transport of AA in adrenocortical cells, but the K_m is increased in leukocytes at elevated concentrations (TABLE 4). The inhibition of AA transport by DEA and DAA is even more pronounced in FMLP stimulated than in resting granulocytes (not shown).

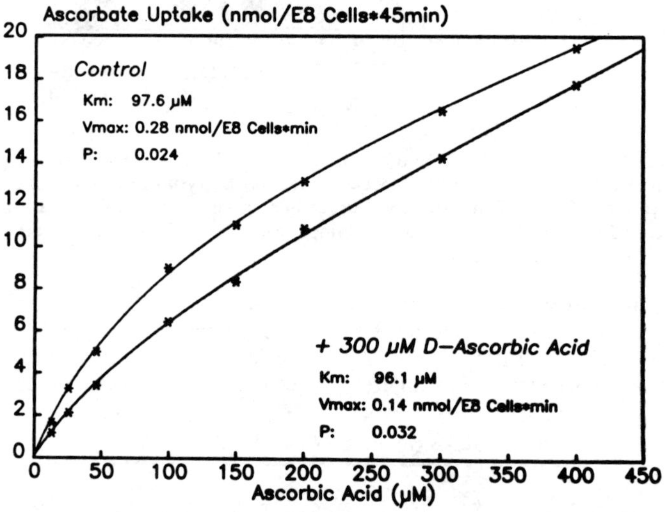

FIGURE 12. Inhibition of ascorbate uptake by D-ascorbate in mononuclears. See legend to FIGURE 4. All incubations of the lower curves contained 300 μM D-ascorbate. Kinetic parameters have been calculated according to equation **2**.

CONCLUSION

Data from our experiments and from the literature suggest that ascorbic acid is transported into several organs by an active transport mechanism. Human leukocytes accumulate ascorbic acid in the cytosol up to a concentration that exceeds the plasma level over 80-fold.[2] The uptake occurs by a stereoselective active transport that is temperature dependent, the maximal velcoity being at 40°C. At elevated extracellular concentrations passive diffusion becomes most effective, but under physiological conditions this may account only for 15-35% of total transport. Glucose inhibits the uptake in neutrophils (K_i = 3.7 mM) but not in adrenocortical cells. The glucose analogues 2-deoxy-D-glucose and 3-*O*-methyl-D-glucose inhibit the transport of de-

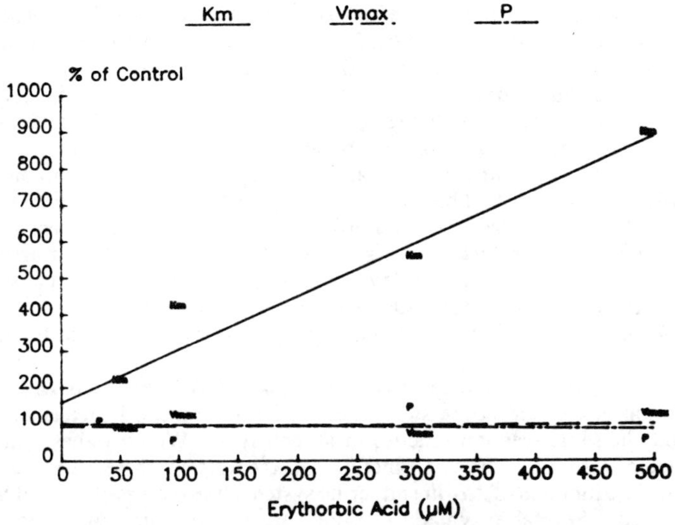

FIGURE 13. Effect of erythorbate on ascorbate uptake in granulocytes. Kinetic parameters obtained from experiments as shown in FIGURE 11 are expressed as percentage of the values measured in a control experiment, *i.e.*, without erythorbate (FIG. 3).

hydroascorbic acid in neutrophils.[9] The addition of glucose impairs the uptake of dehydroascorbate in erythrocytes[20] too, but there is also a glucose-independent uptake.[28] The transport of AA in rat alveolar macrophages is saturable and sodium dependent,[29] as in adrenal cells.[4-8] In granulocytes, however, only the increased uptake of AA is dependent on a cation, Ca^{2+}, during stimulation with *N*-formyl peptides, ionophore A28187, or bacteria. Sodium, potassium, and calcium are without effect in resting cells.

TABLE 3. Effect of D-Erythorbic Acid and D-Ascorbic Acid on Kinetic Parameters of Ascorbate Uptake in Adrenals and Mononuclear Leukocytes

Isomer	Conc. (mM)	Cell Type	K_m	V_{max} (% of Control)	P
D-Erythorbic Acid	0.1	adrenals	148	100	—
	0.5		201	97	—
	0.1	mononuclears	99	77	94
	0.3		111	71	87
	0.5		409	126	46
D-Ascorbic Acid	0.2	adrenals	114	94	—
	0.1	mononuclears	47	41	151
	0.3		99	49	135
	0.5		38	26	174

Adrenocortical cells and mononuclear leukocytes have been incubated as mentioned in METHODS. The parameters have been calculated according to equation **1** (adrenals) or equation **2** (mononuclear cells). *P*: Permeability constant.

The two isomers D-ascorbic acid and D-erythorbic acid are poor substrates of the carrier; only trace amounts have been found in leukocytes after an incubation period of 45 minutes. In adrenocortical cells the uptake of the isomers occurs at elevated concentrations in the medium at a 60–70% lower rate than AA. Both isomers inhibit the transport of AA into leukocytes; in adrenal cells only D-erythorbic acid acts as an inhibitor. D-Erythorbic acid increases the K_m of the uptake of AA in granulocytes without changing V_{max} and the permeability constant P, thus the inhibition is competitive with regard to AA. These results are in good agreement with an earlier study on the distribution of labeled D-erythorbic acid in guinea pigs, where it was found that the isomer is transported into tissues less effectively than AA.[1] Furthermore, D-erythorbic acid inhibits the transport of AA in brush border membrane vesicles from guinea pig small intestine[30] (K_i = 20 mM) and lowers the AA content in liver, adrenal gland, spleen, and kidney of guinea pigs fed 100 mg erythorbic acid in addition to 5 mg AA.[31]

It can therefore be concluded that the uptake of AA into cells with a high AA concentration is mediated by a stereospecific, active transport mechanism that does not exhibit the same reaction kinetics in all cell types. Whether these variations can be correlated with the different functions AA exerts in these cells remains to be elucidated. The function of AA in endocrine systems like the adrenal gland has recently been reviewed.[32] Several enzymatic reactions are known where AA acts as cofactor, like dopamine β-hydroxylase and peptide amidation. In leukocytes several functions have been described to be dependent on AA: stimulation of chemotactic migration,[23] inhibition of 3-hydroxy-3-methylglutaryl coenzyme A reductase activity,[33] and stimulation of the hexose monophosphate shunt (HMP).[34] No differences could be found between control and scorbutic guinea pigs regarding H_2O_2 production and killing of *Staphylococcus aureus.*[35] Increasing the intracellular level of AA in neutrophils by preincubation with dehydroascorbate only slightly enhanced the HMP shunt (TABLE 5). A possible site of action of AA could be the regulation of the elastase activity, since the α-proteinase inhibitor is inactivated by oxidation of methionine residues in or near the active site, which increases the activity of the elastase.[36] In addition, AA as well as aqueous humor inhibit the myeloperoxidase extracted from inflamed rabbit iris-ciliary bodies, indicating a function as antiinflammatory agent.[37] Furthermore, AA inhibits platelet lipoxygenase[38] and corneal lipoxygenase,[39] and regulates neutrophil 5-lipoxygenase.[40] These interactions of AA with the prostaglandin metabolism emphasize the role of AA as immunomodulator. Thus population groups with known lower AA levels in neutrophils such as chronic hemodialysis patients,[41,42] psychogeriatric patients,[43] institutionalized elderly,[44] pregnant women, neonates, and their mothers[45] may be at increased risks of infections and inflammation.

TABLE 4. Effect of D-Ascorbate on Kinetic Parameters of L-Ascorbate Uptake in Granulocytes

	Control	+300 μM D-Ascorbate
K_m (μM)	23.0	353.8
V_{max} (nmol/10^8 cells × min)	0.14	0.23
P (μl/10^8 cells × min)	1.59	1.35

4 × 10^6 granulocytes have been incubated in presence of 10–400 μM L-[1-^{14}C]ascorbate ± 300 μM D-ascorbate. The curves have been fitted according to equation **2.**

TABLE 5. Stimulation of Glucose Oxidation by Ascorbic Acid

	$[^{14}C]CO_2$ (dpm) Derived from			
	$[1\text{-}^{14}C]$Glucose		$[6\text{-}^{14}C]$Glucose	
AA in Cells (μg/10^8) / Zymosan (μg)	12.9	82.3	9.2	56.8
0	1337	2178	10	16
7	1465	2369	11	23
70	4704	5803	—	—
700	8799	11,105	95	75

3×10^6 cells were preincubated with or without 0.5 mM dehydroascorbate for 10 minutes, washed, and reincubated for 30 minutes in the presence of 100,000 dpm $[1\text{-}^{14}C]$glucose or $[6\text{-}^{14}C]$glucose and zymosan as indicated.

REFERENCES

1. Hornig, D. 1975. Ann. N.Y. Acad. Sci. **258:** 103-118.
2. Evans, R. M., L. Curie & A. Campbell. 1982. Br. J. Nutr. **47:** 473-482.
3. Levine, M. 1986. N. Engl. J. Med. **314:** 892-902.
4. Finn, F. M. & P. A. Johns. 1980. Endocrinology **106:** 811-817.
5. Moser, U. & F. Weber. 1984. Experientia **40:** 610.
6. Levine, M. A. & H. B. Pollard. 1983. FEBS Lett. **158:** 134-138.
7. Diliberto, E. J., Jr., G. D. Heckman & A. J. Daniels. 1983. J. Biol. Chem. **258:** 12886-12894.
8. Daniels, A. J., G. Dean, O. H. Viveros & E. J. Diliberto, Jr. 1982. Science **216:** 737-739.
9. Bigley, R., M. Wirth, D. Layman, M. Riddle & L. Stankova. 1983. Diabetes **32:** 545-548.
10. Stankova, L., M. Riddle, J. Larned, K. Burry, D. Menashe, J. Hart & R. Bigley. 1984. Metabolism **33:** 347-353.
11. Chen, M. S., M. L. Hutchinson, R. E. Pecoraro, W. Y. L. Lee & R. F. Labbé. 1983. Diabetes **32:** 1078-1081.
12. Moser, U. & F. Weber. 1984. Int. J. Vit. Nutr. Res. **54:** 47-53
13. Winterbourn, Ch. C. & M. C. M. Vissers. 1983. Biochim. Biophys. Acta **763:** 175-179.
14. Hjorth, R., A. K. Jonsson & P. Vretblad. 1981. J. Immunol. Methods **43:** 95-101.
15. Böyum, A. 1968. Scand. J. Clin. Lab. Invest. **21**(Suppl. 97): 77-89.
16. Fredlund, P., S. Saltman & K. J. Catt. 1975. Endocrinology **97:** 1577-1586.
17. Role, L. W. & R. L. Perlman. 1980. J. Neurosci. Methods **2:** 253-265.
18. Zannoni, V., M. Lynch, S. Goldstein & P. Sato. 1974. Biochem. Med. **11:** 41-48.
19. Beutler, H. O. & G. Beinstingl. 1980. Dtsch. Lebensm.-Rdsch. **76:** 69-75.
20. Mann, G. V. & P. Newton. 1975. Ann. N.Y. Acad. Sci. **258:** 243-252.
21. Schmidt, K. & U. Moser. 1985. Int. J. Vit. Nutr. Res. Suppl. **27:** 363-379.
22. Schiffmann, E., B. A. Cocoran & S. M. Wahl. 1975. Proc. Natl. Scad. Sci. U.S.A. **72:** 1059-1062.
23. Anderson, R. & P. T. Jones. 1982. Clin. Exp. Immunol. **47:** 487-496.
24. Chirico, G., A. Chiara, A. Lanfranchi, P. Malfa, D. Rondena, A. G. Ugazio & G. Rondini. 1985. Biol. Neonate **48:** 368.
25. Anderson, R. 1981. Clin. Exp. Immunol. **43:** 180-188.

26. PATRONE, F., F. DALLEGRI, E. BONVINI, F. MINERVINI & C. SACCHETTI. 1982. Acta Vitaminol. Enzymol. **4:** 163-168.
27. HEMILÄ, H., P. ROBERTS & M. WIKSTRÖM. 1984. FEBS Lett. **178:** 25-30.
28. BIANCHI, J. & R. C. ROSE. 1986. Proc. Soc. Exp. Biol. Med. **181:** 333-337.
29. CASTRANOVA, V., J. R. WRIGHT, H. D. COLBY & P. R. MILES. 1983. J. Appl. Physiol. Respirat. Environ. Exercise Physiol. **54:** 208-214.
30. SILIPRANDI, L. P. VANNI, M. KESSLER & G. SEMENZA. 1979. Biochim. Biophys. Acta **552:** 129-142.
31. ARAKAWA, N., E. SUZUKI, T. KURATA, M. OTSUKA & C. INAGAKI.. 1986. J. Nutr. Sci. Vitaminol. **32:** 171-181.
32. LEVINE, M. & K. MORITA. 1985. Vitam. Horm. **42:** 1-64.
33. HARWOOD JR., H. J., Y. J. GREENE & P. W. STACPOOLE. 1986. J. Biol. Chem. **261:** 7127-7135.
34. DE CHATELET, L. R., M. R. COOPER & C. E. MCCALL. 1972. Antimicrob. Agents Chemother. **1:** 12-16.
35. STANKOVA, L., N. B. GERHARDT, L. NAGEL & R. BIGLEY. 1975. Infect. Immun. **12:** 252-256.
36. CARP, H. & A. JANOFF. 1980. J. Clin. Invest. **66:** 987-995.
37. WILLIAMS, R. N., C. A. PATERSON, K. E. EAKINS & P. BHATTACHERJEE. 1984. Exp. Eye Res. **39:** 261-265.
38. CORDOVA, C., A. MUSCA, F. VIOLI, C. ALESSANDRI, A. GHISELLI & L. IULIANO. 1984. Prostaglandins **27**(Suppl): 103.
39. WILLIAMS, R. N. & C. A. PATERSON. 1986. Exp. Eye Res. **43:** 7-13.
40. STEINHILBER, D., U. MOSER, H.-J. ROTH & K. H. SCHMIDT. 1987. Interaction between ascorbate and the 5-lipoxygenase pathway of arachidonic acid in human PMNL. This volume.
41. SULLIVAN, J. F. & A. B. EISENSTEIN. 1972. J. Am. Med. Assoc. **222:** 1697-1699.
42. PÖNKÄ, A. & B. KUHLBÄCK. 1983. Acta Med. Scand. **213:** 305-307.
43. HONTELY, S., J. VOBECKY, D. SHAPCOTT & J. S. VOBECKY. 1983. Nutr. Rep. Int. **27:** 1101-1111.
44. SCHORAH, C. J. 1981. Vitamin C status in population groups. *In* Vitamin C (Ascorbic Acid). J. N. Counsell & D. H. Hornig, Eds.: 23-47. Applied Science. London.
45. COCCHI, P., A. RAVINA, U. PIERRO & M. PARADISO. 1985. Haematologica **70:** 263-265.

DISCUSSION OF THE PAPER

E. J. DILIBERTO (*Wellcome Research Labs, Research Triangle Park, N.C.*): Your distribution data would suggest that you were dealing mainly with the adrenal cortex.

U. MOSER (*Hoffmann-La Roche, Basel, Switzerland*): Yes, that's whole adrenal gland.

E. J. DILIBERTO: That's the whole adrenal gland which is predominantly adrenal cortex. Otherwise you would have seen more in the mitochondrial fraction.

Also with regard to your glucose inhibition data, do you know whether or not it may be due to the oxidation of ascorbic acid, and that in fact you are showing inhibition because of dehydroascorbic acid going through the glucose transporter?

U. MOSER: I don't know, we tried to measure the dehydroascorbic acid in several experiments and we never could find any.

E. J. DILIBERTO: But you didn't try that experiment that you did with the DTT to prevent oxidation of the ascorbic acid on the outside?

U. MOSER: No, we didn't do this experiment with glucose.

B. LADU (*University of Michigan Medical School, Ann Arbor, Mich.*): I understand that years ago it was reported that the absolute configuration of L-ascorbic acid and the D-glucose were supposed to be the same, and I'm wondering whether the glucose transport system is actually being employed here in the cells that you've studied; this has been proposed also for the absorption of L-ascorbic acid in the kidney. Would you have some comments on that?

U. MOSER: I doubt that the transport system is the same. We can fit the glucose inhibition to a model very well where only it's V_{max} and its diffusion is involved. That would mean that the glucose inhibits noncompetitively with regard to ascorbic acid. In adrenal cells there is no inhibition by glucose. Perhaps the dehydroascorbic acid may slip through like glucose, but I don't know whether it's true for the reduced form.

B. L. HORECKER (*Cornell University Medical College, New York, N.Y.*): In that connection I wonder if you tried 2-deoxyglucose, because that would give you a pretty clear picture if it were inhibited by the glucose transport system.

U. MOSER: Yes, the deoxyglucose gives the same result as glucose.

B. L. HORECKER: Then it might be the glucose transport.

R. C. ROSE (*Pennsylvania State University College of Medicine, Hershey, Pa.*): Was the uptake in the presence of an ionophore a saturable mechanism? You didn't show us data on that as opposed to simply busting open holes in the membranes.

U. MOSER: That is one thing we have to do now.

R. C. ROSE: Your early data on saturation in the adrenal gland was quite convincing, but then you went on to conclude that it was an active transport process and to really say that it's an active transport process one needs to demonstrate that the substrate moves against an electrochemical potential difference. Are you relying on other data in addition to saturation to propose a real active transport process?

U. MOSER: Of course this could be also facilitated diffusion, however the uptake is temperature dependent. From the temperature curve you can calculate an activation energy of about 30 kcal per mole.

R. C. ROSE: Facilitated diffusion can also be temperature dependent.

E. J. DILIBERTO: What was the K_i for inhibition of phlorizin?

U. MOSER: It's about the same as glucose.

E. J. DILIBERTO: So that means it's in the millimolar range. So that would be also suggestive that it would be the glucose transporter.

U. MOSER: That's what we concluded before, but I'm not so sure because in other tissues it is not the same.

B. TOLBERT (*University of Colorado, Boulder, Colo.*): In the literature there are approximately seven papers which say erythorbic acid will completely replace ascorbic acid at about 20 times its level, and there are another six papers which say that it won't, so it's a 50-50 result. Do you have any feeling as to whether sufficiently high levels of erythorbic acid can replace ascorbic acid in the guinea pig?

U. MOSER: Probably to some extent if you give really high doses. It may fulfill functions which depend on the reducing property only, if you have enough, but getting enough inside the cell would be a problem.

S. L. ROMNEY (*Albert Einstein College of Medicine, Bronx, N.Y.*): Is the leukocyte concentration of ascorbic acid a good indicator of tissue concentration?

U. MOSER: It certainly is, but it's very hard to measure. I think it is still better if you have the plasma level taken several times. It is certainly much easier than to have leukocyte measurements, because in leukocytes it depends very much on how well you separate the cells and then it depends also on the method you use to measure the ascorbic acid.

Effect of Vitamin C on Tubulin Tyrosinolation in Polymorphonuclear Leukocytes

JAYASREE NATH[a] AND JOHN I. GALLIN[b]

[a]*Department of Hematology*
Walter Reed Army Institute of Research
Washington, D. C. 20307-5100

[b]*Laboratory of Clinical Investigation*
National Institute of Allergy and Infectious Diseases
National Institutes of Health
Bethesda, Maryland 20892

INTRODUCTION

In recent years, studies from several laboratories have indicated significant alterations in the cytoskeletal organization of human polymorphonuclear leukocytes (PMN), coincident with its orientation or migration in response to chemotactic stimuli.[1–6] The ability of PMN to orient and migrate in gradients of chemotactic factors and to phagocytize foreign particles essentially depends on the cytoskeletal apparatus and its dynamic properties. During the past decade, efforts were directed toward the elucidation of two different, though related, problems: to identify the components of the contractile machinery for ameboid locomotion and phagocytosis, and to determine the role of the microtubule system in neutrophil activation, migration, phagocytosis, and degranulation. Although much remains to be learned about the precise role of microtubules, their integrity and specific organization is important for a number of neutrophil responses.[1–7] The function of the microtubule apparatus in directed migration is not completely understood. Studies by Malech *et al.*[1] and others[2–4] strongly suggest that migrating neutrophils have a polarized internal organization, with centrioles and associated microtubules located between the advancing pseudopod and nucleus. These studies provide evidence to indicate that the position of the centrosome and the organization of the microtubule system may be important for the directed migration of PMN[1–4] and for their secretory responses.[5–7] Recent high-voltage electron microscopic studies have revealed that upon activation of PMN, not only is there an increase in the total number of centriole-associated microtubules, there is also a substantial increase in the average length of microtubules.[4] Therefore, it may be assumed that activation not only results in a reorganization of the PMN microtubule system, but also induces changes in the total number of microtubules and the polymer state of tubulin.

Tubulin, a heterodimer consisting of α- and β-chains, is the major subunit protein of microtubules.[8] The α-chain of tubulin can be enzymatically modified by a reversible addition of a tyrosine residue at its carboxy terminus.[9–12] A tubulin-tyrosine ligase catalyzes the ATP-dependent addition of tyrosine to the carboxy terminus of tubulin α-chain.[9–12] A distinctly different carboxypeptidase-type enzyme, which releases the tyrosine, has also been identified and characterized.[13] The latter appears to be the initial modification, since tyrosine also is the coded terminal residue.[14,15] Although the presence or absence of the tyrosine has not been found to affect the assembly of microtubules, we have observed significant changes in the state of tyrosinolation of tubulin in cells undergoing cytoskeletal reorganization due to differentiation[16] or during mitosis[17] During studies involving a variety of neuronal and nonneuronal cells,[16–19] a specific, dose-dependent stimulation of tubulin tyrosinolation was noted in rabbit peritoneal neutrophils[20] and also in human peripheral blood PMN,[21] as induced by the synthetic peptide chemoattractant *N*-formyl-methionyl-leucyl-phenylalanine (FMLP). In order to better understand this reaction in PMN, we also studied and compared tubulin tyrosinolation in PMN obtained from normal donors and those from patients with the Chediak-Higashi syndrome (CHS)[21] and with chronic granulomatous disease (CGD).[22] Ascorbate and other reducing agents were found to have profound effects on PMN tubulin tyrosinolation, suggesting a correlation between PMN redox state and modulation of tubulin tyrosinolation in these cells.[21] Some of the pertinent results observed during the course of these studies are discussed in the present monograph.

EXPERIMENTAL PROCEDURES

Since tubulin tyrosinolation (or detyrosinolation) is a posttranslational modification and does not require de novo protein synthesis,[9,10] the reaction can be studied in intact functional cells by measuring the incorporation of radiolabeled tyrosine into trichloroacetic-acid-insoluble fraction, in the complete absence of protein synthesis.[16] The latter can be achieved by prior incubation of cells with a mixture of cycloheximide, puromycin, and chloramphenicol, a condition that inhibits [^{14}C]leucine incorporation by more than 99%.[16–22] The details of this method have been published.[20] Specificity of the reaction, *i.e.,* incorporation of radiolabeled tyrosine into α chain of tubulin, was documented by SDS-urea polyacrylamide gel electrophoresis and subsequent fluorography.[21,22] Under the experimental conditions used, radioactivity was detected only in a single protein band comigrating with authentic [^{14}C]tyrosinolated α-tubulin.[18–21] For quantitation of the amount of radioactivity in tubulin bands, the fluorogram was scanned and the corresponding peak areas were measured. Procedures for the determination of the specific radioactivity of the intracellular pool of tyrosine in various PMN samples were as previously described.[20,21] Other details of the experimental procedures used in these studies have been reported.[20–22]

RESULTS

Comparison of Resting Levels of Tubulin Tyrosinolation in Normal and Patient PMN

As reported earlier,[20–22] there was a 2-3-fold stimulation of tubulin tyrosinolation when resting PMN were stimulated with the peptide chemoattractant FMLP. This stimulation was receptor mediated and dose dependent.[20,21] In order to further understand the mechanism of stimulation, we also studied the resting and FMLP-stimulated levels of tubulin tyrosinolation in PMN isolated from normal donors and compared with those from patients with CHS and CGD. The striking differences in the resting levels of posttranslational incorporation of tyrosine in normal PMN and in PMN from patients with CHS and CGD are shown in FIGURE 1. Since many of the functional abnormalities of PMN from CHS patients have been associated with impaired microtubule function,[23-25] we decided to study these cells in more detail. The results of these experiments are discussed below.

Tubulin Tyrosinolation in PMN of CHS Patients

We studied two CHS patients (CHS_1 and CHS_2) who have been extensively studied at the NIH.[26] We also had occasion to study a third (CHS_3) patient, whose clinical course and laboratory findings have been reported.[27] The resting levels of tubulin tyrosinolation in PMN of all three CHS patients were 2-3-fold higher than in PMN from normal donors. As shown in FIGURE 2, the higher tyrosinolation in resting PMN of CHS patients could be further stimulated in the presence of 10^{-7} M FMLP. The elevated levels of tyrosinolation in CHS cells was not attributable to a parallel increase in the specific radioactivity of the intracellular pool of tyrosine, and was not due to an increase in ligase or carboxypeptidase activity or amount of cellular tubulin. Polyacrylamide slab gel electrophoresis of [^{14}C]tyrosine-labeled PMN and subsequent fluorography demonstrated the specific nature of the posttranslational incorporation of tyrosine, both in normal and CHS cells. FIGURE 3 demonstrates the specific incorporation of tyrosine into tubulin of normal and CHS_1 PMN and also shows a quantitative increase in the amount of [^{14}C]tyrosine incorporated in CHS cell tubulin.

Effect of Ascorbic Acid on PMN Tubulin Tyrosinolation

The rationale for studying the effect of ascorbic acid on tubulin tyrosinolation in PMN were twofold: first, ascorbic acid has been reported to stimulate neutrophil chemotaxis and to promote microtubule assembly, both in *in vitro* and in *in vivo* studies;[28,29] second, although reports on the efficacy of ascorbate administration in correcting some of the functional abnormalities of PMN related to CHS have been controversial,[30,31] it has been reported that ascorbate improved defective adherence, chemotaxis, and degranulation of PMN from some CHS patients.[30] Since the reaction which we are studying specifically involves cellular tubulin-microtubules, it was of interest to examine the effect of ascorbate on tubulin tyrosinolation in PMN of CHS patients.

As shown in FIGURE 4, prior incubation with 10^{-4} M ascorbate had a profound effect both on the rate and the final extent of posttranslational incorporation of tyrosine in PMN of CHS patient and the incorporation was very similar to the resting levels observed in normal PMN. Addition of ascorbate also had a small but reproducible effect on the resting levels of tyrosine incorporation in normal cells. Perhaps the most significant result obtained from these earlier studies on PMN tubulin tyrosinolation that provided valuable insight into the possible mechanism of modulation of this reaction in PMN came from the observation of complete inhibition of the FMLP-induced response in the presence of 10^{-4} M ascorbate (FIGURE 4), both in normal PMN and in PMN obtained from CHS patients.

In view of the results presented in FIGURE 4, we also studied the effect of oral administration of a large dosage (8 g) of ascorbate to the CHS patients on the rate

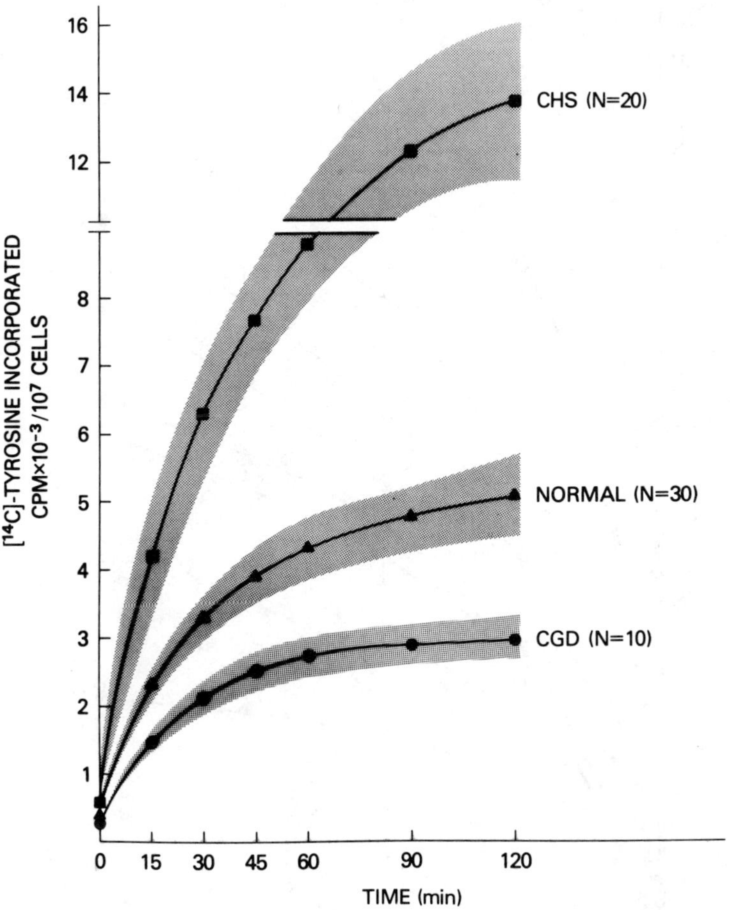

FIGURE 1. Comparison of resting levels of posttranslational incorporation of tyrosine in normal, CHS, and CGD PMN. Cells at 10^7/ml were incubated with inhibitors of protein synthesis for 30 min at 37°C, prior to the addition of [^{14}C]tyrosine (5 μCi = 0.01 μmol/ml) at 0 min. Trichloroacetic-acid-insoluble radioactivity was determined at indicated times. *N* represents the number of different experiments and the *shaded regions* represent the range of values (From Nath and Gallin.[33])

and extent of posttranslational incorporation of tyrosine in their PMN. Absorption of ascorbic acid from the gastrointestinal tract was documented in each patient. Serum ascorbate levels increased from 70 μM/liter to 205 μM/liter in CHS_1 and from 55 μM/liter to 115 μM/liter in CHS_2 while they were taking ascorbate. FIGURE 5A demonstrates the levels of posttranslational incorporation of [^{14}C]tyrosine in PMN

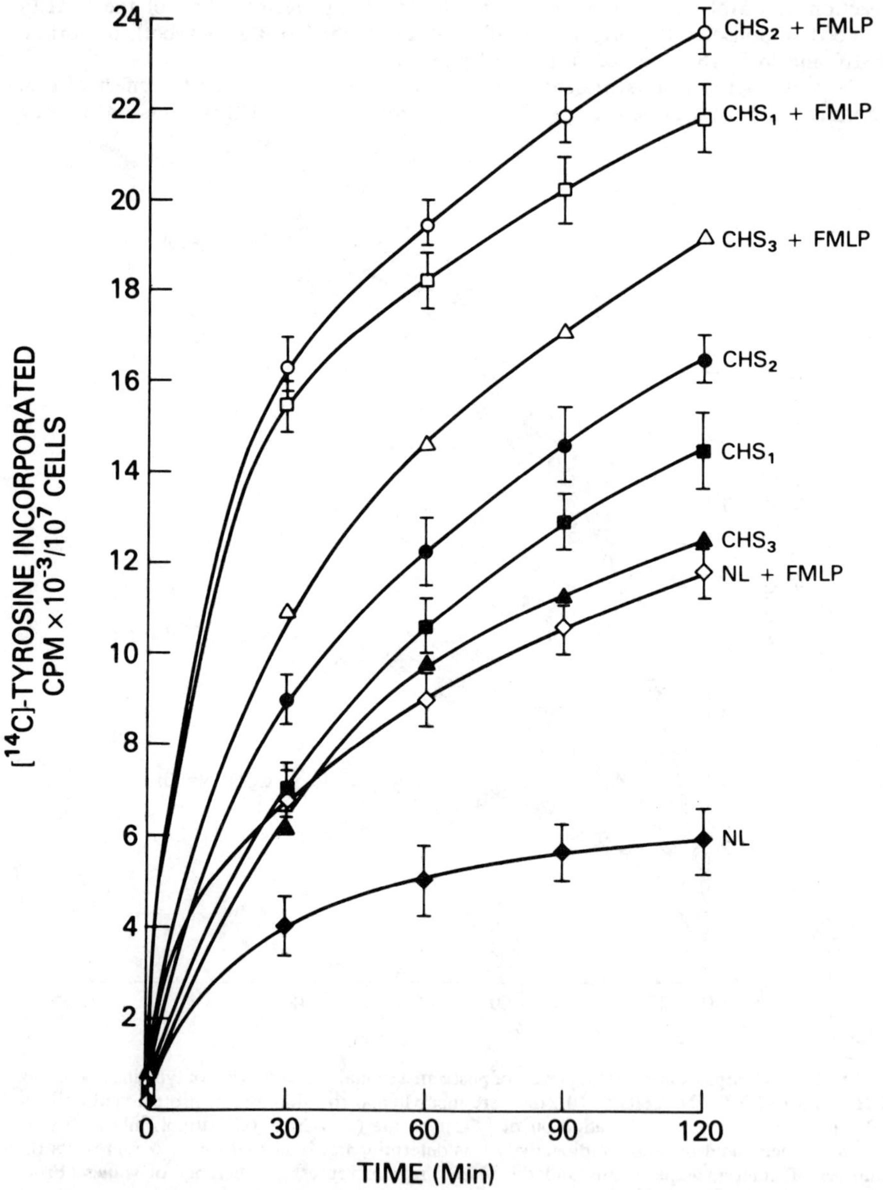

FIGURE 2. Resting and FMLP-stimulated levels of posttranslational incorporation in PMN from normal (*NL*) and three Chediak-Higashi syndrome patients (*CHS_1, CHS_2, and CHS_3*). Experimental conditions were identical to those described in FIGURE 1. Where indicated, 10^{-7} M FMLP was added just before addition of [^{14}C]tyrosine. The results are means ± SEM for three experiments each with CHS_1 and CHS_2, and for six experiments with normal individuals. The values for CHS_3 are means of duplicate determinations obtained from a single experiment. (From Nath *et al.*[21])

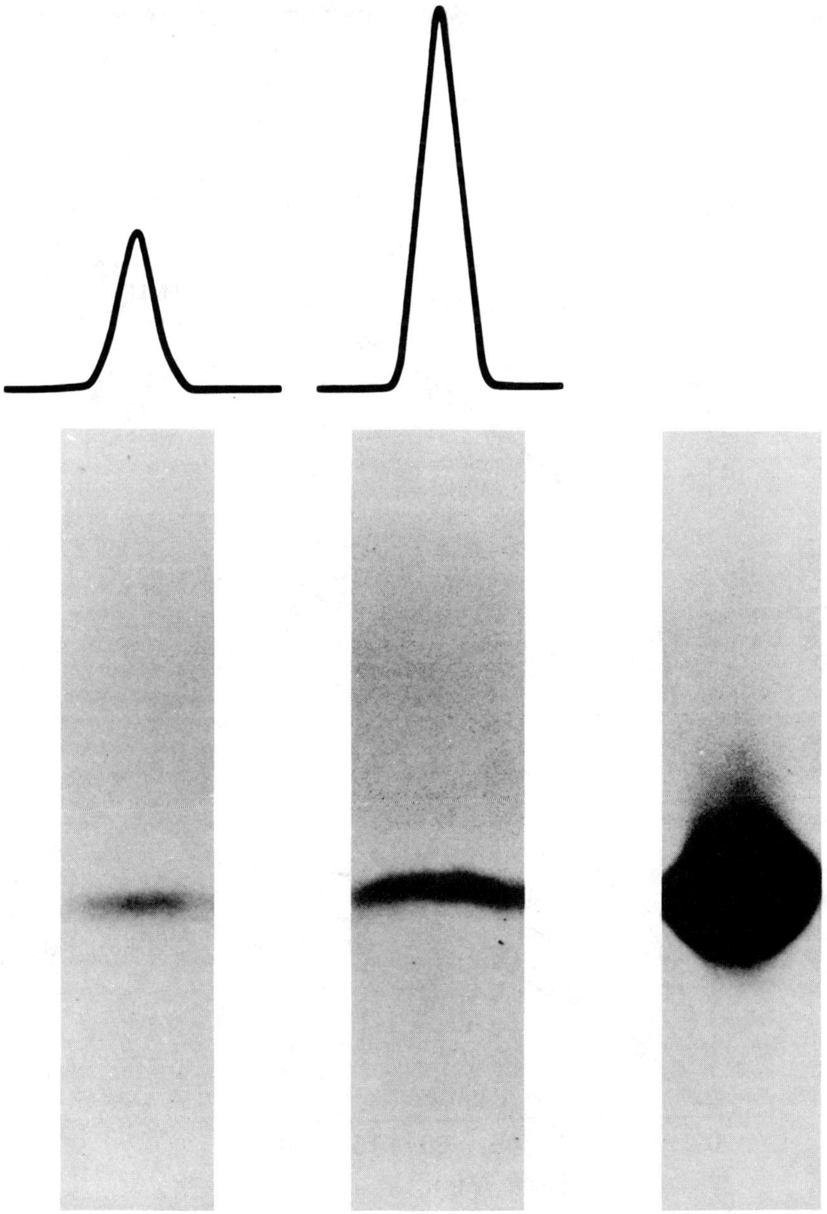

FIGURE 3. Quantitative demonstration of the specific incorporation of [^{14}C]tyrosine into tubulin α-chains of normal and CHS_1 PMN. Equal amounts of [^{14}C]tyrosinolated PMN samples as obtained from the experiment described in FIGURE 2 were subjected to slab gel electrophoresis and the gel was processed for fluorography. Details of this procedure are given in refs. 20 and 21. Lane A, normal PMN; lane B, CHS_1 PMN; lane C, an overexposed sample of purified [^{14}C]tyrosinolated tubulin preparation. The upper panel (in **A** and **B**) represents the densitometric scans of the corresponding radioactive bands in the lower panel. (From Nath *et al.*[21])

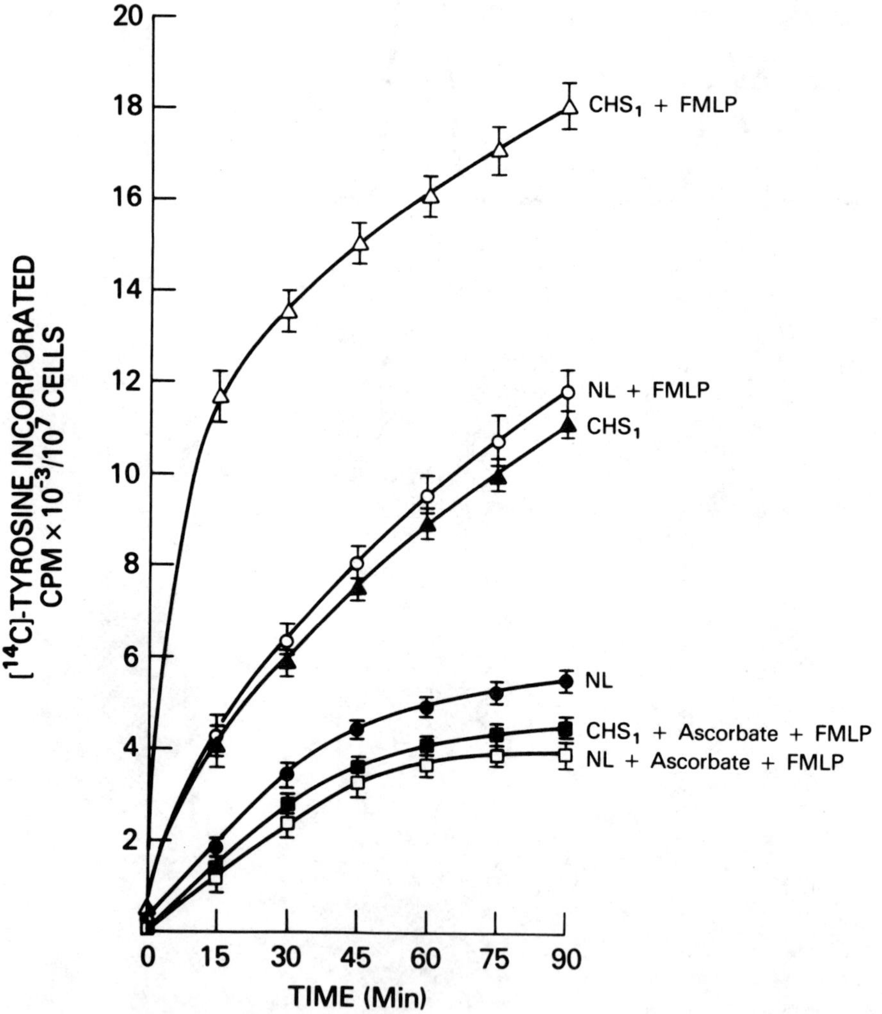

FIGURE 4. Effect of addition of ascorbate *in vitro* on the resting and FMLP-stimulated levels of tyrosine incorporation in normal and CHS_1 PMN. Where indicated, cells were incubated for 30 min with 10^{-4} M ascorbate prior to the addition of 10^{-7} M FMLP, and [^{14}C]tyrosine at 0 min. Other experimental conditions were the same as in FIGURE 1. The results are means ± SEM of three separate experiments. (From Nath and Gallin.[33])

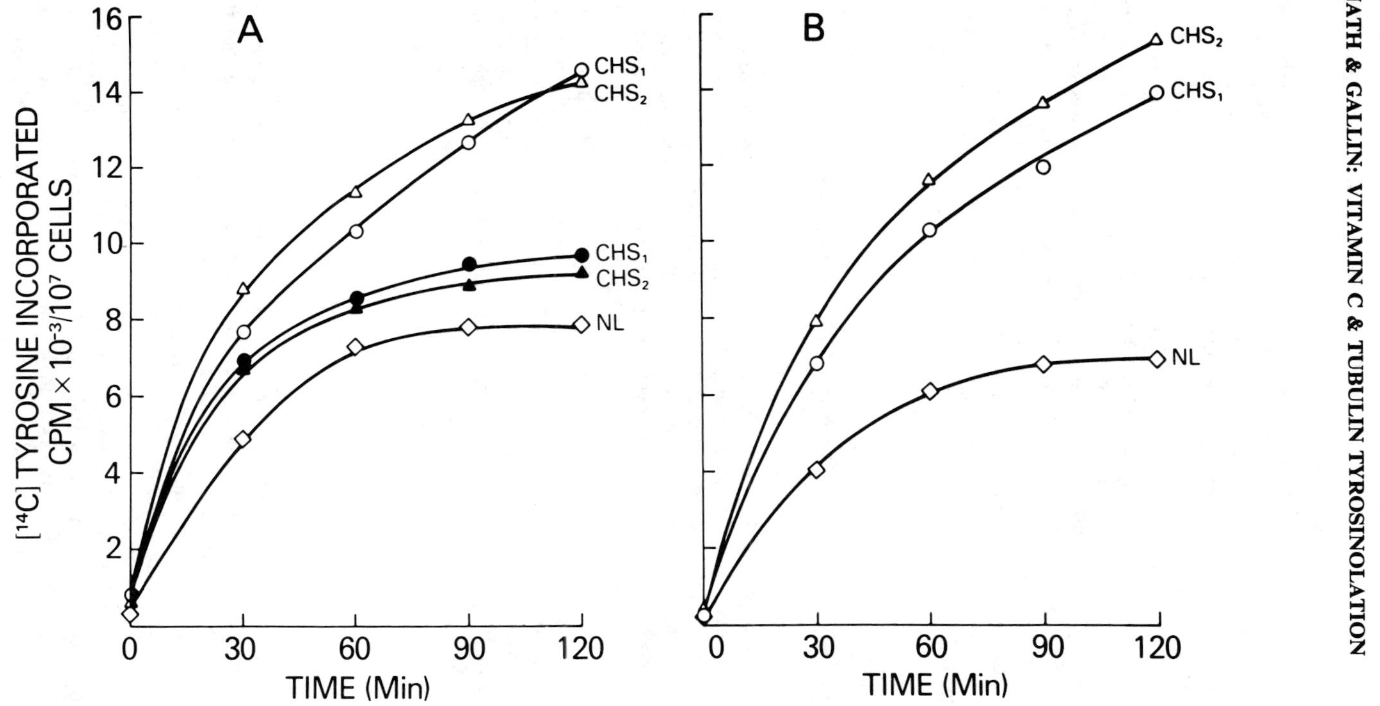

FIGURE 5. Effect of administration of ascorbate *in vivo* on the resting levels of posttranslational tyrosine incorporation into PMN of CHS patients. PMN at 10^7/ml were preincubated with antibiotics, and [^{14}C]tyrosine incorporation was measured at indicated times. (**A**) ○ and △ represent preascorbate levels of tyrosine incorporation in CHS_1 and CHS_2; ● and ▲ represent the corresponding 2-h postascorbate values; ◇ shows resting levels of incorporation in PMN of a normal donor. (**B**) The corresponding resting levels of tyrosine incorporation in CHS_1 and CHS_2 1 week after withdrawal of ascorbate therapy compared to the normal levels. The results are the means of duplicate determinations. (From Nath *et al.*[21])

of CHS_1 and CHS_2 before and 2 hours after administration of ascorbate. A significant effect of ascorbate was noted that was very similar to the results shown in FIGURE 4, where 10^{-4} M ascorbate was added *in vitro* to the cell incubation medium. FIGURE 5B shows tyrosine incorporation in PMN of the same CHS patients a week after ascorbate was withdrawn and demonstrates the typical high levels of incorporation as shown in FIGURE 4 or FIGURE 5A. We also performed a control experiment in which PMN were incubated for 1 hr in the presence of 10^{-4} M ascorbate and then centrifuged and resuspended in either fresh incubation medium without ascorbate or in medium containing 10^{-4} M ascorbate. When tubulin tyrosinolation was studied in these PMN, the effect of ascorbate (as shown in FIGS. 4 and 5) persisted for at least another 90 min after removal of ascorbate from the incubation medium, with the resting levels of posttranslational incorporation of tyrosine in the absence of ascorbate resembling the incorporation in the presence of added ascorbate. This was true in both normal and CHS_1 PMN (data not shown).

Effect of Various Reducing Agents on PMN Tubulin Tyrosinolation

Since ascorbate is a powerful reducing agent and undergoes reversible oxidation/reduction in granulocytes,[32] we wanted to assess whether the effect of ascorbate on PMN tubulin tyrosinolation was specific or was mediated via its reducing equivalents. As shown in FIGURE 6, prior incubation of normal or CHS_1 PMN with other reducing agents, such as reduced glutathione (GSH), cysteine or dithiothreitol (DTT), produced effects on posttranslational incorporation of tyrosine similar to that of ascorbate (FIGS. 4 and 5), a result true for both resting and FMLP-stimulated PMN. These results suggested a link between cellular oxidative-reductive reactions and tubulin tyrosinolation in PMN. In subsequent studies, by further manipulation of the redox state in normal PMN and by studies in PMN from patients with CGD, a strong correlation between changes in PMN redox state and the modulation of tubulin tyrosinolation has been established.[22,33]

DISCUSSION

Although the mechanism of modulation of tubulin tyrosinolation and its possible functional role in PMN has not been elucidated, our studies clearly indicate that the activation of the membrane-associated NADPH oxidase-mediated respiratory burst and the related changes in the cellular redox state in FMLP-stimulated PMN are coupled to the stimulation of tubulin tyrosinolation in these cells.[21,22,33] The PMN oxidative-reductive changes are also coupled to intracellular changes in the free Ca^{2+} concentration[34] and we have recently demonstrated a requirement for optimal Ca^{2+} concentration for proper modulation of PMN tubulin tyrosinolation.[35] It is likely, therefore, that the regulation of tubulin tyrosinolation in PMN is quite complex and rests on the delicate equilibrium between a number of physiological and biochemical processes of the cell. Tubulin contains essential sulfhydryl groups[36] that are reportedly critical for its proper assembly and function.[36,37] In *in vitro* studies, there are, however, no available data to suggest a role of the tubulin sulfhydryl groups in tyrosinolation

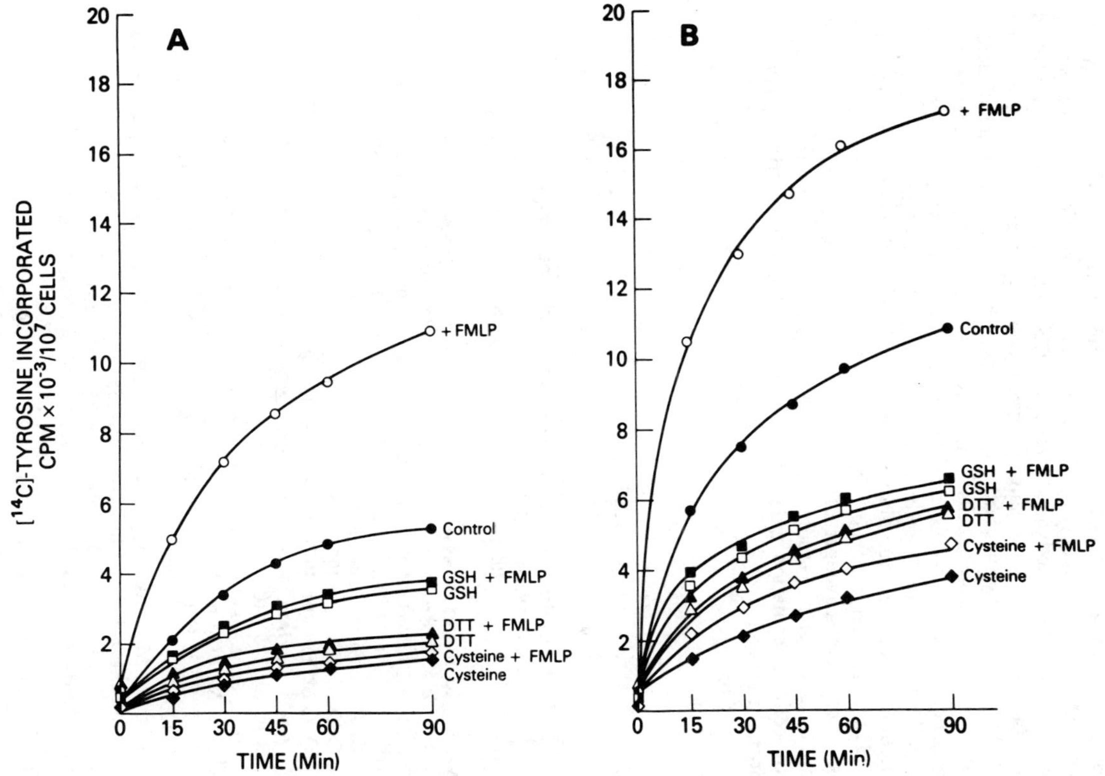

FIGURE 6. Effects of various reducing agents on the resting and FMLP-stimulated levels of tyrosine incorporation in normal and CHS_1 PMN. Where indicated, PMN were incubated for 30 min with 10^{-3} M GSH, cysteine, or DTT, prior to the addition of 10^{-7} M FMLP at 0 min. Other experimental conditions were the same as in FIGURE 1. (**A**) Normal PMN. (**B**) CHS_1 PMN. The results are the means of duplicate determinations (From Nath *et al.*[21])

or detyrosinolation of its α chain. Although the reaction was discovered more than a decade ago, nothing definitive is known about the cellular regulation or the functional role of this intriguing reaction.

Ascorbate is a powerful reducing agent and it has been proposed that in conjunction with GSH it constitutes a functional part of the redox-active components of PMN[38,39] and therefore, may function to protect cell constituents from denaturation by the oxidants generated during phagocytosis.[39] There is also reported evidence suggesting that ascorbate and GSH are effective reductants and free radical scavengers that protect PMN from damages caused by oxidant and free radical injury as caused by ionizing radiation.[40,41] It is noteworthy that both GSH and ascorbate are present in sufficiently high concentrations in white blood cells[42,43] and levels of both are reported to transiently fall (and their corresponding oxidized species to rise) upon activation of PMN.[39] It is conceivable, therefore, that intracellular ascorbate and GSH may play a physiological role in the modulation of PMN tubulin tyrosinolation. Since GSH peroxidase catalyzed reduction of H_2O_2 plays an important functional role in PMN redox changes[44,45] and ascorbate is an effective scavenger of superoxide radicals,[32] such speculation is not totally without basis. Further studies are clearly warranted to understand the complexities of the cellular regulation and possible functional role of this intriguing reaction in PMN and in other cells.

REFERENCES

1. MALECH, H. L., R. K. ROOT & J. I. GALLIN. 1977. J. Cell Biol. **75:** 666-693.
2. OLIVER, J. M. 1978. Am. J. Pathol. **93:** 221-278.
3. ANDERSON, D. C., L. J. WIBLE, B. J. HUGHES, C. W. SMITH & B. R. BRINKLEY. 1982. Cell **31:** 719-729.
4. SCHLIWA, M., K. B. PRYZWANSKY & U. EUTENEUR. 1982. Cell **31:** 705-717.
5. GALLIN, J. I. & A. S. ROSENTHAL. 1974. J. Cell Biol. **62:** 594-609.
6. HOFFSTEIN, S. T., I. M. GOLDSTEIN & G. WEISSMANN. 1977. J. Cell Biol. **73:** 242-256.
7. ALBERTINI, D. F., R. D. BERLIN & J. M. OLIVER. 1977. J. Cell Sci. **26:** 57-75.
8. DUSTIN, D. 1978. Microtubules. Springer-Verlag. New York.
9. RAYBIN, D. & M. FLAVIN. 1977. Biochemistry **16:** 2189-2194.
10. RAYBIN, D. & M. FLAVIN. 1977. J. Cell Biol. **73:** 492-504.
11. ARCE, C. A., M. E. HALLAK, J. A. RODRIGUEZ, H. S. BARRA & R. CAPUTTO. 1978. J. Neurochem. **31:** 205-210.
12. KOBAYASHI, T. & M. FLAVIN. 1981. Comp. Biochem. Physiol. B. **69:** 387-392.
13. KUMAR, N. & M. FLAVIN. 1981. J. Biol. Chem. **256:** 7678-7686.
14. LEMISCHKA, I. R., S. FARMER, V. R. RACANIELLO & P. A. SHARP. 1981. J. Mol. Biol. **151:** 101-120.
15. VALENZUELA, P., M. QUIROGA, J. ZALDIVER, W. J. RUTTER, M. W. KIRSCHNER & D. W. CLEVELAND. 1981. Nature **289:** 650-655.
16. NATH, J. & M. FLAVIN. 1979. J. Biol. Chem. **254:** 1505-1510.
17. NATH, J., J. WHITLOCK & M. FLAVIN. 1978. J. Cell Biol. **79:** 294a.
18. NATH, J. & M. FLAVIN. 1980. J. Neurochem. **35:** 693-706.
19. NATH, J. & M. FLAVIN. 1984. Biochem. Biophys. Acta **803:** 314-322.
20. NATH, J., M. FLAVIN & E. SCHIFFMANN. 1981. J. Cell Biol. **91:** 232-239.
21. NATH, J., M. FLAVIN & J. I. GALLIN. 1982. J. Cell Biol. **95:** 519-526.
22. NATH, J. & J. I. GALLIN. 1983. J. Clin. Invest. **71:** 1272-1281.
23. OLIVER, J. M., R. B. ZURIER & R. D. BERLIN. 1975. Nature **253:** 471-473.
24. OLIVER, J. M. 1976. Am. J. Pathol. **85:** 395-417.
25. OLIVER, J. M. & R. B. ZURIER. 1976. J. Clin. Invest. **57:** 1239-1247.

26. WOLFF, S. M., D. C. DALE, R. A. CLARK, R. K. ROOT & H. KIMBALL. 1972. Am. J. Intern. Med. **76:** 293-306.
27. WHITE, J. G. & C. C. CLAWSON. 1979. Am. J. Hematol. **7:** 349-356.
28. GOETZEL, E. J., S. I. WASSERMAN, I. GIGLI & K. F. AUSTEN. 1974. J. Clin. Invest. **53:** 813-818.
29. BOXER, L. A., B. VANDERBILT, S. BOUSIB, R. JERSILD, H. YANG & R. L. BAEHNER. 1979. J. Cell Physiol. **100:** 119-126.
30. BOXER, L. A., A. M. WATANABE, M. RISTER, H. R. BESCH, J. ALLEN & R. L. BAEHNER. 1976. N. Engl. J. Med. **295:** 1041-1045.
31. GALLIN, J. I., R. J. ELIN, R. T. HUBERT, A. S. FAUCI, M. A. KALINER & S. M. WOLFF. 1979. Blood **53:** 226-234.
32. KRAUT, E. H., E. N. METZ & A. L. SAGONE. 1980. J. Reticuloendo. Soc. **27:** 359-366.
33. NATH, J. & J. I. GALLIN. 1984. *In* White Cell Mechanics. H. J. Meiselman, M. A. Lichtman & P. L. LaCelle, Eds.: 95-110. A. R. Liss, Inc. New York.
34. KORCHAK, H. M., L. E. RUTHERFORD & G. WEISSMANN. 1984. J. Biol. Chem. **259:** 4070-4075.
35. NATH, J. & J. I. GALLIN. 1986. J. Immunol. **136:** 628-635.
36. MELLON, M. G. & L. I. REBHUN. 1976. J. Cell Biol. **70:** 226-238.
37. REBHUN, L. I. & J. NATH. 1976. *In* Cell Motility, Vol. 3. R. Goldman, T. Pollard & J. Rosenbaum, Eds.: 1343-1366. Cold Spring Harbor Laboratory. Cold Spring Harbor, N.Y.
38. BIGLEY, R. H. & L. STANKOVA. 1974. J. Exp. Med. **139:** 1084-1092.
39. STANKOVA, L., N. B. GERHARDT, L. NAGEL & R. H. BIGLEY. 1975. Infect. Immun. **12:** 252-256.
40. TAPPEL, A. L. 1973. Fed. Proc. Fed. Am. Soc. Exp. Biol. **32:** 1870-1874.
41. STANKOVA, L., D. RIGAS, C. HEAD, B. T. GAY & R. BIGLEY. 1979. Rad. Res. **80:** 49-60.
42. ALBANESE, A. A., E. H. WEIN & L. A. MATA. 1975. Nutr. Rep. Internat. **12:** 271-289.
43. CASTELLI, A., G. E. MARTORANA, E. MEUCCI & G. BONETTI. 1982. Acta Vitaminol. Enzymol. **4:** 189-196.
44. OLIVER, J. M., D. F. ALBERTINI & R. D. BERLIN. 1976. J. Cell Biol. **71:** 921-932.
45. WEDNER, H. J., L. SIMCHOWITZ, W. F. STENSON & C. M. FISCHMAN. 1981. J. Clin. Invest. **68:** 535-543.

DISCUSSION OF THE PAPER

J. PATRICK (*Alacer Corp., Buena Park, Calif.*): Would you speculate that deficiency of ascorbate might be in any way involved in Alzheimer's disease?

J. NATH (*Walter Reed Army Institute of Research, Washington, D. C.*): I will not be able to comment on that. All that I can say is there is a sufficiently high level of ascorbate and glutathione in cells and tissues, and it could very well be that they play a physiological role in the modulation of this reaction. But it is not totally without basis to suggest that physiological molecules like ascorbate or glutathione which are present in sufficiently high concentration in these cells play a modulatory role.

E. J. DILIBERTO (*Wellcome Research Labs, Research Triangle Park, N.C.*): What is the percentage of tyrosinolation of tubulin that you would normally find in the resting cells?

J. NATH: In neuronal cells the concentration of tubulin is very high. In nonneuronal cells the concentration ranges from 1-2% of the total cellular protein. Mole for mole if I calculate, the resting level is about 0.3 nanomole of tyrosine per mole of tubulin, so about 30% of the α chains have tyrosine in their carboxy terminals, which goes up 2-3-fold.

E. J. DILIBERTO: The other question is, have you done a chase experiment where you looked at the off rate?

J. NATH: Yes, it is a very interesting question. All those studies are done and published. It's a very intriguing reaction because the reaction has been extensively studied *in vitro* with purified tubulin and purified ligase and carboxypeptidase. But *in vivo* only in neuronal cells can the reaction be chased. In HeLa cells I cannot chase that, in neutrophils I cannot chase that even in large excess of cold tyrosine, suggesting that after the incorporation of tyrosine there is some further modification which prevents the chase type of reactions. It's very interesting that even in neuronal cells only half of the freshly incorporated tyrosine can be chased out. A sample of tubulin isolated has some preexisting tyrosine in α chains (about 10-25%, it varies) and then it can take another 10-15% additional tyrosine which it can accept *in vitro* in presence of enzymes added under appropriate optimal conditions. But in *in vitro* experiments we have never been able to show that 100% of the tubulin α chains can accept tyrosine.

So there is a nonsubstrate moeity as we call it, and this is a big black box still in this area of research.

B. L. HORECKER (*Cornell University Medical College, New York, N.Y.*): If I gather correctly from your observations the ectoform, the proper form for tubulin for proper polymerization, is the detyrosinolated form?

J. NATH: That's not known.

B. L. HORECKER: When you enhance the tyrosinolation you block the response to FMLP or is it the other way? With ascorbic acid do you block the response to FMLP?

J. NATH: I'm blocking the response to the peptide FMLP which enables the cell to accept additional tyrosine.

B. L. HORECKER: But if you block the tyrosinolation with ascorbic acid, then you should enhance the ability of the cells to respond to FMLP.

J. NATH: That depends on whether we can believe that this has any direct correlation with the reaction. The functional significance of this intriguing reaction is still not known.

M. LEVINE (*National Institutes of Health, Bethesda, Md.*): With regard to the inhibition that you see of tyrosine incorporation, do you get a difference if you measure intracellular versus extracellular reducing agent?

J. NATH: No, it's a strictly intracellular reaction. You can add scavengers outside, SOD, or you can generate superoxide radicals like using xanthine, xanthine oxidase extracellularly, either way it's not going to affect the reaction in any way intracellularly. It's a strictly intracellular reaction, so any reducing agent or scavenger you want to use has to go inside the cell.

A Biological Role for Ascorbate in the Selective Neutralization of Extracellular Phagocyte-derived Oxidants

R. ANDERSON AND P. T. LUKEY

Division of Immunology
Department of Medical Microbiology
Institute for Pathology
University of Pretoria
Pretoria, Republic of South Africa

INTRODUCTION

Activation of the phagocyte-membrane-associated enzyme NADPH oxidase leads to the generation of superoxide, which is the precursor of a series of highly reactive antimicrobial oxidants such as hydroxyl radical, singlet oxygen, and especially H_2O_2.[1,2] Interaction of H_2O_2 with the primary granule enzyme myeloperoxidase (MPO) and a halide leads to the generation of a secondary series of oxidants such as hypohalous acids, halides, chloramines, and chloramides.[1,2] These phagocyte-derived reactive oxidants (RO) are indiscriminate and ideally their generation and activities should be confined to the intraphagocytic milieu. Reactive oxidants that are released extracellularly by activated phagocytes are normally neutralized by innate antioxidants such as superoxide dismutase (SOD), catalase, glutathione peroxidase, and low molecular weight agents such as ascorbate, β-carotene, α-tocopherol, and nonprotein sulfhydryls.[2] However, these biological antioxidant systems may be overwhelmed during excessive and chronic activation of phagocytes leading to uncontrolled activity of extracellular RO. Phagocyte-derived RO that are released extracellularly have wide-ranging deleterious activities *in vitro.* They are autotoxic to phagocytes,[2–5] immunosuppressive,[6–9] mutagenic,[10–12] and promote the oxidative inactivation of the elastase-inhibitory capacity of human α-l-protease inhibitor (API).[13,14] Although the *in vivo* significance of these findings has not been firmly established, the sustained release of RO by chronically activated phagocytes has been linked to the pathogenesis of some human diseases. The association of certain cancers and immune attrition with chronic inflammatory disorders of infective and noninfective origin is well-recognized.[15–19] There is also evidence to suggest that cigarette smoking leads to chronic recruitment and activation of both polymorphonuclear leukocytes (PMNL) and macrophages in the lungs.[20–22] Increased numbers of hyperactive macrophages and PMNL in the lungs of smokers with consequent chronic generation of phagocyte-derived RO may be involved in the etiology of smoking-related lung diseases such as emphysema and bronchial carcinoma. With increasing awareness of the deleterious consequences of the extracellular release of RO by phagocytes it is important that safe and effective antioxidants are identified. Ideally these should selectively neutralize extracellular phagocyte-derived RO without compromising the intraphagocytic generation of antimicrobial oxidants. In this report we describe the effects of ascorbate on the extracellular and intracellular generation of RO by activated human PMNL *in vitro.*

METHODS AND RESULTS

Ascorbate Concentrations

For *in vitro* investigations the water-soluble sodium salt of ascorbic acid was used at final concentrations of 5, 10, 20, and 50 μg/ml. The lower concentrations are equivalent to physiological serum concentrations and the higher concentrations can be achieved during oral and intravenous administration of 1 g of ascorbate.[9,23] To measure the effects of oral administration of ascorbate on the generation of RO by PMNL *in vitro,* effervescent tablets containing 1 g of the vitamin were used (Redoxon [R], Hoffmann-La Roche & Co., Basel, Switzerland).

Measurement of the Effects of Ascorbate on Oxidant Generation by Activated PMNL

For these experiments human PMNL were obtained from heparinized venous blood and resuspended to 2×10^7 PMNL/ml in Hepes (*N*-2-hydroxyethylpiperazine-*N*'-2-ethanesulfonic acid) buffered Hanks' balanced salt solution (HBSS) after separation on cushions of Ficoll:metrizoate, gelatin sedimentation, and ammonium chloride (0.85%) lysis of residual erythrocytes.[24] We investigated the effects of ascorbate (5-50 μg/ml) on the extracellular and intracellular generation of RO using luminol (0.1 mM; 5-amino-2,3-dihydro-1,4-phthalazinedione) enhanced chemiluminescence (LECL) of PMNL activated with the leukoattractant FMLP (1 μM; *N*-formyl-methionyl-leucyl-phenylalanine) and cytochalasin B (CB; 1 μg/ml). This is a useful method since it reveals distinct extracellular and intracellular oxidative events during stimulation of PMNL membrane-associated oxidative metabolism.[25,26] PMNL (2×10^5) were preincubated with ascorbate and luminol for 30 min at 37°C and then activated by addition of FMLP/CB and LECL was recorded at 10-sec intervals for 1 min and thereafter at 1 min intervals using a Lumac (R) Biocounter (model 2010, Lumac System Inc., Fla.) These results are expressed as relative light units (r.l.u.).

The results of these experiments for control systems and systems containing 50 μg/ml ascorbate are shown in FIGURE 1. Ascorbate caused significant inhibition of the early-occurring (within 1 min) extracellular LECL response of FMLP/CB-activated PMNL and increased the intensity of the later-occurring intracellular peak. The corresponding results for all the ascorbate concentrations tested are shown in TABLE 1. Ascorbate caused a dose-related selective inhibition and stimulation respectively of the extracellular and intracellular peaks of LECL in FMLP/CB-activated PMNL. These results indicate that the antioxidant activity of ascorbate occurs ex-

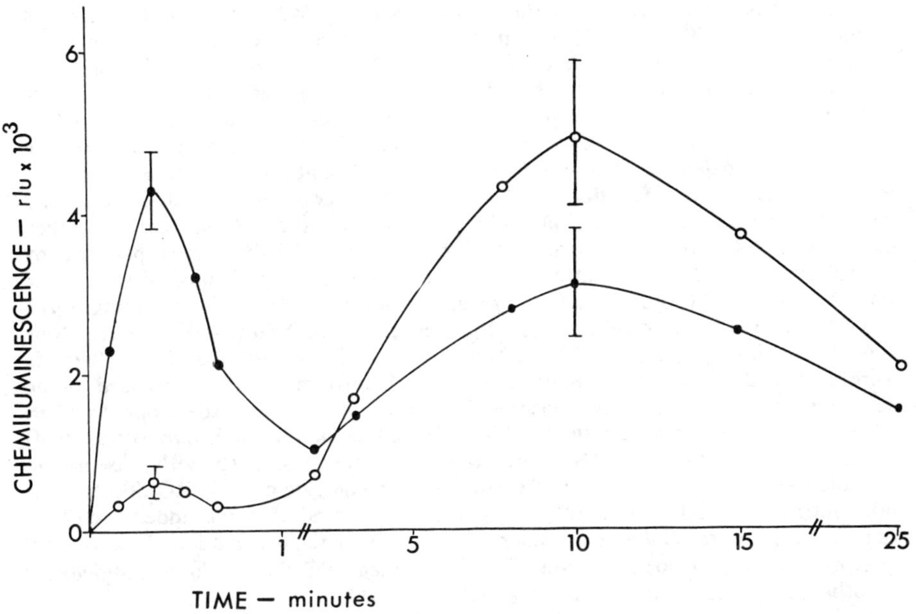

FIGURE 1. FMLP/CB-activated luminol-enhanced chemiluminescence of control PMNL (●——●) and PMNL coincubated with 50 μg/ml of sodium ascorbate (○——○). The results are expressed as the mean value ± SEM in relative light units (r.l.u.) of eight different experiments.

clusively in the extracellular compartment and that intracellular oxidant generation is unaffected and even increased by ascorbate.

Measurement of the Effects of Ascorbate on the Intracellular Killing of S. aureus *by PMNL*

Ideally antioxidants should not interfere with, or at worst minimally compromise, the oxygen-dependent intraphagocytic antimicrobial systems. We have recently de-

TABLE 1

Ascorbate Concentration (μg/ml)	Mean Percentage Inhibition of the Extracellular Peak of FMLP/CB-activated LECL	Mean Percentage Stimulation of the Intracellular Peak of FMLP/CB-activated LECL
5	60 ($p < 0.001$)	18 ($p < 0.025$)
10	75 ($p < 0.001$)	44 ($p < 0.01$)
20	78 ($p < 0.001$)	52 ($p < 0.01$)
50	86 ($p < 0.001$)	56 ($p < 0.01$)

Effects of varying concentrations of ascorbate (5-50 μg/ml) on the early (extracellular) and late (intracellular) peaks of FMLP/CB-activated luminol-enhanced chemiluminescence. The results are expressed as the mean percentage stimulation or inhibition of the early and late values respectively, relative to the ascorbate-free control systems. Statistical analyses were performed using the paired t statistic. For the control systems the peak early and late FMLP/CB-activated LECL responses were 4353 ± 417 and 3131 ± 676 (mean r.l.u. ± SEM of eight experiments).

veloped a sensitive radioassay for the measurement of PMNL intracellular killing of *S. aureus* (strain ATCC 25923). This method has been described in detail elsewhere.[27] Briefly, 5×10^6 PMNL were incubated with 1×10^8 colony-forming units of opsonized *S. aureus* (ratio of 1 PMNL:20 microorganisms) with and without ascorbate (5-50 μg/ml) in 1 ml HBSS in polypropylene test tubes. The tubes were rotated on a turntable for 30 min at 37°C after which 10 μg of lysostaphin was added to each tube and reincubated for 15 min at 37°C. The PMNL and intracellular bacteria were then pelletted by centrifugation, washed, and resuspended in 1 ml HBSS with and without ascorbate. The PMNL/intracellular bacteria mixtures ± ascorbate were then incubated for 4 and 18 hours at 37°C, after which the PMNL were pelletted by centrifugation, washed, and intracellular bacteria released by distilled water lysis of PMNL. The lysate (50 μl) was then inoculated into nutrient broth containing 5 μCi of L-[^{14}C(U)]amino acid mixture (specific activity 55 mCi/matom of carbon; New England Nuclear Corp., Boston, Mass.) and incubated for 8 hours at 37°C. The bacteria were then washed three times and the extent of incorporation of radiolabeled amino acids into bacterial cells was measured in a liquid scintillation spectrophotometer. These results are shown in TABLE 2. The intracellular killing of *S. aureus* was unaffected by coincubation of PMNL with ascorbate. These results show that ascorbate-mediated selective neutralization of extracellular phagocyte-derived RO does not inhibit intrinsic PMNL antimicrobial systems and suggest that the added ascorbate remains in the extracellular environment. To investigate the distribution of the vitamin (*i.e.*, extracellular versus intracellular) we have measured the uptake of radiolabeled ascorbate by resting and FMLP/CB-stimulated PMNL.

Measurement of the Uptake of Radiolabeled Ascorbate by Resting and FMLP-activated PMNL

PMNL were prepared as described above. They were resuspended at 10^7/ml of 0.15 M phosphate-buffered saline containing 1.3 mM $CaCl_2$, 0.8 mM $MgSO_4$, and 1 mg bovine serum albumin (BSA) per milliliter. PMNL (100 μl) were incubated for 30 minutes at 37°C in the presence of 10 μg ascorbate/ml (containing 6×10^5 cpm of ascorbic acid (L-[1-^{14}C] 7.7 mCi/mmol, New England Nuclear Corp., Boston, Mass.) During exposure to ascorbate PMNL were either in a resting (no stimulant)

TABLE 2

Test System	Intraphagocytic Killing (%) of *S. aureus*
PMNL only	69 ± 10
PMNL + 5 μg/ml ascorbate	66 ± 13
PMNL + 10 μg/ml ascorbate	65 ± 13
PMNL + 20 μg/ml ascorbate	67 ± 14
PMNL + 50 μg/ml ascorbate	65 ± 11

Measurement of the effects of coincubation with ascorbate (5-50 μg/ml) on the antimicrobial activity of PMNL for *S. aureus*. The results are expressed as the mean percentage killing of phagocytosed *S. aureus* ± SEM of four different experiments. The growth of surviving intracellular bacteria for ascorbate-free, control PMNL was 6062 ± 1910 counts per minute and 19496 ± 4624 cpm for the PMNL-free system containing bacteria only.

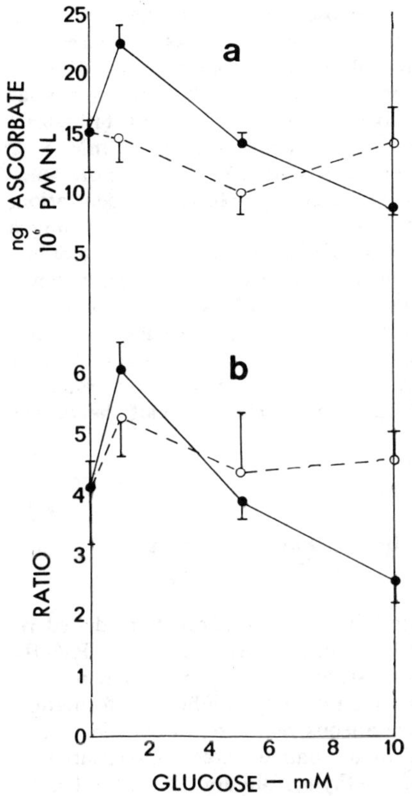

FIGURE 2. Ascorbate uptake by PMNL. PMNL (10^6) were incubated with 10 μg/ml ascorbate (26×10^5 cpm) and varying concentrations of glucose in the presence of 1 μM FMLP and 1 μg/ml cytochalasin B (○- - - -○) or in the absence of stimulant (●——●). After 30 minutes at 37°C PMNL were pelletted through silicone oil and pellets and supernates subjected to liquid scintillation counting. Results are expressed as **(a)** nanograms of ascrobate associated with 10^6 PMNL or as **(b)** the ratio of intracellular ascorbate concentration versus extracellular concentrations.

or activated (1 μM FMLP and 1 μg/ml cytochalasin B) state. The effect of glucose on uptake of ascorbate was also investigated by including glucose at concentrations ranging from 0 to 10 mM in the incubation buffer. The total incubation volume was 200 μl. PMNL together with bound ascorbate were separated from unbound ascorbate by centrifugation through 150 μl silicone oil at 12000 $\times$ g for 3 minutes at 4°C.[28] Tubes were frozen at -70°C and sliced to isolate the cell pellet and its supernatant. The cell pellet was solubilized in Protosol (New England Nuclear Corp.) and subjected to liquid scintillation, as were the supernates. Intracellular volume was calculated using tritiated water[28,29] (New England Nuclear Corp.). PMNL were incubated as above with substitution of 6×10^6 cpm of tritiated water for ascorbate.

Results are expressed both as nanograms of ascorbate associated with 10^6 PMNL and as the ratio of the amount of ascorbate bound to PMNL divided by the amount bound due solely to diffusion (calculated using tritiated water results). This ratio, therefore, if greater than 1 suggests active uptake of ascorbate, if less than 1 active exclusion, and if equal to 1 diffusion. Results are presented in FIGURES 2a and 2b. Ascorbate is actively taken up by resting PMNL in the absence of glucose (FIG. 2b). This uptake is increased at low glucose concentrations (1 mM; $p < 0.025$) but decreased at higher glucose concentrations (10 mM; $p < 0.005$). This latter result is in agreement with previous reports of glucose inhibition of ascorbate uptake due to their sharing the same transport system.[30] The effect of low glucose concentrations may be to activate the transport pathway without reaching saturation of the transport system.

Stimulation of the PMNL resulted in a loss of the glucose effect. However a comparison of ascorbate uptake between resting and stimulated PMNL reveals an increase in ascorbate uptake by stimulated PMNL only at high glucose concentrations (10 mM, $p < 0.01$). This is also in agreement with a previous report.[31] The trend is similar when results are expressed as nanograms ascorbate bound to 10^6 PMNL (FIG. 2a). Low glucose stimulates ascorbate uptake at low glucose concentrations (1 mM; $p < 0.005$), but inhibits at high concentrations (10 mM; $p < 0.005$). Furthermore stimulation of PMNL causes increased ascorbate uptake at high glucose concentrations (10 mM; $p < 0.01$). The actual amounts bound to PMNL are low, reaching a maximum with resting PMNL at 1 mM glucose, viz. 22.34 ± 1.28 ng ascorbate bound to 10^6 PMNL. This may be expressed as a function of intracellular volume, which, by using tritiated water, is calculated as 0.4 μl/10^6 PMNL. This value is in agreement with a previous estimate using cell dimensions to give a value of 32 μl/10^8 PMNL. A maximum value of 22.34 ± 1.8 ng/0.4 μl gives a maximum intracellular ascorbate concentration of 282.1 ± 16.2 μM. This value is in the same range as concentrations previously reported *in vitro*,[31] but is approximately half the *in vivo* estimated concentration for PMNL.[32]

Investigation of the Antioxidant Mechanisms of Ascorbate

We have previously reported that ascorbate inhibits the superoxide-mediated reduction of ferricytochrome *c* (cyt *c*) by FMLP-activated PMNL and by a PMNL-free xanthine/xanthine oxidase superoxide generating system.[33] In agreement with others,[34] ascorbate appeared to scavenge superoxide rather than inhibit its formation. However our results were compounded by spontaneous reduction of cyt *c* by ascorbate.[33] To overcome this problem we have used an alternative method of measurement of superoxide generation, viz. lucigenin (bis-*N*-methylacridinium nitrate) enhanced chemiluminescence (LUECL).[35] By using this method the effects of ascorbate on superoxide generation by FMLP/CB-activated PMNL and by xanthine/xanthine oxidase system have been investigated. PMNL (1×10^6) were preincubated with ascorbate (5-50 μg/ml) and lucigenin (2 mM) for 15 min at 37°C followed by the addition of FMLP/CB. LUECL was recorded at 10-s intervals for 2 min and thereafter at 1-min intervals for 10 min. In the PMNL-free system ascorbate was added to 0.1 mM xanthine in 1 ml HBSS containing lucigenin and reactions initiated by the addition of 100 mU of xanthine oxidase and LUECL recorded. In both systems the addition of superoxide dismutase (SOD, 200 units/ml) eliminated LUECL. The results for the control systems and systems containing 20 μg/ml ascorbate are shown in FIGURES 3a and 3b for FMLP/CB-activated PMNL and the xanthine/xanthine oxidase system respectively. The maximal responses for all the ascorbate concentrations tested in both systems are shown in TABLE 3. Ascorbate caused a dose-dependent inhibition of superoxide-mediated LUECL in PMNL and with the xanthine/xanthine oxidase system. Ascorbate mediates a spontaneous dose-related reduction of cyt *c*,[33] and this property of the vitamin was utilized in a further series of experiments designed to investigate its superoxide scavenging potential. Ascorbate at a fixed concentration of 10 μg/ml was incubated with 0.1 mM xanthine and varying concentrations of xanthine oxidase (25-500 mU/ml) in a final volume of 1 ml of 0.15 M phosphate-buffered saline (PBS), pH 7.4. The reactions were initiated by the addition of the enzyme and terminated after 5 min of incubation at room temperature by the addition of 200 units of SOD. The volume in each tube was made up to 3 ml with PBS followed by the

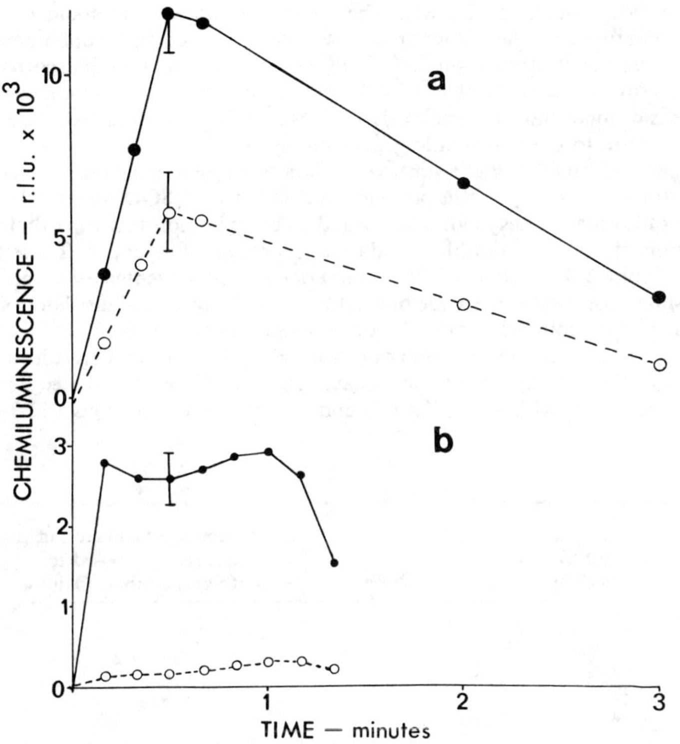

FIGURE 3. (**a**) FMLP/CB-activated lucigenin-enhanced chemiluminescence of control PMNL (●——●) and PMNL coincubated with 20 μg/ml ascorbate (○- - - -○) and (**b**) lucigenin-enhanced chemiluminescence by a xanthine (0.1 mM)/xanthine oxidase (100 mU) system in the presence (○- - - -○) and absence (●——●) of 20 μg/ml ascorbate. The results are expressed as the mean value ± SEM in relative light units (r.l.u.) of three different experiments.

TABLE 3

Ascorbate Concentration (μg/ml)	Percentage Inhibition of Superoxide Generation by:	
	FMLP-activated PMNL	Xanthine/Xanthine Oxidase
5	29 ± 5	62 ± 8
10	44 ± 6	81 ± 3
20	52 ± 5	90 ± 1
50	55 ± 4	97 ± 1

Effects of varying concentrations of ascorbate (5-50 μg/ml) on superoxide-mediated lucigenin-enhanced chemiluminescence by FMLP/CB-activated PMNL (10^6) and by xanthine (0.1 mM)/xanthine oxidase (100 mU) superoxide generating system. The results are expressed as the mean percentage (± SEM) inhibition of LUECL 1 minute after addition of the FMLP/CB or xanthine oxidase for three different experiments. The mean values ± SEM for the control ascorbate-free systems were 11969 ± 2101 and 2452 ± 445 r.l.u. Statistically significant (paired *t* statistic) inhibition of superoxide generation was observed for all concentrations of ascorbate tested with both systems ($p < 0.025$-$p < 0.005$).

addition of 50 μl of cyt *c* (1 mM). The tubes were allowed to stand for 15 min at room temperature and the reduction of cyt *c* was recorded spectrophotometrically at 550 nm. These results are shown in TABLES 4a-4c. Coincubation of ascorbate resulted in loss of reducing activity that was directly proportional to the concentrations of ascorbate, xanthine, and xanthine oxidase. These results show that ascorbate is oxidized during exposure to the superoxide generating system.

To identify the mediator of ascorbate oxidation, 10 μg of the vitamin were incubated with xanthine (0.1 mM) in the presence and absence of SOD and catalase (50-200 units/ml of both enzymes) individually and in combination in 1 ml PBS followed by the addition of 100 mU xanthine oxidase and processed as above. These results are shown in TABLE 5. Inclusion of SOD but not of catalase prevented ascorbate oxidation. These results show that superoxide but not H_2O_2 is the principal mediator of ascorbate oxidation. Indeed coincubation of 10 μg/ml ascorbate with H_2O_2 at concentrations of 0.01-1 mM H_2O_2 for 15 min did not cause ascorbate oxidation (results not shown). In a further series of experiments ascorbate (10 μg/ml) was incubated with 0.1 mM xanthine, xanthine oxidase (100 mU), and various concentrations (12.5-200 mU/ml) of purified myeloperoxidase (MPO) prepared from human PMNL. The reactions were initiated by the addition of xanthine oxidase and the tubes incubated for 15 min, after which the residual reducing activity of ascorbate was measured. These results are shown in TABLE 6. Inclusion of MPO with the xanthine/xanthine oxidase superoxide/H_2O_2 generating system potentiated ascorbate oxidation. These results show that in addition to superoxide, oxidants generated by the MPO/H_2O_2 system are also scavenged by ascorbate.

TABLE 4a

Ascorbate Concentration (μg/ml)	Percentage Oxidation of Ascorbate Exposed to Xanthine/Xanthine Oxidase
1.25	100
2.5	97 ± 2
5	78 ± 1
10	52 ± 1
20	27 ± 2

Measurement of the oxidation of ascorbate (1.25-20 μg/ml) during exposure (15 min/37°C) to a xanthine (0.1 mM)/xanthine oxidase (100 mU) superoxide generating system. Ascorbate oxidation was determined spectrophotometrically at 550 nm according to loss of reducing activity for ferricytochrome *c*. The results of three experiments are expressed as the mean percentage oxidation of ascorbate ± SEM.

Mechanism of Ascorbate-mediated Potentiation of Intracellular Oxidant Generation

The respiratory burst enzyme of human phagocytes is susceptible to autoxidative inactivation by myeloperoxidase (MPO).[4,5] We therefore measured the effects of added

TABLE 4b

Xanthine Oxidase Concentration (mU)	Percentage Oxidation of 10 μg/ml Ascorbate Exposed to Xanthine/ Xanthine Oxidase
25	14 ± 1
50	21 ± 2
100	46 ± 3
250	70 ± 2
500	98 ± 1

Measurement of the oxidation of a fixed concentration of ascorbate by exposure to 0.1 mM xanthine and varying concentrations (100-500 mU/ml) of xanthine oxidase. The results of three experiments are expressed as the mean percentage oxidation of ascorbate ± SEM.

MPO (25-200 mU/ml) and H_2O_2 (0.1-100 μM) or an H_2O_2-generating system of xanthine (0.5 mM) and xanthine oxidase (100 mU) on PMNL superoxide production in the presence and absence of ascorbate (5-50 μg/ml). PMNL (1×10^7) were exposed to MPO and H_2O_2 ± ascorbate for 15 minutes at 37°C, washed, resuspended in HBSS, and the superoxide generating capacity was measured using LUECL with FMLP/CB as the stimulus. In control systems the PMNL were exposed to the individual components of the assay system. Exposure of PMNL to MPO and H_2O_2 individually and in combination at all concentrations tested did not affect superoxide generation on subsequent exposure to FMLP/CB (results not shown). We reasoned that receptor-mediated assembly and activation of the NADPH oxidase complex may be a prerequisite for autoxidative inactivation by MPO. To investigate this possibility, PMNL were preincubated for 1 min with a predetermined, suboptimal concentration of FMLP (0.1 μM) prior to addition of MPO, H_2O_2, and ascorbate, then washed, and LUECL was measured after addition of FMLP/CB to 1×10^6 PMNL in HBSS. Inclusion of FMLP during exposure of PMNL to MPO/H_2O_2 caused a marked inhibition of FMLP/CB-activated LUECL, which was directly related to the MPO and H_2O_2 concentrations. Both MPO and H_2O_2 individually caused a slight inhibition of FMLP/CB-activated LUECL. These results are shown in TABLE 7. Ascorbate at all concentrations tested prevented the oxidative inactivation of the respiratory burst enzyme mediated by MPO/H_2O_2 in FMLP-primed PMNL (TABLE 7).

TABLE 4c

Xanthine Concentration (mM)	Percentage Oxidation of 10 μg/ml Ascorbate Exposed to Xanthine/ Xanthine Oxidase
0.1	55 ± 3
0.25	81 ± 2
0.5	100
1	100

Measurement of the oxidation of ascorbate (10 μg/ml) following exposure to 100 mU of xanthine oxidase and varying concentrations of xanthine (0.1-1 mM). The results of three experiments are expressed as the mean percentage oxidation of ascorbate ± SEM.

TABLE 5

Oxidizing System	Percentage Oxidation of Ascorbate
Xanthine + xanthine oxidase only	59 ± 3
+ 12.5 units of SOD	26 ± 3
+ 25 units of SOD	18 ± 3
+ 50 units of SOD	10 ± 4
+ 100 units of SOD	8 ± 4
+ 200 units of SOD	8 ± 4

Measurement of the protective effects of superoxide dismutase (12.5-200 units) on the oxidation of 10 μg/ml ascorbate by the xanthine (0.1 mM)/xanthine oxidase (100 mU) superoxide generating system. The results are expressed as the mean percentage oxidation of ascorbate ± SEM. Inclusion of SOD caused statistically significant protection of ascorbate oxidation ($p < 0.005$ for all SOD concentrations).

To investigate the effects of cell-associated ascorbate on the activity of intracellular MPO, PMNL (1×10^7) were preincubated with ascorbate (10-50 μg/ml) for 30 minutes at 37°C in HBSS, then washed twice, resuspended to 2 ml in HBSS, and sonicated with 3 × 10 s bursts at an amplitude of 8 μm peak to peak. To 200 μl of sonicate were added 5 mg of bovine serum albumin (BSA), 0.06 μCi of radiolabeled sodium iodide ($Na^{125}I$, 17 Ci/mg, New England Nuclear Corp.) and 40 nmol of carrier NaI. The reactions were initiated by the addition of 100 μM H_2O_2, incubated for 15 minutes at 37°C and terminated by the addition of 10% tricholoracetic acid (TCA). The BSA precipitates were washed three times with TCA. The extent of MPO-mediated iodination of BSA in the protein precipitate was determined in a solid-state gamma counter and the results expressed as nmol ^{125}I/5 mg BSA. In the presence of added H_2O_2 the results were 6.1 ± 0.3, 6.1 ± 0.1, 6.1 ± 0.2, and 5.7 ± 0.2 nmol of ^{125}I/5 mg BSA for MPO activity released from control, untreated PMNL, and PMNL pretreated with 10, 20, and 50 μg/ml ascorbate respectively. The corresponding values for control systems in the absence of added MPO were 0.81 ± 0.3, 0.81 ± 0.1, 0.83 ± 0.5, and 0.86 ± 0.6 nmol ^{125}I/5 mg BSA. These results of three separate determinations are expressed as the mean values ± SEM. No statistically significant differences were observed, which shows that ascorbate does not affect the activity of intracellular MPO.

TABLE 6

Oxidizing System	Percentage Oxidation of Ascorbate
Xanthine + xanthine oxidase only	42 ± 5
+ 12.5 mU of MPO	47 ± 6
+ 25 mU of MPO	49 ± 6
+ 50 mU of MPO	55 ± 5
+ 100 mU of MPO	80 ± 4
+ 200 mU of MPO	100

Measurement of the effects of inclusion of myeloperoxidase (12.5-200 mU) on the oxidation of ascorbate (10 μg/ml) mediated by the xanthine (0.1 mM)/xanthine oxidase (100 mU) superoxide generating system. The results are expressed as the mean percentage oxidation of ascorbate ± SEM. MPO at all concentrations tested caused statistically significant potentiation of ascorbate oxidation ($p < 0.05$-$p < 0.005$).

Modulation by Ascorbate of the Release of Reactive Oxidants by Circulating Phagocytes from Cigarette Smokers

Neutralization of extracellular RO by ascorbate may have biological significance, especially in conditions associated with chronic activation of phagocytes such as chronic inflammatory diseases and cigarette smoking, both of which are associated with depletion of leukocyte and plasma ascorbate.[36,37] We have developed a whole-blood LECL method to measure the generation of RO by FMLP/CB-activated circulating phagocytes from cigarette smokers and nonsmoking control subjects. This method, which has been described in detail elsewhere,[38] eliminates the trauma to PMNL of purification techniques and removal of environmental (plasma) antioxidant systems. As with purified PMNL, activation of whole blood with FMLP/CB resulted in a bimodal pattern of LECL with distinct (within 1 minute) early extracellular and later-occurring intracellular peaks of LECL.

TABLE 7

Test System	FMLP/CB-activated LUECL
PMNL only	11230 ± 358
PMNL + FMLP	7254 ± 666
PMNL + FMLP + MPO + H_2O_2	2964 ± 1136
PMNL + FMLP + MPO + H_2O_2 + 5 μg/ml ascorbate	6950 ± 1152
PMNL + FMLP + MPO + H_2O_2 + 10 μg/ml ascorbate	8936 ± 804
PMNL + FMLP + MPO + H_2O_2 + 20 μg/ml ascorbate	9910 ± 1052
PMNL + FMLP + MPO + H_2O_2 + 50 μg/ml ascorbate	9684 ± 994

Measurement of the protective effects of ascorbate (5-50 μg/ml) on the autoxidative inactivation of the respiratory burst enzyme mediated by added MPO (200 mU) and H_2O_2 (100 μM). PMNL were preincubated with 0.1 μM FMLP for 1 min to promote assembly and activation of the respiratory burst enzyme followed by addition of MPO and H_2O_2 ± ascorbate. After 15 min incubation at 37°C the PMNL were washed, resuspended in HBSS, and superoxide generation stimulated by FMLP/CB and measured using lucigenin-enhanced chemiluminescence. The results of four different experiments are expressed as the mean values ± SEM in r.l.u. for the maximal LUECL response. Ascorbate caused a statistically significant dose-related ($p < 0.025$-$p < 0.005$) protection of the respiratory burst enzyme from inactivation by MPO/H_2O_2. The values for control systems containing FMLP/MPO and FMLP/H_2O_2 pretreated PMNL were 6586 ± 1072 and 7320 ± 2122 r.l.u. respectively.

The spontaneous and FMLP/CB-activated LECL responses of heparinized whole blood taken from 30 asymptomatic cigarette smokers and 30 nonsmoking adult volunteers were measured *in vitro.* No differences in spontaneous LECL responses were detected. However FMLP/CB-activated LECL responses were markedly elevated in smokers' blood relative to nonsmokers. These results for the peak extracellular and intracellular LECL responses as well as serum ascorbate levels and circulating leukocyte counts are shown in TABLE 8. The mean percentage increases in the FMLP/CB-activated LECL responses of smokers' blood (relative to nonsmokers) were 58% ($p < 0.005$) and 75% ($p < 0.005$) for the extracellular and intracellular peaks respectively.

The same investigations were performed on all the cigarette smokers and nonsmoking control subjects before and 2 hours after the oral administration of a single

TABLE 8

Group	Leukocyte Counts	Serum Ascorbate Levels	Chemiluminescence Peaks Extracellular	Chemiluminescence Peaks Intracellular
Nonsmokers	6.2 ± 0.3	11.6 ± 3.8	705 ± 76	707 ± 56
Smokers	7.9 ± 0.4	9.9 ± 2.9	1119 ± 108	1235 ± 100

Comparison of circulating leukocyte counts, serum ascorbate levels, and FMLP/CB-activated peak extracellular and intracellular LECL responses of whole blood in 30 cigarette smokers and nonsmoking control subjects. The results are expressed as leukocytes $\times 10^6$/ml, μg ascorbate/ml serum and as relative light units respectively. The mean ages ± SEM were 29.3 ± 1.5 and 42.5 ± 2 for the nonsmokers and smokers respectively. The corresponding male:female ratios were 16:14 and 19:11 for each group. Relative to the nonsmokers the circulating leukocyte counts and peak extracellular and intracellular FMLP/CB-activated LECL responses of smokers were significantly increased ($p < 0.005$).

1 g dose of ascorbate. These results are shown in TABLE 9. Early occurring extracellular FMLP/CB-activated LECL responses were significantly decreased for smokers (mean inhibition 28%, $p < 0.005$) and nonsmokers (17%, $p < 0.01$). Intracellular LECL responses were unaffected in smokers and increased in nonsmokers (16%) following administration of ascorbate.

These results show that cigarette smoking primes circulating phagocytes to hyperreact to stimuli of membrane-associated oxidative metabolism. Administration of ascorbate to both smokers and nonsmokers is associated with decreased levels of phagocyte-derived RO.

Ascorbate-mediated Prevention of Oxidative Inhibition of Lymphocyte Proliferation

Mitogen-induced lymphocyte proliferation is susceptible to oxidative inhibition by H_2O_2[6–9] only and by H_2O_2 and peroxidases.[9,39] It has been suggested that immune

TABLE 9

Time of Testing	Leukocyte Counts	Serum Ascorbate Levels	Chemiluminescence Peaks Extracellular	Chemiluminescence Peaks Intracellular
Before ascorbate	7.9 ± 0.4	9.9 ± 2.9	1119 ± 108	1235 ± 1000
2 Hours after administration of 1 gram ascorbate	8.3 ± 1.3	17.2 ± 4.8	872 ± 106	1274 ± 120

Measurement of the effects of oral administration of a single 1 g dose of ascorbate on the peak extracellular and intracellular FMLP/CB-activated LECL responses of blood from 30 cigarette smokers measured before and 2 h after ascorbate administration. Measurement of circulating leukocyte counts and serum ascorbate levels were also performed before and 2 h after administration of ascorbate. Serum ascorbate levels increased significantly and peak extracellular FMLP/CB-activated LECL decreased significantly ($p < 0.005$) after the administration of ascorbate.

attrition (anergy) associated with certain cancers and chronic inflammatory disorders may be related to the immunosuppressive activity of RO released by chronically activated phagocytes.[8] To investigate a possible role for ascorbate in the protection of RO-exposed lymphocytes, mononuclear leukocytes (MNL) prepared as described previously[9] were exposed to an experimental oxidizing system (0.1 mM xanthine, 100 mU xanthine oxidase and 50 mU of MPO) ± ascorbate (10-200 μg/ml) washed and tested for responsiveness to the mitogen phytohemagglutinnin (PHA, 2.5 μg/ml) as previously described.[9] In control systems MNL were coincubated with MPO only, ascorbate only, or with the xanthine/xanthine oxidase system only. Mitogen-induced MNL proliferation was measured according to the extent of uptake of radiolabeled thymidine by MNL.[9] These results are shown in FIGURE 4. Exposure of

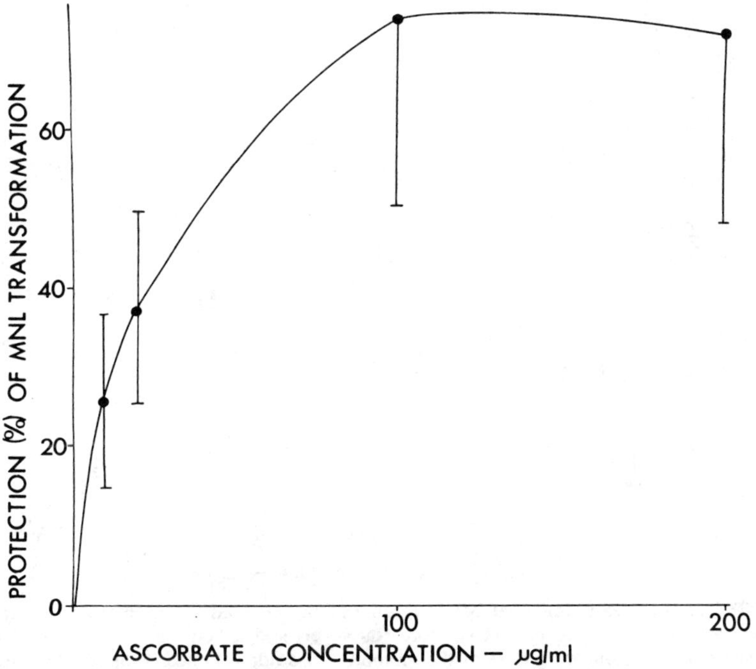

FIGURE 4. Measurement of the protective effects of ascorbate (10-200 μg/ml) on the inhibition of MNL PHA-activated transformation mediated by exposure to xanthine (0.1 mM)/xanthine oxidase (100 mU) and myeloperoxidase (50 mU) for 15 minutes. The results are expressed as mean percentage protection (± SEM) of four separate experiments. The absolute values for the xanthine/xanthine oxidase and oxidant-exposed systems in the absence of ascorbate were 27678 ± 12479 and 9679 ± 3055 counts per minute respectively.

MNL to MPO in the presence of xanthine and xanthine oxidase inhibited PHA-activated MNL proliferation (mean percentage inhibition ± SEM of 50 ± 16). Inclusion of ascorbate caused statistically significant dose-related protection of MNL proliferation (FIGURE 4). Exposure to ascorbate or MPO only or to the xanthine/xanthine oxidase system only did not influence MNL proliferation.

To investigate the possible *in vivo* significance of these findings MNL mitogen-induced transformation was investigated *in vitro* before and 2 hours after the oral administration of 1 gram of ascorbate to six adult human volunteers. To prevent loss of ascorbate during MNL processing procedures a whole blood method was preferred to the use of purified MNL. The mitogen PHA was used at suboptimal, optimal, and supraoptimal concentrations (0.3-20 μg/ml). To control for the effects of serial bleeding a control group of six adult volunteers was included in which PHA-induced MNL transformation was measured using blood taken at time 0 and 2 hours later. These results for the ascorbate and control groups are shown in FIGURES 5a and 5b respectively. Administration of ascorbate was associated with statistically significant

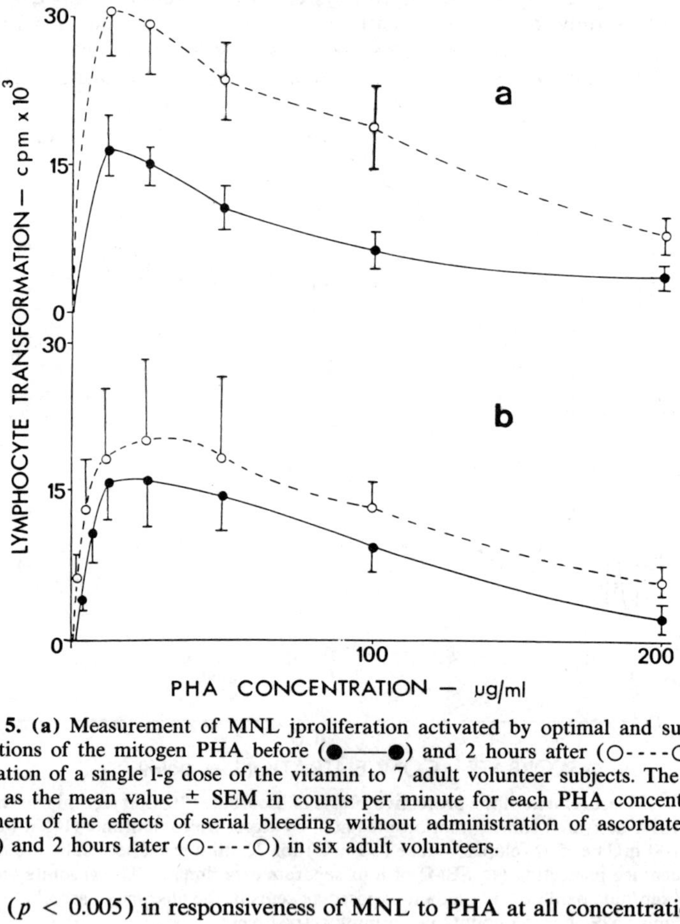

FIGURE 5. **(a)** Measurement of MNL jproliferation activated by optimal and supraoptimal concentrations of the mitogen PHA before (●——●) and 2 hours after (○- - - -○) the oral administration of a single l-g dose of the vitamin to 7 adult volunteer subjects. The results are expressed as the mean value ± SEM in counts per minute for each PHA concentration. **(b)** Measurement of the effects of serial bleeding without administration of ascorbate at time 0 (●——●) and 2 hours later (○- - - -○) in six adult volunteers.

increases ($p < 0.005$) in responsiveness of MNL to PHA at all concentrations of the mitogen.

DISCUSSION

Activation of human phagocytes by soluble and particulate stimuli of membrane-associated oxidative metabolism leads to the generation of a series of stable and unstable

RO. Phagocyte-derived RO are not confined to the intracellular compartment but are also released extracellularly. Since these RO possess wide-ranging deleterious activities, their sustained release in conditions associated with chronic phagocyte activation presents the potential hazard of oxygen toxicity. Antioxidant mechanisms that neutralize phagocyte-derived RO *in vivo* are doubtless of major importance in the maintenance of good health. The results presented here demonstrate that ascorbate mediates efficient neutralization of phagocyte-derived potentially harmful extracellular oxidants, while the intracellular generation of antimicrobial oxidants remains intact.

Ascorbate caused a dose-related inhibition of the early-occurring peak of LECL in FMLP/CB-activated PMNL. The early peak corresponds to the extracellular release of RO by activated PMNL.[25,26] However, ascorbate did not inhibit and indeed potentiated the later-occurring intracellular peak of LECL in FMLP/CB-activated PMNL. This could indicate a predominantly extracellular mode of action for ascorbate, which would correlate with its water solubility. We have shown here, however, that ascorbate is indeed actively taken up by PMNL. This uptake is modulated by glucose concentration and cell activation. At 10 mM glucose (the concentration in HBSS used here), activation of PMNL increases ascorbate uptake. In fact, the average intracellular volume is 0.4 μl (calculated from the uptake of tritiated water into 10^6 PMNL), therefore at 10 mM glucose intracellular ascorbate concentration is 179.8 $\pm$ 34.8 μM in stimulated PMNL and 110.1 $\pm$ 11.6 μM in resting PMNL. As ascorbate at these concentrations extracellularly is able to inhibit LECL significantly, one must assume that their failure to inhibit intraphagocytic LECL is due to the relative inaccessibility of sites of intracellular oxidant generation. Potentiation of intracellular RO generation is due to an extracellular protective effect of ascorbate on PMNL. The RO generating system is susceptible to autoxidative inactivation[40,41] and this is largely prevented by extracellular ascorbate. Alternatively stimulation of intracellular RO production by ascorbate may be due to reduction of the ferryl heme iron of inactive compound II into the ferric heme iron of native MPO and thus increase the activity of the MPO/H_2O_2/halide system.[42] This latter effect takes place at submicromolar concentrations of ascorbate. However in our system intracellular ascorbate concentrations are < 100 μM, and pretreatment of PMNL with ascorbate followed by washing did not increase the activity of intracellular MPO released by sonication and assayed using a protein iodination assay with added H_2O_2. Ascorbate therefore increases intraphagocytic oxidant generation by neutralizing extracellular autoreactive oxidants leading to sustained activity of the respiratory burst enzyme in the PMNL membrane.

In addition to inhibition of LECL, which detects oxidants generated by the MPO/H_2O_2/halide system,[25,26] we also observed that ascorbate caused a dose-related inhibition of lucigenin-enhanced chemiluminescence by FMLP/CB-activated PMNL as well as by a cell-free xanthine/xanthine oxidase superoxide generating system. To investigate the mechanism of ascorbate-mediated inhibition of LUECL (*i.e.,* a true inhibition of superoxide generation or a radical scavenging mechanism) we measured the residual reducing capacity of ascorbate following exposure to xanthine/xanthine oxidase. The reducing activity of ascorbate was diminished by exposure to the superoxide generating system and confirms the results of a similar study by Nishikimi.[43] This shows that ascorbate is oxidized to dehydroascorbate by superoxide and supports a scavenging mechanism according to the following reaction[43]:

$$\text{Ascorbate}^- + 2O_2^{\bullet -} + 3H^+ \rightarrow \text{dehydroascorbate} + 2H_2O_2.$$

Hemila *et al.* have reported that extracellular ascorbate is effectively oxidized by activated PMNL *in vitro* and does not interfere with membrane-associated oxidative metabolism.[34] Superoxide was the principal oxidant of ascorbate with a lesser contribution from the MPO/H_2O_2/halide system. In agreement with this report, we observed that the reducing capacity of ascorbate is protected by SOD and that oxidation of ascorbate exposed to the xanthine/xanthine oxidase system is potentiated by MPO. Exposure of ascorbate to concentrations of up to 1 mM H_2O_2 (in the absence of MPO) did not promote ascorbate oxidation. Similarly, inclusion of catalase with the xanthine/xanthine oxidase superoxide generating system did not prevent ascorbate oxidation. In experiments with catalase the results with enzymatically active enzyme were always compared with heat-treated (80°C for 1 min) enzymatically inactive catalase. From these investigations it can be concluded that ascorbate functions as an antioxidant primarily in the extracellular milieu where it scavenges superoxide and RO generated by the MPO/H_2O_2/halide system. The vitamin is probably of considerable importance in the prevention of toxicity mediated by these phagocyte-derived RO. Since ascorbate does not neutralize H_2O_2 per se it is probably ineffective in protection against H_2O_2-mediated toxicity. Importantly, the protective antioxidant functions of ascorbate are accomplished without compromising the antimicrobial functions of the PMNL.

Increased consumption of ascorbate is a consequence of cigarette smoking[36] and chronic inflammatory disorders such as rheumatoid arthritis,[37] both of which are associated with chronic phagocyte activation. This suggests a biological role for ascorbate in the neutralization of phagocyte-derived RO. We investigated the effects of oral administration of ascorbate on the extracellular reactivity of oxidants released by activated phagocytes from cigarette smokers. Cigarette smoking is a good model to use for this type of investigation since it is uncomplicated by chemotherapy. In agreement with previous reports, we observed that circulating phagocytes from cigarette smokers are primed to release increased amounts of RO.[21] Administration of a single 1-g oral dose of ascorbate was associated with a significant decrease (28%; $p < 0.005$) in the peak of extracellular LECL by FMLP/CB-activated smokers' phagocytes.

Phagocyte-derived RO are also autotoxic for phagocytes[3–5] and inhibit the proliferation of T lymphocytes,[6–9,39] B lymphocytes,[39] and cytotoxic activity of natural killer cells.[39] Oxidants generated by the MPO/H_2O_2/halide systems are potent immunosuppressive agents.[9,39] The immune attrition associated with certain cancers and chronic inflammatory disorders has been attributed to RO released extracellularly by chronically activated phagocytes.[8] To investigate the protective effects of ascorbate on oxidative inactivation of MNL proliferation, we exposed human MNL to an experimental oxidizing system (MPO/xanthine/xanthine oxidase) in the presence and absence of ascorbate (10-200 μg/ml). Exposure to the oxidizing system caused marked inhibition of mitogen-induced proliferative activity. However inclusion of ascorbate with the oxidizing system caused a dose-related protection of proliferative activity. Ascorbate was administered (as a single 1-g oral dose) to seven adult volunteer subjects. Mitogen-activated proliferative responses of blood taken 2 hours after ascorbate administration were compared with the corresponding preascorbate responses. A whole blood method was used to prevent the loss of cell-associated ascorbate by preparative procedures and to include PMNL as a source of immunosuppressive oxidants. Ascorbate administration was associated with significantly increased proliferative responses to optimal and supraoptimal concentrations of the mitogen PHA. These findings suggest that ascorbate may be of value in reducing the level of immunosuppression that may accompany conditions associated with chronic phagocyte activation.

Finally phagocyte-derived RO which are autotoxic,[3–5] immunosuppressive,[6–9] mutagenic,[10,11] and potentially carcinogenic[12] and which enhance proteolysis[13,14] are neutralized by ascorbate *in vitro.* It is proposed that this vitamin is of considerable importance in the neutralization of toxic phagocyte-derived RO and consequently in the maintenance of good health.

REFERENCES

1. BADWEY, J. A. & M. L. KARNOVSKY. 1980. Ann. Rev. Biochem. **49:** 695-726.
2. BABIOR, B. M. 1984. Blood **64:** 959-966.
3. BAEHNER, R. L., L. A. BOXER, J. M. ALLEN & J. DAVIS. 1977. Blood **50:** 327-335.
4. STENDAHL, O., B. I. COBLE, C. DAHLGREN, J. HED & L. MOLIN. 1984. J. Clin. Invest. **73:** 366-373.
5. MATZNER, Y., L. M. BRASS, B. J. MCMURRICH, W. A. PETERS, J. ANDRE-SCHWARTZ & B. M. BABIOR. 1982. Blood **60:** 822-826.
6. ZOSCHKE, D. C. & R. P. MESSNER. 1984. Clin. Immunol. Immunopathol. **32:** 29-40.
7. AUNE, T. M. 1984. J. Immunol. **133:** 899-906.
8. CARSON, D. A., S. SEKO & D. B. WASSON. 1986. J. Exp. Med. **163:** 746-751.
9. ANDERSON, R. 1981. Am. J. Clin. Nutr. **34:** 1906-1911.
10. WEITZMAN, S. A. & T. P. STOSSEL. 1982. J. Immunol. **128:** 2770-2772.
11. WEITBERG, A. B., S. A. WEITZMAN, M. DESTREMPES, S. A. LATT & T. P. STOSSEL. 1983. N. Engl. J. Med. **308:** 26-30.
12. WEITZMAN, S. A., A. B. WEITBERG, E. P. CLARK & T. P. STOSSEL. 1985. Science **227:** 1231-1233.
13. CLARK, R. A., P. J. STONE, A. EL-HAG, J. D. CALORE & C. FRANZBLAU. 1981. J. Biol. Chem. **256:** 3348-3353.
14. CARP, H. & A. JANOFF. 1980. J. Clin. Invest. **66:** 987-995.
15. MANALE, B. L. & T. D. BROWER. 1973. Surg. Gynecol. Obstet. **136:** 63-64.
16. KUNTZ, R. E., A. M. CHEEVER & B. J. MYERS. 1972. J. Natl. Cancer Inst. **68:** 223-235.
17. GILBERT, J. B. 1944. J. Urol. **51:** 296-300.
18. SAEED, W., S. KIM & B. H. HURCH. 1974. Arch. Surg. **108:** 376-379.
19. FOX, H. B., P. DE TOGNI & B. M. BABIOR. 1985. Immunology Today **6:** 327-328.
20. LUDWIG, P. W., B. A. SCHWARTZ, J. R. HOIDAL & D. E. NIEWOEHNER. 1985. Am. Rev. Resp. Dis. **131:** 838-830.
21. HOIDAL, J. R. & D. E. NIEWOEHNER. 1982. Am. Rev. Resp. Dis. **126:** 548-552.
22. LUDWIG, R. W. & J. R. HOIDAL. 1982. Am. Rev. Resp. Dis. **126:** 977-980.
23. ANDERSON, R. 1981. Clin. Exp. Immunol. **43:** 180-188.
24. ANDERSON, R. & A. J. VAN RENSBURG. 1979. Immunology **37:** 15-24.
25. BENDER, J. G. & D. E. VAN EPPS. 1983. Infect. Immun. **41:** 1062-1070.
26. BRIHEIM, G., O. STENDAHL & C. DAHLGREN. 1984. Infect. Immun. **45:** 1-5.
27. ANDERSON, R., G. JOONE & C. E. J. VAN RENSBURG. 1986. J. Infect. Dis. **153:** 593-600.
28. NACCACHE, P. H., H. J. SHOWELL, E. L. BECKER & R. I. SHA'AFI. 1977. J. Cell Biol. **73:** 428-444.
29. JOHNSON, J. D., W. L. HAND, J. B. FRANCIS, N. KING-THOMPSON & R. W. CARWIN. 1980. J. Lab. Clin. Med. **95:** 429-439.
30. BIGLEY, R., M. WIRTH, D. LAYMAN, M. RIDDLE & L. STANKOVA. 1983. Diabetes **32:** 545-548.
31. MOSER, U. & F. WEBER. 1984. Int. J. Vit. Nutr. Res. **54:** 47-53.
32. STELMAZYNSKA, T. & J. M. ZGLICZYNSKI. 1974. Eur. J. Biochem. **45:** 305-312.
33. THERON, A. J. & R. ANDERSON. 1985. Am. Rev. Resp. Dis. **132:** 1049-1054.
34. HEMILA, H., P. ROBERTS & M. WIKSTROM. 1984. FEBS Lett. **178:** 25-30.
35. DAHLGREN, C., H. ANIANSSON & K-E. MAGNUSSON. 1985. Infect. Immun. **47:** 326-328.
36. BROOK, M. & J. J. GRIMSHAW. 1968. Am. J. Clin. Nutr. **21:** 1254-1258.
37. SAHUD, M. A. & R. J. COHEN. 1971. Lancet **1:** 937-938.
38. ANDERSON, R., P. T. LUKEY, A. J. THERON & U. DIPPENAAR. 1986. Agents Actions. In press.

39. El-Hag, A., P. E. Lipsky, M. Bennett & R. A. Clark. 1986. J. Immunol. **136:** 3420-3426.
40. Jandl, R. C., J. Andre-Schwartz, L. Borges-Du Bois, R. S. Kipnes, B. J. McMurrich & B. M. Babior. 1978. J. Clin. Invest. **81:** 1176-1185.
41. Nauseef, W. M., J. A. Metcalf & R. K. Root. 1983. Blood **61:** 483-489.
42. Bolscher, B. G. J. M., G. R. Zoutberg, R. A. Cuperus & R. Wever. 1984. Biochem. Biophys. Acta **784:** 189-191.
43. Nishikimi, M. 1975. Biochem. Biophys. Res. Commun. **63:** 463-468.

DISCUSSION OF THE PAPER

A. B. Kitabchi (*University of Tennessee, Memphis, Tenn.*) These are very nice studies. I have a couple of questions. First, the specificity of ascorbic acid or have you tried any others? Secondly, you mentioned up to 50 micrograms. I'm wondering whether you believe that megadoses of vitamin C could be injurious?

R. Anderson (*University of Pretoria, Pretoria, South Africa*): To answer your first question, the mechanism I demonstrated here is not exclusive to ascorbic acid; other water-soluble reducing agents have the same effect. We've tested cysteine and it does essentially the same although it's a little less impressive than ascorbic acid. We chose not to exceed 50 μg per ml; we wanted to remain within the physiological concentrations and presumably reasonable pharmacological concentrations.

We found that if we went up to 200 μg per ml of ascorbic acid we did observe interference with the phagocyte antimicrobial systems. But up to 100 μg per ml there is no obvious interference.

E. J. Diliberto (*Wellcome Research Labs, Research Triangle Park, N.C.*): I'm a little confused. What would you predict to be the effect of ascorbic acid on the cytotoxic effects of killer lymphocytes on tumor cells? Can you make any extrapolations in that system, where I think various oxygen species have been shown to be the mediator in cytotoxicity?

R. Anderson: I think that has not been confirmed and that it's actually granule factors which mediate the cytotoxic activity. So one has to be very cautious there, it would appear that the role of reactive oxidants in the mediation of cytotoxic activity by NK cells is somewhat dubious.

B. L. Horecker (*Cornell University Medical College, New York, N.Y.*): Has anyone looked at ascorbic acid in that model?

A. Bendich (*Hoffmann-La Roche, Nutley, N.J.*): We haven't published it. We looked at the effect of vitamin C on NK activity in rat and guinea pig cells and the only activity that we were able to discern was in vitamin-C-low or vitamin-C-deficient guinea pigs the NK activity increased. It may be that vitamin C is important in maintaining the membrane structure of the natural killer cells and when it is deficient the cell increases its nonspecific natural killer cell activity.

R. Anderson: It has recently been reported that the cytotoxic activity of NK cells can be modulated by phagocyte-derived oxidants, oxidants released by the myeloperoxidase hydrogen peroxide system.

Unidentified Speaker: Have you looked at macrophages at all and have you done any biological correlates like phagocytic index or bacteriocidal activity?

R. ANDERSON: We haven't looked at macrophage effects, and with regard to functional correlates we believe that its immunostimulatory potential is probably related to these antioxidant mechanisms which I've described here by protecting the phagocyte membrane; the phagocyte function such as migratory phagocytic capacity and antimicrobial activities are probably sustained.

Ascorbic Acid Metabolism in Diabetes Mellitus[a]

ROGER E. PECORARO AND MEI S. CHEN[b]

Seattle VA Medical Center
Seattle, Washington 98108
and
Department of Medicine
University of Washington
Seattle, Washington 98105

Levels of ascorbic acid (AA) are decreased in various tissues of animals with experimental diabetes.[1,2] Altered plasma levels and turnover of AA have been reported in diabetic patients,[3,4] but these observations and their biologic importance have not been explained. Noting the structural similarity between glucose and AA (FIG. 1), Mann suggested in 1974 that glucose and vitamin C might occupy the same membrane transport system.[5] He subsequently reported with Newton that elevated glucose levels interfered with cellular AA transport in erythrocytes.[6] Others have observed inhibition by glucose of AA transport *in vitro* by human lymphocytes[7] and bovine endothelial cells.[8] Bigley *et al.* described competitive inhibition between the *in vitro* uptake of dehydroascorbic acid and glucose analogues by human polymorphonuclear leukocytes (PMN) and fibroblasts, and concluded on the basis of kinetic data that the competing ligands utilized the same membrane carrier.[9]

We recently subjected normal human volunteers to intravenous glucose tolerance testing in order to study the effect of acute hyperglycemia *in vivo* on the concentration of AA in mononuclear leukocytes (MNL).[10] Significant decreases in intracellular AA levels promptly followed the rapid glucose infusions.

The concentrations of AA in polymorphonuclear leukocytes and mononuclear leukocytes reportedly exceed plasma levels 25-fold and 80-fold, respectively, in normal individuals,[11] although the physiologic significance of this concentration gradient is not understood. Several published observations suggest a possible association between intracellular ascorbate, abnormal PMN function, and an increased frequency of infection in subgroups of patients with congenital leukocyte disorders.[12–23] Furthermore, ascorbate may affect neutrophil and monocyte chemotaxis.[15,16,23–25] Leukocyte dysfunction occurs in diabetes mellitus,[26–31] and is particularly associated with hyperglycemia and poor control.[32–35]

In this study we compared the effects of an acute intravenous glucose load on the ascorbic acid levels of polymorphonuclear leukocytes and mononuclear leukocytes in

[a] This work was supported by an ADA feasibility grant, a research grant-in-aid from Hoffmann-La Roche, Inc., and the Veterans Administration. A portion of this work was conducted at the Clinical Research Center of the University of Washington and supported by NIH grant no. RR-37.

[b] Current address: Kaohsiung Adventist Clinic, Kaohsiung 800, Taiwan.

normal subjects. We examined concurrently whether leukocyte chemotaxis changed in association with alterations in intracellular AA. Finally, we investigated the effects of protracted *in vivo* hyperglycemia, maintained for up to four hours by a glucose clamp technique, on leukocyte ascorbic acid concentrations and chemotactic function.

MATERIALS AND METHODS

Normal male volunteers (ages 27-68) were studied according to a protocol approved by the Human Subjects Review Committee of the University of Washington, and after obtaining written informed consent. Subjects were asked to withhold all over-the-counter medications, including vitamins, for at least 24 hours before the study and to maintain a 12-hour overnight fast. Ten subjects received glucose as a bolus infusion, followed by examination of its effects on leukocyte AA concentrations.

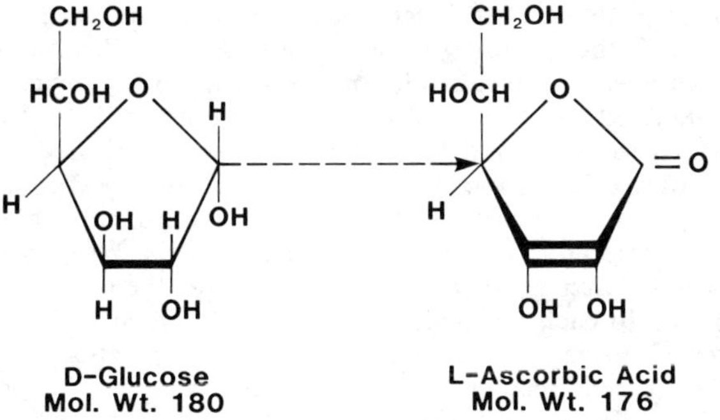

FIGURE 1. Molecular structures of D-glucose and L-ascorbic acid.

Chemotaxis by leukocytes from eight of the subjects was examined as well. Six of the eight returned at a later date for similar studies during continuous glucose infusions to maintain the blood glucose above 250 mg/dl for 4 hours.

Two baseline blood samples (−5, 0 minutes) were initially obtained from an antecubital vein prior to all glucose infusions. During studies with a single acute challenge of i.v. glucose, baseline blood sampling was followed by administration of 0.3 g per kg of 50% glucose given as a bolus to an indwelling butterfly needle in the contralateral antecubital vein. Blood samples were drawn at 5, 15, 20, 30, 60, 90, and 120 minutes after the glucose bolus. Glucose clamp studies were performed on another day several weeks later. Baseline blood samples were obtained, followed by the infusion of a loading dose of 25 g of 50% glucose over 12 minutes into an antecubital vein. In order to maintain the plasma glucose concentration above 250 mg/dl, the loading dose was followed by a continuous infusion of 10% glucose at a rate of 1 g per minute or, if necessary, 20% glucose at a rate of 2 g per minute.[36] Blood samples were subsequently drawn from the contralateral antecubital vein at 5-15 minute intervals to monitor the plasma glucose levels, and at 60, 120, 210, and 240 minutes for plasma and intracellular leukocyte AA levels. Tests of leukocyte chemotaxis were done at baseline and at 210 and 240 minutes.

Plasma was separated from cellular components by centrifuging 20 ml heparin anticoagulated venous blood. An aliquot of plasma was frozen at −20°C for subsequent glucose determination. The remainder of the plasma and aliquots of the MNL and PMN fractions (described below) were processed on the same day of sample collection for ascorbic acid determination as described previously.[37]

After dilution with PBS, the cells were separated by centrifugation over a Ficoll-Hypaque solution with a density of 1.095 g/ml.[38] Two distinct fractions representing predominantly PMN and MNL cells were separately washed and counted using a hemocytometer. Smears were prepared and stained by Diff-Quick (Scientific Products) for differential cell counting. The PMN fraction contained 95-98% PMNs, with mononuclear cells composing the non-PMN cell contaminants. The MNL fraction represented 85 ± 5% lymphocytes, 8 ± 4% monocytes, with the remainder composed of contaminating PMNs. After aliquots of the PMN and MNL fractions were removed for determinations of ascorbic acid, the remaining cells were reserved for tests of chemotactic function.

Plasma glucose was determined by the glucose oxidase method using the Technicon (Tarrytown, N.Y.) Autoanalyzer II. Plasma and cellular ascorbic acid levels were determined by HPLC.[37] The intra-assay coefficient of variation (CV) is 2.4%. The inter-assay CV is 4.3%.

Chemotaxis of PMN and MNL cells was measured in microchemotaxis chambers (Neuroprobe Co., Bethesda, Md.) using the modified Boyden technique.[39,40] PMNs were diluted with Gey's basic salt solution (with 2% bovine serum albumin fraction V, without phenol red, pH 7.2 ± 0.1; Gibco) to contain 1×10^6 /ml, and chemoattractants used were 5% and 10% human zymosan activated serum (ZAS). For MNL, the cells were diluted to contain 1×10^6 /ml myeloperoxidase positive cells; 10% and 20% ZAS were used as chemoattractants. Additionally, chemotaxis for both cell fractions was mesured to the chemoattractant, 10^{-8} M *N*-formylmethionyl-leucyl-phenylalanine-methylester (FMLP). PBS was used for the control. Chemoattractant (25 μl) was added to each chamber's bottom compartment, with 50 μl of cell suspension introduced to the top compartment. For PMN, chemotaxis was quantified as the total number of cells that migrated completely through a 3μ pore size filter and were visible in five adjacent light microscopic fields (450×), after incubation for three hours at 37°C (5% CO_2, 100% humidity). For MNL, the filter pore size was 5 μ; incubation time was 90 minutes at 37°C (5% CO_2, 100% humidity), and chemotaxis was quantified as the total number of cells that migrated completely through the filter and were visible in five adjacent light microscopic fields (970×). Chemotaxis was determined in triplicate for each chemoattractant and the replicates averaged.

Changes from baseline for values of leukocyte intracellular AA concentrations and leukocyte chemotaxis were evaluated using Student's *t*-test. Correlation between leukocyte AA and plasma AA concentrations was tested by least-squares linear regression analysis. The relationship between mean fasting intracellular AA concentrations and the nadir of mean intracellular AA after glucose infusions was examined by one-way analysis of variance. Associations between changes in chemotaxis and decreases in leukocyte AA concentration were evaluated by the Kendall rank order correlation test.

RESULTS

Intracellular AA Leukocyte Levels

The concentration of AA in polymorphonuclear leukocytes in the fasting state was about half that in mononuclear leukocytes, as shown in TABLE 1. In order to permit comparison with plasma AA levels, the intracellular AA concentration was expressed as mg/dl, converted from $\mu g/5 \times 10^6$ cells, using cell volume data from the literature.[41]

We found the average intracellular AA level in PMN was 17-fold concentrated compared with plasma, and MNL levels 43-fold increased. The MNL AA and plasma AA levels were correlated ($r = 0.65, p = 0.04, n = 10$), but no significant correlation was found between PMN AA and plasma AA ($r = 0.35, p = 0.32$).

FIGURE 2 demonstrates the effect of acute hyperglycemia on leukocyte AA levels. Ascorbic acid levels were expressed as the percentage of the basal concentration to facilitate analysis, since we observed a significant correlation between the fasting intracellular AA levels and the nadir of intracellular AA after glucose infusion for both MNL ($r = 0.79, p < 0.01, n = 10$) and PMN ($r = 0.75, p < 0.05, n = 10$). The decline in intracellular AA induced by hyperglycemia was not as dramatic in PMN (nadir $-26 \pm 12\%$ [mean ± SD] at 30 minutes, $n = 8, F = 6.29, p < 0.001$) compared to that in MNL ($-38 \pm 21\%$ at 20 minutes, $n = 8, F = 4.64, p < 0.01$). Similar results were obtained when statistical significance was calculated for the absolute decreases in PMN and MNL mean intracellular AA levels. The effect persisted for 30 minutes in MNL compared with 120 minutes in PMN. Plasma AA levels did not change.

TABLE 1. The Comparative Ascorbic Acid Concentration in Leukocytes and Plasma in Ten Normal Subjects in the Fasting State

	Measured			Calculated	
	MNL ($\mu g/5 \times 10^6$ Cells)	PMN ($\mu g/5 \times 10^6$ Cells)	Plasma (mg/dl)	MNL (mg/dl)	PMN (mg/dl)
Range	0.34-2.55	0.16-1.1	0.12-3.12	18.9-141.7	7.1-48.9
Mean	1.12	0.56	1.43	62.2	24.8
SD	0.66	0.34	0.89	36.8	15.2
Mean cell/plasma ratio				43×	17×
Cell/plasma ratio (range), Evans *et al.*[11]				26-84×	14-33×

Leukocyte Chemotaxis

We initially tested leukocyte chemotaxis before and after the declines in intracellular AA concentrations induced by acute hyperglycemia. Chemotaxis was measured for MNL at −5, 0, 15, and 30 minutes, since the decline in the intracellular AA concentration reached a nadir between 15-20 minutes and persisted for 30 minutes. Chemotactic function was evaluated in PMN at 0, 60, 90, and 120 minutes, since the decline in intracellular AA persisted for 120 minutes. We were not able to demonstrate any acute change in chemotactic function in either MNL or PMN following the single glucose pulse.

Unlike the observations following the single bolus infusion of glucose, chemotaxis by both PMN and MNL was significantly suppressed to all chemoattractants tested after exposure to hyperglycemia *in vivo* for 210 or 240 minutes (TABLES 2 and 3; FIG. 3). During these glucose clamp studies, plasma glucose was successfully maintained throughout the duration of the infusion above 250 mg/dl. We observed a decrease in the plasma AA concentration from a mean baseline concentration of 0.82 mg per dl to 0.56 mg per dl at 240 minutes (FIG. 3). Total urinary AA output was measured during the infusion in four subjects, and varied from 1-20 mg/4 h, which appeared to be sufficient to account for the decrease in plasma AA. Mean intracellular concentrations of AA decreased from the baseline in both MNL and PMN, and remained depressed throughout the period of the hyperglycemic clamp. However, this decrease was not statistically significant at all points due to small sample size.

In both PMN and MNL, the decrements in chemotaxis compared to the mean baseline were similar ($p > 0.05$) for all the experimental chemoattractants. Changes

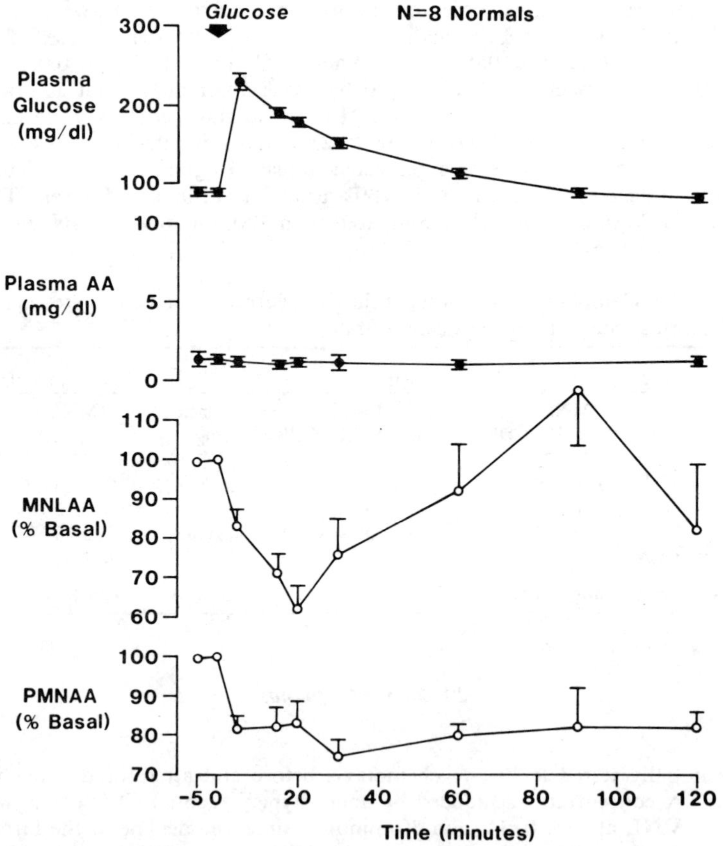

FIGURE 2. The effect of acute hyperglycemia on plasma, mononuclear (MNL), and polymorphonuclear (PMN) leukocyte ascorbic acid (AA) concentrations ($n = 8$, mean ± SEM).

in chemotaxis to all chemoattractants were combined for an analysis of possible association between changes in chemotaxis and the corresponding decreases in intracellular AA concentration. The changes in the various tests of chemotaxis by PMN correlated significantly with decreases in intracellular AA at 210 min ($p = 0.03$) and 240 min ($p < 0.01$). Changes in tests of chemotaxis by MNL correlated with decreases in intracellular AA at 240 min ($p = 0.03$), but the correlation was not significant at 210 min ($p = 0.1$).

TABLE 2. Chemotaxis by MNL from 6 Normals to Various Chemoattractants before and during Hyperglycemic Clamp (Mean ± SD)

	Time		
	Mean Baseline	210 min	240 min
Buffer	78 ± 3	75 ± 5	73 ± 4[a]
FMLP[d]	64 ± 3	59 ± 3[b]	58 ± 3[c]
10% ZAS[e]	104 ± 2	96 ± 3[c]	95 ± 3[c]
20% ZAS[e]	140 ± 4	134 ± 3[b]	128 ± 5[c]

[a] $p < 0.05$.
[b] $p < 0.01$.
[c] $p < 0.001$.
[d] Formyl-methionyl-leucyl-phenylalanine-methylester.
[e] Zymosan activated serum.

DISCUSSION

In these studies using the sensitive HPLC method to measure AA, we have confirmed that both polymorphonuclear leukocytes and mononuclear leukocytes concentrate ascorbic acid *in vivo.* The intracellular AA concentration averaged 17-fold in PMN compared with fasting plasma levels and 43-fold in MNL, comparable to concentrtions calculated by Evans *et al.*[11] Our data revealed a correlation between MNL AA and plasma AA, but a similar relationship could not be shown for PMN.

In order to assess a possible physiologic correlation of the acute decreases in leukocyte intracellular AA observed after i.v. glucose infusion, we examined whether changes occurred simultaneously in leukocyte chemotaxis. A central role for AA in leukocyte function has been implied by previous observations, but has not been clearly defined. Addition of exogenous AA to PMN has been reported to stimulate oxidative metabolism [12] and glycolysis,[13] to enhance microtubule assembly,[14] to stimulate random and directional movement,[15] and to prevent chemotactic deactivation.[16] In one study the AA effect on neutrophil mobility was related to the inhibition of the activity of

TABLE 3. Chemotaxis by PMN from 6 Normals to Various Chemoattractants before and during Hyperglycemic Clamp (Mean ± SD)

	Time		
	Mean Baseline	210 min	240 min
Buffer	25 ± 3	19 ± 1[a]	19 ± 1[a]
FMLP[c]	29 ± 3	23 ± 4[a]	23 ± 4[a]
5% ZAS[d]	64 ± 3	57 ± 3[a]	55 ± 6[a]
10% ZAS[d]	105 ± 3	97 ± 3[b]	95 ± 5[b]

[a] $p < 0.01$.
[b] $p < 0.001$.
[c] Formyl-methionyl-leucyl-phenylalanine-methylester.
[d] Zymosan activated serum.

the myeloperoxidase (MPO) H_2O_2/halide system.[17] Ascorbic acid has been shown to improve abnormal bactericidal activity in several conditions.[18] Ascorbic acid given to patients with Chediak-Higashi syndrome [19,20] and chronic granulomatous disease [12,21] has been reported to improve neutrophil function and clinically to decrease the frequency of infection.

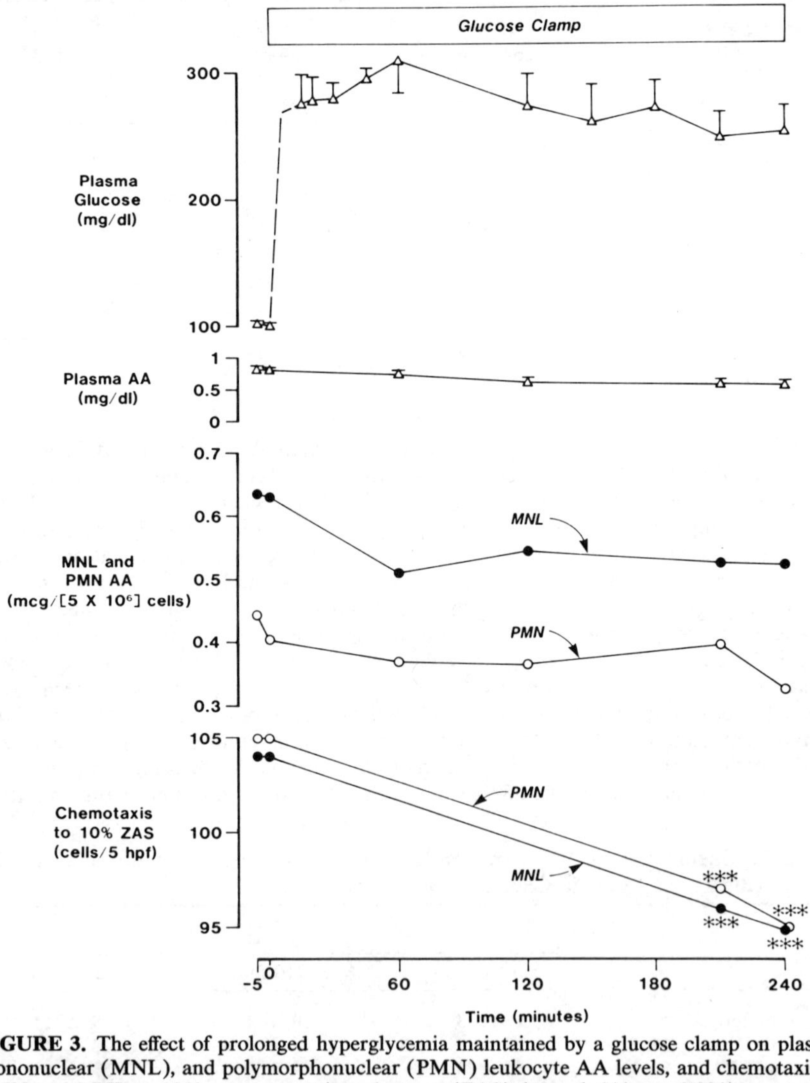

FIGURE 3. The effect of prolonged hyperglycemia maintained by a glucose clamp on plasma, mononuclear (MNL), and polymorphonuclear (PMN) leukocyte AA levels, and chemotaxis by PMN and MNL to 10% zymosan activated serum (ZAS) ($n = 6$; $^{***}p < 0.001$).

Abnormal PMN chemotaxis has been associated with hyperglycemia,[30,31] as have impaired intracellular killing and impaired phagocytosis.[32,33] PMNs from patients with poorly controlled diabetes reportedly failed to kill bacteria as effectively as neutrophils from normal subjects or patients with controlled diabetes.[34,35] The mechanisms responsible for these effects have not been demonstrated.

We elected to measure chemotaxis as a screening test for leukocyte function since it represents an important initial step in the acute inflammatory response, a function known to be affected *in vitro* by ascorbic acid [15,16,24] and reported to be depressed in diabetes mellitus.[23,25–31] We were unable to demonstrate acute changes in chemotactic function of either PMN or MNL from normal men following acute, transient hyperglycemia. However, following prolonged hyperglycemia, maintained *in vivo* by an intravenous glucose clamp for 3½ hours or longer, chemotaxis by both PMN and MNL directed at several experimental chemoattractants was significantly inhibited. Of particular interest, the changes in chemotaxis by both cell types correlated with decreases in the intracellular concentrations of ascorbic acid after 4 hours of continous hyperglycemia. These observations support the hypothesis that chronic hyperglycemia, as in the case of diabetes mellitus, may impair leukocyte chemotactic function by chronically depressing intracellular ascorbate. Continuous increased urinary losses of AA via osmotic diuresis in parallel with chronic glycosuria are likely to deplete the plasma compartment of ascorbic acid. These hypothetical mechanisms could result in a defective acute inflammatory response in diabetes, which in turn might contribute to increased susceptibility to bacterial infections and perhaps to delayed wound healing, which depends on prompt early infiltration by PMNs and macrophages to initiate the normal cellular sequence of wound repair.[42–44]

SUMMARY

Competition for membrane transport between glucose and ascorbic acid (AA) has been shown *in vitro* in human lymphocytes, granulocytes, and fibroblasts. Therefore, we examined the effects of acute administration of i.v. glucose on AA levels in mononuclear (MNL) and polymorphonuclear leukocytes (PMN) and on leukocyte chemotaxis. Plasma glucose and AA, MNL AA, PMN AA, and chemotaxis by MNL and PMN were measured before and after i.v. glucose in fasted normal male volunteers. A decline in AA occurred in PMN as well as MNL, but decreases in AA induced acutely by transient hyperglycemia were not associated with changes in chemotaxis. However, under conditions of prolonged hyperglycemia maintained by a glucose clamp technique, significant changes ($p < 0.01$) in chemotaxis by both PMN and MNL were observed after 210 and 240 min, with changes in chemotaxis to several chemoattractants significantly correlated with decreases in intracellular AA after 240 min ($p < 0.05$). These results are consistent with the hypothesis that chronic hyperglycemia may be associated with intracellular deficits of leukocyte AA, an impaired acute inflammatory response, and altered susceptibility to infection and faulty wound repair in patients with diabetes.

REFERENCES

1. YEW, M. S. 1983. Effect of streptozotocin diabetes on tissue ascorbic acid and dehydroascorbic acid. Horm. Metab. Res. **15:** 158.
2. RIKANS, L. E. 1981. Effect of Alloxan diabetes on rat liver ascorbic acid. Horm. Metab. Res. **13:** 123.
3. SOM, S., D. BASU, S. MUKHERJEE, *et al.* 1981. Ascorbic acid metabolism in diabetes mellitus. Metabolism **30:** 572-577.

4. BANERJEE, A. 1982. Blood dehydroascorbic acid and diabetes mellitus in human beings. Ann. Clin. Biochem. **19:** 65-70.
5. MANN, G. V. 1974. The impairment of transport of ascorbic acid by monosaccharides. Fed. Proc. **53:** 251.
6. MANN, G. V. & P. NEWTON. 1975. The membrane transport of ascorbic acid. Ann. N. Y. Acad. Sci. **258:** 243-252.
7. DAVIS, K. A., W. Y. L. LEE & R. F. LABBE. 1983. Energy dependent transport of ascorbic acid into lymphocytes. Fed. Proc. **42:** 2011.
8. KAPEGHIAN, J. C. & A. J. VERLANGIERI. 1983. The effects of glucose on ascorbic acid uptake in heart endothelial cells: Possible pathogenesis of diabetic angiopathies. Life Sci. **34:** 577-584.
9. BIGLEY, R., M. WIRTH, D. LAYMAN, M. RIDDLE & L. STANKOVA. 1983. Interaction between glucose and dehydroascorbate transport in human neutrophils and fibroblasts. Diabetes **32:** 545-548.
10. CHEN, M. S., L. HUTCHINSON, R. E. PECORARO, W. Y. LEE & R. F. LABBE. 1983. Hyperglycemia induced intracellular depletion of ascorbic acid in human mononuclear leukocytes. Diabetes **32:** 1078-1081.
11. EVANS, R. M., L. CURRIE & C. CAMPBELL. 1982. The distribution of ascorbic acid between various cellular components of blood in normal individuals and its relation to the plasma concentration. Br. J. Nutr. **47:** 473-482.
12. PATRONE, F. 1982. Effects of ascorbic acid on neutrophil function. Studies on normal and chronic granulomatous disease neutrophils. Acta Vitaminol. Enzymol. **4**(1-2): 163-168.
13. ANDERSON, R. & A. THERON. 1979. Effects of ascorbate on leukocytes. Part I. Effects of ascorbate on neutrophil motility and intracellular cyclic nucleotide levels in vitro. S. Afr. Med. J. **56:** 394-400.
14. BOXER, L. A., B. VANDERBILT, S. BONSIB, R. JERSILD, H. H. YANG & R. L. BAEHNER. 1979. Enhancement of chemotactic response and microtubule assembly in human leukocytes by ascorbic acid. J. Cell. Physiol. **199:** 119-126.
15. DALLEGRI, F., G. F. LANZI & F. PATRONE. 1980. Effects of ascorbic acid on neutrophil locomotion. Int. Arch. Allergy Appl. Immunol. **61:** 40-45.
16. PATRONE, F., F. DALLEGRI, G. F. LANZI & C. SACHETTI. 1980. Prevention of neutrophil chemotactic deactivation by ascorbic acid. Br. J. Exp. Pathol. **61:** 486-489.
17. ANDERSON, R. 1981. Ascorbate-mediated stimulation of neutrophil mobility and lymphocyte transformation by inhibition of the peroxidase/H_2O_2/halide system in vitro and in vivo. Am. J. Clin. Nutr. **34:** 1906-1911.
18. REBORA, A., F. DALLEGRI & F. PATRONE. 1980. Neutrophil dysfunction and repeated infections: Influence of levamisole and ascorbic acid. Br. J. Dermatol. **102:** 49-56.
19. WEENING, R. S., E. P. SCHOOREL, D. ROOS, *et al.* 1981. Effect of ascorbate on abnormal neutrophil, platelet and lymphocyte function in a patient with the Chediak-Higashi syndrome. Blood **57:** 856-865.
20. BOXER, L. A., A. M. WATANABE, M. RISTER, H. R. BESCH, JR., J. ALLEN & R. L. BAEHNER. 1976. Correction of leukocyte function in Chediak-Higashi syndrome by ascorbate. N. Engl. J. Med. **295:** 1041-1045.
21. ANDERSON, R. & O. C. DITRICH. 1979. Effect of ascorbate on leukocyte. Part IV. Increased neutrophil function and clinical improvement after oral ascorbate in two patients with chronic granulomatous disease. S. Afr. Med. J. **56:** 476-480.
22. DRATH, D. B. & M. L. KARNOVSKY. 1974. Bactericidal activity of metal-mediated peroxide-ascorbate systems. Infect. Immun. **10:** 1077-1083.
23. SANDLER, J. A., J. I. GALLIN & M. VAUGHAN. 1975. Effects of serotonin, carbomylcholin, and ascrbic acid on leukocyte cyclic GMP and chemotaxis. J. Cell Biol. **67:** 480-484.
24. ANDERSON, R., R. OOSTHUIZEN, R. MARITZ, *et al.* 1980. The effects of increasing weekly doses of ascorbate on certain cellular and humoral immune functions in normal volunteers. Am. J. Clin. Nutr. **33:** 71-76.
25. GANGULY, R., M. F. DURIEUX & R. H. WALDMAN. 1976. Macrophage function in vitamin C deficient guinea pigs. Am. J. Clin. Nutr. **29:** 762-765.
26. MACCUISH, A. C., S. J. URBANIAK, C. J. CAMPBELL, L. J. P. DUNCAN & W. J. IRVINE. 1974. Phytohemagglutinin transformation and circulating lymphocyte subpopulations in insulin-dependent diabetic patients. Diabetes **23:** 708-712.

27. BRODIE, J. I. & K. MERLIE. 1970. Metabolic and biosynthetic features of lymphocytes from patients with diabetes mellitus: Similarities to lymphocytes in chronic lymphocytic leukemia. Br. J. Haematol. **19:** 193-201.
28. RAGAB, A. H., B. HAZLITT & D. H. COWEN. 1972. Response of peripheral blood lymphocytes from patients with diabetes mellitus to phytohemagglutinin and candida ablicans antigen. Diabetes **21:** 906-907.
29. CASEY, J. I., B. J. HESTER & K. A. KLYSHEVICH. 1977. Impaired response of lymphocytes of diabetic subjects to antigen of staphylococcus aureus. J. Infect. Dis. **136:** 495-501.
30. MOWAT, A. G. & J. BAUM. 1971. Chemotaxis of polymorphonuclear leukocytes from patients with diabetes mellitus. N. Engl. J. Med. **284:** 621-627.
31. MOLENAAR, D. M. *et al.* 1976. Lymphocyte chemotaxis in diabetic patients and their non-diabetic first degree relatives. Diabetes **25:** 880-883.
32. NOLAN, C. M., H. N. BEATY & J. D. BAGDADE. 1978. Further characterization of the impaired bactericidal function of granulocytes in patients with poorly controlled diabetes. Diabetes **27:** 889-894.
33. BAGDADE, J. D., R. K. ROOT & R. J. BULGER. 1974. Impaired leukocyte function in patients with poorly controlled diabetes. Diabetes **23:** 9-15.
34. RAYFIELD, E. J. *et al.* 1982. Infection and diabetes: The case for glucose control. Am. J. Med. **72:** 439-450.
35. REPINE, J. E., C. C. CLAWSON & F. C. GOETZ. 1980. Bactericidal evaluation of neutrophils from patients with acute bacterial infections and from diabetics. J. Infect. Dis. **142:** 869-875.
36. DE FRONZO, R. A., J. D. TOBIN & A. REUBIN. 1979. Glucose clamp technique: A method for quantifying insulin secretion and resistance. Am. J. Physiol. **237:** 214-223.
37. LEE, W., P. HAMERNYIK, M. HUTCHINSON, V. A. RAISYS & R. F. LABBE. 1982. Ascorbic acid in lymphocytes: Cell preparation and liquid chromatographic assay. Clin. Chem. **28:** 2165-2169.
38. FERRANTE, A. & Y. H. THONG. 1978. A rapid one-step procedure for purification of mononuclear and polymorphonuclear leukocytes from human blood using a modification of the Hypaque-Ficoll technique. J. Immunol. Methods **24:** 389-393.
39. FALK, W., R. H. GOODWIN, JR. & E. J. LEONARD. 1980. A 48-well micro chemotaxis assembly for rapid and accurate measurement of leukocyte migration. J. Immunol. Methods **33:** 239-247.
40. MALECH, H. L., R. K. ROOT & J. I. GALLIN. 1977. Structural analysis of human neutrophil migration. J. Cell Biol. **75:** 666-693.
41. TIVEY, H., J. G. LI & E. E. OSGOOD. 1951. Average volume of leukemia leulocytes. Blood **6:** 1013-1020.
42. CLARK, R. A. F. 1985. Cutaneous tissue repair. Basic biologic considerations. I. J. Am. Acad. Dermatol. **13:** 701-725.
43. LEIBOVICH, S. J. & R. ROSS. 1975. The role of the macrophage in wound repair. Am. J. Pathol. **78:** 71-91.
44. NORRIS, D. A., R. A. F. CLARK, L. M. SWIGART, *et al.* 1982. Fibronectin fragments(s) are chemotactic for human peripheral blood monocytes. J. Immunol. **129:** 1612-1618.

DISCUSSION OF THE PAPER

B. LANE (*Columbia University School of Public Health, New York, N.Y.*): It's well known that diabetics are at greater risk for glaucoma and ocular hypertension than normals. My 1980 case-control study showed that ascorbic acid and the glucose tolerance factor, chromium, appeared to collaborate in lowering intraocular pressure. One explanation was a putative role of ascorbic acid as an insulin receptor potentiator.

Are you suggesting that your explanation of competition for transport is complementary to this previous explanation? Or do you feel that your explanation displaces the older explanation?

R. E. PECORARO (*Seattle VA Medical Center, Seattle, Wash.*): No, I don't think it's relevant to your observations. The data suggest strongly that transport occurs by the same transport mechanism in these cells. It's very clear from other data that's been discussed today and yesterday that in other cellular systems the transport mechanism appears to be quite different, so I would be very hesitant to extrapolate this information to other tissues.

B. LANE: Are you saying that there may be more a factor here of transport than potentiation of insulin receptors?

R. E. PECORARO: Well the insulin induces the glucose transporter at the cell membrane and this has been shown in fibroblasts both for ascorbic acid and for glucose. There's increased uptake of both of those after induction by insulin for a matter of hours.

E. J. DILIBERTO (*Wellcome Research Labs, Research Triangle Park, N. C.*): Even though you didn't show any fall in the ascorbate levels in plasma, did you by chance measure the dehydroascorbic acid levels as well?

R. E. PECORARO: No, we didn't. I think that would clarify a lot of these questions about uptake if we were able to do that. Our method measures both the dehydroascorbate and ascorbic acid as ascorbic acid.

J. HATHCOCK (*Food and Drug Administration, Washington, D.C.*): Do you think that ascorbic acid supplementation would help remedy the impaired leukocyte status in the hyperglycemic diabetics?

R. E. PECORARO: We have not done the appropriate work to answer that question.

S. TANNENBAUM (*Massachusetts Institute of Technology, Cambridge, Mass.*): Why did you lump monocytes and lymphocytes when they have such different function and origin? You essentially obliterated any interesting information that you might have derived from the monocytes by having lymphocytes there.

R. E. PECORARO: With regard to the chemotaxis?

S. TANNENBAUM: With regard to levels of ascorbic acid or chemotaxis.

R. E. PECORARO: It's a technical problem. Separations of cell types are difficult, and they have to be processed quickly in order to do studies such as chemotaxis without major artifacts. Based on limited data, monocyte levels appear to have much higher intracellular levels than lymphocytes. I would point out though that the chemotaxis results on the mononuclear cell fraction represented only monocyte chemotaxis in that peroxidase positive cells were aliquotted for those studies.

UNIDENTIFIED SPEAKER: When using the Ficoll gradient technique, do you have aggregates of white leukocytes close to the red cell layer and have difficulty in separating them? Do you lose any of your leukocytes in that way?

R. E. PECORARO: You lose some. We've examined carefully the various cellular makeups of these fractions, and red cell contamination is not a problem. It's less than 5%.

Vitamin C and Airways

VAHID MOHSENIN AND ARTHUR B. DuBOIS

John B. Pierce Foundation Laboratory
New Haven, Connecticut 06519

BACKGROUND

Although Reisseissen has been widely quoted for his first observation on the relationship of "convulsive asthma" and severe scurvy, James Lind, who wrote extensively on the treatment and prevention of scurvy, made no mention of any asthmatic conditions in the scurvy-striken sailors.[1] In the first symposium held in 1961 on vitamin C there was no mention of the airway effect of vitamin C.[2] And in the second symposium on this vitamin, the relationship of vitamin C and respiratory viral infection was explored.[3] Even though several studies failed to show a beneficial effect of large doses of vitamin C in airway constriction[4,5] or treatment of asthma,[6] research on the effect of vitamin C in regulation of airway tone and modulation of airway reactivity continues. Scurvy is not only of historical interest, but continues to exist even in the 1980s.[7] With an enlarging at-risk population of elderly individuals with marginal economic resources, clinicians are likely to encounter scurvy more frequently. The possible role of vitamin C in the maintenance of the redox state of the lung and modulation of arachidonic acid metabolism has provided new bases for investigation of the role of vitamin C in the pathophysiology of airway reactions. In this report, we review the studies on animals and man in reference to lung and airway effects of vitamin C and also review our own experience in New Haven.

ANIMAL STUDIES

Most of the animal studies on the effect of vitamin C on airways and lungs can be divided into two categories: (1) effect of administration of ascorbic acid, or its deficiency, on mediator release from lung tissues, and (2) effect of vitamin C administration or its deficiency on the airway responses to antigen or agonist. Based on the earlier observation by Hochwald[8] that large doses of vitamin C protected the guinea pigs from anaphylactic shock, the focus of investigations was directed toward histamine metabolism. Dawson and West[9] showed that during initial phases of ascorbic acid deficiency in the guinea pig the basal urinary excretion of histamine was considerably increased and there was a concomittant increase in histamine content of the lungs. At this time the sensitivity to histamine aerosol, as judged by the time taken to produce the first convulsive cough, was increased. As the scorbutic state developed further, the basal urinary excretion of histamine was considerably reduced, as also the histamine content of the lungs, and the sensitivity of the animal to the histamine aerosol were decreased. After 30 days on the scorbutic diet, there was less histidine converted to histamine and the guinea pig became very insensitive to the histamine aerosol. During the early phases of scurvy, the increased sensitivity to a histamine aerosol was related with the lowered level of circulating corticosteroids, as reported by Prunty *et al.*[10] Conversely, during the advanced stages of vitamin C deficiency the raised levels of circulating corticosteroid may have been responsible for the insensitivity to histamine aerosol.

Later on, it was demonstrated that the pretreatment of guinea pigs with ascorbic acid (200 mg/kg) inhibited the bronchoconstriction induced by intravenous administration of histamine.[11] However, after a short delay the response to histamine was potentiated, and after another 15 minutes, it had reduced to control levels. They concluded that the bronchoconstrictive responses were potentiated by vitamin C in low concentrations and were inhibited by high concentrations of this substance.

The relation between adrenal function and sensitivity to histamine aerosol in scorbutic guinea pigs was investigated by Guirgis.[12] The drop in the resistance of the guinea pig to histamine aerosol after 2 weeks being on scorbutic diet and the marked rise on the 21st day coincided with the stages of adrenal insufficiency and adrenal hyperfunction respectively. Vitamin C treatment at both stages restored to normal not only the impaired adrenal function but also restored the histamine sensitivity to normal. From these studies it was concluded that the functional state of the adrenal cortex in the scorbutic guinea pig determines the resistance of the animal to histamine aerosol. Actively sensitized scorbutic guinea pigs are reported to be less sensitive to antigen aerosols than guinea pigs maintained on normal diets.[9] When sensitization is induced in the scorbutic state, sensitivity to antigen is decreased but is restored to normal by incorporating ascorbic acid in the drinking water. Similarly when the scorbutic diet replaces the complete diet during the period of sensitization there is also a decrease in sensitivity to antigen, but this is restored by ascorbic acid. The release of histamine from guinea pig lungs *in vitro* by the antigen was markedly reduced when the animals were maintained on the scorbutic diet. In a similar study, lungs from actively sensitized scorbutic guinea pigs released lower amounts of histamine after *in vitro* challenge with antigen, when compared with lungs from animals maintained on 10 and 100 mg of vitamin C per day.[13] Differences in the amount of histamine released in normal versus deficient animals could not be explained by differences in histamine content. In an earlier study it was shown that certain antioxidants, such as α-tocopherol and propylgallate (synthetic antioxidant), have inhibitory effects on the biosynthesis of prostaglandins in sheep vesicular glands.[14] However, the effect of exogenous antioxidants in other tissues depends both on the type of antioxidant and the concentration, and also the effects of *in vivo* and *in vitro* antioxidant on arachidonate turnover are not identical.[15] Pugh and coworkers [16] subsequently demonstrated an inhibitory effect of vitamin C on the generation of prostaglandin F from guinea pig uterine homogenates. This inhibitory effect was only observed, at lower concentrations of ascorbic acid in the medium, so long as the pH of the medium remained above 7.17. In other tissues prostaglandin inhibition appears to be pH-dependent.[17] Therefore, it is conceivable that the lack of inhibitory effect of ascorbic acid at higher concentration was due to changes in pH.

The corollary to these previous works was to investigate the effect of endogenous levels of ascorbic acid on regulation of arachidonic acid metabolites in lungs of guinea pigs.[18] The *in vitro* biosynthesis of prostaglandins, prostacyclin and thromboxane were examined in isolated microsomal membranes from control and scorbutic guinea pigs. The total cyclooxygenase products synthesized from lung microsomes of scorbutic animals were significantly higher than those of controls, provided that the arachidonic acid concentration was not excessive in the incubation medium. At higher concentrations of arachidonic acid, both groups synthesize larger amounts of prostanoids.[18] The prostaglandin $F_2\alpha$ increased more in guinea pigs that were deficient in ascorbic acid than in controls. Prostaglandin E_2 was not detectable in either group. In summary, the work by Pugh [16] demonstrated that addition of ascorbic acid to uterine preparations inhibits the generation of $PGF_2\alpha$, whereas the deficiency of ascorbic acid causes an increase in the biosynthesis of $PGF_2\alpha$ in the lungs of guinea pigs.[18] In addition to its bronchoconstrictive effect in the lung, $PGF_2\alpha$ is known to cause a decrease in cAMP levels.[19] This decrease is associated with the release of various mediators, particularly histamine.[20] In a similar study, tracheal chain preparations from scorbutic guinea pigs produced significantly larger amounts of $PGF_2\alpha$ than like tissues from control group.[21] Conversely, the same tissues from scorbutic guinea pigs generated lesser quantities of PGE_2 than the controls.

The same investigators confirmed the previous findings in *in vitro* system that ascorbic acid antagonized in a dose-dependent fashion the airway constriction induced by $PGF_2\alpha$ in the guinea pig. The ascorbic acid enhanced the production of PGE_2 but not $PGF_2\alpha$ in tracheal chain preparations from normal guinea pigs.[22] The protective effect of ascorbic acid on airway constriction is not only against $PGF_2\alpha$, but has been observed against 5-hydroxytryptamine and histamine. Furthermore, the results reported by Dawson and West [9] indicate that the protective effect of ascorbic acid is not mediated by catecholamines and does not involve β-adrenergic receptors. This led us to examine the relationship between ascorbic acid deficiency, airway responses to aerosolized histamine, and prostanoid generation in guinea pigs. To examine the possible direct effect of ascorbic acid deficiency on membrane properties of tracheal smooth muscle cells we also measured the resting membrane potential of airway smooth muscle cells, and the contribution of electrogenic sodium pump to this potential. Three groups of guinea pigs were studied: control, partially deficient in vitamin C, and scorbutic. The airway response to aerosolized histamine was enhanced in scorbutic guinea pigs and animals partially deficient in vitamin C. Indomethacin treatment did not significantly affect airway responses to histamine in control animals, whereas in scorbutic guinea pigs the anti-inflammatory drug caused a slight but significant enhancement of the preexisting airway hyperresponsiveness to histamine.

The scorbutic and partial deficiency states had a profound effect on spontaneous generation of prostanoids from different parts of the lung. There was an increased production of $PGF_2\alpha$ by tracheal tissues from scorbutic guinea pigs, while there were no significant differences in the spontaneous generation of PGE_2, thromboxane B_2 and 6-keto-$PGF_1\alpha$ by tracheal tissues from control and scorbutic guinea pigs. The generation of all four prostanoids by tracheal tissues from animals fed vitamin C deficient food but with vitamin C supplementation (partially deficient) was significantly reduced compared to controls. Vitamin C deficiency whether total or partial had no significant effect on the generation of prostanoids by bronchial tissues except that thromboxane B_2 production was significantly reduced in bronchial tissues from partially vitamin C deficient animals compared to scorbutic animals. Parenchymal tissues from scorbutic and partially vitamin C deficient guinea pigs generated less $PGF_2\alpha$ than like tissues from control animals. There were no differences in the production of PGE_2 and thromboxane B_2 by parenchymal tissues from control, partially deficient in vitamin C, and scorbutic animals. The amount of 6-keto-$PGF_1\alpha$ released from lung parenchymal slices of partially vitamin C deficient guinea pigs was less than that seen in control tissues. No differences between the resting membrane potentials of airway smooth muscle cells of control and scorbutic animals were observed. Further, the contribution of the electrogenic sodium pump to resting membrane potential was not significantly different between the treatment and control groups (TABLE 1). Even though scorbutic animals showed enhanced airway hyperresponsiveness to histamine *in vivo,* we were unable to show any significant differences between the contractile responses induced by histamine in tracheal segments from control and scorbutic animals when compared as groups. However, four of six scorbutic guinea pigs demonstrated higher contractile force by at least 2 standard deviations from mean of control animals. These observations must be interpreted with caution because neither dose-response relationships nor active tensions were determined.

In the present study, we have shown that ascorbic acid deficiency caused an increase in airway responsiveness to histamine aerosol in guinea pigs. Our data also indicate that even an intermediate level of lung ascorbic acid is associated with heightened airway responsiveness. It has been hypothesized that ascorbic acid exerts its action directly on the smooth muscle.[11] Therefore, we examined whether vitamin C deficiency leads to an alteration in the membrane properties of airway smooth muscle cells. If such an alteration occurred, one might expect changes in resting membrane potential and/or ion fluxes. However, our data did not demonstrate any significant change in either the resting membrane potential or the contribution of electrogenic sodium pump to resting membrane potential. Rather surprisingly, our results suggest that the passive

TABLE 1. Resting Membrane Potential and Response of Tracheal Smooth Muscle Isolated from Control and Scorbutic Guinea Pigs to Histamine *in Vitro*[a]

	Em (mV)	Electrogenic Na Pump[b]	After Histamine 10^{-5} M: ΔEm (mV)	After Histamine 10^{-5} M: ΔIF (g)	Lung Vitamin C[c]
Control (n = 6)	−61.7 ±0.4	18.6 ±1.4	10.3 ±1.1	2.8 ±0.3	23.0 ±3.2
Scorbutic (n = 6)	−62.4 ±0.5	21.5 ±2.2	10.5 ±1.0	3.8 ±0.4	0.3 ±0.1

[a] Values are mean ±SE.
[b] Percent contribution of electrogenic sodium ion pump to membrane potential measured by degree of depolarization after ouabain.
[c] μg/100 mg lung wet weight.
Em: resting membrane potential; mV: millivolts; ΔEm: degree of depolarization induced by histamine; ΔIF: change in isometric force induced by histamine.

and active ion fluxes which contribute to the resting membrane potential are probably not influenced by vitamin C deficiency even though, if prolonged, the latter is known to significantly affect tissue structure.[23]

The mechanism of action of ascorbic acid on airway smooth muscle therefore appears to be more subtle than a direct effect on the membrane or antagonism of a specific agonist. Indeed, previously published data has shown that ascorbic acid can antagonize airway constriction to $PGF_2\alpha$, methacholine, and histamine both *in vivo* and *in vitro.*[21] Since these effects are not due to catecholamine release,[9] it seemed possible that ascorbic acid mediated its actions on airway smooth muscle by altered prostanoid synthesis since indomethacin apparently reverses some of the effects of ascorbic acid.[24–26]

HUMAN STUDIES

The use of ascorbic acid in the treatment of asthma has intrigued investigators for over half a century. Initially, the rationale for the use of ascorbic acid in the treatment of asthma was based on the following observations which were outlined by Hunt in 1938.[6] First, ascorbic acid was known to be present in the adrenals and second, asthmatic patients and adrenalectomized dogs were reportedly hypersensitive to proteins and both conditions were associated with eosinophilia. From these observations, it was concluded that asthma or allergy was associated with adrenal hypofunction and the adrenal medulla was chiefly at fault and finally "some special relation exists between ascorbic acid and the adrenals." Hunt studied 25 patients with asthma who were treated with 50 mg of ascorbic acid twice a day for 2 weeks. No improvement in the degree of wheezing, the incidence of attacks, or the general condition was found. Five of the patients were given 500 mg of ascorbic acid intravenously with no improvement in their symptoms within thirty minutes of injection. This study was apparently not in agreement with an earlier study by Hochwald[27] who reported improvement in symptoms of asthma after large intravenous injections of ascorbic acid.

Based on the report of Dawson and West[9] on the protective effect of ascorbic acid against anaphylactic shock, Zuskin and coworkers[28] hypothesized that ascorbic acid might have a therapeutic effect in asthma despite the contradictory clinical reports. Seventeen healthy subjects participated in a randomized controlled study comparing the effect of 500 mg of ascorbic acid taken orally or placebo on histamine-induced bronchoconstriction. Following the administration of ascorbic acid, histamine reduced flow rates less than after administration of placebo. Although ascorbic acid reduced the effect of histamine significantly, the decreases of flow rates were in most instances still significant compared to the control value before histamine inhalation. This indicated that the ascorbic acid only reduced the airway obstruction and did not prevent it. The same investigators tested the efficacy of ascorbic acid in prevention of airway obstructions induced by breathing aerosols of hemp dust extract.[29] Exposure to textile dust is known to cause acute airway constriction and if the exposure is prolonged, workers develop a chronic obstructive lung disease. Pretreatment of 10 nonsmoking healthy subjects with 1 g ascorbic acid significantly diminished the acute reductions in flow rates over the entire 30-min experiment in comparison with those after the placebo. These findings led to another experiment on mild asthmatics by the same investigators.[4] The efficacy of ascorbic acid was examined in a double-bind controlled experiment on seven patients with mild asthma. The histamine-induced airway obstruction was used as the end point. Ascorbic acid given as 500 mg orally four times a day for 3 days prior to histamine challenge had no significant effect on the maximum flow rate at 40% of vital capacity measured from a partial expiratory flow-volume curve. The authors, however did not rule out the potential beneficial effect of ascorbic acid in induced bronchoconstriction and hypothesized that the protective action of ascorbic acid may be temporary, occurring shortly after single high doses (as in the previous studies) but not after 3 days of treatment, as in this study.

Based on some reports claiming that large doses of ascorbic acid are an effective prophylaxis against upper respiratory tract viral infections,[30–32] and respiratory infections are known to precipitate asthmatic attacks in predisposed persons, a group of investigators from Nigeria followed 41 asthmatic patients for 14 weeks in a randomized double-blind trial.[33] One group took 1 g of ascorbic acid once daily and the second group took a matching placebo. At the end of the 14-week trial, the frequency and severity of asthmatic attacks were compared between the two groups. No objective assessment of airflow obstruction was made and the severity and frequency of asthmatic attacks were solely based on patient's description and report. They found that frequency and severity of attacks were significantly lower in the ascorbic acid group compared to the placebo. The dosing issue of ascorbic acid was addressed by a study that examined the effect of a short course of moderately high doses of vitamin C (500 mg four times daily for 3 days and 1 g prior to spirometric evaluation) in 20 subjects with mild asthma.[34] Objective assessment of lung functions consisting of forced vital capacity, timed vital capacity, and measures of flow in small airways were made before and after vitamin C. No significant improvement was noted in the ventilatory parameters. Again authors did not rule out the potential beneficial effect of ascorbic acid in treatment of asthma and argued that the effect might take some time before it could be detected.

Employing exercise to induce bronchoconstriction, Schachter and Schlesinger demonstrated a small but significant diminution in airway constriction (FEV_1) 5 minutes after exercise. This protective action of vitamin C had no effect on flow rates at low lung volumes. This suggested that the effect of ascorbic acid was more prominent in the larger airways.[35]

So far, the aforementioned studies were descriptive in nature and investigators attempted to elicit the effect of ascorbic acid by measuring various parameters of airway function. The following investigations were basically designed to explore the mechanisms of action of ascorbic acid in human airways.

Ogilvy and other investigators at the John B. Pierce Foundation Laboratory studied the airway responsiveness of six healthy subjects to inhaled methacholine after treatment with 1 g ascorbic acid or 1 g ascorbic acid and 50 mg indomethacin.[24] Based on the previous animal studies (see above) it was hypothesized that if ascorbic acid exerts its action through alteration of prostanoid metabolism, one might expect to see diminution on the effect of ascorbic acid once cyclooxygenase enzymes are inhibited by indomethacin. They found that methacholine inhalation induced concentration-dependent airway constriction. Responses to low doses of the methacholine were similar before and after ascorbic acid ingestion. The response to the highest methacholine dose used (the dose which decreased specific airway conductance, SGaw, by 35% from the control value) was attenuated after ascorbic acid ingestion in both magnitude and duration. Ascorbic acid decreased the fall in SGaw by 25%. After administration of ascorbic acid and indomethacin airway responses to inhaled methacholine were equal to or more severe than responses to the challenge without treatment. This study provided indirect evidence that the protective effect of ascorbic acid on methacholine-induced airway constriction is via alteration of prostanoid metabolism.

Mohsenin and coworkers from the same Laboratory examined the mechanism of action of ascorbic acid in asthma.[26] Four general mechanisms had already been invoked to explain ascorbic acid's effect on smooth muscle contractile states. These included (1) the effect of vitamin C in accelerating the metabolism of histamine,[9,36,37] (2) the direct effect of vitamin C on smooth muscle,[11,38,39] (3) vitamin C's effect on cyclic AMP metabolism,[40] and (4) the modulation of prostanoid production by vitamin C.[23,41] Mohsenin and associates examined the effect of 1 g ascorbic acid alone, 50 mg indomethacin alone or of a combination of these two on airway reactivity of 14 asthmatics to inhaled methacholine. They noted a 34% increase in the dose of methacholine necessary to produce decreases in specific airways conductance after vitamin C administration. This effect of vitamin C was small—especially when one notes the 10- 100-fold greater sensitivity to methacholine of asthmatics when compared with nonasthmatic subjects. Nevertheless, this protecive effect of ascorbic acid was abolished when indomethacin was coadministered with the vitamin, implying the involvement of prostanoids in the regulation of airway tone under contractile states. Since indomethacin administration had no effect on baseline lung functions or airway reactivity, this suggests that prostaglandins probably do not play a major role in regulation of airway tone in the resting state. Since ascorbic acid did not reverse the air flow limitation or significantly abolish airway hyperreactivity to agonist, this relatively small influence of vitamin C may explain both the failure of previous studies to elicit a significant effect and the inconclusive nature of reports on the efficacy of vitamin C in the treatment of asthma.

MECHANISM OF ACTION OF VITAMIN C IN REGULATION OF AIRWAY TONE

Ascorbic acid is a ketolactone and a powerful reducing agent; it provides reducing equivalents to other enzymes. Antioxidants such as vitamin C have the potential to affect the activity of both the prostaglandin endoperoxide synthetase and the profile of products that are formed from the endoperoxide intermediates. The mechanism of action of sodium ascorbate was investigated in human lung parenchyma *in vitro.*[42] Ascorbate increased the quantities of the prostacyclin metabolites 6-keto $PGF_1\alpha$ and thromboxane B_2 generated from human lung parenchyma in a dose-dependent fashion. At high concentrations of sodium ascorbate (0.1-1.0 M), production of prostaglandin E_2 and $F_2\alpha$ were also increased. Co-incubation of tissues with indomethacin almost completely abolished the stimulatory effect of ascorbic acid. Ascorbic acid most likely stimulated prostanoid synthesis either by directly increasing cyclooxygenase activity or by increasing substrate availability for the cyclooxygenase. Altered substrate availability is an attractive explanation because ascorbic acid is known to promote lipid peroxidation and to alter hydrogen peroxide formation.[43] A direct modulatory role for the prostanoids on airway muscle appears unlikely because, although human airways generate large quantities of PGE_2,[44] this prostanoid does not appear to directly modulate the contractile responses of human bronchi,[44] as has been demonstrated for guinea pig airways.[45–47] The spontaneous electrical and mechanical activation of tracheal smooth muscle of dogs treated with repeated administration of indomethacin was completely suppressed by PGE_2, indicating a physiological role of prostaglandins in feedback inhibitory mechanisms for acetylcholine release from the nerve terminals during resting and active states.[48] This phenomenon has been recently confirmed in a study using dog tracheal smooth muscle, and has shown that the local generation of PGE_2 can modulate cholinergic neurotransmission.[49] It is, therefore, probable that ascorbic acid modifies airway reactivity by promoting prostanoid synthesis. These products may then modulate neurotransmission in the airway smooth muscle where reflex cholinergic bronchoconstriction is a major regulator of airway tone and reactivity.

REFERENCES

1. Lind, J. 1980. A Treatise on the Scurvy. Gryphon Editions, Ltd. Birmingham, Ala.
2. Burns, J. J., Ed. 1961. Ann. N.Y. Acad. Sci. **92:** 1-382.
3. Coulehan, J. L., L. Kapner, S. Eberhard, F. H. Taylor & K. D. Rogers. 1975. Vitamin C and upper respiratory illness in Navaho children: Preliminary observations (1974). Ann. N.Y. Acad. Sci. **258:** 513-522.
4. Kreisman, H., C. Mitchell & A. Bouhuys. 1977. Inhibition of histamine-induced airway constriction negative results with oxtriphylline and ascorbic acid. Lung **154:** 223-229.
5. Cockcroft, D. W., D. W. Killian, J. J. Adrian-Mellon & F. E. Hargreave. 1977. Protective effect of drugs on histamine induced asthma. Thorax **32:** 429-437.
6. Hunt, H. B. 1938. Ascorbic acid in bronchial asthma. Report of a therapeutic trial on twenty-five cases. Br. Med. J. **1:** 726-727.

7. REULER, J. B., V. C. BROUDY & T. G. COONEY. 1985. Adult scurvy. J. Am. Med. Assoc. **253:** 805-807.
8. HOCHWALD, A. 1935. Allergiefragen und vitamin C. Zentralbl. Med. **56:** 769-771.
9. DAWSON, W. & G. B. WEST. 1965. The influence of ascorbic acid on histamine metabolism in guinea-pigs. Brit. J. Pharmacol. **24:** 725-734.
10. PRUNTY, F. T. G., B. E. CLAYTON & J. E. HAMMANT. 1957. Experiments on the level of blood corticotrophin with particular reference to scurvy. Ciba Found. Colloquia on Endocrinology **11:** 150-160.
11. DAWSON, W., B.A. HEMSWORTH & M. A. STOCKHAM. 1967. Actions of sodium ascorbate on smooth muscle. Br. J. Pharmacol. Chemother. **31:** 269-275.
12. GUIRGIS, H. M. 1965. The regulatory role of vitamin C on the adrenal function and resistance to histamine aerosol in the scorbutic guinea-pig. J. Pharm. Pharmacol. **17:** 674-675.
13. HITCHCOCK, M. 1980. Effect of variation in endogenous levels of ascorbic acid on the *in vitro* immunological release of histamine and slow reacting substance of anaphylaxis from actively sensitized guinea-pig lung fragments. Br. J. Pharmacol. **71:** 539-543.
14. VAN DORP, D. A. 1971. Recent developments in the biosynthesis and the analysis of prostaglandins. Ann. N.Y. Acad. Sci. **180:** 181-199.
15. CARPENTER, M. P. 1981. Antioxidant effects on the prostaglandin endoperoxide synthetase product profile. Fed. Proc. **40:** 189-194.
16. PUGH, D. M., S. C. SHARMA & C. W. M. WILSON. 1975. Inhibitory effect of L-ascorbic acid on the yield of prostaglandin F from the guinea-pig uterine homogenates. Br. J. Pharmacol. **53:** 469P.
17. ROSE, A. J. & A. J. COLLINS. 1974. The effect of the pH on the production of prostaglandin E_2 and $F_2\alpha$ and a possible pH-dependent inhibitor. Prostaglandins **8:** 271-283.
18. ROTHBERG, K. G. & M. HITCHCOCK. 1983. Effects of ascorbic acid deficiency on the *in vitro* biosynthesis of cyclooxygenase metabolites in guinea-pig lungs. Prostaglandins, Leukotrienes and Medicine **12:** 137-147.
19. SHARMA, S. C. & C. W. M. WILSON. 1980. The cellular interaction of ascorbic acid with histamine, cyclic nucleotides, and prostaglandins in the immediate hypersensitivity reaction. Int. J. Vit. Nutr. Res. **50:** 163-170.
20. TAUBER, A. J., M. KALINER, D. J. STECHSCHULTZ & K. F. AUSTEN. 1973. Immunologic release of histamine and slow reacting substances of anaphylaxis from human lung. V. Effect of prostaglandins on release of histamine. J. Immunol. **111:** 27-32.
21. PUGLISI, L., F. BERTI, E. BOSISIO, D. LONGIAVE & S. NICOSIA. 1976. Ascorbic acid and $PGF_2\alpha$ antagonism on tracheal smooth muscle. *In* Advances in Prostaglandins and Thromboxane Research. B. Samuelsson & R. Paoletti, Eds. Raven Press. New York.
22. PUGLISI, L. & F. MAGGI. 1977. Respiratory system. *In* Prostaglandins and Thromboxanes. F. Berti, Ed. Plenum Press. New York.
23. BURNS, J. J.1959. Biosynthesis of L-ascorbic acid: Basic defect in scurvy. Am. J. Med. **26:** 740-748.
24. OGILVY, C. S., A. B. DUBOIS & J. S. DOUGLAS. 1981. Effect of ascorbic acid and indomethacin on the airways of healthy male subjects with and without induced bronchoconstriction. J. Allergy Clin. Immunol. **67:** 363-369.
25. KAWANI, H., R. SOEJIMA, T. MATSUSHIMA, S. YAGI, H. HARA & M. WATANABE. 1985. Effect of ascorbic acid on bronchoconstriction induced by methacholine inhalation challenge in healthy subjects. Kawasaki Med. J. **11:** 87-92.
26. MOHSENIN, V., A. B. DUBOIS & J. S. DOUGLAS. 1983. Effect of ascorbic acid on response to methacholine challenge in asthmatic subjects. Am. Rev. Respir. Dis. **127:** 143-147.
27. HOCHWALD, A. 1936. Die Rolle reduzierender Substanzen bei der hyperergischen Reaktion. Klin. Wochenschr. **15:** 894-898.
28. ZUSKIN, E., A. J. LEWIS & A. BOUHUYS. 1973. Inhibition of histamine-induced airway constriction by ascorbic acid. J. Allergy Clin. Immunol. **51:** 218-226.
29. ZUSKIN, E. & A. BOUHUYS. 1975. Byssinosis: Airway responses in textile dust exposure. J. Occup. Med. **17:** 357-359.
30. WILSON, C. W. M. & M. S. LOH. 1973. Vitamin C and colds. Lancet **1:** 638-641.
31. PAULING, L. 1970. Vitamin C and the Common Cold. Freeman. San Francisco.

32. ANDERSON, T. W., D. B. W. REID & G. H. BEATON. 1972. Vitamin C and the common cold: A double-blind trial. Can. Med. Assoc. J. **107:** 503-508.
33. ANAH, C. O., L. N. JARIKE & H. A. BAIG. 1980. High dose ascorbic acid in Nigerian asthmatics. Trop. Geogr. Med. **32:** 132-137.
34. TING, S., L. E. MANSFIELD & J. YARBROUGH. 1983. Effects of ascorbic acid on pulmonary functions in mild asthma. J. Asthma **20:** 39-42.
35. SCHACHTER, E. N. & A. SCHLESINGER. 1982. The attenuation of exercise-induced bronchospasm by ascorbic acid. Ann. Allergy **49:** 146-151.
36. CHATTERJEE, I. G., S. D. BUPTA, A. K. MAJUMDER, B. K. KAUDI & N. SUBRAMANIAN. 1975. Effect of ascorbic acid on histamine metabolism in scorbutic guinea pigs. J. Physiol. **251:** 271-279.
37. SUBRAMANIAN, N., B. K. NANDI, A. K. MAJUMDER & I. B. CHATTERJEE. 1973. Role of L-ascorbic acid on detoxification of histamine. Biochem. Pharmacol. **22:** 1671-1673.
38. GHOURI, M. K., M. W. RHANA & T. J. HALEY. 1977. Ascorbic acid blockade of muscle contractions by neurotransmitters and 2-aminoethanol. Life Sci. **20:** 213-222.
39. JOINER, P. 1973. Effect of ascorbate on contractile activity of ileal smooth muscle. J. Pharmacol. Exp. Ther. **135:** 84-93.
40. LEWIN, S. 1976. Vitamin C: Its Molecular Biology and Medical Potential. Academic Press. New York.
41. PARKER, C. & D. SNIDER. 1973. Prostaglandins and asthma. Ann. Int. Med. **78:** 963-965.
42. FANN, Y.-D., K. G. ROTHBERG, P. G. TREMML, J. S. DOUGLAS & A. B. DUBOIS. 1986. Ascorbic acid promotes prostanoid release in human lung parenchyma. Prostaglandins **31:** 361-368.
43. POLGAR, P. & L. TAYLOR. 1980. Stimulation of prostaglandin synthesis by ascorbic acid via hydrogen peroxide formation. Prostaglandins **19:** 693-700.
44. BRINK, C., C. GRIMAUD, C. GUILLOT & J. OREHEK. 1980. The interaction between indomethacin and contractile agents on human isolated airway muscle. Br. J. Pharmacol. **69:** 383-388.
45. GRODZINSKA, L., B. PANCZENKO & R. J. GRYGLEWSKI. 1975. Generation of prostaglandin E-like material by the guinea pig trachea contracted by histamine. J. Pharm. Pharmacol. **27:** 88-91.
46. OREHEK, J., J. S. DOUGLAS, A. J. LEWIS & A. BOUHUYS. 1973. Prostaglandin regulation of airway smooth muscle tone. Nature New Biology **245:** 84-85.
47. OREHEK, J., J. S. DOUGLAS & A. BOUHUYS. 1975. Contractile response of the guinea pig trachea in vitro: Modification by prostaglandin synthesis-inhibiting drugs. J. Pharmacol. Exp. Ther. **194:** 554-564.
48. ITO, Y. & K. TAJIMA. 1981. Spontaneous activity in the trachea of dogs treated with indomethacin: An experimental model for aspirin related asthma. Br. J. Pharmacol. **73:** 563-571.
49. SHORE, S. A. & J. G. MARTIN. 1985. Evidence for presynaptic inhibition of cholinergic neurotransmission by endogenous prostaglandins in canine tracheal smooth muscle *in vitro.* Am. Rev. Respir. Dis. **131:** A286.

DISCUSSION OF THE PAPER

E. GUMBPRICHT: (*Center for Air and Environmental Studies, University Park, Pa.*): Since PGE_1 is a more potent bronchodilator than PGE_2, and vitamin C stimulates PGE_1 in platelets, I was wondering if there was an association?

V. MOHSENIN: (*John B. Pierce Foundation Laboratory, New Haven, Conn.*) That's a good question but I'm not aware of measurements of PGE_1 in the lung and airways specifically. Are you aware of PGE_1 being generated in the airways?

E. GUMPRICHT: I'm not really sure. There's a lot of PG synthetase in the lung and the amount of dihomogammalinoleic acid in the body may determine the amount of PGE.

My second quick question is, are there any trials that you know about that would use a combination of vitamin C and vitamin E with bronchoconstriction of histamine? Vitamin E might inhibit the formation of leukotrienes which are very potent, much more potent than $PGF_2\alpha$ in constricting smooth muscle.

V. MOHSENIN: I'm not aware of a study combining these two in regard to the histamine sensitivity.

Evaluation of the Effects of Vitamin C on Ozone-induced Bronchoconstriction in Normal Subjects[a]

MARIE D. CHATHAM,[b] JOHN H. EPPLER, JR., LARRY R. SAUDER, DON GREEN, AND THOMAS J. KULLE

Division of Pulmonary Medicine
Department of Medicine, Critical Care Division
The University of Maryland Hospital
and
Franklin Square Hospital Center
Baltimore, Maryland 27237

INTRODUCTION

The lung, with a surface area approximately that of a football field, is exposed to some 9000 liters of air daily. Thus it is an obvious target organ for pollutant-related injury. A major pollutant, ozone (O_3), represents some 90% of the oxidant stress from smog, a ubiquitous urban environmental hazard.[1] However, ozone exposure also occurs in other settings, such as during high-altitude flights and in proximity to welding or to high-voltage electrical discharges.[2,3]

Ozone has been extensively studied and minor effects such as nasal, pharyngeal, and conjuctival irritation, chest tightness or soreness, and cough are well-documented.[5,6] Of more concern are the bronchoconstrictive effects of ozone, which can be seen in some "sensitive" normals and asthmatics after as little as 2 hours of exposure to levels commonly observed on days of poor air quality.[7–10] Of great concern as well is the recognition that ozone exposure increases nonspecific airways reactivity,[8,11–13] which may be a sign of risk for future chronic lung disease.[14,15]

The precise mechanisms whereby ozone adversely affects the lungs are unclear, but some generalizations regarding toxin inhalation are better understood and will be reviewed briefly. Basically, toxicity of inhaled pollutants is related both to solubility and chemical reactivity.[16] Of the three most commonly encountered pollutants, ozone, nitrogen dioxide, and sulfur oxides, the first two are considerably less soluble; this

[a] Supported by grants from the American Lung Association of Maryland and Hoffmann-La Roche Inc.

[b] Present address: Department of Medicine, Franklin Square Hospital Center, 9000 Franklin Square Dr., Baltimore, Maryland 21237.

results in a higher percentage of inhaled ozone (or nitrogen dioxide) reaching the alveolar surface, accounting for the fact that damage may occur at any point from the nasopharynx down to the pulmonary parenchyma. The chemical reactivity of ozone is also quite great; only fluorine has a more electronegative oxidation potential. Since it contains two unpaired electrons, it also has a free radical potential.[17]

Theories on the mechanism of ozone toxicity include lipid peroxidation,[18–20] inactivation of sulfhydryl groups of small compounds as reduced glutathione,[21] activation of irritant receptors or vagal pathways,[22] or involvement of inflammatory mediators.[23–25]

Support for the oxidation theory is founded on evidence of the ability of water- or lipid-soluble antioxidants to reduce ozone toxicity. Ascorbic acid, a water-soluble vitamin, is concentrated intracellularly by the lung, and tissue levels are decreased after ozone exposure; this suggests that it may function as an expendable antioxidant.[26] Indeed ascorbate has been reported to protect mice against ozone.[27] Likewise, animals deficient in α-tocopherol, a lipid-soluble antioxidant, show increased toxicity to ozone.[28]

The purpose of this study was to evaluate the ability of vitamin C and E to alter the response to ozone in normal subjects as reflected by changes in symptoms, pulmonary function, and nonspecific airways reactivity.

METHODS

Subjects

Twenty young, nonsmoking volunteers with normal baseline pulmonary function who gave informed consent were studied. All subjects were instructed to refrain from any drugs, including vitamin supplements, during the time of study.

Protocol

The protocol consisted of a screening day on which histories, physical examinations, EKGs, and spirometry were done. A treadmill exercise test, which consisted of 14 minutes of walking at a load sufficient to produce a minute ventilation of 15 times the initial FEV_1, was also conducted in order to give approximate settings for the study days.

On the study days, the subjects reported 1 hour prior to actual exposure to ozone and received in random double-blind fashion either placebo or vitamin C (1 g ascorbic acid po). Just prior to entry into the O_3 exposure room of the chamber, physiologic measurements of flow and lung volume were made. During the 2-hour chamber exposure, the volunteers repeatedly exercised four times for 14 minutes at a load sufficient to attain the target ventilation; each 14-minute exercise stint was followed by 16 minutes of rest. Minute ventilation was checked during the 7th and 13th minute of each exercise stint and minor adjustments made as necessary to achieve the target ventilation. Spirometry was checked at the end of the 2-hour exposure, which was also 8 minutes after the last exercise stint body plethysmograph was done immediately after leaving the exposure chamber. Subjects then underwent methacholine challenge.

Lastly, subjective data in the form of symptoms were collected via a questionnaire consisting of five questions regarding the presence and severity of cough, eye irritation, headache, chest discomfort or tightness, and palpitations. The subjects classified their response to each symptom as none, mild, moderate, or severe. For statistical analysis a score of 0 was assigned to none, 1 to mild (present, but not annoying), 2 to moderate (annoying), and 3 to severe (debilitating).

One week later, in order to avoid the effects of adaptation of ozone,[29] this same sequence was repeated using the other medication.

During the last stage of the study, subjects took 800 IU of vitamin E daily. They came in every 2 weeks to receive additional medication and were called at least once a week to emphasize the importance of compliance. The last day of study, the subjects again reported 1 hour prior to ozone exposure, and were given a single gram dose of vitamin C.

Subjects were tested at the same time of day in order to minimize the effects of diurnal variation in pulmonary function.[30]

Physiologic Measurements

Spirometric measurements (FVC, FEV_1, FEV_3, FEF_{25-75}, and IC) were made with a 10-l Stead-Wells Spirometer interfaced with an Eagle II Microprocessor (Warren E. Collins, Braintree, Mass.). The specifications for measurements exceeded American Thoracic Society criteria.[31] The two maneuvers with the highest sum of FVC and FEV_1 were used to determine the mean FVC, FEV_1, and IC. All lung volumes were expressed in liters at BTPS.

Airway resistance and volume of thoracic gas (Vtg) were measured by the whole-body pressure plethysmographic technique of DuBois and associates[32] with certain modifications[33] in a variable pressure constant volume body plethysmograph (Warren E. Collins, Braintree, Mass.). The plethysmographic data were analyzed by a computer and the actual data represented by an average of 15-20 panting breaths. Airways resistance (Raw) was measured between flows of +0.5 liter/sec inspiratory to −0.5 liter/sec expiratory. Raw was computed and converted to specific airway conductance (SGaw), using the Vtg at which each Raw was measured.

During exercise, minute ventilation was measured using a 120-l Gasometer (Warren E. Collins, Braintree, Mass.). Methacholine challenge was performed by the method of Chai *et al.*[34] with modifications previously described.[29] The logarithm dose of methacholine that decreased the specific airways conductance (SGaw) by 35% (PD_{35}) was derived by linear regression.

Chamber and Ozone Generation

The exposures were conducted in an environmental chamber (2.1 × 4.3 × 2.4 m) utilizing air conditioned by passage through high efficiency particulate absolute (HEPA) filters as well as activated charcoal filters. The ozone was produced by passing 100% O_2 through a water-cooled, silent-arc, ozone generator and introduced via stainless steel tubing into the air supply duct of the exposure room, thus ensuring that pollutant exposure was limited only to ozone. Ozone concentration was continuously monitored by both an ultraviolet absorption photometer (Dasibi Model 1008-PC, Glendale, Calif.) that has its own self-contained calibrator and a Mast Microconlomb Ozone Monitor (Davenport, Iowa). The Dasibi analyzer (Monitor Labs Model 8410E Ozone Analyzer, San Diego, Calif.) was calibrated by comparison of their values to those obtained from similar units at the State Air Management Administration (Baltimore, Md.); agreement was within 4 ppb in all instances in the range of ozone concentration used in this study.

Data Analysis

The data were analyzed by comparing the difference of post- to preexposure measurements on each day of study by the ANOVA technique. The statistical significance of these standardized differences was determined by Tukey's Test.

RESULTS

Of the twenty recruits, 14 completed the study. Of these, 5 were "insensitive" to ozone, having less than a 7% decrease in FEV_1 after exposure. These subjects were not included in the results, since we could not expect to see "protection" by intervention in subjects who had so little worsening (in pulmonary function) with ozone. The remaining nine subjects are described in TABLE 1 and were young, with a mean age of 23.7 years, and had normal pulmonary function.

During the 3 days of study, concentration of ozone (0.3 ppm ± 0.001 ppm), temperature (72 ± 2°F), and humidity (50 ± 3%) were similar. The mean minute ventilations were also not different (64.1 ± 10.2 with placebo, 63.2 ± 11.5 with C, and 64.6 ± 9.5 with C + E).

As shown in FIGURE 1, the decrease in FVC after placebo for the group was greater than with C or C + E. Although the decrease was lesser with vitamin C, this did not reach statistical significance. However, the effect with C + E was significantly different than with placebo and achieved significance.

Results for changes FEV_1 or SGaw are presented in TABLE 2 and did not reach statistical significance. Another way of looking at the effect of the medications is to calculate the percent protection by the following formula:

$$\%\ \text{protection} = \frac{\%\ \text{decrease placebo} - \text{decrease drug}}{\%\ \text{decrease placebo}}$$

The level of nonspecific reactivity as measured by methacholine was also not statistically different. The PD_{35} for the group on the placebo day was 1.346, with vitamin C 1.357, and with C + E 1.354. Differences in symptom scores shown in TABLE 3 did not reach statistical significance.

There was a large range in individual response for the different days of the study as shown in FIGURE 2. Individual responses varied, some showed protection with C and more with C + E, others had greater protection with C alone than with C + E; still others had more protection with vitamin C + E and no protection with C alone.

TABLE 1. Subject Characteristics

Subject	Sex	Height	Age	FEV_1	Pred.	% Pred.	FVC	Pred.	% Pred.
1	F	64	23	3.59	3.19	1.13	4.06	3.96	1.03
2	M	66	29	3.08	3.88	0.79	3.63	4.8	0.76
3	F	65	25	3.71	3.23	1.15	3.94	4.03	0.98
4	M	70	21	4.39	4.51	0.97	5.91	5.6	1.06
5	M	70	23	4.77	4.44	1.07	5.51	5.55	0.99
6	F	64	23	3.28	3.19	1.03	3.97	3.96	1.0
7	M	70	22	5.1	4.48	1.14	6.5	5.57	1.17
8	M	71	23	3.81	4.54	0.84	4.86	5.69	0.85
9	M	71	24	4.0	4.5	0.89	4.73	5.67	0.83
Means ± SD		67.9 ± 2.9	23.7 ± 2.2	3.97 ± 0.63	4.0 ± 0.59	1.0 ± 0.13	4.79 ± 0.94	4.98 ± 0.75	0.96 ± 0.12

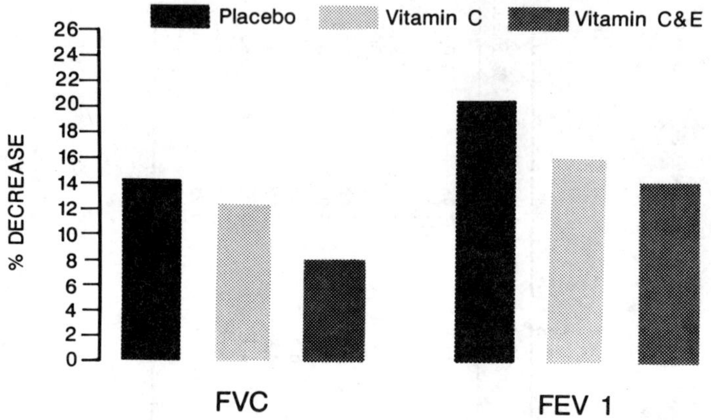

FIGURE 1. Group response to premedication with placebo, vitamin C, and vitamins C + E in normal subjects. Results are expressed as a percent fall from baseline FVC and FEV_1. See text for discussion of these responses.

DISCUSSION

The results of this study suggest that there is partial protection against ozone-induced bronchoconstriction in normal subjects by the combination of vitamins E and

TABLE 2. Mean Changes in Pulmonary Function in Normals

	Placebo	Vit. C	Vit. C + E
FEV_1			
Pre-exposure	3.95	3.91	3.95
Postexposure	3.16	3.25	3.38
% Decrease ± SE	20.7 ± 3.5	16.4 ± 1.7	14.1 ± 3.6
% Protection		20.4	31.8
FEF_{25-75}			
Pre-exposure	4.10	4.05	4.11
Postexposure	2.78	3.02	3.17
% Decrease ± SE	31.8 ± 5.9	24.2 ± 4.9	20.7 ± 7.9
% Protection		24.0	35.0
FVC			
Pre-exposure	4.73	4.73	4.75
Postexposure	4.09	4.16	4.35
% Decrease ± SE	14.1 ± 2.5	12.3 ± 2.2	8.2 ± 1.8
% Protection		12.9	41.8[a]
SGaw			
Pre-exposure	0.184	0.197	0.183
Postexposure	0.149	0.158	0.157
% Decrease ± SE	18.4 ± 2.9	18.7 ± 2.0	13.4 ± 3.8
% Protection		−1.66	27.3

[a] $p < 0.05$.

TABLE 3. Normals: Mean Symptom Scores[a]

	Placebo	Vit. C	Vit. C + E
Chest discomfort and tightness	0.78	0.44	0.44
Cough	1.00	0.78	0.33
Nose or throat irritation	0.11	0.22	0.33
Eye irritation	0.11	0.11	0.11
Headache	0.00	0.00	0.11
Heart palpitation	0.00	0.00	0.00

[a] Symptom scores: 0 = none, 1 = mild, 2 = moderate, 3 = severe.

C. Trends were seen toward protection with only vitamin C, but did not reach statistical significance. With a larger number of subjects, this may have reached significance; a prior study at this institution using one gram of vitamin C 1 hour prior to ozone exposure (0.3 ppm for 2 hours) in normals did show statistically significant amelioration of bronchoconstriction for general measures of pulmonary function (FEV_1, FVC, FEF_{25-75}).[35]

The effect of vitamin C and vitamin C + E showed individual variation, which has also been observed by others.[35] Known variables such as diurnal rhythm,[30] temperature, humidity, minute ventilation during exercise,[36] and ozone concentration were carefully controlled and thus probably did not account for the individual variability. It is unlikely that direct effects by vitamin C on airway tone influenced the results since pulmonary function 1 hour after vitamin C was similar to that 1 hour after placebo, and one would have expected near maximum serum levels during this time period.[37,38]

Adaptation to ozone is associated with chronic exposure and should not have affected the results given that at least 7 days were allowed between exposures.[29]

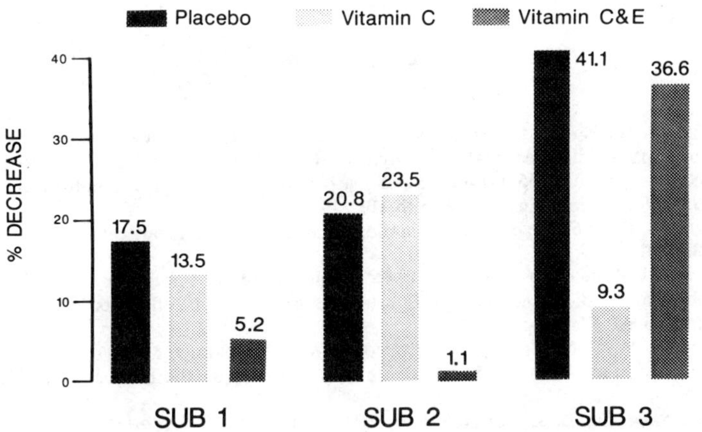

FIGURE 2. Different patterns of individual response to pretreatment with placebo, vitamin C, and vitamin C + E in terms of percent decrease in FEV_1 after ozone exposure (see text).

Changes in methacholine reactivity were insignificant, suggesting that the protection seen with vitamin C + E was not due to shifts in nonspecific reactivity. Ogilvy and colleagues found that ascorbic acid (500 mg) decreased both the duration and intensity of bronchoconstriction from methacholine challenge, and that these effects were blocked by indomethacin.[39] They suggested that vitamin C might have acted by modulation of prostaglandin metabolism. Further support for this hypothesis was published in a later study using asthmatics.[40] We think the lack of shift in reactivity after vitamin C and ozone exposure argue against this being the major mechanism in the model of ozone-induced bronchoconstriction.

The apparent synergism of vitamins C and E supports the hypothesis that they act as antioxidants. Although we are unaware of similar studies in man, animal studies have demonstated protective effects of dietary antioxidants in models of ozone toxicity.[27,28] Thus, while we cannot define the precise mechanism whereby vitamins C + E work, we find it quite plausible that they are acting as expendable antioxidants. It is also probable that prostaglandins are involved to some extent in these reactions, as both vitamins have been shown to have complex effects on arachidate metabolism.[41,42]

To summarize, ascorbic acid and vitamin E when given concurrently appear to ameliorate ozone-induced bronchoconstriction in normal subjects. This study supports results of previous animal studies documenting the efficacy of dietary antioxidants in providing protection against ozone toxicity. Since ozone exposure is ubiquitous and often unavoidable, pharmacologic protection is important for public health. Further studies aimed at examining other population groups (such as asthmatics) as well as elucidating precise mechanisms will be of great interest and relevance.

ACKNOWLEDGMENTS

We express our gratitude to Dr. Rebecca Bascon for her assistance. In addition, we wish to thank Bonnie Berman and Barbara Tuma for their assistance in measurements and typing.

REFERENCES

1. Mueller, P. K. & M. Hitchcock. 1969. Air quality criteria—Toxicological appraisal for oxidants, nitrogen oxides and hydrocarbons. J. Air Pollut. Control Assoc. **19:** 670.
2. Stokinger, H. E. 1965. Ozone toxicology. A review of research and industrial experience, 1954-1964. Arch. Environ. Health **10:** 719.
3. Bennett, G. 1962. Ozone contamination of high altitude aircraft cabins. Aerospace Med. **33:** 969.
4. National Research Control. 1977. Ozone and Other Photochemical Oxidants. Committee on Medical and Biologic Effects of Environmental Pollutants. National Academy of Sciences. Washington, D.C.
5. Bates, D. V., G. M. Bell, C. D. Burnham, M. Hazucha, J. Mantha, L. D. Pengelly & F. Silverman. 1972. Short-term effects of ozone on the lung. J. Appl. Physiol. **32:** 176.
6. Silverman, F., L. J. Folinsbee, J. Barnard & R. J. Shephard. 1976. Pulmonary function changes in ozone-interaction of concentration and ventilation. J. Appl. Physiol. **41:** 859.

7. McDonnell, W. F., D. H. Horstman, M. J. Hazucha, E. Seal, Jr., E. D. Haak, S. A. Salaam & D. E. House. 1983. Pulmonary effects of ozone exposure during exercise: Dose-response characteristics. J. Appl. Physiol. **54**(5): 1345.
8. Golden, J. A., J. A. Nadel & H. A. Boushey. 1978. Bronchial hyperirritability in healthy subjects after exposure to ozone. Am. Rev. Respir. Dis. **118:** 287.
9. Folinsbee, L. J., B. L. Drinkwater, J. F. Bedi & S. M. Horvath. 1978. The influence of exercise on the pulmonary function changes due to exposure to low concentrations of ozone. *In* Environmental Stress: Individual Human Adaptations. L. J. Folinsbee, J. A. Wagner, J. F. Borgia, B. L. Drinkwater, J. A. Gilner & J. F. Bedi, Eds.: 111-124. Academic Press. New York.
10. Schoettlin, C. E. & E. Landau. 1961. Air pollution and asthmatic attacks in the Los Angeles area. Public Health Rep. **76:** 545.
11. Dimeo, M. J., M. G. Glenn, M. J. Holtzman, J. R. Sheller, J. A. Nadel & H. A. Boushey. 1981. Threshold concentration of ozone causing an increase in bronchial reactivity in humans and adaptation with repeated exposures. Am. Rev. Respir. Dis. **124:** 245-248.
12. Holtzman, M. J., J. H. Cunningham, J. R. Sheller, G. B. Irsigler, J. A. Nadel & H. A. Boushey. 1979. Effect of ozone on bronchial reactivity in atopic and nonatopic Subjects. Am. Rev. Respir. Dis. **120:** 1059.
13. Holtzman, M. J., L. M. Fabbri, P. M. O'Byrne, B. D. Gold, H. Aizawa, E. H. Walters, S. E. Alpert & J. A. Nadel. 1983. Importance of airway inflammation for hyperresponsiveness induced by ozone. Am. Rev. Respir. Dis. **127:** 686.
14. Barter, C.E. & A. H. Campbell. 1976. Relationship of constitutional factors and cigarette smoking to decrease in 1-second forced expiratory volume. Am. Rev. Respir. Dis. **113:** 305-314.
15. Britt, E. J., B. Cohen, H. Menkes, E. Bleecker, S. Permutt, *et al.* 1980. Airways reactivity and functional deterioration in relatives of COPD patients. Chest **77:** 260-261.
16. Health Efects of Air Pollution. 1978. American Thoracic Society Medical Section of the American Lung Association.
17. Mustafa, M. G. & D.F. Tierney. 1978. Biochemical and metabolic changes in the lung with oxygen, ozone, and nitrogen dioxide toxicity. Rev. Respir. Dis. **118:** 1061.
18. Cross, D.E., A. J. De Lucia, A. K. Reddy, M. Z. Hussain, C-K. Chow & M. G. Mustafa. 1976. Ozone interactions with lung tissue. Am. J. Med. **60**(7): 929.
19. Mustafa, M. G. & C. E. Cross. Efefcts of short-term ozone exposure on lung mitochondrial oxidative and energy metabolism. Arch. Biochem. Biophys. **162:** 585.
20. Chow, C. L., C. J. Dillard & A. L. Tappel. 1974. Glutathione peroxidase system and lysozyme in rats exposed to ozone or nitrogen dioxide. Environ. Res. **7:** 311.
21. Mentzel, D. B. Toxicity of ozone, oxygen, and radiation. Ann. Rev. Pharmacol. **10:** 379-394.
22. Lee, L-Y, E. R. Bleecker & J. A. Nadel. 1977. Effects of ozone on bronchomotor response to inhaled histamine aerosol in dogs. J. Appl. Physiol. **43:** 626-31.
23. Drazen, J. M., K. F. Austen, R. A. Lewis, *et al.* 1980. Comparative airway and vascular activities of leukotrienes C-1 and D in vivo and in vitro. Proc. Natl. Acad. Sci. **77:** 4354-4358.
24. Perez, H. D., I. M. Goldstein & B. B. Weksler. 1980. Generation of a biologically active lipid from arachidonic acid by exposure to a superoxide-generating system. AAS **7:** 109-114.
25. O'Byrne, P. M., E. H. Walters, B. D. Gold, H. A. Aizawa, L. M. Fabbri, S. E. Alpert, J. A. Nadel, M. J. Holtzman. 1984. Neutrophil depletion inhibits airway hyperresponsiveness induced by ozone exposure. Am. Rev. Respir. Dis. **130:** 214-219.
26. Kratzing, C. C. & R. J. Willis. 1980. Decreased levels of ascorbic acid in lung following exposure to ozone. Chem. Biol. Interactions **30:** 53-56.
27. Matzen, R. N. 1957. Effect of vitamin C and hydrocortisone on the pulmonary edema produced by ozone in mice. J. Appl. Physiol. **11:** 105-109.
28. Mustafa, M. G. 1975. Influence of dietary vitamin E on lung cellular sensitivity to ozone in rats. Nutr. Rep. Int. **11:** 473-6.
29. Kulle, T. J., L. R. Sauder, H. D. Kerr, B. P. Farrell, M. S. Bermel & D. M.

SMITH. 1982. Duration of pulmonary function: Adaptation to ozone in humans. Am. Ind. Hyg. Assoc. J. **43**(11): 832-837.
30. KERR, H. D. 1973. Diurnal variation of respiratory function independent of air quality: Experience with an environmentally controlled exposure chamber for human subjects. Arch. Environ. Health **26**: 144-152.
31. AMERICAN THORACIC SOCIETY STATEMENT. 1979. Snowbird workshop on standardization of spirometry. Am. Rev. Respir. Dis. **119**: 831-838.
32. DUBOIS, A. B., S. Y. BOTELHO & J. H. COMROE, JR. 1956. A new method for measuring airway resistance in man using a body plethysmograph: Values in normal subjects and in patients with respiratory disease. J. Clin. Invest. **35**: 327-335.
33. SAUDER, L. 1982. Computer analysis versus technician analysis of body plethysmographic analogue recordings of airway resistance and thoracic gas volume. Respir. Care **27**: 62-69.
34. CHAI, H., F. S. FARR, L. A. FROEHLICH, *et al.* 1975. Standardization of bronchial inhalation challenge procedures. J. Allergy Clin. Immunol. **56**: 323-327.
35. CHATHAM, M. D., L. R. SAUDER & T. J. KULLE. 1984. Evaluation of the effect of vitamin C on ozone-induced bronchoconstriction in normal subjects. Am. Rev. Respir. Dis. **129**: 145.
36. DEAL, E. C., E. R. MCFADDEN, R. H. INGRAM & J. J. JAEGER. 1979. Hypernea and heat flux: Initial reaction sequence in exercise-induced asthma. J. Appl. Physiol. **46**: 476-483.
37. WRIGHT, I. S., A. LILIENFLED & E. MACLENATHEN. 1937. Determination of vitamin C saturation. Arch. Intern. Med. **60**: 264.
38. GOODMAN, L. S. & A. GILMAN. 1965. The Pharmacological Basis of Therapeutics. 3d ed. Macmillan Co., New York. p. 1665.
39. OGILVY, C. S., A. B. DUBOIS & J. S. DOUGLAS. 1981. Effects of ascorbic acid and indomethacin on the airways of healthy male subjects with and without induced bronchoconstriction. C. V. Mosby Co. **67**(5): 363-369.
40. MOHSENIN, V., A. B. DUBOIS & J. S. DOUGLAS. 1983. Efect of ascorbic acid on response to methacholine challenge in asthmatic subjects. Am. Rev. Respir. Dis. **127**: 143-147.
41. PUGLISI, L. & F. MAGGI. 1977. Respiratory system. *In* Prostaglandins and Thormboxanes. F. Berti, Ed.: 150-154. Plenum Press. New York.
42. HOPE, W. L., C. DALTON, L. J. MACHLIN, R. J. FILIPSKI & F. M. VANE. 1975. Prostaglandins **10**: 557.

DISCUSSION OF THE PAPER

UNIDENTIFIED SPEAKER: Some work that we have done with vitamin C shows that there is a tremendous difference in the oral absorption of the vitamin and there's some speculation that there might even be poor absorbers and good absorbers. I was wondering if your results have something to do with that?

M. D. CHATHAM: (*University of Maryland Hospital, Baltimore, Md.*) At this point I can't answer that question. I agree that knowing that might help explain some of the individual variability.

C. ROMNEY (*Dartmouth College, Hanover, N.H.*): How consistent were your groupings socioeconomically?

M. D. CHATHAM: They were generally young, first- and second-year medical students.

A. KALLNER (*Karolinska Hospital, Stockholm, Sweden*): Is 0.3 ppm ozone a concentration which anybody is exposed to?

M. D. CHATHAM: Yes it is. Many of the studies actually look at anywhere up to 0.6 ppm, which produces more dramatic effects but is not a very realistic exposure. Levels of 0.3 ppm to 0.4 p ppm are seen on days with bad air quality.

D. HORNIG (*Hoffmann-La Roche, Basel, Switzerland*): What was the vitamin status of the placebo group before they started these experiments?

M. D. CHATHAM: The medical students were asked to refrain from taking any vitamin supplements from the point we did the initial screening, which was at least two weeks, and they were also asked to abstain from foods known to be rich in vitamin C.

D. HORNIG: You did not measure vitamin C?

M. D. CHATHAM: We did but those results are unavailable at this point.

C. BOREK (*Columbia University, New York, N.Y.*) As controls would it not have been advisable to use pure oxygen since ozone is more soluble than oxygen, and because ozone's not a free radical, it's a producer of free radicals?

M. D. CHATHAM: Ozone was used since we were interested in studying, in the environmental chamber, a constituent of smog and air pollution.

D. HORNIG: Do you have any figures on the exposure of ozone in a transatlantic flight?

M. D. CHATHAM: It depends on where you're sitting. There's a concentration gradient and, if you're sitting in the rear of the plane, levels of 10 to 15 parts per million have been reported.

UNIDENTIFIED SPEAKER: Medical students are occasionally a highly stressed population. Did you take into account the stress situation?

M. D. CHATHAM: This was a randomized study so that hopefully we negated those effects. Some were given a placebo first, some were given vitamin C first.

Ascorbic Acid and the Eye with Special Reference to the Lens

SHAMBHU D. VARMA

Department of Ophthalmology
University of Maryland School of Medicine
Baltimore, Maryland 21201

The concentration of ascorbic acid is substantially high in various ocular tissues. As TABLE 1 indicates, the cornea and the lens are next only to adrenals and liver in this regard. The aqueous and vitreous humors of the eye are also substantially rich in their ascorbate content. The level of this substance in these two intraocular fluids is the highest among all the extracellular fluids of the body. It has previously been shown that the high concentration of ascorbic acid in the aqueous humor is due to an active transport of the nutrient by the ciliary epithelium across the blood aqueous barrier.[1] The aqueous humor can then perhaps act as a source of ascorbate to all the other ocular fluids and tissues including the lens, cornea, vitreous humor, and retina. The various ocular tissues are thus well fortified with this nutrient. Why nature has developed this mechanism of maintaining a high concentration of ascorbate in the aqueous humor, lens, and the other ocular tissues is as yet an unsolved mystery. Since ascorbic acid is a relatively strong reducing agent, it is quite possible that the role of this nutrient in such high concentrations may be to protect the tissues of the eye against the deleterious effects of the photochemical or the ambient oxidation reactions involving oxygen and its radicals. Previous studies have shown that exposure of animals to hyperbaric oxygen leads to the development of cataracts.[2] This has been found to be true in the case of humans also.[3] There now exists a substantial body of literature to suggest that the development of several oxygen-induced pathologic manifestations is initiated by conversion of the relatively inactive dioxygen to the superoxide radical anion and its consequent derivatization to other active species of oxygen.[4] The formation of the superoxide is considered imperative in many of the autoxidative and enzymatic reactions. An indirect proof of the concept that superoxide generation is toxic to the tissues is derived from the fact that most tissues are endowed with superoxide dismutase, an enzyme meant to effectively remove this free radical at a rate faster than would happen in its absence.[4] Thus the first task before us was to examine if the lens has superoxide dismutase (SOD). As summarized in FIGURE 1, that was found to be true.[5] The formation of adrenochrome from epinephrine, an O_2^--dependent reaction,[6] is inhibited by the dialyzed extracts of the rat lens, the inhibition being proportionate to the protein concentration. Similar reports on the presence of SOD in the lens were made simultaneously by other laboratories.[7,8] That the inhibition of adrenochrome formation by the lens extract is indeed specific to superoxide dismutase was subsequently proven by us following electrophoretic separation of the lens proteins on polyacrylamide gel and negative staining of the gel[9] for superoxide dismutase per se (FIG. 2).[5] Definitive evidence on the presence of SOD led us to postulate that superoxide generation (FIG. 3) in the aqueous humor might be a factor in the pathogenesis of senile cataracts and that ascorbate may be an element

TABLE 1. Concentration of Ascorbic Acid in Various Tissues and Fluids

	Man	Ox	Rabbit	Rat
Cornea	—	310	550	100
Aqueous humor	160	190	346	0
Lens	190	360	130	0
Vitreous	360	80	70	14
Ciliary body	—	87	196	
Retina		98	425	
Choroid		49	177	
Liver	70	290	170	240
Pancreas	11	125		
Adrenals	4000	1285	340	
Brain		170		
Cardiac muscle	—	38	—	—
Kidney	130	130	—	150
Saliva	0			
Seminal vesicle secretion	50	140		40
Urine	25	0		
Plasma	10	4.5	4.1	
GSF	12			
Leukocytes	165			

Values are expressed as mg/kg wet weight of the tissue or as mg/l of the fluid. The data have been compiled from those given in *Biochemists' Handbook*, D. Van Nostrand Company, Inc., New York (1961).

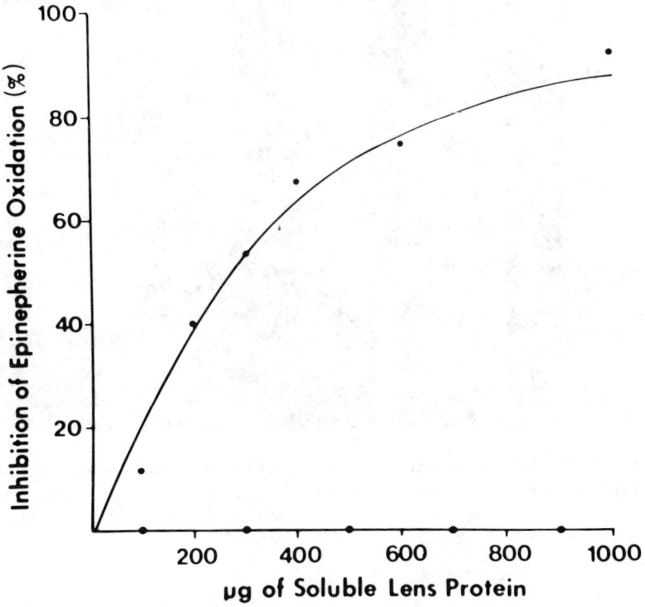

FIGURE 1. Inhibition of O_2^- dependent oxidation of epinephrine by dialyzed extract of rat lens homogenates. The rate of epinephrine oxidation was monitored by determining ΔOD_{460} nm. The reaction mixture consisted of sodium carbonate 0.05 M, pH 10.2, and epinephrine, 3×10^{-4} M. The reation was conducted in the presence and absence of the lens extract so that the values are expressed as the percentage of epinephrine oxidized in the absence and presence of the lens extract. The reaction was followed for a period of 3 minutes.[5]

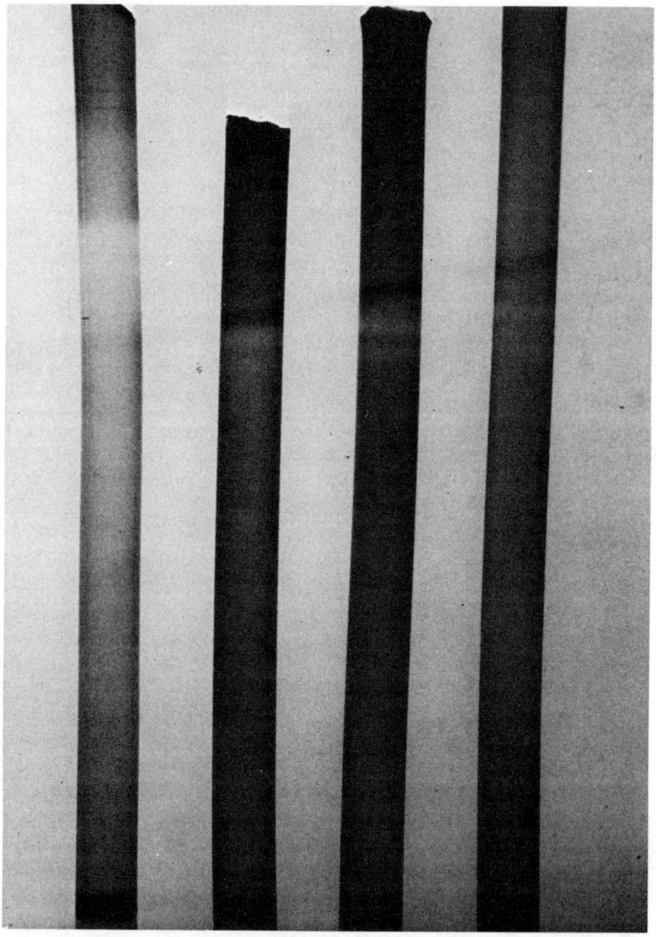

FIGURE 2. Identification of superoxide dismutase activity in lens. Soluble fraction of the lens homogenates representing about 200 μg of protein was electrophoresed on a 10% polyacrylamide gel and SOD activity, represented here by achromatic bands, was localized by negative staining according to the procedure of Beauchamp and Fridovich.[9]

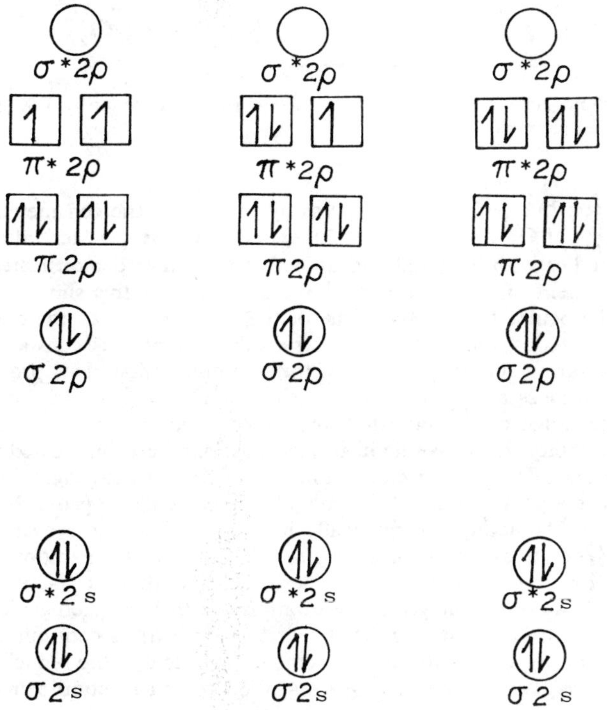

O_2	O_2^-	O_2^{2-}
oxygen	superoxide ion	peroxide ion
bond order, 2	bond order, 1½	bond order, 1
unpaired electrons,2	unpaired electrons,1	unpaired electrons,0

FIGURE 3. Electronic structure of oxygen, superoxide anion, and the peroxide anion. The orbitals have been indicated according to standard chemical textbooks. Electrons only of the principal quantum $n = 2$ have been taken into consideration.

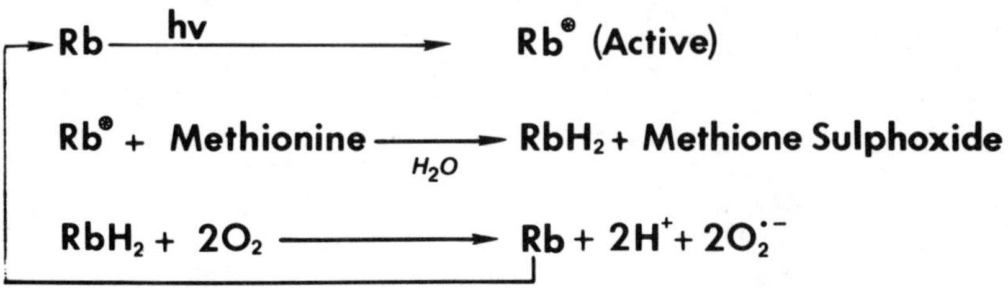

FIGURE 4. Photochemical generation of O_2^-: *Rb* = riboflavin, *hv* = light. Methionine can be replaced by other electron donors.

acting to thwart this pathological process by protecting the cell membranes against the toxic effects of O_2^-, hydrogen peroxide, and other oxygen derivatives.[5]

Several studies with bacterial and animal tissues have demonstrated that O_2^- is an important agent of oxygen toxicity.[4] We thought that this should be more so in the case of the ocular tissues. By virtue of the transparency of the cornea, aqueous humor, lens, vitreous, and retina, the intraocular chamber is constantly filled with light, at least during the long periods of the photopic vision. This provides a unique situation for an incessant photochemical generation of O_2^- and its derivatives in the vicinity of the tissue membrane sites, and makes oxidation of their components an imminent possibility. Excessive light, in fact, has long been implicated in the genesis of senile cataracts. The incidence of cataracts is higher in the geographic areas of the world that have a greater amount of sunlight. Thus, we developed a theory that light and oxygen may be acting synergistically in contributing to the overall cataractous process; the synergic action involves *in situ* photochemical generation of O_2^- in the aqueous and vitreous humors and its derivatization to other active oxidants.[10,11] The function of ascorbate may then be to attenuate this cataractogenic process.[5] An example of a reaction that may yield O_2^- and its derivatives continuously under the influence of light, as long as the system remains aerobic and the reducing equivalents are available, has been shown in FIGURE 4, the detailed reaction being described in FIGURE 5.

Experiments were, therefore, designed to test the hypothesis that such a photocatalytic system, like the one described in FIG. 4, would damage the lens physiologically. Such a test has been done by *in vitro* organ culture techniques. Briefly, freshly isolated rat lenses were cultured in riboflavin-containing medium in the absence and presence of the fluorescent daylight and their ability to actively transport rubidium against an electrochemical gradient was measured (FIG. 6). The results have been described as the distribution ratio of the ion between lens water and the medium

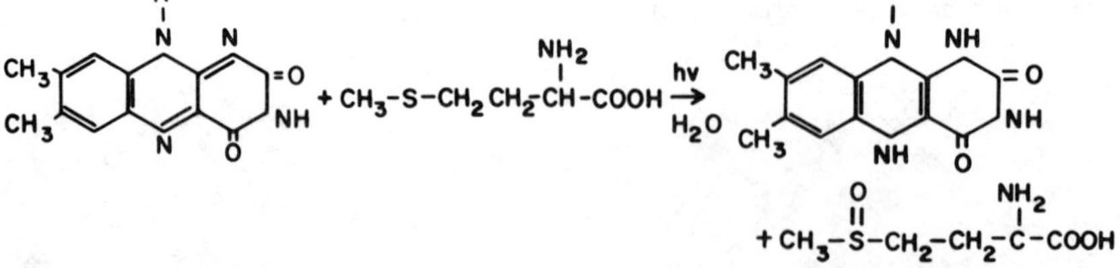

FIGURE 5. Reduction of riboflavin by methionine. The reaction is photocatalyzed.

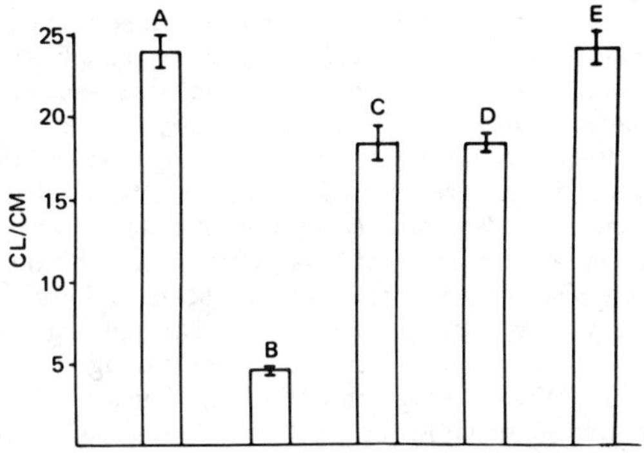

FIGURE 6. Uptake of rubidium by rat lens in culture. Lenses isolated from rats were incubated in TC 199 medium containing 2.5 mM Ca^{2+} and 27 mM HCO_3^- and pulsed with Rb-86. After 20 hours, the distribution ratio was determined by measuring the radioactivity in the lens and the medium. *A,* Control with or without riboflavin in the dark and without riboflavin in the light; *B,* with riboflavin in the light; *C,* B + SOD, 15 units; *D,* B + ascorbate (2.5 mM); *E,* B + catalase (25 units).[11]

TABLE 2. Uptake of Rb-86 by the Rat Lens Incubated in Light in Medium Containing Riboflavin: Effect of Various Scavengers

Conditions	n	× CL/CM	× E/C × 100
Blank control	20	22.0 ± 3.0	
−SOD (control)	20	5.06 ± 2.27	
+8 Units Mn SOD	9	9.19 ± 2.98	166.9 ± 26.9[a]
+20 Units Mn SOD	11	10.04 ± 2.75	198.2 ± 34.8[a]
−Catalase (control)	20	6.5 ± 2.5	
+4 Units catalase	10	20.0 ± 3.00	307.0 ± 80.0[a]
+16 Units catalase	10	21.0 ± 2.00	323.0 ± 80.0[a]
$-Na_3Fe(CN)_6$ (control)	16	7.44 ± 1.43	
+5 μM $Na_3Fe(CN)_6$	8	20.44 ± 2.35	213.2 ± 39.8[a]
+10 μM $Na_3Fe(CN)_6$	8	20.88 ± 3.97	280.2 ± 18.4[a]
$-Na_4Fe(CN)_6$ (control)	10	7.00 ± 1.47	
+5 μM $Na_4Fe(CN)_6$	5	19.55 ± 2.00	
+10 μM $Na_4Fe(CN)_6$	5	19.97 ± 3.86	223.3 ± 15.1[a]
−Mannitol (control)	7	6.02 ± 3.20	
+100 μM Mannitol	8	11.05 ± 3.00	138.9 ± 34.5[a]
$+FeSO_4$ (5 μM) (control)	8	8.33 ± 2.19	
$+FeSO_4$ (5 μM) + 100 μM Mannitol	10	12.09 ± 3.07	156.6 ± 31.3[a]

The blank control consisted of lenses incubated in the dark in medium containing 50 μM riboflavin. In other cases, the controls and corresponding experimentals were incubated in light for a period of 18-20 hours and CL/CM determined. The values have been expressed as mean ± standard deviation.

[a] Values are significantly different from the contralateral controls, the p values being < 0.001.[12]

attained following overnight incubation. The uptake of this ion by the lenses incubated exposed to light was substantially lower; no effect on rubidium accumulation was observed in the lenses incubated in the dark. The lower accumulation of the ion represents physiological damage. The damage is clearly photochemical in nature. The addition of catalase to the medium protects the lens against this damage completely, the uptake of rubidium in the presence of this enzyme being identical to that in the dark. Thus, hydrogen peroxide generated by the dismutation of the photochemically produced O_2^- is an important agent of lens injury. The addition of superoxide dismutase and ascorbate also protected the lens significantly. These experiments, therefore, demonstrate very simply the effectiveness of ascorbate against the induction of oxygen-dependent photoinjury to the lens. That the damage to the lens membranes may involve $OH^\bullet$, in addition to O_2^- and H_2O_2, is indicated by more recent experiments wherein the photodamage to the lens cation pump could also be prevented by ferricyanide, ferrocyanide, and mannitol, in addition to superoxide dismutase and catalase (TABLE 2). The protective effect of ferrocyanide has been explained to be due to its interaction with O_2^-, thus removing the latter from the medium of incubation. The interaction follows the following sequence:

	$\Delta E^{\circ\prime}$ volts
$O_2^- \rightarrow O_2 + e$	$+0.33$
$Fe(CN)_6^{3-} + e \rightarrow Fe(CN)_6^{4}$	$+0.46$
$O_2^- + Fe(CN)_6^{3-} \rightarrow O_2 + Fe(CN)_6^{4-}$	$+0.79$

The protective effect of the ferrocyanide appears to be due to its reaction with H_2O_2, as well as with $OH^\bullet$. This was indicated from the DMPO-dependent inhibition of the oxidation of ferrocyanide to ferricyanide by H_2O_2, as well as by the protection of the pump damage by mannitol, a hexitol known to scavenge $OH^\bullet$ more widely.[12]

The observed protective effect of ascorbate against the photosensitized damage to the lens cation pump led us to investigate the significance of ascorbate in relation to oxygen damage to the lens in greater detail. In experiments described in FIGURE 7, lens incubations were conducted in the medium TC 199 without any additional amount of riboflavin and the photodamage was assessed in terms of lipid peroxidation[13] as apparent by the level of malonaldehyde (MDA) formed by the series of reactions shown in FIGURE 7. As summarized in FIGURE 8, the physiological level of this aldehyde in the lens is substantially low. Incubation in the dark did not affect this level. However, MDA content of the lenses incubated in the presence of light was observed to be substantially greater than that of the lenses incubated in the dark. Since the elevation in the level of MDA took place in the medium without any additional amount of photosensitizers over and above that normally present in the culture medium, peroxidative degradation of the lens membrane lipids as a contributing factor in the genesis of cataracts appears quite feasible. The composition of TC 199 is close to that of the aqueous humor and the light intensity used in these experiments is within physiological limits.

Again, addition of superoxide dismutase, catalase, and physiological amounts of ascorbate to the culture system prevented the light-catalyzed formation of MDA. Since superoxide dismutase and catalase would not enter the cell, the damage to the lens lipids appears to be localized at the membrane level. The protective effect of ascorbic acid is apparently due to its interaction with O_2^-, H_2O_2, as well as $OH^\bullet$, all of which are supposed to react with ascorbate with different rate constants. (TABLE 3).

A tissue organ cultured in the presence of light can, however, be structurally altered irrespective of the generation of the active species of oxygen because of the

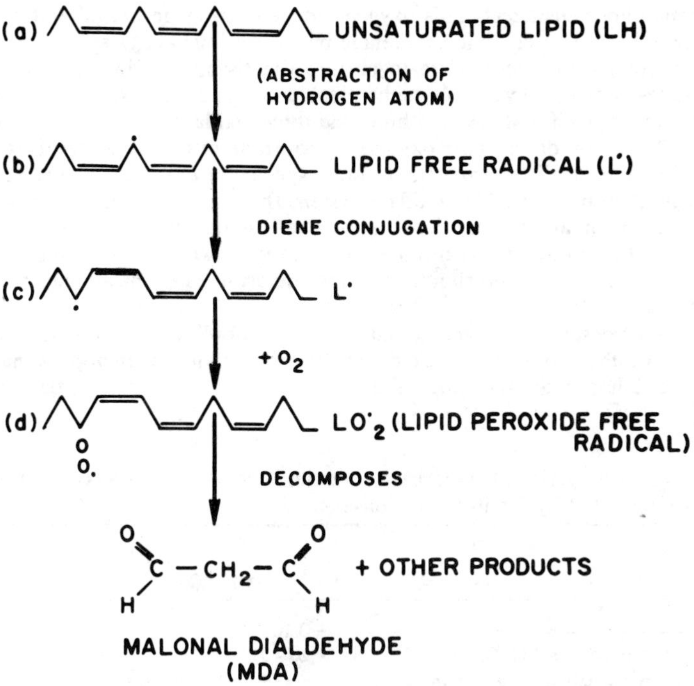

FIGURE 7. Peroxidative degradation of unsaturated lipids. Malonaldehyde is one of the final products.

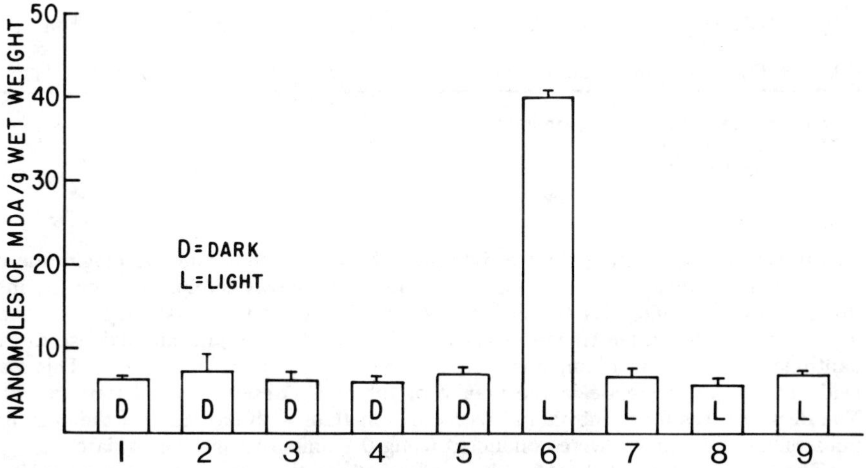

FIGURE 8. Malonaldehyde content of rat lenses freshly dissected and following incubation. The incubation conditions were similar to those described in FIGURE 6, except that the medium lacked riboflavin. MDA was determined in the trichloroacetic acid lens extract by treating the latter with 2-thiobarbituric acid.[13] 1, Freshly dissected unincubated lens; 2, basal medium (BM) in dark (D); 3, BM + SOD (D); 4, BM + catalase (D); 5, BM + ascorbate (D); 6, BM + light (L); 7, BM + SOD (L); 8, BM + catalase (L); 9, BM + ascorbate (L).

endogenous photosensitizers. This oxygen-independent structural alteration may render the tissue more susceptible to damage by the various active species of oxygen. It was, thus, considered appropriate to culture the tissue also in a system producing active species of oxygen by nonphotochemical means and then examine the moderating effect of ascorbate, if any. A xanthine/xanthine oxidase system was used for this purpose. The action of xanthine oxidase on xanthine in the presence of oxygen leads to the formation of all the important active species of oxygen, *i.e.*, O_2^-, H_2O_2, and $OH^\bullet$.[14] The xanthine/xanthine oxidase system thus offers the opportunity of testing the efficacy of an antioxidant in a more challenging situation. Lenses were thus incubated in the presence of xanthine and xanthine oxidase generating H_2O_2 at the rate shown in FIGURE 9, and rubidium uptake determined in the absence and presence of ascorbate.[15]

FIGURE 10 describes the accumulation of the rubidium by the lens as a function of time. In the absence of xanthine oxidase, the uptake follows an approximately linear course until 2 hours. Subsequently, the process becomes curvilinear, tending towards

TABLE 3. The Energetics of Ascorbate Reaction with Xanthine Oxygen Interaction Products Generated by Xanthine Oxidase

	Δn E° (Volts)	$\Delta G^{\circ\prime}$ (kcal/mol)	Rate Constant
$HA^- \rightarrow A + H^+ + 2e$	−0.108		
$H_2O_2 + 2H^+ + 2e \rightarrow 2H_2O$	+2.640		
$HA^{-a} + H_2O_2 + H^+ \rightarrow A + 2H_2O$	2.532	−58.4	$8\ M^{-1}s^{-1}$
$HA^- \rightarrow A^{\bullet -} + H^+ + e$	−0.282		
$OH^\bullet + H^+ + e \rightarrow H_2O$	+2.180		
$HA^{-\,b} + OH^\bullet \rightarrow H_2O + A^{\bullet -}$	1.898	−43.8	$1.3 \times 10^{10}\ M^{-1}s^{-1}$
$HA^{-\,c} + O_2^- \rightarrow$ products		−15.2	$5 \times 10^4\ M^{-1}s^{-1}$

[a] Determined in the author's laboratory.[12]
[b] Farhataziz and Ross.[22]
[c] Cabelli and Bielski.[23]

an asymptote approaching the steady state. The time course and the magnitude of accumulation remained virtually unaffected if xanthine was eliminated from the medium. Thus, xanthine, over the time period, did not affect the physiology of the lens as reflected by its cation transport activity. The addition of xanthine oxidase to the xanthine-containing medium, however, exerted a toxic effect on the lens. This was evidenced by the decrease in the uptake of rubidium, observed at all time periods. The decrease was more substantial with 0.1 unit than with 0.01 unit of the enzyme. Subsequent experiments were conducted using 0.1 units of xanthine oxidase.

The incorporation of 2 mM sodium ascorbate in the xanthine/xanthine oxidase containing medium led to a greater accumulation of rubidium as compared to that in the contralateral control lenses incubated in medium without any ascorbate (FIG. 11). At the end of a 3-hour incubation experiment, the distribution ratio attained in the presence of ascorbate was approximately twice that in its absence. The overall rate of accumulation was also greater in the presence of ascorbate. The effect of varying

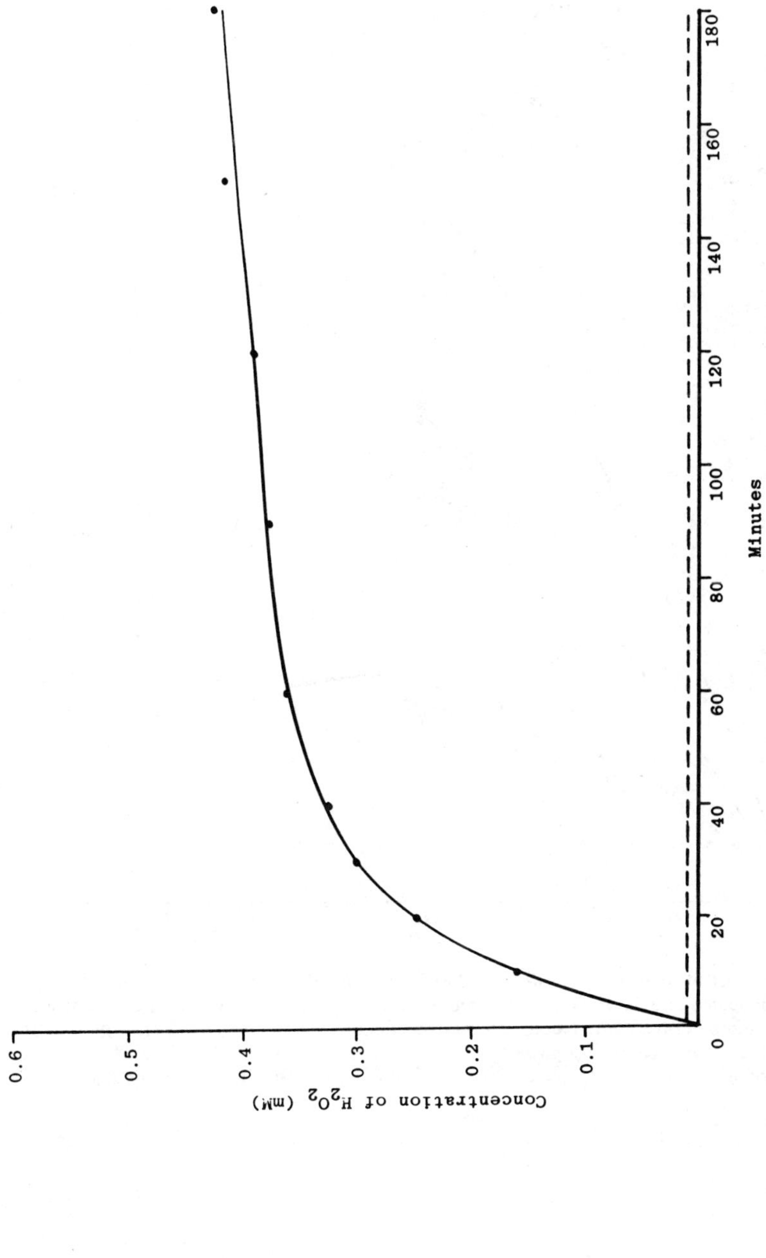

FIGURE 9. Hydrogen perioxide generation in the incubation medium: xanthine (1 mM), xanthine oxidase (0.1 unit/4 ml). The incubation medium was tyrode with Mg^{2+} and contained 5.5 mM glucose. The dotted line close to the abscissa indicates traces of H_2O_2 in the medium incubated with ascorbate (2 mM). H_2O_2 was determined iodometrically; 0.5 ml of the medium was mixed with 0.1 M acetate buffer pH 3.5, 1 ml of 0.6 M KI, 0.5 ml of 1.7×10^{-3} M, ammonium molybdate, and 0.5 ml of 1% starch. The liberated iodine was titrated with 0.5 mM $Na_2S_2O_3$. In the medium incubated with ascorbate, the peroxide determination was done after reacting 0.5 ml of the medium with 20 units of ascorbate oxidase for 15 minutes.

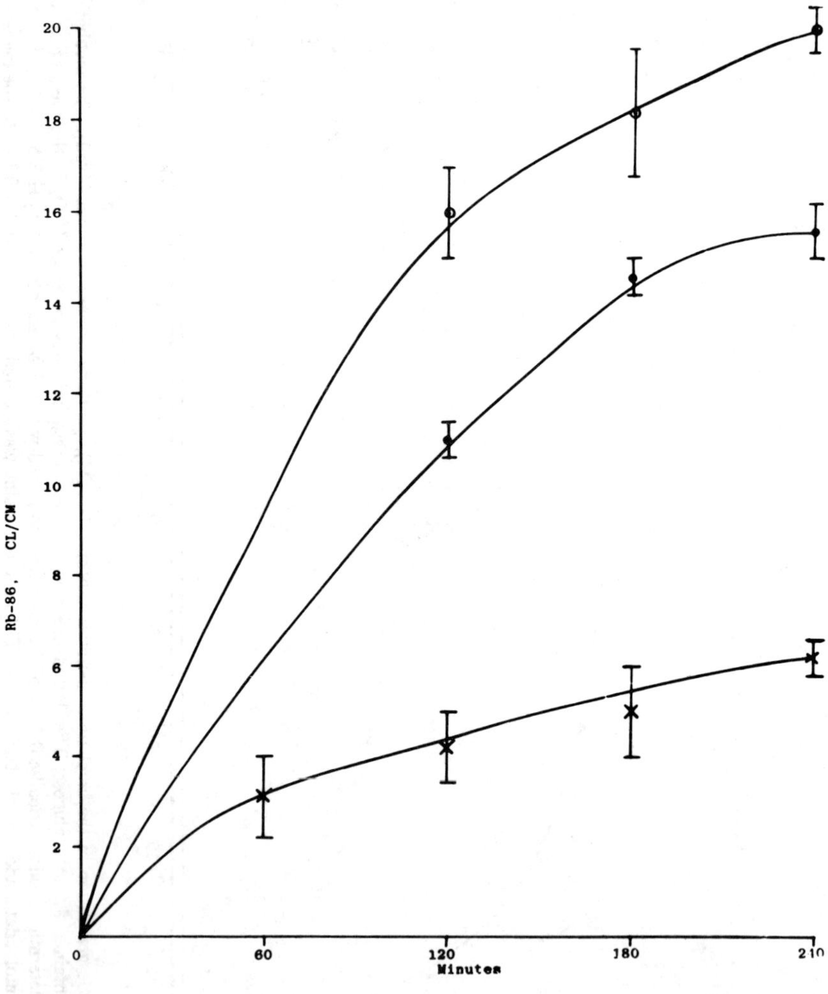

FIGURE 10. Effect of xanthine/xanthine oxidase system on Rb-86 accumulation by rat lens. Lenses were incubated at 36°C in tyrode medium without Mg^{2+} containing 5.5 mM glucose and 1 mM sodium xanthine. Xanthine oxidase/4 ml of incubation medium; ○ = 0 unit; ● = 0.01 unit; × = 0.1 unit. Incubator was gassed with oxygen/air mixture (95/5). CL/CM = concentration of rubidium in the lens/concentration of rubidium in the medium of incubation.

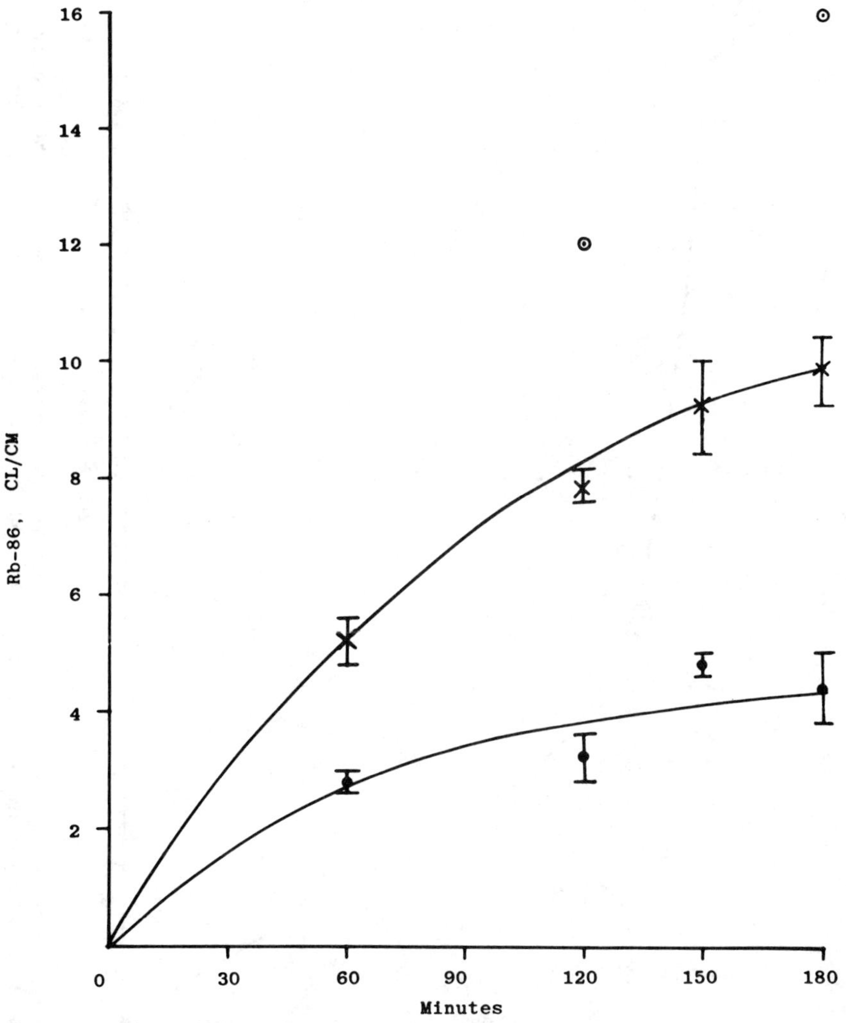

FIGURE 11. Uptake of rubidium in rat lenses incubated in medium containing xanthine (1 mM) and xanthine oxidase (0.1 unit/4 ml of medium): effect of ascorbate. Ascorbate concentration: ● = 0, × = 2 mM. Blank controls (⊙) consisted of lenses incubated without xanthine oxidase and ascorbate. The values corresponding to 60, 120, and 150 minutes represent mean ± standard deviation of three experiments. The values at 3 hours represent mean ± standard deviation of six experiments. Contralaterally paired lenses were used.

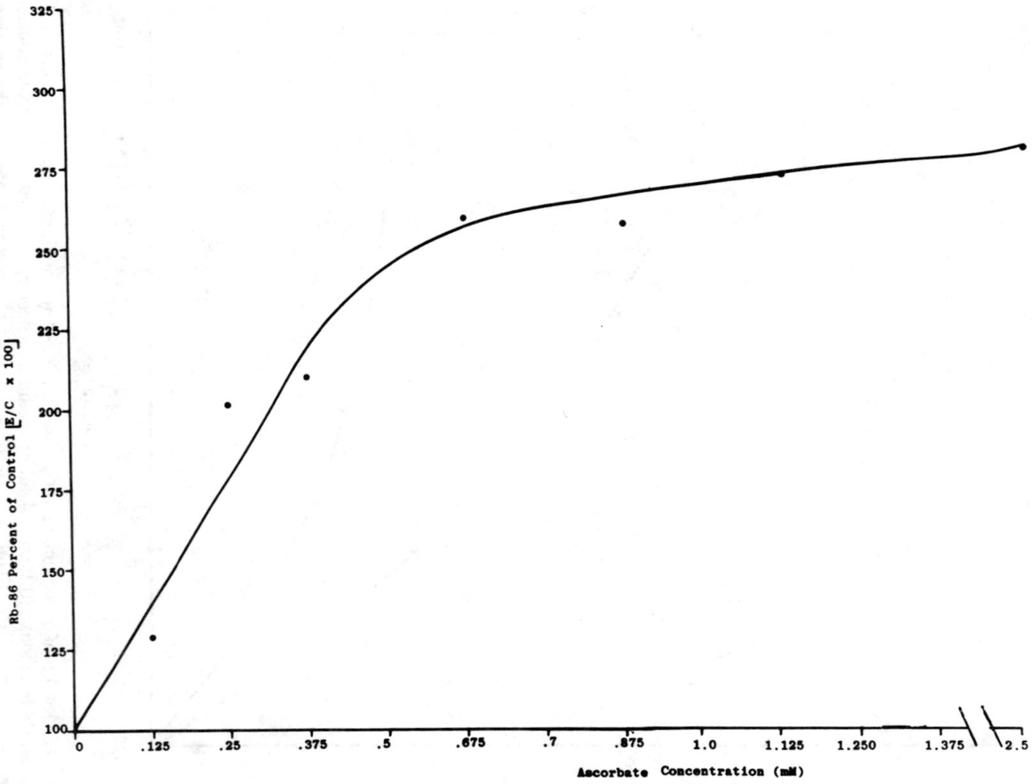

FIGURE 12. Effect of varying concentrations of ascorbate on Rb-86 uptake. Rat lenses were incubated in the presence of medium containing xanthine (1 mM) and xanthine oxidase (0.1 unit/4 ml) as described in FIGURE 10. The results are expressed as E/C × 100, E being the uptake by the lens in the presence and C being the uptake in the absence of ascorbate.

concentrations of ascorbate on the uptake of rubidium by the lens has been described in FIG. 12. Here the results are expressed as the ratio between the rubidium contents of the lenses incubated in the presence and absence of ascorbate multiplied by 100 (E/C × 100). A value above 100 indicated the protective effect. The protective effect of ascorbate was approximately linear up to 0.5 mM. Subsequent increases in ascorbate did not cause a commensurate increase in the protective effect.

The time courses of α-aminoisobutyric acid transport from the medium containing xanthine and xanthine oxidase in the absence (control) and presence (experimental)

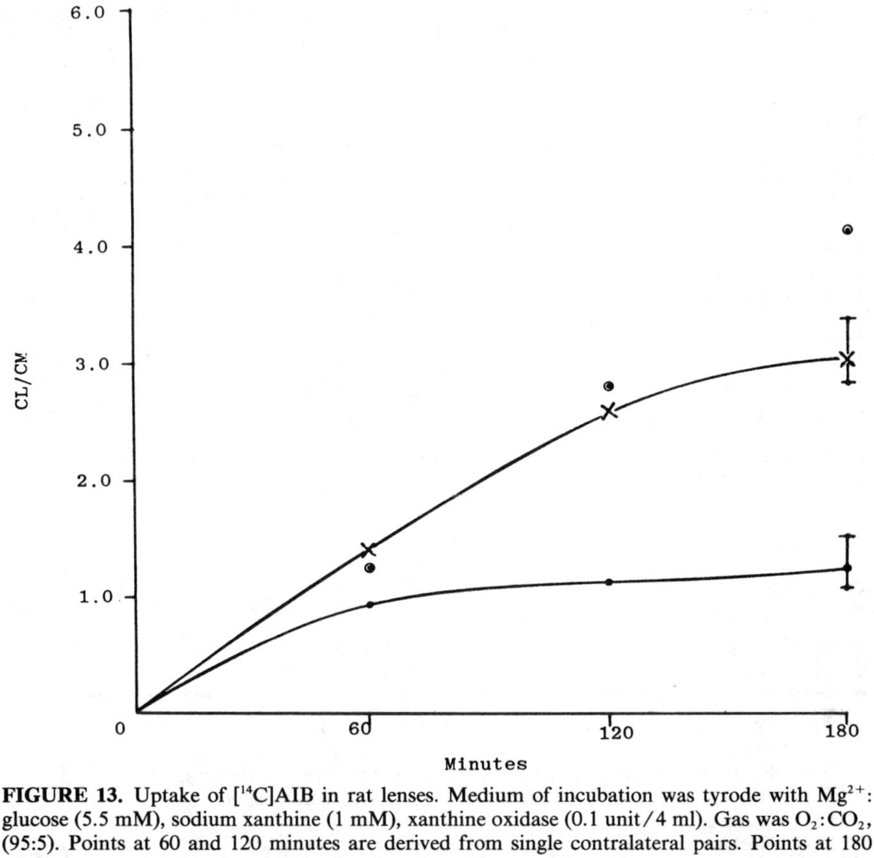

FIGURE 13. Uptake of [^{14}C]AIB in rat lenses. Medium of incubation was tyrode with Mg^{2+}: glucose (5.5 mM), sodium xanthine (1 mM), xanthine oxidase (0.1 unit/4 ml). Gas was O_2:CO_2, (95:5). Points at 60 and 120 minutes are derived from single contralateral pairs. Points at 180 minutes represent three contralateral pairs. ● = No ascorbic acid; × = ascorbate (2 mM); ⊙ = blank controls, representing duplicate incubation experiments without xanthine oxidase and without ascorbate.

of sodium ascorbate along with the blank controls have been described in FIGURE 13. Akin to the case of rubidium, the magnitude as well as the rate of AIB uptake was greater in the presence of ascorbate and compared favorably with that observed in the blank control experiments conducted in the medium without xanthine oxidase.

Data for comparison between various groups were obtained by 3-hour incubation experiments. As described in TABLE 4, ascorbate by itself has no effect on Rb^+ uptake by the lenses incubated in the basal medium without xanthine/xanthine oxidase. Also, dehydroascorbate had no protective effect, either in terms of rubidium or AIB uptake. Ascorbate is indeed able to protect the lens even in the oxidizing medium generated by xanthine/xanthine oxidase, and it is quite possible that it may be able to perform a similar antioxidant function *in vivo* also.

Ascorbate, however, should not be an exception in protecting the lens against oxidative damage. It is most effective probably because of the existence of the ascorbate pump and consequent maintenance of its high concentration in the aqueous and the lens. Vitamin E, another nutritional antioxidant, though not known to be actively transported across the blood aqueous barrier, should be equally effective. As summarized in FIGURE 14, low concentrations of this nutrient also inhibit the photoperoxidative degradation of lens lipids quite effectively.[16] Higher concentrations were not prooxidative.

TABLE 4. Effect of Ascorbic Acid on Rubidium and α-Aminoisobutyric Acid (AIB) Uptake[a, d]

Additions to the Medium	CL/CM	E/C × 100	n[c]
	RUBIDIUM UPTAKE		
C[e] none	21 ± 1.5		6
E[f] Na ascorbate (1.0 mM)	21 ± 2.0	100	2
E Na ascorbate (2.0 mM)	21.5 ± 0.5	102	3
C xanthine (1 mM)	19 ± 1.5	24	6
E xanthine + XO[b] (0.1 unit)	4.5 ± 1.0		
C XO (0.1 unit)	5.2 ± 0.5	200	10
E XO + Na ascorbate (2mM)	10.4 ± 0.4		
C XO (0.1 unit)	4.54 ± 0.4	105	5
E XO + DHA (2.5 mM)	4.78 ± 0.28		
	AIB UPTAKE		
xanthine (1 mM)	5.5 ± 0.5		4
C xanthine + XO (0.1 unit)	1.48 ± 0.38	203	7
E xanthine + Na ascorbate (2 mM)	3.0 ± 0.47		
C XO (0.1 unit)	0.92 ± 0.31	108	5
E XO + DHA (2.5 mM)	1.0 ± 0.5		

[a] Paired lenses were incubated as C and E in the medium with additions indicated.
[b] XO = xanthine oxidase.
[c] n = Number of contralateral pairs used in each group of experiments.
[d] Rubidium and AIB were used as ^{86}RbCl and C-14 AIB respectively.
[e] C = Control.
[f] E = Contralateral lenses incubated as experimental.[12]

The final test for involvement of oxidation of membrane lipids, membrane proteins, or membrane-bound enzymes in the formation of cataracts must come from a study of the animal model system. Unfortunately, there is a general lack of models of senile cataracts. One that has been recently developed is known as the Emory mouse. These animals are known to develop cataracts by the ninth to eleventh month. More recently,

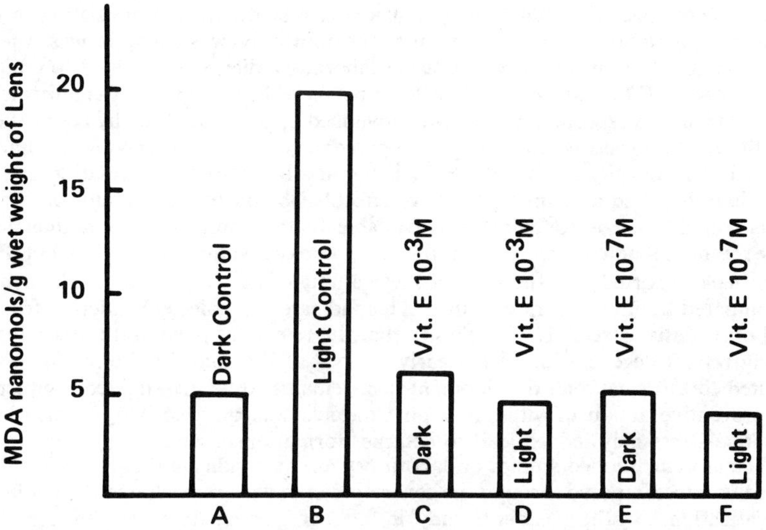

FIGURE 14. Photoperoxidation of rat lens lipids under organ culture conditions: prevention by vitamin E.[16] Period of incubation was 20 hours.

studies from this laboratory demonstrate that an oxidative stress may indeed be one of the contributing factors involved in the development of cataracts in these animals. Six-week-old mice were housed in individual cages and given Purina lab chow ad lib. Animals in one group[17] were given weekly 0.1 ml of coconut oil intraperitoneally as a placebo and designated as the control group. The members of the other group, designated as the experimental group, received intraperitoneally 0.1 ml of the coconut oil containing 10 mg of DL-α-tocopherol acetate. The progress of cataracts, as monitored by an ophthalmoscope and Scheimpflug photography, using 1% homatropine as a pupillary dilator, is depicted in FIGURE 15. The animals were also weighed at

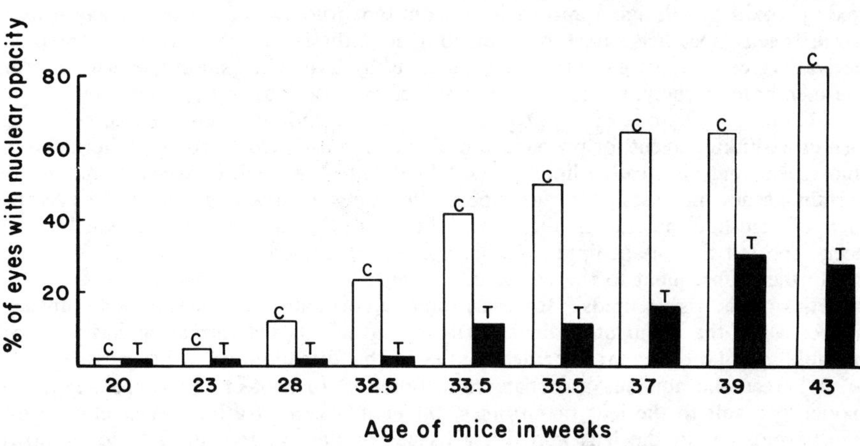

FIGURE 15. Effect of vitamin E on the progress of cataracts in emory mice[17]; *C*, treated with placebo, *T*, treated with vitamin E.

frequent intervals. If the animals were kept on a synthetic diet consisting of casein, starch, coconut oil, vitamins, and salts, the mortality was high. It was, therefore, necessary to maintain them on the usual laboratory diet which is said to contain 30 mg of vitamin E (*dl*-α-tocopherol acetate) per kg. The majority of the animals in the control group, as reported, did develop advanced opacity (nuclear) by about the tenth month. In the experimental group, the percentage of animals having a comparable degree of lens opacity was substantially lower at all periods of examination, indicating that vitamin E had a definite protective effect.[17] Obviously, the amount of vitamin E present in the diet is somehow not available to the animal, or the animal has, for some reason unknown at the present time, an increased vitamin E requirement. Bhuyan *et al.*[7] have reported that the liver of these animals has a higher malonaldehyde level as compared to other laboratory mice. This indicates the mice are suffering from some kind of oxidative stress. The possibility that the release of polyunsaturated fatty acids during retinal degeneration in the early periods of the animal's life might have contributed towards cataract development also remains to be investigated. Nevertheless, the preventive action of vitamin E on cataracts was quite striking, indicating that oxidative stress may indeed lead to cataract formation *in vivo.*

The aforementioned studies on laboratory animals indicate the possibility that the oxidative stress imposed through the generation of oxygen radicals and their subsequent derivatization to other oxidants may initiate or participate in the process of age-dependent cataractogenesis (senile cataracts) in human beings also, light and oxygen, as explained earlier, acting synergistically through some photosensitizers. As mentioned before, these radicals can induce oxidation of proteins, lipids, and other chemical constituents of metabolic and structural importance to the tissue. The oxidation of protein sulfhydryl associated with human cataracts appears now to be well established. Organ culture studies reported above and elsewhere indicate the additional possibility of membrane lipid damage due to the generation of oxidizing radicals and their active derivatives. That such may be the case in human beings is quite possible. Preliminary evidence of lipid damage associated with human cataracts has been obtained by us and others. The content of malonaldehyde was found to be higher in the cataractous lenses as compared to the noncataractous lenses. Furthermore, the level is higher in the brunescent cataracts than in the nonbrunescent cataracts (TABLE 5).[17] Since malonaldehyde is derived by the degradation of the lipid peroxides, the formation of such toxic peroxides with aging and their concomitant role in membrane damage leading towards lens opacification is an interesting hypothesis to consider. These peroxide derivatives can initiate oxidation of protein sulfhydryls while simultaneously altering the membrane structure. Tryptophan residues are also known to be involved.

Malonaldehyde, one of the products of lipid peroxidation, has been shown to act as a cross-linking agent for proteins and phospholipids, yielding complex lipid-soluble fluorescent materials called lipofuscins.[18] The content of such fluorescent material in certain tissues increases with age. Spectrofluorometric studies indicate the presence of lipofuscin-like material in the human lens also. In addition, the human lens has been reported to contain lipid material containing conjugated dienes.[17] These dienes are formed subsequent to the peroxidative degradation of polyunsaturated fatty acid moieties of the tissue lipids. However, data to correlate the contents of lipofuscins, as well as of the conjugated dienes, with aging or cataract formation has not been possible to obtain thus far. Further studies in this direction are in progress.

Whereas the aqueous ascorbate may function to protect the tissue against an oxidative insult to the lens membranes, the significance of a high intracellular level of this nutrient in the lens and other tissues of the eye remains as yet unknown. Mapson and Moustafa[19] observed that in germinating pea seedlings, the dehydroascorbate-ascorbate redox couple is able to facilitate electron transfer from NADPH to oxygen enzymatically, the GSSG/GSH couple acting as an intermediate electron carrier operating between NADPH and dehydroascorbate (DHA); see FIGURE 16. The possibility of a similar role of the DHA-AA couple also exists in the lens and other tissues of the eye where there is excessive GSH.[20] Therefore, one of the ultimate effects of such an electron transfer may be a stimulation of the hexose monophosphate shunt (HMP shunt). We have recently found this to be true. The activity of the HMP shunt was measured by incubating lenses in the balanced salt medium containing [I-^{14}C]glucose and measuring the CO_2 liberated from utilization of C-1 of glucose. The time course of CO_2 production through the shunt in the presence and the absence of ascorbate and the dehydroascorbate has been described in FIGURE 17. The rate of CO_2 production was higher in the case of the lenses incubated with ascorbate, the enhancement in the rate becoming significant after thirty minutes of incubation. The amount of CO_2 produced by incubation of the lenses for 60 minutes in the presence of ascorbate was approximately twice that produced in its absence. This increase in shunt activity brought about by ascorbate was not abolished by the addition of catalase in the medium. This suggests that the shunt stimulation does not involve, to any

TABLE 5. Malonaldehyde Content in Human Cataracts

Cataract Type	Malonaldehyde (nmol) Per Lens	Malonaldehyde (nmol) Per g Wet Lens
Yellow cataract	0.566 ± .04	2.181 ± 0.10 (20)
Brown cataract	0.915 ± .05	3.715 ± 0.30 (18)

Intracapsularly removed cataracts were used. Following surgical extraction, they were quickly weighed and transferred to homogenizing tubes containing 1 ml of 10% trichloroacetic acid solution; 0.5 ml of the protein free supernatant obtained after centrifugation was mixed with 0.5 ml of 0.71% L-thiobarbituric acid solution. The mixture was heated in boiling water bath for 15 minutes and MDA values calculated from the optical density of the resulting solution at 532 nm. Tetramethoxy propane was used as a standard. The number of samples is given in parentheses. The values are expressed as mean ± standard error; p value between brown and yellow cataracts is < 0.01.[17]

significant extent, hydrogen peroxide that might be generated in the incubation medium during the oxidation of ascorbate. The concentration of the ascorbate in the medium following 1 hour of incubation remained approximately 95% of the starting concentration. No peroxide could be detected. The possibility of its decomposition to water because of the excess of ascorbate cannot be ruled out. An alternate possibility, that the stimulation of the shunt may be due to the dehydroascorbate derived by ascorbate oxidation, was examined next. In accordance with this prediction, addition of DHA to the medium stimulated the shunt very substantially. The amount of CO_2 liberated by the lens when incubated in the presence of DHA was approximately 4 to 10 times that in its absence (FIG. 17). It was also about 4 to 5 times over that produced by the lenses incubated in the presence of similar levels of ascorbate.

Subsequent experiments were designed to investigate whether the stimulation of the shunt by the oxidized and the reduced forms of the vitamin C is related to the coupling of the DHA-AA pair with GSSG-GSH pair on the one hand and the oxygen-H_2O pair on the other hand, oxygen serving as the final acceptor of the reducing equivalents. The concentration of glutathione in the tissue is maintained constant due

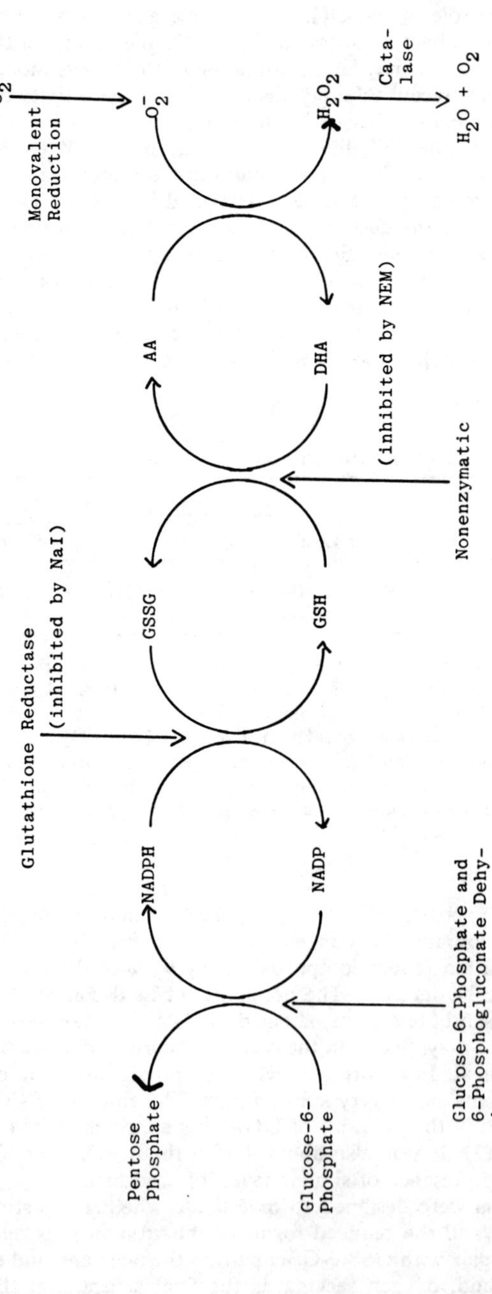

FIGURE 16. Integration of ascorbate (*AA*)/dehydroascorbate (*DHA*) with the transport of hydrogen from glucose-6-phosphate to oxygen through NADPH.

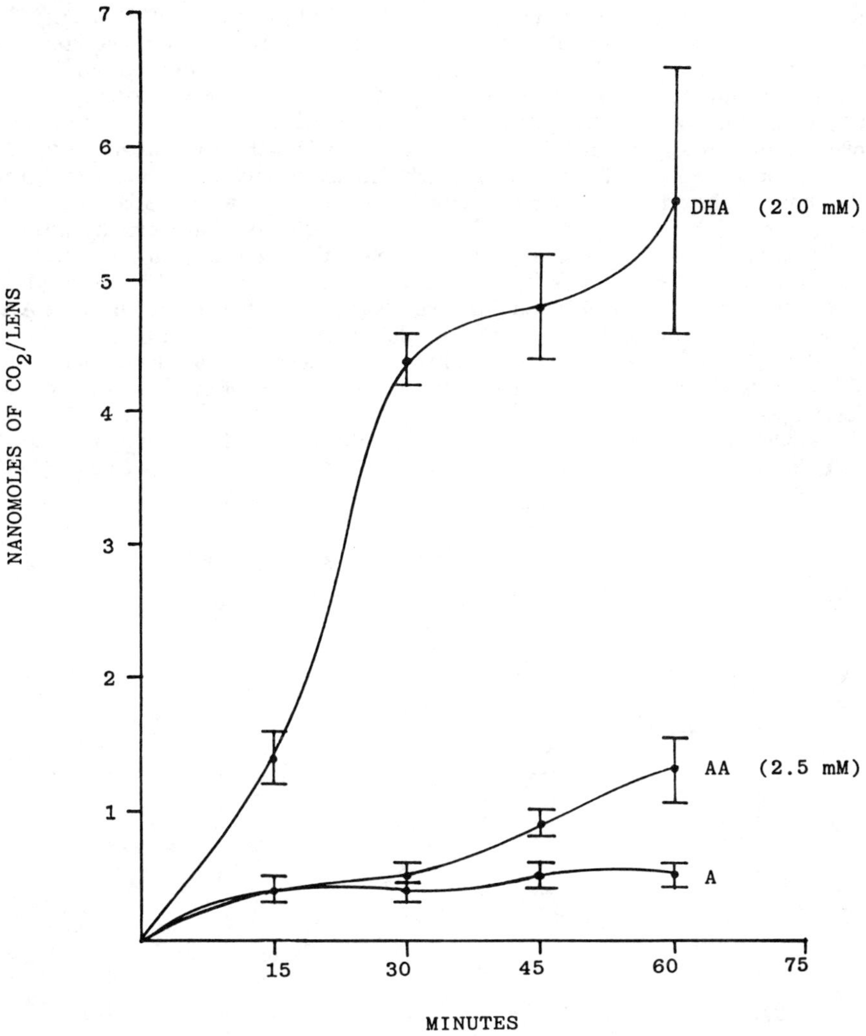

FIGURE 17. Hexose monophosphate shunt in rat lens: effect of ascorbate and dehydroascorbate. The utilization of glucose by the rat lenses through the hexose monophosphate shunt was measured in terms of CO_2 produced by isolated lenses incubated at 37°C in the medium pulsed with [1-^{14}C]glucose. *A,* basal medium; *AA,* medium containing ascorbic acid; *DHA,* medium containing dehydroascorbic acid. Each point represents a mean of four experiments ± standard deviation. The basal medium consisted of tyrode without Mg^{2+}. The glucose concentration was 5.5 mM. The specific activity of ^{14}C was 3.5 μCi/μmol of glucose. Incubations were conducted in test tubes containing a sidearm closed with a rubber stopper. After introducing the lens, the test tube was closed with a stopper through which was hung a plastic cup containing filter paper soaked in hyamine hydroxide (0.1 N). At the end of the incubation, 100 μl of a 10% solution of trichloroacetic acid was injected through the sidearm and CO_2 collected in the hyamine-containing cup by incubating the tubes for another 30 minutes. The plastic cup was then transferred to LSC fluid and radioactivity determined.

to the continued reduction of GSSG by NADPH produced by the HMP shunt. The reduction reaction is catalyzed by glutathione reductase. HMP shunt measurements were thus conducted in the presence of sodium iodide. This compound is a known inhibitor of glutathione reductase.[19] Addition of the iodide to the incubation medium in concentrations ranging from 0 to 4 mM was observed to antagonize the stimulation of the shunt brought about by ascorbate, as well as by dehydroascorbate (FIG. 18). The accelerating effect of ascorbate and dehydroascorbate can thus be attributed to the ability of the cell to continue to maintain glutathione at an appropriate level. The importance of the reduced glutathione was additionally apparent from the effect of NEM on the HMP shunt (FIG. 19). The addition of this compound, in concentrations ranging from 0 to 4 mM, also led to an abolition of the ascorbate and dehydroascorbate stimulating effects. The iodide and NEM effects thus indicate that the DHA- or AA-dependent stimulation of the shunt involves glucose, NADPH, GSSH/GSH, and very likely oxygen. Direct oxygen consumption measurements by the tissue due to the small size of the rat lenses incubated in microliter volumes of medium could not be feasible.

As summarized in the APPENDIX, the potential difference of the reduction of DHA by GSH at pH 7 and 37° C is approximately 0.284 V. The free energy change ($\Delta G°$)

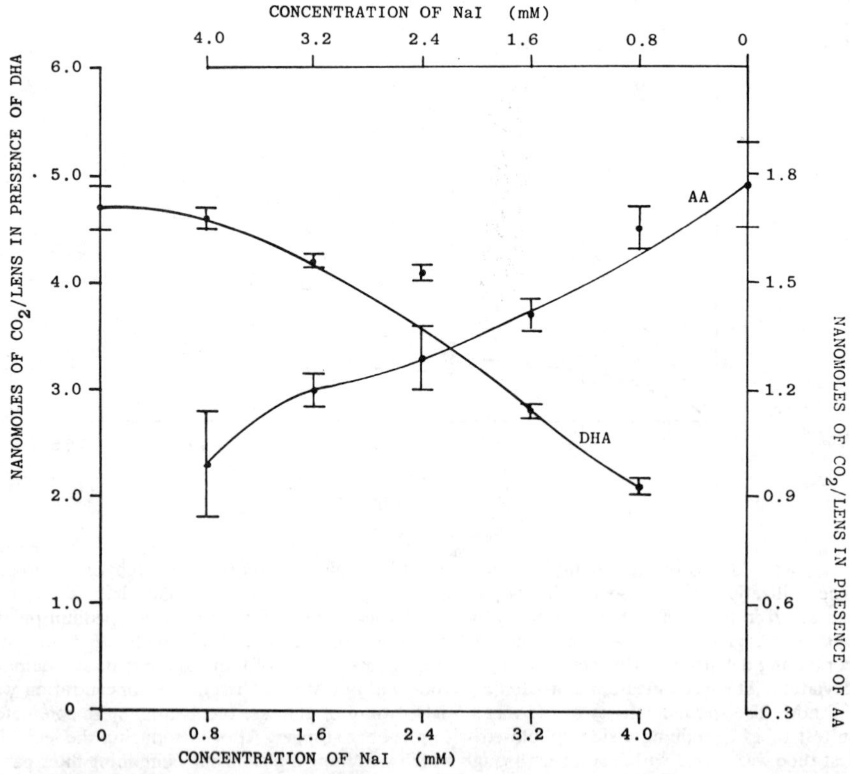

FIGURE 18. Effect of NaI on the production of CO_2 from [1-^{14}C]glucose in the presence of DHA (2 mM) and AA (2.5 mM). Experimental techniques were similar to those described in the legend of FIGURE 17. Each point represents a mean of three experiments ± standard deviation.

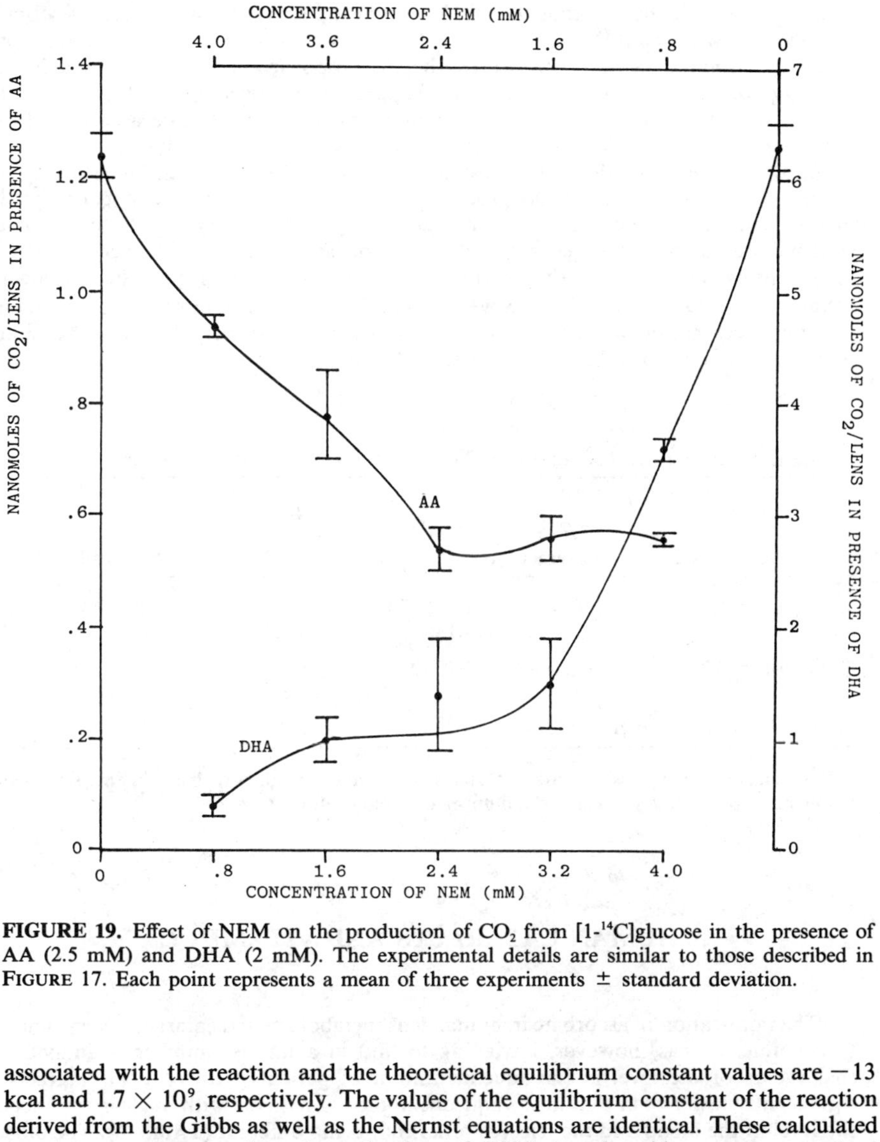

FIGURE 19. Effect of NEM on the production of CO_2 from [1-^{14}C]glucose in the presence of AA (2.5 mM) and DHA (2 mM). The experimental details are similar to those described in FIGURE 17. Each point represents a mean of three experiments ± standard deviation.

associated with the reaction and the theoretical equilibrium constant values are -13 kcal and 1.7×10^9, respectively. The values of the equilibrium constant of the reaction derived from the Gibbs as well as the Nernst equations are identical. These calculated parameters point out that the reduction of DHA by glutathione under standard conditions is thermodynamically quite feasible. The observed acceleration of the shunt by DHA is thus in accordance with the energy considerations of the overall reaction. Based on the equilibrium constant calculations under standard conditions, the ratio of AH_2/DHA is 1.7×10^8. Although the precise relationship between the steady-state concentrations and equilibrium concentrations remains somewhat disputed for most reactions in the biological milieu, the values derived form the equilibrium constant attainable under standard conditions do provide a reasonably good measure of approximation. In the present case, the AH_2/DHA ratio, as calculated under standard states, agrees with the fact that the DHA level in the lens is below the ordinary limits of detection, whereas AH_2 is about 3 mg/kg wet weight of the tissue. The reactions in the direction indicated in FIGURE 16, therefore, appear kinetically so favorable that the suggested electron flux may proceed in many types of animal cells, despite the absence of dehydroascorbate reductase and ascorbate oxidase. The ascorbate-dehydroascorbate couple may thus be of significant importance in the dynamic regulation of glucose metabolism through the shunt. The physiological significance of the ascorbate-driven electron transport process described above is not quite apparent at the present time, except perhaps in the prevention against oxidative stress to the tissue concomitant to an *in situ* generation of active species of oxygen. The acceleration of the shunt is another way to get rid of the peroxide that is generated following dismutation of O_2^-. It should, however, be mentioned that ascorbate can act in certain circumstances as an oxidant and the possibility of its participation in cataractous process, although unlikely, cannot be entirely ruled out.

TABLE 6. Ascorbic Acid Content of Blood, Aqueous Humor, and Lens

Age in Years	Type of Cataract	Blood (mg/100 ml)	Aqueous Humor (mg/100 ml)	Lens (mg/100 ml)	Group
55.28 (64) ± 10.16	cortical	1.735 (60) ±1.136	14.0 (52) ±8.05	31.81 (64) ±10.8	A
63.75 (20) ± 9.3	nuclear	2.10 (20) ±1.43	28.3 (12) ±5.26	30.73 (20) ±7.55	B
p values between A and B		<1.22	<0.001*	<0.5	

Values have been expressed as mean ± standard deviation; *p* values with *asterisk* are significant. Numbers in parentheses indicate the number of samples in each case.[20]

HUMAN CATARACTS AND VITAMIN C

The implication of ascorbate in human lens metabolism and cataract is very difficult to examine. It was, however, intriguing to find in a limited number of individuals that the ascorbate level in the aqueous humor of patients with cortical cataracts is significantly lower than that in the aqueous humor of patients with nuclear cataracts,[20] the level in the blood and the lens (cataract) not being different (TABLE 6). The blood-to-aqueous ratio of ascorbate in the case of patients with cortical cataracts is approximately 1:7, whereas in patients with nuclear cataracts it was 1:14. This lowering of the ratio in the case of cortical cataracts points out that blood aqueous ascorbate pump may be more sluggish in patients who develop cortical cataracts. It may also be coincidental with the incidence of cortical cataract. The final answer to the question of the role of ascorbate in fighting cataracts must, therefore, await further experimental studies. The importance of ascorbate in another part of the eye, the retina, also must be studied. Recent studies from this laboratory demonstrate that the structures of the neural retina undergo severe degeneration in scurvy.

APPENDIX

$$\begin{array}{lc} & \Delta E^{\circ\prime}\ (\text{pH } 7.0) \\ 2\ \text{GSH} \longrightarrow \text{GSSG} + 2\text{H}^+ + 2\text{e} & 0.23\ \text{V}^{23} \\ \text{DHA} + 2\text{H} + 2\text{e} \longrightarrow \text{AH}_2 & 0.054\ \text{V}^{24} \\ \hline \text{DHA} + 2\text{GSH} \longrightarrow \text{GSSG} + \text{AH}_2 & \Delta E^{\circ\prime} = 0.284 \end{array}$$

$$\begin{aligned} \Delta G^{\circ\prime} &= -\,(n)\,(\text{F})\,(\Delta E^{\circ\prime}) \\ &= -\,(2)\ 23.06\ \frac{\text{kcal}}{\text{mol °K V}} \times 0.284 \text{ volts} \qquad \text{Gibbs equation} \\ &= -\ 13.1 \text{ kcal/mol of DHA reduced} \end{aligned}$$

$$\begin{aligned} \Delta G^{\circ} &= -\ \text{RT}\ \ln K \\ \frac{-13.1 \text{ kcal}}{\text{mol}} &= -\ 1.987 \times 10^{-3}\ \frac{\text{kcal}}{\text{mol °K}} \times 310\text{°K}\ \ln K \qquad \text{Gibbs equation} \\ \ln K &= 21.26 \\ K &= 1.7 \times 10^9 \end{aligned}$$

$$\begin{aligned} \text{Log } K &= \frac{(\Delta E^{\circ}) \times n}{\text{RT/F}} \\ &= \frac{(0.284)\ (2)}{0.0615} = 9.23 \qquad \text{Nernst equation} \\ K &= 1.7 \times 10^9 \end{aligned}$$

$$K = \frac{[\text{GSSG}]\ [\text{AH}_2]}{[\text{GSH}]^2\ [\text{DHA}]}, \quad \left.\begin{array}{l} \text{GSH} = 10^{-2}\ \text{M} \\ \text{GSSG} = 0.05 \times 10^{-2}\ \text{M} \end{array}\right\} \text{ in rat lens}$$

$$1.7 \times 10^9 = \frac{[0.05 \times 10^{-2}]\ [\text{AH}_2]}{[10^{-4}]\ [\text{DHA}]}$$

$$\frac{[\text{AH}_2]}{[\text{DHA}]} = \frac{1.7 \times 10^9 \times 10^{-4}}{0.05 \times 10^{-2}} = 3.4 \times 10^8$$

In the rat lens, ascorbate is $\approx$ 50 mg/l = 2.8×10^{-4} M (ref. 13)

$$\frac{[\text{AH}_2]}{[\text{DHA}]} = 3.4 \times 10^8$$

$$\text{Approximate [DHA]} = \frac{2.8 \times 10^{-4}}{3.4 \times 10^{8}} \text{ M} = 0.82 \times 10^{-12} \text{ M}$$

$\Delta E^{\circ\prime}$ = Sum of the reduction potentials of both the half reactions
$\Delta G^{\circ\prime}$ = Gibbs energy of the reaction
n = Number of electrons involved in the redox reaction
F = Faraday, expressed in terms of kcal/mol °K volt
R = Gas constant in terms of kcal/mol °K
T = Absolute temperature in Kelvin
K = Equilibrium constant
In rat lens, [GSH] = 10^{-2} M, [GSSG] = 0.05×10^{-2} M
Ascorbate = 2.8×10^{-4} M
(For conventions, see ref. 21.)

REFERENCES

1. BECKER, B. 1967. Ascorbate transport in guinea pig eyes. Invest. Ophthalmol. Vis. Sci. **6:** 410-415.
2. SCHOCKET, S. S., J. ESTERSON, B. BRADFORD, M. MICHAELIS & R. D. RICHARDS. 1972. Induction of cataracts in mice by exposure to oxygen. Isr. J. Med. Sci. **8:** 1596-1601.
3. PALMQUIST, B., B. PHILIPSON & P. BARR. 1984. Nuclear cataract and myopia during hyperbaric oxygen therapy. Brit. J. Ophthalmol. **68:** 113-117.
4. FRIDOVICH, I. 1978. The biology of oxygen radicals. Science **201:** 875-880.
5. VARMA, S. D., T. K. ETS & R. D. RICHARDS. 1977. Protection against superoxide radicals in rat lens. Ophthal. Res. **9:** 421-431.
6. MISRA, H. P. & I. FRIDOVICH. 1972. The role of superoxide anion in the autooxidation of epinephrine and a simple assay of superoxide dismutase. J. Biol. Chem. **247:** 3170-3175.
7. BHUYAN, D. K. & K. C. BHUYAN. 1977. Superoxide dismutase of calf lens. Invest. Ophthalmol. Vis. Sci. **16**(ARVO Suppl.): 15.
8. KUCK, J. F. R., JR. 1977. Activity of superoxide dismutase in rat lens epithelium and in fiber. Invest. Ophthalmol. Vis. Sci. **16**(ARVO Suppl.): 14.
9. BEAUCHAMP, C. & I. FRIDOVICH. 1971. Superoxide dismutase: Improved assays and an assay applicable to acrylamide gels. Anal. Biochem. **44:** 276-287.
10. VARMA, S. D. 1981. Superoxide and lens of the eye: A new theory of cataractogenesis. Int. J. Quantum Chem. **20:** 479-484.
11. VARMA, S. D., S. KUMAR & R. D. RICHARDS. 1979. Light-induced damage to ocular lens cation pump: Prevention by vitamin C. Proc. Natl. Acad. Sci. U.S.A. **76:** 3504-3506.
12. VARMA, S. D. & J. M. MOONEY. 1986. Photodamage to the lens in vitro: implications of the Haber-Weiss reaction. J. Free Rad. Biol. Med. In press.
13. VARMA, S. D., V. K. SRIVASTAVA & R. D. RICHARDS. 1982. Photoperoxidation in lens and cataract formation: Preventive role of superoxide dismutase, catalase, and vitamin C. Ophthal. Res. **14:** 167-175.
14. FRIDOVICH, I. 1970. Quantitative aspects of the production of superoxide anion radical by milk xanthine oxidase. J. Biol. Chem. **245:** 4053-4057.
15. VARMA, S. D., S. M. MORRIS, S. A. BAUER & W. KOPPENOL. 1986. In vitro damage to rat lens by xanthine/xanthine oxidase: Protection by ascorbate. Exp. Eye Res., in press.
16. VARMA, S. D., N. BEACHY & R. D. RICHARDS. 1982. Photoperoxidation of lens lipids: Prevention by vitamin E. Photochem. Photobiol. **36:** 623-626.
17. VARMA, S. D., D. CHAND, Y. R. SHARMA, J. F. KUCK, JR. & R. D. RICHARDS. 1984. Oxidative stress on lens and cataract formation: Role of light and oxygen. Current Eye Res. **3:** 35-57.

18. TAPPEL, A. L. 1980. *In* Free Radicals in Biology, Vol. 4. W. Pryor, Ed.: 2-44. Academic Press. New York.
19. MAPSON, L. W. & E. M. MOUSTAFA. 1956. Ascorbic acid and glutathione as respiratory carriers in the respiration of pea seedlings. Biochem. J. **62:** 248-259.
20. CHANDRA, D. B., R. VARMA, A. SUHAIL & S. D. VARMA. 1986. Vitamin C in the human aqueous humor and cataracts. Int. J. Vit. Nutr. Res. **56:** 165-169.
21. KOPPENOL, W. H. & J. BUTLER. 1985. Energetics of interconversion of reaction of oxygen radicals. Adv. Free Rad. Biol. Med. **1:** 99-131.
22. FARHATAZIZ & A. B. ROSS. 1977. Selected specific rates of reactions of transients from water in the aqueous solution. III. Hydroxyl radicals and their ions. N.S.R.D.S.-N.B.S. Washington, D.C. p. 59.
23. CABELLI, D. E. & J. H. BIELSKI. 1983. Kinetics and mechanism of oxidation of ascorbate by H_2O_2/O_2 radicals. A pulse radio-isotope and stopped flow study. J. Phys. Chem. **87:** 1809-1832.

DISCUSSION OF THE PAPER

L. MACHLIN (*Hoffmann-La Roche Inc., Nutley, N.J.*): In your last slide, there was quite a substantial difference in the aqueous humor between the cortical and nuclear cataracts but no difference in the lens. I thought you said that it's the uptake in the aqueous humor that is the critical factor?

S. D. VARMA (*University of Maryland School of Medicine, Baltimore, Md.*): The lenses are not different because they are both cataracts taken by intracapsular extraction. When you estimate ascorbic acid you don't know what you're measuring because those lenses are porous. The aqueous humor estimations are more reliable. We hypothesize that the difference between cortical and nuclear cataract ascorbic acid is related to the transport of ascorbic acid from the blood to the aqueous humor. This needs confirmation. The number of cases is not that large, but to get aqueous humor is very difficult.

UNIDENTIFIED SPEAKER: Most of your data was with the rat. Nocturnal animals don't have high concentrations of ascorbate in the aqueous humor and might not need this kind of mechanism.

S. D. VARMA: We have said in a very early paper that the concentration of ascorbic acid is higher in the diurnal animals compared to the nocturnal animals. Also, the concentration of ascorbic acid is lower in the animals in utero than postnatally. So there could be some component involving their adaptation to the light and that's why there is apparently a higher concentration of ascorbic acid in diurnal animals.

H. SPRINCE (*Jefferson Medical College, Philadelphia, Pa.*): From your slide showing the riboflavin liberation of oxygen, may I infer that the high doses of riboflavin could generate the formation of cataracts?

S. D. VARMA: Possibly.

H. SPRINCE: Are people at risk with high doses of riboflavin?

S. D. VARMA: Well, any photosynthetizer at a high dose will be bad because it will initiate unwanted photochemical reactions.

A. TAYLOR (*Tufts University, Boston, Mass.*): You made a distinction between nuclear and cortical opacities in the Emory mice and I think that you showed animals on higher E diets have delayed onset of that cataract. We have shown in the same mice that dietary restriction delays the onset of cataracts.

About 15 years ago we carried out studies on the distribution of [^{14}C]ascorbic acid in guinea pigs. At that time we were very much surprised that the lens accumulated the radioactivity to a large extent. In addition we were surprised that the half-life of retention of lens ascorbic acid was very similar to the half-life we found in brain and in testes.

S. D. VARMA: It is very difficult to deplete the lens of ascorbic acid.

Ascorbic Acid and Cataract[a]

W. LOHMANN

Institut für Biophysik
Universität Giessen
Giessen, Federal Republic of Germany

In the preceding paper, Dr. Varma has pointed out that the total (reduced and oxidized) ascorbic acid concentration (ASC_{total}) decreases with progressing cataract. We have confirmed this result. In addition, we were able to show that this decrease is, however, not due to a reduction of the supply of ascorbic acid but rather to a rapid oxidation of vitamin C.[1]

Ascorbic acid is present in biological systems mainly in its reduced form (~90-95%). During cataractogenesis it is oxidized to its oxidative decay products with a transient increase and decrease of the semioxidized (ascorbyl radical) and the completely oxidized form (dehydroascorbic acid). Because we cannot consider these oxidative decay products (diketogulonic acid up to methylglyoxal) to be vitamin C, the total concentration of ascorbic acid in the lens decreases during cataractogenesis.

This we were able to confirm[1] as can be seen by the values given in TABLE 1. A total of 451 lenses have been investigated with five control lenses and 86, 354, and 6 respectively for the other groups mentioned (incipiens, matura, hypermatura). Very recently we obtained a very dark lens from a patient with a very advanced cataract. Its vitamin C values are shown in the last column. As can be seen in TABLE 1, the total ascorbic acid concentration (ASC_{tot}) decreases with progressing cataract, while the DHA fraction increases considerably (in the case of the dark lens up to ~68%). For comparison, the Na, K, Ca, and Mg concentrations are also shown.

Since the ascorbyl radical is the first oxidation product of ascorbic acid, it should be possible to detect this transient species by electron spin resonance (ESR) measurements during cataractogenesis.[1] The ESR spectra obtained for lenses with different states of cataract are shown in FIGURE 1. As can be seen, a young healthy lens (control) exhibits a very small signal only. At an early state of the disease (cataracta nuclearis incipiens), a signal at $g = 2.005$ can be recorded that we could assign to the ascorbyl radical.[2] As expected, this signal disappears again with progression of the disease. At an advanced state (hypermatura), another signal with an increased spin concentration can be detected, which cannot as yet be identified. It seems to belong to some radical state of the oxidative decay products of ascorbic acid, *i.e.,* a state beyond dehydroascorbic acid.

During autoxidation of ascorbic acid solutions, their color changes from transparent to dark brown.[3] These color changes are very similar to those observed in lenses during cataractogenesis. Despite intensive studies, not much is known about the chemical nature of these chromophores formed within the cataractous lens.

[a] Parts of this work were supported by grants from the Bundesministerium für Forschung und Technologie and from the Fonds der Chemischen Industrie.

TABLE 1. Concentrations of Total (Reduced and Oxidized) Ascorbic Acid (ASC_{tot}), Dehydroascorbic Acid (DHA), and of a Few Electrolytes in Human Lenses without (Control) and with Different States of Nuclear Cataract; SD ≤ 10%

		Cataracta Nuclearis			
	Control	Incipiens	Matura	Hypermatura	
Na (μmol/100 mg dry mass)	7.7	15.0	42.3	46.5	
K (μmol/100 mg dry mass)	27.8	21.7	8.2	9.0	
Ca (μmol/100 mg dry mass)	0.1	1.1	2.4	2.9	
Mg (μmol/100 mg dry mass)	0.87	0.88	0.92	1.05	
ASC_{tot} (μg/100 mg fresh weight)	18.5	18.2	15.1	10.5	2.2
DHA (μg/100 mg fresh weight)	2.2	3.1	3.0	2.8	1.5
DHA (% of ASC_{tot})	11.8	16.8	19.8	26.6	68

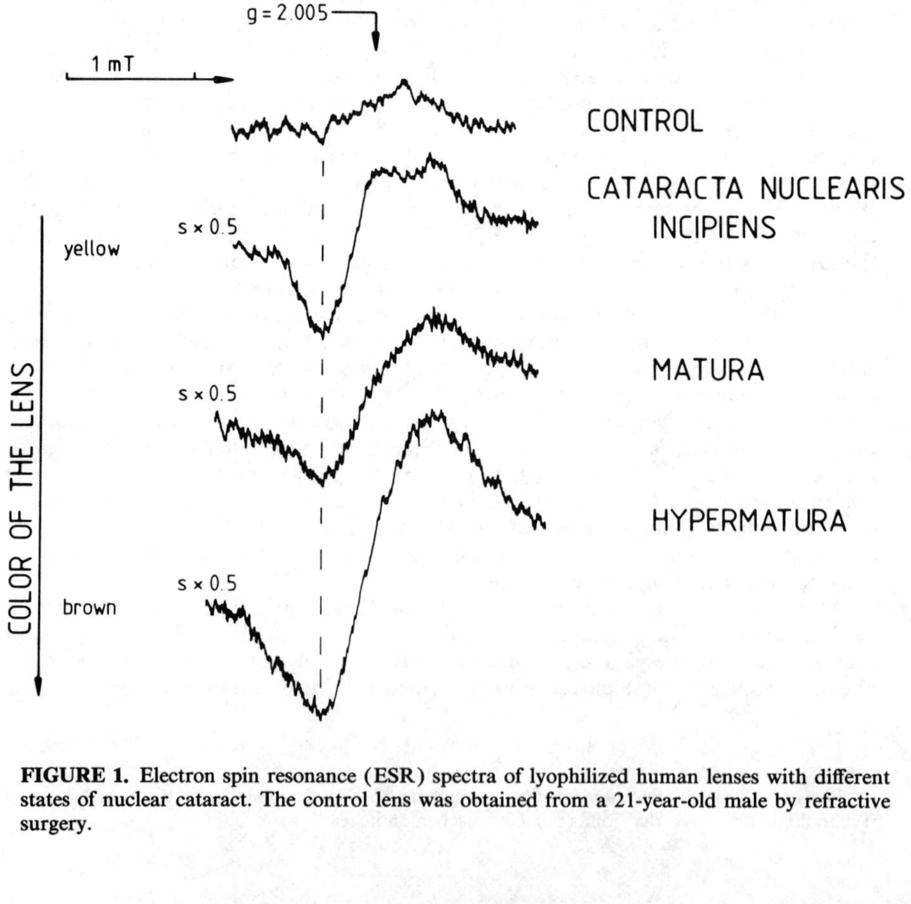

FIGURE 1. Electron spin resonance (ESR) spectra of lyophilized human lenses with different states of nuclear cataract. The control lens was obtained from a 21-year-old male by refractive surgery.

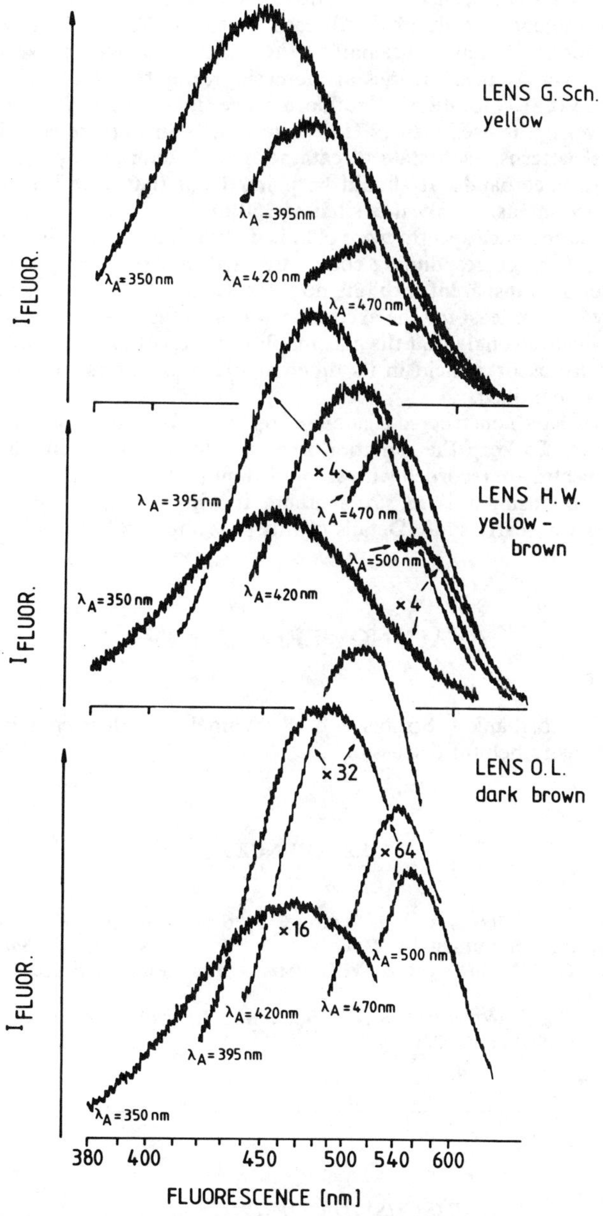

FIGURE 2. Fluorescence spectra obtained with three different-colored lenses after excitation at 350, 395, 420, 470, and 500 nm. In order to compare the intensity of the spectra, they must be multiplied by the factors given.

After monochromatic excitation of extracted lenses at either 350, 395, 420, 470, or 500 nm, typical fluorescence spectra are obtained which can be used for describing quantitatively the actual state of the disease[4] (FIG. 2). At the onset of the disease, only the shorter excitation wavelengths (FIG. 2, lens G.Sch.) cause a fluorescence response with decreasing intensities at increasing excitation wavelengths. With progressing disease (*i.e.*, coloration), the fluorescence light is more intense at the longer excitation wavelengths used (lenses H.W. and O.L.). Since there is a gradual change between these patterns, each state of cataract is characterized by a typical intensity ratio of fluorescence bands. It should be pointed out that a healthy lens exhibits a low fluorescence intensity only if excited at 350 nm.

Identical fluorescence spectra were obtained with "aged," that is, colored ascorbic acid solutions, if the corresponding colors are used. A freshly prepared ascorbic acid solution, which is transparent, exhibits no fluorescence at all. This again corresponds to the healthy lens, at least for the excitation wavelengths at and above 395 nm. Thus, the results indicate strongly that the chromophores present in cataractous lenses might be identical with ascorbic acid in its different oxidation states that are colored, but still have to be identified.

Recently we have constructed a suitable apparatus for determining the fluorescence response *in vivo.* To keep the radiation exposure to the eye as low as possible, the fluorescence spectra are recorded with an optical multichannel analyzer (OMA) within 10 to 50 ms. By using this *in situ* method, it will be possible to detect cataract formation at a very early state. Details of this procedure will be published elsewhere.

ACKNOWLEDGMENTS

I would like to thank J. Strobel and W. Schmehl for their contributions to this work and for many helpful discussions.

REFERENCES

1. LOHMANN, W., W. SCHMEHL & J. STROBEL. 1986. Exp. Eye Res. **43:** 859-862.
2. LOHMANN, W., J. SCHREIBER & W. GREULICH. 1979. Z. Naturforsch. **34c:** 550-554.
3. BENSCH, K. G., J. E. FLEMING & W. LOHMANN. 1985. Proc. Natl. Acad. Sci. U.S.A. **82:** 7193-7196.
4. LOHMANN, W., M. WUNDERLING, W. SCHMEHL, J. STROBEL & K. G. BENSCH. 1986. Naturwissenschaften **73:** 266-267.

DISCUSSION OF THE PAPER

S. D. VARMA (*University of Maryland School of Medicine, Baltimore, Md.*): When the membrane integrity is compromised, all kinds of oxidation reactions take place in the lens. Just because of the fluorescence spectra, you say that these are oxidation products of ascorbic acid. However, there are other things which will oxidize preferentially.

W. LOHMANN (*University of Giessen, Giessen, FRG*): We have checked other substances too and it turns out that only ascorbic acid after aging can be excited between 350 and 500 nm. Freshly prepared ascorbic acid solution, which is transparent, does not show any fluorescence spectra at all. Nobody knows what the structure of these colored ascorbic acid solutions is.

S. TANNENBAUM (*Massachusetts Institute of Technology, Cambridge, Mass.*): The problem with the interpretation of these spectras is that they are simply typical spectra of products of browning reactions. If either sugars or other carbonyl compounds, such as malonaldehyde, are reacted with proteins you get very similar pigments whether you start with malonaldehyde, or ascorbic acid, or glucose. There's been some very good work showing that in people with cataracts there are adducts, amadori compounds which form from the reaction of glucose with proteins. So the fluorescence spectra don't really prove what the cause is, all they prove is that browning reactions may be important in cataract formation.

Effect of Ascorbic Acid on Male Fertility

EARL B. DAWSON, WILLIAM A. HARRIS,
WILLIAM E. RANKIN, LEONARD A. CHARPENTIER,
AND WILLIAM J. McGANITY

Department of Obstetrics and Gynecology
The University of Texas Medical Branch at Galveston
Galveston, Texas 77550

INTRODUCTION

Lindsay and Medes showed in 1926 that a diet deficient in ascorbic acid (AA) for one month caused a massive degeneration of the germinal epithelium with desquamation into the lumina of the tubules in the testes of guinea pigs.[1] In 1941, it was reported that the seminal plasma of normal man contains 12.8 ± 1.6 mg/dl of AA, which is actively concentrated and secreted by the seminal vesicles during ejaculation.[2-4] This level is many times the 0.2-1.40 mg/dl level normally found in serum, and suggests a vital metabolic role for AA in the seminal plasma. In 1960, Lindahl and Kihlstrom described a proteinaceous substance in seminal plasma, which appears in two forms (oxidized or reduced), and which is reversibly transformed from one form to the other.[5] The reduced form has antiagglutinic activity—that is, it prevents clumping and immotility of sperm in seminal plasma. This form is attached to the surface of the sperm and loses its ability to be fixed to the cell surface when oxidized. Ascorbic acid, a biological reducing agent, has been shown to reduce the oxidized protein in several *in vitro* studies.[6] In 1986, the studies of Lindsay and Medes were duplicated when male guinea pigs with proven fertility were placed on an avitaminotic C diet for 3 weeks and sacrificed. The most severe histological damage was observed in the testicular tissue. The most pronounced damage was in the cauda epididymus, wherein sperm undergo maturation, with thickened peritubular muscle layer and connective tissue, decreased tubular diameter and epithelial cell height, and nuclear pycnosis. Similarly affected was the vas deferens and accessory sex glands, although to a lesser degree. Sperm analysis of semen samples collected from testis, cauda epididymus, and vas deferens showed decline in total sperm count and sperm motility, as well as an increase in the number of abnormal sperm.[7]

Over a decade ago, the obstetricians of our clinical service commenced prescribing AA to husbands when it was apparent from fertility testing of both husband and wife that infertility was possibly due to sperm agglutination in excess of 25%. The practice is to prescribe 1 g per day for 10 days and then 500 mg per day through two menstrual cycles (60 days). This practice has proved successful. Subsequently, we reported a comparison of the differences in sperm qualities between a group of 20 men with sperm agglutination over 25% who followed the AA supplementation protocol and a control group of 10 men (who received no supplement or placebo), for 60 days. At the end of 60 days, all of the group who had received AA had impregnated their wives, and none of the control group reported pregnancies. Comparison of the sperm qualities before and after the 60-day period showed marked improvement in all of the measured properties in the group receiving AA supplementation and no change in the control group. Most notably, there was observed a significant increase in sperm count and motility, and decrease in abnormal sperm, sperm precursors ($p \leq 0.05$), and agglutination ($p \leq 0.001$). The group mean ± SEM of the seminal plasma AA levels of the control group before the study was 1.41 ± 0.4 mg/dl and 1.25 ± 0.32 mg/dl at the end. The group mean ± SEM seminal plasma levels of AA of the supplemented group was 2.29 ± 0.96 mg/dl prior to supplementation and 6.75 ± 0.43 mg/dl after 60 days of AA supplementation.[8]

Dietary AA is assimilated and distributed throughout the body within a few hours. Additionally, the metabolic turnover time for assimilated AA in basal man is approximately 20 days, which is much longer than the 3-day turnover time in the guinea pig.[9] Therefore, the apparent beneficial effects of increasing seminal plasma AA levels with increased supplementation on sperm agglutination, motility, and abnormal morphology would possibly be observed within several weeks. Additionally, a daily dietary AA supplementation of 1.0 g is larger than the 60 mg per day recommended by the Nutrition Research Council and has been considered a therapeutic level when continued for long periods of time.[10] The purpose of this study is to monitor the effect of graded levels of AA supplementation on these sperm qualities during a 3-week study in men with sperm agglutination in excess of 25%.

MATERIALS AND METHODS

The study subjects were 30 men between the ages of 25 and 45 with sperm agglutination over 25% when their sperm was studied by microscopic analysis. Sperm antibody studies, according to the method of Franklin and Dukes, were negative in each subject.[11,12] All were considered physically normal and showed no signs of inflammatory disease. The research protocol was described to the subjects, human consent forms signed, and 10 ml of blood were obtained for the measurement of serum AA. An aliquot of seminal plasma was prepared for AA measurement. By random selection, three groups of ten subjects each were provided one month's supply of a daily supplement of 1000 or 200 mg of AA, or a placebo. The pure L-ascorbic acid used for supplement and placebo were provided by Hoffmann-LaRoche, Inc., Nutley, New Jersey. Each subject was told he was receiving AA and expected improvement in sperm quality. The subjects returned weekly and were requested to refrain from sexual activity during the length of the study.

Routinely, all samples were obtained within the laboratory. The sperm analysis was with an American Optical Binocular Microscope (AO Scientific Instruments, Buffalo, N. Y.) and was completed within ten minutes. The sperm qualities determined were: total count as millions of sperm; viability as percent of sperm count showing movement; motility as percentage of sperm count showing progressive forward movement; sperm precursors as millions per milliliter; abnormal sperm morphology as percentage of sperm count; and, sperm agglutination as percentage of sperm count clumped together with no forward movement, but obviously viable. All of these counts were in triplicate of separate fields with a high-powered lens (FIGS. 1 and 2). Sperm

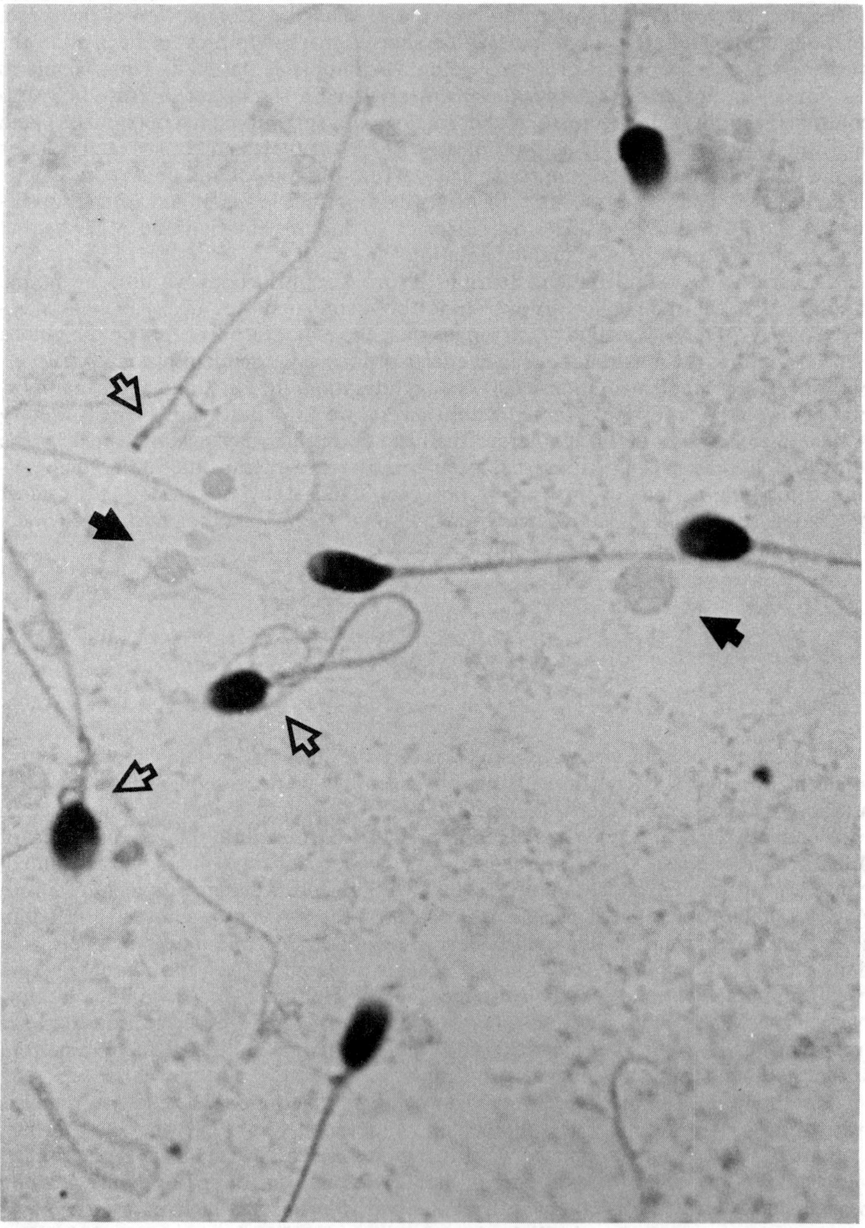

FIGURE 1. *Open arrowheads*: abnormal sperm; *closed arrowheads*: sperm precursors. High-power magnification.

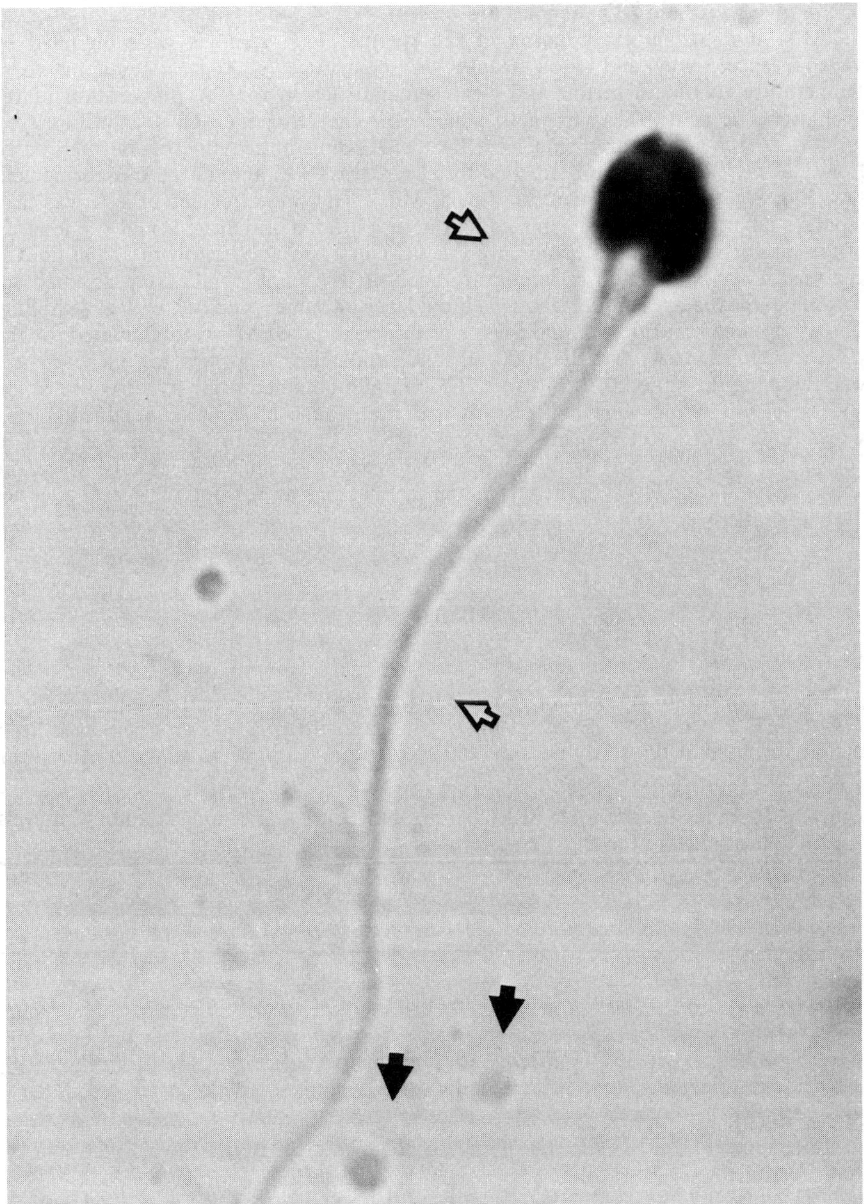

FIGURE 2. *Open arrowheads*: agglutinated normal mature sperm; *closed arrowheads*: sperm precursors. High-power magnification.

precursors are easily discernable: (1) they are larger in size than white blood cells, which are usually present, (2) the nucleus appears as a small tip on the outer perimeter of the precursor, and (3) vacuoles are present.[13]

After microscope examination of the sperm, blood samples were obtained by venapuncture. Semen and blood samples were centrifuged (300 g × 10 min) at room temperature to obtain serum and clear seminal plasma for the preparation of the trichloroacetic acid (TCA) extracts, which were then frozen (−60° C) until assayed for AA. All AA measurements were by the 2,4-dinitrophenylhydrazine method of Lowry *et al.*,[14] and the standard used for comparison was anhydrous L-ascorbic acid (A-7506, Sigma Chemical Co., St. Louis, Mo.). The measurement of AA was in a separate laboratory after all samples had been collected and stored, and the identity of the numbered test tubes containing the TCA extracts was unknown at the time of analysis. After completion of laboratory analysis, the sample data were grouped according to the dietary supplementation provided each subject and the week of sampling. The group mean and the standard error of the mean (±SEM) was calculated for the placebo, 200 mg AA, and the 1000 mg AA supplemented groups for each of the 4 weeks during the study: 0, 1, 2, and 3. The samples obtained prior to subject selection and supplementation were at 0 weeks, and the groups of samples obtained during weeks 1, 2, and 3 were during supplementation. The Student's *t* test was used to determine the statistical significance of the weekly group mean change from the presupplementation levels in the serum and seminal plasma AA and the sperm qualities during supplementation.

RESULTS

The group mean ± SEM of AA levels for serum and seminal plasma are listed in TABLE 1. The mean ± SEM serum AA levels for the placebo group showed little change throughout the study period: 0.50 ± 0.16 mg/dl prior to study, 0.61 ± 0.11

TABLE 1. Changes in Mean ± SEM Serum and Seminal Plasma Vitamin C Levels (mg/dl) during Study for the Three Groups

	Study Weeks			
	0	1	2	3
Serum				
Placebo (n = 10)	0.57 ± 0.16	0.61 ± 0.11	0.65 ± 0.08	0.65 ± 0.09
Supplemented				
200 mg (10)	0.40 ± 0.12	0.90 ± 0.20[a]	1.80 ± 0.14[d]	2.20 ± 0.10[d]
1000 mg (10)	0.20 ± 0.01	1.30 ± 0.30[c]	2.50 ± 0.15[d]	2.40 ± 0.15[d]
Seminal Plasma				
Placebo (10)	3.7 ± 1.9	2.9 ± 2.0	3.0 ± 1.1	2.0 ± 1.1
Supplemented				
200 mg (10)	2.0 ± 0.5	3.8 ± 0.4[b]	6.3 ± 0.3[d]	8.1 ± 0.5[d]
1000 mg (10)	4.2 ± 0.7	12.7 ± 1.1[d]	11.2 ± 1.3[d]	13.1 ± 1.3[d]

[a] $p \leq 0.05$.
[b] $p \leq 0.02$.
[c] $p \leq 0.01$.
[d] $p \leq 0.001$.

TABLE 2. Changes in Mean ± SEM Percent Sperm Viability, Motility, and Agglutination for the Three Groups

	Study Weeks 0	1	2	3
Viability, %				
Placebo (n = 10)	53 ± 9	56 ± 8	58 ± 3	60 ± 4
Supplemented				
200 mg (10)	45 ± 8	59 ± 2	68 ± 3[b]	68 ± 3[b]
1000 mg (10)	37 ± 3	55 ± 3[d]	52 ± 3[c]	61 ± 3[d]
Motility, %				
Placebo (10)	40 ± 10	46 ± 10	51 ± 3	49 ± 8
Supplemented				
200 mg (10)	28 ± 7	44 ± 2[a]	54 ± 4[c]	51 ± 7[a]
1000 mg (10)	38 ± 6	86 ± 12[c]	87 ± 13[c]	94 ± 10[d]
Agglutination				
Placebo (10)	35 ± 4	34 ± 2	30 ± 3	29 ± 5
Supplemented				
200 mg (10)	34 ± 5	16 ± 4[b]	15 ± 4[c]	10 ± 4[c]
1000 mg (10)	36 ± 3	14 ± 5[c]	13 ± 4[d]	11 ± 5[d]

[a] $p \leq 0.05$.
[b] $p \leq 0.02$.
[c] $p \leq 0.01$.
[d] $p \leq 0.001$.

mg/dl after 1 week, 0.65 ± 0.08 mg/dl after 2 weeks, and 0.65 ± 0.09 mg/dl after 3 weeks. There was no significant difference between the weekly mean ± SEM serum AA levels before and after receiving placebos. The mean ± SEM seminal plasma AA levels for the placebo group were: 3.7 ± 1.9 mg/dl prior to study, 2.9 ± 2.0 mg/dl after 1 week, 3.0 ± 1.1 mg/dl after 2 weeks, and 2.0 ± 1.1 mg/dl after 3 weeks. There was no significant difference between the weekly mean ± SEM seminal plasma AA levels before and after receiving placebos. The study group supplemented with 200 mg AA per day showed progressively higher group mean ± SEM serum AA levels during the three weeks of supplementation. The group mean ± SEM prior to supplementation was 0.4 ± 0.12 mg/dl, 0.9 ± 0.2 mg/dl after 1 week ($p \leq 0.05$), 1.80 ± 0.14 mg/dl after 2 weeks ($p \leq 0.001$), and 2.20 ± 0.10 mg/dl after 3 weeks of supplementation ($p \leq .001$). The group mean ± SEM seminal plasma AA levels were significantly higher following supplementation than the initial 2.0 ± 0.5 mg/dl; 3.8 ± 0.4 mg/dl after 1 week ($p \leq 0.02$), 6.3 ± 0.3 mg/dl after 2 weeks ($p \leq 0.001$), and 8.1 ± 0.5 mg/dl after 3 weeks ($p \leq 0.001$). The group mean ± SEM serum AA levels of those subjects receiving 1000 mg AA per day showed a rapid elevation from the 0.20 ± 0.01 mg/dl level prior to supplementation: 1.3 ± 0.3 mg/dl after 1 week ($p \leq 0.01$), 2.50 ± 0.05 mg/dl after 2 weeks ($p \leq 0.001$), and 2.40 ± 0.05 mg/dl after 3 weeks ($p \leq 0.001$). The group mean ± SEM seminal plasma AA level of these subjects showed similar rapid elevation from the presupplementation level of 4.2 ± 0.7 mg/dl: 12.7 ± 1.1 mg/dl after 1 week ($p \leq 0.001$), 11.2 ± 1.3 mg/dl after 2 weeks ($p \leq 0.001$), and 13.1 ± 1.3 mg/dl after 3 weeks of supplementation ($p \leq 0.001$).

The changes in sperm viability, motility, and agglutination of the three study groups are depicted in TABLE 2. The placebo group showed very little variation in mean ± SEM viability throughout the study. The initial group mean ± SEM viability was 53 ± 9%, 56 ± 8% after 1 week, 58 ± 3% after 2 weeks, and 60 ± 4% after 3 weeks. None of these weekly group viability levels were significantly different from the initial levels. The group mean ± SEM percent viability of the group receiving 200 mg AA per day supplementation showed significant improvement during the study. The initial group means ± SEM percent viability was 45 ± 8%, 59 ± 2% after 1 week of supplementation (not significant), 68 ± 3% after 2 weeks ($p < .02$), and 68 ± 3% after 3 weeks ($p \leq 0.02$). The group of subjects receiving 1000 mg of AA per day supplementation showed a more significant improvement in sperm viability during the period of supplementation. Initially, the group mean viability was 37 ± 3%, 55 ± 3% after 1 week of supplementation ($p \leq 0.001$), 54 ± 3% after 2 weeks ($p \leq 0.01$), and 61 ± 3% after 3 weeks ($p \leq 0.001$). After 1 week of supplementation there was a mean average of 31% improvement in the sperm viability of the group of subjects receiving 200 mg AA per day and 49% improvement in the group receiving 1000 mg AA per day (see TABLE 4).

The placebo group showed no significant change in weekly group ± SEM percent sperm motility during the study (TABLE 2). The initial mean ± SEM motility was 40 ± 10%, 46 ± 10% after 1 week, 51 ± 3% after 2 weeks, and 49 ± 8% after 3 weeks. The 200 mg AA per day supplementation group showed significant improvement in sperm motility after 1 week and each week thereafter. The initial mean ± SEM motility was 28 ± 7%, 44 ± 2% after one week ($p \leq .05$), 54 ± 4% after 2 weeks ($p \leq .01$), and 51 ± 7% after 3 weeks ($p \leq 0.05$). The 1000 mg AA per day supplementation group showed a more rapid improvement in sperm motility during the study than the 200 mg AA supplemented group. The group mean ± SEM motility prior to supplementation was 38 ± 6%, 86 ± 12% after 1 week ($p \leq 0.01$), 87 ± 13% after 2 weeks ($p \leq 0.01$), and 94 ± 10% after 3 weeks ($p \leq 0.001$). After 1 week of supplementation, there was an average of 52% improvement in sperm motility in the 200 mg AA supplemented group and 126% improvement in the group receiving 1000 mg AA per day (see TABLE 4).

The level of the percent sperm agglutination for the placebo group showed slight variation throughout the study (TABLE 2). The initial mean ± SEM of percent sperm agglutination was 35 ± 4%, 34 ± 2% after 1 week of placebo supplementation, 30 ± 3% after 2 weeks, and 29 ± 5% after 3 weeks. The 200 mg AA group showed significant improvement following supplementation. The group mean ± SEM for sperm agglutination was 34 ± 5% prior to supplementation, 16 ± 4% after 1 week ($p \leq 0.02$), 15 ± 4% after 2 weeks ($p \leq 0.01$), and 10 ± 4% after 3 weeks ($p \leq 0.01$). There was a mean average of 53% decrease in sperm agglutination after 1 week of 200 mg AA per day supplementation. The group of subjects receiving 1000 mg AA per day showed significant improvement throughout the period of supplementation with a higher level of statistical confidence than the group receiving 200 mg AA per day. Initially, the group mean ± SEM sperm agglutination was 36 ± 3%, 14 ± 5% after 1 week ($p \leq 0.01$), 13 ± 4% after 2 weeks ($p \leq 0.001$), and 11 ± 5% after 3 weeks ($p < 0.001$). There was a mean average of 61% decrease in sperm agglutination after 1 week supplementation with 1000 mg AA per day, which is 8% lower than the decrease noted for the 200 mg AA supplemented group (see TABLE 4).

The group mean ± SEM weekly changes in sperm precursors as million per ml, percent abnormal morphology, and total sperm count in million per ml for the three study groups are listed in TABLE 3. The group of subjects receiving placebo supplementation showed no significant variation in the group mean ± SEM sperm precursor count throughout the study. Initially, the sperm precursor count was 2.0 ± 1.0 million/ml, 1.3 ± 1.0 million/ml after 1 week, 0.3 ± 0.3 million/ml after 2 weeks, and 1.2 ± 1.0 million/ml after 3 weeks. The 200 mg AA supplemented group showed a significant reduction in the group mean ± SEM sperm precursor count after 1 week. The group mean ± SEM sperm precursor count prior to supplementation was 5.0 ± 1.0 million/ml, 2.2 ± 0.5 million/ml after one week of supplementation ($p \leq .05$), 1.5 ± 0.7 million/ml after 2 weeks ($p \leq 0.01$), and 0.5 ≤ 0.3 million/ml after 3 weeks ($p \leq 0.001$). There was a 56% reduction in the group mean sperm precursor count after 1 week of supplementation with 200 mg AA per day. The subjects receiving 1000 mg AA per day supplementation showed a more significant reduction in the sperm precursor count than the subjects receiving 200 mg AA supplementation. The initial mean ± SEM precursor count for the 1000 mg AA supplemented group was 6.5 ± 0.4 million/ml, 2.2 ± 0.7 million/ml after one week ($p \leq 0.001$), 2.3 ± 0.6 million/ml after 2 weeks ($p \leq 0.001$), and 3.0 ± 1.0 million/ml after 3 weeks ($p \leq 0.01$). There was a 66% reduction in the group mean precursor count after one week of supplementation with 1000 mg AA per day. This was a 10% greater reduction than observed in the 200 mg AA per day group (see TABLE 4).

TABLE 3. Changes in Mean ± SEM Sperm Precursors, Abnormal Morphology, and Total Count for the Three Groups (million/ml)

	Study Weeks			
	0	1	2	3
Precursors				
Placebo (n = 10)	2.0 ± 1.0	1.3 ± 1.0	0.3 ± 0.3	1.2 ± 1.0
Supplemented				
200 mg (10)	5.0 ± 1.0	2.2 ± 0.5[a]	1.5 ± 0.7[b]	0.5 ± 0.3[c]
1000 mg (10)	6.5 ± 0.4	2.2 ± 0.7[c]	2.3 ± 0.6[c]	3.0 ± 1.0[b]
Abnormal Morphology, %				
Placebo (10)	45 ± 4	40 ± 5	41 ± 3	41 ± 8
Supplemented				
200 mg (10)	64 ± 5	43 ± 3[b]	39 ± 4[b]	36 ± 6[b]
1000 mg (10)	62 ± 5	42 ± 4[b]	43 ± 4[b]	35 ± 5[b]
Count				
Placebo (10)	153 ± 84	195 ± 84	132 ± 51	177 ± 68
Supplemented				
200 mg (10)	83 ± 48	176 ± 59	205 ± 92	216 ± 121
1000 mg (10)	99 ± 37	238 ± 71	215 ± 80	216 ± 84

[a] $p \leq 0.05$.
[b] $p \leq 0.01$.
[c] $p \leq 0.001$.

There was little variation in percent abnormal sperm morphology during the study in the placebo group. Prior to the study, the mean ± SEM for abnormal sperm morphology was 45 ± 4%, 40 ± 5% 1 week later, 41 ± 3% after 2 weeks, and 41 ± 8% after 3 weeks. None of these weekly changes were significantly different from the prestudy mean ± SEM. The group mean ± SEM abnormal morphology for the 200 mg AA supplemented subjects showed a significant decrease in the percentage of abnormal sperm following supplementation. The group mean ± SEM abnormal morphology was 64 ± 5% before supplementation, 43 ± 3% after 1 week of supplementation ($p \leq 0.01$), 39 ± 4% after 2 weeks ($p \leq 0.01$), and 36 ± 6% after three weeks ($p \leq 0.01$). There was a 33% decrease in the mean percent of abnormal sperm after 1 week of 200 mg AA supplementation. Similar results were observed in the group receiving 1000 mg AA per day supplementation. The group mean ± SEM abnormal sperm morphology was 62 ± 5% prior to supplementation, 42 ± 4% after 1 week of supplementation ($p \leq 0.01$), 43 ± 4% after 2 weeks ($p \leq 0.01$), and 35 ± 5% after 3 weeks ($p \leq 0.01$). There was a 32% decrease in percent abnormal sperm morphology after one week of supplementation with 1000 mg AA per day, which is comparable to the 33% decrease observed in the 200 mg AA per day group (see TABLE 4).

There was a nonsignificant increase observed in the total sperm count during supplementation of all the groups studied (TABLE 3). The group ± SEM of the placebo group was 153 ± 84 million prior to the study, 195 ± 84 million after 1 week, 132 ± 5 million after 2 weeks, and 177 ± 68 million after 3 weeks. Although a 27% increase in mean sperm count is apparent after 1 week, this increase was not

TABLE 4. Mean Percent Improvement in Sperm Qualities following One Week of Ascorbic Acid Supplementation

Quality	Placebo	200 mg AA	1000 mg AA
Viability	6	31	49
Motility	15	52	126
Agglutination	3	53	61
Precursors	35	56	66
Morphology	13	32	33
Count	22	112	140

statistically significant. The 200 mg AA per day group showed no significant difference in sperm count during the study. The group mean ± SEM count was 83 ± 48 million prior to supplementation, 176 ± 59 million after one week, 205 ± 92 million after 2 weeks, and 216 ± 121 million after 3 weeks. However, there was a 112% increase in group mean total sperm count after 1 week of supplementation. The increases in mean sperm count observed during the weeks of supplementation were not statistically significant due to the large SEMs. The 1000 mg AA per day group showed similar changes during the 3 weeks of supplementation. The group mean ± SEM was 99 ± 37 million prior to supplementation, 238 ± 71 million after one week, 215 million after two weeks, and 216 million after three weeks. None of these weekly changes in sperm count during supplementation were significantly different from presupplementation mean due to the large SEM of each mean. There was a 140% increase in group mean sperm count observed after one week of supplementation with 1000 mg per day, which was 28% higher than the 112% increase observed following one week of 200 mg AA supplementation (see TABLE 4).

DISCUSSION

The previous studies in the scorbutic guinea pig have shown that the most severely affected tissue was the cauda epididymus, wherein sperm precursors mature, are nurtured into fertile sperm, and assume normal morphology. Secondarily, the germinal epithelium showed shrinkage, nuclear pycnosis, and desquamation.[1,7] Several *in vitro* studies with normal mature sperm of man have shown that agglutination of normal sperm will occur in the absence of AA in the medium of seminal plasma and will reverse when AA is added.[5,6] Thus, These studies suggest that low or deficient AA levels in man are associated with semen containing an increase in sperm precursors and sperm with abnormal morphology, and decreased sperm count due to the scorbutic affect on the testes. Further, the scorbutic effect in seminal plasma is decreased viability and motility, as well as increased agglutination of mature sperm, and is due to decreased AA secretion by the seminal vesicles during ejaculation.

Much variation exists in these semen qualities between individuals and between samples. For this reason, the most rapid and dependable technique for identifying the scorbutic effect on fertility in man is the microscopic measurement of sperm agglutination. The wide range of normal variations of the other sperm qualities preclude their use for this purpose and the measurement of semen or seminal plasma AA levels is a lengthy laboratory procedure. Furthermore, a scorbutic condition in man does not appear to affect the libido, so that no physical signs or symptoms suggest sperm defects.[7] For this reason, we may not have received the only weekly semen sample from some of our subjects, and both the quality and the quantity of the sample may be so affected.

The AA levels of semen and seminal plasma showed progressive weekly increases, with more rapid and higher increases in the 1000 mg AA group than the 200 mg AA group. This suggests that the subjects were supplementing daily according to protocol. It is noteworthy that the serum and seminal plasma levels of AA were elevated significantly in both supplemented groups within one week ($p \leq 0.05$, ≤ 0.02, ≤ 0.02, ≤ 0.001) and indicates that possible beneficial effects of higher levels of AA on the studied sperm qualities would be anticipated.

The results show significant improvement after 1 week for the 1000 mg AA per day subjects in viability ($p \leq 0.001$), motility ($p \leq 0.01$), agglutination ($p \leq 0.001$), immature precursors ($p \leq 0.001$), abnormal morphology ($p \leq 0.01$). Also, apparently, total count improved 140%. The results for the 200 mg AA per day group after one week were similar, but of lower significance. The sperm viability was improved 31%, but the improvement was not statistically significant. However, significant improvement was observed in motility ($p \leq 0.05$), agglutination ($p \leq 0.02$), immature precursors ($p \leq 0.05$), and abnormal morphology ($p \leq 0.01$). The total sperm count was also apparently improved 112%. The significance of the improvements in sperm qualities were very nearly identified for both AA supplementation groups after two weeks of supplementations. Therefore, it is possible to conclude that daily supplementations with 200 mg of AA may be as effective as supplementation with 1000 mg AA in the improvement of sperm qualities within one week, and certainly after two weeks of supplementation. However, it is important to note that the mean ± SEM seminal plasma AA level of the 1000 mg AA group rose from the presupplementation level of 4.2 ± 0.7 mg/dl to 12.7 ± 1.1 mg/dl after only 1 week of supplementation, and this level has previously been reported to be the average level in normal men.[2–4] In contrast, this level was never reached in the 200 mg/day AA group throughout the study. The presupplementation level was 2.0 ± 0.5 mg/dl and was progressively elevated to 8.1 ± 0.5 mg/dl after 3 weeks.

The results of this study appear to agree with previous reports and provide direct evidence of the importance of AA in maintaining the physiological integrity of testis, epididymus, and accessory glands. Ascorbic acid is readily oxidized to dehydroascorbic acid (DHA). Dehydroascorbic acid, in turn, is readily reduced back to AA, enabling AA conservation. Therefore, the AA/DHA ratio may reflect one form of control on cell biosynthetic processes. Ascorbic acid is required for the hydroxylation of essential amino acids, and several oxidase enzymes, some of which function in the catabolism of organo-chlorine pesticides, heavy metals, and steroids.[15-17] The essential role of AA in the physiology of the testes appears to be associated with the role of AA in protein metabolism. In contrast, the role of AA in the qualities of the mature sperm within seminal plasma appears to be the effects of the AA/DHA ratio on antiagglutination and motility of sperm, and possibly of AA oxidase activity on improved sperm viability.

REFERENCES

1. LINDSAY, B. & G. MEDES. 1926. Histological changes in the testis of the guinea pig during scurvy and inanition. Am. J. Anat. **37:** 213-230.
2. BERG, O. C., C. HUGGINS & C. V. HODGES. 1941. Concentration of ascorbic acid and the phosphatases in secretion of the male genital tract. Am. J. Phys. **133:** 82-87.
3. PAZ, G., A. SOTER, Z. T. HOMMONAI & P. F. KRAICER. 1977. Human semen analysis: Seminal plasma and prostatic fluid compositions and their interrelations with sperm quality. Int. J. Fertil. **22:** 140-147.
4. LI, T. K. & B. L. VALLEE. 1973. *In* Modern Nutrition in Health and Disease (chap. 8B). R. S. Goodhart & M. E. Shils, Eds. Lea and Febiger. Philadelphia.
5. LINDAHL, P. E. & J. E. KIHLSTROM. 1960. Some factors influencing the biological activity of sperm agglutinins. J. Reprod. Fertil. **1:** 3-22.
6. LINDAHL, P. E. 1966. Sperm agglutinating and anti-agglutinating factors in normal follicular fluid from cattle. Int. J. Fertil. **11:** 297-305.
7. CHINOY, N. J., R. P. BUCH-NEE, R. R. MELITA, L. SEETHALAKSHMI, J. D. SHARMA & M. R. CHINOY. 1986. Effects of vitamin C deficiency on physiology of male reproductive organs of guinea pigs. Int. J. Fertil. **31:** 232.
8. HARRIS, W. A., T. E. HARDEN & E. B. DAWSON. 1979. Apparent effect of ascorbic acid medication on semen metal levels. Fertil. Steril. **32**(4): 455-459.
9. TOLBERT, B. M. 1979. Ascorbic acid metabolism and physiological function. Int. J. Vit. Nutr. Res. Suppl. **19:** 127-142.
10. ALHADETT, L., T. GUALTIERI & M. LIPTON. 1984. Toxic effects of water-soluble vitamins. Nutr. Rev. **42:** 33-40.
11. FRANKLIN, R. R. & C. D. DUKES. 1964. Antispermatozoal antibody and unexplained infertility. Am. J. Obstet. Gynecol. **89:** 6-9.
12. FRANKLIN, R. R. 1964. Further studies on sperm agglutination antibody and unexplained fertility. J. Am. Med. Assoc. **190:** 602.
13. AMELAR, R. D. 1966. The Semen Analysis: Infertility in Men (chap. 2). C. E. Heaton, Ed. F. A. Davis. Philadelphia.
14. LOWRY, O. H., O. A. BESSEY, M. J. BROCK & J. A. LOPEZ. 1946. The interrelationship of dietary, serum, white blood cell, and total body ascorbic acid. J. Biol. Chem. **166:** 111-119.
15. AGRAWAL, N. K., C. J. JUNEJA & C. L. MAHAJAN. 1978. Protective role of ascorbic acid in fishes exposed to organochlorine pollution. Toxicology **11:** 369-375.
16. EICHBAUM, F. W., A. O. GUEDES, J. P. NETO & F. V. CARVALHO. 1977. Protecting effect of ascorbic acid in strychnine poisoning and in tetanus. Int. J. Vit. Nutr. Res. Suppl. **16:** 31-39.
17. TUCKER, B. W. & J. E. HALVER. 1984. Ascorbate-2-sulfate metabolism in fish. Nutr. Rev. **42:** 173-179.

DISCUSSION OF THE PAPER

E. HOPES (*SUNY at Stony Brook, Stony Brook, N.Y.*): Did you look at vitamin E concurrently while you were evaluating the ascorbic acid responses with these sperm studies, or did you just measure vitamin C in your study?

E. B. DAWSON (*University of Texas Medical Branch, Galveston, Tex.*): Just vitamin C. We have been using vitamin C solely for many years in our clinics. This was based on the earlier studies, where agglutination was titrated against the amount of ascorbic acid. This study was to determine how fast this reaction will take place in reducing the agglutination and how small a dosage of vitamin C could be used. We did a correlation study between the blood serum level of ascorbic acid and the percentage of agglutination in the seminal plasma. We came out with a correlation factor of 0.88, very highly significant. We also found that when the serum plasma level of ascorbic acid was 0.8 mg/dl there was no agglutination. On the other hand when the serum plasma level was 0 mg/dl there was 45% agglutination of the sperm.

A. KALLNER (*Karolinska Hospital, Stockholm, Sweden*): I wonder whether you have reported any pregnancy in the groups studied?

E. B. DAWSON: In this particular group we were not looking for pregnancy. We did the study about 10 years ago where we compared a group of 20 infertile men with at least 25% agglutination who took vitamin C supplements with a similar group of 10 men who had 25% agglutination and received no supplement. All 20 men who received ascorbic acid supplement had impregnated their wives and the group which had received nothing had no pregnancies at all, nor was there any improvement in any of the sperm qualities.

S. L. ROMNEY (*Albert Einstein College of Medicine, New York, N.Y.*): The findings are interesting, but I think you must be able to provide some pregnancy rates in terms of follow-up in order for anybody to really assess the significance of ascorbic acid.

E. B. DAWSON: We didn't follow it up in this study because it had already been published.

Is There a Physiological Role of Vitamin C in Iron Absorption?

LEIF HALLBERG, MATS BRUNE, AND
LENA ROSSANDER-HULTHÉN

Department of Medicine
University of Göteborg
Sahlgren's Hospital
S-413 45 Göteborg, Sweden

INTRODUCTION

Iron deficiency is very common all over the world in both industrialized and developing countries, especially in children and women at childbearing age. Iron deficiency is more prevalent and more severe in developing countries. Its main cause is nutritional, *i.e.,* absorption of iron from the diet is insufficient to cover physiological needs. Of the two kinds of iron in the diet with respect to mechanism of absorption, heme and nonheme iron, the latter forms the main part, usually more than 90%. The absorption of the nonheme iron is influenced by several factors in the diet, some enhancing the absorption, for example vitamin C (ascorbic acid, AA), and meat or fish, others inhibiting the absorption, for example phytates and tannins (polyphenols); for a review, see ref. 1. It has long been known that AA increases the absorption of iron,[2] but in recent years with the introduction of new methodology to study iron absorption it has become quite evident that AA is the most potent promotor of the absorption of dietary iron.[3–11] The question I would like to discuss is the following: Is the effect of AA on iron absorption a physiological one, in other words, is AA a nutritional factor that is essential for the absorption of iron in man?

Ascorbic acid is not essential in the sense that the presence of AA in a meal is a condition for the absorption of iron. This fact does not, however, exclude the possibility that AA may be essential for the iron absorption and that AA has a physiological role in the absorption. The crucial point is whether AA, in the amounts that normally may occur in the diet of man, has a determining, critical influence on the iron balance. What should be considered normal in this context will be discussed later. What we need to know and what the main question is concerned about is basically the relationship between the amount of AA in a meal and its effect on the absorption of iron.

EFFECT OF DIFFERENT AMOUNTS OF ASCORBIC ACID ON IRON ABSORPTION FROM DIFFERENT TYPES OF MEALS

In an early study we found that in a simple maize meal there was a strong semiexponential relationship between the amount of ascorbic acid added to the meal and the induced absorption increase.[6] Another laboratory reported very similar findings with a semisynthetic meal.[7] Giving the same amount of ascorbic acid (50 mg) in different meals, we consistently found a significant increase of the absorption but the effect varied for different meals.[12] The relationship between the amount of ascorbic acid added to a meal and the absorption increase can be illustrated by regression lines with different slopes (FIG. 1).

It should also be mentioned that the same enhancement of iron absorption is obtained when it is administered in its native form as an integral component of a

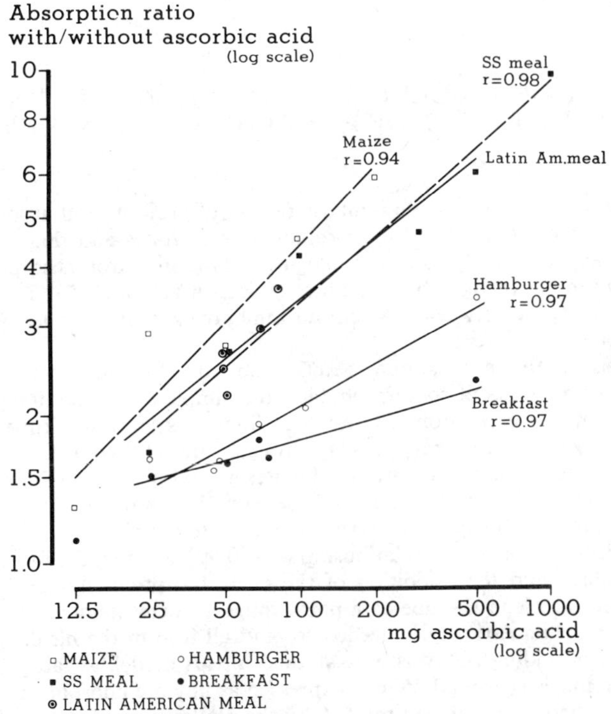

FIGURE 1. The relationship between the amount of ascorbic acid added and the ratio of the absorption of iron when different meals are given with or without ascorbic acid. (From Hallberg and Rossander.[13])

food, as when synthetic ascorbic acid is given,[12] which suggests that ascorbic acid is rapidly and completely released from vegetables and fruits. It is probable that differences in the effect of AA on iron absorption from different meals is in some way related to varying contents of factors enhancing and inhibiting iron absorption. Besides AA, meat and fish are the other main promotors of iron absorption. The effect of AA, however, seems to be about the same whether or not meat or fish is present in a meal. This was noted in studies on the effect of AA in rice meals served with or without fish [12] and in studies on the effect of meat and fish in vegetarian meals with a high content of AA.[13]

The mechanism of action of AA is usually related either to its ability to form soluble iron complexes or its ability to reduce ferric to ferrous ions. Both mechanisms will reduce the probability that iron ions will be strongly bound to other ligands in the intestinal content, such as hydroxyl ions, and that iron will be prevented from being absorbed. Ascorbic acid will thus counteract the influence of ligands that bind iron ions and inhibit iron absorption. The most potent dietary inhibitors known are phytates and polyphenols. The amounts in different foods of such polyphenols that inhibit iron absorption are not precisely known. Phytates can be more easily quantified but the interplay between phytates and AA in determining the bioavailability of iron is not known.

EFFECT OF ASCORBIC ACID IN COUNTERACTING THE INHIBITORY EFFECT OF PHYTATES ON IRON ABSORPTION

Recent studies show that the inhibiting effect of bran on iron absorption [14,15] can be almost completely explained by its content of phytates and that sodium phytate can be used experimentally as a physiological inhibitor of iron absorption.[16]

To get more knowledge about the mechanism of action of AA in the absorption of iron, the interplay between AA and an inhibitor, sodium phytate, was studied in a simple model experiment.

As a basis for the investigation the effect on iron absorption was first studied by adding various amounts of sodium phytate to a simple meal containing no phytates. Bread baked of low extraction, white wheat flour was used and known amounts of sodium phytate were added after baking. In each subject a comparison was made of the iron absorption from bread with no phytates and with a known amount of phytate. The iron in the two kinds of bread was labeled with two different radioiron isotopes and the two kinds of bread were served on alternate days.

A very strong, exponential relationship ($r = 0.99$) was found between the amount of phytate added and the inhibition of the iron absorption. As seen in FIGURE 2, there is a clear break in the line at a point roughly corresponding to the amount of phytate that stochiometrically is needed to bind all iron in the meal.

The effect of adding AA was now studied in this model. A fairly common level of phytates in meals is around 25 mg (expressed as phytate phosphorus). At this level the absorption had been reduced to 35% by the phytates compared with a meal with no phytates. The addition of 50 mg AA at this phytate level increased the absorption about twice and the addition of 100 mg AA increased it three or four times, to 119% of the original. This means that in this meal about 80 mg AA would probably completely neutralize the inhibiting effect of the phytates.

Another level studied was 250 mg of phytate P. This is a high but realistic level in meals containing phytate rich cereals, beans, etc. The effect of 50 mg of AA increased the absorption three times. The absolute increase of the iron absorption, however, was less at this higher level of phytate inhibition. This observation implies that the more the absorption is inhibited by phytates the more AA is needed to counteract the inhibition. A rough estimate would suggest that about 200 mg AA should be needed at this level of phytate content to completely counteract the inhibition of the absorption.

The presence of a strong dietary inhibitor of iron absorption such as a phytate is not a condition for obtaining an effect of ascorbic acid on the absorption of iron. This was noted when giving 50 mg AA in the same meal with no known inhibitor composed of only white wheat bread. A 60% increase was seen (see FIG. 2).

This finding implies that the absorption of iron may be inhibited also by a binding of iron ions to ligands in foods other than phytates or polyphenols or by a binding to ligands, *e.g.*, hydroxyl ions normally present in the intestinal tract, and that AA interferes with such bindings. This interpretation would also imply that if by some means, in a meal with no known inhibitors, the native content of AA would be reduced, for example by destroying AA by cooking, a reduction of the absorption of iron from that meal would be expected. This was actually observed when comparing the absorption from two initially identical meals containing meat, vegetables, and potatoes.[17] One meal was kept warm for 4 hours, which led to a reduction of both the content of AA and the absorption of iron. A restitution of the content of AA

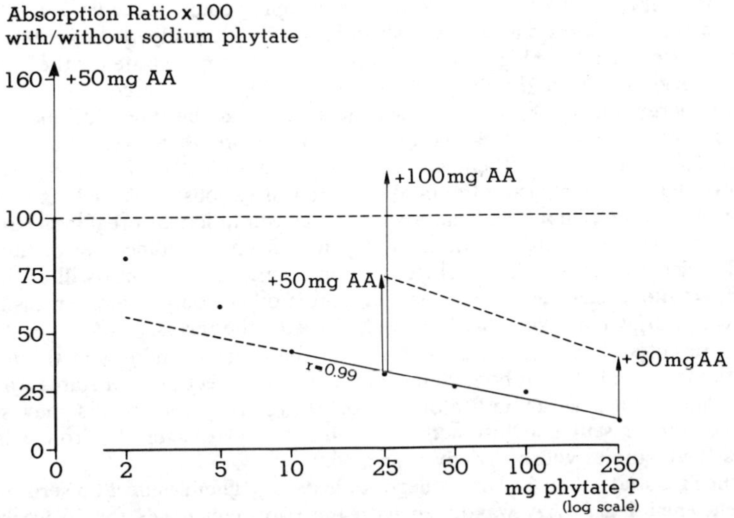

FIGURE 2. Effect of different amounts of sodium phytate (expressed as phytate phosphorus) added to white wheat bread on the absorption of iron (solid regression line) and the effect of adding different amounts of ascorbic acid to bread with different phytate content (0, 25, 250 mg phytate phosphorus). Each bread contains 3.8 mg iron, mainly from the addition of ferrous sulfate (3.7 mg) to the white wheat flour (extraction 72%).

normalized the absorption, thus proving that the reduction of the iron absorption was not caused by destruction of some other heat labile factor in the meal.

EVALUATION OF THE PHYSIOLOGICAL ROLE OF ASCORBIC ACID IN THE ABSORPTION

The studies reported above show that AA enhances the absorption of iron from all kinds of meals, also those containing no known inhibitor of nonheme iron absorption. They also show that the relative effect of AA is higher if a meal contains potent inhibitors of iron absorption such as phytates, but that more AA is needed in such a meal to counteract the inhibition of iron absorption and that more AA is needed the higher the content of phytates. In a diet with a high content of cereals, pulses, and vegetables more AA is thus needed to balance its higher content of phytates and tannins.

Early man had a diet with no refined foods and its contents of phytates and tannins from various vegetables, fruits, nuts, etc. were probably high. On the other hand the content of foods enhancing iron absorption was probably considerably higher than today. It has been estimated that as much as 35% of the energy came from meat. The intake of ascorbic acid has also been estimated to be high, amounting to about 400 mg daily (for a review see ref. 18).

Marked changes in dietary habits have occurred over just a few hundred generations, a period that is probably too short to allow major adaptations to diets in the last ten or even the last few generations. In developing countries today the intakes of enhancers of iron absorption (meat or fish and AA) are generally low. In industrialized countries these intakes are higher but usually too low to compensate for the low intake of both energy and iron. In industrialized countries, as a consequence of the low intake of energy and foods, we have become aware that the intake of fiber has also become too low for normal intestinal functions. Therefore, there is at present a strong recommendation to increase the fiber content of the diet. One of the few feasible ways to achieve this is by increasing the intake of bran in various ways such as increasing the extraction rate of flour and increasing the consumption of breakfast cereals. In turn this will lead to a higher intake of phytates. Such a modification of the diet is probably scientifically well founded from certain points of view, but will at the same time impair the already low absorption of iron if other compensatory measures are not taken, such as increasing the intake of AA with the meals.

Two recent studies have been reported on the effect of increasing the intake of AA with main meals.[19,20] In both studies no significant effect on iron status was found using serum ferritin as an indicator of iron stores. As these results may seem to contradict the present conclusion that AA has a key role for the iron balance in humans these studies will be briefly discussed.

In the first study,[19] on 17 U.S. college students, a negligible effect on serum ferritin was seen when 1 g of AA was taken with the two main meals for 16 weeks. Iron absorption studies with a semisynthetic meal showed that the enhancing effect of AA was not reduced after the 16 weeks of AA supplementation. An intestinal adaptation to the high doses was thus excluded.

The iron absorption from these semisynthetic meals is much lower than from the actual diet and the effect of AA much more pronounced. The students considered their diet to be well balanced and to contain generous quantities of meat and poultry. It is thus probable that the intakes of both AA and meat were high. The findings that no effect was seen by giving extra AA do not contradict the present conclusion that AA has a key physiological role in the absorption of iron as the bioavailability of the dietary iron in the supplementation study was probably near maximum already before giving the extra AA.

In the other study,[20] Irish student nurses were given 100 mg AA with their three main meals for 8 weeks. Changes in serum ferritin were compared in 25 treated and 23 control nurses. There was a slight (16%) increase in the AA group and no change in the control group, but the difference was not statistically significant. There may be different explanations for this negative result, such as the short supplementation period and the fact that serum ferritin, because of its fairly marked normal variation, is not a good measure of small changes of the iron stores in the normal range. An important factor to consider in evaluating the average effect of AA is the influence of those women who have not only exhausted iron stores but also suboptimal hemoglobin values at the start of the study. Theoretically, no effect on serum ferritin is expected until the hemoglobin is normalized, which may take a long time. In this study no information was given about the prevalence of iron deficiency anemia, but 25% had subnormal serum ferritin values initially. This Irish study had a good double blind design, but just as in iron fortification trials a great number of subjects must be studied for a long time to obtain statistically significant differences.

These two studies have been discussed in greater detail as there is a risk that the results might be misinterpreted by those who are not quite familiar with the many pitfalls in iron intervention studies. One problem, based on recent World Health Organization (WHO) studies, is the slow response expected. It usually takes 2 years or more to obtain statistically significant changes in fairly large groups. A special problem is related to the effect of the initial hematological status on the rate of change of different iron parameters. In groups with a high prevalence of anemia, for example, the response of serum ferritin can be expected to be low and slow, and in groups with a low prevalence of anemia the hemoglobin response will be low and slow.

CONCLUDING REMARKS

There is no doubt that AA is of great importance for the absorption of dietary iron. Even if iron can be absorbed from the diet in the absence of AA, its effect is so marked and so regular that it must be considered as a physiological factor that is essential for dietary iron absorption.

Ascorbic acid has several important functions in the body. It is involved in the metabolism of several tissues and in a multitude of biochemical processes. In setting the nutritional requirements for AA, its key role in the absorption of iron from the diet has not been sufficiently considered.

Our calculations, based on several hundred absorption studies in man by several investigators in recent years, strongly indicate that each main meal should preferably contain at least 25-50 mg of AA. If meals have a higher content of factors inhibiting iron absorption, such as phytates and tannins, a content of AA higher than those mentioned should be considered.

SUMMARY

Nonheme iron usually constitutes more than 90% of the dietary iron. Its absorbability is a resultant of the balance between factors enhancing and inhibiting the absorption. Ascorbic acid is the most potent enhancer, and is the same for native and synthetic AA. The enhancing effect is strongly dose related (log dose/effect), and is different for different meals probably mainly due to varying content of inhibitors in the meals. AA also increases the iron absorption from simple meals with no known inhibitor, probably because AA impairs the formation of unavailable iron complexes with ligands normally present in the gastrointestinal lumen. The effect of AA is so unequivocal and marked that it must be considered as a physiological factor essential for the absorption of dietary iron.

REFERENCES

1. HALLBERG, L. 1981. Bioavailability of dietary iron in man. Ann. Rev. Nutr. **1:** 123-147.
2. MOORE, C. V. & R. DUBACH. 1951. Observations on the absorption of iron from foods tagged with radioiron. Trans. Assoc. Am. Physic. **64:** 245-256.
3. SAYERS, M. H., S. R. LYNCH, P. JACOBS, R. W. CHARLTON, T. H. BOTHWELL, R. B. WALKER & F. MAYET. 1973. The effects of ascorbic acid supplementation on the absorption of iron in maize, wheat and soy. Br. J. Haematol. **24:** 209-218.
4. SAYERS, M. H., S. R. LYNCH, R. W. CHARLTON, T. H. BOTHWELL, R. B. WALKER & F. MAYET. 1974. Iron absorption from rice meals cooked with fortified salt containing ferrous sulphate and ascorbic acid. Br. J. Nutr. **31:** 367-375.
5. SAYERS, M. H., S. R. LYNCH, R. W. CHARLTON, T. H. BOTHWELL, R. B. WALKER & F. MAYET. 1974. The fortification of common salt with ascorbic acid and iron. Br. J. Haematol. **28:** 483-495.
6. BJÖRN-RASMUSSEN, E. & L. HALLBERG. 1974. Iron absorption from maize. Effect of ascorbic acid on iron absorption from maize supplemented with ferrous sulphate. Nutr. Metabol. **16:** 94-100.
7. COOK, J. D. & E. R. MONSEN. 1977. Vitamin C, the common cold and iron absorption. Am. J. Clin. Nutr. **30:** 235-241.
8. GILLOOLY, M., T. H. BOTHWELL, J. D. TORRANCE, A. P. MAC PHAIL, D. P. DERMAN, W. R. BEZWODA, W. MILLS, R. W. CHARLTON & F. MAYET. 1983. The effects of organic acids, phytates and polyphenols on the absorption of iron from vegetables. Br. J. Nutr. **49:** 331-342.
9. MORCH, T. A., S. R. LYNCH & J. D. COOK. 1982. Reduction of the soy-induced inhibition of non-heme iron absorption. Am. J. Clin. Nutr. **36:** 219-223.
10. DERMAN, D. P., T. H. BOTHWELL, A. P. MAC PHAIL, J. D. TORRANCE, W. R. BEZWODA, R. W. CHARLTON & F. G. H. MAYET. 1980. Importance of ascorbic acid in the absorption of iron from infant foods. Scand. J. Haematol. **25:** 193-201.
11. HALLBERG, L. & L. ROSSANDER. 1984. Improvement of iron nutrition in developing countries: comparison of adding meat, soy protein, ascorbic acid, citric acid and ferrous sulphate on iron absorption from a simple Latin American-type of meal. Am. J. Clin. Nutr. **39:** 577-583.
12. HALLBERG, L. & L. ROSSANDER. 1986. Effect of ascorbic acid on iron absorption from different types of meals. Studies with ascorbic-acid-rich foods and synthetic ascorbic acid given in different amounts with different meals. Human Nutrition: Applied Nutrition **40A:** 97-113.
13. HALLBERG, L. & L. ROSSANDER. 1982. Absorption of iron from Western-type lunch and dinner meals. Am. J. Clin. Nutr. **35:** 502-509.
14. BJÖRN-RASMUSSEN, E. 1974. Iron absorption from wheat bread. Influence of various amounts of bran. Nutr. Metabol. **16:** 101-110.
15. SIMPSON, K. M., E. R. MORRIS & J. D. COOK. 1981. The inhibitory effect of bran on iron absorption in man. Am. J. Clin. Nutr. **34:** 1469-1478.
16. HALLBERG, L., L. ROSSANDER & A. B. SKÅNBERG. Phytates and the inhibitory effect of bran on iron absorption in man. Am. J. Clin. Nutr. In press.
17. HALLBERG, L., L. ROSSANDER, H. PERSSON & E. SVAHN. 1982. Deleterious effects of prolonged warming of meals on ascorbic acid content and iron absorption. Am. J. Clin. Nutr. **36:** 846-850.
18. EATON, B & M. KONNER. 1985. Paleolithic nutrition. A consideration of its nature and current implications. N. Engl. J. Med. **312:** 283-289.
19. COOK, J. D., S. S. WATSON, K. M. SIMPSON, D. A. LIPSCHITZ & B. S. SKIKNE. 1984. The effect of high ascorbic acid supplementation on body iron stores. Blood **64:** 721-726.
20. MALONE, H. E., J. P. KEVANY, J. M. SCOTT, S. D. O'BROIN & G. O'CONNOR. 1986. Ascorbic acid supplementation: Its effects on body iron stores and white blood cells. Irish J. Med. Sci. **155:** 74-79.

DISCUSSION OF THE PAPER

A. KITABCHI (*University of Tennessee, Memphis, Tenn.*): How important is the source, preparation, and storage of various foods?

L. HALLBERG (*University of Göteborg, Göteborg, Sweden*): Certainly what really matters is the amount of ascorbic acid that is on the plate when you eat it. There's a major source of error in just the storage of the food and the cooking. For instance, potatoes in canteens often are cooked in the morning and served at noon time and there's very little ascorbic acid left in the potato. We did a screening in school lunch programs in Sweden and investigated several meals and found almost zero content of ascorbic acid in the meals because they have been stored for several hours.

J. BERGER (*Tufts University, Boston, Mass.*): Does the vitamin C have to be consumed in the same meal as the iron in order to promote the absorption? If so, a single bolus of vitamin C in the morning which some people take as supplements is not going to be effective.

L. HALLBERG: That's right.

P. J. GARRY (*University of New Mexico School of Medicine, Albuquerque, N.M.*): Does the vitamin C have the same effect on an individual who is iron deficient as it would on a person who is iron sufficient?

L. HALLBERG: Yes and no. The percentage increase would be the same, but the absolute increase in absorption would be higher in iron deficient than in normal ones.

B. DEFAVIO (*New York Hospital, New York, N.Y.*): I've had several patients suffering from iron-deficiency anemia. I have had them have a half a cup to a cup of orange juice with each meal and there has been a very dramatic increase in their iron nutriture within a very short time (2-3 weeks). Is there anything related to orange juice, preferably fresh, aside from ascorbic acid, that may be helpful?

L. HALLBERG: We have studied that and found it to be identical with exactly the same amount of ascorbic acid.

As to the rapidity in response, you have to consider that the normal absorption in an iron-deficient woman with a fairly good diet would be something around 2 or 3 mg a day. You can't achieve much more than doubling that. If there's a hemoglobin deficit of let's say 500 mg of hemoglobin iron, it would take fairly long time in order to cover this deficit, so your rapid response may be due to something else. But it's worthwhile to say people should have orange juice or vegetables with a meal.

D. J. MORRÉ (*Purdue University, West Lafayette, Ind.*): I'm just delighted to see this type of relationship, because the biochemical dogma now is that iron would have to be reduced as it crossed the membrane and the enzyme system plasma redox that drives it uses ascorbate very nicely as a substrate. So there may be a mechanistic connection here that would be very nice to pursue.

A. KITABCHI: Do you hypothesize that with large dose of vitamin C you may have an iron overload?

L. HALLBERG: I would not think so because...

A. KITABCHI: Because of the limiting factor for iron absorption?

L. HALLBERG: Exactly, and the more iron you have in the stores the lower the extra iron is being absorbed so you're flattening it out. It may be, however, that you could diagnose iron overload perhaps one year earlier, but that wouldn't induce iron overload in more subjects.

A. KITABCHI: Is there one good method to determine whether in fact the vitamin C is immediately effective for iron absorption?

L. HALLBERG: I think you should not use a good diet to treat iron deficiency anemias, you should use a good diet in order to prevent the development of iron-deficient anemia. You need iron tablets to treat iron-deficient anemia, otherwise it would take too long.

It would take a year or two to treat iron-deficient anemia by a good diet, which you can do in six weeks with iron treatment. But in order to *prevent* the development, and to reduce the probability of development of an iron deficiency, a good diet is important.

UNIDENTIFIED SPEAKER: There's been considerable concern in this country that we have too high an RDA for iron and that the ascorbic acid recommendations are also enhancing the possibility of iron overload. Now I'm not convinced myself but apparently you're not either.

L. HALLBERG: We have been studying homozygotic patients with hemochromatosis to determine the effect of vitamin C, and the effect is not more pronounced there than in other subjects. Actually they are so overloaded that the vitamin C effect is fairly marginal.

UNIDENTIFIED SPEAKER: The most significant losses of vitamin C and other nutrients in food occur in the home and in food service practices. The food service practices, as you mentioned regarding school lunches, is a very good example of that as well as institutional feedings in hospitals and nursing homes.

B. LANE (*Columbia University School of Public Health, New York, N.Y.*): As I recall from the literature, there's a lesser uptake of iron from vegetables as compared to meat, except in people who need the iron and then the value for vegetables comes up more, comparable to what it is in meat. Is that correct?

L. HALLBERG: If I understand what you are asking, there is little relationship of heme iron absorption to iron status, it is usually about 25%, whereas nonheme iron absorption is very much related to the iron status of the subject.

In some meals you can get as much as 45% iron absorption. The highest figure in our record is from a meal containing sauerkraut. Sauerkraut has certain factors which we are studying at the moment which increased absorption very dramatically. Another such meal is Russian borscht soup and it's not just related to pH. We are looking for certain acids which may also promote iron absorption in a way similar to ascorbic acid.

Experimental Vitamin C Depletion and Supplementation in Young Men

Nutrient Interactions and Dental Health Effects

ROBERT A. JACOB,[a] STANLEY T. OMAYE,[a]
JAMES H. SKALA,[a] PENELOPE J. LEGGOTT,[b]
DAVID L. ROTHMAN,[b] AND PATRICIA A. MURRAY[b]

[a] *United States Department of Agriculture*
ARS Western Human Nutrition Research Center
Presidio of San Francisco, California 94129

[b] *School of Dentistry*
University of California at San Francisco
San Francisco, California 94143

INTRODUCTION

A number of interactions of ascorbic acid (AA) with other nutrients have been reported, particularly in human subjects consuming large intakes of AA. Interactions with water-soluble vitamins (B_6 and B_{12}), fat-soluble vitamins (A and E), and minerals (iron and copper) have been postulated.

Evidence of interaction between vitamin C and vitamin B_6 was first suggested by Russian investigators who found that excess vitamin C intake increased urinary excretion of 4-pyridoxic acid, the major metabolite of vitamin B_6.[1] These results suggested a possible higher requirement of vitamin B_6 during elevated ascorbate intakes. Human experimental deficiency studies of vitamins B_6 and C by Baker *et al.* also suggested an interaction between the two vitamins, the precise interrelation being unknown.[2,3] A recent study, however, showed no significant effect of short-term supplementation of up to 1 g/day ascorbic acid on urinary 4-pyridoxic acid or plasma pyridoxal phosphate in human subjects.[4]

The initial suggestion of a metabolic relation between ascorbic acid and vitamin B_{12} came from observations of unexplained ascorbate deficiency in vitamin-B_{12}-deficient patients.[5] Concern over the effect of excess ascorbic acid intakes on vitamin B_{12} nutriture was generated from reports of the ascorbate-related destruction of vitamin B_{12} in foods[6] and lowered serum vitamin B_{12} levels in patients taking ascorbate supplements.[7,8] Later studies claimed the apparent ascorbate-related destruction of vitamin B_{12} in food or serum was only an artifact of the method of vitamin B_{12}

analysis.[9,10] Two recent studies showed no effect of supplemental ascorbic acid on vitamin B_{12} nutriture in humans.[11,12]

Since both vitamins C and E are natural antioxidants, the possible synergism of the two vitamins as regards tissue antioxidant potential has been investigated. Results from animal and free radical chemistry studies suggest that excess vitamin C spares metabolic utilization of vitamin E and results in higher plasma vitamin E levels.[13,14] However, excess vitamin C supplements were found to lower tissue antioxidant potential during marginal vitamin E nutriture, thus suggesting an increased vitamin E requirement during high vitamin C intakes.[15,16] Other studies have shown a synergistic effect on antioxidant activity when both vitamins are present.[17,18] Excess vitamin C was shown to affect antioxidant activity in vitamin-A-stressed rats.[19]

Supplemental AA has been shown to affect mineral absorption, possibly by its action as a reductant or metal chelator. The vitamin has clearly been shown to increase nonheme iron absorption,[20] however its effect on copper absorption is not nearly so well documented. An antagonistic effect of high AA intakes on copper absorption and status has been postulated from limited animal[21,22] and human studies.[23]

Much of the evidence for AA-nutrient interactions comes from animal or human studies that were not specifically designed to study such interactions. Since AA is the most frequently taken single nutrient supplement,[24,25] it is important that these postulated interactions be studied further and in a focused way.

We have studied AA depletion and supplementation in young men residing for 14 weeks in a live-in metabolic unit, focusing on biochemistries related to AA[26,27] and AA-nutrient interactions. Since the role of AA in periodontal disease continues to be controversial,[28] we also studied dental effects of the varying AA intakes and have previously reported these results.[29,30] The controlled nature of this study and the in-depth biochemical determination of AA status provided a unique investigation of AA nutriture and periodontal health as compared to the usual free-living studies of this question. We present here the results of the AA-nutrient interaction studies and summarize the dental effects. The results on the effects of varying AA intakes on copper absorption and status have been reported separately.[31]

SUBJECTS AND METHODS

Eleven nonsmoking male volunteers aged 19-32 were admitted to the metabolic unit of the USDA-ARS Western Human Nutrition Research Center after medical and psychological screening. Subjects selected had relatively good existing oral health, a natural dentition of at least 20 teeth, minimal gingival calculus, and no mucosal disease. The study protocol and Informed Consent given was approved by the Committee for the Protection of Human Subjects of the University of California, San Francisco. For the 14-week study duration the subjects lived and ate all meals in the Metabolic Unit under close supervision. Because of personal reasons, two subjects left the study, at 10 and 12 weeks respectively. Subjects were confined to the Metabolic Unit at all times except for supervised outings. Subjects ate a 7-day rotating diet of natural foods that excluded fruits and vegetables and was calculated from food composition tables to provide an average of 4.9 mg/day (range 4.8-5.0 mg/day) of AA (food items were chosen to avoid intakes of isoascorbic acid). The specific food items that constitute the 7-day rotating menu are listed in a separate report.[26] The diet provided a calculated average of 2652 kcal/day, 56% as carbohydrate (8% simple carbohydrate), 28% as fat ($P/S = 0.95$) and 16% as protein. Individual caloric adjustments were made by supplementing the basal diet with soda, sucrose, or Polycose[c] sweetener (Ross Laboratories, Columbus, Ohio). Water was consumed ad libitum and consistent amounts of soda pop, coffee, and tea with no appreciable AA content were consumed. No tobacco or alcoholic beverages were allowed. No drugs were given except for occasional aspirin as an analgesic.

The basal diet was shown from food nutrient composition table calculations to contain at least 80% of the RDA of nutrients except for AA, copper, magnesium, and folic acid. The latter three micronutrients were supplemented into the diet.

Appropriate caloric intakes were estimated initially from dietary records and basal metabolic rate measurements. Individual adjustments in caloric intake to maintain constant weight were made during the first 2-week equilibration period and at two points during the study. Complete urine collections were made throughout the study. Physical activity regimens were prescribed on an individual basis and included walks, calesthenics, or jogging. Subjects who missed elements of their regular exercise regimen were given treadmill assignments as makeup.

The same basal diet (5 mg AA/day) was fed throughout the study. The experimental periods I-V of varying AA intakes are shown in TABLE 1. The initial 2-week baseline period allowed the individual subjects to adjust to the metabolic unit and the basal diet with an RDA intake of AA. Ascorbic acid intake was varied by feeding at each meal a grape juice drink that contained 0, 20, or 200 mg of supplemental AA powder (Hoffmann-La Roche, Nutley, N.J.) corresponding to the experimental period. A stock grape juice drink was prepared each day with Polycose sweetener and the proper amount of AA added so that a 200 ml aliquot at each of three meals would provide the desired amount of daily AA supplementation. Prestudy blind taste testing showed that test subjects were unable to predict which grape juice drinks had the higher or lower levels of AA, hence, study subjects were blind to the amount of AA being supplemented throughout the study. The stock grape juice solutions were kept at 4°C throughout the day so as to minimize oxidation of the AA. During each of the first four study periods several replicates of the daily supplemental AA dose were prepared by combining the three grape juice aliquots served at each meal. After the AA was stabilized by the addition of an equal volume of 10% metaphosphoric acid, the grape juice composite was analyzed for AA by a modification of a published HPLC procedure.[32] Analyzed vs. calculated daily intakes of AA were 65 ± 1 mg vs. 60 mg, 599 ± 34 mg vs. 600 mg, and <0.02 mg vs. zero ascorbate.

Fasting blood (45 ml) was taken before breakfast by venipuncture weekly throughout the study for biochemical determinations. Eight additional mid-week draws of 3.5 ml were taken during the two 4-week depletion periods for additional determinations of plasma AA. An extra 10 ml aliquot of blood was taken at the end of each metabolic period for the determination of total leukocyte AA. Whole blood anticoagulated with EDTA was kept on ice until the red cells were centrifuged at 4°C. The AA in plasma, blood cells, urine, and saliva was stabilized by the addition of an equal volume of 10% trichloroacetic acid (TCA). Aliquots of nonacidified 24-h urine collections kept at 4°C were taken for determination of AA and creatinine by the Jaffe method as a measure of completeness of collection. Whole mixed saliva was collected by expectoration after stimulation by paraffin chewing, and kept on ice until the AA was stabilized by addition of TCA. Total biologically active vitamin C (ascorbic and dehydroascorbic acids) was determined in plasma, urine, and saliva by spectropho-

[c]Mention of a trademark or proprietary product does not constitute a guarantee or warranty of the product by the U. S. Department of Agriculture, and does not imply its approval to the exclusion of other products that may also be suitable.

TABLE 1. Experimental Design

Experimental Period	Duration (Weeks)	AA Intake (mg/Day)
I	2	65
II	4	5
III	3	605
IV	4	5
V	1	605/65

tometric procedures using 2,4-dinitrophenylhydrazine as chromogen.[33] Analytical variability (standard deviation), as measured by repeated analysis of frozen aliquots of a specimen pool, was ± 0.05 mg/dl for serum, urine, and saliva. Leukocytes were isolated by a Percoll density gradient procedure and analyzed for AA by HPLC after deproteinization with 6% metaphosphoric acid.[33] Leukocyte AA determinations were done less frequently then plasma AA determinations because of the large amounts of blood required. Details of the leukocyte AA methodology are reported separately.[26] Urinary oxalate was measured by a colorimetric procedure using HCl acidified aliquots of 24-hour urine collections.[34]

As a measure of vitamin B_6 excretion, total urinary pyridoxine was determined by microbiological assay.[35] Serum vitamin B_{12} was determined using a commercially available radioassay kit, Immophase Vitamin B_{12}/Folate Radioassay (Corning Medical, Medfield Mass.). Serum retinol and α-tocopherol were determined by a reversed-phase HPLC method.[36] Serum ferritin was determined using a commercially available enzyme immunoassay, Ferrizyme (Abbott Laboratories Diagnostics Division, North Chicago, Ill.). The percent saturation of transferrin was calculated as the ratio of serum iron to total iron binding capacity (TIBC) after serum iron and TIBC were determined colorimetrically with ferrozine.[37]

Weekly oral examinations beginning the second week were conducted throughout the study. Biweekly reinforcement of an oral hygiene program of toothbrushing and interproximal cleaning with floss were also conducted throughout the study period beginning the second week. Oral indices were assessed by a single examiner who was unaware of the ascorbic acid supplementation schedule. Indices were obtained on the subset of six teeth described by Ramfjord[38,39] and included the plaque index,[40] gingival index (GI),[41] gingival bleeding index (BI),[42] and probing depth measured to the nearest millimeter from the gingival margin. Measures of plaque accumulation and gingival inflammation for each patient were expressed both as means (± SD) and as frequency distributions that showed the percentage of total sites scored for each index criterion.

Biochemical data were analyzed by a computerized statistical package (Minitab statistical computing system, Dept. of Statistics, Pennsylvania State University, University Park, Pa.) using linear regression and t-statistics. Oral index values were analyzed by the Wilcoxon Signed-Rank test and repeated measures analysis of variance. Both the parametric and the nonparametric tests gave similar results, and in no case disagreed on rejection of the null hypothesis of no difference from baseline to subsequent examinations.

RESULTS AND DISCUSSION

Ascorbic Acid Status

Group mean plasma and leukocyte AA levels are shown in FIGURE 1. The levels clearly reflect the changes in AA intake through the four experimental periods. The subjects were in a state of biochemical AA deficiency by the end of both depletion periods (5 mg/day), as indicated by the mean plasma AA and leukocyte levels, which were below the lower limits of normal for these indices, 0.3 mg/dl for plasma AA and 20 $\mu g/10^8$ cells for leukocyte AA.[33] No clinical effects of AA deficiency, *i.e.*, scurvy, were observed at any time during the study, hence the subjects were not severely depleted of body AA. The lower leukocyte AA values at the end of the second depletion period compared to the first depletion period suggests a conditioned deficiency or "rebound" effect after high AA intakes. These results have been discussed in a previous report.[26]

During the high AA intake period (605 mg/day) mean plasma AA levels rose abruptly and plateaued at a maximal mean value of 1.4 mg/dl. The mean leukocyte AA at the end of this period was 135% of the week 2 baseline value. Mean urinary AA excretion rose to 400 mg/day or two thirds of the AA intake. Hence, subjects were fully repleted with AA at the end of the high AA intake period. The biochemical indices of AA status show clearly that the subjects were depleted and repleted of AA body reserves to varying degrees as the study was designed to do.

Urinary oxalate levels ranged from 12-38 mg/day throughout the study. These levels were not affected by the varying AA intakes and were within the normal limit of 40 mg/day.[43] Concern has been raised that high AA intakes may increase body levels of oxalate, an end product of AA metabolism, thereby contributing to renal stone formation.[44] Whereas excess AA intake is largely excreted into the urine unchanged, or is not absorbed, only a small amount is metabolized to oxalate regardless

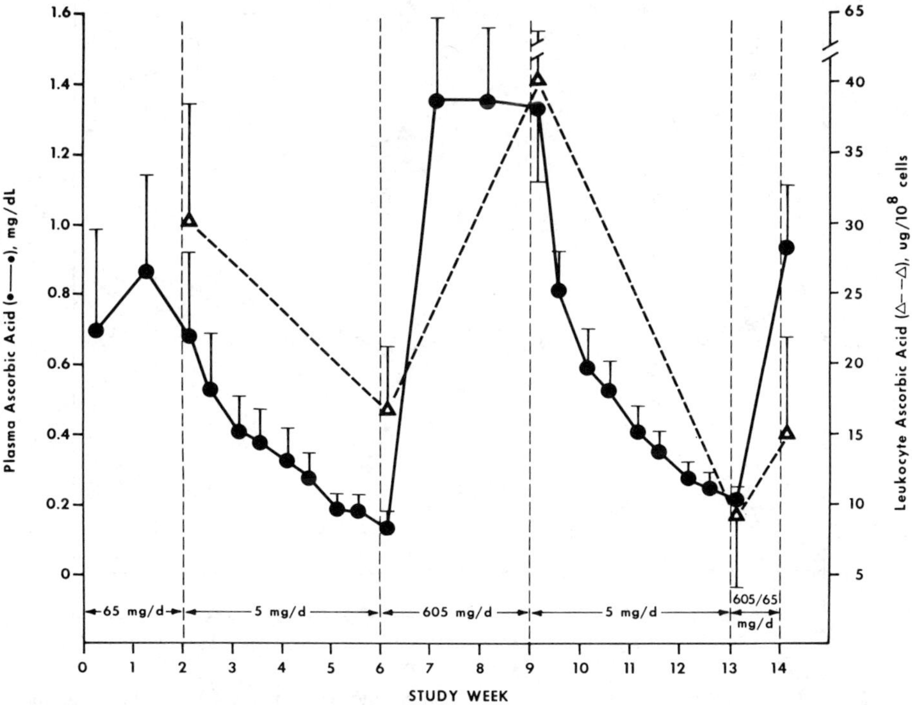

FIGURE 1. Mean + SD plasma and leukocyte ascorbic acid (AA) levels in 11 young men receiving varying AA intakes. (From Jacob *et al.*[27] Reproduced by permission of the *American Journal of Clinical Nutrition.*)

of the AA intake.[45] Most studies have shown that increased AA intakes do not increase body oxalate levels.[46] Results of the present study affirm this, and show that moderate supplemental AA intakes do not increase urinary oxalate levels in healthy young men even relative to low AA intakes.

Ascorbic Acid-Nutrient Interactions

Markers of vitamins B_6 (total urinary B_6) and B_{12} (serum cobalamin) are shown in FIGURE 2. We measured urinary B_6 levels because an increased excretion of 4-pyridoxic acid, the major urinary metabolite of pyridoxine, has been reported in humans consuming supplemental AA.[1]

As FIGURE 2 shows, we found no effect of the various AA intakes on pyridoxine excretion in our young men. This affirms the results of Schultz and Leklem,[4] who found no effect of AA supplementation on 4-pyridoxic acid excretion or pyridoxine status in human subjects. Likewise, our finding that various AA intakes did not influence serum cobalamin levels is consistent with two recent reports of human studies that found no effect of supplemental AA on human vitamin B_{12} status.[11,12]

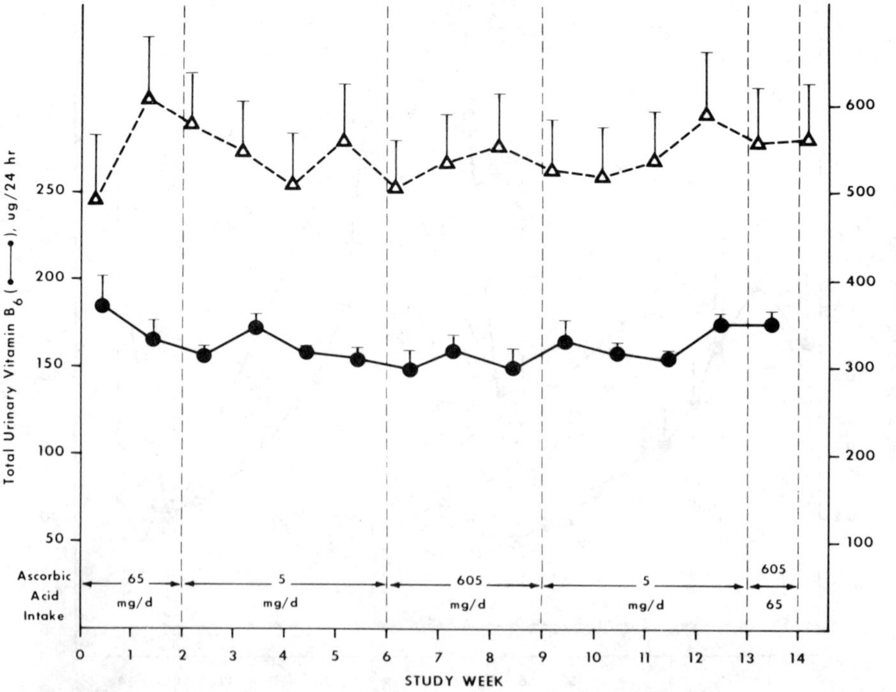

FIGURE 2. Mean + SE of total urinary vitamin B_6 and serum vitamin B_{12} in 11 young men receiving varying AA intakes.

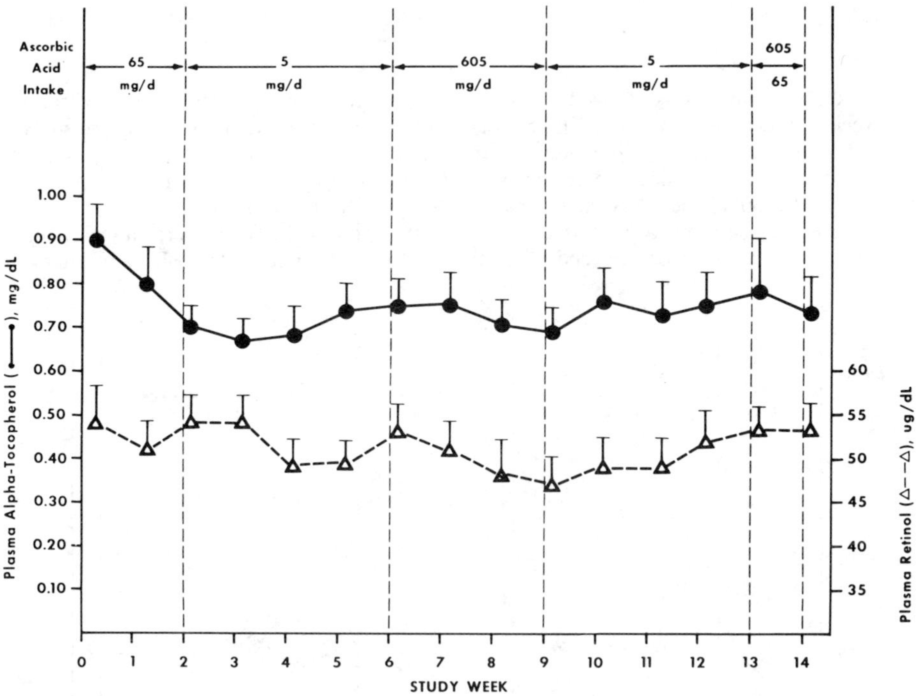

FIGURE 3. Mean + SE of plasma α-tocopherol and plasma retinol in 11 young men receiving varying AA intakes.

Markers of fat-soluble vitamin status, plasma α-tocopherol and plasma retinol, are shown in FIGURE 3. Neither vitamin level was affected by AA intake. Mixed results and conclusions have been reported from previous studies of vitamin-C-tocopherol interactions. Generally, supplemental AA has been found to spare vitamin E and/or augment the antioxidant activity of vitamin E.[13,14,17,18] Niki has reported (in this volume) that AA augments the antioxidant potential of membrane tocopherol by reducing the tocopherol radical to regenerate the protective vitamin E form.[47] Our results show no effect of AA intake on circulating tocopherol levels, possibly because we fed moderate supplements of AA (605 mg/day) and/or our subjects were in no antioxidant stress.

Blood hemoglobin and hematocrit levels did not change throughout the study. More sensitive indices of iron deficiency are shown in FIGURE 4. The percent saturation of transferrin did not change. Serum ferritin declined gradually throughout the study due to phlebotomy blood loss of approximately 50 cc/week. Although AA is known to facilitate iron absorption and the diet contained an adequate amount of iron (20 mg/day average), we found no effect of the high AA intake period of 605 mg/day on ameliorating the gradual decline in serum ferritin levels. This may be due to several reasons: (1) the diet was relatively high in animal protein and hence a majority of the dietary iron was as heme rather than nonheme form; (2) high intakes of AA were not fed for a long enough period to see an effect on changes in iron stores; (3) the enhancing effect of AA on iron absorption is not marked in subjects with adequate iron status.

Dental Health Effects

Oral examinations showed no severe periodontal disease during either of the 4-week AA depletion periods. The results of the four oral indices measured are shown in FIGURE 5. No changes in plaque accumulation or pocket depth were seen throughout the study. The bleeding indices are scored from 0 to 3 after probing at the gingival margin (gingival index) or at the base of the pocket (bleeding index).

Bleeding sites are scored as 2 or 3. FIGURE 5 shows the mean percentage of total sites that did not bleed, *i.e.,* score less than 2, for both indices. The effects of changes

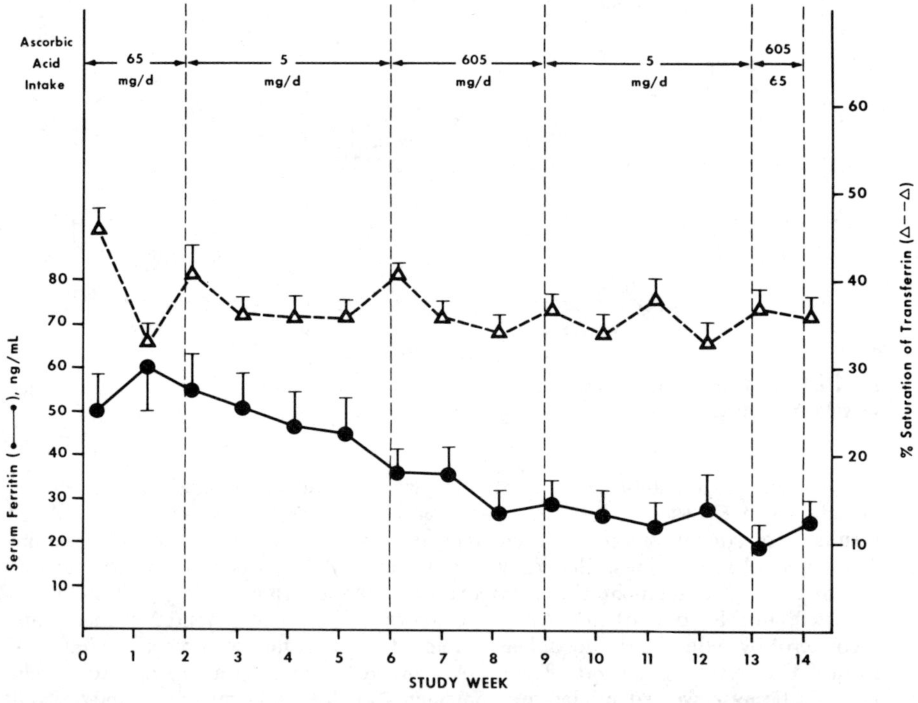

FIGURE 4. Mean + SE of serum ferritin and percent saturation of serum transferrin in 11 young men receiving varying AA intakes.

in AA intakes were similar for both bleeding indices. Compared to the baseline values, the percentage of nonbleeding sites decreased during the first depletion period, increased significantly ($p < 0.01$) during the AA supplementation, and decreased significantly during the second depletion period. The proportion of total sites with gingival index scores of 0, 1, and 2 or 3 at the end of each experimental period is shown in FIGURE 6. After the AA supplementation period the proportion of healthy sites increased to 30% while the proportion of bleeding sites decreased below 5%. After the second depletion period the proportion of bleeding sites was 60% and some 20% greater than at the end of the first depletion period. This latter result is consistent

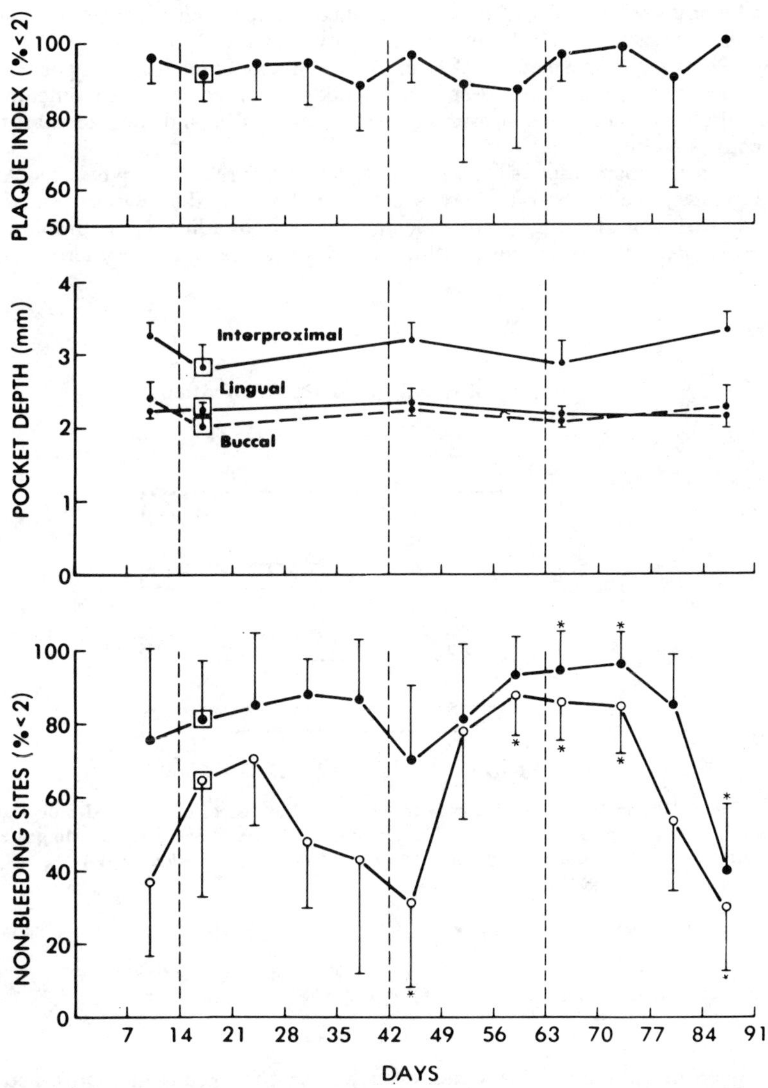

FIGURE 5. Oral indices (plaque index, pocket depth, gingival index [GI], and bleeding index [BI]) during periods of varying AA intakes (separated by *vertical dashed lines*) of 65, 5, 605, and 5 mg/d (left to right). *Asterisks* mark levels significantly ($p < 0.01$) different from the baseline levels shown inside squares. *Bars* indicate standard deviations. (From Leggott *et al.*[29] Reproduced by permission of the *Journal of Periodontology.*)

with a lower leukocyte AA level and AA body status at the end of the second depletion as compared to the first, *i.e.*, a conditioned deficiency effect.

Overall, the propensity of the gingiva to bleed, a measure of gingival inflammation, was found to be inversely related to AA intakes. The supplemental AA intake of 605 mg/day showed a benefit to gingival integrity over the RDA level intake of 65 mg/day. The lack of a benefit of AA intakes on periodontal health reported in some previous studies may have been due to poor study control, experimental design, nonobjective or insensitive measures of periodontal health, and/or inaccurate measures of body AA status.

The mechanism responsible for the observed direct relationship between AA intake and gingival health observed here is not clear, however the relatively rapid changes in sensitivity of the gingiva to inflammation and bleeding upon probing suggests changes in early inflammatory pathways and/or the periodontal microvasculature.

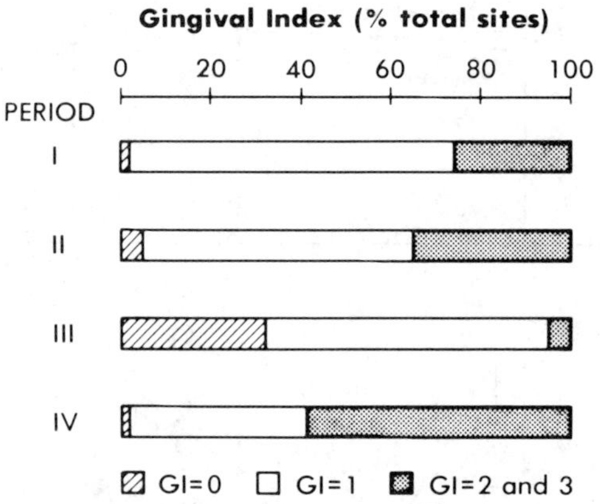

FIGURE 6. Proportion of total sites with gingival index scores of 0, 1, and 2 or 3 at the end of each study period. Scores of 0 and 1 are nonbleeding sites; 2 and 3 are bleeding sites. (From Leggott *et al.*[29] Reproduced by permission of the *Journal of Periodontology.*)

SUMMARY

Biochemical indices of AA clearly showed that the young men in this study were brought into various states of AA depletion and repletion according to their dietary AA intakes.

While previous studies have postulated that supplemental intakes of AA may adversely affect body status of vitamins B_6 and B_{12}, we found no changes in the B vitamin status of the young men receiving varying AA intakes. Moderate AA supplementation (605 mg/day) showed no antagonistic effect on markers of vitamins B_6 and B_{12}. Blood markers of fat-soluble vitamins A and E and iron status were not affected by AA intakes.

The propensity of the gingiva to become inflamed or bleed on probing was reduced after normal (65 mg/day) AA intakes as compared to deficient (5 mg/day) intakes and upon supplementary (605 mg/day) AA intakes as compared to normal intakes. The results suggest that AA status may influence early stages of gingival inflammation and crevicular bleeding, and warrant further study of the relationship between AA and periodontal health.

REFERENCES

1. SELIVANOVA, V. M., V. K. AGASHIN & I. N. POLYAKOVA. 1963. The effect produced by ascorbic acid on the 4-pyridoxic acid passage with urine in healthy humans. Vopr. Pitan. **22:** 55-57.
2. BAKER, E. M., J. E. CANHAM, W. T. NUNES, H. E. SAUBERLICH & M. E. MCDOWELL. 1964. Vitamin B_6 requirement for adult men. Am. J. Clin. Nutr. **15:** 59-66.
3. BAKER, E. M., R. E. HODGES, J. HOOD, H. E. SAUBERLICH, S. C. MARCH & J. E. CANHAM. 1971. Metabolism of ^{14}C and ^{3}H labeled L-ascorbic acid in human scurvy. Am. J. Clin. Nutr. **24:** 444-454.
4. SCHULTZ, T. D. & J. E. LEKLEM. 1982. Effect of high dose ascorbic acid on vitamin B_6 metabolism. Am. J. Clin. Nutr. **35:** 1400-1407.
5. COX, E. V. 1968. The anemia of scurvy. Vitam. Horm. **26:** 635.
6. HERBERT, V. & E. JACOB. 1974. Destruction of vitamin B_{12} by ascorbic acid. J. Am. Med. Assoc. **230:** 241-242.
7. HINES, J. D. 1975. Ascorbic acid and vitamin B_{12} deficiency. J. Am. Med. Assoc. **234:** 24.
8. COX, E. V., M. J. MEYNELL & B. E. NORTHAM. 1967. The anemia of scurvy. Am. J. Med. **42:** 220.
9. NEWMARK, H. L., J. M. SCHEINER, M. MARCUS & M. PRABHUDESAI. 1979. Ascorbic acid and vitamin B_{12}. J. Am. Med. Assoc. **242:** 2319-2320.
10. MARCUS, M., M. PRABHUDESAI & S. WASSEF. 1980. Stability of vitamin B_{12} in the presence of ascorbic acid in food and serum: Restoration by cyanide of apparent loss. Am. J. Clin. Nutr. **33:** 137-143.
11. EKVALL, S., I. W. CHEN & R. BOZIAN. 1981. The effect of supplemental ascorbic acid on serum vitamin B_{12} levels in myelomeningocele patients. Am. J. Clin. Nutr. **34:** 1356-1361.
12. WATSON, W. S., B. D. VALLANCE, M. M. MUIR & R. HUME. 1982. The effect of megadose ascorbic acid ingestion on the absorption and retention of vitamin B_{12} in man. Scott. Med. J. **27:** 240-243.
13. CHEN, L. H., M. S. LEE, W. F. HSING & S. H. CHEN. 1980. Effect of vitamin C on tissue antioxidant status of vitamin E deficient rats. Int. J. Vitam. Nutr. Res. **50:** 156-162.
14. PACKER, J. E., T. F. SLATER & R. L. WILSON. 1979. Direct observation of a free radical interaction between vitamin E and vitamin C. Nature **278:** 737-738.
15. CHEN, L. H. & H. M. CHANG. 1979. Effects of high level of vitamin C on tissue antioxidant status of guinea pigs. Int. J. Vitam. Nutr. Res. **49:** 87-91.
16. CHEN, L. H. 1981. An increase in vitamin E requirement induced by high supplementation of vitamin C in rats. Am. J. Clin. Nutr. **34:** 1036-1401.
17. BARCLAY, L. R. C., S. J. LOCKE & J. M. MACNEIL. 1983. The autoxidation of unsaturated lipids in micelles. Synergism of inhibitors vitamins C and E. Can. J. Chem. **61:** 1288-1290.
18. LEUNG, H. W., M. J. VANG & R. D. MAVIS. 1981. The cooperative interaction between vitamin E and C in suppression of peroxidation of membrane phospholipids. Biochim. Biophys. Acta **664:** 266-272.
19. GEORGE, T., N. J. BAI & S. KRISHNAMURTHY. 1978. Ascorbic acid effect on hypervitaminosis A in rats. Int. J. Vitam. Nutr. Res. **48:** 233-239.
20. LYNCH, S. R. & J. D. COOK. 1980. Interaction of vitamin C and iron. Ann. N. Y. Acad. Sci. **355:** 32-44.

21. VAN CAMPEN, D. & E. GROSS. 1968. Influence of ascorbic acid on the absorption of copper by rats. J. Nutr. **95:** 617-622.
22. MILNE, D. B. & S. T. OMAYE. 1980. Effect of vitamin C on copper and iron metabolism in the guinea pig. Int. J. Vitam. Nutr. Res. **50:** 301-308.
23. FINLEY, E. B. & F. L. CERKLEWSKI. 1983. Influence of ascorbic acid supplementation on copper status in young adult men. Am. J. Clin. Nutr. **37:** 553-556.
24. GRAY, G. E., A. PAGANINI-HILL & R. K. ROSS. 1983. Dietary intake and nutrient supplement use in a Southern California retirement community. Am. J. Clin. Nutr. **38:** 122-128.
25. SCHUTZ, H. G., M. READ, R. BENDEL, V. S. BHALLA, I. HARRILL, J. E. MONAGLE, E. T. SHEEHAN & B. R. STANDAL. 1982. Food supplement usage in seven western states. Am. J. Clin. Nutr. **36:** 897-901.
26. OMAYE, S. T., J. H. SKALA & R. A. JACOB. 1986. Plasma ascorbic acid in adult males: Effect of depletion and supplementation. Am. J. Clin. Nutr. **44:** 257-264.
27. JACOB, R. A., J. H. SKALA & S. T. OMAYE. 1987. Biochemical indices of human vitamin C status. Am. J. Clin. Nutr. In press.
28. WOOLFE, S. N., E. B. KENNEY & W. R. HUME. 1980. Ascorbic acid and periodontal disease: A review of the literature. J. West. Soc. Periodontol. **28:** 44.
29. LEGGOTT, P. J., P. B. ROBERTSON, D. L. ROTHMAN, P. A. MURRAY & R. A. JACOB. 1986. The effect of controlled ascorbic acid depletion and supplementation on periodontal health. J. Periodontol. **57:** 480-485.
30. LEGGOTT, P. J., P. B. ROBERTSON, D. L. ROTHMAN, P. A. MURRAY & R. A. JACOB. 1986. Response of lingual ascorbic acid test and salivary ascorbate levels to changes in ascorbic acid intake. J. Dent. Res. **65**(2): 131-134.
31. JACOB, R. A., S. T. OMAYE, J. H. SKALA & J. R. TURNLUND. The effect of vitamin C intakes on human copper absorption and status. J. Nutr. In press.
32. KUTNINK, M. A., J. H. SKALA, H. E. SAUBERLICH & S. T. OMAYE. 1985. Simultaneous determination of ascorbic acid, isoascorbic acid (erythorbic acid) and uric acid in human plasma by high performance liquid chromatography with amperometric detection. J. Liq. Chromatogr. **8:** 31-46.
33. OMAYE, S. T., J. P. TURNBULL & H. E. SAUBERLICH. 1979. Selected methods for the determination of ascorbic acid in animal cells, tissues, and fluids. *In* Methods in Enzymology, Vol. 62 (chap. 1). Academic Press. New York. pp. 3-11.
34. HODGKINSON, A. & A. WILLIAMS. 1972. An improved colorimetric procedure for urine oxalate. Clin. Chim. Acta **36:** 127-132.
35. SAUBERLICH, H. E. 1967. Vitamin B_6. *In* The Vitamins: Chemistry, Physiology, Pathology, Methods, Vol. VII. P. Gyorgy & W. N. Pearson, Eds.: 169-177. Academic Press. New York.
36. BIERI, J. G., T. J. TOLLIVER & G. L. CATIGNANI. 1979. Simultaneous determination of alpha-tocopherol and retinol in plasma or red cells by high pressure liquid chromatography. Am. J. Clin. Nutr. **32:** 2143-2149.
37. WHITE, J. M. & H. A. FLASHKA. 1973. An automated procedure, with use of ferrozine for assay of serum iron and total iron-binding capacity. Clin. Chem. **19:** 526-528.
38. RAMFJORD, F. P. 1959. Indices for prevalence and incidence of periodontal disease. J. Periodontol. **30:** 51.
39. GETTINGER, G., M. R. PATTERS, M. A. TESTA, *et al.* 1982. The use of six selected teeth in population measures of periodontal status. Periodontol. **54:** 155.
40. SILNESS, J. & H. LOE. 1964. Periodontal disease in pregnancy. II. Correlation between oral hygiene and periodontal condition. Acta Odontol. Scand. **22:** 121.
41. LOE, H. & T. SILNESS. 1963. Periodontal disease in pregnancy. I. Prevalence and severity. Acta Odontol. Scand. **21:** 533.
42. ARMITAGE, G. C., W. R. DICKINSON, R. J. JENDERSECK, *et al.* 1982. Relationship between the percentage of subgingival spirochetes and the severity of periodontal disease. J. Periodontal. **53:** 550.
43. TIETZ, N. W., ED. 1983. Clinical Guide to Laboratory Tests. W. B. Saunders. Philadelphia. p. 336.
44. MCMICHAEL, A. J. 1978. Kidney stone hospitalization in relation to change in vitamin C consumption in Australia 1966-1976. Community Health Studies **2:** 9-13.

45. SAUBERLICH, H. E. 1981. Ascorbic acid (vitamin C). *In* Symposium on Laboratory Assessment of Nutritional Status. Clinics in Laboratory Medicine. R. F. Labbe, Ed.: 1967. W. B. Saunders. Philadelphia.
46. TSAO, C. S. & S. L. SALIMI. 1984. Effect of large intakes of ascorbic acid on urinary and plasma oxalic acid levels. Int. J. Vitam. Nutr. Res. **54:** 245-249.
47. NIKI, E. 1987. Interaction of ascorbate and α-tocopherol. This volume.

DISCUSSION OF THE PAPER

B. LANE (*Columbia University School of Public Health, New York, N.Y.*): When you administered the 600-mg doses for the 3 weeks and then for the final week, how was that divided during the day?

R. A. JACOB (*USDA, San Francisco, Calif.*): The 600 mg was divided three ways, 200 mg in each meal. It was in grape juice drink they consumed with each meal. The 60 mg. was also divided three ways: 20 mg in each meal.

B. N. LADU (*University of Michigan Medical School, Ann Arbor, Mich.*): When Sandy Miller was at MIT they were trying to use buccal cells as an index of tissue ascorbic acid concentrations. With your interesting findings on the gingivitis relationship it might be interesting at some time to use buccal cell analyses. As you know, the plasma is a quick reflection of a dietary intake, and the white cells seem to take longer to establish a level. It would be interesting to know what the cells in that general area might show as an index of tissue concentration.

R. A. JACOB: I agree wholeheartedly. We're very interested in what buccal cells might show as markers of nutritional status for vitamin A, for instance. They're accessible and we're interested in noninvasive markers that are easy to analyze for. We looked at salivary ascorbate, and we didn't find an effect of the intakes of ascorbate on the salivary ascorbate levels. However, we took whole saliva, we didn't take parotid saliva which would probably be a better way to do that.

H. SPRINCE (*Jefferson Medical College, Philadelphia, Pa.*): In your drug interactions studies did you consider the interaction between ascorbic acid and riboflavin?

R. A. JACOB: No, we didn't.

B. TOLBERT (*University of Colorado, Boulder, Colo.*): It begins to look like there might be two separate measurements of nutritional status. One of them would be the immediate concentration of ascorbic acid available for things such as your plasma ascorbic acid measurements would indicate. The second one might be the body pool in tissues such as the brain which have been shown to have very long turnover times. The long-term time to scurvy would be affected by the deep pool, and the other one would be your immediate effects. Would you comment?

R. A. JACOB: I agree that a more informative or accurate measurement of nutritional status would be a tissue that would reflect the body pool. The leukocyte we think is a tissue that reflects the body pool closely. There is a lot of variation in plasma levels for several reasons, one of which is recent intake. If you wanted to pick out in a population those that are suffering from low intake, and in fact may have low body status, then plasma ascorbate should be a pretty good tool to do that because the variation of plasma ascorbate in those body pools and those plasma values would be consistently low.

B. TOLBERT: Yes, but I'm just a little bit worried about your concept that the leukocytes or the plasma ascorbic acid level reflect the body pool. For instance, with the leukocytes you showed that in a period of 4 weeks these things go down by a factor of two to three and a half and yet it seems a terribly short half-time to consider that the body pool would go down by a factor of two to three and a half in 4 weeks. I mean that's quite inconsistent with Baker's data and the rest of the data that suggests that this half-time is nearer to sixty days.

R. A. JACOB: Well it's well known that nutrients compartmentalize in various tissues and are refractory to depletion, to varying degrees. It's well known for instance for B_6 that it's very protected in the brain, though not so much in the muscle or liver. These leukocyte values could go down further than they are.

The Effects of Vitamin C Supplementation on Blood and Hair Levels of Cadmium, Lead, and Mercury

EDWARD J. CALABRESE, ANNE STODDARD, DENISE A. LEONARD, AND SALVATORE R. DINARDI

Division of Public Health
University of Massachusetts
Amherst, Massachusetts 01003

SUMMARY

Fifty-two adult male subjects were randomly assigned to one of three possible treatment groups: supplemental ascorbic acid at one of two levels, 500 mg or 1000 mg, or a placebo. Hair and blood samples were taken before vitamin C or placebo supplementation was started and at monthly intervals thereafter for three months. Samples were analyzed for cadmium, lead, and mercury. Results indicate that vitamin C did not significantly affect levels of these metals in either hair or blood samples.

INTRODUCTION

There is considerable debate concerning the capacity of vitamin C nutritional status to affect heavy metal toxicity and body burden. In both animal and human studies various investigators have found that ascorbic acid influences mineral metabolism and there is evidence it also may affect heavy metal exposure or may ameliorate heavy metal toxicity. For example, vitamin C therapy alone and in combination with iron reduced cadmium toxicity in both Coturnix (Japanese Quail) and rats.[1–6] In some animal models, vitamin C was shown to increase the body burden of mercury,[7,8] whereas in others it was found to have a beneficial effect.[9,10] Vitamin C and iron were found to be protective but not curative against lead toxicity.[11,12] Vitamin C has been reported as a possible chelator of lead, being as effective as EDTA.[13] Studies on human subjects provide conflicting results on the efficacy of vitamin C treatment for prevention and reduction of lead toxicity.[14–17]

This study was designed to determine the effect of ascorbic acid on heavy metal body burden. Body burden of heavy metals such as Cd, Pb, and Hg in a healthy adult population may be inferred by an assessment of blood and/or hair levels. Several studies have demonstrated that hair levels of heavy metals are a good indicator of environmental exposure,[18–30] and blood lead levels have long been used as an indicator of lead exposure.[31,32]

METHODOLOGY

Fifty-two (52) adult male subjects aged 29 to 64 were recruited for this study and randomly assigned to one of three groups: fructose placebo control, 500 mg vitamin C, and 1000 mg vitamin C. The subjects were blind as to which treatment group they were assigned. Subjects were given two vials of pills, one labeled "morning" and one labeled "night," and were instructed to take one pill twice a day, morning and night, from the respective vials. Subjects in the placebo group were given placebos for morning and night. Subjects in the 500 mg group were given 500 mg vitamin C in the morning and a fructose placebo at night, while those in the 1000 mg group were given vitamin C for both morning and night. Subjects were given enough pills for a 1-month period and then given the rest of their pills at the second data-collecting time. The vitamin C and placebo pills were prepared and supplied by Hoffmann-La Roche, Inc., Nutley, New Jersey.

Baseline data were collected immediately before subjects were started on the vitamin regimen. Samples collected for baseline data were 20 ml whole blood, 0.5 g hair (cut from the nape of the neck), and a urine specimen. Blood and urine specimens were collected three more times at 4-week intervals and hair was again collected at 3 months.

Urine samples were tested for vitamin C using the Ames C Stix (Ames Division, Miles Laboratories, Inc., Elkhart, Ind.). The dip stick uses a colorimetric test to indicate vitamin C levels in the urine. This was used to test subject compliance.

Lead content in blood was analyzed by a procedure adapted from the method of Subramanian *et al.*[33] on a Perkin Elmer Model 460 atomic absorption spectrophotometer with a UGA-2100 graphite furnace.

Instrument conditions were as follows:
Background corrector on
Integration time: 8 s
Interrupt mode on
Sample size: 10 μl
Drying temperature: 100°C
Drying time: 50 s
Charring temperature: 700°C
Charring time: 30 s
Atomization temperature: 2300°C
Atomization time: 8 s

Blood samples were collected in heparinized tubes suitable for trace mineral analysis. Samples were prepared for analysis in 1.5 ml Eppendorf tubes. Into each tube was placed the following: (1) 0.1 ml of a 5% Triton X-100 solution, (2) 0.6 ml of distilled, deionized water, (3) 0.2 ml of whole blood, and (4) 0.1 ml of a 5% $(NH_4)_2HPO_4$ solution. All samples were run in duplicate and well shaken before analysis.

Standards were made by dilution of a 1000 ppm stock (Sigma Chemical Company, St. Louis, Mo.) to 0.1 ppm and 0.05 ppm. The 1000 ppm lead stock was first diluted to 1 ppm in 0.5% HNO_3. The 0.1 ppm lead standard was prepared in a 10 ml volumetric flask by adding (1) 1.0 ml of 1 ppm lead stock, (2) 1 ml of a 5% Triton X-100 solution, and (3) 1 ml of a 5% $(NH_4)_2PO_4$ solution and bringing to volume with 0.5% HNO_3. The 0.05 ppm lead standard was made by a 1:2 dilution of the 0.1 ppm lead standard. These standards corresponded to 25.0 μg/dl (0.05 ppm lead standard) and 50.0 μg/dl (0.1 ppm lead standard) of lead in whole blood. This conversion accounts for the dilution of the blood in the sample preparation. A blank was prepared the same way as the blood samples with the substitution of distilled, deionized water for the whole blood.

Quality control was maintained by running controls obtained from the CDC Lead Proficiency Testing Program with each run of samples.

Cadmium and mercury levels in the blood were analyzed by Doctor's Data (West Chicago, Ill.). Five (5) ml of whole blood were mixed with 9.5 ml of nitric:perchloric acid (1:2) and digested on a hot plate until the perchloric acid fumed. The resulting volume was 2 ml and was diluted to 10 ml with distilled water. This sample was analyzed on a vacuum inductively coupled argon plasma emission spectrograph Model ARL 34000. Mercury levels were not quantitatively measured as the levels in blood were below the level of detection for the methodology and instrumentation used.

All three metal levels in hair were also analyzed by Doctor's Data on a vacuum inductively coupled argon plasma emission spectrograph Model ARL 34000. Samples were weighed, washed, and wet digested in nitric:perchloric acid with a final dilution for analysis of 1:25. Quality control was maintained by the use of internal standards and a spiked, pooled sample.

Data were transformed by a log transformation because sample values were skewed and variances were not equal. Analysis of Variance (ANOVA) was performed on the transformed data. Tukey's HSD test was used to determine individual treatment effects.

RESULTS

Of the 52 participants, 7 dropped out of the study principally because of complicating vacation schedules. However, two participants dropped out because of perceived health concerns.

Based on the qualitative analysis of vitamin C in the urine and the subjects' reports of pill usage, all participants in the vitamin C groups had complete compliance. The subjects in the 1000 mg/day group displayed 40 mg ascorbic acid/dl urine while those in the 500 mg/day group displayed between 20-40 mg/dl. The control or placebo group typically had less than 20 mg/dl.

TABLES 1 through 5 list the mean levels of each metal in hair and blood samples for each time period. For all the metals tested, there was no significant treatment effect for either the blood samples or the hair samples. There was no significant treatment-over-time effect except for the cadmium levels in hair, which significantly increased ($p = 0.0188$) over time. TABLE 2 lists the mean hair cadmium values; in all three treatment groups the mean level of Cd in hair from the three-month sample rose from baseline. Increase in mean hair Cd was greatest in the 1000 mg group and lowest in the 500 mg group.

There was a significant time effect in all groups except the hair mercury levels. Mean values of lead in hair increased over time, but only in the placebo and 1000 mg treatment group.

TABLE 1. Mean Values of Cadmium in Blood (ppm)

	Placebo	500 mg	1000 mg
Baseline	0.088 ± 0.028	0.071 ± 0.003	0.077 ± 0.020
1 Month	0.098 ± 0.019	0.096 ± 0.022	0.107 ± 0.028
2 Months	0.090 ± 0.015	0.094 ± 0.016	0.085 ± 0.012
3 Months	0.109 ± 0.018	0.113 ± 0.021	0.114 ± 0.020

TABLE 2. Mean Values of Cadmium in Hair (ppm)

	Placebo	500 mg	1000 mg
Baseline	0.506 ± 0.425	0.444 ± 0.421	0.378 ± 0.401
3 Months	1.015 ± 0.756	0.651 ± 0.700	0.969 ± 1.255

TABLE 3. Mean Values of Lead in Blood (ppm)

	Placebo	500 mg	1000 mg
Baseline	0.142 ± 0.571	0.120 ± 0.061	0.128 ± 0.056
1 Month	0.104 ± 0.393	0.090 ± 0.058	0.088 ± 0.055
2 Months	0.117 ± 0.346	0.101 ± 0.043	0.112 ± 0.043
3 Months	0.125 ± 0.697	0.096 ± 0.070	0.111 ± 0.068

TABLE 4. Mean Values of Lead in Hair (ppm)

	Placebo	500 mg	1000 mg
Baseline	12.400 ± 11.407	19.769 ± 47.912	10.875 ± 17.914
3 Months	18.133 ± 12.322	19.308 ± 30.322	21.313 ± 29.992

TABLE 5. Mean Values of Mercury in Hair (ppm)

	Placebo	500 mg	1000 mg
Baseline	1.247 ± 0.481	1.054 ± 0.380	1.219 ± 0.341
3 Months	0.847 ± 0.311	1.000 ± 0.248	1.050 ± 0.354

Analysis of the time effect on the results of the blood samples by Tukey's HSD test indicated no trends common to all the metals.

DISCUSSION

Megadoses of vitamin C have been shown to increase deposition of orally ingested mercury in the tissues of guinea pigs.[7,8] In a study by Chatterjee and Pal,[9] vitamin C supplementation did not have any beneficial effect on mercury intoxication in rats, although tissue levels of mercury were not measured so that it is not known whether tissue levels were affected due to ascorbic acid supplementation. Similar results were obtained by Hill [10] in chicks. In the current study, there was no statistically significant increase of mercury in hair in the vitamin-C-dosed groups. If hair mercury levels are a good indicator of tissue Hg levels, this finding suggests that in humans ascorbic acid supplementation of up to 1000 mg per day did not cause an increase in mercury concentration in tissues.

In the study by Murray and Hughes[8] a short exposure period was used and guinea pigs were sacrificed after five days. Similarly, the exposure period was 15 days in the Blackstone *et al.*[7] study. In addition, in both the latter studies, the guinea pigs were challenged orally with mercury in levels which would be in excess of environmental and occupational exposure levels. It is not possible to resolve the apparent differences between the Murray and Hughes[8] and Blackstone[7] reports with the present study. Possible differences in the site of biomonitoring (hair vs. tissue) and level and duration of exposure contributed to the divergent findings as well as possible interspecies variation.

Studies on cadmium toxicity in young Coturnix (Japanese Quail)[2–5] and rats[1,6] have indicated that a combination treatment of ascorbic acid and iron reduced cadmium toxicity. In addition, studies have shown that supplemental dietary ascorbic acid alone had both a curative and preventative effect on cadmium toxicity in young Coturnix.[2,4,5] The present study found no significant change in cadmium levels in either hair or blood due to vitamin C supplementation.

There were no significant changes in blood or hair lead due to ascorbic acid supplementation. In contrast, Sohler *et al.*[14] found that a therapeutic regimen of zinc and ascorbic acid significantly decreased the blood lead levels in psychiatric outpatients. However, these findings are not in agreement with Lauwerys *et al.*,[15] who gave occupationally exposed workers supplements of zinc and vitamin C and found no effect on blood lead levels. It should be emphasized that both the Lauwerys *et al.*[15] report and the present study utilized external controls as well as having proper baseline data, and as such offered a stronger study design than the earlier report of Sohler *et al.*[14]

CONCLUSION

The present findings indicate that dietary supplementation of ascorbic acid up to 1000 mg/kg for three months did not affect the body burdens of three heavy metals—Pb, Cd, and Hg—in a nonoccupationally exposed population of healthy adult Caucasian males using blood and hair as indicators of exposure.

REFERENCES

1. SUZUKI, I. & A. YOSIDA. 1978. Long-term effectiveness of dietary iron and ascorbic acid in the prevention and cure of cadmium toxicity in rats. Am. J. Clin. Nutr. **31:** 1491-1498.
2. FOX, M. R. S., B. E. FRY, JR., B. F. HARLAND, M. D. SCHERTEL & C. E. WEEKS. 1971. Effect of ascorbic acid on cadmium toxicity in the young Coturnix. J. Nutr. **101:** 1295-1306.
3. FOX, M. R. S., R. M. JACOBS, A. O. L. JONES, B. E. FRY, JR. & C. L. STONE. 1980. Effects of vitamin C and iron on cadmium metabolism. Ann. N.Y. Acad. Sci. **355:** 249-261.
4. FOX, M. R. S. 1975. Protective effects of ascorbic acid against toxicity of heavy metals. Ann. N.Y. Acad. Sci. **258:** 144-150.
5. RICHARDSON, M. E., M. R. S. FOX & B. E. FRY, JR. 1974. Pathological changes produced in Japanese quail by ingestion of cadmium. J. Nutr. **104:** 323-338.
6. MAJI, T. & A. YOSHIDA. 1974. Therapeutic effect of dietary iron and ascorbic acid on cadmium toxicity of rats. Nutr. Rep. Int. **10**(3): 139-149.
7. BLACKSTONE, S., R. J. HYRLEY & R. E. HUGHES. 1974. Some interrelationships between vitamin C (L-ascorbic acid) and mercury in the guinea pig. Food Cosmet. Toxicol. **12:** 511-516.
8. MURRAY, D. R. & R. E. HUGHES. 1976. The influence of dietary ascorbic acid on the concentration of mercury in guinea pig tissues. Proc. Nutr. Soc. **35**(3): 118A-119A.
9. CHATTERJEE, G. C. & D. R. PAL. 1975. Metabolism of L-ascorbic acid in rats under in vivo administration of mercury: Effect of L-ascorbic acid supplementation. Int. J. Vitam. Nutr. Res. **45:** 284-292.
10. HILL, C. H. 1980. Interactions of vitamin C with lead and mercury. Ann. N.Y. Acad. Sci. **355:** 262-265.
11. SUZUKI, T. & A. YOSHIDA. 1979. Effect of dietary supplementation of iron and ascorbic acid on lead toxicity in rats. J. Nutr. **109:** 982-988.
12. SUZUKI, T. & A. YOSHIDA. 1979. Effectiveness of dietary iron and ascorbic acid in the prevention and cure of moderately long-term lead toxicity in rats. J. Nutr. **109:** 1974-1978.
13. GOYER, R. A. & M. G. CHERIAN. 1979. Ascorbic acid and EDTA treatment of lead toxicity in rats. Life Sci. **24:** 433-438.
14. SOHLER, A., M. KRUESI & C. C. PFEIFFER. 1977. Blood lead levels in psychiatric outpatients reduced by zinc and vitamin C. J. Othomol. Psych. **6**(3): 1-5.
15. LAUWERYS, R., H. RUELS, J.-P. BUCKET, A. A. BERNARD. L. VERHOEVEN & J. KONINGS. 1983. The influence of orally-administered vitamin C on the absorption of and biological response to lead. J. Occup. Med. **25**(9): 668-678.
16. HOLMES, H. N., K. CAMPBELL & E. J. AMBERG. 1939. The effect of vitamin C on lead poisoning. J. Lab. Clin. Med. **24:** 1119-1127.
17. MARCHMONT-ROBINSON, S. W. 1940. Effect of vitamin C on workers exposed to lead dust. J. Lab. Clin. Med. **26:** 1478-1481.
18. CHATTOPADHYAY, A., T. M. ROBERTS & R. E. JERVIS. 1977. Scalp hair as a monitor of community exposure to lead. Arch. Environ. Health. pp. 226-236.
19. CORRIDAN, J. P. 1974. Head hair samples as indicators of environmental pollution. Environ. Res. **8:** 12-16.
20. CREASON, J. P., T. A. HINNER, J. E. BUMGARNER & C. PINKERTON. 1975. Trace elements in hair, as related to exposure in metropolitan New York. Clin. Chem. **21:** 603-612.
21. HAMBRIDGE, K. M. 1982. Hair analysis: Worthless for vitamins, limited for minerals. Am. J. Clin. Nutr. **36:** 943-949.
22. HAMMER, D. I., J. F. FINKLEA, R. H. HENDRICKS, C. M. SHY & R. J. M. HORTON. 1971. Hair trace metal levels and environmental exposure. Am. J. Epidemiol. **93**(2): 84-92.
23. HANSEN, J. C., H. C. WULF, N. KROMANN & K. ALBOGE. 1983. Human exposure to heavy metals in East Greenfield. I. Mercury. Sci. Total Environ. **26:** 233-243.
24. HEFFEREN, J. J. 1976. Usefulness of chemical analysis of head hair for exposure to mercury. J. Am. Dent. Assoc. **92:** 1213-1216.
25. HOPPS, H. C. 1977. The biologic bases for using hair and nail for analysis of trace elements. Sci. Total Environ. **7:** 71-89.

26. KOPITO, L., A. M. BRILEY & H. SCHWACHMAN. 1969. Chronic plumbism in children. J. Am. Med. Assoc. **209**(2): 243-248.
27. KOPITO, L., R. BYERS & H. SCHWACHMAN. 1967. Lead in hair of children with chronic lead poisoning. N. Engl. J. Med. **276**(17): 949-953.
28. OLERU, U. G. 1975. Epidemiological implication of environmental cadmium. I. The probable utility of human hair for occupational trace metal (cadmium) screening. Am. Ind. Hyg. Assoc. J. pp. 229-233 (March).
29. PUESCHEL, S. M., L. KOPITO & H. SCHWACHMAN. 1972. Children with an increased lead burden: A screening and follow-up study. J. Am. Med. Assoc. **222**(4): 462-466.
30. HOWIE, R. A. & H. SMITH. 1967. Mercury in human tissue. Forensic Sci. **7**: 90-96.
31. CHISOLM, J. J. & D. M. O'HARA. 1982. Lead Absorption in Children: Management, Clinical and Environmental Aspects. Urban & Schwarzenberg. Baltimore.
32. WALDBOTT, G. L. 1973. Health Effects of Environmental Pollutants. Mosby. St. Louis.
33. SUBRAMANIAN, K. S. & J. C. MERANGER. 1981. A rapid electrothermal atomic absorption spectrophotometric method for cadmium and lead in human whole blood. Clin. Chem. **27**: 1866-1871.

DISCUSSION OF THE PAPER

B. LANE (*Columbia University School of Public Health, New York, N.Y.*): I see by your data that the baseline values for mercury in hair were 1.2, 1.0, and 1.2 ppm for the three groups that were studied. This suggests that these people were not eating much large fish since persons eating tuna fish twice a week for a year or more will show values well above 2 parts per million and if they've been eating fish three or more times a week will show 3-7 parts per million. It would be interesting to see the effect of ascorbic acid on a population that are consuming large fish and see if the ascorbic acid doesn't make a meaningful difference then.

E. J. CALABRESE (*University of Massachusetts, Amherst, Mass.*): I think you make a good point that one of the limitations of our study here is the nature of the population. We were disappointed from an experimentalist's point of view that the inherent backgrounds in terms of lead was as low as it was and as you suggest the mercury may also not be as high as we would have liked for such a study.

A. BENDICH (*Hoffmann-LaRoche, Nutley, N.J.*): Are we in fact looking at data in your study on normal levels of lead, mercury, and cadmium and the effect of vitamin C maintaining normal levels? I might also comment that in reviewing the literature recently it came to my attention that in numerous studies where high mercury and lead levels were found in children and adults, when 200 mg of vitamin C or more was taken with chelating agents, body lead burdens were lowered.

E. J. CALABRESE: You're right, it was a population that had what appeared to have "normal" levels of lead, mercury, and cadmium.

Inhibition of Nitrosamine Formation by Ascorbic Acid[a]

STEVEN R. TANNENBAUM AND JOHN S. WISHNOK

Department of Applied Biological Sciences
Massachusetts Institute of Technology
Cambridge, Massachusetts 02139

INTRODUCTION

Nitrosation is now known to occur under a wide variety of conditions by reaction of virtually all types of amines with any of a large number of nitrosating species.[1–7] The latter includes a group of equilibrated nitrosating agents arising from nitrite ion (NO_2^-) under conditions commonly found in the human stomach.[8] Nitrite can be formed *in vivo* via bacterial reduction of nitrate[9] and by the biochemical effects of activated macrophages,[10] and is therefore typically present in the stomach even with low-nitrate diets. Nitrosatable amines are found in many types of foods and some, *e.g.*, dimethylamine, are synthesized in the human body.[11] Precursors of *N*-nitroso compounds are thus almost constantly present together in the body under favorable reaction conditions and there is, consequently, considerable interest in the question of whether or not there is a human health risk arising from exposure to this class of compounds via endogenous formation.

The assumption, that this risk is indeed real, suggests that research dealing with endogenous nitrosation should probably be complemented with research into safe methods of inhibiting this process.

Although many nitrosation inhibitors are known, attention in this area has often focused on ascorbic acid, which is known to react with many nitrosating agents[12,13] and which has been long known to be virtually nontoxic to humans. This paper will review the chemistry of ascorbic acid inhibition of nitrosation reactions, the background of its application as a nitrosation inhibitor in man, and the clinical and epidemiological evidence for its potential importance in health intervention.

The effect of ascorbic acid on the nitrosation of an amine is conceptually very simple: ascorbic acid reacts faster than the amine with the nitrosating agent and the amine is therefore not measurably nitrosated. The reality, however, is not so simple, especially in a reactor as complex as the human stomach. There is not, first of all, a single nitrosating agent. Nitrite, in the presence of protons, gives rise to at least two nitrosating species, and in the presence of halogens, or pseudo-halogens such as thiocyanate (SCN^-), can yield an additional nitrosating species for each. All of these are in equilibrium as shown schematically below (SCHEME 1), with the halogens and pseudohalogens designated with an "X"; the known nitrosating agents are shown in italics.

[a]This investigation was supported by PHS grant no. P01-CA26731, awarded by the National Cancer Institute, DHHS.

$$NO_2^- + H^+ \rightleftarrows HNO_2$$
$$HNO_2 + H^+ \rightleftarrows \mathit{H_2NO_2^+} \rightleftarrows H_2O + \mathit{NO^+}$$
$$2HNO_2 \rightleftarrows \mathit{N_2O_3} + H_2O$$
$$HNO_2 + HX \rightleftarrows \mathit{NOX} + H_2O$$

SCHEME 1

The relative concentrations of the various nitrosating species depend on the pH and on the concentrations of halogens and pseudohalogens. In general, for a typical dialkylamine in a straightforward chemical situation, the maximum initial rate for nitrosation will occur at about pH 3.4, resulting from a combination of increased concentrations of nitrosating species, and decreased concentration of the free amine (the protonated amine does not react; see ref. 14) as the pH is lowered. Catalysts such as thiocyanate generally shift the position of the pH maximum and cause the rate to fall off more slowly as the pH decreases.

A detailed treatment of the intervention of ascorbic acid in nitrosation must therefore consider the interaction of ascorbic acid with each of these nitrosating species along with the competing reactions of the amine being studied. The situation is further complicated by the fact that ascorbic acid is in a pH-dependent equilibrium with the ascorbate anion, and finally, by the reaction of ascorbic acid with oxygen.

CHEMISTRY OF ASCORBIC ACID

Some relevant aspects of this topic have been discussed in greater detail in an earlier report,[15] but a summary is appropriate here. Ascorbic acid, under anaerobic conditions, can usually react with N_2O_3, $H_2NO_2^+$, and NOX with rates higher in each case than the corresponding nitrosation rates for amides or dialkyl amines. Ascorbic acid can therefore generally inhibit the *in vitro* nitrosation of these classes of compounds.[12,13] This property has been exploited to prevent nitrosamine formation in foods, and to some extent, to inhibit *in vivo* nitrosation.[16,17]

Ascorbic acid thus appears to have potential importance as an *in vivo* nitrite scavenger. This potential, however, has yet to be completely realized; not only are the reactions that occur in live organisms more complex than might have been expected, but the *in vitro* model systems themselves, as noted earlier, are not straightforward. The equilibria shown in SCHEME 1, for example, involve several nitrosating species that can react with ascorbic acid to form NO, which does not nitrosate amines. Under anaerobic conditions these reactions can then exhaust the nitrosating capacity of the system. Oxygen, however, can react with NO (as shown in SCHEME 2) to form N_2O_3 and N_2O_4, both of which are capable of nitrosation.

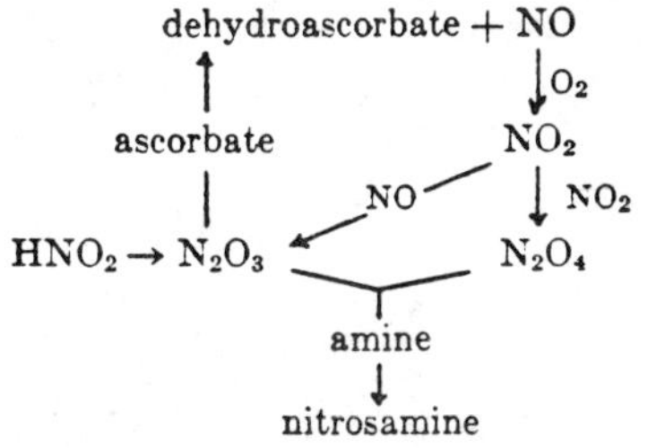

SCHEME 2

Under these circumstances, the ascorbic acid may be effectively removed from the system without significantly affecting the concentration of nitrosating species. In addition to these processes, ascorbic acid itself can also react directly with oxygen, undergoing conversion to dehydroascorbate.[18,19] FIGURE 1 illustrates some reactions of ascorbic acid with nitrite under aerobic and anaerobic conditions, and demonstrates that a combination of oxygen and nitrite can completely exhaust the ascorbic acid; nitrite alone reacts stoichiometrically.

EXPOSURE TO NITROSATING AGENTS

FIGURE 2 summarizes some of the interlocking pathways of human exposure to nitrosating agents. The bacterial reduction of nitrate to nitrite in saliva in the hypoclorhydric stomach and in the infected bladder are well-documented phenomena that have been adequately described in the past.

The major entry point for nitrate and nitrite into the body is food and water, except in cases of infection or inflammation. This latter situation has been shown to involve activated macrophages,[10] and most importantly, to lead to the direct formation of nitrite. A limited number of quantitative studies on infected people has indicated that hundreds of millimoles of nitrite might be formed in an individual in a single day (ref. 20 and unpublished observations). We have, additionally, recently presented evidence that some of this nitrite can be converted to nitrosamines.[21] The contribution of this process to the net endogenous synthesis of *N*-nitroso compounds is unknown. There is also a background level of endogenous nitrate formation, comprising perhaps as much as 2000 μmol of nitrate and/or nitrite per day.[22,23] This source of nitrate and/or nitrite may involve an oxidative process unrelated to the stimulation of macrophages, but the mechanism is currently not well understood.

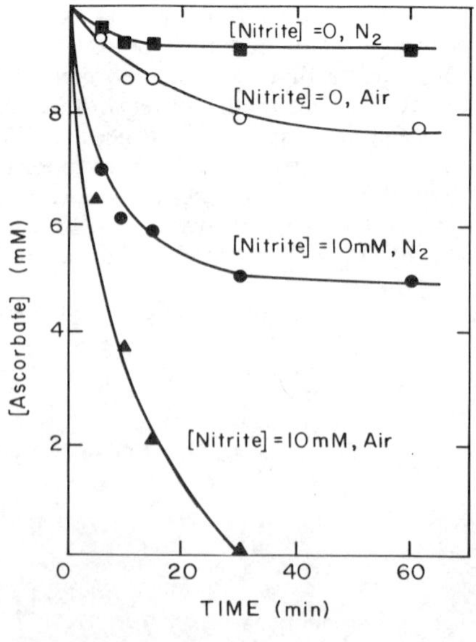

FIGURE 1. Reaction of ascorbic acid with nitrite and air at pH 4 (25°C). (From Garland *et al.*[24])

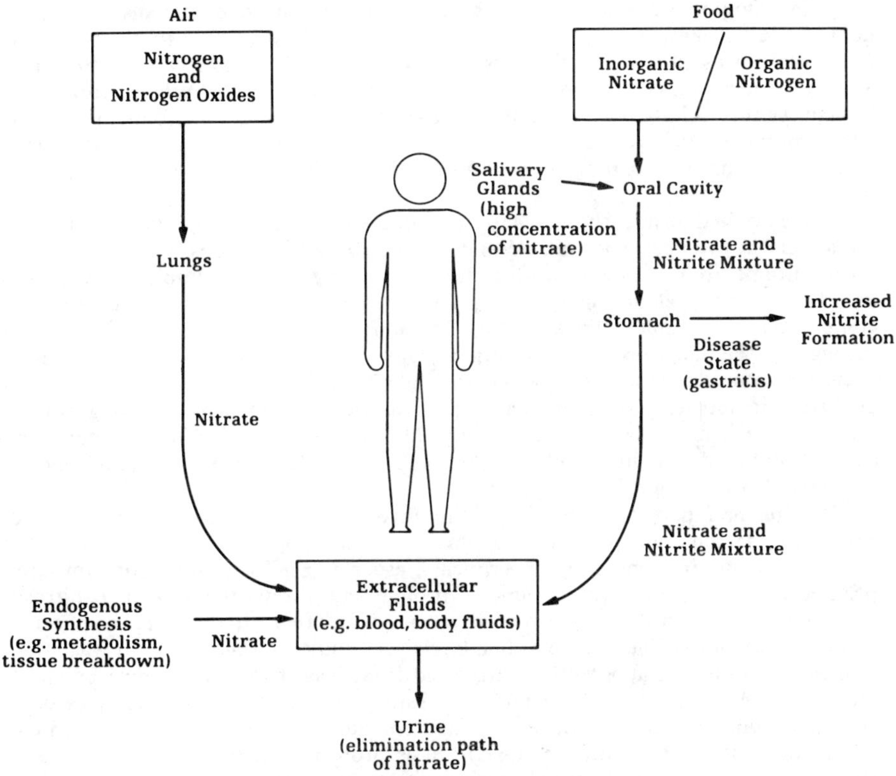

FIGURE 2. Nitrate distribution in humans.

Another potentially important, but also poorly understood, mechanism of exposure to nitrosating agents is via nitrogen oxides in air. This may be particularly significant for individuals who are exposed to combustion gases or polluted air. Although estimations of the contributions of this source to the body burden of nitrosating agents suggest that it is small relative to nitrate and nitrite exposure, there is evidence that it may be directly related to nitrosamine exposure as shown recently by Garland and coworkers,[24] who found a positive correlation between atmospheric NO_2 levels and urinary excretion of *N*-nitrosodimethylamine over a period of time.

EXPOSURE TO ENDOGENOUSLY FORMED *N*-NITROSO COMPOUNDS

Ohshima and Bartsch, in 1981, developed the "nitrosoproline test" for endogenous nitrosation, in which proline and/or nitrate can be given to humans and the resulting *N*-nitrosoproline can be detected in the urine.[25] *N*-nitrosoproline is not metabolized in humans and is consequently nontoxic and noncarcinogenic. Proline is a naturally occurring amino acid and can thus be administered safely to humans. Nitrate is nontoxic to humans, but can be reduced endogenously by bacteria to nitrite. Levels of urinary nitrosoproline are then assumed to be related to endogenous nitrosation, and experiments based on this test have been carried out in many laboratories. Although the results have often been difficult to interpret because of confounding factors, one immutable conclusion has emerged, *i.e.,* virtually all humans tested thus far have been found to have *N*-nitrosoproline in their urine, and it is undoubtedly of endogenous origin.

In the original form of the test, and the one subject to the least number of variables, the system is swamped with nitrate and proline so that variations in the amount of nitrosoproline are dependent on other factors. Among the most important variables are the temporal relationships among the reactants and catalysts in the reaction compartment (presumably the stomach). The administered nitrate must be distributed throughout the body, taken up into saliva, converted to nitrite, and swallowed, prior to the reaction. Individual differences in rates of nitrate distribution, rates of conversion of nitrate to nitrite, and other factors such as rates of stomach emptying will be important determinants of nitrosoproline yield. These and additional parameters, *e.g.,* concentrations of inhibitors and catalysts, will finally determine the overall amount of excreted *N*-nitrosoproline.

Despite the inherent complexity of the nitrosoproline test, it has been used successfully to demonstrate partial *in vivo* inhibition of endogenous nitrosation by ascorbic acid in humans. In some early experiments, ascorbic acid was simply administered prior to or concurrently with proline or proline and nitrate; the yield of *N*-nitrosoprolone was lower in the presence of vitamin C.[26] In other experiments, the effect of ascorbic acid on baseline nitrosoproline levels was studied. In these cases it was found that nitrosoproline is apparently synthesized in at least two physiological compartments, one that utilizes an endogenous nitrosating agent and is unaffected by ascorbic acid, and another that can utilize exogenous nitrite and in which ascorbic acid can inhibit nitrosation. The latter compartment is probably the stomach; the nature of the former is unknown. TABLE 1 summarizes an experiment of this type in which ^{15}N-labeled nitrate was given either along with or in the absence of ascorbic acid, and the resulting nitrosoproline was analyzed by mass spectrometry for the incorporation of the labeled nitrogen. There is complete inhibition of proline nitrosation from nitrosating species arising from the nitrate, but no inhibition of nitrosation by other nitrosating agents.[20]

In a second form of the test, currently being used in some epidemiological studies,[27] individual capacity for nitrosation of amines, including exogenous proline, is tested under loading conditions of proline and vitamin C. The other amines, including

TABLE 1. Incorporation of [^{15}N]Nitrate into Nitrosoproline: Effect of Ascorbic Acid[a]

	% Incorporation of ^{15}N into Nitrosoproline	
Subject	Low C Diet	High C Diet
1	13	6
2	20	1
3	52	0
4	30	17
5	24	1
6	47	0

[a] Adapted from Wagner *et al.*[26]

TABLE 2. Urinary Nitrosoproline in Lin-Xian County[a]

Subjects	*n*	Nitrosoproline[d]	Total NAA[c, d]
Undosed	44	5.7	21.2
Proline (3 × 100 mg)[b]	50	8.3	18.0
Proline + C (3 × 100 mg)[b]	48	2.4	6.6

[a] From Lu *et al.*[27]
[b] Proline and C dosed 1 hour after meals.
[c] NAA = Nitrosoamino acids.
[d] Nitrosoproline and NAA measured as μg per person per day.

thiazolidine carboxylic acid, are endogenous or ingested as part of typical diets. This form of the test is of course subject to all of the variables noted above for the original form. TABLE 2 lists some results from this study. Note that some nitrosoproline, as well as other nitrosoamino acids, is excreted even in the absence of exogenous proline. Added proline leads to increased nitrosoproline excretion, and ascorbic acid administered concurrently with the proline leads to a marked decrease in nitrosoproline excretion. The objective of this series of experiments was to determine if people living in an area of increased cancer risk showed higher rates of endogenous nitrosation than did people from areas of normal cancer risk. There did appear to be some correlation of risk with nitrosating ability (data not shown). The experiments, however, may be somewhat inconclusive since the molar ratio of ascorbic acid to proline was only 0.7; Leaf *et al.*[28] have shown that the effectiveness of ascorbic acid as an *in vivo* nitrosation inhibitor is as much as tenfold lower than would be predicted from *in vitro* anaerobic experiments and that at least a twofold molar excess of ascorbic acid over proline is required to reduce endogenously formed nitrosoproline to near-baseline levels.

Despite these considerations concerning the details of the effect, however, the overall conclusion arising from these experiments is that vitamin C can at least partially inhibit gastric nitrosation of proline (and presumably other amines) in humans.

The ultimate question, however, remains whether or not endogenous nitrosation represents a human health risk and whether or not this risk can be significantly lowered by vitamin C. There is currently no direct answer to either aspect of this question. The uncertainty arises largely from the facts that human exposure to nitrosamines, whether endogenous or exogenous, generally involves chronic low-dose situations and is superimposed on countless other factors (*e.g.*, exposure to other types of toxic substances or to modifiers of metabolism) that may enhance, inhibit, or mask the effects of the *N*-nitroso compounds. *N*-nitroso compounds have, nonetheless, been implicated in several human epidemiological situations, including elevated risk toward gastric cancer in some well-defined geographical areas. FIGURE 3 shows schematically some of the characteristics of gastric cancer, along with some etiological hypotheses, based on available epidemiological and biochemical evidence. In virtually all cases, a well-defined progressive change in stomach physiology and morphology from chronic gastritis through atrophic gastric to apparently precancerous intestinal metaplasia is observed. As the stomach begins increasingly to resemble the intestine, the pH rises, leading to conditions that support or favor increased bacterial growth. This in turn leads to higher levels of gastric nitrate from reduction of nitrate by some of these bacteria. This could conceivably result in increased formation of *N*-nitroso compounds from amines present in the stomach either via the diet or from endogenous synthesis. If this is indeed the case, then it is also conceivable that the process could be interrupted at this point by the ingestion of ascorbic acid.

As noted earlier in this report, however, the interactions of ascorbic acid with nitrosating agents in the stomach are far from straightforward, and the clinical effectiveness of vitamin C in reducing a real human cancer risk still remains to be demonstrated.

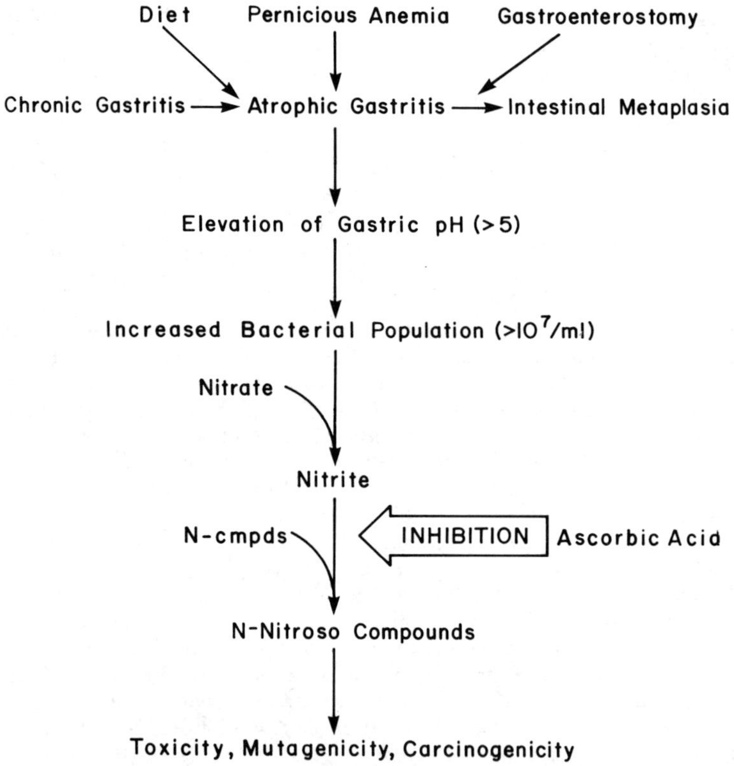

FIGURE 3. Model for gastric cancer etiology.

REFERENCES

1. WISHNOK, J. S. 1977. J. Chem. Educ. **54:** 440-442.
2. SANDER, J. & G. BURKLE. 1969. Z. Krebsforsch. **73:** 54-66.
3. GREENBLATT, M., S. S. MIRVISH & B. T. SO. 1971. J. Natl. Cancer Inst. **46:** 1029-1034.
4. MYSLIWY, T. S., E. L. WICK, M. C. ARCHER, R. C. SHANK & P. M. NEWBERNE. 1974. Br. J. Cancer **30:** 279-283.
5. TANNENBAUM, S. R., A. J. SINSKEY, M. WEISMAN & W. W. BISHOP. 1974. J. Natl. Cancer Inst. **53:** 79-84.
6. TANNENBAUM, S. R., M. C. ARCHER, J. S. WISHNOK & W. W. BISHOP. 1978. J. Natl. Cancer Inst. **60:** 251-253.

7. Tannenbaum, S. R., D. Fett, V. R. Young, P. D. Land & W. R. Bruce. 1978. Science **200:** 1487-1489.
8. Mirvish, S. S. 1975. Toxicol. Appl. Pharmacol. **31:** 325-351.
9. Tannenbaum, S. R., M. Weisman & D. Fett. 1976. Food Cosmet. Toxicol. **14:** 549.
10. Stuehr, D. J. & M. A. Marletta. 1985. Proc. Natl. Acad. Sci. U.S.A. **82:** 7738-7742.
11. Zeisel, S. H., K.-A. DaCosta & J. G. Fox. 1985. Biochem. J. **232:** 403-408.
12. Mirvish, S. S., L. Wallcave, M. Eagen & P. Shubik. 1972. Science **177:** 65-68.
13. Archer, M. C., S. R. Tannenbaum, T. Y. Fan & M. J. Weisman. 1975. J. Natl. Cancer Inst. **54**(5): 1203-1205.
14. Fan, T. Y. & S. R. Tannenbaum. 1973. J. Agric. Food Chem. **21:** 237-240.
15. Kim, Y.-K., S. R. Tannenbaum & J. S. Wishnok. 1982. *In* Ascorbic Acid: Chemistry, Metabolism and Uses. P. A. Seib & B. M. Tolbert, Eds.: 571-585. American Chemical Society. Washington, D.C.
16. Greenblatt, M. 1973. J. Natl. Cancer Inst. **50:** 1055-1056.
17. Fiddler, W., J. W. Pensabene, E. G. Piotrowski, R. C. Doerr & A. E. Wasserman. 1973. J. Food Sci. **38:** 1084-1085.
18. Fennema, O. R. 1976. Principles of Food Science, Part 1: Food Chemistry. Marcel Dekker. New York. pp. 360-364.
19. Bunton, C. A., H. Dahn & L. Loewe. 1959. Nature. **183:** 163-165.
20. Wagner, D. A., D. E. G. Shuker, C. Bilmazes, M. Obiedzinski, I. Baker, V. R. Young & S. R. Tannenbaum. 1985. Cancer Res. **45:** 6519-6522.
21. Miwa, M., D. J. Stuehr, M. A. Marletta, J. S. Wishnok & S. R. Tannenbaum. 1987. Carcinogenesis. In press.
22. Dull, B. J. & J. H. Hotchkiss. 1984. Carcinogenesis **5:** 1161-1164.
23. Saul, R. L. & M. C. Archer. 1984. Carcinogenesis **5:** 77-81.
24. Garland, W. A., W. Keunzig, F. Rubio, E. P. Norkus & A. H. Conney. 1986. Cancer Res. **46:** 5392-5400.
25. Ohshima, H. & H. Bartsch. 1981. Cancer Res. **41:** 3658-3662.
26. Wagner, D. A., D. E. G. Shuker, C. Bilmazes, M. Obiedzinski, V. R. Young & S. R. Tannenbaum. 1985. *In* N-Nitroso Compounds: Occurrence, Biological Effects and Relevance to Human Cancer. I. K. O'Neill, R. C. von Borstel, J. E. Long, C. T. Miller & H. Bartsch, Eds.: 223-230. International Agency for Research on Cancer. Lyon, France.
27. Lu, S.-H., H. Ohshima, H.-M. Fu, Y. Tian, F.-M. Li, M. Blettner, J. Wahrendorf & H. Bartsch. 1986. Cancer Res. **46:** 1485-1491.
28. Leaf, C. D., A. J. Vecchio, D. A. Roe & J. H. Hotchkiss. 1987. Ninth International Symposium on N-Nitroso Compounds: Relevance to Human Cancer. International Agency for Research on Cancer. Lyon, France. In press.

DISCUSSION OF THE PAPER

M. Levine (*National Institutes of Health, Bethesda, Md.)*: Could you comment on the specificity of ascorbic acid in inhibiting both nitrosamine formation, the reaction you showed at the beginning, and also perhaps in inhibiting nitrosamine or nitrated compound formation in your macrophage models?

S. Tannenbaum (*Massachusetts Institute of Technology, Cambridge, Mass.*): In the case of gastric endogenous synthesis any substance that would effectively compete for the nitrosating agent would act as an inhibitor. Now another class of compounds that's been used effectively has been some of the phenols, particularly some of the substituted phenols. But the situation there can be very complex because none of them actually are catalysts and the role is not very clear. However, it is true that the role of vitamin C in that situation is not its role as a vitamin, but its role as an extremely good reactant for the nitrosating agent.

M. LEVINE: So agents with similar redox potentials then should be expected to have similar effects?

S. TANNENBAUM: It is not simply a redox effect, it is the ability of ascorbic acid and its anion to react effectively with N_2O_3 or with other various forms of nitrosating agents. It's not a redox effect, it's actually a reaction pathway effect.

In the case of the macrophages, one of the compounds that we were very interested in looking at was uric acid, because that also would have the capacity of acting in a similar fashion. In fact it does not. With the very limited amount of data we have, we think ascorbic acid is playing a very specific role. And I think that from other work that's been presented at this meeting it's clear that the active transport system, which makes it possible to get elevated levels of vitamin C and vitamin E into cells, is obviously playing an important role.

E. J. DILIBERTO (*Wellcome Research Labs, Research Triangle Park, N.C.*): Are nitroso compounds themselves catalysts for formation of nitroso compounds? And what are some of the other catalysts that you referred to very briefly?

S. TANNENBAUM: Nitroso compounds are not. This is not an autocatalytic process. Another catalyst is, for example, thiocyanate, which is an important anion in gastric juice. It depends on the system. Compared to water, chloride would be a catalyst. There are more potent nitrosating agents than N_2O_3. Under proper conditions citrate could be a catalyst. Catalysts are generally things which form a more reactive nitrosating species.

S. L. ROMNEY (*Albert Einstein College of Medicine, Bronx, N.Y.*): Would you say a word about ascorbic acid RDA food processing, food storage, and nitrosamine formation?

S. TANNENBAUM: That's just too broad a question, so I'll just say something about requirements, which will probably be a very controversial point of view.

I'm beginning to think that people have been looking at the question of ascorbic acid requirements incorrectly. I think that ascorbic acid is a substance that's been around a long time that in a single molecule has a multipotential role for many kinds of cells. It can act as an antioxidant, it can act as a chelating agent, and it can act to combine with such substances as nitrites. If you look at the role that ascorbic acid plays in the process which I described, which is the biosynthesis of nitrite by macrophages, in fact macrophages can make both nitrites and oxygen radicals very nicely without ascorbic acid. So, what ascorbic acid does, as I concluded, is to protect the host against its own defense mechanisms for processes which are intended to kill bacteria, viruses, and tumor cells. I think that's why the concentrations in those kinds of cells is so high. Now what that suggests is that there is in fact no optimum level for ascorbic acid, that under certain conditions maybe the more the better, although not without limit. I mean the idea that tens of milligrams is going to suffice for a condition where the host is challenged is absolute nonsense in my opinion.

C. BOREK (*Columbia University, New York, N.Y.*): What is the time course in the interrelation between the vitamin C and the nitrates and the nitrosamine production in your macrophages? Do you have to precondition the cells with the vitamin C, or can you add it after you've stimulated the cells with the LPS?

S. TANNENBAUM: If you're working with mouse peritoneal macrophages, they already have a high level of vitamin C. If you're working with cell lines and there is no ascorbic acid in the medium, you have to add ascorbic acid and wait for a period of time in which uptake takes place. Then since macrophages are adherent cells, you can just pour off the medium, add a medium which contains no ascorbic acid, and then you'll get exactly the same results. So it's the ascorbic acid in the cell and not the ascorbic acid in the medium which is playing a role here.

C. BOREK: So if you add it 96 hours after you've stimulated the cells it's not going to have an effect?

S. TANNENBAUM: No, it will have no effect at all.

K. F. GEY (*Hoffmann-La Roche, Basel, Switzerland*): You have shown that vitamin E pretreatment in comparison to vitamin C treatment did not reduce the nitrosamine formation. Would you like to comment on the findings of other authors, who found that vitamin E has an effect on nitrosamine reduction?

I should like to comment that in the Basel Prospective Study, we found a synergistic effect of vitamins C and E with regard to gastrointestinal cancers. High levels of both vitamins showed the highest protection. Do you mind speculating on the synergism or the cooperative action on these compounds and whether there is a similar action on nitrosamine formation?

S. TANNENBAUM: We have proposed several times that they would have a synergistic effect, predominantly because vitamin C is a water-soluble substance and vitamin E is a fat-soluble substance. The data in that one slide that I showed are a little bit misleading. What in fact is the case of vitamin E is that it was quite effective for some people and quite ineffective for others. Let me just warn people, this is free α-tocopherol, not tocopherol acetate. So this is not just vitamin E, this is specifically α-tocopherol. I think that it probably does have a role to play, but maybe more so in the lung than in the stomach. I think to prevent nitrosamine formation in the stomach vitamin C is perfectly adequate. You don't need vitamin E. To prevent nitrosamine formation in the skin (which has been shown from nitrogen oxides) or in the lung, vitamin E may be much more effective. However, no one has yet done experiments to definitively test this hypothesis.

UNIDENTIFIED SPEAKER: You have on the one hand epidemiological evidence for protection from gastric cancer by vitamins, and on the other hand the fact that nitrosamines are carcinogenic. What direct evidence is there that nitrosamines are in fact a major contributor to gastric cancer?

S. TANNENBAUM: The problem with dealing with any environmental carcinogen is that you can demonstrate the carcinogenicity of those substances in animals but you can't obviously do those kind of experiments in humans. What you have to do is develop animal models and then test out those hypotheses on human populations in epidemiological studies. There's very little risk to continuing to undertake studies, in particular doing interventions in high-risk populations with large amounts of vitamin C. By large amounts of vitamin C, I mean 1-2 g a day and as far as I can see there's extremely small downside risk and a lot to be gained.

Ascorbic Acid, Alcohol, and Environmental Chemicals[a]

V. G. ZANNONI, J. I. BRODFUEHRER, R. C. SMART,[b]
AND R. L. SUSICK, JR.[c]

Department of Pharmacology
University of Michigan Medical School
Ann Arbor, Michigan 48109

INTRODUCTION

It is apparent through the efforts of a number of investigators that ascorbic acid is involved in the metabolism and detoxification of numerous xenobiotics. Interestingly, the vitamin participates at a variety of levels, including the important hepatic electron transport systems, *i.e.*, cytochrome P-450 mixed function oxygenase (MFO)[1–21] and flavin-containing monooxygenase (FMO)[22–24]; protection against covalent binding of "reactive intermediates" to macromolecular proteins[25–49]; and more recently involvement in the metabolism and toxicological consequences of a most commonly used and abused drug, alcohol.[51–61,63,66] Although the precise biochemical mechanism of the vitamin's participation at these levels warrants further investigation, the role of ascorbic acid in xenobiotic metabolism may have important consequences.

RESULTS

Ascorbic Acid and Hepatic Electron Transport Systems

It has been previously established and well documented that the hepatic microsomal cytochrome P-450 mixed function oxygenase (MFO) electron transport system is markedly reduced in vitamin C deficiency.[3,7,15,20] In addition, more recently, it has been demonstrated that the flavin-monooxygenase transport system (FMO),[22–24] responsible for the metabolism of many nitrogen- and sulfur-containing xenobiotics, is also jeopardized in vitamin C deficiency (TABLE 1). Guinea pigs maintained on an

[a]This work supported in part by grant no. 23007 from Hoffmann-La Roche, Inc., Nutley, N.J. and grant nos. 5M01-RR00042 and 2 P60 AM20572 from the National Institutes of Health.

[b]Present address: Toxicology Program, North Carolina State University, Raleigh, N.C. 27695.

[c]Present address: Department of Pathology and Experimental Toxicology, Warner-Lambert/Parke-Davis, Ann Arbor, Mich. 48105.

TABLE 1. Effect of Ascorbic Acid Deficiency on Hepatic Transport Systems

Condition	FMO Dimethylaniline *N*-Oxidation (nmol/min/mg Protein)	Cytochrome P-450 (nmol/100 mg Protein)	Cytochrome b_5 (nmol/100 mg Protein)	Cytochrome b_5 Reductase (nmol/min/μg Protein)
Deficient diet plus ascorbate	1.20 ± 0.60 (20)	9.3 ± 3.4 (20)	25.0 ± 5.6 (7)	0.97 ± 0.18 (7)
Deficient diet	0.66 ± 0.30[a] (20)	4.7 ± 1.6[a] (20)	22.0 ± 4.6 (7)	1.0 ± 0.15 (7)

Dimethylaniline and cytochrome P-450 were assayed in the 12,000 g supernatant fraction and cytochrome b_5 and cytochrome b_5 reductase were assayed in the 100,000 × g microsomal pellet. The average hepatic level of ascorbic acid in supplemented animals was 29 ± 13 mg/100 g wet liver weight and 1.8 ± 1.3 mg in the deficient animals. The numbers in parentheses equal the number of animals, and the data are means ± SD. (From Brodfuehrer and Zannoni.[22])

[a] $p < 0.001$.

ascorbic-acid-free diet had a significant reduction (45%) in the *N*-oxidation of dimethylaniline, a substrate for the FMO. Cytochrome P-450 was also reduced, whereas the microsomal heme-containing cytochrome b_5 and the flavin-containing b_5 reductase were not affected. The decrease in FMO could be differentiated from the decrease in the MFO system by the use of specific inhibitors, phenobarbital pretreatment, thermal treatment, and pH activity profiles. The effect of inhibitors is shown in FIGURES 1A and 1B. The *N*-oxidation of dimethylaniline (DMA), a FMO substrate, was in fact stimulated (twofold) by 3 mM *n*-octylamine (FIG. 1A), while aniline, an MFO substrate, was inhibited 90%. Also, SKF-525A, another MFO inhibitor, caused a decrease of 60% in the metabolism of p-NO_2 anisole, an MFO substrate, and did not affect the *N*-oxidation of DMA (FIG. 1B). Phenobarbital pretreatment of guinea pigs did not induce FMO, while the MFO was induced twofold in both ascorbic-acid-supplemented and ascorbic-acid-deficient guinea pigs. The FMO could be selectively inactivated at 50°C (95%), while the MFO was not significantly altered. In addition, FMO activity was reduced by 50% at pH 7.0 in preparations from both ascorbate-supplemented and ascorbic-acid deficient animals while no significant decrease of MFO activity was observed at this pH. In addition to the decrease in FMO due to the deficiency of the vitamin, there was a further decrease in FMO activity in the deficient guinea pigs who had lost up to 10-15% of their body weight compared to ascorbic-acid-supplemented animals (TABLE 2). Guinea pigs on an ascorbic-acid deficient diet that had no significant weight loss had a 40% reduction in FMO activity, while the ascorbic acid deficient animals that had lost 10-15% of their body weight had a reduction of 83%.

Purification of the FMO from ascorbic-acid-deficient and ascorbic-acid-supplemented guinea pig livers with DEAE cellulose and Blue-2-Agarose column chromatography resulted in two FMO fractions (A and B) in both cases. Fraction A was eluted with 0.35 M KCl, pH 7.8, and required exogenous FAD (0.05 mM) for optimal activity. The recovery of enzyme activity in the fraction from ascorbic-acid-deficient microsomal preparations was four times greater than microsomal preparations from ascorbic-acid supplemented guinea pigs. There was, however, a marked reduction in the quantity of FMO activity in the B fraction isolated from the ascorbic-acid-deficient microsomes: a 5% recovery compared to a 25% recovery for the ascorbic-acid-supplemented animals. Furthermore, subjection of purified preparations (30- to 120-fold; DEAE, Blue-2-Agarose, and 2′, 5′-ADP Sepharose) to polyacrylamide gel electrophoresis resulted in a marked decrease in the quantity of protein banding at 56,000 daltons in enzyme prepared from vitamin-C-deficient animals.[24]

Purified FMO isolated from ascorbic-acid-deficient animals was unstable to freezing at −20°, whereas comparable preparations from ascorbic-acid-supplemented animals maintained over 90% of their activity up to 4 weeks. Furthermore, marked substrate inhibition occurred with enzyme prepared from the ascorbic-acid-deficient animals. At 2 mM DMA, the initial rate of the reaction as measured for 100 seconds was linear with time and comparable to enzyme purified from the ascorbic-acid-supplemented animals. However, this initial rate decreased at 400 seconds, a more than 95% decrease with enzyme prepared from ascorbic-acid-deficient animals compared to a decrease of 40% with enzyme prepared from ascorbic-acid-supplemented animals.[24]

Kinetic experiments with purified preparations from ascorbic-acid-deficient or supplemented animals indicated no significant difference in the affinity of either DMA or NADPH. The K_m for DMA is 3.1×10^{-4} M for enzyme prepared from ascorbic-acid-deficient animals and 3.2×10^{-4} M for enzyme prepared from ascorbic-acid-supplemented animals. The K_m for NADPH is 7.7×10^{-5} M for ascorbic-acid-deficient and 3.8×10^{-5} M for ascorbic-acid-supplemented animals for FMO activity associated with the B fraction. The apparent affinity for FAD with purified FMO fraction A isolated from ascorbic-acid-deficient animals was 8.0×10^{-7} M.[24]

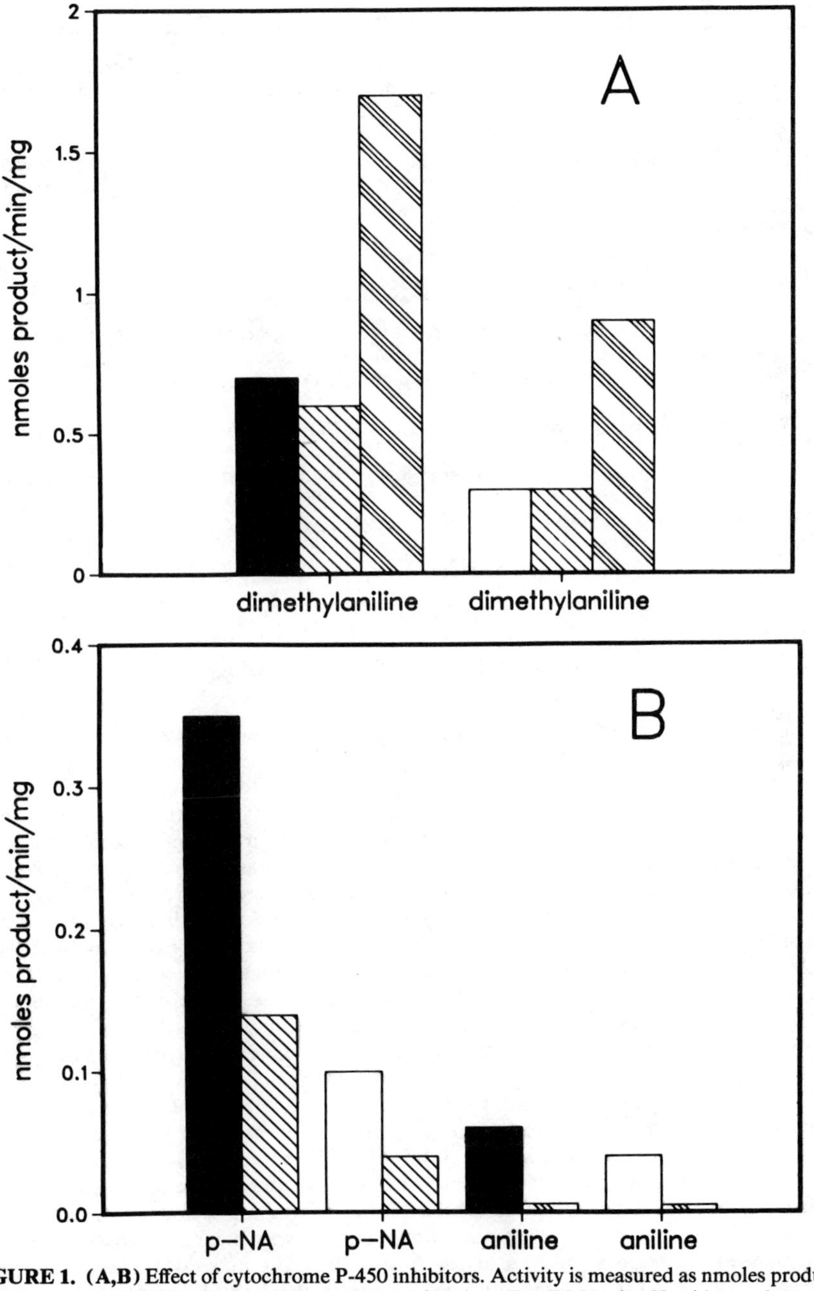

FIGURE 1. (A,B) Effect of cytochrome P-450 inhibitors. Activity is measured as nmoles product per min per mg 12,000 $\times g$ hepatic supernatant fraction. For DMA the *N*-oxide product was determined. *p-Na,* p-nitroanisole. The average hepatic level of ascorbic acid in supplemented animals was 29 ± 13 mg/100 g wet liver and 1.8 ± 1.3 in the deficient animals. The data presented in the figure represent a typical experiment: ■ = plus ascorbate diet; □ = no ascorbate diet; ▧ = SKF-525A, 1 mM; and ◪ = *n*-octylamine, 3 mM. (From Brodfuehrer and Zannoni.[22] Reprinted by permission of *Biochemical Pharmacology*.)

TABLE 2. Effect of Ascorbic Acid on Hepatic Flavin Monooxygenase[a]

Diet	FMO (nmol/min/mg Protein)	Liver Ascorbate (mg/g Wet Liver wt)
10-15% Weight Loss		
Deficient diet plus ascorbate	2.3 ± 0.5 (6)	20.9 ± 4.5 (6)
Deficient diet, no ascorbate	0.4 ± 0.08[b] (5)	1.9 ± 1.4 (5)
No Weight Loss		
Deficient diet plus ascorbate	1.3 ± 0.4[d] (7)	22.0 ± 6.6 (7)
Deficient diet, no ascorbate	0.8 ± 0.3[c,e] (7)	1.3 ± 0.5 (7)

[a] Dimethylaniline *N*-oxidation was assayed in the 12,000 g hepatic supernatant fraction. The numbers in parentheses are the number of animals and the data are means ± SD.

[b] The mean value is significantly different ($p < 0.001$) compared to the corresponding plus ascorbate group.

[c] The mean value is significantly different ($p < 0.02$) compared to the corresponding plus ascorbate group.

[d] The mean value is significantly different ($p < 0.01$) compared to the corresponding group that lost weight.

[e] The mean value is significantly different ($p < 0.02$) compared to the corresponding group that lost weight.

(Modified from Brodfuehrer and Zannoni.[23])

Ascorbic Acid and Covalent Binding

Several laboratories have demonstrated a protective effect of ascorbic acid against covalent binding of reactive xenobiotic intermediates to macromolecular protein.[25–36] Benzene and its metabolites exemplify this protective property of the vitamin. Upon chronic exposure, benzene produces hemopoietic toxicity resulting in reported cases of aplastic anemia.[37,38] It has also been implicated as a human leukemogen.[39] The toxicological consequences are thought to be via metabolic activation of benzene presumably through covalent binding of its metabolites to protein.[40–49] A possible metabolic pathway of benzene or its metabolite, phenol, leading to intermediates capable of covalent binding to protein is given in FIGURE 2. Benzene is oxidized to phenol, via the P-450 system, and phenol is further metabolized to catechol or hydroquinone. Subsequent oxidation of phenol through $1e^-$ steps lead to semiquinone intermediates and quinoid products. Peroxidases can catalize the oxidative steps; diaphorases are capable of reducing the quinoid products to hydroquinone and catechol.

The ability of ascorbic acid to inhibit covalent binding of benzene or phenol metabolites is shown in TABLE 3. The inhibition by ascorbic acid was 75% with benzene. Glutathione caused marked inhibition of binding (95%). When phenol was incubated with hepatic microsomes, ascorbic acid or glutathione inhibited over 95%. Ascorbic acid can reduce quinoids while GSH can form adducts with them. The effect of ascorbic acid on covalent binding when phenol was incubated with horseradish peroxidase or myeloperoxidase isolated from guinea pig marrow is shown in TABLE 4. HRP increased the covalent binding of phenol metabolites 30-fold and myeloperoxidase increased it sixfold compared to the binding that occurred in their absence (TABLES 3 and 4). The binding was decreased over 95% by 1 mM ascorbic acid (TABLE 4).

The effect of DT diaphorase, ascorbic acid, or glutathione on the covalent binding of benzene or phenol metabolites is shown in FIGURES 3A and 3B. With benzene as the substrate, DT-diaphorase inhibited binding by only 18%, ascorbic acid inhibited it by 55% and GSH by 95% (FIG. 3A). In contrast, with phenol as the substrate, DT-diaphorase inhibited binding by 70% and ascorbic acid or GSH inhibited binding over 95% (FIG. 3B). Ascorbic acid or GSH had no significant effect on MFO metabolism of phenol or hydroquinone. However, ascorbic acid decreased the formation of phenol from benzene by 35%.[35]

The effect of dietary ascorbic acid on the covalent binding of phenol metabolites with bone marrow, the target tissue for the toxicological events, was also determined. Covalent binding in bone marrow homogenates isolated from guinea pigs on an ascorbic-acid-free diet was fourfold higher than the bone marrow isolated from animals receiving 0.5 mg ascorbic acid per ml of drinking water (TABLE 5). In addition, the bone marrow ascorbic acid concentration was tenfold higher in the ascorbic-acid-treated animals while the GSH concentration in the bone marrow was not significantly different between the two groups. In view of these results, the effect of ascorbic acid

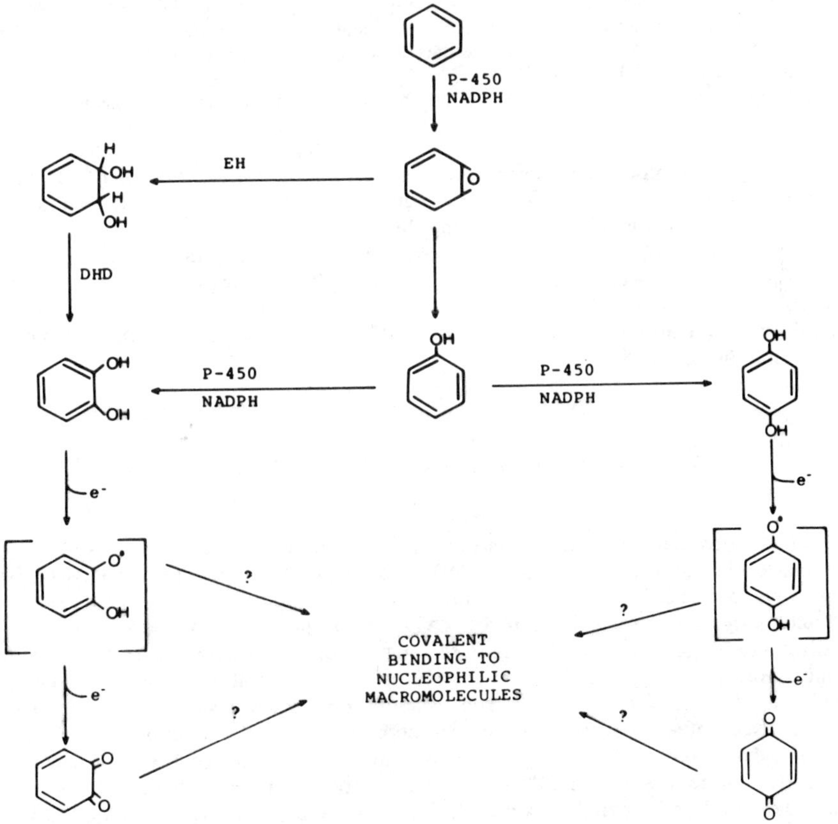

FIGURE 2. Proposed metabolic pathway of benzene or phenol leading to intermediates capable of covalent binding. (From Smart and Zannoni.[34] Reprinted by permission of *Molecular Pharmacology.*)

TABLE 3. Effect of Ascorbic Acid and Glutathione on Covalent Binding of [^{14}C]Benzene Metabolites to Hepatic Microsomal Protein *in Vitro*

Additions	Covalent Binding[a] (pmol Phenol or Benzene Equivalents Bound/min/mg Microsomal Protein)
With [^{14}C]Benzene	
None (control with NADPH)	104.4 ± 18.8 (3)[b]
Ascorbate (1 mM)	29.9 ± 3.4 (3)
Glutathione (1 mM)	3.3 ± 2.2 (3)
With [^{14}C]Phenol	
None (control with NADPH)	252.7 ± 12.3 (4)
Ascorbate (1 mM)	13.4 ± 1.7 (4)
Glutathione (1 mM)	1.0 ± 1.9 (4)

[^{14}C]Phenol (1 mM) or [^{14}C]benzene (1 mM) were incubated with 0.7-1.0 mg of washed hepatic microsomal protein isolated from phenobarbital-pretreated guinea pigs, NADPH (1 mM), and sodium phosphate buffer (100 mM, pH 7.4). The incubation was carried out for 10 or 15 min at 37°C. The total volume was 1 ml. (Modified from Smart and Zannoni.[35])

[a] Values are expressed as $\bar{x}$ ± SD.

[b] The number in parentheses equals the number of experiments done in duplicate.

on covalent binding of benzene metabolites after i.p. administration of [^{14}C]benzene to guinea pigs placed on different dietary intakes of the vitamin was determined. In both liver and bone marrow, the binding decreased with higher tissue concentrations of ascorbic acid. In the liver the decrease was 50% with animals with 2.63 μmol ascorbate per g liver compared to animals with 0.28 μmol ascorbic acid per g (FIG. 4). In the marrow the decrease was on the order of 30% with animals with 1.84 μmol ascorbic acid per g marrow compared to animals with 0.12 μmol ascorbic acid per g marrow (FIG. 5).

Ascorbic Acid and Alcohol

The involvement of ascorbic acid in alcohol metabolism and toxicity is of current interest and has been studied in several laboratories.[51-63,66] The principal enzymatic pathways for the oxidation of methanol or ethanol to their aldehydes include cytosolic alcohol dehydrogenase, which utilizes NAD^+; cytosolic alcohol dehydrogenase, which utilizes H_2O_2; and the cytochrome P-450 mixed function oxygenase system (MFO), which requires NADPH and O_2. *In vitro* studies in our laboratory indicate that ascorbic acid can promote alcohol oxidation via a catalase-mediated reaction.[57] A comparison of this ascorbate-dependent alcohol oxidizing system to the alcohol dehydrogenase and the cytochrome P-450 microsomal system is shown in TABLE 6. The ascorbate-dependent system was over 200 times more active than the microsomal system for both methanol and ethanol; it was over 500 times more active for methanol and over 100 times more active for ethanol compared to alcohol dehydrogenase. The effect of catalase inhibitors on the ascorbate-dependent alcohol oxidation is given in TABLE 7. Sodium azide and 3-amino-1,2,4-triazole were effective inhibitors with hepatic supernatant, 12,000 × g hepatic pellet fractions, and when purified catalase was used as the enzyme source. When a peroxide generating system was used in place of ascorbic acid, the amount of inhibition with azide or triazole was comparable to the ascorbate-dependent system. A comparison of oxygen consumption, H_2O_2 production, and alcohol oxidation utilizing ascorbic acid or a H_2O_2 generating system is further evidence for an ascorbic-acid-catalase mediated reaction (TABLE 8). Hydrogen peroxide was produced and O_2 consumed with either ascorbic acid or a H_2O_2 generating system. However, more oxygen was consumed with the ascorbic acid system than could be accounted for by its incorporation into H_2O_2. The active oxidizing species of catalase and H_2O_2 is catalase-H_2O_2 complex I, which can be measured spectrophotometrically.[50] A comparison of alcohol oxidation and complex I formation utilizing either an H_2O_2 generating system or ascorbic acid resulted in the formation of complex I. Furthermore, upon addition of alcohol the quantity of complex I was reduced in both cases.[57]

In view of the *in vitro* participation of ascorbic acid in alcohol metabolism, a determination of any *in vivo* protective effect of the vitamin against the toxicity of ethanol was of interest. The vitamin has been previously assessed, but to a limited degree, and in the main, in species that can synthesize it.[51,55,56] Yunice and coworkers did, however, find protection against hepatic steatosis by ascorbic acid in guinea pigs, a species which cannot synthesize the vitamin. In this study the animals were chronically infused with ethanol.[61] The data presented in TABLE 9 are the result of an acute study in our laboratory in which serum enzymes, triglycerides, and liver weight to body weight ratios were determined after a single dose of ethanol given to guinea pigs on various regimens of ascorbic acid. Animals with hepatic ascorbic acid concentrations

TABLE 4. Effect of HRP and Myeloperoxidase on Covalent Binding of [^{14}C]Phenol and Metabolites to Microsomal Protein *in Vitro*

Condition	Covalent Binding[a] (pmol Phenol Equivalents Bound/min/0.7 mg Microsomal Protein)
HRP[b] (0.12 unit), H_2O_2 (8.8 mM)	8287.8 ± 995.5
HRP (0.12 unit), H_2O_2 (8.8 mM), ascorbate (1 mM)	50.8 ± 23.3[c]
Myeloperoxidase[b] (0.17 unit), H_2O_2 (8.8 mM)	1458.6 ± 57.2
Myeloperoxidase (0.17 unit), H_2O_2 (8.8 mM), ascorbate (1 mM)	70.6 ± 20.1[c]
H_2O_2 (4.4 mM)	114.1 ± 32.1[c]

[^{14}C]Phenol (1 mM) was incubated with 0.7 mg of washed hepatic microsomal protein isolated from PB-pretreated guinea pigs and sodium phosphate buffer (100 mM, pH 7.4). The incubation was carried out for 5 min at 37°C. The total volume was 1 ml. (Modified from Smart and Zannoni.[34])

[a] Values are means ± standard deviation, $n = 4$, 1 unit = 1 μmol of guaiacol oxidized/min/mg; HRP = 126 units/mg; guinea pig myeloperoxidase = 1.3 units/mg.

[b] Control.

[c] Significantly different from control ($p < 0.01$).

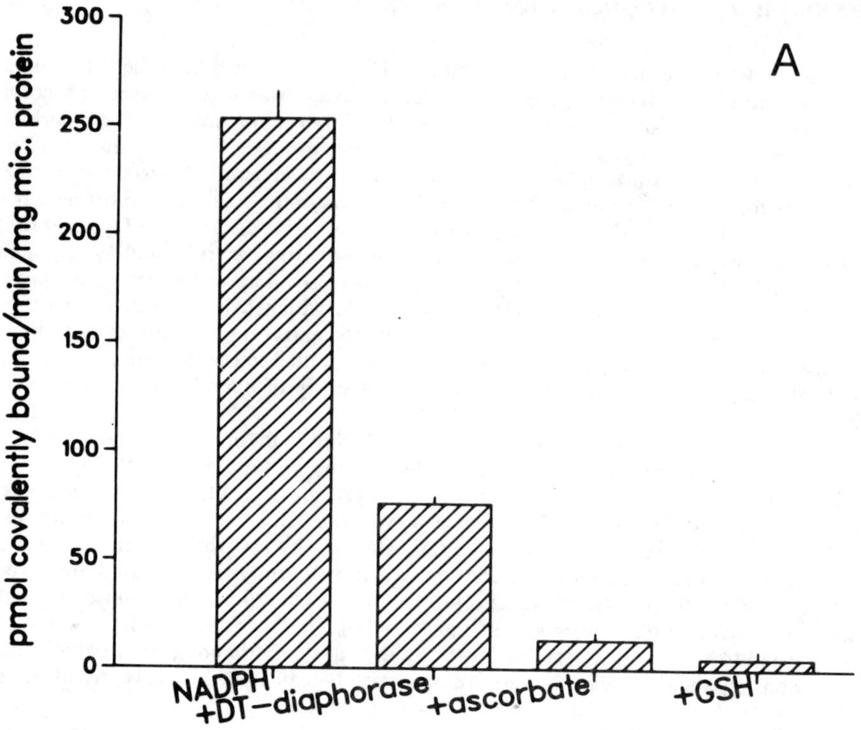

A
pmol covalently bound/min/mg mic. protein
300
250
200
150
100
50
0
NADPH
+DT-diaphorase
+ascorbate
+GSH

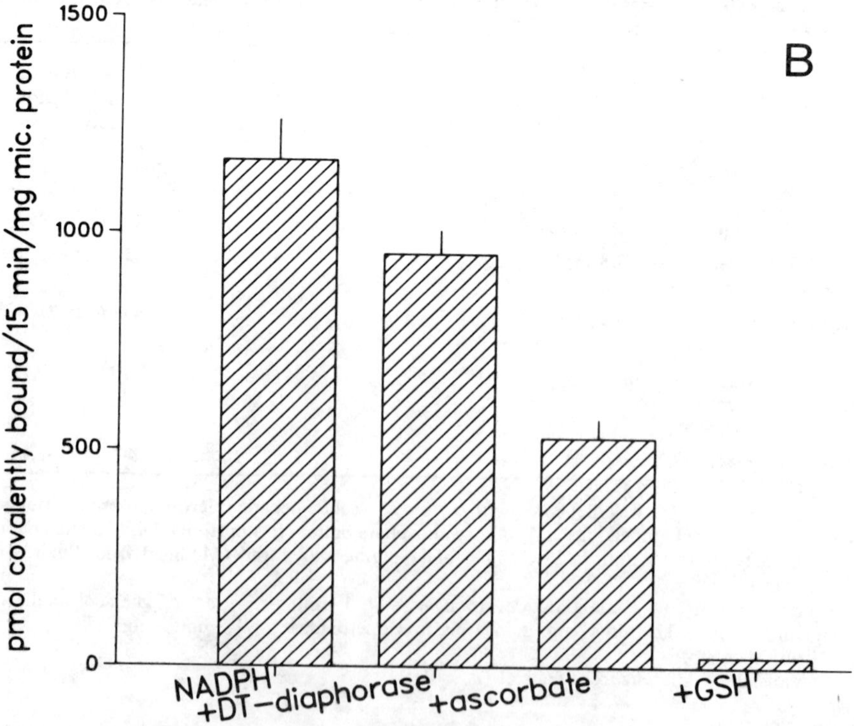

B
pmol covalently bound/15 min/mg mic. protein
1500
1000
500
0
NADPH
+DT-diaphorase
+ascorbate
+GSH

from 1 to 16 mg per 100 g liver had a 12-fold increase in SGOT levels. In contrast, animals with hepatic concentrations of the vitamin from 17 to 36 mg per 100 g liver had only a fivefold increase. The significance between the SGOT levels of these groups was $p < 0.01$. No correlation was found between the concentrations of the vitamin and levels of SGPT or serum triglycerides. In addition, there was no significant increase in ethanol clearance in animals with 1 to 16 mg ascorbic acid per 100 g liver (773 ± 104 μmol per 100 ml per h) compared to animals with liver ascorbic acid concentrations above 16 mg per 100 g liver (755 ± 55 μmol per 100 ml per h).

The data in TABLE 10 are the result of a study to determine the effect of ascorbic acid on chronic alcohol consumption in the guinea pig. Levels of SGOT and SGPT were significantly elevated in animals on the low ascorbic acid diet that had received alcohol: 120% and 250%. FIGURES 6A and 6B give a representative example of guinea pigs on a low or high ascorbic acid diet that received alcohol for 14 weeks. The development of hepatic steatosis is shown in the figure. Half of the animals on the low ascorbic acid diet also developed necrosis of their hepatocytes. None of the animals on the high ascorbic acid diet that received alcohol manifested these changes.

Based on the *in vivo* animal studies demonstrating that ascorbic acid gave some protection against alcohol toxicity, and the *in vitro* studies demonstrating ascorbic acid involvement in alcohol oxidation, a determination was made on the effect of the vitamin on the consequences of acute alcohol consumption in man.[59] The study included ethanol clearance, toxicity, and behavioral patterns. Ascorbic acid pretreatment resulted in a significant increase in blood ethanol clearance. Ten subjects had a greater than 10% difference in clearance after vitamin pretreatment. Nine of these ten subjects had an 11-74% increase in clearance; one had a 15% decrease. Of the remaining ten subjects, four had an increase in clearance of 1-10%, five had a decrease of 1-10%, and one had no change. In general, those individuals who had the largest increase in clearance with ascorbic acid pretreatment were those with the slowest clearance with the placebo (FIG. 7). In addition, ascorbic acid pretreatment resulted in a significantly higher level of serum triglycerides (FIG. 8). Eleven subjects had a greater than 20% difference in serum triglyceride levels after ascorbic acid pretreatment. Ten of these eleven subjects had a 36-133% increase in triglyceride levels; one subject had a 30% decrease. Of the remaining nine subjects, four had a 10-20% increase, four had a 1-20% decrease, and one had no change. Ascorbic acid pretreatment did not significantly influence blood lactate : pyruvate ratios.

FIGURE 3. (A) Effect of DT-diaphorase, ascorbate, or GSH on covalent binding of benzene metabolites with microsomes isolated from benzene-pretreated guinea pigs. The incubation contained [^{14}C]benzene (1 mM), 2 units DT-diaphorase, ascorbate (1 mM), or GSH (1 mM). One unit of DT-diaphorase is defined as 1 μmol of *p*-benzoquinone reduced/min. Each value represents the mean of three experiments done in duplicate ± SD. **(B)** Effect of DT-diaphorase, ascorbate, or GSH on covalent binding of phenol metabolites with microsomes isolated from phenobarbital-pretreated guinea pigs. The incubation contained [^{14}C]phenol (1 mM), 2 units DT-diaphorase, ascorbate (1 mM), or GSH (1 mM). One unit of DT-diaphorase is defined as 1 μmol *p*-benzoquinone reduced/min. Each value represents the mean of three experiments done in duplicate ± SD. (From Smart and Zannoni.[35] Reprinted by permission of *Toxicology and Applied Pharmacology*.)

TABLE 5. Effect of Dietary Ascorbate on Covalent Binding of [^{14}C]Phenol with Isolated Bone Marrow Preparations from Pair-Fed Guinea Pigs[a,b]

Diet	pmol Phenol Equivalents Covalently Bound/5 min[c]	mg% Ascorbate (Hepatic)[c]
0.5 mg ascorbate per ml drinking water; ascorbate-free guinea pig chow	882.3 ± 105.5	34.7 ± 17.9
0.0 mg ascorbate per ml drinking water; ascorbate-free guinea pig chow	3108.7 ± 159.2	3.1 ± 1.6

[a] $n = 5$ individual guinea pigs per group.
[b] All incubations contained [^{14}C]phenol (1 mM), sodium phosphate buffer (1 mM, pH 7.4), H_2O_2 (8.8 mM), 35 mg bone marrow tissue, and 0.7 mg microsomal protein.
[c] Values are expressed as the means of five individual values ± SD.
(From Smart and Zannoni.[35])

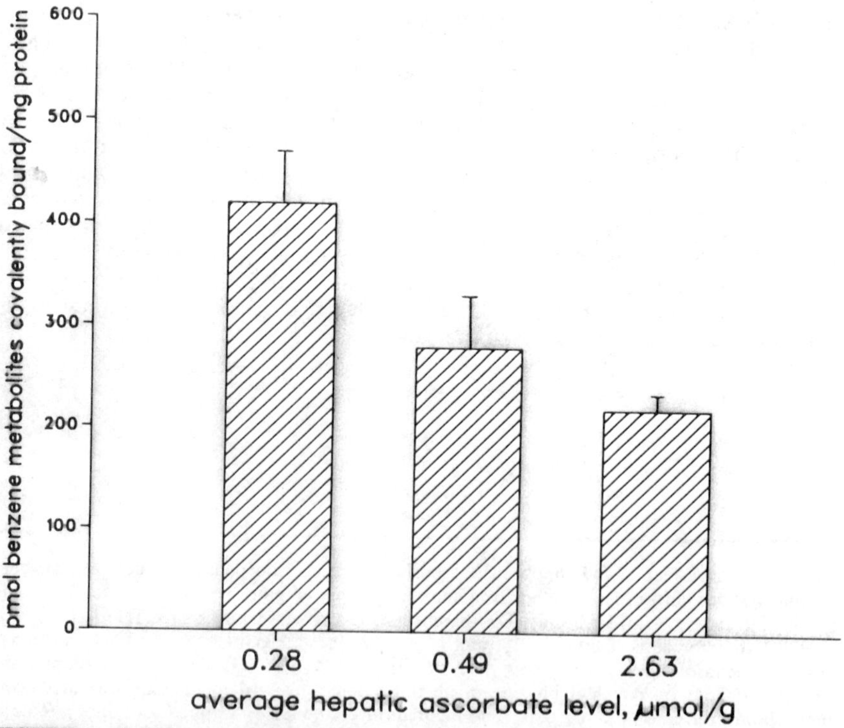

FIGURE 4. Covalent binding of the metabolites of benzene in the liver after i.p. administration of [^{14}C]benzene. [^{14}C]benzene (660 mg/kg) was administered 12 and 6 hours before termination to three groups of guinea pigs on different dietary intakes of ascorbate. Ascorbate and binding values were determined for each guinea pig and the values are expressed as the mean ± standard deviation; $p < 0.01$ with respect to 0.28 to 2.63 μmol/g ascorbate. (From Smart and Zannoni.[36] Reprinted by permission of *Biochemical Pharmacology.*)

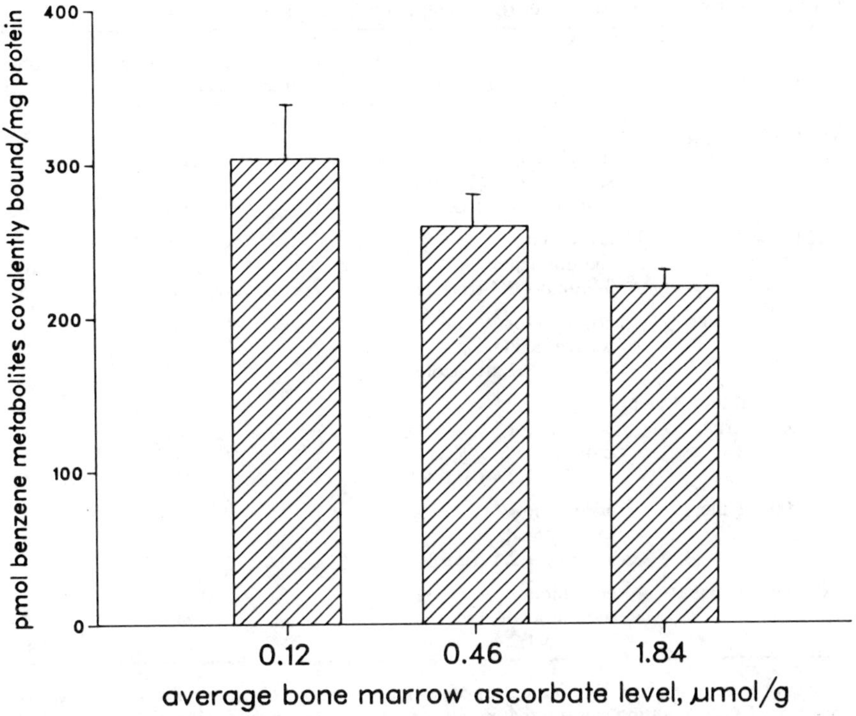

FIGURE 5. Covalent binding of the metabolites of benzene in the bone marrow after i.p. administration of [^{14}C]benzene. [^{14}C]Benzene (660 mg/kg) was administered 12 and 6 hours before termination to 3 groups of guinea pigs on different dietary intakes of ascorbate. Ascorbate and binding values were determined for each guinea pig and the values are expressed as the mean ± standard deviation; $p < 0.01$ with respect to 0.12 to 1.84 µmol/g ascorbate. (From Smart and Zannoni.[36] Reprinted by permission of *Biochemical Pharmacology*.)

TABLE 6. Activities of Alcohol Dehydrogenase, Microsomal Ethanol-oxidizing System and the Ascorbate-Dependent Alcohol-oxidizing System with Methanol and Ethanol[a]

	Specific Activity[b] (nmol CH_2O or CH_3CHO/min/mg Protein)		
Substrate	Alcohol Dehydrogenase	Microsomal Ethanol-oxidizing System	Ascorbate-Dependent Alcohol-oxidizing System
Methanol	3 ± 0.6	6 ± 1	1720 ± 200
Ethanol	11 ± 1.0	6 ± 1	1570 ± 140

[a] Assay conditions were as follows. Alcohol dehydrogenase: glycine buffer, 0.1 M, pH 9.5; NAD^+, 1 mg/ml; guinea pig hepatic 10^5 g supernatant fraction, 1.0 mg protein; substrate, 12 mM; total volume, 1.0 ml, 37°C. Microsomal ethanol-oxidizing system: sodium phosphate buffer, 0.1 M, pH 7.5; NADPH, 2 mM; guinea pig hepatic microsomes, 1.0 mg protein; substrate, 200 mM; total volume 1.0 ml, 37°C. Ascorbate-dependent alcohol-oxidizing system: Tris-HCl buffer, 0.1 M + 10 mM 1,10-phenanthroline, pH 8.5; ascorbate, 4 mM; guinea pig hepatic 10^5 g supernatant fraction, 15 µg protein; substrate, 800 mM; total volume, 1.0 ml, 37°C.

[b] Values are means of at least three experiments.

(From Susick and Zannoni.[57])

TABLE 7. Inhibition of the Ascorbate-Dependent Alcohol Oxidation[a]

Assay Condition	Inhibitor	Specific Activity (nmoles CH_2O/min/mg)	Inhibition (%)
Ascorbate			
10^5 g Supernatant	no inhibitor	1,484	
	sodium azide (2 mM)	nd[b]	100
	3-amino-1,2,4-triazole (100 mM)	547	63
12,000 g Pellet	no inhibitor	760	
	sodium azide	nd	100
	3-amino-1,2,4-triazole	258	66
Purified catalase	no inhibitor	45,203	
	sodium azide	nd	100
	3-amino-1,2,4-triazole	12,804	72
H_2O_2 Generating System			
10^5 g Supernatant	no inhibitor	867	
	sodium azide	89	90
	3-amino-1,2,4-triazole	328	62
12,000 g Pellet	no inhibitor	634	
	sodium azide	68	89
	3-amino-1,2,4-triazole	144	77
Purified catalase	no inhibitor	30,009	
	sodium azide	1,944	94
	3-amino-1,2,4-triazole	5,468	82

[a] Assay conditions: Tris-HCl buffer, 0.1 M + 1,10-phenanthroline, 10 mM, pH 8.5; ascorbate, 4 mM or H_2O_2 generating system, 5.4 mg glucose + 0.1 mg glucose oxidase; guinea pig hepatic fraction, 15 μg protein, or purified beef liver catalase, 0.4 μg protein; methanol, 800 mM, total volume 1.0 ml, 37°C.

[b] Not detected.

(From Susick and Zannoni.[57])

TABLE 8. Comparison of Methanol Oxidation, H_2O_2 Production, and O_2 Consumption with an H_2O_2 Generating System and the Ascorbate Oxidation System[a]

Assay Conditions	Methanol Oxidation (nmol/min/mg)	H_2O_2 Production (nmol/min)	O_2 Consumption (nmol/min)
H_2O_2 generating system	223 ± 14 (3)	3.9 ± 0.9 (3)	4.3 ± 0.6 (5)
Ascorbate	247 ± 22 (3)	6.9 ± 0.4 (4)	23.9 ± 1.0 (6)

[a] Assay conditions were as follows. Methanol oxidation: Tris-HCl buffer, 0.1 M + 1,10-phenanthroline, 10 mM, pH 8.5; ascorbate, 1 mM, or H_2O_2 generating system, 30 mM glucose + 0.02 mg glucose oxidase; guinea pig hepatic 10^5 g supernatant fraction, 15 μg protein; methanol, 800 mM; total volume, 1.0 ml, 25°C. H_2O_2 production: Tris-HCl buffer, 0.1 M + 1,10-phenanthroline, 10 mM, pH 8.5; ascorbate, 1 mM, or H_2O_2 generating system, 30 mM glucose + 0.04 mg glucose oxidase; boiled guinea pig hepatic 10^5 g supernatant fraction, 30 μg protein; total volume 2.0 ml, 25°C. O_2 consumption: Tris-HCl buffer, 0.1 M + 1,10-phenanthroline, 10 mM, pH 8.5; ascorbate, 1 mM, or H_2O_2 generating system, 30 mM glucose + 0.03 mg glucose oxidase; boiled guinea pig hepatic 10^5 g supernatant fraction, 22.5 μg protein; total volume, 1.5 ml, 25°C. Data are means ± SD. Numbers in parentheses represent number of experiments.

(Modified from Susick and Zannoni.[57])

TABLE 9. Effect of Hepatic Ascorbic Acid Concentration on Indices of Acute Ethanol Toxicity[a]

	SGOT (Units/ml[b])	SGPT (Units/ml)	Serum Triglycerides (mg/100 ml)	Liver Weight : Body Weight (mg/g)
Hepatic Ascorbate (mg/100 g liver)				
1 to 16	281 ± 140 (14)	64 ± 24 (14)	200 ± 86 (12)	46 ± 7 (14)
17 to 36	111 ± 26 (7)	53 ± 12 (7)	213 ± 101 (7)	40 ± 2 (7)
p Value[c]	< 0.01	ns[d]	ns	< 0.05
Baseline before Alcohol	22 ± 9 (21)	17 ± 8 (21)	121 ± 37 (19)	

[a] Values represent means ± SD, with number of animals in parentheses, after 4.0 g ethanol/kg i.p. Baseline values taken before ethanol administration were not affected by ascorbate concentration and were combined.
[b] 1 unit = 4.82×10^{-4} μmoles glutamate formed/min.
[c] *p* values calculated from a two-sided Student's *t*-test.
[d] not significant, $p > 0.05$.
(From Susick and Zannoni.[58])

TABLE 10. Serum Enzymes and Liver Weight to Body Weight Ratios after Chronic Alcohol Consumption[a]

	Low Ascorbic Acid Diet[b]		High Ascorbic Acid Diet[c]	
	No Alcohol	Alcohol	No Alcohol	Alcohol
SGOT (units/ml)[d]	42 ± 8 (5)	92 ± 34 (8)**	32 ± 12 (6)	47 ± 12 (12)*
SGPT (units/ml)[d]	15 ± 9 (5)	52 ± 29 (8)*	14 ± 5 (6)	24 ± 11 (12)
Liver Weight / Body Weight (mg/gram)	41 ± 3 (6)	42 ± 10 (9)	43 ± 7 (6)	38 ± 10 (13)

[a] Values represent the mean ± SD with number of animals in parentheses.
[b] Mean hepatic ascorbic acid concentration: no alcohol, 30 ± 9 mg/100 g protein; alcohol, 20 ± 6 mg/100 g protein.
[c] Mean hepatic ascorbic acid concentration: no alcohol, 130 ± 34 mg/100 g protein; alcohol, 120 ± 28 mg/100 g protein.
[d] 1 unit = 4.82×10^{-4} μmoles glutamate formed/min.
* $p < 0.05$ from nonalcohol control group.
** $p < 0.01$ from nonalcohol control group.
(From Susick *et al.*[60])

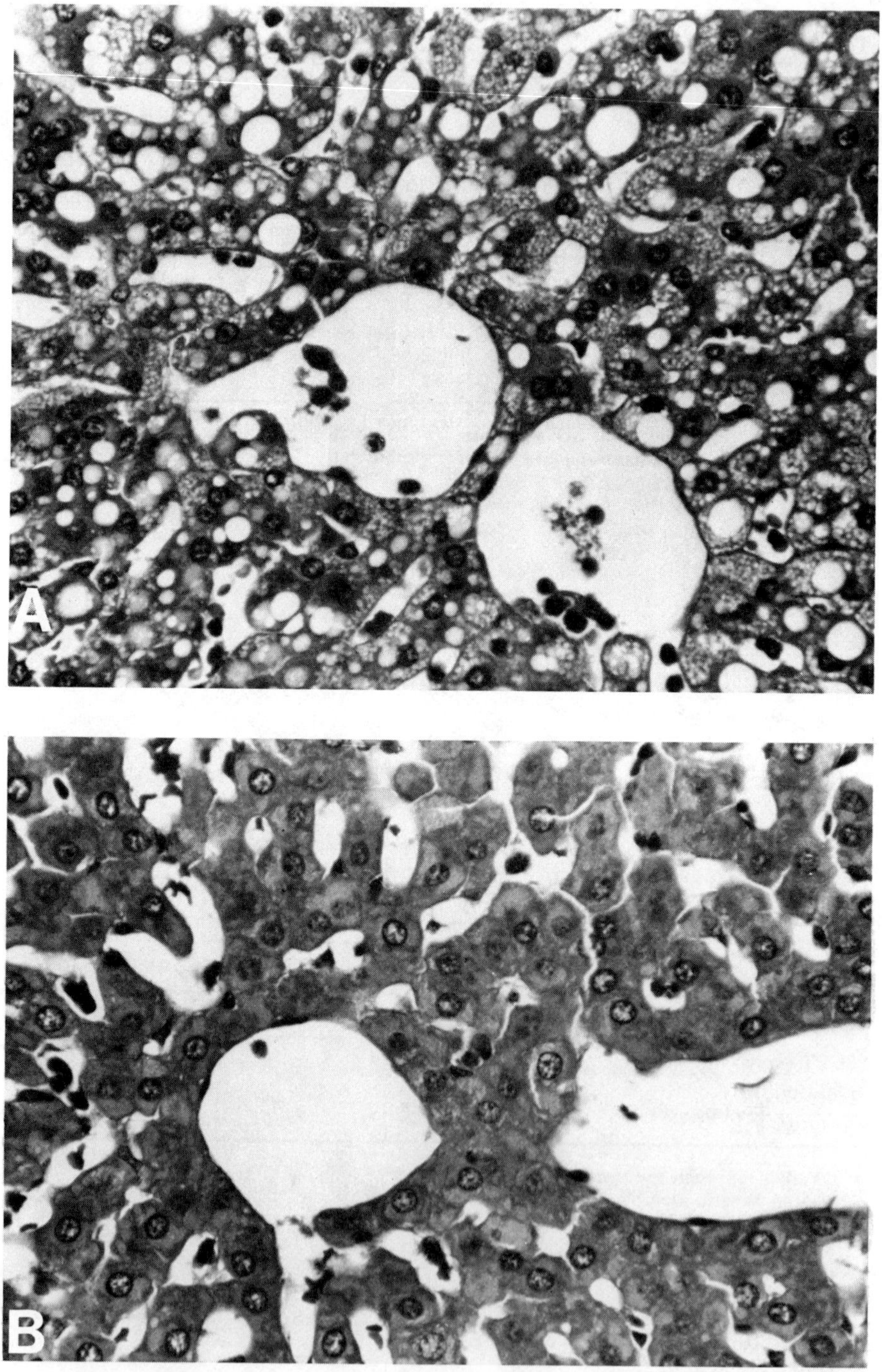
A
B

With regard to behavioral tests, ethanol consumption impaired motor coordination and the vitamin significantly reduced this impairment (FIG. 9). Thirteen subjects had a greater than 20% difference in motor coordination after ascorbic acid pretreatment. Eleven of these thirteen subjects had a 21-50% improvement after vitamin pretreatment; two had a decrease of 48% and 56%. Of the remaining seven subjects, one had an improvement in motor coordination of 14% and six had a decrease from 1% to 20%. Pretreatment with the vitamin also protected against decrease in color discrimination after ethanol (FIG. 10). Ten subjects had a greater than 20% difference in color discrimination after ascorbic acid pretreatment. Nine of these had a 20-100% improvement. The remaining three subjects had a 10-20% decrease in color discrimination. There was no statistically significant effect of ascorbic acid, however, on intellectual function.

DISCUSSION

The participation of ascorbic acid in the major xenobiotic metabolizing systems requiring endoplasmic reticulum electron transport is rather specific. The depletion of the vitamin does not involve an impairment leading to a general membrane alteration. Cytochrome P-450, a heme protein, and the flavin monooxygenase, a nonheme, FAD-requiring enzyme, are significantly decreased in deficiency.[1-24] On the other hand, microsomal b_5 reductase, a FAD-requiring enzyme, and microsomal cytochrome b_5, a heme protein, are not affected by the depletion of the vitamin. With regard to the cytochrome P-450 mixed function monooxygenase and the flavin monooxygenase enzyme, the participation of ascorbic acid shares some similarities but also some differences. With either system there are no significant alterations between enzyme prepared from vitamin C or supplemented animals in the apparent affinity of drug substrates or cofactors.[6,7,15,22,24] In addition, with purification there is a significant decrease in the quantity of cytochrome P-450 as well as the flavin monooxygenase.[24] With regard to cytochrome P-450, the decrease involves three specific heme isozymes while others are unaltered.[19] Also, inhibitor studies with ferrous iron chelators and substrate binding information indicate that the vitamin's participation, most likely, is at the heme-ferrous ion level, and specifically with those isozymes most affected by the deficiency.[16]

To date, the observed decrease in the quantity of FMO could be more general, or for that matter involve a disaggregation of monomeric forms of FMO resulting in a loss of recoverable enzyme protein.[22,24] In addition to a decrease in the quantity of FMO protein, vitamin C deficiency also resulted in purified FMO fractions, which had an obligatory need of exogenous FAD for optimal activity. Perhaps the binding site of FAD to apo-enzyme is jeopardized, leading to disaggregation with subsequent decrease in viable enzyme. In keeping with this, FMO purified from vitamin-C-deficient animals is highly unstable and vulnerable to marked substrate inhibition. This is not the case with purified cytochrome P-450 isolated from vitamin-C-deficient animals.

FIGURE 6. Photomicrographs of the liver, hematoxylin and eosin, magnification 470×. **(A)** From animal on the low ascorbic acid diet, after 14 weeks of ethanol feeding. Note the marked, diffuse vacuolar change in hepatocytes, indicative of steatosis. **(B)** From animal on the high ascorbic acid diet, after 14 weeks of ethanol feeding. In contrast to **A,** there is no steatosis. (From Susick *et al.*[60] Reprinted by permission of *Toxicology and Applied Pharmacology.*)

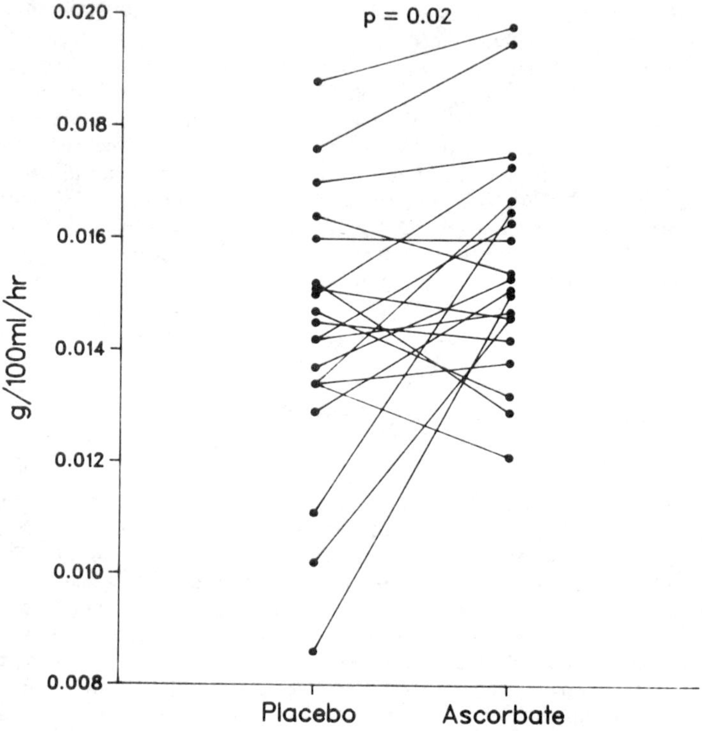

FIGURE 7. Effect of ascorbic acid pretreatment on blood ethanol clearance after an acute dose of ethanol in 20 subjects. Ascorbic acid or placebo (lactose) was given for 2 weeks prior to the alcohol; 1.0 gram of ascorbic acid or placebo was taken five times a day. On the test day 0.95 gram ethanol/kg body weight was taken orally as a 30% solution in ginger ale over a 2½-h period. (From Susick and Zannoni.[59] Reprinted by permission of *Clinical Pharmacology and Therapeutics.*)

Furthermore, the decrease in FMO activity in vitamin C deficiency was enhanced by a concomitant loss of 10-15% body weight, which was not found with cytochrome P-450.[23] The *in vivo* consequences of a decreased capacity of these two major electron transport systems will obviously depend upon the particular xenobiotic under investigation, that is, the potential toxicity of the substrate itself or its subsequent metabolites. In addition, since both electron transport systems are affected in vitamin C deficiency, their contribution to metabolism and the possibility of forming toxic intermediates will rely heavily on the degree to which particular species of either transport system becomes rate limiting. Although it is difficult to predict the ultimate consequences of jeopardized electron transport metabolic systems by a biochemical *in vitro* analysis, an *in vivo* model is at hand whereby a variety of xenobiotics could be tested under depletion of the vitamin and an examination of any toxicological consequences documented.

In addition to the effect of ascorbic acid on the important electron transport systems, MFO and FMO, the vitamin partakes via its reductive property in preventing quinoid-type reactive intermediates from covalent binding to macromolecular proteins.[25,28,30,32-36,48] This participation was illustrated in the metabolic transformation of benzene or phenol, which results in electrophilic reactive quinone species, the latter undergoing reduction by the vitamin.[34] Although the precise damaging intermediate of benzene or phenol resulting in myelotoxicity remains to be elucidated—it may be a hydroxyl cyclohexadienyl radical formed from benzene[49] or a semiquinone or quinone type metabolite of phenol—the vitamin affords *in vitro* protection.[35] In keeping with these findings, when the dietary ascorbic acid was varied in the guinea pig *in vivo*, the ensuing covalent binding of benzene metabolites to hepatic or bone marrow tissue was inversely related to the concentration of the vitamin found in these tissues.[36] Both the *in vitro* and *in vivo* effect of varying amounts of ascorbic acid on covalent binding with benzene, however, was less than when phenol was used as a substrate. For example, ascorbate is capable of inhibiting the *in vitro* covalent binding of phenol by over 90% compared to 35% with benzene as a substrate. Similarly, the *in vivo* covalent binding with an i.p. administration of benzene decreased 30% in the bone marrow of animals on a high intake of the vitamin.[36] This is suggestive that the most potent species of benzene transformation may not be oxidized forms of hydroquinone, ca-

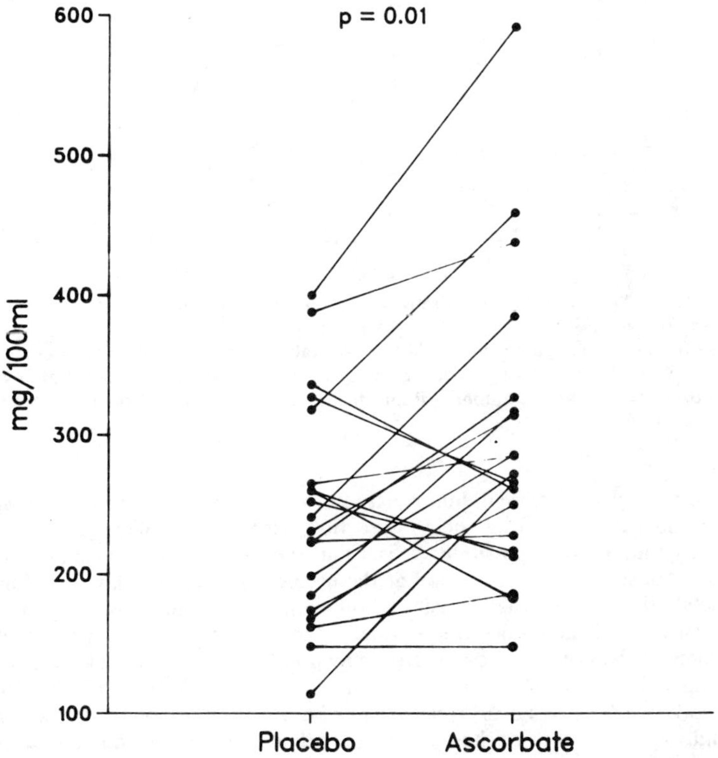

FIGURE 8. Effect of ascorbic acid pretreatment on serum triglyceride levels after an acute dose of ethanol in 20 subjects. Ascorbic acid or placebo (lactose) was given for 2 weeks prior to the alcohol; 1.0 gram of ascorbic acid or placebo was taken five times a day. On the test day 0.95 gram ethanol/kg body weight was taken orally as a 30% solution in ginger ale over a 2½-h period. (From Susick and Zannoni.[59] Reprinted by permission of *Clinical Pharmacology and Therapeutics.*)

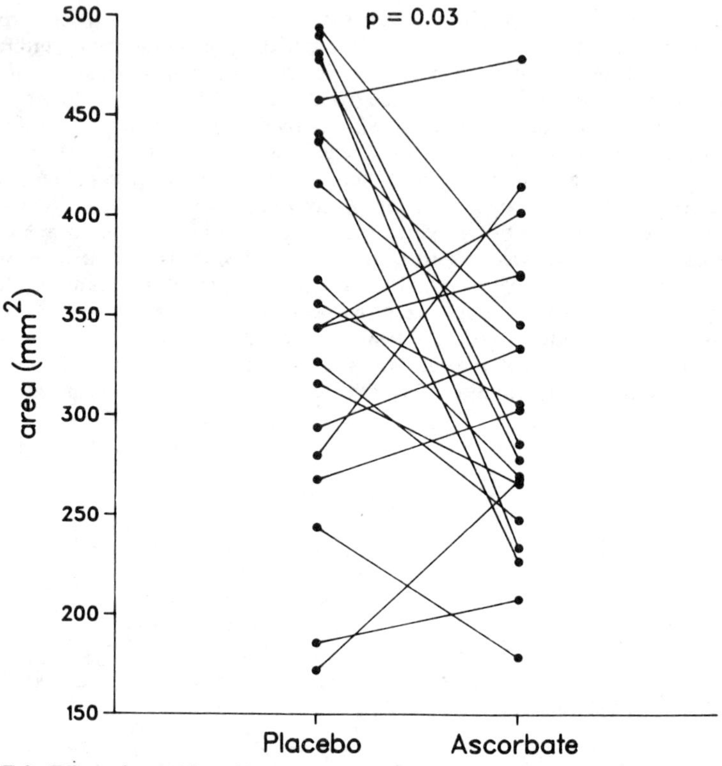

FIGURE 9. Effect of ascorbic acid pretreatment on motor coordination after an acute dose of ethanol in 20 subjects. Ascorbic acid or placebo (lactose) was given for 2 weeks prior to the alcohol; 1.0 gram of ascorbic acid or placebo was taken five times a day. On the test day 0.95 gram ethanol/kg body weight was taken orally as a 30% solution in ginger ale over a 2½-h period. (From Susick and Zannoni.[59] Reprinted by permission of *Clinical Pharmacology and Therapeutics.*)

techol, or phenol since the vitamin is extremely effective in reducing these metabolites and consequently preventing their interaction with macromolecular tissue sites. On the other hand, another possibility exists that the cellular distribution of the vitamin is not at a substantial concentration at the precise macromolecular site of interaction with metabolites of benzene. Consideration should be given, however, to the findings that modulation of the intake of the vitamin can alter the covalent binding of benzene metabolites in the bone marrow *in vivo,* which is the target tissue of benzene toxicity.

In the United States at least 10 million people suffer from alcohol abuse. An estimated 205,000 deaths every year are attributable to that abuse, the causes ranging from highway fatalities to cirrhosis.[64] In large urban areas of the United States, cirrhosis has become the third most common cause of all deaths in the age group 25 to 65 years.[65] The ability of any agent to influence the effect of alcohol on the body is of importance. In this regard, vitamin C has been demonstrated to influence ethanol metabolism, toxicity, and behavioral impairment in man and animals.[51,55–56,58–61,66]

Ascorbic acid pretreatment enhanced ethanol clearance from the blood in healthy human subjects.[59,66] We have described and characterized an ascorbic-acid-dependent alcohol oxidizing system that is catalyzed by catalase.[57] In this system, ascorbic acid generated a peroxide that is utilized by catalase in a peroxidatic oxidation of the alcohol. The ability of ascorbic acid pretreatment to enhance blood ethanol clearance may be due to its ability to supply peroxide and thus allow catalase to contribute to ethanol oxidation.

It is well established that ethanol affects lipoprotein metabolism and that an acute dose increases levels of serum and hepatic triglycerides.[67–69] Ethanol causes a number of events to occur that lead to increased hepatic fat levels, *e.g.*, a stress-induced mobilization of fat from adipose tissues, a decrease in the rate of fatty acid oxidation, an increase in the rate of fatty acid synthesis, and an increase in the esterification of fatty acids to triglycerides. In addition, ethanol inhibits export of fat from the liver by impairing the synthesis and secretion of very low-density lipoproteins (VLDL), the export form of triglycerides. Ascorbic acid pretreatment in human subjects caused an increase in serum triglyceride levels.[59] The possibility that the vitamin increases the rate of fat transport out of the liver should be considered. This could be accomplished by a stimulation of the synthesis or secretion of the VLDL or by prevention

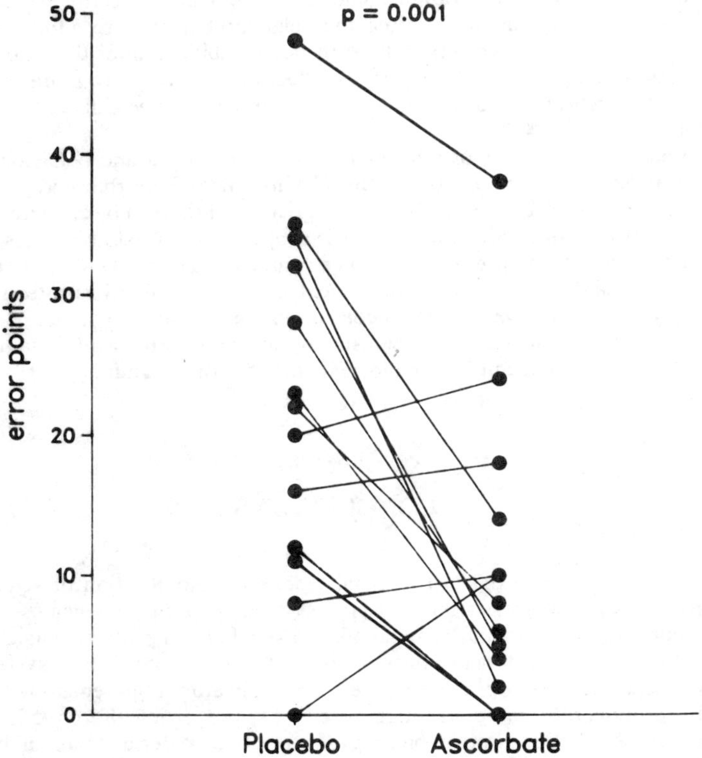

FIGURE 10. Effect of ascorbic acid pretreatment on color discrimination after an acute dose of ethanol in 13 subjects. Ascorbic acid or placebo (lactose) was given for 2 weeks prior to the alcohol; 1.0 gram of ascorbic acid or placebo was taken five times a day. On the test day 0.95 gram ethanol/kg body weight was taken orally as a 30% solution in ginger ale over a 2½-h period. (From Susick and Zannoni.[59] Reprinted by permission of *Clinical Pharmacology and Therapeutics.*)

of alcohol's impairment of these processes. This mechanism would result in less fat accumulation in the liver. In keeping with this, *in vivo* guinea pig studies have demonstrated that ascorbic acid pretreatment results in less hepatic steatosis after chronic ethanol consumption.[60] Ascorbic acid may also protect the liver against direct ethanol toxicity, as evidenced in acute and chronic *in vivo* guinea pig studies where ascorbic acid protected against elevated SGOT and SGPT levels, indices of increased permeability and/or necrosis of hepatocytes.[59,60]

The participation of ascorbic acid in neurochemical events is of current interest. Its involvement in Na^+ and K^+ transport and metabolism of neurotransmitters, including the synthesis of catecholamines, storage of norepinephrine, and regulation of dopaminergic transmission, have been reported.[70–72] In addition, the vitamin has been implicated in modulating the behavioral effects of antipsychotic drugs by influencing their binding to the dopamine receptor.[73] With regard to ethanol's impairment of behavior, ascorbic acid has been shown to prevent impaired swimming in mice caused by an intoxicating dose of ethanol.[55] The protective effect of ascorbic acid pretreatment on impaired motor coordination demonstrated in the clinical study (FIG. 9) supports a neurochemical role for the vitamin. The protective effect is not due to altered blood ethanol concentrations, since ascorbic acid pretreatment had no effect on ethanol concentrations at the time the behavioral tests were administered. The mechanism of the protective action, be it through stabilization of the neuronal membrane, an effect on Na^+/K^+ transport, an effect on the synthesis, storage, release or binding of the neurotransmitters, or an as yet unknown mechanism, remains to be determined.

Ascorbic acid is capable of influencing both biochemical and behavioral changes caused by an acute dose of ethanol in man. The influence of ascorbic acid pretreatment, although not observed in every subject, was greatest in those subjects with the largest alcohol impairment of behavior and in those subjects with the slowest rates of ethanol elimination after placebo pretreatment. The results of the clinical study coupled with the animal studies suggest that ascorbic acid may be of value in protecting against some harmful consequences of alcohol consumption in certain individuals. In addition, the determination of the exact mechanisms by which ascorbic acid acts may lead to the elucidation of important new biological roles for the vitamin.

CONCLUSIONS

Ascorbic acid is involved in the metabolism and detoxification of xenobiotics, in the hepatic microsomal electron transport system, in protection against binding of reactive intermediates, and in the metabolism of alcohol and protection against alcohol toxicity. Its deficiency results in a decrease of the flavin monooxygenase system (50%). There is a decrease in recoverable enzyme and an obligatory requirement for exogenous FAD. In addition, the vitamin, through its reductive properties, affords 75% protection against macromolecular protein binding of "reactive" intermediates of benzene or phenol. Furthermore, ascorbic acid participates in alcohol metabolism by generating a peroxide that is utilized by catalase to oxidize alcohol. The vitamin protects against acute and chronic alcohol toxicity. SGOT levels are lowered (50%) and fatty infiltration and necrosis of hepatocytes are minimized. In clinical studies with an acute dose of ethanol, vitamin pretreatment enhanced ethanol clearance (11-74%), increased serum triglycerides (33-133%), and was beneficial in behavioral responses of motor coordination and color discrimination in 50% of the subjects.

REFERENCES

1. RICHARDS, R. K., K. KUETER & T. I. KLATT. 1941. Effect of vitamin C deficiency on action of different types of barbiturates. Proc. Soc. Exp. Biol. Med. **48:** 403-409.
2. AXELROD, J., S. UDENFRIEND & B. B. BRODIE. 1954. Ascorbic acid in aromatic hydroxylation. III. Effect of ascorbic acid on hydroxylation of acetanilide, aniline and antipyrine *in vivo.* J. Pharmacol. Exp. Ther. **111:** 176-181.
3. CONNEY, A. H., G. A. BRAY, C. EVANS & J. J. BURNS. 1961. Metabolic interactions between L-ascorbic acid drugs. Ann. N.Y. Acad. Sci. **92:** 115-127.
4. KATO, R. A., TAKANAKA & T. OSHIMA. 1969. Effect of vitamin C deficiency on metabolism of drugs and NADPH-linked electron transport system in liver microsomes. Jpn. J. Pharmacol. **19:** 25-33.
5. LEBER, H., E. DEGKWITZ & H. STAUDINGER. 1969. Untersuchungen zum Einfluss der Ascorbinaure auf die Aktivitat und die Biosynthese mischfunktioneller Oxygenasen sowie den Gehalt an Hamoproteiden in der Mikrosomenfraktion der Meerschweinchenleber. Hoppe-Seyler's Z. Physiol. Chem. **350:** 439-455.
6. ZANNONI, V. G., E. J. FLYNN & M. LYNCH. 1972. Ascorbic acid and drug metabolism. Biochem. Pharmacol. **21:** 1377-1392.
7. ZANNONI, V. G. & M. M. LYNCH. 1973. The role of ascorbic acid in drug metabolism. Drug Metab. Rev. **2:** 57-69.
8. GUNDERMANN, K., E. DEGKWITZ & H. STAUDINGER. 1973. Mischfunktionelle Oxygenierung von (+) und (−) Hexobarbital und spektrale Anderungen des Cytochrome P-450 in der Leber ascorbinsaurefrei ernahrter Meerschweinchen. Hoppe-Seyler's Z. Physiol. Chem. **354:** 238-242.
9. GINTER, E. 1973. Cholesterol: Vitamin C controls its transformation to bile acids. Science **179:** 702-704.
10. GUNDERMANN, K., E. DEGKWITZ & H. STAUDINGER. 1973. Mixed function oxygenation of (+) and (−) hexobarbital and spectral changes of cytochrome P-450 in liver of guinea pigs fed without L-ascorbic acid. Hoppe-Seyler's Z. Physiol. Chem. **354:** 238-242.
11. DEGKWITZ, E. & K. S. KIM. 1973. Comparative studies on the influence of L-ascorbic acid, D-arabino-ascorbate and 5-x-D-gluconate on the amounts of cytochrome P-450 and b_5 in liver microsomes of guinea pigs. Hoppe-Seyler's Z. Physiol. Chem. **354:** 555-561.
12. ZANNONI, V. G., M. M. LYNCH & P. H. SATO. 1974. *In* Perinatal Pharmacology: Problems and Priorities. J. Dancis & J. C. Hwang, Eds.: 131-147. Raven Press. New York.
13. DEGKWITZ, E., S. WALSCH & M. DUBBERSTEIN. 1974. Influence of L-ascorbate on the concentrations of microsomal cytochrome P-450 and cytochrome b_5 in adrenals, kidney and spleen of guinea pigs. Hoppe-Seyler's Z. Physiol. Chem. **355:** 1152-1158.
14. SATO, P. H. & V. G. ZANNONI. 1974. Stimulation of drug metabolism by ascorbic acid in weanling guinea pigs. Biochem. Pharmacol. **23:** 3121-3127.
15. ZANNONI, V. G. & P. H. SATO. 1975. Effects of ascorbic acid on components of liver microsomal drug metabolizing enzymes. Ann. N.Y. Acad. Sci. **258:** 119-131.
16. SATO, P. H. & V. G. ZANNONI. 1976. Ascorbic acid and hepatic drug metabolism. J. Pharmacol. Exp. Ther. **198:** 295-307.
17. RIKANS, L. E., C. R. SMITH & V. G. ZANNONI. 1977. Ascorbic acid and heme synthesis in deficient guinea pig liver. Biochem. Pharmacol. **26:** 797-799.
18. KUENZIG, W., V. TKACZEVSKI, J. J. KAMM, A. H. CONNEY & J. J. BURNS. 1977. The effect of ascorbic acid deficiency on extrahepatic microsomal metabolism of drugs and carcinogens in the guinea pig. J. Pharmacol. Exp. Ther. **201:** 527-533.
19. RIKANS, L. E., C. R. SMITH & V. G. ZANNONI. 1978. Ascorbic acid and cytochrome P-450. J. Pharmacol. Exp. Ther. **204:** 702-713.
20. ZANNONI, V. G., E. HOLSZTYNSKA & S. S. LAU. 1982. Biochemical functions of ascorbic acid in drug metabolism. *In* Ascorbic Acid: Chemistry, Metabolism and Uses. P. A. Seib & B. M. Tolbert, Eds.: 349-368. American Chemical Society. Washington, D.C.
21. ZANNONI, V. G., R. L. SUSICK & R. C. SMART. 1984. Ascorbic acid as it relates to the metabolism of drugs and environmental chemicals. *In* Nutrition in the 20th Century, Vol. 13. M. Winick, Ed.: 21-35. John Wiley & Sons. New York.
22. BRODFUEHRER, J. I. & V. G. ZANNONI. 1987. Ascorbic acid deficiency and the flavin-containing monooxygenase. Biochem. Pharmacol. **35:** 637-644.

23. BRODFUEHRER, J. I. & V. G. ZANNONI. 1987. Modulation of the flavin-containing monooxygenase by ascorbic acid and food restriction. J. Nutrition. **117:** 286-291.
24. BRODFUEHRER, J. I. 1986. Ascorbic acid and the flavin-containing monooxygenase. Ph.D. thesis, University of Michigan.
25. HOUSTON, J. B. & G. LEVY. 1976. Drug biotransformation interaction in man. VI. Acetaminophen and ascorbic acid. J. Pharm. Sci. **65:** 1218-1221.
26. HINSON, J. A., S. D. NELSON & J. R. MITCHELL. 1977. Studies on the microsomal formation of arylating metabolites of acetaminophen and phenacetin. Mol. Pharmacol. **13:** 625-633.
27. HINSON, J. A., L. R. POHL & J. R. GILLETTE. 1979. N-Hydroxyacetaminophen: A microsomal metabolite of N-hydroxyphenacetin but apparently not of acetaminophen. Life Sci. **24:** 2133-2138.
28. CORCORAN, G. B., J. R. MITCHELL, Y. N. VAISHNAV & E. C. HORNING. 1980. Evidence that acetaminophen and N-hydroxyacetaminophen form a common arylating intermediate, N-acetyl-p-benzoquinoneimine. Mol. Pharmacol. **18:** 536-542.
29. HINSON, J. A., L. R. POHL, T. J. MONKS, J. R. GILLETTE & F. P. GUENGERICH. 1980. 3-Hydroxyacetaminophen: A microsomal metabolite of acetaminophen. Evidence against an epoxide as the reactive metabolite of acetaminophen. Drug Metab. Dispos. **8:** 289-294.
30. LAKE, G., R. A. HARRIS, J. C. PHILLIPS & S. D. GANGOLLI. 1981. Studies on the effects of L-ascorbic acid on acetaminophen-induced hepatotoxicity. Toxicol. Appl. Pharmacol. **60:** 229-240.
31. GILLETTE, J. R., S. D. NELSON, G. J. MULDER, D. J. JOLLOW, J. R. MITCHELL, L. R. POHL & J. A. HINSON. 1982. Formation of chemically reactive metabolites of phenacetin and acetaminophen. *In* Biological Reactive Intermediates. II. Chemical Mechanisms and Biological Effects (part B). R. Snyder, D. V. Parke, J. J. Kocsis, G. Gibson & D. J. Jollow, Eds.: 931-950. Plenum Press. New York.
32. ZANNONI, V. G., E. G. MARKER & S. S. LAU. 1982. Hepatic bromobenzene epoxidation and binding: Their prevention by ascorbyl palmitate. J. Drug Nutrient Interact. **1:** 193-204.
33. MILLER, M. G. & D. J. JOLLOW. 1983. Effect of ascorbic acid on acetaminophen-induced hepatotoxicity and covalent binding in hamsters. Drug Metabol. Dispos. **12:** 271-279.
34. SMART, R. C. & V. G. ZANNONI. 1984. DT-diaphorase and peroxidase influence the covalent binding of the metabolites of phenol, the major metabolite of benzene. Mol. Pharmacol. **26:** 105-111.
35. SMART, R. C. & V. G. ZANNONI. 1985. Effect of ascorbate on covalent binding of benzene and phenol metabolites to isolated tissue preparations. Toxicol. Appl. Pharm. **77:** 334-343.
36. SMART, R. C. & V. G. ZANNONI. 1986. Effect of dietary ascorbate on covalent binding of benzene to bone marrow and hepatic tissue *in vivo.* Biochem. Pharmacol. **35:** 3180-3182.
37. SNYDER, R. & J. J. KOCSIS. 1975. Current concepts of chronic benzene toxicity. CRC Crit. Rev. Toxicol. **3:** 265-288.
38. SNYDER, R. S., S. L. LONGACRE, C. M. WITMER, J. J. KOCSIS, L. S. ANDREWS & E. W. LEE. 1981. Biochemical toxicology of benzene. *In* Reviews in Biochemical Toxicology, Vol. 3. E. Hodgson, J. R. Bend & R. M. Philpot, Eds.: 123-153. Elsevier/North-Holland. New York.
39. GOLDSTEIN, B. D. 1977. Hematotoxicity in humans. *In* Benzene Toxicity, a Critical Evaluation. J. Toxicol. Environ. Health (Suppl.) **2:** 69-105.
40. LEE, E. W., J. J. KOCSIS & R. SNYDER. Dose dependent inhibition of ^{59}Fe incorporation into erythrocytes after a single dose of benzene. Res. Commun. Chem. Pathol. Pharmacol. **5:** 547-550.
41. ANDREWS, L. S., E. W. LEE, C. M. WITMER, J. J. KOCSIS & R. SNYDER. 1977. Effects of toluene on the metabolism, disposition and hemopoietic toxicity of ^{14}C-benzene. Biochem. Pharmacol. **26:** 293-300.
42. NOMIYAMA, K. 1962. Studies on the poisoning by benzene and its homologues: Oxidation rate of benzene and benzene poisoning. Med. J. Shinshu Univ. **7:** 41-48.
43. SAMMETT, D., E. W. LEE, J. J. KOCSIS & R. SNYDER. 1979. Partial hepatectomy reduces both metabolism and toxicity of benzene. J. Toxicol. Environ. Health **5:** 785-792.
44. BOLCSAK, L. E. & D. E. NERLAND. 1983. Inhibition of erythropoiesis by benzene and benzene metabolites. Toxicol. Appl. Pharmacol. **69:** 363-368.

45. TUNEK, A., K. L. PLATT, P. BENTLEY & F. OESCH. 1978. Microsomal metabolism of benzene to species irreversibly binding to microsomal protein and effects of modifications of this metabolism. Mol. Pharmacol. **14:** 920-929.
46. RICKERT, D. E., T. S. BAKER, J. S. BUS, C. S. BARROW & R. D. IRONS. 1979. Benzene disposition in the rat after exposure by inhalation. Toxicol. Appl. Pharmacol. **49:** 417-423.
47. GREENLEE, W. F., J. D. SUN & J. S. BUS. 1981. A proposed mechanism of benzene toxicity: Formation of reactive intermediates from polyphenol metabolites. Toxicol. Appl. Pharmacol. **59:** 187-195.
48. SAWAHATA, T. & R. A. NEAL. 1983. Biotransformation of phenol to hydroquinone and catechol by rat liver microsomes. Mol. Pharmacol. **23:** 453-460.
49. JOHANSSON, I. & M. INGELMAN-SUNDBERG. 1983. Hydroxyl radical mediated cytochrome P-450-dependent metabolic activation of benzene in microsomes and reconstituted enzyme systems from rabbit liver. J. Biol. Chem. **258:** 7311-7316.
50. CHANCE, B. & A. C. MAEHLY. 1955. Assay of catalases and peroxidases. *In* Methods in Enzymology, Vol. II. S. P. Colowick & N. O. Kaplan, Eds.: 764. Academic Press. New York.
51. VANHA-PERTTULA, T. P. J. 1960. The influence of vitamin C on eosinophil response to acute alcohol intoxication in rats. Acta Endocrinol. **35:** 585-593.
52. DILUZIO, N. R. 1964. Prevention of the acute ethanol-induced fatty liver by the simultaneous administration of antioxidants. Life Sci. **3:** 113-118.
53. TEPHLY, T. R., M. ATKINS, G. J. MANNERING & R. E. PARKS, JR. 1965. Activation of a catalase peroxidative pathway for the oxidation of alcohols in mammalian erythrocytes. Biochem. Pharmacol. **14:** 435-444.
54. PAWAN, G. S. L. 1968. Vitamins, sugars and ethanol metabolism in man. Nature **220:** 374-376.
55. BUSNEL, R. G. & A. G. LEHMANN. 1980. Antagonistic effect of sodium ascorbate on ethanol-induced changes in swimming of mice. Behav. Brain Res. **1:** 351-356.
56. YUNICE, A. A. & R. D. LINDEMAN. 1977. Effect of ascorbic acid and zinc sulfate on ethanol toxicity and metabolism. Proc. Soc. Exp. Biol. Med. **154:** 146-150.
57. SUSICK, R. L., JR. & V. G. ZANNONI. 1984. Ascorbic acid and alcohol oxidation. Biochem. Pharmacol. **24:** 3963-3969.
58. SUSICK, R. L., JR. & V. G. ZANNONI. 1987. Ascorbic acid and elevated SGOT levels after an acute dose of ethanol in the guinea pig. Alcoholism: Clin. Exp. Res. **11** (3).
59. SUSICK, R. L., JR. & V. G. ZANNONI. 1987. Effect of ascorbic acid on the consequences of acute alcohol consumption in man. Clin. Pharmacol. Ther. **41:** 502-509.
60. SUSICK, R. L., JR., G. D. ABRAMS, C. A. ZURAWSKI & V. G. ZANNONI. 1986. Ascorbic acid and chronic alcohol consumption in the guinea pig. Toxicol. Appl. Pharmacol. **84:** 329-335.
61. YUNICE, A. A., J. M. FAHMY & S. HENRY. 1984. Ethanol-ascorbate interrelationship in acute and chronic alcoholism in the guinea pig. Proc. Soc. Exp. Biol. Med. **177:** 262-271.
62. OHEL, G., R. KISSELEVITZ, E. J. MARGALIOTH & J. G. SCHENKER. 1985. Ascorbate-dependent lipid peroxidation in the human placenta and fetal membranes. Gynecol. Obstet. Invest. **19:** 73-77.
63. SPRINCE, H. 1981. Ascorbic acid, sulfur compounds, and antiadrenergic agents as protectants against acetaldehyde toxicity: implications in alcoholism and smoking. Brit. J. Alcohol Alcoholism **16:** 5-9.
64. MAJCHROWICZ, E. & E. P. NOBLE. 1979. Biochemistry and Pharmacology of Ethanol, Vol. 2. Plenum Press. New York.
65. LIEBER, C. 1982. Medical disorders of alcoholism: Pathogenesis and treatment. Major Health Problems in Internal Medicine, Vol. 22. W. B. Saunders Company. Philadephia.
66. KRASNER, N., M. R. MOORE, J. DOW & A. GOLDBERG. 1974. Ascorbic acid and ethanol metabolism. Lancet **21:** 693-695.
67. BARAONA, E. & C. S. LIEBER. 1979. Effect of ethanol lipid metabolism. J. Lipid Res. **20:** 289-315.
68. HANSSON, P. & P. NILSSON-EHLE. 1983. Acute effects of ethanol and its metabolism on plasma lipids and lipoprotein lipase activity. Ann. Nutr. Metab. **27:** 328-337.
69. CROUSE, J. R. & S. M. GRUNDY. 1984. Effects of alcohol on plasma lipoproteins and cholesterol and triglyceride metabolism in man. J. Lipid Res. **25:** 486-496.

70. SUBRAMANIAN, N. 1977. On the brain ascorbic acid and its importance in metabolism of biogenic amines. Life Sci. **20:** 1479-1484.
71. LOHMANN, W. 1984. Structure of ascorbic acid and its biological function. VI. Its importance for Na^+/K^+ transport. Biophys. Struct. Mech. **10:** 205-210.
72. REBEC, G. V., J. M. CENTORE, L. K. WHITE & K. D. ALLOWAY. 1985. Ascorbic acid and the behavioral response to haloperidol: Implications for the action of antipsychotic drugs. Science **227:** 438-440.
73. TOLBERT, L. C., T. N. THOMAS, L. D. MIDDAUGH & J. W. ZEMP. 1979. Effect of ascorbic acid on neurochemical, behavioral, and physiological systems mediated by catecholamines. Life Sci. **25:** 2189-2195.

DISCUSSION OF THE PAPER

B. LANE (*Columbia University School of Public Health, New York, N.Y.*): Did it look like there was a worsening of color vision following the pretreatment?

V. G. ZANNONI (*University of Michigan Medical School, Ann Arbor, Mich.*): Color vision is improved.

M. LEVINE (*National Institutes of Health, Bethesda, Md.*): Will your effects in the guinea pig studies on alcohol be specific for ascorbic acid in terms of increasing alcohol metabolism? Did you try other reducing agents or was this effect specific for alcohol for ascorbic acid?

V. G. ZANNONI: We didn't try other reducing agents *in vivo* but we did measure the glutathione levels in the two groups and they were exactly the same. However, in the *in vitro* experiments as far as alcohol metabolism is concerned, it's relatively specific for ascorbic acid or analogues of ascorbic acid; other reducing agents have absolutely no effect.

B. N. LADU (*University of Michigan Medical School, Ann Arbor, Mich.*): Didn't the involvement of the catalase system as a peroxidase require the generation of peroxide, which is coming from ascorbic acid, so theoretically anything that would generate peroxide in a similar way should bring the catalases into account.

S. L. ROMNEY (*Albert Einstein College of Medicine, Bronx, N.Y.*): This is not meant to be facetious. On basis of your work would you recommend that liquor manufacturers supplement with some form of ascorbic acid?

V. G. ZANNONI: No actually I'd recommend that people drink less and have a few oranges a day.

Measurement of Vitamin C in Blood Components by High-Performance Liquid Chromatography

Implication in Assessing Vitamin C Status

STANLEY T. OMAYE,[a] ELLEN E. SCHAUS,
MARK A. KUTNINK, AND WAYNE C. HAWKES

Biochemistry Research Unit
Western Human Nutrition Research Center
U. S. Department of Agriculture, ARS
Presidio of San Francisco, California 94129

The search for an understanding of the biochemical mechanism of the action of vitamin C remains dynamic. Even the relationship between the deficiency disease scurvy and the biochemical mode of action is not clear,[1-3] and as a consequence we are uncertain what clinical or chemical measures best reflect the vitamin's status. Likewise for this reason, the development of methodologies for assessing vitamin C status has remained a dynamic process that has been hampered by the issues of (1) what form(s) of the vitamin should be measured, (2) what measurable body tissue best reflects vitamin C status, and (3) what functional process is specific for vitamin C and reflects the vitamin's status.

There are now many published techniques and procedures for the determination of ascorbic acid that are applicable to many different types of biological materials. In general, the chemical methods used[4,5] are time-consuming, cumbersome, and often inaccurate due to the presence of interfering substances. Even more important, these methods usually lack the sensitivity to deal with the limited sample sizes that are usually available from preparations of blood components. We have chosen to concentrate our efforts on the development of a high-performance liquid chromatographic (HPLC) method for quantifying ascorbate in blood components, since HPLC is sensitive enough to measure very small amounts of the vitamin.

It is almost certain that vitamin C acts intracellularly by modifying or assisting enzyme or biochemical reactions. Therefore, it might well be argued that the best measurement of vitamin C status would be a measure of its biological activity at the site of its action.[6-8] However, until the fundamental role of vitamin C can be determined, a test for the vitamin based on its function is unlikely. Like others,[9,10] we have focused our resources on cellular analysis in an attempt to provide a better understanding of whether or not white blood cell components reflect the body's vitamin C status.

[a] Present address: Toxicology Division, Letterman Army Institute of Research, Presidio of San Francisco, Calif. 94129.

PLASMA, WHOLE BLOOD, RBC, OR URINE ASCORBIC ACID

During the past decade there has been an increasing awareness of the limitations of measuring nutrients in serum and urine for the purpose of assessing nutrient status.[6–8] Plasma and serum are readily available and have been used by many in an attempt to assess body reserves of the vitamin. The theory is that the ascorbate in these biological fluids will equilibrate between the various organs of the body. But plasma and serum ascorbic acid concentrations are usually more reflective of recent intakes than of total body stores.[11] After consuming ascorbic acid, plasma values rise rapidly within 1 or 2 hours,[12] reaching a peak concentration in about 3 to 6 hours. When the vitamin intake is poor, the plasma concentration can become depleted even when tissue reserves remain adequate.[13] Plasma ascorbic acid is also rapidly affected by acute illness, environmental pollutant exposures, and physiological-endocrine changes.[2,4]

Whole blood and/or red blood cell (RBC) ascorbic acid values seem to reflect loss of body stores only until a certain degree of deficiency is reached. A relationship between whole blood levels of ascorbic acid and the vitamin's body pool exists only when the body pool is in excess of 300 mg (about 20% of saturation), and as the body pool decreases below that, no further decrease in whole blood levels is noted.[14]

The use of urinary levels of ascorbic acid has limited value because the excretion of ascorbic acid in the urine ceases when the body pool size of ascorbic acid drops to only slightly below normal.[15] Urinary ascorbic acid values at best reflect recent dietary intakes of the vitamin and often, because of degradation to other compounds,[5] can lead to low estimates.

LEUKOCYTE ASCORBIC ACID

As previously stated, during the past few years there has been an increasing awareness of the limitations of using plasma or serum to assess nutrient status. Investigators have subsequently turned to cellular analyses to provide a better understanding of the total body status of a nutrient. Selecting a cell whose ascorbic acid concentration could be used as an index of status does not necessarily require knowledge of the role of vitamin C in that cell. Other considerations may be equally important. For example, among blood cells it is likely that leukocytes have some advantages compared to red cells. They are arguably more representative of lean tissue in that they have a nucleus, mitochondria, and all the other organelles, and they are not dedicated to the transport of one unique protein as red blood cells are.

In TABLE 1 we have listed the human tissue distribution of ascorbic acid. The lowest concentrations of ascorbic acid appear to be in the plasma. This table supports the concept, demonstrated in other publications,[16,17] that transfer of the vitamin from blood to tissue must occur against a concentration gradient, most likely through active transport mechanisms. Although no data have been found to clearly show that any one tissue or organ acts as a storage reservoir for vitamin C, leukocytes must be regarded as likely candidates. Correlations between leukocyte and plasma ascorbic acid levels suggest that leukocyte levels reflect the amount of ascorbic acid available for storage while plasma ascorbate levels reflect the metabolic turnover rate of the vitamin.[5] Recently, in a study with young adult males, it was shown that during

TABLE 1. Ascorbic Acid Content of Adult Human Tissues[a]

Tissue	Concentration of Ascorbic Acid: mg/100 g Wet Tissue	mg/Total Tissue	% of Total Pool
Pituitary gland	40-50	0.22-0.28	0.01
Adrenal glands	30-40	3-4	0.17
Eye lens	25-31		
Brain	13-15	185-215	10.3
Liver	10-16	158-253	8.8
Spleen	10-15	16-24	0.9
Kidneys	5-15	15-24	0.8
Heart muscle	5-15	16-48	0.9
Lungs	7	52-155	2.9
Skeletal muscle	3	1200	66.6
Testes	3	1	0.06
Thyroid	2	0.55	0.03
Leukocytes	35	63	3.5
Plasma	0.4-1.0	93	5.2

[a] Calculations based on previously published report[4] and assumptions based on a 70 kg adult standard.

vitamin C repletion leukocyte ascorbic acid responded to dietary intake of the vitamin, but not as rapidly as plasma ascorbic acid.[18] The implication is that tissue levels of ascorbic acid are replenished at a slower rate than plasma ascorbic acid. In addition, it has been shown (TABLE 2) that measurement of the ascorbate present in a mixed leukocyte population has the best correlation to liver stores of the vitamin ($r = 0.683$) and to the total body pool ($r = 0.923$) in monkeys.[3,9] This has been confirmed in the guinea pig.[16]

Compared to plasma, the cells contributing to the leukocyte population, mononuclear (MN), polymorphonuclear (PMN), and platelets (PL), show a marked ability to concentrate the vitamin.[19] In addition, mixed leukocyte ascorbic acid levels drop slowly during repletion, only reaching zero just before the onset of clinical symptoms

TABLE 2. Relationship between Blood Ascorbic Acid and Liver or Body Pool Ascorbic Acid in the Subhuman Primate

Blood Component	r	p
LIVER		
Whole blood	< 0.15	ns
Plasma	0.477	< 0.05[a]
Leukocytes	0.683	< 0.01
BODY POOL		
Whole blood	< 0.1	ns
Red blood cells	< 0.1	ns
Plasma	< 0.1	ns
Leukocytes	0.923	< 0.001

[a] When plasma ascorbic acid was > 0.1 μg/ml.[9]

of scurvy.[13,20] For these reasons, leukocyte ascorbic acid levels have often been used as an easy-to-sample "best" indicator of vitamin C status in humans.

But measuring the vitamin in a mixed population of cells is fairly crude since MN, PMN, and PL contain very different amounts of ascorbic acid.[19,21–24] In addition, the relative number of each cell type can vary between individuals and can be influenced by the number of circulating leukocytes[25] and by the method used to isolate them.[19] Leukocyte ascorbic acid content has also been shown to be influenced by many other factors such as stress,[26] surgery,[27] myocardial infarction,[28] tobacco smoking,[29] and exercise.[30]

For these reasons, it seems necessary to separate the leukocyte population into MN, PMN, and PL fractions. Even if ascorbic acid can routinely be measured in different cell types, understanding how the ascorbic acid content of the cell components varies under different states of health or vitamin C intake will still provide us with a "static" index but will not address the fundamental question of what role vitamin C plays in these different cells. In order to investigate functional roles that ascorbic acid plays within leukocyte cell types, methodology must be developed to measure ascorbic acid within reasonably pure cell populations with limited sample sizes.

ISOLATION OF BLOOD PLATELETS AND LEUKOCYTES

Fifteen ml blood is drawn from volunteers into heparinized tubes. To minimize cell damage, plastic equipment is used throughout the procedure and cell pellets are resuspended with minimum delay by gently drawing the cells in and out of plastic pasteur pipettes. FIGURE 1 illustrates the steps of cell isolation. Whole blood is immediately centrifuged at 90 × g for 15 minutes and the upper platelet-rich plasma (PRP) layer is removed. An aliquot of PRP is used for counting MN, PMN, red blood cells (RBC), and platelets, and the remaining PRP is centrifuged at 800 × g for 10 minutes to obtain the platelet pellet. The upper platelet-poor plasma (PPP) layer is removed, differential cell counts are performed, and the cells are preserved for vitamin analysis. Platelets are resuspended in Hanks' balanced salt solution (HBSS) without Ca^{2+} or Mg^{2+} and recentrifuged at 800 × g for 10 minutes two times to remove contaminating plasma. The washed platelet pellet is resuspended in 50 μl of 5% metaphosphoric acid (MPA) containing 0.54 mM disodium EDTA and stored at −70° C until analysis.

Mononuclear leukocytes and polymorphonuclear leukocytes are isolated by a modification of several established procedures.[31–35] The blood remaining after removal of the PRP is diluted 1:1 with HBSS and kept at 4°C until layered onto a discontinuous Percoll gradient with a top density of 1.085 g/l and a bottom density of 1.095 g/l. The layered cells are centrifuged at 700 × g for 30 minutes at room temperature. The top layer of MN leukocytes and the bottom layer of PMN leukocytes are removed with a plastic pasteur pipette. Mononuclear leukocytes are washed and centrifuged once with HBSS and brought to a known volume, and differential cell counts are performed. The remaining MN leukocytes are centrifuged into a pellet and extracted with 5% MPA and 0.01% disodium EDTA and stored at −70°C until analysis. The PMN leukocytes are washed once with HBSS, centrifuged, and then brought up to a known volume; aliquots are taken for differential cell counts. The remaining cells are extracted with 5% MPA containing EDTA and stored at −70°C until analysis.

The erythrocytes remaining after removing the MN and PMN leukocytes are washed twice with HBSS and centrifuged at 800 × g for 15 minutes at room temperature. Aliquots are used for cell counting and 1 ml of erythrocytes is extracted with an equal volume of 10% MPA and centrifuged at 800 × g for 20 minutes. The supernatant is stored at −70°C until analysis.

Mononuclear and polymorphonuclear leukocytes are counted manually using a hemocytometer. Erythrocytes and platelets in the cell fractions are measured with an automated cell counter. Microscopic examination of the various cell fractions is carried out using standard hematologic techniques. Cell viability is determined by the Trypan blue exclusion test. Results indicate more than 85% of the cells are viable.

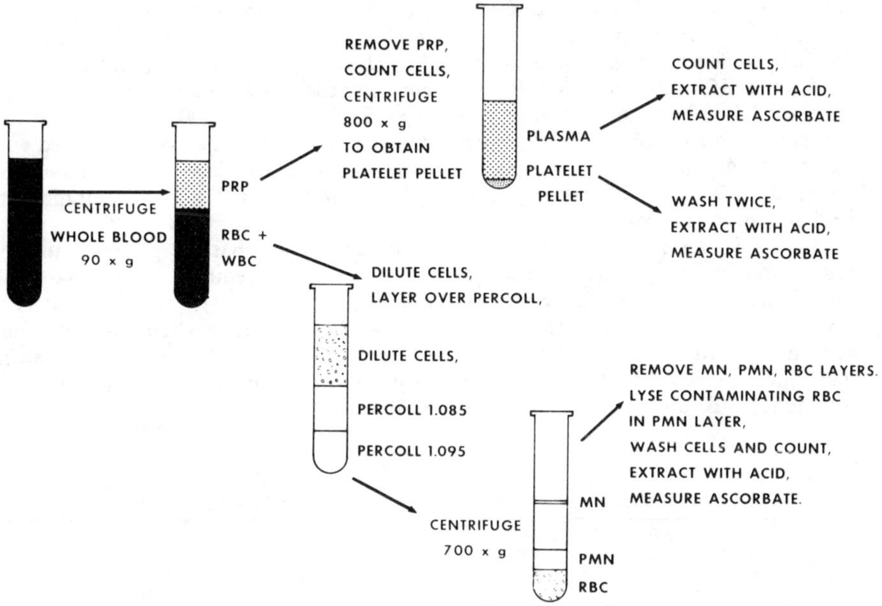

FIGURE 1. Diagrammatic representation of discontinuous gradient centrifugation procedure for the differential isolation of PMN, MN, platelets, and RBC.

HPLC DETERMINATION OF ASCORBIC ACID

Several HPLC methods have been reported for the determination of ascorbic acid in various types of samples.[5,36–38] They employ a variety of columns and elution conditions to separate ascorbic acid from interfering substances, isomers, and metabolites. A desirable HPLC method for ascorbic acid analysis should be simple and, if it cannot be rapid, it should be automated so that excessive operator time is not required. A simple method to determine plasma and leukocyte ascorbic acid is described in the following section. As with most analytical methods for biological materials, sample preparation consumes the major portion of the time and effort. It is worthwhile to note that we are very interested in coupling the automated procedure described here with some modification of the robotic extraction procedure recently published by Vanderslice and Higgs.[39]

Cell ascorbic acid is extracted into cold 5% MPA containing 0.54 mM disodium EDTA. Plasma is mixed with an equal volume of cold (4°C) 10% MPA containing 0.54 mM disodium EDTA. The precipitated plasma proteins are removed by centrifugation and the supernatants frozen at −70°C until analysis. Upon thawing, samples are mixed with an equal volume of cold 5% MPA containing 0.54 mM disodium EDTA and the internal standard, isoascorbic acid. The resulting solutions are diluted 25-fold with cold 1.04 mM cysteine-HC1 containing 0.54 mM disodium EDTA and filtered through 0.2-micron nylon filters into glass HPLC vials. Final concentrations of MPA, EDTA, and cysteine HC1 in both standards and samples are 0.2% (wt/vol), 0.54 mM, and 1.0 mM respectively. The internal standard concentration is 0.2 μg/ml. Under refrigerated conditions, samples and standards can be held up to at least 24 hours with negligible change in the calculated ascorbic acid values.

A Perkin-Elmer chromatography system composed of a Series 4 solvent delivery system, LCI 100 integrator tied on line with a Perkin-Elmer 7500 computer for data processing, and an ISS 100 refrigerated autosampler is used. The system is fitted with a 25 cm × 4.6 mm Altex Ultrasphere ODS (C18) reversed-phase analytical column, spherical particle size of 5 microns, and a 3 cm Brownlee RP18 guard column with 5-micron spherical particles.

TABLE 3 summarizes the chromatographic conditions that are presently used in our laboratory. They have been adapted from earlier procedures[37,38] to allow unattended analysis of up to 70 samples within a 24-hour period. The mobile phase is 40 mM sodium acetate, 0.54 mM disodium EDTA, 1.5 mM dodecyltriethylammonium phosphate (paired-ion agent) and 7.5% methanol, taken to pH 4.75 with glacial acetic acid. Elution is isocratic at ambient temperature and a 0.8 ml/min flow rate. Injection volume is 50 μl and the analysis time is 20 minutes.

TABLE 3. HPLC Parameters for the Analysis of Blood Ascorbic Acid

Column	Altex Ultrasphere ODS (C18): 5 micron, 250 × 4.6 mm Brownlee Guard Column (C18): 5 micron, 30 × 4.6 mm
Mobile Phase	sodium acetate 40 mM disodium EDTA 0.54 mM dodecyltriethylammonium phosphate 1.5 mM methanol 7.5% pH adjusted to 4.75 with glacial acetic acid filtered through a 0.2 micron nylon filter isocratic elution, 0.8 ml/minute
Detector	LC4B Amperometric Controller (Bioanalytical Systems) glassy-carbon working electrode stainless steel auxiliary electrode Ag/AgCl reference electrode applied potential + 0.5 V, range 50 nA
Chromatograph	Series 4 solvent delivery system (Perkin-Elmer) ISS-100 autosampler (with refrigeration) LCI-100 computing integrator-recorder injection volume 50 μl
Samples and Standards	all contain 10 ng isoascorbic acid as internal standard per 50 μl injection in: 0.2% (wt/vol) metaphosphoric acid, 0.54 mM disodium EDTA, 1.0 mM cysteine HCl

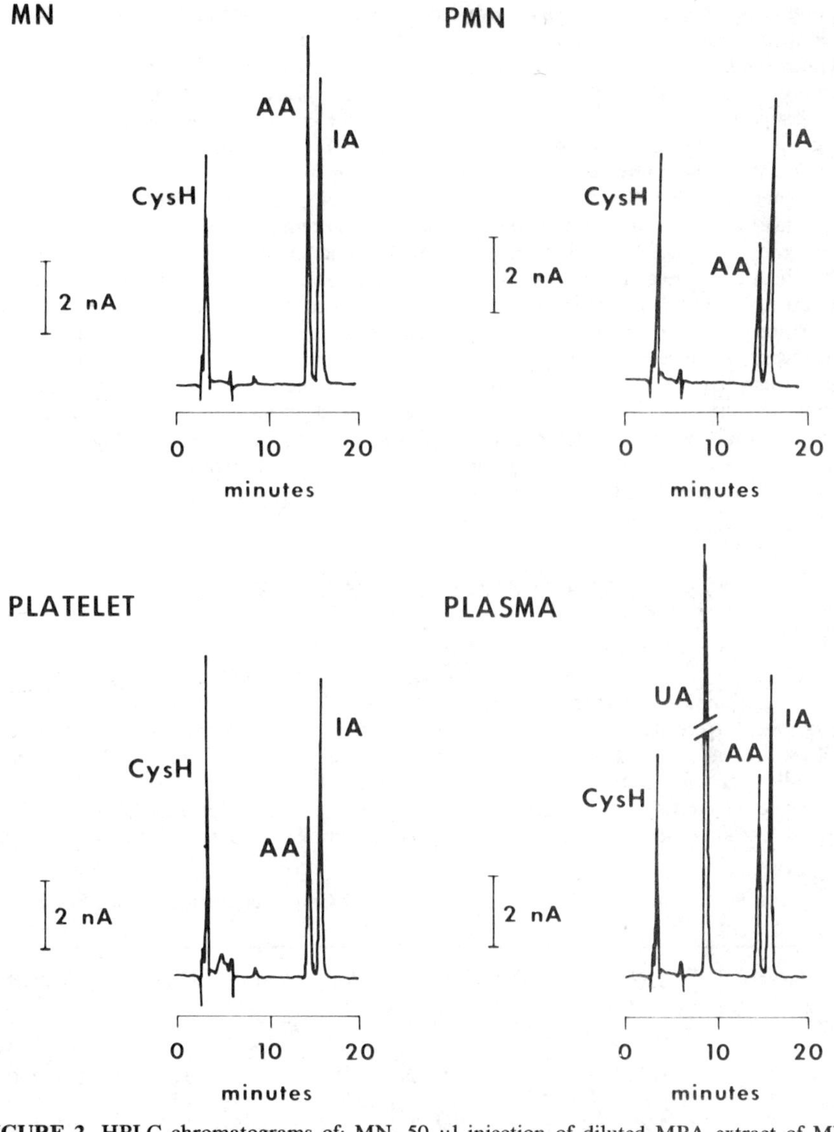

FIGURE 2. HPLC chromatograms of: MN, 50 μl injection of diluted MPA extract of MN cells (10.8 ng ascorbic acid, 10 ng isoascorbic acid); PMN, 50 μl injection of diluted MPA extract of PMN cells (4.8 ng ascorbic acid, 10 ng isoascorbic acid); platelets, 50 μl injection of diluted MPA extract of platelets (5.2 ng ascorbic acid, 10 ng isoascorbic acid); and plasma, 50 μl injection of diluted MPA extract of plasma (6.5 ng ascorbic acid, 10 ng isoascorbic acid).

An amperometric detection system (BAS, West Lafayette, Ind.), which includes an LC4B controller and an electrode flowcell consisting of a glassy-carbon electrode, a stainless steel auxiliary electrode, and an Ag/AgC1 reference electrode is used, with an applied potential of +0.5 V (oxidative) and a sensitivity range of 50 nA. Quantitation is based on graded levels of ascorbic acid standards containing a constant amount of isoascorbic acid as an internal standard. A least-square linear regression calibration plot of ascorbic acid/isoascorbic acid standard peak area ratio is used to determine sample ascorbic acid concentrations. The calibration curves are linear over a range of 0.5 to at least 30 ng ascorbic acid in the presence of 10 ng isoascorbic acid per 50 μl injection volume. On a routine basis we can detect as little as 0.02 ng and quantitate as little as 0.2 ng per 50 μl injection. Typical chromatograms for plasma, PMN, MN, and platelet ascorbic acid are shown in FIGURE 2. We observe a broad asymmetric peak following the cysteine (CysH) peak in extracts of platelets or cells with heavy platelet contamination. Further work is required to determine the chemical nature of this peak and whether it may serve as an index of platelet contamination.

The direct measurement of UV absorbance at 245 nm is relatively insensitive, requiring as much as 20 times the concentration of ascorbic acid.[4,40] Amperometric detection has been shown to be very sensitive for ascorbic acid measurement.[41] The disadvantage of electrochemical detection is that the oxidation product of ascorbic acid, dehydroascorbic acid, cannot be detected in the oxidative mode.[421] Vitamin C occurs naturally in two forms, most in the reduced form as ascorbic acid and minor quantities in the oxidized form as dehydroascorbic acid. Since most biological samples have a much greater content of ascorbic acid than dehydroascorbic acid, methods that determine the reduced form are useful.[4,5,43]

EXPERIMENTAL OBSERVATIONS

The ascorbic acid concentration in plasma, platelets, MN, PMN and RBC from six healthy males was measured by HPLC and differential cell counts in each blood component were used to quantify cross-contamination. The means of four replicate values for each individual's blood components are shown in TABLE 4 along with the overall mean values obtained for the six subjects. HPLC replicates agree fairly well, with the variability among plasma values being less than for PL, MN, and PMN replicates. This most likely reflects the variability inherent to cell counting techniques,

TABLE 4. Ascorbic Acid Content in the Various Cell Preparations from Six Healthy Males[a]

Subject No.	Plasma (mg/dl)	Polymorphonuclear (μg/10^8 Cells)	Mononuclear (μg/10^8 Cells)	Platelets (μg/10^8 Cells)
1	0.45 ± 0.007	2.60 ± 0.08	12.3 ± 1.56	0.302 ± 0.007
2	0.72 ± 0.007	1.47 ± 0.01	11.1 ± 0.04	0.214 ± 0.057
3	0.87 ± 0.007	2.37 ± 0.19	7.3 ± 1.55	0.235 ± 0.012
4	1.28 ± 0.007	5.66 ± 0.03	19.5 ± 1.20	0.294 ± 0.005
5	1.31 ± 0.007	1.17 ± 0.27	23.2 ± 0.30	0.173 ± 0.057
6	1.26 ± 0.007	4.57 ± 0.88	21.5 ± 4.71	0.299 ± 0.006
Mean ± SD	0.98 ± 0.357	2.97 ± 1.78	15.8 ± 6.44	0.253 ± 0.054
CV (%)	36.4	59.9	40.8	21.3
Mean %SD of replicates	0.83	9.11	10.61	11.79

[a] Mean ± SD of replications (not corrected for ascorbic acid contribution from other cell types).

TABLE 5. Mean Percentage Contribution by Different Cell Types to the Total Ascorbic Acid Measured in Various Blood Components[a]

Blood Component	Polymorphonuclear PMN	Mononuclear MN	Platelets	Erythrocytes
plasma	0	0	0.5 (0.3-0.9)	0
platelets	0.1 (0-0.3)	0.1 (0-0.5)	99.8 (99.3-100)	0
MN	2 (0-6)	96 (92-99)	2 (0.1-4.8)	0
PMN	80 (57-97)	20 (4-43)	0	0

[a] Values in parentheses are the range. Percentage contributions are based on predicted ascorbic acid concentration per pure cell type, and numbers of cells contaminating each blood component.

although different levels of ascorbic acid within cells of a fraction, while unlikely, cannot be discounted. In agreement with reports of others,[18,23-26] MN leukocytes have the highest concentration of ascorbic acid when compared to PL or PMN fractions (MN > PMN > PL). Also confirming previous reports,[4,5] our HPLC methodology, which measures only the reduced form of the vitamin, failed to detect any significant levels in RBC.

Simultaneous equations were solved in order to estimate PL, MN, and PMN ascorbic acid per cell for each subject. These values were multiplied by the number of cross-contaminating cell types per fraction in order to calculate the percentage of total ascorbic acid measured that could be attributed to contaminating cell types. As shown in TABLE 5, the percentage of total ascorbic acid contributed by cross-contaminating cells in plasma, PL, and MN was quite small ($< 6\%$). The largest percentage of contaminating ascorbic acid was from MN contamination in the PMN fraction (range = 4-43%). The amount of ascorbic acid per cell derived from simultaneous equations showed a mean ($\pm$ SD) of $1.52 \pm 0.63 \times 10^{-4}$ ng for MN, $0.24 \pm 0.16 \times 10^{-4}$ ng for PMN, and $0.03 \pm 0.005 \times 10^{-4}$ ng for PL (MN > PMN > PL).

TABLE 6 shows our data corrected for ascorbate contributed by cross-contamination and literature values for measurements in the same types of samples using various methods of analysis. Very few investigators[19] have measured all the blood fractions in one study and none have identified or corrected for cross-contamination in all fractions. The most complete report is that of Evans *et al.*,[19] which utilized similar separation methodology but coupled the separation methodology with the colorimetric analysis for ascorbic acid.

SUMMARY

The fact that platelets, PMN leukocytes, and MN leukocytes concentrate ascorbic acid suggests that vitamin C has an important role in their physiological functions.

TABLE 6. Comparison of HPLC Leukocyte Ascorbic Acid Values with the Literature

		Ascorbic Acid Concentration (Mean ± SD)					
Method	Ref.	Plasma (mg/dl)	Platelets ($\mu g/10^8$ Cells)	MN ($\mu g/10^8$ Cells)	PMN ($\mu g/10^8$ Cells)	RBC ($\mu g/10^8$ Cells)	Corrected for Contamination
HPLC, EC		0.97 ± 0.34	0.25 ± 0.05	15.2 ± 6.28	2.43 ± 1.63	nd	yes[a]
HPLC, EC	21	0.53-1.06 (range)	—	5.3-10.6 (range)	2.6-4.4 (range)	—	no
DNPH[c]	19	0.78 ± 0.33	0.55 ± 0.28	24.5 ± 5.80	9.3 ± 2.99	0.058 ± 0.021	yes[b]
DNPH	22	—	—	88.5 ± 29.2	10.9 ± 3.83	—	no
DNPH	23	—	0.34 ± 0.23	—	—	—	no
DNPH	24	0.69 ± 0.06	0.29 ± 0.02	—	—	0.055 ± 0.004	no

[a] Mean values from TABLE 4 corrected for ascorbic acid contribution from other cell fractions (calculated by simultaneous equations).
[b] MN leukocyte fraction corrected for platelet contamination.
[c] DNPH = 2,4-dinitrophenylhydrazine.

The question still remains as to which one of the cells best reflects vitamin C status. The ascorbic acid content of PMNs and platelets correlates positively with plasma concentration and supplementation with vitamin C, as shown in Evans *et al.*[19] They also found that MN leukocytes, in contrast, do not show any such relationship[19]; however, MN leukocytes maintain the highest levels of ascorbic acid and play a very important function in immunocompetence. We have found that with a limited number of subjects, ascorbic acid content of MN and PMN leukocytes correlates positively with plasma ascorbic acid, but there was no correlation between platelets and plasma ascorbic acid (unpublished results). Therefore, further work is necessary to evaluate these three blood components for the best cellular marker of vitamin C status.

We have developed a reversed-phase HPLC method for ascorbic acid that can be used in conjunction with our cellular differential centrifugation technique for the determination of ascorbic acid in relatively pure blood cell fractions. The chromatographic method is simple, sensitive, and automated. It clearly resolves ascorbic acid, which is the major form of the vitamin found *in vivo* and is not prone to interference by sugars, carbohydrates, or nucleotides.

REFERENCES

1. CHOJKIER, M., R. SPANHEIMER & B. PETERKOFSKY. 1984. Specifically decreased collagen biosynthesis in scurvy dissociated from an effect on proline hydroxylation and correlated with body weight loss: In vitro studies in guinea pig calvarial bones. J. Clin. Invest. **72:** 826.
2. LEVINE, M. 1986. New concepts in the biology and biochemistry of ascorbic acid. N. Engl. J. Med. **314:** 892.
3. OMAYE, S. T., J. A. TILLOTSON & H. E. SAUBERLICH. 1982. Metabolism of L-ascorbic acid in the monkey. *In* Advances in Chemistry. P.A. Seib & B. M. Tolbert, Eds.: 318-334. American Chemical Society. Washington, D.C.
4. OMAYE, S. T., J. D. TURNBULL & H. E. SAUBERLICH. 1979. Selected methods for the determination of ascorbic acid in animal cells, tissues and fluids. Methods Enzymol. **62:** 1.
5. SAUBERLICH, H. E., M. D. GREEN & S. T. OMAYE. 1982. Determination of ascorbic acid and dehydroascorbic acid. *In* Advances in Chemistry. P. A. Seib & B. M. Tolbert, Eds.: 199-221. American Chemical Society. Washington, D.C.
6. SOLOMONS, N. W. & L. H. ALLEN. 1983. The functional assessment of nutritional status: Principles, practice and potential. Nutr. Rev. **41:** 33.
7. RUSSELL, R. M., S. F. KRASINSKI & B. DAWSON-HUGHES. 1984. Indices of fat-soluble vitamin states. Clin. Nutr. **3:** 161.
8. PATRICK, J. & C. DERVISH. 1984. Leukocyte zinc in the assessment of zinc status. Crit. Rev. Clin. Lab. Sci. **20:** 95.
9. TURNBULL, J. D., J. H. SUDDUTH, H. E. SAUBERLICH & S. T. OMAYE. 1981. Depletion and repletion of ascorbic acid in the Rhesus monkey: Relationship between ascorbic acid concentration in blood components with total body pool and liver concentrations of ascorbic acid. Int. J. Vitam. Nutr. Res. **51:** 47.
10. ELIN, R. J. 1983. The status of cellular analysis. J. Am. Coll. Nutr. **4:** 329.
11. SAUBERLICH, H. E., R. P. DOWDY & J. H. SKALA. 1974. *In* Laboratory Tests for the Assessment of Nutritional Status. CRC Press. Boca Raton, Fla.
12. COULEHAN, J. L., S. EBERHARD, L. KAPNER, F. TAYLOR, K. ROGERS & P. GARRY. Vitamin C and acute illness in Navajo schoolchildren. 1976. N. Engl. J. Med. **295:** 973.
13. CRANDON, J. H., C. C. LUND & D. B. DILL. 1940. Experimental human scurvy. N. Engl. J. Med. **223:** 353.

14. BAKER, E. M., R. E. HODGES, J. HOOD, H. E. SAUBERLICH, S. C. MARCH & J. E. CANHAM. 1971. Metabolism of ascorbic-1-^{14}C acid in experimental human scurvy. Am. J. Clin. Nutr. **24:** 444.
15. HODGES, R. E., E. M. BAKER, J. HOOD, H. E. SAUBERLICH & S. C. MARCH. 1969. Experimental scurvy in man. Am. J. Clin. Nutr. **22:** 535.
16. STEVENSON, N. R. & M. K. BRUSH. 1969. Existence and characteristics of Na^+-dependent active transport of ascorbic acid in the guinea pig. Am. J. Clin. Nutr. **22:** 318.
17. HORNIG, D. 1975. Metabolism of ascorbic acid. World Rev. Nutr. Diet. **23:** 225.
18. OMAYE, S. T., J. H. SKALA & R. A. JACOB. 1986. Plasma ascorbic acid in adult males: Effects of depletion and supplementation. Am. J. Clin. Nutr. **44:** 257.
19. EVANS, R. M., L. CURRIE & A. CAMPBELL. 1982. The distribution of ascorbic acid between various cellular components of blood in normal individuals, and its relation to the plasma concentration. Br. J. Nutr. **47:** 473.
20. BASU, T. K. & C. J. SCHORAH. 1982. *In* Vitamin C in Health and Disease. AVI Publ. Co., Inc. Westport, Conn.
21. LEE, W., P. HAMERNYIK, M. HUTCHINSON, V. A. RAISYS & R. F. LABBE. 1982. Ascorbic acid in lymphocytes: Cell preparation and liquid chromatographic assay. Clin. Chem. **28:** 2165.
22. IKEDA, T. 1984. Comparison of ascorbic acid concentrations in granulocytes and lymphocytes. Tohoku J. Exp. Med. **142:** 117.
23. ATTWOOD, E. D., E. D. ROBEY, J. ROSS, F. BRADLEY & J. J. KRAMER. 1974. The determination of platelet and leukocyte vitamin C and the levels found in normal subjects. Clinica Chimica Acta **54:** 95.
24. BARKHAN, P. & A. N. HOWARD. 1958. Distribution of ascorbic acid in normal and leukaemic blood. Biochem. J. **70:** 163.
25. VALLANCE, S. 1979. Leukocyte vitamin C and the leukocyte count. Br. J. Nutr. **41:** 409.
26. LOH, H. S. 1972. The relationship between dietary ascorbic acid intake and buffy coat and plasma ascorbic acid concentrations at different ages. Int. J. Vitam. Nutr. Res. **42:** 80.
27. VALLANCE, S. 1986. Platelets, leucocytes and buffy layer vitamin C after surgery. Human Nutr.: Clin. Nutr. **40C:** 35.
28. HUME, R., E. WEYERS, T. ROWAN, D. S. REID & W. S. HILLIS. 1972. Leucocyte ascorbic acid levels after acute myocardial infarction. Br. Heart J. **40:** 64.
29. PELLETIER, O. 1970. Vitamin C status of cigarette smokers and nonsmokers. Am. J. Clin. Nutr. **23:** 520.
30. BODDY, K., R. HUME, P. C. KING, E. WEYERS & T. ROWAN. 1974. Total body, plasma and erythrocyte potassium and leucocyte ascorbic acid in "ultra-fit" subjects. Clin. Sci. Mol. Med. **46:** 271.
31. BOYUM, A. 1968. Separation of leukocytes from blood and bone marrow. Scand. J. Clin. Lab. Invest. **21:** 1.
32. GIUDICELLI, J. P., J. M. PHILIP, P. DELQUE & P. A. SUDAKA. 1982. A single-step centrifugation method for separation of granulocytes and mononuclear cells from blood using discontinuous density gradient of Percoll. J. Immunol. Methods **54:** 43.
33. HAUFFMAN, H. F., P. R. LEVERING & K. DEVRISS. 1983. A single centrifugation step method for the isolation of erythrocytes, granulocytes, and mononuclear cells on continuous density gradients of Percoll. J. Immunol. Methods **57:** 1.
34. ENGLISH, D. & B. R. ANDERSON. 1974. Single-step separation of red blood cells, granulocytes and mononuclear leukocytes on discontinuous density gradients of Ficoll-Hypaque. J. Immunol. Methods **5:** 249.
35. SEGAL, A. W., A. FORTUNATO & T. HERD. 1980. A rapid single centrifugation step method for the separation of erythrocytes, granulocytes, and mononuclear cells on continuous density gradients of Percoll. J. Immunol. Methods **32:** 203.
36. BUI-NGUYEN, M. H. 1985. Ascorbic acid and related compounds. *In* Modern Chromatographic Analysis of the Vitamins. A. P. DeLeenheer, W. E. Lambert & M. G. M. DeRuyter, Eds: 267-301. Marcel Dekker. New York.
37. KUTNINK, M. A., J. H. SKALA, H. E. SAUBERLICH & S. T. OMAYE. 1985. Simultaneous determination of ascorbic acid, isoacorbic acid (erythorbic acid) and uric acid in human plasma by high-performance liquid chromatography with amperometric detection. J. Liq. Chromatogr. **8:** 31.

38. KUTNINK, M. A. & S. T. OMAYE. 1987. Determination of ascorbic acid, erythorbic acid, and uric acid in cured meats by high-performance liquid chromatography. J. Food. Sci. **52:** 53.
39. VANDERSLICE, J. T. & D. J. HIGGS. 1985. Robotic extraction of vitamin C from food samples. J. Micronutr. Anal. **1:** 143.
40. PELLETIER, O. 1985. Vitamin C (L-ascorbic and dehydro-L-ascorbic acids). *In* Methods of Vitamin Assay. J. Augustin, B. P. Klein, D. A. Becker & P. B. Venugopal, Eds.: 303-347. John Wiley & Sons. New York.
41. PACHLA, L. A. & P. T. KISSINGER. 1979. Analysis of ascorbic acid by liquid chromatography with amperometric detection. Methods Enzymol. **62:** 15.
42. FARBER, C. M., S. KANENGISER, R. STAHL, L. LIEBES & R. SILBER. 1983. A specific high-performance liquid chromatography assay for dehydroascorbic acid shows an increased content in CLL lymphocytes. Anal. Biochem. **134:** 355.
43. SCHAUS, E. E., M. A. KUTNINK, D. K. O'CONNOR & S. T. OMAYE. 1986. A comparison of leukocyte ascorbate levels measured by the 2,4-dinitrophenylhydrazine method with high-performance liquid chromatography using electrochemical detection. Biochem. Med. **36:** 369.

DISCUSSION OF THE PAPER

P. J. GARRY (*University of New Mexico School of Medicine, Albuquerque, N.M.*): Are you planning to measure the dehydroascorbic acid in these cells?

S. T. OMAYE (*USDA, San Francisco, Calif.*): There was an interesting poster yesterday that described how the authors converted the material with homocysteine to dehydroascorbic acid and subtracted. That's very easily applied to our system, so we're probably going to do that too.

W. LOHMANN (*University of Giessen, Giessen, FRG*): Have you been able to determine separately the vitamin C concentration of T and B lymphocytes?

S. T. OMAYE: No, we haven't gotten to that point yet, that's very important.

R. SHAEFFER (*National Bureau of Standards, Gaithersburg, Md.*): More of an announcement than a question. I'm from the National Bureau of Standards. We have an external quality assurance program for plasma levels of vitamin C, and any laboratory interested in the accuracy of their measurements are invited to join us. We also do as part of that study measurements of selenium, vitamin A, vitamin E, and β-carotene, so if you're interested contact me, Robert Schaeffer, at The Department of Chemistry, National Bureau of Standards.

Ascorbic Acid and Cancer[a]

W. LOHMANN

Institut für Biophysik
Universität Giessen
Giessen, Federal Republic of Germany

There has been considerable discussion on vitamin C and cancer during the last few years. Although a few authors believe that vitamin C can protect against cancer or, at least, against some types of cancer, the majority seems to be inclined to doubt such a function for this vitamin.

Like any other biochemical system, the vitamin C redox system is important for maintaining normal, optimum biological functions. The electron flux caused by coupling the vitamin C redox system with other redox systems as well as its kinetic equilibrium maintained by the different components can be influenced by cancerogenic substances acting as electron donors or acceptors. These modifications can change the intensity (spin concentration) of the free radicals present normally in healthy biological systems or might produce new ones. These can be detected by means of the electron spin resonance (ESR) method. In this way, an elucidation of molecular changes that occur during the formation and progression of certain diseases might be possible. Only by knowing these molecular mechanisms, especially if they are specific for certain diseases, can an unequivocal diagnosis and an optimum therapy be possible. Also, such a specificity permits the use of these radicals as tumor markers and as an indicator for the success of the therapy applied.

Recently we have shown that the ESR spectra of lyophilized erythrocytes of patients with acute lymphatic leukemia (ALL) exhibit an increase in spin concentration and an additional signal at about g = 2.005 not present in control samples[1] (FIG. 1). By comparing both of these spectra it should be noticed that the "ALL" spectrum has been registered with about one sixth of the sensitivity of the control spectrum. With progression of the disease, the spin concentration increases, reaches an optimum, and decreases again, yet the typical ALL shape of the spectrum is maintained throughout. A completely different ESR spectrum is obtained with erythrocytes from patients with an acute myeloic leukemia (AML), which seems to be characteristic for this disease[2] (FIG. 1, lower spectrum). This latter spectrum will be discussed later on.

The question of how informative, indicative, and cognitive the "ALL" erythrocyte ESR spectrum is will be discussed first.

The "ALL" spectrum could never be observed in erythrocytes from patients with other types of leukemia or other diseases of the hematopoetic or lymphatic system.[3] Furthermore, it could not be observed in the erythrocytes of patients with other malignant diseases of other organ systems. It should also be emphasized that the presence and intensity of the "ALL" signal is independent of the leukocyte count, that is, patients with a low leukocyte count (1,000 to 3,000 cells/μl) also exhibit the typical "ALL" ESR spectrum while in cases of nonleukemic leukocytosis this spectrum

[a] This work was supported in part by grants from the Bundesministerium für Forschung und Technologie (nos. 01Z0014/9 and 01Z0062/8) and from the Fonds der Chemischen Industrie.

can never be observed. If the patients are treated successfully with prednisolone, vincristine, and doxorubicine, the signal intensity as well as its shape change with treatment and appear to become almost normal concomitantly with an almost complete remission of the bone marrow and the peripheral blood (FIG. 2). Thus, the "ALL" ESR spectrum seems to be indicative for ALL.

To assign this signal, the substance and its interacting partner that transforms it to the radical reflecting the spectrum shown in FIG. 1 must be identified. Moreover, the location of this interaction is of special importance since it will indicate one possible defect in ALL erythrocytes.

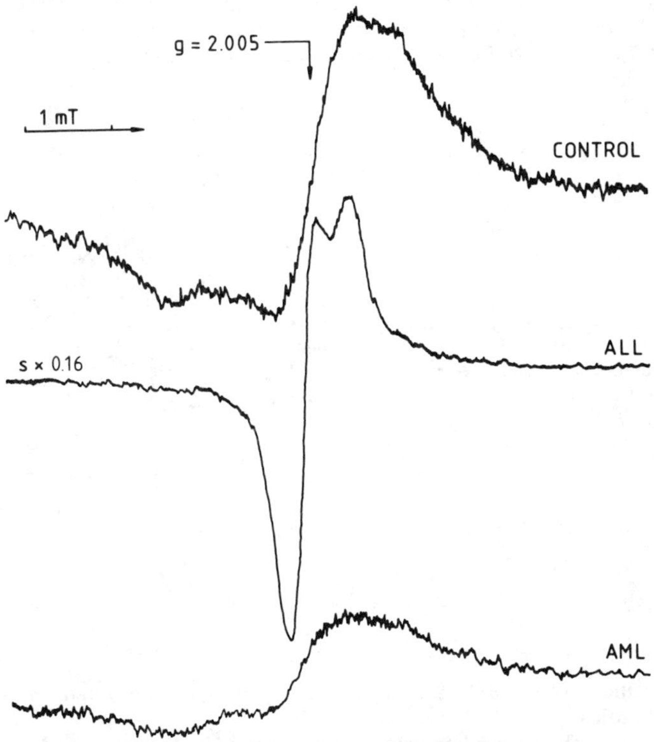

FIGURE 1. ESR spectra of lyophilized erythrocytes of healthy persons (control) and of patients with acute lymphatic leukemia (ALL) or acute myeloic leukemia (AML); *s* ≙ rel. sensitivity. (From Lohmann *et al.*[2] Reprinted by permission from *Blut.*)

Since it could be assumed that a one-electron oxidation step of an antioxidant will result in the radical form shown in FIGURE 1, more than 100 different organic and inorganic antioxidants were added to healthy erythrocytes. Of all the antioxidants tested,[1,3] only vitamin C resulted in a spectrum identical to the "ALL" spectrum (FIG. 3). Other vitamins, *e.g.,* E and A, did not give this typical spectrum. Thus, the semidehydroascorbate (= ascorbyl, SDA) radical seems to be typical for ALL. It is produced, as shown in FIGURE 4, by a one-electron oxidation of ascorbic acid which is present normally in biological systems in its reduced form. This step is obviously incomplete, since two additional protons have to be removed also.[4]

Generally, it is assumed that only dehydroascorbic acid (DHA), the completely oxidized form of ascorbic acid, can permeate the cytoplasm membrane. If so, the radical formation must occur at the surface of or just within the membrane. For that reason, scanning electron microscopical investigations were conducted.[1] The shape of healthy erythrocytes (FIG. 5A) looks like erythrocytes obtained from ALL patients (FIG. 5C) if 1 mM of vitamin C has been added to the healthy erythrocytes (FIG. 5B).

From this it can be concluded that a membrane defect must be present in "ALL" erythrocytes, that is that the membrane must be the location at which vitamin C interacts with some partner. For that reason, white ghosts (erythrocyte membranes) were isolated. Only an addition of vitamin C (1 mM) to these white ghosts resulted in the "ALL" erythrocyte ESR spectrum[5] (FIG. 6, compare with FIG. 1). From this,

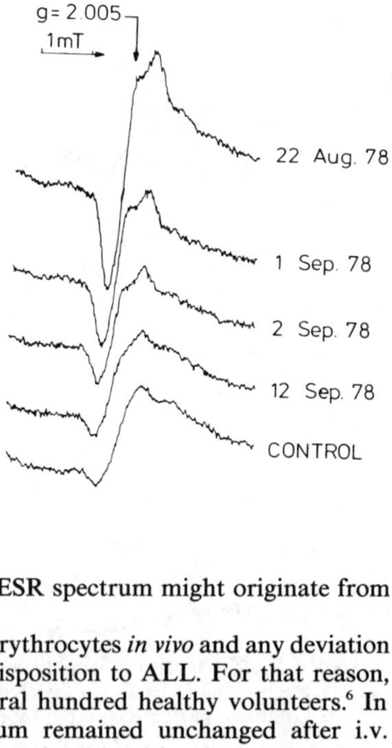

FIGURE 2. ESR spectra of lyophilized erythrocytes of a female patient with ALL taken after treatment with several cytostatic agents (see text). The blood was drawn on the dates given. Upper spectrum: untreated blood.

it can be concluded that the "ALL" erythrocyte ESR spectrum might originate from the ascorbyl radical.

Such a reaction should not happen with intact erythrocytes *in vivo* and any deviation from that can be taken as an indicator for a predisposition to ALL. For that reason, vitamin C experiments were conducted with several hundred healthy volunteers.[6] In most of the cases, the erythrocyte ESR spectrum remained unchanged after i.v. injection of 1 g of vitamin C. As expected, the vitamin C level in plasma was about 17 mg/100 ml 2 min after injection and decreased exponentially. After several hours, it reached the control level of about 1 mg/100 ml. The vitamin C level of the erythrocytes increased slightly within the first few hours.

In the course of these investigations, two interesting observations were made: in one volunteer, the vitamin C level of the plasma was 76 mg/100 ml 2 min after i.v. injection. The erythrocyte ESR spectrum was very similar to an "ALL" spectrum[7] (FIG. 7). After several hours, the vitamin C level of the plasma as well as the erythrocyte ESR spectrum were normal again. Additional investigations are needed

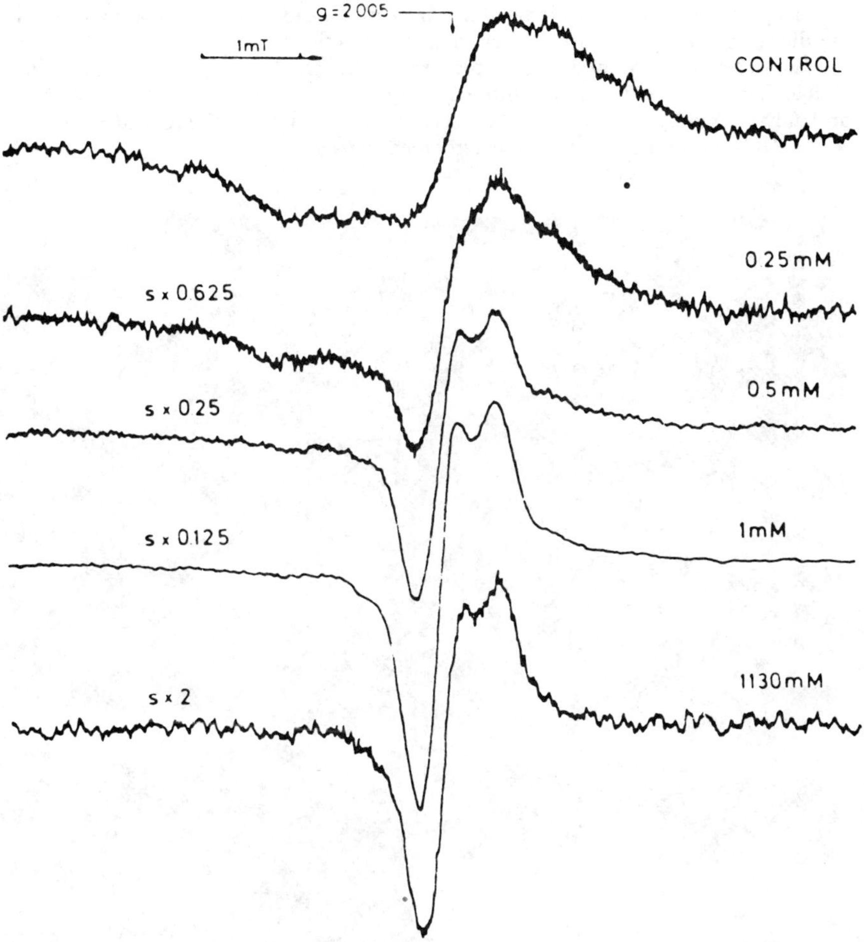

FIGURE 3. The effect of different concentrations of ascorbic acid on the ESR spectrum of erythrocytes; *s* ≙ rel. sensitivity. (From Lohmann *et al.*[1] Reprinted by permission from *Blut.*)

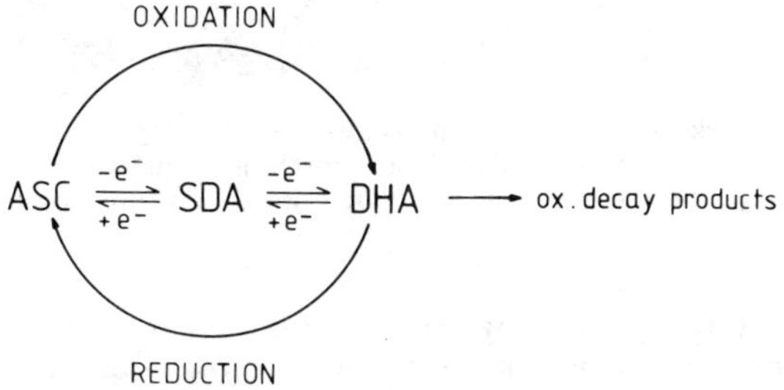

FIGURE 4. The vitamin C redox system. *ASC* ≙ ascorbic acid (reduced form); *SDA* ≙ semidehydroascorbic acid (ascorbyl radical); *DHA* ≙ dehydroascorbic acid (oxidized form). *ox. decay products* = oxidized decay products of vitamin C.

to clarify if this volunteer might be a prospective candidate for ALL, in which case the illness might appear after oxidative stress reactions.

The other observation concerns the response of the leukocyte count after vitamin C injection. A few volunteers exhibited a slightly increased level (12-18 $\times 10^3/\mu l$) probably caused by an infection. After vitamin C injection, this count dropped to 5-6 $\times 10^3/\mu l$ within a few hours and remained constant.

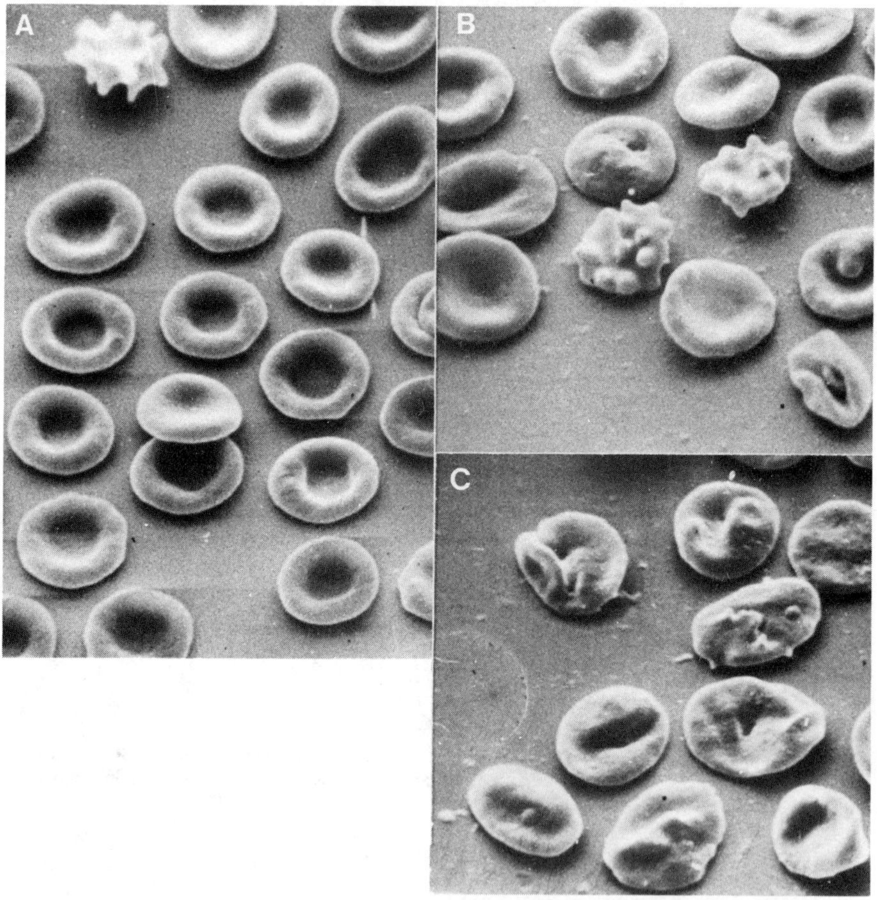

FIGURE 5. Scanning electron microscopical pictures of erythrocytes of healthy volunteers (**A**) before and (**B**) after treatment with 1 mM of ascorbic acid and (**C**) of patients with ALL. Magnification: 1650 ×. (From Lohmann *et al.*[1] Reprinted by permission from *Blut.*)

These *in vivo* investigations suggest that the formation of the ascorbyl radical occurs at the outer membrane surface. It is caused by an interaction partner which is protected in the case of healthy erythrocytes. If this protective effect is missing, this interaction results in an oxidation of vitamin C, that is in its radical form (SDA) or its completely oxidized form (DHA). This results in a facilitated transport across

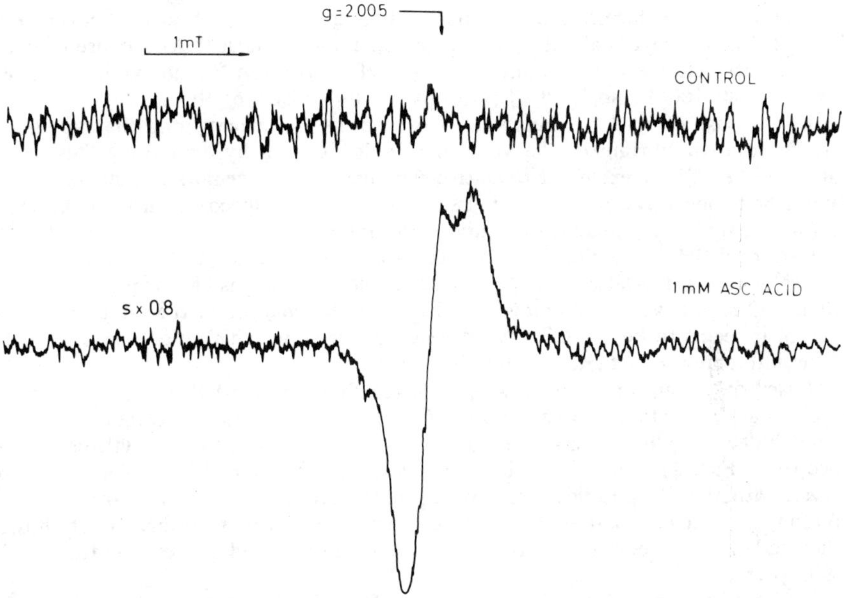

FIGURE 6. The effect of 1 mM of ascorbic acid on the ESR spectrum of white ghosts of erythrocytes; *s* ≙ rel. sensitivity.

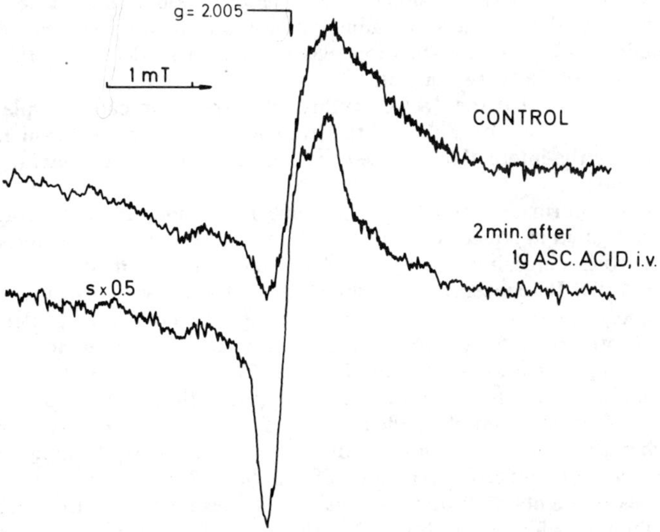

FIGURE 7. ESR spectra of lyophilized erythrocytes of a healthy volunteer immediately prior to and 2 min after i.v. injection of 1 g of ascorbic acid; *s* ≙ rel. sensitivity.

the membrane; the permeation rate (permeation coefficient ca. 4×10^{-7} cm/s at pH 7 and 37°C) of DHA is about 100 times larger than for SDA.[4,8] As a consequence of this, the intracellular total vitamin C level as well as its DHA fraction will be increased (by about 80-100% and 7-18%, respectively) in "ALL" erythrocytes.

In the case of ALL, a metabolic defect seems to be prevailing that apparently disturbs the equilibrium of the vitamin C redox system by oxidation. This can be caused either by an extended oxidative damage (cancerogenic substances) or by a diminished concentration of reducing substances (*e.g.,* reduced glutathione, GSH). It is, thus, tempting to assume that, under the influence of certain enzymes, the concentration of SH containing compounds, especially cysteine (CySH) and GSH, decreases, whereby the latter redox system is especially responsible for maintaining the vitamin C redox system. Whatever might cause the changes in the vitamin C redox system, it seems to be an oxidative damage or stress which results in a disturbance of the interaction of vitamin C and glutathione.

Based on certain indications, we postulated that the ascorbyl radical might be the result of an interaction between vitamin C and Cu^{2+} proteins.[9] Several Cu^{2+} proteins tested indeed produced ESR spectra very similar to the "ALL" erythrocyte ESR spectrum (FIG. 8). This differs, however, considerably from the doublet usually observed with the SDA radical (see spectrum at bottom). In addition, other copper-containing proteins, such as ascorbate oxidase (FIG. 8) or tyrosinase (not shown), when added to an ascorbate solution and then lyophilized, did not result in the "ALL" ESR spectrum.

It should be pointed out that addition of Fe^{2+} (as $FeSO_4$) to either ALL erythrocytes or healthy erythrocytes treated with 0.2 mM of ascorbic acid resulted in an ESR spectrum of healthy erythrocytes, suggesting that Fe^{2+} causes a reduction of the ascorbyl radical to ascorbic acid with a concomitant increase of the $g = 4.3$ signal.[10] This latter signal is indicative of a ferric high spin iron complexed presumably with ascorbic acid and histidine.[11]

Recently we have shown that the signal is produced only when the oxygen atom attached to C-3 of the ascorbate anion is electrostatically bound to a partner molecule.[12] There was some evidence that an alkaline cation might be involved. For this reason, a more detailed molecular investigation was conducted to determine the formation and the structure of the SDA radical.

It could be shown that the "ALL erythrocyte" spectrum can be explained by an interaction between Na^+ or K^+ and the anionic ascorbyl radical which has been produced first by the interaction between ascorbic acid and some oxidant, *e.g.,* Cu^{2+}-containing proteins.[13]

To get more information on the involvement of Na^+ and K^+, the ascorbyl radical has been investigated in aqueous solution at room temperature. These investigations[13] revealed that with increasing pH the ascorbyl radical is formed, showing a hyperfine structure (FIG. 9). Its concentration exhibits an optimum at a pH of around 7 and decreases rapidly at larger values. Since the coupling constants of the triplets produced by the CH_2-6 protons differ between ascorbic acid and isoascorbic acid, it was concluded that a ring formation must exist between C(3)-O^- and C(6)-OH.

Additional ^{1}H-NMR investigations revealed that with increasing pH only the H-4 proton is shifted considerably upfield, while the H-5 and CH_2-6 protons remain almost unchanged in their positions, indicating an increased shielding of the H-4 proton.[14] A very pronounced downfield shift of the C-3 resonance was observed by ^{13}C-NMR measurements, indicating a considerable deshielding of this carbon atom.

The IR findings which have shown that the two O-H stretching bands originating from the OH groups attached to C-3 and C-6 are missing at pH 7 also favor the existence of a cyclic structure of the ascorbic acid side chain.[15] This ring closure occurs,

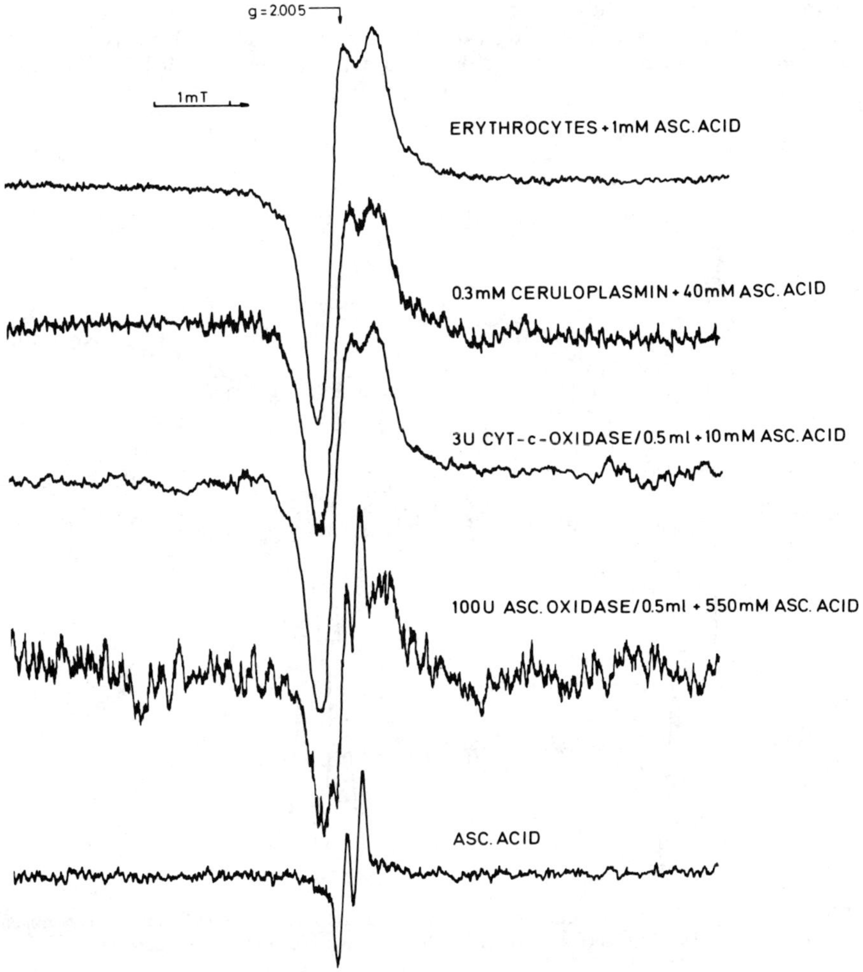

FIGURE 8. The effect of ascorbic acid on erythrocytes and different copper proteins. Spectrum at the bottom represents the ascorbyl radical obtained after lyophilizing ascorbic acid. Note: different relative sensitivities were used for the different samples. Spectra demonstrate qualitative response only. (From Lohmann and Winzenburg.[8] Reprinted by permission from *Zeitschrift für Naturforschung.*)

presumably, via a hydrogen bond formation between C(3)-O^- and C(6)-OH resulting, *e.g.*, in a deshielding of the C-3 atom. The change in electron densities at C-3 and C-6 is also expressed by a shift of the C = O and C = C stretching bands towards smaller wave numbers. This result is confirmed by UV absorption spectra, which exhibit a red shift (245 → 264 nm) of the π-π^* excitation of the C=C double bond with the formation of the radical.

Mass spectrometrical investigations have shown that the Na-ascorbyl radical formed is electroneutral (probably due to a 1:1 ASC^-:Na^+ ratio).[16] This and the

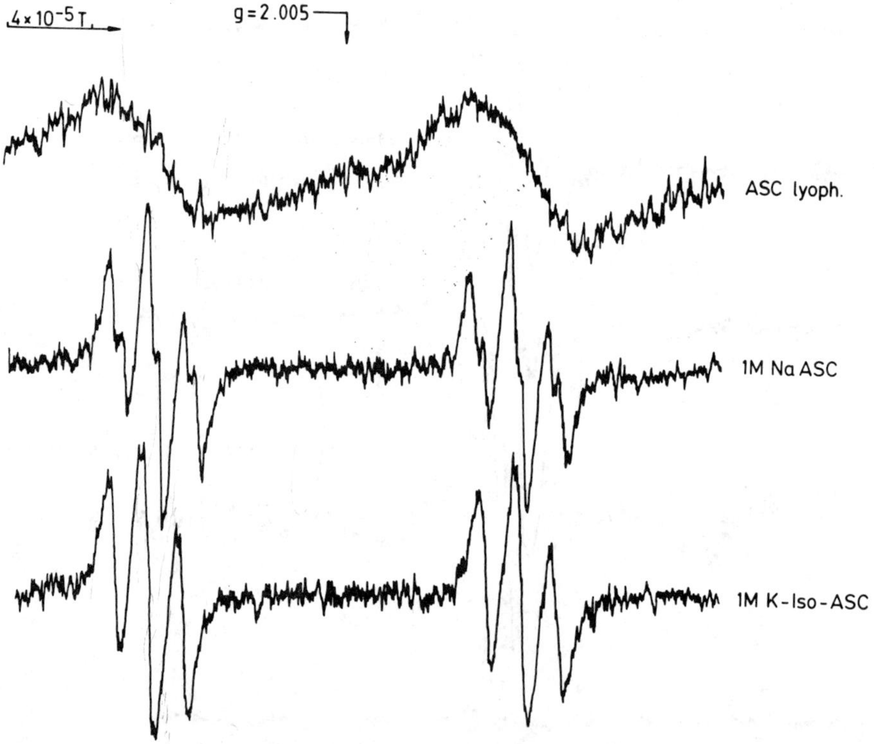

FIGURE 9. ESR spectra of aqueous solutions of Na ascorbate (*Na-ASC*) and K isoascorbate (*K-Iso-ASC*) at pH 7.2. Upper spectrum: lyophilized ascorbic acid for comparison.

cyclic side chain structure should enable the Na-ASC radical to permeate membranes even at room temperature. This could be confirmed by experiments on dipalmitoyl-phosphatidylcholine (DPPC) vesicles, in which the radicals reduced the spin label I[1,14] located at the apolar end of the CH_2 chain.[8] Since there is no reduction at pH < 4, it can be concluded that the hydrophilic free side chain prevents the permeation of ASC across membranes. When, however, the bicyclic structure is formed (radical) and the total charge is zero, permeation is possible. Such a transport has its highest efficacy at around 37°C and pH 7 (the physiological temperature and pH value).[13]

Furthermore, it should be pointed out that the ascorbyl anion radical prefers to form a complex with Na^+ instead of with K^+.

These results could be confirmed recently[17] by determining the disproportionation rate of the ascorbyl radical in the presence of Li^+, Cs^+, K^+, and Na^+. While Li^+ did not influence the disproportionation rate at all, the influence of the other ions tested was of the following order: $Na^+ > K^+ > Cs^+$.

Furthermore, an experiment on frog rectus muscles[18] showed that a pretreatment with ascorbate enhanced the slow transient changes of extracellular Na^+ activity during contractures stimulated by acetylcholine. The inverse effect was obtained with K^+.

In summary, it might be concluded that the ascorbyl doublet is the result of an interaction between an unpaired electron (OH), produced by the cleavage of a hydrogen atom belonging to a water molecule that forms a hydrogen bond with C(5)-OH, and the proton attached to C-4.[4] This cleavage can be produced, for example, by lyophilization. In aqueous neutral solution, an additional proton is removed from C(3)-OH, allowing the suggested ring closure of the side chain.

These observations confirm the assumption expressed in the text to FIGURE 4, according to which the formation of the ascorbyl radical occurs by the removal of one proton and one H atom each.

In the presence of O_2, then, the H atom can react with oxygen resulting in HO_2, and finally, in H_2O_2. Such a mechanism might explain the formation of H_2O_2 during the oxidation of ascorbic acid. This had been assumed and has now been detected by quite a few investigators. H_2O_2 can inhibit Na^+/K^+-ATPase and, thus, can affect the active Na^+ and K^+ transport across the membrane. The ascorbyl-radical-Na^+ complex, however, will facilitate the passive Na^+ transport.

The change in intracellular Na^+ and K^+ (and Ca^{2+}) concentrations, caused, therefore, by the formation of the ascorbyl radical might be a consequence of or might even act as a promoter for a great variety of diseases. It will also result in a change in intracellular pH. A few preliminary investigations have shown drastic changes in the concentration of these ions as well as the pH value within the erythrocytes without detecting any change in the plasma in some diseases.

While in the case of ALL the disturbance of the vitamin C redox system resulting in the formation of the ascorbyl radical seems to be of importance, the SH/SS redox system seems to be modified in the case of acute myeloic leukemia (AML).[2] In addition, a considerable decrease in the total vitamin C concentration in plasma as well as in erythrocytes of AML patients has been observed (16% and 46%, respectively, of the controls). Here, too, most of the vitamin C is present in its oxidized form (dehydroascorbic acid).

This should be reflected in the ESR spectra of AML erythrocytes as can be seen in FIGURE 1, lower curve. The ascorbyl radical does not appear. The spin concentration, in comparison to the control, has been considerably reduced. It should be pointed out that the "AML" spectrum seems to be indicative for AML and can be used, therefore, as a tumor marker. In the case of a successful treatment of the patients with alexan, thioguanine, and daunoblastin the gradual change of the ESR spectrum from "typical AML" to "healthy" erythrocytes can be observed (FIG. 10). The vitamin C concentration of the erythrocytes returns to about that of the controls while the plasma values remain low (30-40% of the controls).

Since we have observed recently a considerable increase in cysteine ($\times$ 1.5) and glutamic acid ($\times$ 2.5) concentrations in plasma of patients with AML,[19] the reduction in spin concentration of AML erythrocytes might be due to the interaction between cysteine or glutathione (GSH) and the radical species that are normally present in healthy erythrocytes but have not yet been identified.

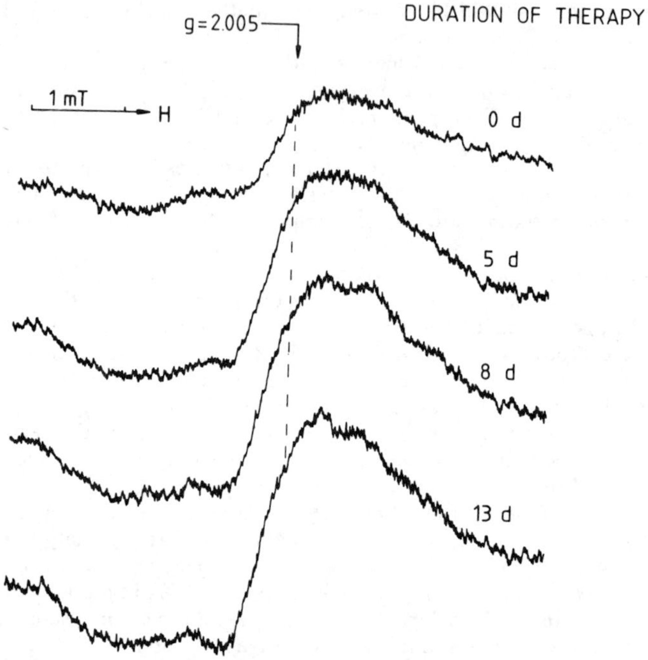

FIGURE 10. Modification of the ESR spectra of lyophilized erythrocytes of a female patient with acute myeloic leukemia (AML) under therapy. The blood at day zero (*0 d*) was drawn just prior to therapy. (From Lohmann *et al.*[2] Reprinted by permission from *Blut.*)

An addition of GSH (or cysteine) to healthy erythrocytes results, indeed, in a reduction of the spin concentration and, at about 70 mM GSH, resembles very closely the ESR spectrum of erythrocytes obtained from AML patients (FIG. 11).

Since the permeation rate of GSH across cell membranes is very low, it might be assumed that GSH, the plasma concentration of which is supposedly high in AML patients, interacts with some compounds at the surface of the membrane. Whether or not this interaction will influence the ascorbic acid transport across membranes or is a consequence of concentration changes of the latter has still to be elucidated. At any rate, these interactions will cause membrane defects in erythrocytes of AML patients resulting in gross changes of the cell shape. This can be seen in FIGURE 12, showing a scanning electron microscopical picture of AML erythrocytes.

For a better understanding of the molecular events occurring during the initiation and progression of ALL and AML (ALL: formation of the ascorbyl radical; AML: SH effect, vitamin C concentration too low in both plasma and erythrocytes), the membrane defects must be elucidated. In both types of leukemia the formation of glycoproteins seems to be diminished. This is supported by the findings that the concentrations of threonine, serine, asparagine, and lysine are considerably reduced in the plasma of patients with AML and ALL.[19] This might cause certain disturbances within the cell membrane resulting in modifications observed in erythrocytes of AML and ALL patients. Since the concentrations of the four amino acids mentioned above are changed to about the same extent in both ALL and AML, the diminished glycoprotein formation is, obviously, not specific for the formation of either one of the two types of leukemia investigated. Thus, other factors might be of importance for the development of the acute leukemias, *e.g.,* modifications in the formation of glycolipids and in the vitamin E redox system.

The results obtained show a modification of the vitamin C metabolism in the case of ALL. Since in healthy biological systems vitamin C is present mainly in its reduced form, the appearance of the ascorbyl radical in erythrocytes obtained from patients with ALL indicates the direct or indirect oxidation of a certain portion of ascorbic acid by an endogenous or exogenous oxidizing (cancerogenic) species, the structure of which is still unknown. It should be emphasized that the ascorbyl radical can be detected only in the erythrocytes from patients with ALL. Because of its specificity, it can serve as a tumor marker for ALL. Furthermore, the total concentration of vitamin C (reduced and oxidized form) is increased in both plasma and erythrocytes. In the case of AML, the total vitamin C concentration is diminished considerably and consists mainly of the oxidized form of vitamin C. In addition, the GSH/GSS redox system is modified.

It is tempting to reconcile these two seemingly different results (obtained with ALL and AML) in a model that will show that both types of leukemia might be caused by the same agent depending on its concentration and, thus, on the extent of the disturbance produced.

In biological systems, ascorbic acid is present mainly in its reduced form, which is maintained by other redox systems. Any deviation from this will disturb the redox equilibria. The presence of relatively small concentrations of an oxidizing cancerogenic agent will cause, directly or indirectly, an oxidation of ascorbic acid. Its extent depends on the concentration of this agent. Such an oxidation will modify, of course, the ASC/DHA redox equilibrium.

As a consequence of this, the biological system tries to supply ascorbic acid by mobilizing the defense system that is specific for restoring the original vitamin C redox

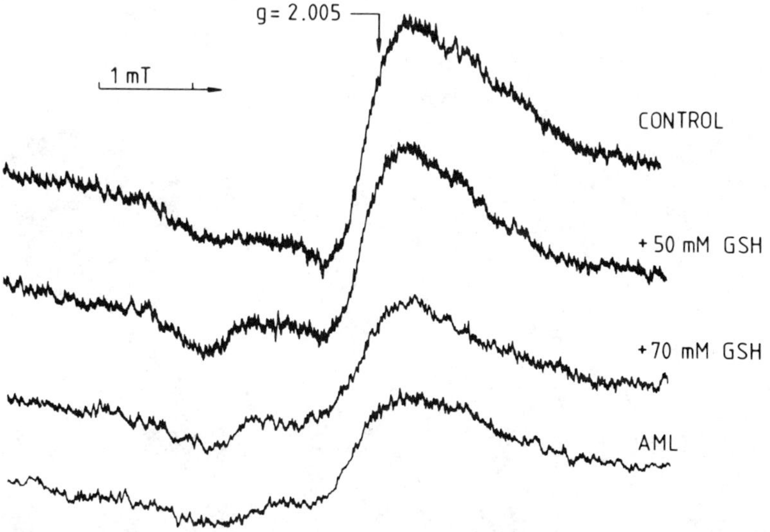

FIGURE 11. Effect of reduced gluthathione (*GSH*) on the ESR spectrum of healthy erythrocytes. For comparison, the ESR spectrum of erythrocytes obtained from an AML patient is shown at the bottom. (From Lohmann *et al.*[2] Reprinted by permission from *Blut.*)

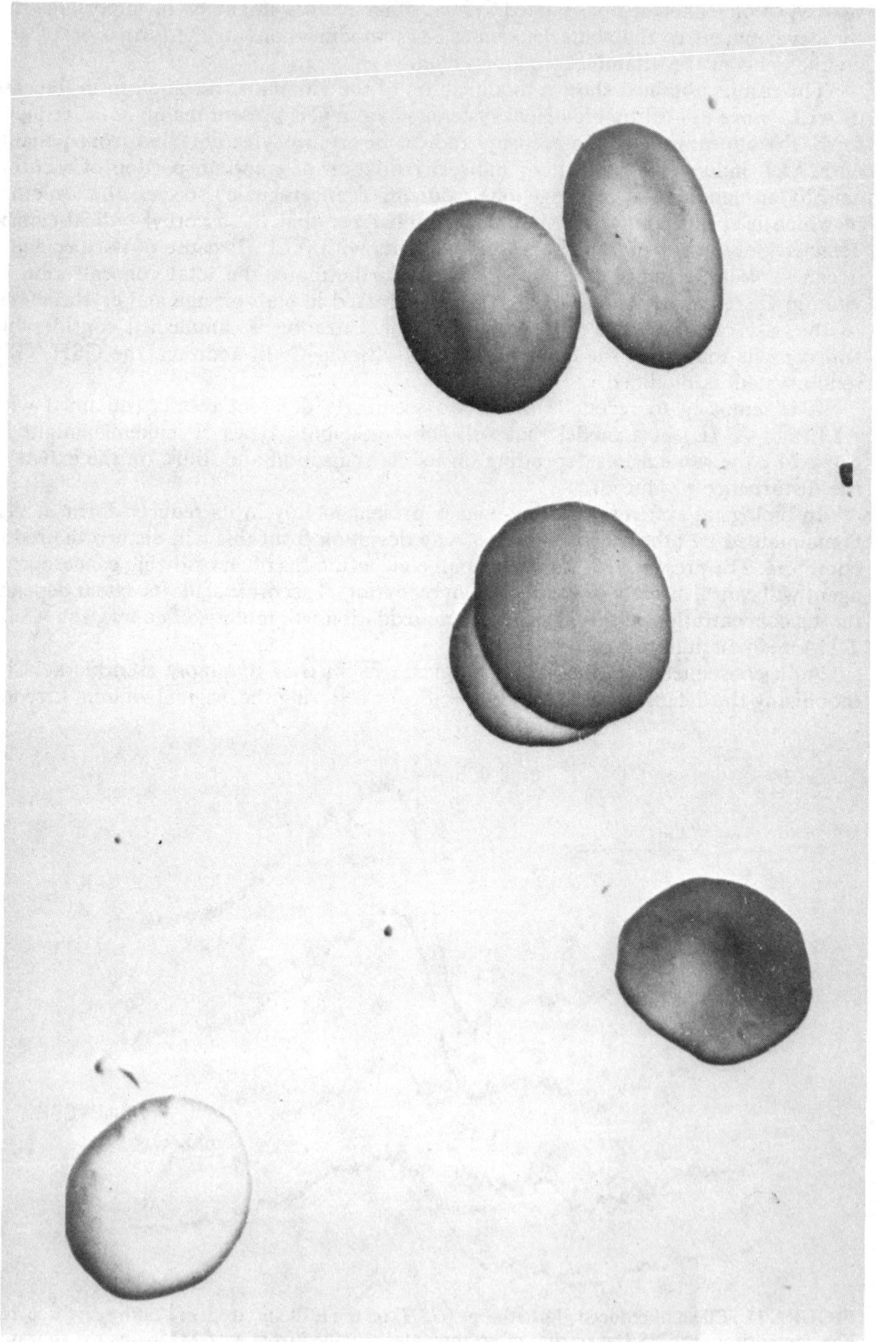

FIGURE 12. Scanning electron micrograph of erythrocytes obtained from a patient with acute myeloic leukemia (AML). Magnification: 3400 ×. (Picture taken by G. Magdowski.)

equilibrium. This can be achieved by supplying lymphocytes, which are the carriers of ascorbic acid as is well-known. Their vitamin C concentration exceeds that of plasma by about 100 times. As a consequence of this response, there will be an overcompensation of the loss of ASC resulting in an increase in the total vitamin C concentration, as has been observed in patients with ALL.

With the progression of the disease, that is with an increase in the concentration of the oxidizing cancerogenic agent, most of the ascorbic acid available will be oxidized to DHA, and subsequently, to its oxidized decay products, *e.g.*, diketogulonic acid and up to methylglyoxal. These oxidative decay products are no longer considered to be vitamin C. Thus, the total vitamin C concentration is diminished, which is caused by a rapid oxidation of ASC to its oxidized decay products and which is not due to a reduced supply of ASC. Most of the remaining portion of vitamin C is present in its oxidized form DHA. At this stage of the disease, the biological system cannot cope any more with the deficiency of ascorbic acid. The vitamin C redox system will be inefficient and, finally, will break down.

At this stage, the GSH/GSS redox system, which interacts with and regulates the vitamin C redox system, will be affected next and will be modified either directly or indirectly by the oxidizing agent. Since the specific defense system mentioned above is unable to compensate these modifications (damage), a more general defense system is mobilized by supplying granulocytes.

According to this model, irregularities in the vitamin C redox system in the blood result first in ALL. With an increase in the oxidizing power of the cancerogenic agent and with the gradual breakdown of the vitamin C redox system, the GSH/GSS redox system will also be affected. This more severe stage of leukemia might be characterized as AML. These considerations do not imply that each AML patient has to pass through an ALL stage first. If the concentration of the oxidizing species is already initially high enough to affect, at least, both of the ASC/DHA and GSH/GSS redox systems, AML will prevail.

This model is a working hypothesis. Future experiments have to prove (or disprove) its validity.

Only a complete and thorough elucidation of the molecular reactions occurring and prevailing in the formation of cancer—as in other diseases, too—will allow an unequivocal diagnosis and, most of all, an optimum therapy.

ACKNOWLEDGMENTS

I would like to thank numerous collaborators and physicians and colleagues, especially from the University Hospital in Giessen, Stanford University, and Academia Sinica, Peking, for conducting the experiments and for many helpful discussions.

REFERENCES

1. LOHMANN, W., J. SCHREIBER, W. STROBELT & CH. MÜLLER-ECKHARDT. 1979. Blut **39:** 317-326.
2. LOHMANN, W., W. SCHMEHL, D. HOLZ & M. EVERZ. 1986. Blut **53:** 437-441.
3. LOHMANN, W., J. SCHREIBER, W. GREULICH, W. STROBELT, E. MÜLLER, H. LÖFFLER, H. PRALLE, H. FEUSTEL, K. SCHWEMMLE & R. D. FILLER. 1981. Collective Phenomena **3:** 245-258.

4. LOHMANN, W., 1984. Biophys. Struct. Mech. **10:** 205-210.
5. LOHMANN, W., W. GREULICH & G. DÖLL. 1979. Blut **39:** 327-332.
6. LOHMANN, W., K. G. BENSCH, E. MÜLLER & S. O. KANG. 1981. Z. Naturforsch. **36c:** 1-4.
7. LOHMANN, W. 1981. Z. Naturforsch. **36c:** 804-808.
8. LOHMANN, W. & J. WINZENBURG. 1983. Z. Naturforsch. **38c:** 923-925.
9. LOHMANN, W., J. SCHREIBER & W. GREULICH. 1979. Z. Naturforsch. **34c:** 550-554.
10. LOHMANN, W., B. KIEFER, D. HOLZ, D. SCHMIDT & J. SCHREIBER. 1983. Z. Naturforsch. **38c:** 862-863.
11. LOHMANN, W., D. HOLZ, B. KIEFER & D. SCHMIDT. 1983. Z. Naturforsch. **38c:** 90-93.
12. BENSCH, K. G., O. KÖRNER & W. LOHMANN. 1981. Biochem. Biophys. Res. Commun. **101:** 312-316.
13. LOHMANN, W. & D. HOLZ. 1984. Biophys. Struct. Mech. **10:** 197-204.
14. LOHMANN, W., K. BEINHAUER & H. SAPPER. 1984. Naturwissenschaften **71:** 477-478.
15. LOHMANN, W., D. PAGEL & V. PENKA. 1984. Eur. J. Biochem. **138:** 479-480.
16. LOHMAN, W., F. HILLENKAMP, J. ROSMARINOWSKY, D. BACHMANN & M. KARAS. 1984. Fresenius' Z. Anal. Chem. **317:** 129-130.
17. WIECZOREK, P. & T. OGONSKI. Personal communication.
18. PUPPI, A., I. WITTMANN & M. DELY. 1986. Gen. Physiol. Biophys. **5:** 187-192.
19. SCHREIBER, J., W. LOHMANN, F. BERTHOLD & F. LAMPERT. 1986. Fresenius' Z. Anal. Chem. **325:** 476-477.

DISCUSSION OF THE PAPER

S. D. VARMA (*University of Maryland School of Medicine, Baltimore, Md.*): It looks to me that the defect in the membrane in the ALL is really at the metabolic level because you need a good regenerating system of the thiols to reduce ascorbic acid and the block is probably much earlier to the thiol. Did you measure these or the hexose monophosphate shunt activity of the red blood cells?

W. LOHMANN (*University of Giessen, Giessen, FRG*): When we had observed these results we injected 1 gram of ascorbic acid into healthy volunteers. Their leukocyte count was between 12,000 and 18,000 and when 1 g of ascorbic acid was injected it dropped the leukocyte count down to about 4,000 to 5,000 and it remained low. When 1 g of ascorbic acid was injected into healthy volunteers there generally was no change in the erythrocyte ESR spectrum. If you inject 1 g of ascorbic acid in a healthy human who has 5 liters of blood, you should expect to get about 20 mg of ascorbic acid for 100 ml of plasma, and this is what we always observed only 2 minutes after the injection. If we inject ascorbic acid into a patient with AML the vitamin C concentration in the plasma does not increase to 17 to 20 mg per 100 ml of plasma.

S. D. VARMA: Your membranes are kind of contorted. In the red blood cells there is a mechanism whereby the oxidized glutathione is pumped out and it looks like there is accumulation of oxidized glutathione there, in such a way that the redox cycle is probably not operating.

W. LOHMANN: You are right. Other redox systems seem to be involved. For that reason, we are investigating the interaction between ascorbate radical and either the NADPH redox system or the glutathione redox system.

H. SPRINCE (*Jefferson Medical College, Philadelphia, Pa.*): Since ascorbic acid and sulfhydryl compounds are known to be powerful protectants against aldehyde toxicity, I'm just wondering what the status of metabolic aldehyde in your system is.

W. LOHMANN: We have not investigated it yet.

UNIDENTIFIED SPEAKER: Does your erythrocyte membrane have a protein that you would refer to as ascorbate oxidase?

W. LOHMANN: This is the question which we are investigating with the Department of Biochemistry in Tübingen, to find out whether or not some copper-containing protein or peptide is in the erythrocyte membrane. Our atomic absorption study did show that copper is in erythrocyte membrane. A biochemist at Tübingen is trying to pinpoint it now.

Requirement for Vitamin C Based on Metabolic Studies

ANDERS KALLNER

Department of Clinical Chemistry
Karolinska Hospital
Stockholm, Sweden

Human requirements of vitamin C have been debated almost ever since ascorbic acid was recognized as the antiscorbutic factor in the thirties. As testified by history and verified by many experiments, very small amounts of vitamin C are required to cure the only indisputable symptom of vitamin C deficiency, scurvy. The classical study[1] performed during World War II is frequently cited and regarded to give the final and conclusive information on the amount required to cure scurvy. This amount is in the range of 10 to 15 mg/day for a normal man. However, before the war it had already been suggested that ascorbic acid has effects on the body other than preventing or curing scurvy. For example, the following statement has been ascribed to Szent-Györgyi: "The medical profession itself took a very narrow view. Lack of ascorbic acid caused scurvy, so if there was no scurvy, there was no lack of ascorbic acid. Nothing could be clearer than this. The ony trouble is that scurvy is not the first symptom of ascorbate deficiency but a final collapse, a premortal syndrome and there is a very wide gap between scurvy and full health." Nowadays it is suggested that ascorbic acid participates in intermediary metabolism, in the biosynthesis and metabolism of certain compounds, in the immune system, and in the absorption of iron. Furthermore, ascorbic acid has been tentatively identified as a radical scavenger *in vivo.* Ascorbic acid has also been described as a hormone[2] participating in the regulation of the production of corticosteroids and thyroid hormones.

It has been and still is difficult to identify the symptoms of marginal deficiencies of vitamin C.[3] The reported signs of reduced intakes of vitamin C may be compared with general signs of nutritional deficiency (TABLE 1). Unless these early symptoms are defined, the effects of supplementation cannot be adequately measured. Apart from scurvy, symptoms of vitamin C deficiency are vague and unspecific. The benefits or effects of large intakes are also difficult to demonstrate in an objective manner. Typically, in an epidemiological approach to the study of vitamin C effects on common cold, it was concluded that large samples had to be studied in order to prove statistically a difference between treated and untreated groups.[4] This illustrates the general observation that effects of supplementation above the minimum amounts to avoid scurvy are never particularly dramatic and it has not been possible to use them to establish the minimum requirements of vitamin C. On the other hand, administration of megadoses has been advocated. Very few if any adverse or toxic effects have been reported.[5] The tolerance range, coined in analogy with therapeutic range, for ascorbic acid is thus considerable, ranging from daily intakes of 10 mg to more than 10 g, a factor of a thousand or more.

Considering the wide variety of claims for benefits, it is understandable if the assumed requirements or recommended intake may vary depending on the effect that

TABLE 1. Comparison of General Signs of Nutritional Deficiency and Vitamin C Deficiency

Clinical	General Nutritional Deficiency	Vitamin C Deficiency
normal or reference	normal pool size and tissue, blood concentration; normal metabolite levels and enzyme activities	normal pool size and tissue, blood, etc. concentration; intakes and losses equal; tissue saturation
normal	reduction in body pool size	reduction in body pool size; decreased turnover
normal	lowering of nutrient concentration in blood, urine and tissue; lowered metabolite in urine	lowered blood concentration
behavioral changes	lowering of activity of vitamin- or mineral-dependent enzymes; early signs of metabolic disturbances	fatigue and skin lesions
	severe metabolic disturbances	scurvy

the user wants to demonstrate or achieve. Many attempts have been made over the years to find objective means to define an optimal intake of vitamin C based on biochemical studies. Thus, Baker *et al.*[6–9] designed a study in which volunteers were depleted of vitamin C and then repleted. This is a kinetic approach in which the behavior of ascorbate in the whole system is studied during *changing* conditions. It is possible to extrapolate information from which the pool size and turnover can be estimated. The results might only be valid under the experimental conditions and therefore not necessarily apply to nonexperimental conditions. Essentially these studies indicated that the pool size was in the order of 1500 mg, with a daily turnover of about 3%. If this were true an intake of about 45 mg/day could be satisfactory to maintain the pool size.

Kinetic studies have also been performed in animals such as rats, mice, and guinea pigs. These studies have indicated that at least three different types of tissues might be identified with regard to the uptake and retention of labeled material in the animal body (TABLE 2). In accordance with these findings, Hornig *et al.*[10] used a three-compartment model to describe the events in the guinea pig. In this study guinea pigs on a steady-state diet were given ^{14}C-labeled ascorbate as a single dose. The distribution and disappearance of the label from various tissues was measured in plasma and in tissues after sacrificing and dissecting the animals at different times after administration of the labeled ascorbate. The model allows the estimation of ascorbic acid turnover in various tissues. The disappearance rate of radioactivity from plasma would indicate the total turnover. A comparison between the different sets of data indicated that the distribution equilibrium between labeled ascorbate and available tissue ascorbate had not been reached or that part of the compartmental pool of ascorbate cannot be

TABLE 2. Kinds of Tissues with Different Uptakes of Radioactively Labeled Ascorbate

1. Tissues with a *low* retention capacity, *e.g.*, liver, lungs and kidney.
2. Tissues exhibiting a *high* retention capacity and a high and rapid accumulation of radioactivity, *e.g.*, adrenals and pituitary gland.
3. Tissues having a *very strong* retention capacity and a long-lasting continuing uptake of labeled material, *e.g.*, cerebrum, bulbus olfactorius, and testes.

exchanged. Therefore the turnover calculated from the tissue data was somewhat larger than that calculated from plasma data. The calculated intake was even larger, indicating a bioavailability of the ingested ascorbate of less than 100%.

Kipp and Rivers[11] and Holloway[12] have applied a two-compartment model and compared the results obtained by an isotope dilution method and an isotope excretion method. They identified a rapidly exchangable pool and a slowly exchangable pool. The half-lives of these pools always appear shorter if estimated by the isotope dilution technique. Results of the latter agree closely with those of Hornig and Hartmann obtained by a slightly different technique. It is suggested that the differences between the results might be explained by removal of label into a very slowly exchangable pool which is not sampled by the excretion studies. For the guinea pig different values of turnover or half-lives are thus obtained, depending on whether they are estimated from tissue dissection, isotope excretion, isotope dilution, or from the calculated intakes. The discrepancies the groups find between turnover rates using different experimental approaches clearly demonstrate the difficulties encountered in evaluating these parameters. It seems likely that the isotope dilution approach, however, suffers from the least errors and many studies lead to results in the same order of magnitude.

The experience with the isotope dilution technique described by Hornig *et al.*[9] led to a similar approach to estimate the pool sizes and turnover rates of the various pools in man by Kallner *et al.*[13] The design of these studies aimed at standardizing the ascorbate intake of volunteers with minimal influence on their normal activities or lives. The participants therefore had to follow certain limitations in the diet and were given supplementary ascorbate ranging from 30 to 180 mg per day in order to achieve differentiated intakes. After a period of equilibration the participants were given a single oral dose of ^{14}C-labeled ascorbic acid. The distribution and excretion of the label were followed in the plasma and the excretion of total activity and unchanged ascorbate in the urine was also measured. The design, from a kinetic point of view, is thus an isotope dilution model. A three-compartment model was used in the evaluation of the data. This model assumed a central pool into which the ascorbate is absorbed from the gastrointestinal tract. Two adjacent pools were assumed, one serving as a slow, "deep pool," the other as a pool where the metabolism of ascorbate takes place.

In this approach, the time course of the decline of the radioactivity in plasma is assumed to be described accurately enough by the sum of four exponential terms. The fitting of the experimental data to this equation and resolution of the various constants allow estimation of the total intake, total turnover, and the apparent rate constants for the exchange between the various pools. Other parameters such as apparent volume of distribution, renal threshold, and relation between the total turnover and plasma steady-state concentrations could also be calculated.

Another assumption in the kinetic study was that the main excretion pathway of ascorbate and its metabolites in humans is via the urine. It has been shown several times[14] that the amounts excreted in feces are neglible, but in the past it has been much debated if ascorbate can also be metabolized to carbon dioxide by humans. Conflicting data appear in the literature, but recently Kallner *et al.*[15] showed that the metabolism of ascorbate is influenced by the size of the intakes. Thus, if man ingests much above 200 mg of 1-^{14}C-labeled ascorbate at one time, a substantial amount of the radioactivity will be found as carbon dioxide in the expired air. Experimental evidence indicates, however, that this is due to a "presystematic" degradation of the ascorbate. With the doses used in the kinetic study, the production of carbon dioxide is no source of error.

The results of the kinetic study in man indicate that the relation between the steady-state plasma concentration and the total turnover—or well controlled intake in other studies—is so strong that the plasma concentration of ascorbate may well be used to estimate the total turnover. Garry[16] has presented data describing the relation between intakes and plasma ascorbate concentrations in the elderly that agree well with previous findings.[13] The apparent volume of distribution in steady state is related to the plasma concentration and thus the total turnover. The lower the turnover, the higher the apparent volume of distribution, which is suggested to reflect a varying degree of tissue binding of the ascorbate. In this connection it should be pointed out that at a turnover rate of 180 mg per day and more, about 20% of the absorbed ascorbate is excreted as metabolites whereas at a lower turnover up to 90% of the absorbed amount is metabolized. Even at low calculated turnover rates ascorbic acid is excreted unchanged in the urine. This finding is particularly important to emphasize since it has been claimed that depleted individuals are repleted when excretion of ascorbate in the urine occurs. This has been thought to indicate that the body pools have been saturated.[17] A leveling off of the tissue binding occurs at a total turnover of 60-80 mg per day. At that level, the renal threshold also seems to be exceeded and at higher values the renal turnover of ascorbate increases rapidly.

The overall half-life of ascrobic acid seems to depend on the total turnover. Thus at a turnover of about 15 mg per day a half-life of about 40 days was found, whereas at about 80 mg per day and above, the half-life was estimated to be only about 10 days, indicating a remarkable ability of the body to economize with available vitamin.

The model used can differentiate between renal and metabolic turnover. As discussed above, a renal threshold is found in accordance with the saturable reabsorption described by, for instance, Berger *et al.*[18] An increased risk for ascorbate deficiency in renal disease does not seem to have been described in the literature, but dialysis patients require substitution with large quantities or the addition of ascorbate to the dialysis fluid.[19]

The metabolic turnover displays saturation kinetics indicating that the body is able to increase the metabolism of ascorbate up to a level of about 40 mg per day. Thus, the body at large behaves similarly in this respect to the cells and subcellular entities studied and where ascorbate is assumed to exert its effects. This occurs at a total turnover of about 60 to 80 mg per day, and at a higher total turnover the metabolic turnover levels off and thus approaches a saturation level. The increase in metabolic turnover would correspond to a slope above 1, which could have a biological significance and be interpreted as a "yield" of the absorbed ascorbate below 100%. It even explains why unchanged ascorbate is found in the urine also at a low turnover. It could be argued that this is explained by roles of ascorbate other than being metabolized to other compounds in the sense of being consumed. The reversible oxidation of ascorbate to dehydroascorbate and the intermediate formation of the ascorbate free radical are examples of such a type of reaction in which ascorbate is known to participate.

Another important implication of the leveling off of the metabolic turnover is that man is not able indefinitely to increase the metabolism of ascorbate. Since one of the major end products of ascorbate in man is oxalate, a consequence is that the excretion of this compound due to ascorbate metabolism will be limited. This is the biochemical explanation of the finding by several groups that excessive ascorbate intakes do not lead to a proportional increase in the excretion of oxalate.[20]

The model also allows the estimation of the size of the three assumed pools and their turnover rates. Several interesting features can be highlighted. Thus, the rapidly exchangeable central pool is smaller than the metabolic and the deep, slowly exchangeable pools, which are of about equal sizes. The total pool size varies with the total turnover as discussed above. When normalized to the weight of the participants in the study, the size of the total pools seem to level off at about 20 mg per kg body weight. This would indicate a total pool of about 1500 mg, which is in very good agreement with other studies. The kinetic approach, however, also allows a further evaluation of the data and the apparent principles of distribution and emptying of the pools. Thus, the central and the metabolic pools varied in size depending on the total turnover, whereas the deep pool seemed practically unchanged in relation to the turnover. The participants were never depleted, but these findings may indicate that the rate of degradation of ascorbic acid in the depleted individuals might only or mainly reflect the turnover or emptying of the deep pool. An extrapolation to normal conditions from data collected during depletion might therefore not be adequate. The calculated turnover of 3%, frequently cited from the studies of Baker *et al.*, might therefore be erroneously low.

It is important to discuss if any of the observed features of the kinetics of the ascorbate metabolism could be used in the determination of a biochemical basis for the ascorbate intake. In summary, therefore, the isotope dilution technique forming the basis of the evaluation has been used and proved reliable in animal studies. Many important events in the metabolism of ascorbic acid in man, *i.e.*, tubular reabsorption, tissue binding, metabolic turnover, and apparent half-lives all seem to change at one and the same total turnover. This occurs at a level of about 60 to 80 mg per day. At an intake of this magnitude, the absorption is below 100%—in the range of 85%—and in addition a statistical uncertainty is inherent in the figures. This would indicate that intakes on the order of 100 mg per day would be needed to reach the point where these events occur, and a sound basis for a recommendation of an optimal intake of ascorbate could be based on these figures.

The discussed study only deals with normal, healthy, nonsmoking men. In another study[27] the same protocol was used for smokers and essentially the same results were obtained. The metabolic turnover, however, was increased about 1.4 times and the point where the discussed events were observed was accordingly raised to higher intakes. Similar differences between smokers and nonsmokers have been reported by others in studies based on different experimental approaches.[22] Other risk groups, such as pregnant women, children, diseased, and elderly have not been studied, mainly because of ethical implications of the use of radioisotopes. The technique of using stable isotopes, is therefore of great importance and potential value in furthering our understanding of ascorbate requirements for these groups.

REFERENCES

1. Barnes, A. E., W. Bartley, I. M. Frankau, G. A. Higgins, J. Pemberton, G. L. Roberts & H. R. Vickers. 1953. Medical Research Council Special Report Series No. 280. H. M. Stationary Office. London.
2. Degkwitz, E. 1985. Umschau **10:** 622.
3. Hornig, D. 1982. J. Jap. Soc. Clin. Nutr. **3:** 127.
4. Anderson, T. W. 1977. Acta Vitam. Enzymol. (Milano) **31:** 43.
5. Rivers, J. M. Safety of high-level vitamin C ingestion. This volume.
6. Sauberlich, H. E. 1984. Clin. Biochem. **17:** 132.
7. Baker, E. M., J. C. Saari & B. M. Tolbert. 1966. Am. J. Clin. Nutr. **19:** 371.
8. Tolbert, B. M., A. W. Chen, E. M. Bell & E. M. Baker. 1967. Am. J. Clin. Nutr. **20:** 250.
9. Baker, E. M., R. E. Hodges, J. Hood, H. E. Sauberlich & S. C. March. 1969. Am. J. Clin. Nutr. **22:** 549.

10. HORNIG, D. & D. HARTMANN. 1982. Kinetic behavior of ascorbic acid in guinea pigs. *In* Ascorbic Acid: Chemistry, Metabolism, and Uses. P. A. Seib & B. M. Tolbert, Eds.: 293.
11. KIPP, D. E. & J. M. RIVERS. 1984. J. Nutr. **114:** 1386.
12. HOLLOWAY, B. F. 1981. Kinetic analysis of ascorbic acid in the guinea pig. M.N.S. thesis, Cornell University, Ithaca, N.Y.
13. KALLNER, A., D. HARTMANN & D. HORNIG. Am. J. Clin. Nutr. **32:** 530.
14. HORNIG, D., J. P. VUILLEUMIER & D. HARTMANN. 1980. Int. J. Vitam. Nutr. Res. **50:** 309.
15. KALLNER, A., D. HORNIG & R. PELLIKA. 1985. Am. J. Clin. Nutr. **41:** 609.
16. GARRY, P. J., D. J. VANDERJAGT & W. C. HUNT. 1987. Ascorbic acid intakes and plasma levels in healthy elderly. This volume.
17. HARPER, A. 1975. Ann. N.Y. Acad. Sci. **258:** 491.
18. BERGER, L., C. D. GERSON & T. YU. 1977. Am. J. Med. **62:** 71.
19. STEIN, G., H. SPERSCHNEIDER & S. KOPPE. 1985. Blood Purif. **3:** 52.
20. MOSER, U. & D. HORNIG. 1982. Trends Pharmacol. **3:** 480.
21. KALLNER, A., D. HARTMANN & D. HORNIG. 1977. Int. J. Vitam. Nutr. Res. **4:** 383.
22. SMITH, J. & R. E. HODGES. 1987. Serum levels of vitamin C in relation to dietary and supplemental intake of vitamin C in smokers and nonsmokers. This volume.

DISCUSSION OF THE PAPER

M. SILVER: Does that radioactivity mentioned represent ascorbic acid or is it total radioactivity?

A. KALLNER (*Karolinska Hospital, Stockholm, Sweden*): If you refer to the graph with plasma data that is total radioactivity only.

M. SILVER: That would seriously jeopardize your model parameter estimation, because if you have metabolites then the excretion pattern is going to be different for various species.

A. KALLNER: This method of course was tested in the animal setup, where we also would have that type of metabolism. In addition the model and the results are not calculated from plasma data alone since we have the other type of information as well.

M. SILVER: Are you suggesting that the tissue data you looked at represent unchanged vitamin C?

A. KALLNER: Yes. In the animals, yes.

M. SILVER: Yes, but you can't get volume or distribution parameters using urine data. I'm only taking issue with the kinetic parameters.

S. D. VARMA (*University of Maryland School of Medicine, Baltimore, Md.*): I'd like to ask a philosophical question. Why did nature make us lack gulonate oxidase if ascorbic acid is so essential?

A. KALLNER: That's a good question. I think you should ask somebody else about that, higher up in the hierarchy.

Ascorbic Acid: The Concept of Optimum Requirements[a]

MARK LEVINE AND WILLIAM HARTZELL

Laboratory of Cell Biology and Genetics
National Institute of Diabetes, Digestive and Kidney Diseases
National Institutes of Health
Bethesda, Maryland 20892

INTRODUCTION

Although minimum requirements for many vitamins are known, optimum requirements for vitamins have remained elusive and may be of profound importance.[1,2] By providing for maximal cell function, optimum requirements may have implications for disease prevention.[1] Therefore, our work has focused on both defining the problem of optimum requirements for vitamins and working towards a solution. We have used vitamin C, or ascorbic acid, as a model vitamin, because of its nearly universal presence in the animal kingdom, its absolute requirement by humans and nonhuman primates, its known role as a cofactor in specific reactions, and its relative lack of toxicity *in vivo.*[1-5] Although we have used ascorbic acid as a model, the concept of optimum requirements should be applicable to many other vitamins, particularly cofactors.

OPTIMUM REQUIREMENTS

Optimum requirements may be thought of as part of a spectrum. At one end of the spectrum is a minimum requirement. For vitamins, the minimum requirement is that amount which just prevents a deficiency syndrome.[6] At the other end of the spectrum is that amount which produces toxicity. An optimum amount, within the limits set by minimum and toxic amounts, represents the best or most suitable amount of a vitamin, without toxicity.

But what in turn is meant by the "best amount," and how is it measured? We have addressed this question by first understanding how minimum and toxic amounts of ascorbic acid are determined. For a minimum amount, investigators have measured the amount of ascorbic acid necessary to prevent overt scurvy in animals and humans.[6,7] Since ascorbic acid is known to be required by eight isolated enzymes for maximal activity[1] (TABLE 1), it is also possible to measure that amount of ascorbic acid which produces minimal enzyme activity above baseline. These assays can be performed for

[a] This work was supported in part by the Foundation for Nutritional Advancement and the Rodale Press Foundation.

both isolated enzymes and enzymes *in situ* (enzymes in cells or tissues). In addition, ascorbic acid might be specifically necessary for nonenzymatic reactions,[1,8] and the minimum amount of ascorbic acid for these reactions can be quantitated.

For determining toxic amounts of ascorbic acid, a similar approach is possible using animals, humans, isolated enzymes, and enzymes *in situ*. Toxicity has been difficult to demonstrate with ascorbic acid.[5,9] However, toxicity for other vitamins can be quantitated, as can be seen from vitamins A, D, or B_6. Organ toxicity from excesses of all three vitamins and death from vitamins A and D have been documented in animals and humans.[10–12] For isolated enzymes, amounts of ascorbic acid that are toxic can be measured theoretically. These amounts of ascorbic acid would produce a decrease in the activities of the enzymes in TABLE 1. Diminished enzyme activities could be measured using either preparations of isolated enzymes or enzymes *in situ*. Ascorbic acid toxicity for other nonenzymatic events can also be quantitated, for example by measuring protein inactivation induced by ascorbic acid.[13]

In contrast to the straightforward approach to measuring minimum and toxic amounts, it has been a very difficult problem to measure an optimum amount of ascorbic acid because there has been no definitive assay in animals and humans.[1,2]

TABLE 1. Enzymes Dependent on Ascorbic Acid for Maximal Activity[a]

4-Hydroxyphenylpyruvate dioxygenase (EC 1.13.11.27)
γ-Butyrobetaine, 2-oxoglutarate 4-dioxygenase (EC 1.14.11.1)
Proline hydroxylase (EC 1.14.11.2)
Lysine hydroxylase (EC 1.14.11.4)
Procollagen-proline 2-oxoglutarate 3-dioxygenase (EC 1.14.11.7)
Trimethyllysine-2-oxoglutarate dioxygenase (EC 1.14.11.8)
Dopamine β-monooxygenase (EC 1.14.17.1)
Peptidyl glycine α-amidating monooxygenase

[a] Modified from Levine.[1]

THE CONCEPT OF OPTIMUM REQUIREMENTS: A HYPOTHESIS

We suggest that it is possible to quantitate an optimum amount of ascorbic acid by measuring specific reaction rates as a direct function of ascorbic acid concentration. This should be feasible for the isolated enzymes that are dependent on ascorbic acid for maximal activity (TABLE 1), as well as for nonenzymatic isolated reactions where ascorbic acid might specifically increase product formation.[8] However, the behavior of isolated enzymes and isolated reactions is not necessarily the same as enzymes *in situ* in cells and tissues (refs. 14 and 15, see below). Thus, we propose that optimum amounts of ascorbic acid can be quantitated by measuring specific reaction rates *in situ* of biochemical events dependent on ascorbic acid for maximal activity as a direct function of ascorbic acid concentration *in situ*. These measurements of *in situ* product formation should be performed first in cells, to learn *in situ* reaction behavior and to be able to predict the best conditions for subsequent animal experiments. Animal experiments should utilize the same kinds of product measurement, since the reactions and products should be specific for ascorbic acid.[1] For cells and animals, the data that need to be obtained can be represented graphically by formation of a specific product, or occurrence of a specific biochemical event, as a function of ascorbic acid concentration. As shown in FIGURE 1, these data are similar to a dose-response curve. The concentration of ascorbic acid for this kind of determination may be that found in cells, in the tissue, in plasma, and/or perhaps that total amount given to animals. Toxicity might also be detected by measuring a decrease in product formation at high ascorbic acid concentrations, as indicated by the dashed part of the line in the right side of FIGURE 1.

However, a dose-response curve for one reaction can not suffice to indicate an optimum requirement of ascorbic acid. Therefore, dose-response curves must be generated for several reactions that are specific for ascorbic acid,[1] as shown in FIGURE 2. From these curves, a concentration of ascorbic acid can be chosen that produces maximum *in situ* enzyme activity without toxicity or a fall in reaction velocity (FIG. 2). Thus, *in situ* reaction kinetics can be used to determine an optimum requirement of ascorbic acid, by a union of techniques in biochemistry, cell biology, physiology, and whole animal biology.

TESTING THE HYPOTHESIS OF OPTIMUM REQUIREMENTS

The concept of optimum requirements is straightforward. But can the concept be achieved, and by what approach? We must emphasize that we have not determined

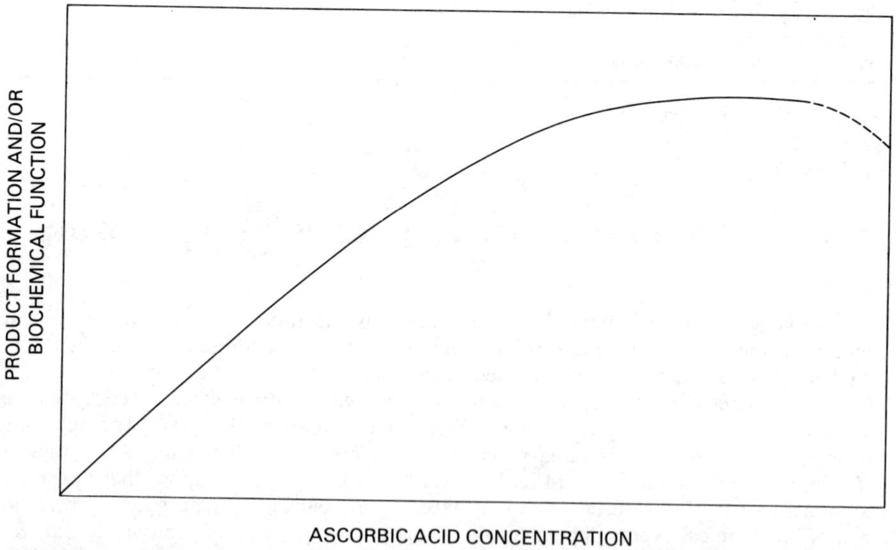

FIGURE 1. Product formation, enzyme activity, and/or a specific biochemical event as a function of ascorbic acid concentration. The *dashed line* represents decreasing product formation at high concentrations of ascorbic acid, perhaps indicative of ascorbic acid toxicity. See text for details.

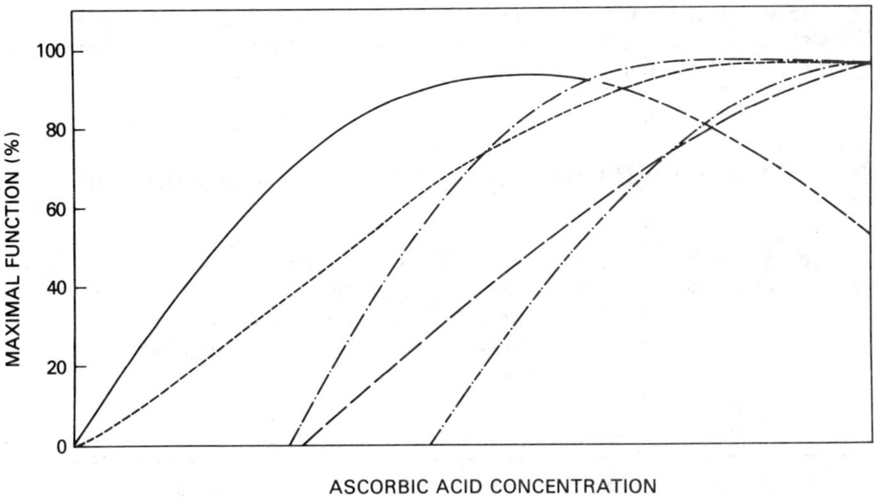

FIGURE 2. *In situ* enzyme activities as a function of ascorbic acid concentration. Each line represents a different enzyme activity or biochemical event. Toxicity is indicated by a fall in enzyme activity, as seen for one enzyme on the right side of the figure. See text for details.

the optimum requirement for ascorbic acid, and indeed the optimum requirement will probably not be determined for many years. However, we have shown that the problem is beginning to be solved by studying *in situ* behavior of an enzyme which when isolated is dependent on ascorbic acid for maximum activity. It is our hope that this approach will be an example of how to begin to solve the problem of optimum requirements.

We began by recognizing that there are eight enzymes dependent on ascorbic acid for maximum activity (TABLE 1). In addition, there are other nonenzymatic reactions that may be dependent on ascorbic acid.[8] We focused our interest on *in situ* enzymatic reactions in which ascorbic acid is a participant, since these reactions are likely to be specific for ascorbic acid in whole animals as well as in cells, and since activity of these enzymes should be regulated by ascorbic acid concentration.

It was necessary, then, to select an enzyme whose activity we could study *in situ* as a function of ascorbic acid concentration. We selected an enzyme by comparing the tissue distribution of enzymes listed in TABLE 1 to the concentration of ascorbic acid in different tissues (TABLES 2 and 3).[16] The tissue with the highest content of ascorbic acid per weight of tissue in rats is the adrenal gland. In humans, the contents of ascorbic acid in adrenals and pituitary are nearly equivalent, and are higher than in all other human tissues. (The uniformly higher ascorbic acid content in rat tissue compared to human tissue may be secondary to delay in analysis of human tissue due to procurement of human tissue postmortem.[17]) In adrenals, ascorbic acid has a similar distribution per wet weight in cortex and medulla.[17] An enzyme dependent on ascorbic acid for maximal activity (TABLE 1) is found in adrenal medulla. The enzyme, dopamine β-monooxygenase (also known as dopamine β-hydroxylase), is essential for biosynthesis of norepinephrine,[18–20] as shown in FIGURE 3. Norepinephrine is a catecholamine hormone of great importance for maintaining homeostasis of the cardiovascular system, and for neurotransmission in the central nervous system.[21] Thus, study of the regulation of dopamine β-monooxygenase activity *in situ* by ascorbic acid appeared to be an attractive model system, since the isolated enzyme was known to require ascorbic acid for maximum activity,[18-20] the enzyme and high concentrations of ascorbic acid colocalize to one tissue,[22] and the product, norepinephrine, has a critical physiologic role.[21]

THE CULTURED CHROMAFFIN CELL AS A MODEL SYSTEM

To test the hypothesis it was first necessary to obtain cells. Adrenal medullary cells, or chromaffin cells, can be readily isolated from animal tissue and cultured for as long as several weeks.[23] Bovine chromaffin cells were selected because large amounts of tissue were available. As examples of their properties, cultured bovine chromaffin

TABLE 2. Tissue Concentrations of Ascorbic Acid in Rats[a]

Tissue	Ascorbic Acid (mg/100 g Tissue)
Adrenal glands	280-400
Pituitary gland	100-130
Liver	25-40
Spleen	40-50
Lungs	20-40
Kidneys	15-20
Testes	25-30
Thyroid	22
Thymus	40
Brain	35-50
Eye lens	8-10
Skeletal muscle	5
Heart muscle	5-10
Bone marrow	12
Plasma	1.6
Blood	0.9

[a] Adapted from Hornig.[16]

cells secrete catecholamines, enkephalins, and ascorbic acid in response to a wide variety of secretagogues.[23-27] These cells thus seemed suitable for studying biosynthesis of catecholamines, particularly ascorbic acid regulation of dopamine β-monooxygenase activity *in situ.*

To study potential ascorbic acid regulation of norepinephrine biosynthesis in cells, it was necessary to be able to vary the concentration of intracellular ascorbic acid. When initially isolated, chromaffin cells contain approximately 100 ng ascorbic acid per 10^5 cells.[28] When the cells are cultured for several days without ascorbic acid, they lose more than 90% of intracellular ascorbic acid. If the cells are then incubated in ascorbic acid, they are able to reaccumulate ascorbic acid 10-fold, to approximately the amount found in the cells originally.[28,29] As seen in FIGURE 4, 3-day-old chromaffin cells contain 12 ng ascorbic acid per 10^5 cells at time 0, before incubation with 200 μM ascorbic acid. After 3 hours in ascorbic acid, ascorbic acid was accumulated 10-

TABLE 3. Ascorbic Acid Content of Adult Human Tissues[a]

Tissue	Ascorbic Acid (mg/100 g Wet tissue)
Adrenal glands	30-40
Pituitary gland	40-50
Liver	10-16
Spleen	10-15
Lungs	7
Kidneys	5-15
Testes	3
Thyroid	2
Heart muscle	5-15
Skeletal muscle	3-4
Brain	13-15
Pancreas	10-15
Eye lens	25-31
Plasma	0.4-1.0
Saliva	0.07-0.09

[a] Adapted from Hornig.[16]

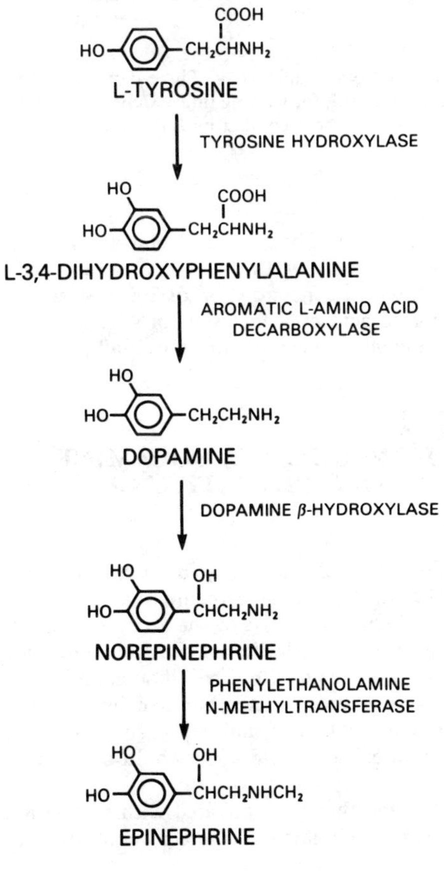

FIGURE 3. Catecholamine biosynthesis from L-tyrosine.

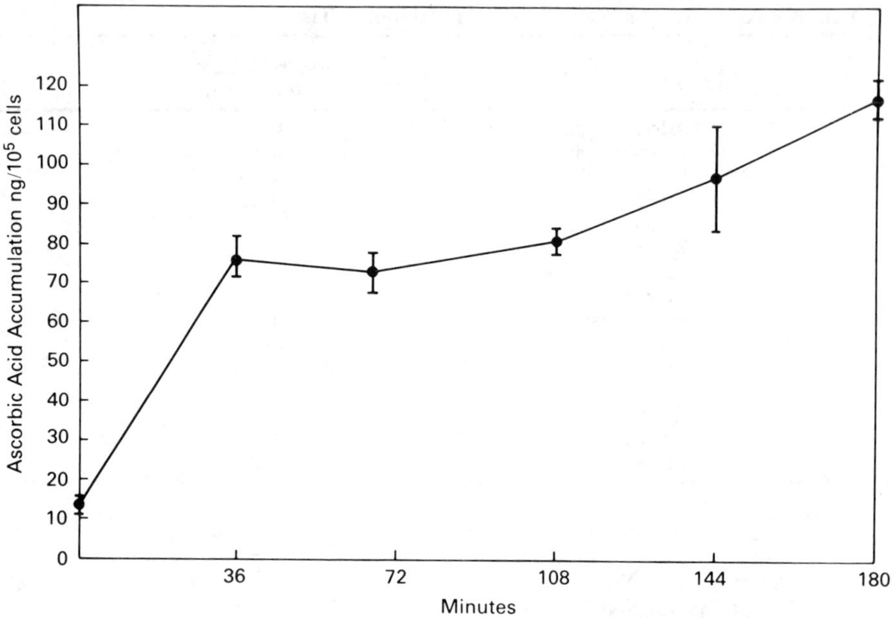

FIGURE 4. Ascorbic acid accumulation in adrenal chromaffin cells. Three-day-old cultured chromaffin cells were incubated with ascorbic acid, 200 μM, for varying times. Cells were washed, lysed by freezing and thawing, and analyzed for ascorbic acid content by HPLC; see ref. 28 for details.

fold by the cells, to more than 120 ng per 10^5 cells. Thus, intracellular ascorbic acid concentration can be regulated in chromaffin cells by the presence or absence of ascorbic acid in the culture medium. The external concentration of 200 μM ascorbic acid was selected to approximate the ascorbic acid concentration in adrenal venous effluents, and to be above the K_m for ascorbic acid transport into chromaffin cells.[29,30]

ASCORBIC ACID SPECIFICALLY REGULATES DOPAMINE β-MONOOXYGENASE ACTIVITY IN CHROMAFFIN CELLS

We selected two conditions to study the effects of ascorbic acid on catecholamine biosynthesis (see FIG. 4). In the first condition, ascorbic acid incubation for 3 hours, there was 10-fold more intracellular ascorbic acid compared to the second condition, no ascorbic acid incubation at time 0. Using these two conditions, we then determined if ascorbic acid increased norepinephrine biosynthesis *in situ* at all, if such an increase was specific for ascorbic acid as a reducing substance, if ascorbic acid increased the activity of one enzyme alone or several enzymes, whether such an increase was truly due to new biosynthesis, and whether such an ascorbic acid effect could occur under a wide variety of conditions.[14]

Chromaffin cells were cultured for 6 days and then preincubated with and without ascorbic acid, to produce the two different intracellular ascorbic acid concentrations noted above. The cells were then incubated at time 0 with [^{14}C]tyrosine, 10 μM, for different times. The biosynthesis of products from [^{14}C]tyrosine was measured (see FIG. 5). [^{14}C]Tyrosine content with and without ascorbic acid preincubation was similar.[14] [^{14}C]Dihydroxyphenylalanine content was identical in either condition and

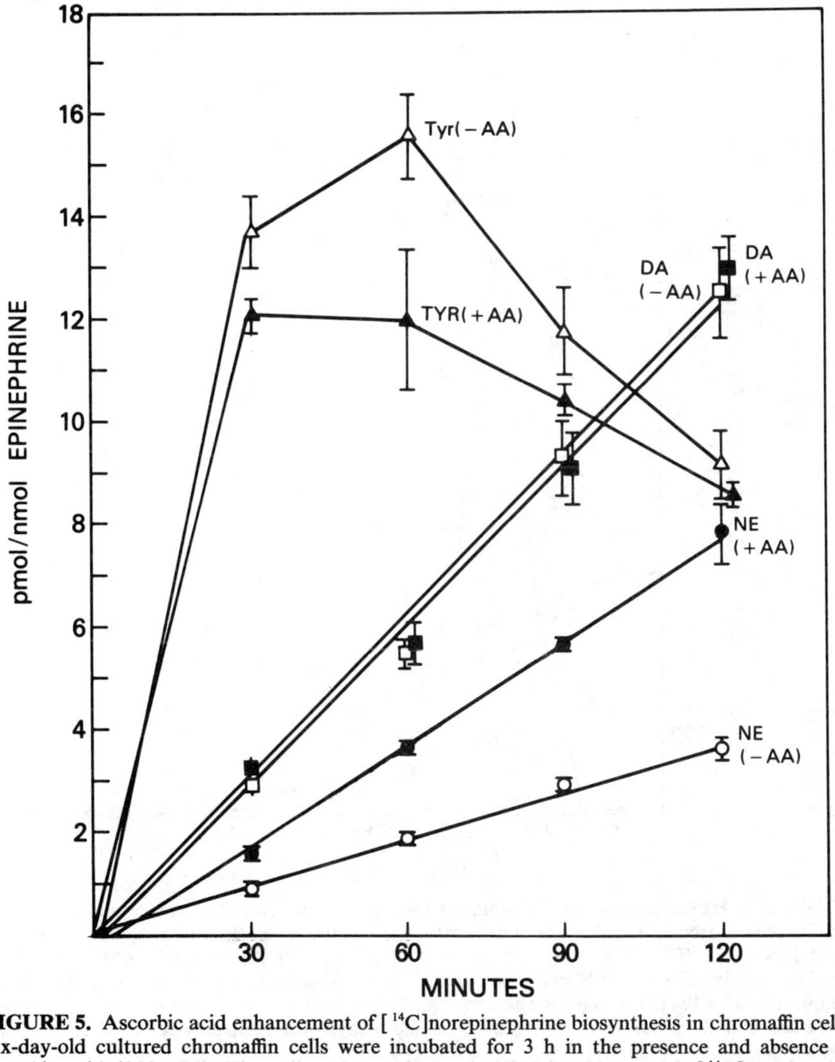

FIGURE 5. Ascorbic acid enhancement of [^{14}C]norepinephrine biosynthesis in chromaffin cells. Six-day-old cultured chromaffin cells were incubated for 3 h in the presence and absence of ascorbic acid (250 μM). The cells were washed twice and incubated with [^{14}C]tyrosine (10 μM) for varying times with (*closed symbols*) and without (*open symbols*) ascorbic acid (188 μM). At the times indicated cells were lysed and analyzed for catecholamines and radiolabeled compounds by HPLC. Symbols are as follows: ●, NE (+AA), [^{14}C]norepinephrine biosynthesis, ascorbic acid incubation; ○, NE (−AA), [^{14}C]norepinephrine biosynthesis, no ascorbic acid incubation; ■, DA (+AA), [^{14}C]dopamine content, ascorbic acid incubation; □, DA (−AA), [^{14}C]dopamine content, no ascorbic acid incubation; ▲, TYR (+AA), [^{14}C]tyrosine content, ascorbic acid incubation; △, TYR (−AA), [^{14}C]tyrosine content, no ascorbic acid incubation; see ref. 14 for details.

was barely detectable, as virtually all [^{14}C]dihydroxyphenylalanine was converted to [^{14}C]dopamine (data not shown, see ref. 28). [^{14}C]Dopamine content was identical whether or not ascorbic acid was present (FIG. 5). By contrast, [^{14}C]norepinephrine biosynthesis was clearly enhanced in the cells preincubated with ascorbic acid. Subsequent [^{14}C]epinephrine biosynthesis did not occur during this time course of

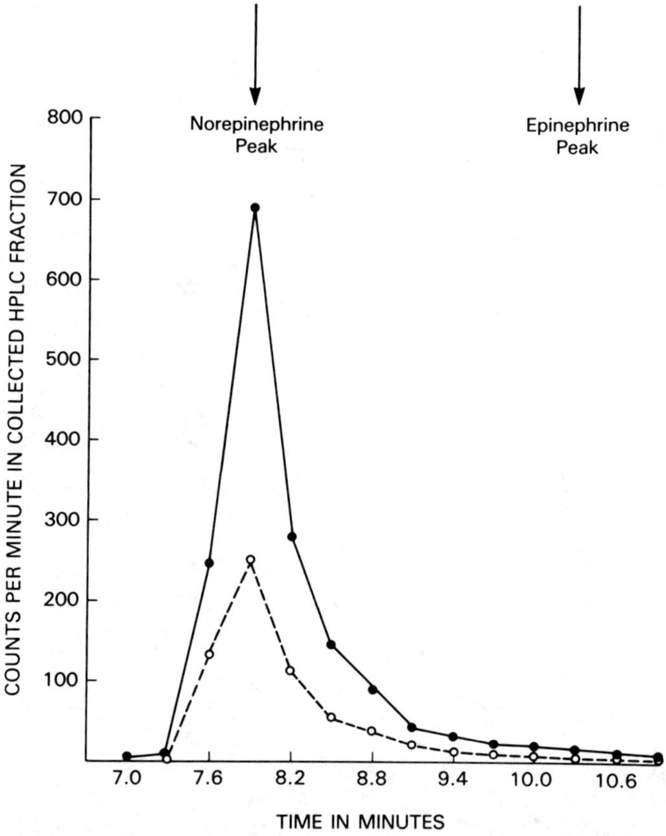

FIGURE 6. HPLC analysis of [^{14}C]epinephrine and [^{14}C]norepinephrine biosynthesis in chromaffin cells. Chromaffin cells were preincubated for 3 hours with (●) and without (○) ascorbic acid (200 μM). The cells were washed, incubated with [^{14}C]tyrosine (20 μM) for 1 hour, washed, and lysed with 0.4 N perchloric acid. Catecholamines were analyzed by HPLC. At the known times for the norepinephrine and epinephrine peaks (indicated by *arrows*), HPLC fractions were sequentially collected every 0.3 minutes and assayed for radioactivity; for details, see ref. 28.

[^{14}C]tyrosine incubation, whether or not ascorbic acid was present (FIG. 6). Thus, ascorbic acid appears to enhance biosynthesis of only norepinephrine in the catecholamine biosynthetic pathway.

If this system were to be useful for measuring product formation as a function of ascorbic acid concentration, it was important to test whether the effect of ascorbic acid was specific for increasing the activity of only one enzyme, dopamine β-monooxygenase. Tyrosine 3-monooxygenase is another enzyme in the catecholamine biosynthetic pathway that requires reducing equivalents. These reducing equivalents are provided by tetrahydrobiopterin.[31] Oxidized tetrahydrobiopterin must be reduced for continued tyrosine 3-monooxygenase activity. Since ascorbic acid can increase isolated tyrosine 3-monooxygenase activity, ascorbic acid might provide reducing equivalents to indirectly regenerate tetrahydrobiopterin, and ascorbic acid has been proposed to provide reducing equivalents indirectly to tyrosine 3-monooxygenase *in vivo* in adrenal medulla.[32–35]

We therefore investigated whether the effect of ascorbic acid was specific for increasing the activity of dopamine β-monooxygenase by inhibiting this enzyme and measuring biosynthesis of [^{14}C]dopamine and [^{14}C]norepinephrine, and [^{14}C]tyrosine content.[14] We used the inhibitor diethyldithiocarbamate, which inhibits dopamine β-monooxygenase, probably by chelating copper moieties.[36] As shown in TABLE 4, in the presence of the inhibitor with and without ascorbic acid, there were no differences in [^{14}C]tyrosine content, total [^{14}C]tyrosine uptake, [^{14}C]dopamine content, and total [^{14}C]dopamine biosynthesis. [^{14}C]Norepinephrine biosynthesis with and without ascorbic acid was inhibited by diethyldithiocarbamate more than 70%. These data indicate that when dopamine β-monooxygenase activity is blocked, ascorbic acid has no other effect on other enzymes in the catecholamine biosynthetic pathway. Thus, ascorbic acid is specific for enhancing the activity of only one enzyme in the norepinephrine biosynthetic pathway. Furthermore, the effect of ascorbic acid with isolated tyrosine 3-monooxygenase activity is not seen under our experimental conditions in chromaffin cells. These experiments emphasize the importance of studying the effect of ascorbic acid on enzyme activity *in situ*.

We then studied the specificity of ascorbic acid as a reductant for enhancing norepinephrine biosynthesis in chromaffin cells. Chromaffin cells were incubated with a variety of reducing agents and [^{14}C]tyrosine content, [^{14}C]dopamine content, and [^{14}C]norepinephrine biosynthesis were measured. As shown in TABLE 5, [^{14}C]norepinephrine biosynthesis was enhanced maximally only by ascorbic acid. These findings can not be explained by lack of uptake of the reducing substances into chromaffin cells,[15] nor by concomitant changes in [^{14}C]dopamine and [^{14}C]tyrosine processing (TABLE 5). Thus, ascorbic acid alone increases norepinephrine biosynthesis.

All of these experiments could have another interpretation: what appeared to be increased norepinephrine biosynthesis might actually be decreased norepinephrine metabolism induced by ascorbic acid. We investigated this possibility and found that virtually no catabolism of norepinephrine occurred under the conditions of these experiments, whether or not ascorbic acid was present.[14] In addition, experiments with isolated chromaffin granules, described below, also strongly suggest that the effect of ascorbic acid is specific for norepinephrine biosynthesis.[15]

For ascorbate regulation of norepinephrine biosynthesis in chromaffin cells to be a model system for studying *in situ* enzyme activity, it was important to test whether the ascorbate effect occurred under varied conditions found in animals. Since chromaffin cells help to maintain homeostasis in animals by secreting catecholamines, we tested whether ascorbic acid enhanced norepinephrine biosynthesis when the cells were induced to secrete. We incubated chromaffin cells with and without ascorbic acid as described, and added the secretagogue carbachol along with [^{14}C]tyrosine at time 0. As shown in FIGURE 7, [^{14}C]norepinephrine biosynthesis was enhanced approximately threefold by ascorbic acid in the presence of carbachol, without other effects on the biosynthetic pathway. Ascorbic acid enhanced norepinephrine biosynthesis in the presence of every secretagogue we have tested.[14] Thus, ascorbic acid enhances norepinephrine biosynthesis under a wide variety of conditions in chromaffin cells.

TABLE 4. Tyrosine 3-Monooxygenase Activity in the Presence and Absence of Ascorbic Acid

	Diethyldithio-carbamate Control[a]	Diethyldithio-carbamate Plus Ascorbic Acid[b]	% Diethyldithio-carbamate Control
pmol [^{14}C]norepinephrine/nmol epinephrine	0.35 ± 0.03	0.65 ± 0.13	186
[^{14}C]norepinephrine-specific activity	0.00042 ± 0.00005	0.00074 ± 0.00015	176
pmol [^{14}C]dopamine/nmol epinephrine	1.78 ± 0.10	1.43 ± 0.02	80
[^{14}C]dopamine-specific activity	0.053 ± 0.001	0.046 ± 0.003	87
total [^{14}C]dopamine biosynthesis (pmol)/nmol epinephrine	2.13 ± 0.10	2.06 ± 0.17	97
pmol [^{14}C]tyrosine/nmol epinephrine	16.25 ± 0.58	16.90 ± 0.64	104
total [^{14}C]tyrosine uptake (pmol)/nmol epinephrine	18.38 ± 0.66	18.95 ± 0.64	103

Six-day-old cultured chromaffin cells were preincubated for 3 h with and without ascorbic acid (250 μM). The cells were washed and incubated for 30 minutes with diethyldithiocarbamate (1 mM) in the presence and absence of ascorbic acid (188 μM). [^{14}C]Tyrosine (10 μM) was then added for an additional 30 minutes. The cells were washed and lysed, and the lysate analyzed for catecholamines and radiolabeled compounds by HPLC; see ref. 14 for details.

[a] DDC control: 74% inhibition of NE biosynthesis.

[b] DDC + AA: 72% inhibition of NE biosynthesis.

TABLE 5. Specificity of Ascorbic Acid for Increasing [^{14}C]Norepinephrine Biosynthesis

	pmol [^{14}C]Norepinephrine/ nmol Epinephrine/h	% Control	pmol [^{14}C]Dopamime/ nmol Epinephrine/h	% Control	pmol [^{14}C]Tyrosine/ nmol Epinephrine/h	% Control
Additions						
None (control)	2.31 ± 0.49	100	5.44 ± 0.44	100	15.49 ± 0.59	100
Ascorbic acid	4.62 ± 0.22	200	6.37 ± 0.22	117	16.08 ± 0.38	104
Glutathione	2.54 ± 0.36	110	5.44 ± 0.29	100	15.38 ± 0.77	99
NADH	2.90 ± 0.47	126	4.71 ± 0.06	87	16.45 ± 1.37	106
Homocysteine	3.11 ± 0.52	135	6.99 ± 0.31	128	6.13 ± 0.11	39
Thiourea	2.94 ± 0.38	127	4.61 ± 0.33	85	16.90 ± 0.64	109

Chromaffin cells were preincubated for 3 hours with one of the reducing agents indicated, at 250 μM. The cells were washed and incubated for 60 minutes with 188 μM of the same reducing agent and [^{14}C]tyrosine (10 μM). Cells were then washed, lysed, and analyzed for catecholamines and radiolabeled compounds by HPLC. Control cells were incubated in medium without any reducing agent. See ref. 14 for details.

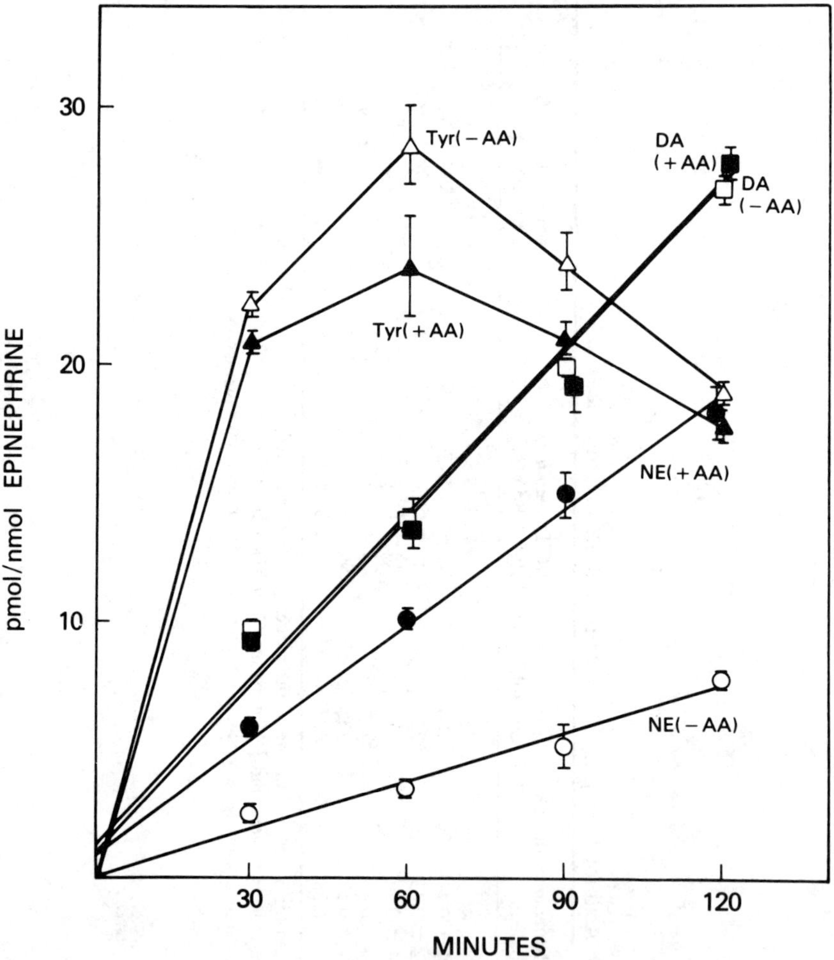

FIGURE 7. Ascorbic acid enhancement of [^{14}C]norepinephrine biosynthesis in chromaffin cells stimulated with the secretagogue carbachol. Cultured chromaffin cells were incubated with and without ascorbic acid as in FIGURE 5. At time 0, carbachol (300 μM) was added with [^{14}C]tyrosine (10 μM), in the presence (*closed symbols*) and absence (*open symbols*) of ascorbic acid (188 μM). At the indicated times the cells were washed, lysed, and analyzed for catecholamines and radiolabeled compounds by HPLC. Catecholamine secretion without ascorbic acid was 18% of total and with ascorbic acid 15% of total throughout the incubation. Symbols are the same as in FIGURE 5; see ref. 14 for details.

All of these data indicate that ascorbic acid increases only dopamine β-monooxygenase activity in the catecholamine biosynthetic pathway *in situ,* that this effect is specific for ascorbic acid, that this effect is not due to metabolism of catecholamines, and that ascorbic acid increases norepinephrine formation under a wide variety of conditions in cells. This system, therefore, appeared to be an ideal model system in which varying concentrations of ascorbic acid would regulate specific product formation.

THE MECHANISM OF ASCORBIC ACID ACTION IS DIFFERENT IN CHROMAFFIN CELLS COMPARED TO ISOLATED DOPAMINE β-MONOOXYGENASE

From all of these experiments, we envisioned a dose-response curve for norepinephrine biosynthesis as a function of ascorbic acid content in chromaffin cells. However, from the present data there are only two points on this curve, one at a relatively high content of ascorbic acid and one at a relatively low content (FIGURE 4). At first, it would appear that the next experiments would be to fill in the dose-response curve (FIGURE 2) by using chromaffin cells with varying contents of ascorbic acid.[14,28]

However, ascorbic acid regulation of norepinephrine synthesis may be more complex than is readily apparent. Dopamine β-monooxygenase is the only catecholamine biosynthetic enzyme that is localized within chromaffin granules, the catecholamine storage vesicles (for review, see ref. 24). The enzyme is present exclusively within granules in both a membrane bound and soluble form, with none of the enzyme on the outer surface of chromaffin granules.[37,38] By contrast, nearly 90% of ascorbic acid is extragranular in bovine cultured chromamaffin cells and in homogenates of bovine adrenal medullae.[27,30,39] This extragranular ascorbic acid appears to be cytosolic and not associated with another vesicle.[39] Chromaffin granules themselves do contain ascorbic acid, at a concentration estimated to be as high as 22 mM,[22] although it may be severalfold lower. However, ascorbic acid is unable to enter isolated chromaffin granules,[17,28] and it is unknown how ascorbic acid enters granules in chromaffin cells.[17]

These observations, then, introduce more complexity to the chromaffin cell model for ascorbate regulation of dopamine β-monooxygenase activity. Though ascorbic acid and the enzyme dopamine β-monooxygenase localize to the same cell, there is not intracellular colocalization, since most of the ascorbic acid is localized outside of granules whereas all of the enzyme is intravesicular. Therefore, since both extragranular and intragranular compartments may affect norepinephrine biosynthesis, and since both compartments contain ascorbic acid, it seemed important to understand the role of ascorbic acid in both compartments. Indeed, we could not assume that the concentration of ascorbic acid in one compartment alone regulated biosynthesis. Furthermore, if ascorbic acid does not enter isolated chromaffin granules, then why did our cell experiments show an effect of ascorbic acid on norepinephrine biosynthesis at all? Thus, the mechanism of ascorbic acid action *in situ* did not appear to be as simple as the mechanism of ascorbic acid action with isolated dopamine β-monooxygenase, where electrons are transferred from ascorbic acid to the enzyme. Before we could determine the dose-response curve for ascorbic acid and norepinephrine biosynthesis in chromaffin cells, it was essential to understand the mechanism of ascorbic acid action in chromaffin tissue.

A critical part of the problem was how dopamine β-monooxygenase received reducing equivalents from ascorbic acid. Since ascorbic acid does not enter isolated chromaffin granules, one possibility is that ascorbic acid can enter granules within cells, and that necessary cytosolic factor(s) are lost when the granules are isolated. We recognized another possibility from the data from homogenized adrenal medullae, where nearly 90% of medullary ascorbate is extravesicular.[39] This distribution suggested that ascorbic acid itself did not have to enter chromaffin granules acutely for norepinephrine biosynthesis. Indeed, all that was required was for reducing equivalents from extragranular ascorbic acid to be transferred across the granule membrane, eventually to dopamine β-monooxygenase.[15] Such an electron transfer system was also independently postulated by Wakefield[41,42] and Njus and coworkers.[43]

We tested the possibility of electron transfer from external ascorbic acid to dopamine β-monooxygenase by studying norepinephrine biosynthesis in isolated chromaffin granules.[15] In the presence and absence of ascorbic acid, dopamine transport into chromaffin granules was identical (FIG. 8). Ascorbic acid did not enter chromaffin granules under these conditions.[15,28] By contrast, norepinephrine biosynthesis was

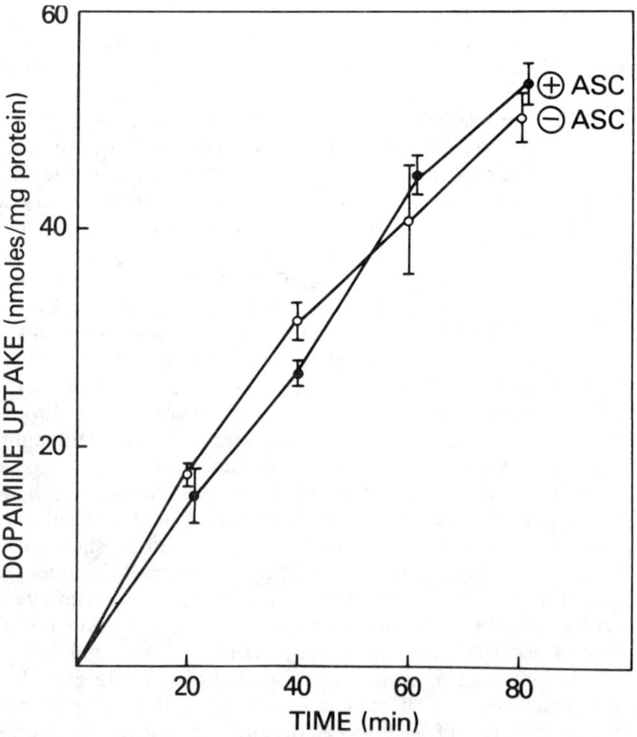

FIGURE 8. [^{3}H]Dopamine uptake into isolated chromaffin granules in the presence and absence of ascorbic acid. Isolated chromaffin granules were incubated with [^{3}H]dopamine (50 μM) and Mg-ATP (2.5 mM), with (●) and without (○) ascorbic acid (2 mM) for varying times. [^{3}H]Dopamine uptake represents the sum of newly transported [^{3}H]dopamine plus newly synthesized [^{3}H]norepinephrine. [^{3}H]Dopamine uptake was determined using a cation exchange resin; see ref. 15 for details.

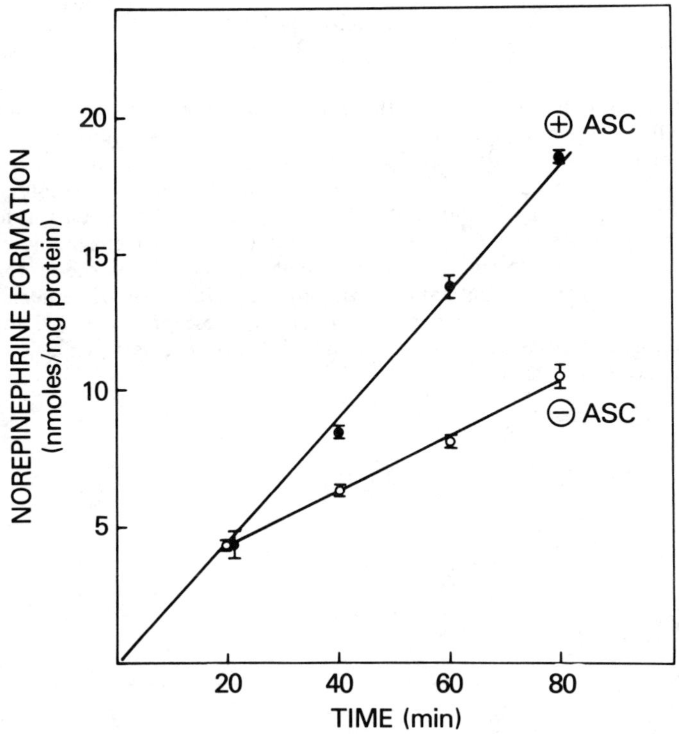

FIGURE 9. [^{3}H]Norepinephrine biosynthesis in isolated chromaffin granules incubated with and without ascorbic acid. [^{3}H]Norepinephrine biosynthesis was determined in the same chromaffin granules shown in FIGURE 8. Isolated chromaffin granules were incubated with [^{3}H]dopamine (50 μM) and Mg-ATP (2.5 mM) with (●) or without (○) ascorbic acid (2 mM). [^{3}H]Norepinephrine biosynthesis was determined using a cation exchange resin; see ref. 15 for details.

clearly increased by external ascorbic acid (FIG. 9). External ascorbic acid enhanced norepinephrine biosynthesis in chromaffin granules in a concentration-dependent manner; the K_m of external (cytosolic) ascorbic acid for norepinephrine biosynthesis was approximately 200 μM (FIG. 10).[15,44] Thus, it is possible to quantitate the K_m of cytosolic ascorbic acid for norepinephrine biosynthesis in chromaffin tissue.

These experiments with isolated granules indicate that it is important to determine the K_m of intragranular ascorbic acid for norepinephrine biosynthesis, with dopamine as the substrate for dopamine β-monooxygenase. We must also learn what other intermediates (if any) participate in electron transfer in addition to ascorbic acid, what energetically drives electron transfer, and what the complete mechanism of electron transfer is. From these experiments to date, it is clear that the interaction between ascorbic acid and dopamine β-monooxygenase *in situ* is strikingly different from the interaction of ascorbic acid and the isolated enzyme.

SUMMARY

Experiments with enzymes *in situ* that are dependent on ascorbic acid for maximum activity will provide critical information about ascorbic acid requirements. Our work with chromaffin tissue as a model system eventually will result in the determination of two dose-response curves for norepinephrine biosynthesis, representing cytosolic ascorbic acid and intragranular ascorbic acid (FIG. 11). These curves for norepinephrine biosynthesis can be combined with curves for other enzymatic events that are also dependent on ascorbic acid for maximal activity. These dose-response curves (FIG. 2) will allow determination of optimum ascorbic acid requirements based on specific product formation and minimum toxicity. These principles are adaptable to other vitamins as well as ascorbic acid, and could form the basis for a new approach to vitamin requirements.

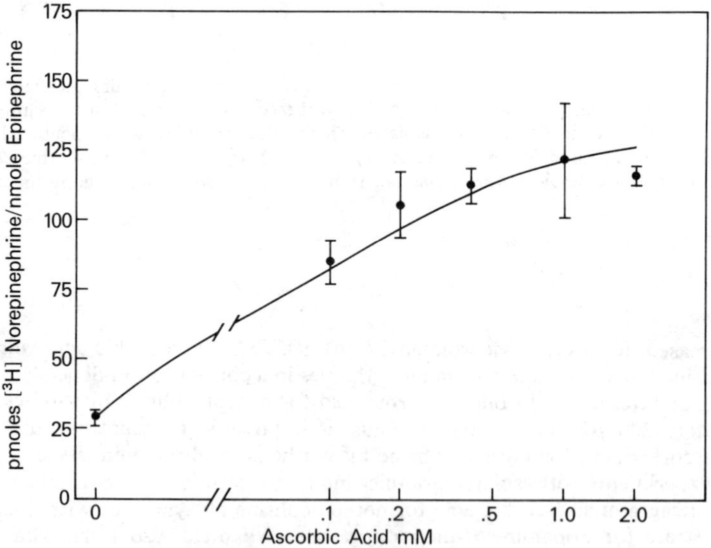

FIGURE 10. Ascorbic acid regulation of [^{3}H]norepinephrine biosynthesis in isolated chromaffin granules. Isolated chromaffin granules were incubated with Mg-ATP (2 mM), [^{3}H]dopamine (10 mM), and varying concentrations of ascorbic acid for 1 hour. [^{3}H]Dopamine uptake was similar regardless of ascorbic acid concentration (not shown). Catecholamine analysis was by HPLC; see refs. 14 and 44 for details.

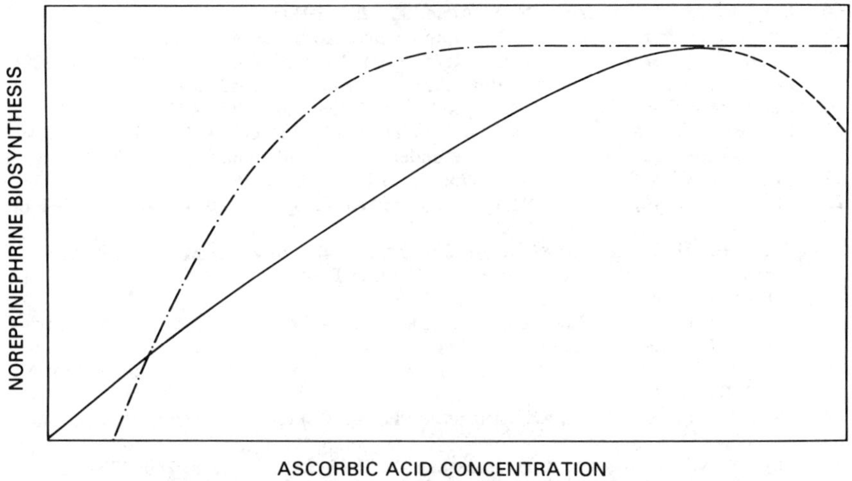

FIGURE 11. Norepinephrine biosynthesis as a function of ascorbic acid concentrations in two different compartments in chromaffin cells. The lines may represent the following: norepinephrine biosynthesis as a function of the concentration of cytosolic ascorbic acid (— • — •), and norepinephrine biosynthesis as a function of the concentration of intragranular ascorbic acid (—). The *dashed line* at the right side of the figure may indicate decreased norepinephrine biosynthesis at high concentrations of ascorbic acid.

REFERENCES

1. LEVINE, M. 1986. N. Engl. J. Med. **314:** 892-902.
2. IRWIN, M. I. & B. K. HUTCHINS. 1976. J. Nutr. **106:** 821-879.
3. BURNS, J. J., P. PEYSER & A. MOLTZ. 1956. Science **124:** 1148-1149.
4. CHATTERJEE, I. B. 1970. *In* Methods in Enzymology, Vol. 18. Part A. D. B. McCormick & L. D. Wright, Eds.: 28-34. Academic Press. New York.
5. HORNIG, D. & U. MOSER. 1981. *In* Vitamin C: Ascorbic Acid. J. N. Counsell & D. H. Hornig, Eds.: 225-248. Applied Science Publishers. London.
6. HODGES, R. E., E. M. BAKER, J. HOOD, H. E. SAUBERLICH & S. C. MARCH. 1969. Am. J. Clin. Nutr. **22:** 535-548.
7. BAKER, E. M., J. C. SAARI & B. M. TOLBERT. 1966. Am. J. Clin. Nutr. **19:** 371-378.
8. BATES, C. J. 1981. *In* Vitamin C: Ascorbic Acid. J. N. Counsell & D. H. Hornig, Eds.: 1-22. Applied Science Publishers. London.
9. RIVERS, J. 1986. Safety of high level vitamin C ingestion. This volume.
10. LUI, N. S. T. & O. A. ROELS. 1980. *In* Modern Nutrition in Health and Disease. R. S. Goodhart & M. E. Shils, Eds.: 142-159. Lea and Febiger. Philadelphia.
11. DELUCA, H. F. 1980. *In* Modern Nutrition in Health and Disease. R. S. Goodhart & M. E. Shils, Eds.: 160-169. Lea and Febiger. Philadelphia.
12. SCHAUMBURG, H., J. KAPLAN, A. WINDEBANK, N. VICK, S. RASMUS, D. PLEASURE & M. J. BROWN. 1983. N. Engl. J. Med. **309:** 445-448.
13. LEVINE, R. 1983. J. Biol. Chem. **258:** 11828-11833.
14. LEVINE, M. 1986. J. Biol. Chem. **261:** 7347-7356.
15. LEVINE, M., K. MORITA, E. HELDMAN & H. B. POLLARD. 1985. J. Biol. Chem. **260:** 15598-15603.

16. HORNIG, D. H. 1975. Ann. N. Y. Acad. Sci. **258:** 103-118.
17. LEVINE, M. & K. MORITA. 1985. Vitamins and Hormones **42:** 1-64.
18. LEVIN, E. Y., B. LEVENBERG & S. KAUFMAN. 1960. J. Biol. Chem. **235:** 2080-2086.
19. LEVIN, E. Y. & S. KAUFMAN. 1961. J. Biol. Chem. **236:** 2043-2049.
20. FRIEDMAN, S. & S. KAUFMAN. 1965. J. Biol. Chem. **240:** 4763-4773.
21. LANDSBERG, L. & J. B. YOUNG. 1985. *In* Textbook of Endocrinology. J. D. Wilson & D. W. Foster, Eds.: 891-965. W. B. Saunders Co. Philadelphia.
22. TERLAND, O. & T. FLATMARK. 1975. FEBS Lett. **59:** 52-56.
23. FENWICK, E. M., P. B. FAJDIGA, N. B. S. HOWE & B. G. LIVETT. 1978. J. Cell Biol. **76:** 12-30.
24. POLLARD, H. B., R. ORNBERG, M. LEVINE, K. KELNER, K. MORITA, R. LEVINE, E. FORSBERG, K. W. BROCKLEHURST, L. DUONG, P. LELKES, E. HELDMAN & M. YOUDIM. 1985. Vitamins and Hormones **42:** 109-196.
25. POLLARD, H. B., C. J. PAZOLES, C. E. CREUTZ, J. H. SCOTT, O. ZINDER & A. HOTCHKISS. 1984. J. Biol. Chem. **259:** 1114-1121.
26. LIVETT, B. G., D. M. DEAN, L. G. WHELAN, S. UDENFRIEND & J. ROSSIER. 1981. Nature **289:** 317-319.
27. LEVINE, M., A. ASHER, O. ZINDER & H. B. POLLARD. 1983. J. Biol. Chem. **258:** 13111-13115.
28. LEVINE, M., K. MORITA & H. B. POLLARD. 1985. J. Biol. Chem. **260:** 12942-12947.
29. DILIBERTO, E. J., JR., G. D. HECKMAN & A. J. DANIELS. 1983. J. Biol. Chem. **258:** 12886-12894.
30. LEVINE, M. & H. B. POLLARD. 1983. FEBS Lett. **158:** 134-138.
31. KAUFMAN, S. 1974. *In* Aromatic Amino Acids in the Brain. CIBA Foundation. Elsevier. Amsterdam. pp. 81-115.
32. STONE, K. J. & B. H. TOWNSLEY. 1973. Biochem. J. **131:** 611-613.
33. LERNER, P., P. HARTMAN, M. M. AMES & W. LOVENBERG. 1977. Arch. Biochem. Biophys. **182:** 164-170.
34. LERNER, P., P. NOSE, M. M. AMES & W. LOVENBERG. 1978. Neurochem. Res. **3:** 641-651.
35. NAKASHIMA, Y., R. SUZUE, H. SANADA & S. KAWADA. 1970. J. Vitaminol. (Kyoto) **16:** 276-280.
36. NAGATSU, T. 1973. Biochemistry of Catecholamines (chap. 1). University Park Press. Baltimore.
37. HORTNAGL, H., H. WINKLER & H. LOCHS. 1972. Biochem. J. **129:** 187-195.
38. LADURON, P. 1975. FEBS Lett. **52:** 132-134.
39. MORITA, K., M. LEVINE, E. HELDMAN & H. B. POLLARD. 1985. J. Biol. Chem. **260:** 15112-15116.
40. TIRRELL, J. & E. WESTHEAD. 1977. Neuroscience **4:** 181-186.
41. WAKEFIELD, L. M., A. E. G. CASS & G. K. RADDA. 1986. J. Biol. Chem. **261:** 9739-9745.
42. WAKEFIELD, L. M., A. E. G. CASS & G. K. RADDA. 1986. J. Biol. Chem. **261:** 9745-9752.
43. NJUS, D., J. KNOTH, C. COOK & P. M. KELLEY. 1983. J. Biol. Chem. **258:** 27-30.
44. LEVINE, M. & W. O. HARTZELL. Manuscript in preparation.

DISCUSSION OF THE PAPER

S. D. VARMA (*University of Maryland School of Medicine, Baltimore, Md.*): You are determining the effect of various reducing agents for norepinephrine biosynthesis. The glutathione you have used will not get into the cell because it's a negatively charged tripeptide.

M. LEVINE (*National Institutes of Health, Bethesda, Md.*): With the isolated vesicles the data if anything are clearer than this, where there is not a problem with penetration; in fact the data are identical using either granules or cells.

S. D. VARMA: You're using a cell line which is capable of synthesizing ascorbic acid from glucose. Therefore how do you make the calculation for endogenous ascorbic acid?

M. LEVINE: These cells do not make ascorbic acid. Ascorbic acid is made by the animal but it's made in the liver. We're taking these cells out of an animal and culturing them, and we've measured by HPLC ascorbic acid content, as I showed at the beginning, over several days. There is no biosynthesis whatsoever of ascorbic acid in these cells.

E. J. DILIBERTO (*Wellcome Research Labs, Research Triangle Park, N. C.*): We have tried to look at synthesis in these cells and we agree it does not occur. In your first slide, where you're looking at the effects or the requirements for ascorbic acid, I think you should separate out those enzymes where it has been shown stoichiometrically that ascorbic acid is utilized in those systems as opposed to the dioxygenases where stochiometric utilization of ascorbic acid is not required.

Secondly, I would like to clarify a few points that you made. Tyrosine hydroxylase is a tetrahydrobiopterin-requiring enzyme and does not require ascorbic acid. Why was there a need for doing those experiments with ascorbic acid?

M. LEVINE: I'm quite cognizant that tetrahydrobiopterin has been reported to be the cofactor for tyrosine hydroxylase or tyrosine monooxygenase. But it's been postulated that ascorbic acid was part of that regenerating system, and not that ascorbic acid was active directly with the enzyme. In our cell experiments we wanted to exclude the possibility that ascorbic acid had any effect on the activity of tyrosine hydroxylase if we were looking for a specific biosynthetic effect.

E. J. DILIBERTO: I think if you were to do the experiments to be able to show what the precise concentration of ascorbic acid in the vesicle is, you can show exactly what the enzymic activity is and correlate those with the vesicular concentration. The regenerating system is obviously required to keep the cofactor in the reduced form.

M. LEVINE: We have reported the concentration of ascorbic acid in crude preparations of vesicles in the *Journal of Biological Chemistry;* it is between 8 and 11 mM. However, the problem with all of the reported concentrations of ascorbate, which have ranged up to 22 mM, is that it's not clear what's actually available, what is free, what's bound, what's complexed, and what the enzyme actually sees.

B. TOLBERT (*University of Colorado, Boulder, Colo.*): I note you list eight enzymes and I'm very hopeful that by the time the next decennial meeting comes to pass they will be multiplied a great deal. There's a great deal of metabolism of ascorbic acid which is not completely understood, especially those involving C5 metabolism which has been reported both in animal systems and in plant systems. The extent that this metabolism of ascorbic acid and the roles of ascorbic acid, and the enzymes related to it, can be amplified are very great. I somehow wonder if your approach will take care of the complexity that will ultimately evolve when we really know what's going on with ascorbic acid.

M. LEVINE: I think that you're right. I think that when we go to animals there's no question that there's going to be a distribution problem and there's going to be a metabolism problem of ascorbic acid. But I think that we have to have an idea of what concentrations in tissues are even going to give us any kind of dose-response information before we can even begin to make guesses.

You're right, I think when we get to animals that there will be a large compartmentalization and metabolism problem but I don't think it will be insurmountable.

H. SPRINCE (*Jefferson Medical College, Philadelphia, Pa.*): As to this concept of optimal requirement of vitamin C, it seems to me that there are many imponderables here which have to be considered, for example, whether people are drinking heavily, whether they're smoking heavily, whether they're exposed to pollutants and so on. This raises the question of whether approaches other than the biochemical approach might not be considered, namely the toxicological approach, the effect of toxic substances and protection therefrom by ascorbic acid. For example, it would appear to me that the optimum requirement of vitamin C might be higher for smokers than people who are nonsmokers.

A. TAYLOR (*Tufts University, Boston, Mass.*): I have a similar concern, though I must say your approach is close to my heart as a chemist and I never knew what a whole animal was until recently. But the fact of the matter is that the whole is much more than the sum of the parts. It's very useful to dissect on a molecular level, but they interelate on yet a whole other level, and then the cells themselves, once they're organized into an organ, similarly interrelate on a yet more complex and interrelated level. It seems that you really have to look for organismic effects.

M. LEVINE: What I'm proposing here is that we use specific functions of ascorbate to help us look for effects in the animal. As one example, norepinephrine biosynthesis is appropriate because no other reducing substance will regulate this synthesis.

A. TAYLOR: But you don't know whether maximizing norepinephrine synthesis is a useful assay, and it seems to me that maybe you're looking at the wrong assay. One should get on to others.

M. LEVINE: That's exactly what I was trying to say: that one assay in and of itself is not going to give the answers. When we have dose-response curves for several of the enzymes that we heard about at the conference, it's the superimposition of those respective curves that is going to give us the answer for optimal ascorbate requirements.

S. L. ROMNEY (*Albert Einstein College of Medicine, Bronx, N. Y.*): I almost have a corollary to that question. Is your position that ascorbic acid is central to stress phenomena? And if that's so, how are you going to define steady state?

M. LEVINE: I really am unable to say that it's essential to stress or not, we're nowhere near being able to answer that.

E. J. DILIBERTO: The nervous system and a number of the endocrine tissues have a high capacity to retain the vitamins. The expression of any deficiencies in those systems is going to be very hard to realize and I think that's why you have to choose your systems well to really understand what's optimal and minimal in terms of the requirements for ascorbic acid. In terms of the communication of neurons and other cells, that is something that's required for survival of the animal and ascorbic requirements are obvious in those cases. But, on the other hand, evolution has given those particular cells a high capacity to retain the vitamin for synthesis of the communicators.

M. LEVINE: I must emphasize again that I do not mean to stress that one biosynthetic pathway is going to provide us with an answer to what is an optimal amount. I think that we have to superimpose several of them. Yes, it turns out that ascorbic acid not only is concentrated in the tissue like the adrenal medulla but there's actually a portal system there in the pituitary as well. But other systems which don't have that portal system are also dependent on ascorbic acid, norepinephrine in peripheral neurons, biosynthesis of the gut hormones, for example, or other amidated peptides which are not in the pituitary and subject to portal regulation.

The point is that we need specific markers of ascorbate function.

Safety of High-level Vitamin C Ingestion

JERRY M. RIVERS[a]

Cornell University
Ithaca, New York 14853

Vitamin C is widely consumed as a dietary supplement, ingested either as a single nutrient or in combination with other vitamins and minerals. Stewart *et al.*[1] reported that 35.1% of the adult U.S. population ingested a vitamin C supplement, with a median intake of 333% of the Recommended Dietary Allowance (RDA) and 28 times the RDA at the 95th percentile. Further, gram amounts of ascorbic acid are suggested for treatment and/or prevention of a wide array of health aberrations.[2–4] Concern about the safety of these practices has been addressed in recent reviews.[5–7]

ABSORPTION, METABOLISM, AND EXCRETION

Ascorbic acid is absorbed in the intestine by an energy-requiring, Na^+-dependent, carrier-mediated transport system.[8–10] *In vivo* intestinal perfusion of vitamin C at concentrations ranging from physiologic (0.85 mM) to pharmacologic (11.36 mM) levels demonstrated saturation kinetics of absorption with a K_m of 5.44 mM.[11] Kubler and Gehler,[12] in a pharmacokinetic study on the absorption of 1.5, 3, 6, and 12 g doses of ascorbic acid, demonstrated an inverse relationship between the size of the dose and the percentage of the dose absorbed. Fifty percent of the 1.5 g dose was absorbed, in contrast to only 16% of the 12 g dose. From these data an average absorption of 71% was extrapolated for physiological doses up to 180 mg. Kallner *et al.*[13] estimated the absorption of 90 and 180 mg daily dietary intakes of ascorbic acid by using [1-^{14}C]ascorbic acid as a marker and measuring radioactivity in the urine over a period of 10 days. Absorption ranged from 78-88% of the dose. Daily urinary excretion of ascorbic acid following loading with high doses was studied by Schmidt *et al.*[14] Ingestion of a single 5 g daily dose resulted in the excretion of approximately 1600-1900 mg, whereas 10 g a day given in 2 doses of 5 g each led to the excretion of 2300-2700 mg, again demonstrating the limited absorption of ascorbic acid at high dose levels. From these results the absorptive capacity in the intestine appears to be reached with oral intakes of about 3 g per day.

Information on tissue concentration of ascorbic acid in humans ingesting large quantities of the vitamin is not available. Tissue levels of ascorbic acid have been compared in guinea pigs fed massive quantities (86 g/kg diet) and control levels (2 g/kg diet) of ascorbic acid for 275 days.[15] The massive intake resulted in a slight

[a] Present address: Graduate Division of Nutrition, University of Texas, Austin, Texas 78712.

elevation of ascorbic acid in all ten tissues analyzed, but this was significant only in the testis. Since the tissues were not perfused before analysis, the high concentration of ascorbate in the extracellular fluid may explain the small increase. These results support the suggestion that tissue levels of ascorbic acid cannot be increased appreciably by ingesting large doses of the vitamin. Studies on tissue transport of ascorbic acid in experimental animals, organs, and isolated cells have demonstrated an active, saturable transport system into the central nervous system,[16,17] lung cells,[18] adrenal cortical cells,[17,19] placenta[20,21] and eye retina.[22] The consistent finding of a total body pool size of about 20 mg/kg body weight in human subjects,[23-25] even when intake of the vitamin is in excess of that required to achieve this pool size,[24,25] confirms that the body pool size is limited by factors other than dietary intake.

In humans ascorbic acid and its metabolites are eliminated in the urine. The quantity of ascorbate filtered by the glomeruli is a function of glomerular filtration rate and plasma ascorbate concentration. Reabsorption of filtered ascorbate in the renal tubules is an active, saturable process with an average maximal reabsorptive capacity of 2.16 mg/min.[26] A kinetic study of ascorbic acid in humans[27] demonstrated a marked increase in renal turnover of unmetabolized ascorbic acid at plasma concentrations of 0.8 to 0.9 mg/dl. At these plasma levels the reabsorption mechanism was saturated. Absorbed ascorbic acid in excess of that required to maintain plasma levels at about 1 mg/dl is, therefore, efficiently eliminated by the kidney.

The metabolic turnover of ascorbic acid is also a saturable process. Kallner *et al.*[24] reported saturation when metabolic turnover reached 40 to 50 mg per day. This quantity of metabolites was associated with a total daily turnover (metabolized plus nonmetabolized) of about 60 mg ascorbic acid. The 40 to 50 mg per day metabolic turnover corresponded to a plasma concentration of 0.8 to 0.9 mg/dl. Metabolic turnover was not increased at higher levels of total turnover, which indicates that large doses of ascorbic acid will not further increase the quantity of metabolites formed.

Physiological mechanisms of ascorbic acid absorption, tissue uptake, metabolism, and elimination by the kidney support the theory that an overload of ascorbic acid is unlikely to occur in man. Reference to these physiological mechanisms will be made in discussing individual topics related to the safety of vitamin C.

OXALATE

Oxalate is a major metabolite of ascorbic acid in man.[23] Ascorbic acid accounts for 35-50% of the 30 to 40 mg of oxalate excreted daily; the remainder arises primarily from glycine degradation (about 40%) and from food (5-10%).[27] Theoretically, large doses of ascorbic acid should not result in increased oxalate formation since the metabolic turnover of the vitamin is limited.[24]

Studies on the effect of large doses of ascorbic acid on urinary oxalate have produced contradictory results. Part of the confusion on the relationship between high intakes of ascorbic acid and urinary oxalate excretion may be explained by assay procedures that result in either erroneously high or low oxalate estimations.[28,29] Evaluation of the literature is difficult because different methodologies are used, and in some cases information on sample collection, storage, extraction, and assay procedures is incomplete.

Early studies revealed no appreciable increase in urinary oxalate when up to 4 g of ascorbic acid was ingested daily, but a daily intake of 9 g increased oxalate excretion by about 50-60 mg.[30] Similar results were reported with oral intakes of 2 or 3 g ascorbic acid for time periods of up to 6 months[31-34] and with 9 g intakes for 3 days.[32] Urinary oxalate excretion was increased only 4 to 11 mg per day in subjects ingesting 1 g ascorbic acid daily.[35,36] However, others[37,38] have reported that high doses of ascorbic acid markedly increase urinary oxalate excretion.

Recently, urinary oxalate excretion after 5 and 10 g daily intakes of ascorbic acid was measured by a new method[39] which is more specific than methods used previously. In this study,[14] 5 g ascorbic acid given daily in five doses of 1 gram each increased the average urinary oxalate excretion by 14.8 mg above baseline values. With intakes of 10 g daily (5 doses of 2 g each) for 5 days urinary oxalate increased from about 50 mg to 87 mg. The 15 to 37 mg per day average increase in oxalate excretion resulting from these large doses of ascorbic acid is similar to the change in urinary content of oxalate that results from consuming normal diets.[27]

An interesting observation, which has been reported in several studies,[14,34,40] is the occasional individual who excretes considerably more oxalate than the other subjects ingesting the same dose of ascorbic acid. The reason for this is unknown but it suggests that some apparently healthy persons have an abnormality in either oxalate absorption or ascorbate metabolism to oxalate. However, it seems safe to conclude that ingestion of large quantities of the vitamin does not constitute a risk factor for calcium oxalate stone formation in most healthy persons.

The role of large ascorbic acid intakes in persons who have a tendency to form stones is less clear. Intakes of 1 g ascorbic acid per day resulted in an average increase of only 5 to 11 mg of oxalate per day in stone-forming subjects, an increase no greater than that observed in the healthy subjects.[35,36] In contrast, Chalmers *et al.*[41] reported that stone formers excrete significantly more oxalate than do normals following 2 g oral daily intakes of ascorbic acid. Another finding was depressed ascorbic acid excretion in the stone formers. Following intravenous infusion of 500 mg of ascorbate, oxalate excretion did not differ between normals and stone formers, although ascorbate excretion was again lower in stone formers. The authors postulated that in stone formers most of the oxalate is derived from malabsorbed ascorbate in the gastrointestinal tract. The decreased excretion of ascorbate by stone formers compared with controls after both oral and intravenous ascorbic acid administration was interpreted as suggesting depleted ascorbate stores in the stone formers. The methodology utilized in this study for specific determination of oxalate in urine samples yields reliable results.[28] This study raises interesting questions about the ascorbate status of stone formers and the source of the increased urinary oxalate observed here and in a few apparently healthy persons as previously discussed.

Results of the study by Chalmers *et al.*[41] indicate that recurrent stone formers should avoid high-dose ascorbate intake. Patients with renal impairment and patients on chronic hemodialysis[42] should also be advised not to ingest large quantities of the vitamin.

URIC ACID EXCRETION

Uric acid and ascorbic acid are both reabsorbed in the proximal tubule. If the two compounds share a common transport system, then some have reasoned that the increased tubular load of ascorbic acid following large intakes could decrease uric acid reabsorption due to competitive inhibition. Underlying this reasoning is the assumption that tubular reabsorption of ascorbic acid can be increased by increasing the tubular load. This assumption appears invalid since the tubular reabsorption of ascorbic acid is a saturable process.[26] However, several studies have investigated the influence of high ascorbic acid intakes on the excretion of uric acid.

Stein *et al.*[43] gave single graded doses of ascorbic acid (0.5, 2.0, and 4.0 g) to three, four, and nine patients respectively. They reported a 70-90% increase in the fractional clearance of uric acid with the 4.0 g dose, but no increase was observed with the 0.5 g and 2.0 g doses. Serum uric acid was not changed. In another test, three patients were given 8.0 g of ascorbic acid daily in four divided doses of 2 g for 3 to 7 days. Fractional clearance of uric acid increased to 174% ± 24% of control values and serum uric acid declined by 1.2 to 3.1 mg/dl. The validity of this study cannot be determined. The patient population consisted of five with gout, three with asymptomatic hyperuricemia and six with normouricemia. Initial laboratory values for these patients showed a three- to fourfold range in serum and urine uric acid and a twofold range in creatinine clearance. A description of the patients actually used in the experiments is not given.

Results of the study by Stein *et al.*[43] have not been confirmed by others. Berger *et al.*[44] reported results of renal clearance studies on five nongouty and six gouty men. The subjects had comparable renal function. A priming dose of ascorbic acid was given intravenously, followed by a sustaining infusion at rates varying from 2.5 to 10 mg/min. The resulting plasma ascorbic acid values varied from 2.9 to 12.4 mg/dl. In eight studies with plasma ascorbic acid ranging between 3.5 and 5.6 mg/dl, the mean Curate : GFR did not differ significantly from the control value. In 10 other studies with plasma ascorbic acid greater than 6 mg/dl, the mean Curate : GFR increased significantly although only to a moderate degree, from a control of 0.081 ± 0.20 to 0.116 ± 0.026. The relatively small increase in urate excretion at extremely high, nonphysiological plasma ascorbate levels led the authors to suggest that either urate has a preferential affinity for the transport mechanism or an additional secretory transport system not shared with ascorbic acid.

Studies on healthy subjects[14,45] have also shown that large intakes of ascorbic acid do not influence uric acid excretion. Mitch *et al.*[45] conducted studies on normal subjects, four men and two women. The ingestion of 4 or 12 g ascorbic acid daily, taken in four divided doses of 1 or 3 g, had no effect on serum uric acid concentration, urine uric acid, or uric acid clearance. Similar results were reported by Schmidt *et al.*[14] Four healthy male subjects ingested 10 g ascorbic acid daily in five divided doses of 2 g for five days. Ascorbic acid loading had no effect on excretion of uric acid.

The evidence does not support claims for an ascorbic-acid-induced uricosuria. Even in patients with gout or hyperuricosuria, it appears doubtful that large doses of ascorbic acid would lead to increased uric acid excretion.

IMPAIRED VITAMIN B_{12} STATUS

In 1974, Herbert and Jacob[46] reported that increasing levels of ascorbic acid added to homogenized test meals before incubation at 37°C for 30 minutes produced increasing destruction of vitamin B_{12}. This study was repeated in another laboratory,[47] except in this case the diets were assayed either (1) directly, (2) after extraction by the Association of Official Analytical Chemists (AOAC) procedure,[48] or (3) by the procedure published by the British Analytical Methods Committee.[49] Vitamin B_{12} was determined by both microbiological assays and radioassays. The results differed sharply from those in the previous report.[46] No loss of vitamin B_{12} was observed in the test meals after incubation with 0.5 g of ascorbic acid. Additional studies on the interaction of vitamins B_{12} and C[50,51] confirmed this finding. When cyanide in the extraction step was adequate to liberate protein-bound cobalamins and convert them to stable cobalamins, the vitamin B_{12} content of serum and food was not decreased, even with ascorbic acid concentrations as high as 1.0 g/dl.[51] Hogenkamp[52] reviewed the stability of cobalamins under varying conditions. Among the naturally occurring cobalamins, only aquocobalamin is readily reduced and subsequently destroyed by ascorbic acid. Aquocobalamin is not one of the major cobalamins in biological materials. Thus, it is highly unlikely that megadoses of ascorbic acid will induce vitamin B_{12} deficiency.

Results of studies on human subjects support the conclusion that a large intake of ascorbic acid will not induce vitamin B_{12} deficiency. Altronz *et al.*[53] reported no deleterious effect of 4 g per day or more intake of ascorbic acid on serum vitamin B_{12} levels in spinal cord injury patients. In another study,[54] long-term ingestion of daily mean doses of 1.65 g of supplemental ascorbic acid by 20 myelomeningocele children did not impair vitamin B_{12} status. These children showed neither deficient serum B_{12} levels, anemia, nor elevated mean corpuscular volume. Herbert *et al.*[55] obtained low serum vitamin B_{12} levels in four of 18 patients with spinal cord injury. These patients had received 2 g of ascorbic acid daily for varying periods of time up to 29 months. Bone marrow examination of the two patients who had serum vitamin B_{12} levels below 100 pg/ml revealed a normoblastic pattern of erythropoiesis and a normal deoxyuridine suppression test. Repeat sera from these two subjects were not available after the authors discovered the protective effect of KCN in the assay procedure. They suggest that the low serum levels were probably an artifact of the assay method.

Neither theoretical considerations nor experimental results provide support for concern that large daily intakes of ascorbic acid will induce vitamin B_{12} deficiency. The evidence has consistently demonstrated that vitamin B_{12} in food and the body is not destroyed by ascorbic acid.

IRON OVERLOAD

Ascorbic acid enhances absorption of dietary nonheme iron[56–58] and fortification iron from food.[59] Ascorbic acid supplementation has been used as a means of reducing the prevalence of iron deficiency in populations whose diets are low in sources of heme iron.[59] Enhancement of iron absorption by the vitamin does not occur in a linear fashion. An almost optimal absorption promoting effect is obtained with 25-50 mg ascorbic acid per meal.[60] Therefore, it appears highly unlikely that large doses of the vitamin would lead to excessive iron accumulation in the body.

Cook *et al.*[61] examined the influence of high ascorbic acid supplementation on body iron stores. Healthy subjects took 1.0 g doses of ascorbic acid with each of the two largest meals of the day for 4 to 24 months. Serum ferritin determinations indicated no significant effect of the vitamin C on iron stores. The authors concluded that the regulatory mechanisms that control body iron stores override any pronounced alterations in food iron availability.

Concern that massive doses of ascorbic acid might lead to progressive iron accumulation in healthy, iron-replete individuals appears unwarranted. Persons who are genetically susceptible to iron overload may be adversely affected by long-term ingestion of large doses of vitamin C.

SYSTEMIC CONDITIONING

Concern that scurvy could result upon cessation of high doses of ascorbic acid has resulted from uncontrolled observation on humans. Cochran[62] reported two cases of scurvy in infants who apparently received adequate vitamin C intakes. He proposed that an "ascorbic acid dependency" was prenatally induced since diet histories revealed a high intake of ascorbic acid by their mothers during pregnancy. Another poorly documented observation was the occurrence of scurvy in two men who presumably decreased ascorbic acid intake after prolonged ingestion of large quantities of the vitamin.[63] A more recent report on conditioned oral scurvy due to megavitamin C withdrawal in an adult male suffers from lack of vitamin C analysis of serum and uncontrolled vitamin C intake.[64] A report[65] on guinea pigs suggested systemic conditioning in utero due to high intakes of the vitamin during pregnancy. Weanling guinea pigs from dams fed a high level of ascorbic acid during pregnancy and then fed a vitamin-C-free diet died sooner than did controls. These results are questionable. Normally, weanling guinea pigs fed a vitamin-C-free diet die of scurvy within 18 to 21 days, whereas in this study some of the control animals lived 40 to 50 days. Others[66] found no increase in the mortality of newborn guinea pigs whose mothers had been injected with 400 mg ascorbic acid per kg daily. Another study[15] on guinea pigs is frequently cited as confirming systemic conditioning, when, in fact it demonstrates the exact opposite. In guinea pigs fed massive quantities of ascorbic acid (86 g/kg diet) for 275 days and then fed diets containing either 3 mg/kg for 68 days or 0 mg/kg for 44 days, there was no evidence to suggest that they would develop scurvy sooner than guinea pigs initially fed control levels (2 g/kg diet). In fact, after feeding the diets to induce chronic and acute deficiency, tissue levels of ascorbic acid in the experimental group were consistently higher than in the control group. Others[67] have reported similar results in pups from mothers fed diets high in ascorbic acid during pregnancy and then switched to a diet providing the minimal daily requirement.

The claim that abrupt cessation of large doses of ascorbic acid will lead to scurvy because of conditioning is not supported by the evidence.

MUTAGENICITY

The idea that ascorbic acid was mutagenic and, therefore, harmful if ingested in large doses originated from a paper by Stitch *et al.*[68] published in 1976. Unfortunately, the results of this *in vitro* study were erroneously applied to *in vivo* systems. Cultured human fibroblasts and microbial cells treated with a mixture of ascorbic acid and Cu^{2+} exhibited DNA fragmentation, DNA-repair synthesis, and chromosome aberrations including chromatid breaks and exchanges. No mutagenic action was observed with pure ascorbic acid. Other investigations,[69,70] using *Salmonella typhimurium* as the test organism, demonstrated that ascorbic acid does not have intrinsic mutagenic activity. If the bacteria were treated with a freshly mixed solution of 5 mM ascorbic acid and 1 or 5 μM cupric ion or if Cu^{2+} ions were added to the bacterial medium containing ascorbic acid,[71] mutagenic action was extensive.

The evidence indicates that ascorbic acid breaks DNA when hydroxyl radicals are produced in the presence of oxygen, a reaction that is stimulated by Cu^{2+} ion[71] and inhibited by calatase.[72] The mutagenicity of the ascorbic acid Cu^{2+} system *in vitro* probably results from ascorbate reduction of oxygen to H_2O_2 and Cu^{2+} to Cu^+. Presumably the Cu^+ then reacts with H_2O_2 to produce hydroxyl radicals. Mammalian organisms seem well protected against damage induced by H_2O_2 and radicals. Glutathione peroxidase catalyzes reactions of glutathione with H_2O_2 and other peroxides; catalase and superoxide dismutase prevent the destructive action of peroxides and free radicals. Furthermore, copper does not exist as free Cu^{2+} ions in the body, rather it occurs only in bound forms.

The inapplicability of *in vitro* studies on ascorbate mutagenicity to *in vivo* systems was clearly demonstrated in studies on sister-chromatid exchanges (SLEs).[73,74] Ascorbate caused a dose-dependent increase in SCEs in Chinese hamster ovary cells and in human lymphocytes.[73] In contrast, oral and intraperitoneal administration of ascorbic acid ranging from 0.2 to 10 g/kg body weight caused no induction of SCEs in the bone marrow of Chinese hamsters.[74]

Concern about a potential mutagenic effect of ascorbic acid is surprising in view of the extensive work on the antimutagenic activity of the vitamin. Presently there is no evidence that high intakes of ascorbic acid will be mutagenic in man.

CONCLUSION

An attempt has been made in this review to select papers that represent opposing views and to present a critical nonbiased interpretation of the results. This has led to the conclusion that the practice of ingesting large quantities of ascorbic acid will not result in calcium-oxalate stones, increased uric acid excretion, impaired vitamin B_{12} status, iron overload, systemic conditioning, or increased mutagenic activity in healthy individuals.

The interaction of ascorbic acid with dietary essential mineral elements other than iron is not included in this review. Research on this topic is revealing interesting results, but is insufficient at this time to formulate valid conclusions.

REFERENCES

1. Stewart, M. L., J. T. McDonald, A. S. Levy, R. E. Schucker & D. P. Henderson. 1985. J. Am. Diet. Assoc. **85:** 1585-1590.
2. Erden, F., S. Gulenc, M. Torun, Z. Kocer, B. Simsek & S. Nebioglu. 1985. Acta Vitaminol. Enzymol. **7:** 131-138.
3. Schrauzer, G. N. 1979. *In* International Review of Biochemistry, Vol. 27. Biochemistry of Nutrition IA. A. Neuberger & T. W. Jukes, Eds. University Park Press. Baltimore.
4. Hoffer, A. 1971. N. Engl. J. Med. **285:** 635-636.
5. Hanck, A. 1982. Int. J. Vit. Nutr. Res. (Suppl. 23). pp. 221-239.
6. Hornig, D. H. & U. Moser. 1981. *In* Vitamin C: Ascorbic Acid. J. N. Counsell & D. H. Hornig, Eds.: 225-248. Applied Science Publishers. London.
7. Sestili, M. A. 1983. Seminars in Oncology **10:** 299-304.

8. STEVENSON, N. 1974. Gastroenterology **67:** 952-956.
9. MELLORS, A. J., D. D. NAHRWOLD & R. C. ROSE 1977. Am. J. Physiol. **233:** E374-E379.
10. TOGGENBURGER, G., M. LANDOLDT & G. SEMENZA. 1979. FEBS Lett. **108:** 473-476.
11. NELSON, E. W., H. LANE, P. J. FABRI & B. SCOTT. 1978. J. Clin. Pharmacol. **18:** 325-335.
12. KUBLER, W. & J. GEHLER. 1970. Internat. Z. Vitam. Forschung **40:** 442-453.
13. KALLNER, A., D. HARTMANN & D. HORNIG. 1977. Int. J. Vitam. Nutr. Res. **47:** 383-388.
14. SCHMIDT, K.-H., V. HAGMAIER, D. H. HORNIG, J.-P. VUILLEUMIER & G. RUTISHAUSER. 1981. Am. J. Clin. Nutr. **34:** 305-311.
15. SORENSEN, D. I., M. M. DEVINE & J. M. RIVERS. 1974. J. Nutr. **104:** 1041-1048.
16. SPECTOR, R. & A. V. LORENZO. 1973. Am. J. Physiol. **225:** 757-763.
17. SHARMA, S. K., R. M. JOHNSTONE & J. H. QUASTEL. 1963. Can. J. Biochem. Physiol. **41:** 597-604.
18. WRIGHT, J. R., V. CASTRANOVA, H. D. COLBY & P. R. MILES. 1981. J. Appl. Physiol: Respirat. Environ. Exercise Physiol. **51:** 1477-1483.
19. FINN, F. M. & P. A. JOHNS. 1980. Endocrinology **106:** 811-817.
20. NORKUS, E. P., J. BASSI & P. ROSSO. 1979. J. Nutr. **109:** 2205-2212.
21. STREETER, M. L. & P. ROSSO. 1981. Am. J. Clin. Nutr. **34:** 1706-1711.
22. HEATH, H. & R. FIDDICK. 1966. Exp. Eye Res. **5:** 156-163.
23. HELLMAN, L. & J. J. BURNS. 1958. J. Biol. Chem. **230:** 923-930.
24. KALLNER, A., D. HARTMANN & D. HORNIG. 1979. Am. J. Clin. Nutr. **32:** 530-539.
25. BAKER, E. M., J. C. SAARI & B. M. TOLBERT. 1966. Am. J. Clin. Nutr. **19:** 371-379.
26. RALLI, E. P., G. J. FRIEDMAN & S. H. RUBIN. 1938. J. Clin. Invest. **17:** 765-770.
27. HAGLER, L. & R. H. HERMAN. 1973. Am. J. Clin. Nutr. **26:** 758-765, 882-889.
28. CHALMERS, A. H., D. M. CROWLEY & B. C. MCWHINNEY. 1985. Clin. Chem. **31:** 1703-1705.
29. CRAWFORD, G. A., J. F. MAHONY & A. Z. GYORY. 1985. Clin. Chim. Acta **147:** 51-57.
30. LAMDEN, M. P. & G. A. CHRYSTOWSKI. 1954. Proc. Soc. Exp. Biol. Med. **85:** 190-192.
31. TAKIGUCHI, H., S. FURUYAMA & N. SHIMAZONO 1966. J. Vitaminol. **12:** 307-312.
32. TAKENOUCHI, K., L. ASO, K. KAWASE, H. ICHIKAWA & T. SHIONI. 1966. J. Vitaminol. **12:** 49-58.
33. ERDEN, F., A. HACISALIHOGLU, Z. KOCER, B. SIMSEK & S. NEBIOGLU. 1985. Acta Vitaminol. Enzymol. **7:** 123-130.
34. BRIGGS, M. H. 1976. Lancet **1:** 154.
35. TISELIUS, H. G. & L. E. ALMGARD. 1977. Eur. Urol. **3:** 41-46.
36. BUTZ, M., H. HOFFMANN & H. KOHLBECKER. 1980. Urol. Int. **35:** 309-315.
37. KALLNER, A. 1977. Acta Med. Scand. **20:** 283-287.
38. YAMAZOE, J. 1966. Eiyo To. Shokuryo **18:** 342-347.
39. SCHMIDT, K. H., V. HAGMAIER, G. BRUCHELT & G. RUTISHAUSER. 1980. Urol. Res. **8:** 177-180.
40. TSAO, C. S. & S. L. SALIMI. 1984. Int. J. Vitam. Nutr. Res. **54:** 245-249.
41. CHALMERS, A. H., D. M. COWLEY & J. M. BROWN. 1986. Clin. Chem. **32:** 333-336.
42. BALCKE, P., P. SCHMIDT, J. ZAZGORNIK, H. KOPSA & A. HAUBENSTOCK. 1984. Ann. Intern. Med. **101:** 344-345.
43. STEIN, H. B., A. HASAN & I. H. FOX. 1976. Ann. Intern. Med. **84:** 385-388.
44. BERGER, L., C. D. GERSON & T. YU. 1977. Am. J. Med. **62:** 71-76.
45. MITCH, W. E., M. W. JOHNSON, J. M. KIRSHENBAUM & R. E. LOPEZ. 1981. Clin. Pharmacol. Ther. **29:** 318-321.
46. HERBERT, V. & E. JACOB 1974. J. Am. Med. Assoc. **230:** 241-242.
47. NEWMARK, H. L., J. SCHEINER, M. MARKUS & M. PRABHUDESAI. 1976. Am. J. Clin. Nutr. **29:** 645-649.
48. OFFICIAL METHODS OF ANALYSIS. 12TH ED. 1975. Association of Official Analytical Chemists. Washington, D. C. pp. 842-843.
49. ANALYTICAL METHODS COMMITTEE. 1956. Analyst **81:** 132-136.
50. NEWMARK, H. L., J. M. SCHEINER, M. MARKUS & M. PRABHUDESAI. 1979. J. Am. Med. Assoc. **242:** 2319-2320.
51. MARKUS, M., M. PRABHUDESAI & S. WASSEF. 1980. Am. J. Clin. Nutr. **33:** 137-143.
52. HOGENKAMP, H. P. C. 1980. Am. J. Clin. Nutr. **33:** 1-3.
53. AFRONZ, M., B. BOTHINARD, J. ETZKORN, S. HORENSTEIN & J. MCGARRY. 1975. J. Am. Med. Assoc. **232:** 246-248.

54. EKVALL, S., I-WEN CHEN & R. BOZIAN. 1981. Am. J. Clin. Nutr. **34:** 1356-1361.
55. HERBERT, V., E. JACOB, K. -T. J. WONG, J. SCOTT & R. D. PFEFFER. 1978. Am. J. Clin. Nutr. **31:** 253-258.
56. CALLENDER, S. T. & G. T. WARNER. 1968. Am. J. Clin. Nutr. **21:** 1170-1174.
57. HALLBERG, L. 1981. *In* Vitamin C: Ascorbic Acid. J. N. Counsell and D. H. Hornig, Eds.: 49-61. Applied Science Publishers. London.
58. COOK, J. D. & E. R. MONSEN. 1977. Am. J. Clin. Nutr. **30:** 235-241.
59. STEKEL, A., F. OLIVARES, M. PIZARRO, M. AMAR, P. CHADUD, M. CAYAZZO, S. LLAGUNO, V. VEGA & E. HERTRAMPF. 1985. Int. J. Vit. Nutr. Res. (Suppl. 27). pp. 167-175.
60. HALLBERG, L. 1985. Int. J. Vit. Nutr. Res. (Suppl. 27). pp. 177-187.
61. COOK, J. D., S. S. WATSON, K. H. SIMPSON, D. A. LIPSCHITZ & B. S. SKIKNE. 1984. Blood **64:** 721-726.
62. COCHRANE, H. A. 1965. Can. Med. Assoc. J. **93:** 893-899.
63. RHEAD, W. A. & G. N. SCHRAUZER. 1971. Nutr. Rev. **29:** 262-263.
64. SIEGEL, C., B. BARKER & M. KUNSTADTER. 1982. J. Periodontol. **53:** 453-455.
65. NORKUS, E. P. & P. ROSSO, 1975. Ann. N. Y. Acad. Sci. **258:** 401-409.
66. ALLEVA, F. R., J. J. ALLEVA & T. BALAZS. 1976. Toxicol. Appl. Pharmacol. **35:** 393-395.
67. NORKUS, E. P. & P. ROSSO. 1981. J. Nutr. **111:** 624-630.
68. STITCH, H. F., J. KARIM, J. KOROPATNICK & L. LO. 1976. Nature **260:** 722-724.
69. NORKUS, E. P., W. KUENZIG & A. H. CONNEY. 1983. Mutat. Res. **117:** 183-191.
70. OMURA, H., K. SHINOHARA, H. MAEDA, M. NOMAKA & H. MURAKAMI. 1978. J. Nutr. Sci. Vitaminol. **24:** 185-194.
71. STITCH, H. F., L. WEI & R. F. WHITING. 1979. Cancer Res. **39:** 4145-4151.
72. STITCH, H. F., L. WEI & P. LAM. 1978. Cancer Lett. **5:** 199-204.
73. GALLAWAY, S. M & R. B. PAINTER. 1979. Mutat. Res **60:** 321-327.
74. SPEIT, G., M. WOLF & W. VOGEL. 1980. Mutat. Res. **78:** 273-278.

DISCUSSION OF THE PAPER

E. HOPE (*SUNY Stony Brook, Stony Brook, N.Y.*): May I ask you for a working definition of large doses? Are we talking about greater than 1 gram or perhaps less than 10 grams?

J. M. RIVERS (*University of Texas, Austin, Tex.*): I think we're talking in the range of 5 g and greater.

E. HOPE: Have there been any studies indicating less than 5 g in relation to these?

J. M. RIVERS: Oh, yes, many studies have been done: 4-g, 3-g, 2-g, and 1-g dose levels. There is no effect.

E. HOPE: Greater than the 5 g to what top limit are we talking about?

J. M. RIVERS: In my review of the literature the highest level used in any of these studies was around 12 g per day, except the i.v. infusion studies.

E. HOPE: These studies were based on otherwise healthy subjects?

J. M. RIVERS: Yes, definitely. I think in some of the studies, for example on oxalate, there were stone formers as well as healthy non-stone formers in the studies; they had gouty men and nongouty men in some of the uric acid studies, but for the most part healthy individuals.

A. TAYLOR (*Tufts University, Boston, Mass.*): I'd like to just follow up on the observations in your review by one comment about a paper that was published by Dr. Fleming and coworkers from the Pauling Institute last year, and I think many people who read that paper would have thought that elevated ascorbate might be procataractogenic. From some of the work that was described by Prof. Varma, and which has been corroborated by J. Blondin and some other people in my lab, it's clear that elevated ascorbate would indeed delay the onset of protein damage caused by ultraviolet radiation in a cataract rather than be procataractogenic.

B. LANE (*Columbia University School of Public Health, New York, N.Y.*): I don't want to take time to comment on the role of ascorbic acid, for example in enhancing ocular accommodation, that's eye focusing, and I don't want to take time to comment on the role of high amounts of ascorbic acid megadoses in depressing calcium metabolism transport or in affecting chromium uptake or in helping to chelate with too much vanadium in the system.

But I would like to mention a paper I presented last year to the National Eye Institute at their symposium on eye disease epidemiology. We found that supplemental ascorbic acid greater than 1500 mg led to an eightfold increased risk for specific disorders of the vitreous.

Furthermore, when the enzyme superoxide dismutase measured in red blood cells was depressed below normal this led to a 24-fold increase in vitreous disorders.

UNIDENTIFIED SPEAKER: There's been a controversy on the different ascorbates being taken internally, like calcium ascorbate, and sodium ascorbate. Is there a problem in one over the other? For instance, would there be more uric acid calculi formed with calcium ascorbate?

J. M. RIVERS (*University of Texas, Austin, Tex.*): I can't answer the question. I know there has been concern about the effect of ingesting, for example, high quantities of sodium along with sodium ascorbate and with calcium and calcium ascorbate but to my knowledge I'm not aware of the studies that have been done on various forms, various salts of the vitamin and whether or not you would get differences.

S. D. VARMA (*University of Maryland School of Medicine, Baltimore, Md.*): A lot of people say that if you take ascorbic acid in large quantities and then stop it, there are withdrawal symptoms. Could you comment on that?

J. M. RIVERS: It seems pretty far-fetched to me that you could get withdrawal symptoms taking large quantities of vitamin C and then going back to a physiological intake level. I'm not aware of any reports in the literature of that occurring. I do think, though, that it's pretty safe to assume that you're not going to increase the metabolism of ascorbate with large quantities.

W. LOHMANN (*University of Giessen, Giessen, FRG*): All you said is probably true for oral intake, but I think it might not be true if you infuse it intraperitoneally.

J. M. RIVERS: That is correct. I hope that I have not implied that massive quantities of ascorbic acid are safe. I chose specific topics on which we have data to support the lack of an adverse effect only for these conditions. I think the work that we need to do on minerals is tremendous. There may be many adverse effects of taking very massive quantities of the vitamin for long periods of time, but at least for those I discussed on which we have considerable data I think it's safe to conclude that large doses will not produce adverse effects.

Early Influence of Testosterone Administration to Normal Intact Male Albino Rats on Hepatic Ascorbic Acid Content

P. M. AMBADKAR, N. F. GANGARAMANI,
AND K. J. DERASARI

Division of Reproductive Endocrinology
Department of Zoology
M. S. University of Baroda
Baroda 390 002, Gujarat, India

It is evident that the testicular hormones significantly influence tissue ascorbic acid (AA) levels in male rats.[1-3] Involvement of AA in steroidogenesis and in various metabolic activities is well established. According to Ambadkar and Gangaramani,[1] hepatic AA levels registered a significant rise after castration and showed a time-dependent fall thereafter. Replacement therapy, in the case of 24-hour castrates, led to results contrary to expected reparative effects. Hence, it was thought necessary to reinvestigate the immediate influence of exogenously administered testosterone propionate (TP) to intact male rats.

Adult male albino rats (*Rattus norvegicus albinus*) were administered TP in doses of 0.1, 0.25, and 0.5 mg as single intramuscular injections per animal. Animals were sacrificed under light ether anesthesia at 30, 60, 90, and 120 min after hormone administration. Hepatic tissue from the median and Spigelian lobes and blood were collected and assayed separately for AA content using the method of Roe.[4]

The results obtained during the present study are presented in TABLE 1.

A number of factors are likely to influence AA metabolism; therefore, age and diet of experimental animals were taken into account.

Depletion in the AA content after a few weeks of orchidectomy has been reported by many workers.[2,3,6,7] Contrary to this, our earlier findings[8,9] indicated an increase within a few hours after gonadectomy, whereas within a few hours of TP replacement subnormal levels of AA could be observed. It is apparent that early responses to hormonal disturbances are markedly different from those observed in previous work after a few weeks of gonadectomy as well as after hormonal replacement. It is desirable to recheck immediate hepatic response of intact animals to exogeneous TP administration. Results indicated reduction of hepatic AA levels within the first 30 minutes. Later fluctuations in blood and hepatic AA levels amply indicated that slightly higher levels of circulating androgens could lead to undesirable alterations in AA metabolism. The only exception was obvious recovery of hepatic AA level at 60 min after 0.5 mg TP, although here also blood AA level was distinctly very low.

In this context, recent work of Muddeshwar *et al.*[7] is noteworthy in that testosterone and other androgenic compounds were shown to exert a depressor type of action at

TABLE 1. Ascorbic Acid Content of Hepatic Tissue and Blood under Influence of TP Administration to Normal Rats (mg/100 g of Tissue or mg/100 ml of Blood)

Normal AA Content			Time Interval (min)	Normal Animal Injected with 0.1 mg TP			Normal Animal Injected with 0.25 mg TP			Normal Animal Injected with 0.5 mg TP		
	Hepatic Tissue											
Blood	Median (M)	Spigelian (Sp)		Blood	M	Sp	Blood	M	Sp	Blood	M	Sp
1.16 ±0.23	31.66 ±3.02	34.53 ±3.33	30	1.68 ±0.3	28.70 ±0.24	29.8[b] ±0.33	0.62[e] ±0.07	24.08[c] ±3.4	19.20[e] ±1.4	0.50[e] ±0.02	24.98[c] ±1.37	27.90[c] ±1.7
			60	0.99 ±0.09	25.42[b] ±2.73	26.23[d] ±0.02	0.65[e] ±0.07	16.53[e] ±2.4	24.97[e] ±2.3	0.82[a] ±0.04	36.35 ±4.0	35.60 ±3.0
			90	1.06 ±0.12	26.0[b] ±1.83	26.79[c] ±2.68	1.1 ±0.1	26.23[c] ±2.6	26.80[e] ±0.7	0.86[a] ±0.09	28.13[c] ±2.1	31.45 ±2.4
			120	1.45 ±0.22	19.64[e] ±1.98	21.83[e] ±2.18	0.47[e] ±0.06	17.12[e] ±1.3	17.31[e] ±1.70	0.61[e] ±0.05	24.08[c] ±3.1	31.12 ±3.53

[a] $p < 0.05$.
[b] $p < 0.01$.
[c] $p < 0.005$.
[d] $p < 0.001$.
[e] $p < 0.0005$.

gene level, after undergoing 5-α reductase action in cytosol that influenced production of specific mRNAs. They reported that an important AA-synthesizing enzyme and a catabolizing enzyme are reciprocally affected by replacement therapy and orchidectomy, respectively. From this, it could be said that circulating testicular androgen levels exert a regulatory effect on the rate of AA synthesis and its concentration in the hepatic tissue. Exogenous TP injection brings about either reduction of hepatic AA synthesis or increases its rate of peripheral utilization. Secondly, intrahepatic utilization of AA might also be enhanced under the influence of TP in excess of normal physiological level, which also may contribute to decrease of hepatic AA levels. It is suggested that hepatic mechanism of AA synthesis and its retention are influenced in different ways and probably independently of each other.

ACKNOWLEDGMENT

One of us (KJD) is grateful to the University Grants Commission (UGC), New Delhi, for the award of a Junior Research Fellowship, which made it possible to undertake this work.

REFERENCES

1. AMBADKAR, P. M. & N. F. GANGARAMANI. 1981. J. Anim. Morphol. Physiol. **28**(1,2): 195-205.
2. STUBBS, D. W. & J. B. MCKERNAN. 1967. Proc. Soc. Exp. Biol. Med. **125:** 1326.
3. STUBBS, D. W., J. B. MCKERNAN & D. B. HAUFREC. 1967. Proc. Soc. Exp. Biol. Med. **126:** 464.
4. ROE, J. H. 1954. Methods of Biochemical Analysis, Vol. 1. D. Glick, Ed. Interscience. New York.
5. EHMKE, D. A., B. L. DAVEY & E. N. TODHUNTER. 1956. J. Nutr. **58:** 281-290.
6. KHANDWEKAR, R. V., M. K. DESHPANDE, N. NATH & M. C. NATH. 1973. Metabolism **22:** 1485-1489.
7. MUDDESHWAR, M. G., N. NATH & S. N. CHARI. 1984. Indian J. Exp. Biol. **22:** 397-400.
8. AMBADKAR, P. M. & N. F. GANGARAMANI. 1980. J. Anim. Morphol. Physiol. **27**(1,2): 209-219.
9. AMBADKAR, P. M., N. F. GANGARAMANI & K. J. DERASARI. 1987. Indian J. Exp. Biol. In press.

A Procedure for the Determination of Ascorbic and Dehydroascorbic Acid in Biological Fluids, Tissues, and Foods

WILLY A. BEHRENS AND RENE MADERE

Bureau of Nutritional Sciences
Food Directorate
Health Protection Branch
National Health and Welfare
Ottawa, Ontario, Canada K1A OL2

Several high-performance liquid chromatography (HPLC) methods have been developed for the analysis of ascorbic acid (AA). These methods utilize various column materials, mobile phases, and detectors, and are based on UV or electrochemical properties of the vitamin.[1] But, so far, no sensitive and reproducible HPLC assay has been reported that allows the determination of both AA and dehydroascorbic acid (DHA). We present here an HPLC procedure that permits the estimation of AA and DHA in biological fluids, tissues, and foods. This procedure is based on the work of Hughes,[2] who showed that incubation of DHA with homocysteine could rapidly and completely produce AA.

Samples are homogenized in the presence of sufficient 17% metaphosphoric acid and water to give a final concentration of 0.85% acid.[3] After centrifugation and dilution, a 20-μl aliquot is injected in a chromatograph equipped with a C18 reversed-phase 5-μm column (250 × 4.6 mm). A mobile phase of 80 mM sodium acetate buffer, pH 4.8, containing 1 mM of *N*-octylamine as the paired-ion reagent, 0.015% metaphosphoric acid, and 15% methanol at a flow rate of 0.9 ml/minute is used. Ascorbic acid is detected with an electrochemical detector preset at + 0.7 V and current of 50 nA. Under these conditions the retention time for AA is 3.0 minutes. Therefore this HPLC procedure is very rapid, analysis being completed within 10 min after sample preparation. Ascorbic acid concentration of the sample extracts is calculated by interpolation on a standard curve obtained after analyzing serial dilutions of AA (50-500 pg/20 μl). Dehydroascorbic acid is indirectly estimated by converting it to AA after reduction with 1.0% homocysteine at pH 7.0-7.2 for 30 min at 25°C.[4] After dilution with mobile phase, a 20 μl aliquot is injected in the chromatograph to obtain total vitamin C (AA + DHA). The concentration of DHA is calculated by subtraction. Ascorbic acid can be reproducibly quantified at concentrations as low as 50 pg/20 μl of sample extract. The assay is suitable for plasma, other biological fluids, tissues, and foods. TABLES 1 and 2 present values for selected foods and rat tissues. These values correlate very well with those obtained with the 2,4-dinitrophenylhydrazine assay (foods, $r = 0.9987$; tissues, $r = 0.9963$).

TABLE 1. Ascorbic and Dehydroascorbic Acid Content of Selected Foods[a]

Food	Total AA	AA	DHA
Broccoli	94.14 ± 5.51	76.89 ± 9.23	17.25 ± 4.39
Carrot	4.07 ± 0.41	2.71 ± 0.25	1.36 ± 0.26
Celery	3.66 ± 0.15	2.06 ± 0.09	1.60 ± 0.20
Cucumber	4.65 ± 0.71	1.50 ± 0.27	3.15 ± 0.70
Lettuce	2.98 ± 0.31	1.96 ± 0.17	1.02 ± 0.34
Tomato	17.54 ± 0.60	12.49 ± 0.74	5.05 ± 1.05
Infant formula	100.03 ± 5.24	85.34 ± 3.60	14.68 ± 4.79
Milk	0.47 ± 0.04	0.23 ± 0.02	0.24 ± 0.06
Apple juice	29.78 ± 2.06	25.91 ± 1.19	3.87 ± 2.68
Orange juice	29.57 ± 2.49	26.60 ± 1.95	2.97 ± 1.86

[a] Results are expressed in mg/100 g of food (or 100 ml of juice); mean ± SD (n = 4).

TABLE 2. Ascorbic and Dehydroascorbic Acid Content of Tissues of Rats Fed 76-AIN Diet

Tissue	Total AA	AA	DHA
Plasma	6.39 ± 1.62	4.31 ± 1.32	2.08 ± 0.37
Heart	61.67 ± 4.92	33.31 ± 2.32	28.36 ± 5.26
Brain	308.22 ± 35.91	284.76 ± 23.00	23.46 ± 15.95
Liver	305.96 ± 30.71	285.49 ± 31.41	20.47 ± 6.98
Spleen	461.31 ± 45.10	313.52 ± 48.98	147.80 ± 13.82
Kidney	97.41 ± 15.89	77.38 ± 13.39	20.03 ± 4.86
Adrenal	3592.00 ± 308.17	3316.00 ± 228.87	276.00 ± 104.79

Results are expressed in μg/g of tissue (or ml of plasma); mean ± SD (n = 5).

REFERENCES

1. Pachla, L. A., D. L. Reynolds & P. T. Kissinger. 1985. Analytical methods for determining ascorbic acid in biological samples, food products, and pharmaceuticals. J. Assoc. Off. Anal. Chem. **68:** 1-12.
2. Hughes, R. E. 1956. The use of homocysteine in the estimation of dehydroascorbic acid. Biochem. J. **64:** 203-208.
3. Pelletier, O. & R. Brassard. 1977. Determination of vitamin C (L-ascorbic acid and dehydroascorbic acid) in food by manual and automated photometric methods. J. Food Sci. **42:** 1471-1477.
4. Dennison, D. B., T. G. Brawley & G. L. K. Hunter. 1981. Rapid high-performance liquid chromatographic determination of ascorbic acid and combined ascorbic acid-dehydroascorbic acid in beverages. J. Agr. Food. Chem. **29:** 927-929.

Dietary Vitamin C Delays UV-induced Eye Lens Protein Damage[a]

J. BLONDIN, V. BARAGI, E. R. SCHWARTZ, J. A. SADOWSKI, AND A. TAYLOR[b]

USDA Human Nutrition Research Center on Aging
at Tufts University
and
Department of Orthopedic Surgery at Tufts University
Boston, Massachusetts 02111

In order to pass light to the retina the eye lens must remain clear throughout life. Eye lens opacification or cataract is due in part to the extensive accumulation, aggregation, and eventual precipitation of postsynthetically modified proteins from the clear lens milieu. One source of damage is the ultraviolet (UV) light and/or oxygen to which the lens is exposed.[1,2] The young lens is protected against this damage by active antioxidant systems and by proteases that in other tissues or cell types are thought to remove damaged proteins.[2,3,4] The observed attenuation of these antioxidant and protease editing systems in old lens tissue is thought to be involved in cataractogenesis.

Recent *in vitro* tests suggested that such photooxidative damage to bulk proteins and proteases can be delayed by enhanced vitamin C (vit C).[2,5] In order to examine the effect of vit C on lens protein integrity under more physiological conditions, guinea pigs were fed 2.0 mg/day (low, L) or 50 mg/day (high, H) vit C for 21 weeks.[6] Guinea pigs, like humans, do not synthesize vit C and hence require vit C in their diet.[7,8] Lenses were excised and analyzed by HPLC for vit C content.[9]

Lens vit C levels were 0.11 and 0.49 mM from animals fed L vit C and H vit C, respectively. In nonirradiated lenses no lens protein pattern differences were observed by SDS-PAGE[10] and exopeptidase activity was similar (substrate: leucyl-*p*-nitroanilide [LpNA]; assay conditions: 2.5 mM LpNA in 0.05 M phosphate, pH 7.0, 37°). When lens soluble proteins were exposed to 2.5×10^{-3} watt cm^{-2} UVA, 0.4×10^{-4} watt cm^{-2} UVB for 4 h, exopeptidase activity decreased in both H vit C and L vit C soluble proteins (FIG. 1). As in *in vitro* tests, protease inactivation is noted before aggregate formation is observed. However, at each of the times monitored the percentage of dark control activity in L vit C samples was significantly lower ($p > 0.05$) than in the H vit C preparations. SDS-PAGE revealed that after UV exposure high molecular weight protein aggregate formation in H vit C homogenates is delayed compared to L vit C homogenates (FIG. 2). These results indicate that dietary vit C

[a]This work was supported by USDA/ARS, the Massachusetts Lions Eye Research Fund, Inc., and the Daniel and Florence Guggenheim Foundation.

[b]Please address inquiries to: Dr. A. Taylor, USDA Human Nutrition Research Center on Aging at Tufts University, 711 Washington St., Boston, Mass. 02111.

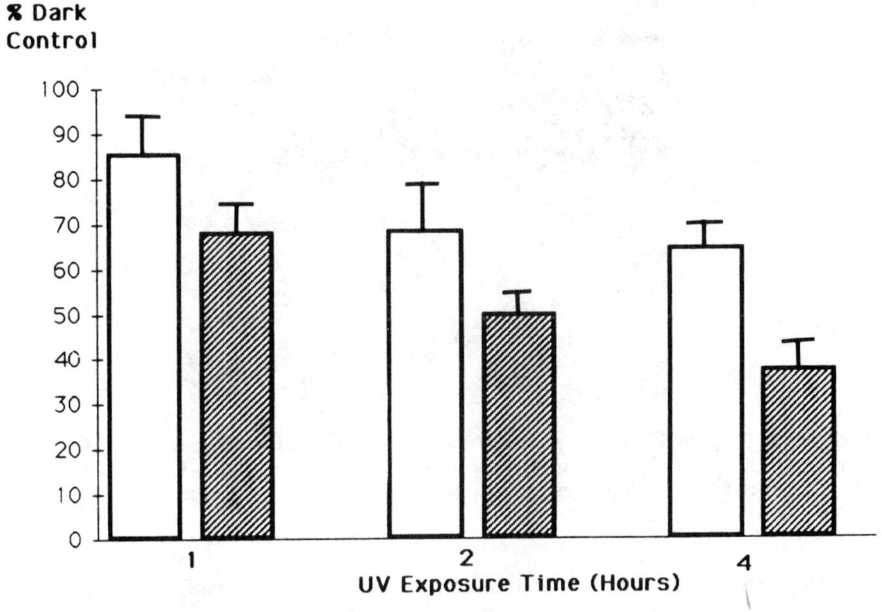

FIGURE 1. Effect of UV exposure on exopeptidase activity in lens homogenates from guinea pigs on 50 and 2 mg/animal/day ascorbate diets. Activity is expressed as percentage of non-UV-exposed (dark) homogenates. *Open bars* represent high ascorbate diet. *Hatched bars* represent low ascorbate diet. Data are from four experiments. *Lines* above bars indicate the standard deviation.

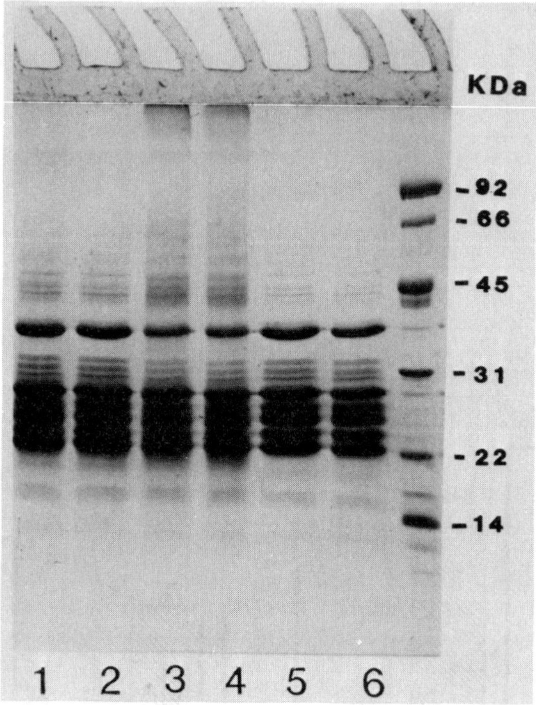

FIGURE 2. SDS-PAGE of lens proteins from guinea pigs fed low (L) and high (H) dietary ascorbate, before and after exposure to UV or dark control. Lanes 1,3,5, L vit C; lanes 2,4,6, H vit C; lanes 1 and 2, 2-h UV exposure; lanes 3 and 4, 4-h UV exposure; lanes 5 and 6, 4-h dark exposure.

protects lens structural proteins and proteases against photooxidation and may be useful in delaying age and cataractous related changes.

These data suggest that optimal levels of nutrients be determined for each stage of life. They also indicate that optimal levels of nutrient intake be considered as recommended standards rather than or in addition to nutrient levels that prevent disease states.

REFERENCES

1. ZIGLER, J. S. & J. D. GOOSEY. 1984. Curr. Eye Res. **3:** 59-65.
2. VARMA, S. D., O. CHAND, Y.R. SHARMA, J. F. KUCK & K. D. RICHARDS. 1984. Curr. Eye Res. **3:** 35-57.
3. JAHNGEN, J., A. L. HAAS, A. CIECHANOVER, J. BLONDIN, D. EISENHAUER & A. TAYLOR. 1986. J. Biol. Chem. **261:** 13760-13767.
4. TAYLOR, A., T. SURGENOR, D. K. R. THOMSON, R. J. GRAHAM & H. C. OETTGEN. 1984. Exp. Eye Res. **38:** 217-229.
5. BLONDIN, J. & A. TAYLOR. 1986. Fed. Proc. **45:** 1599.
6. REID, M. E. & G. M. BRIGGS. 1953. J. Nutrition **51:** 341-354.

7. SVIRBELY, J. L. & A. SZENT-GYÖRGYI. 1932. Biochem. J. **26:** 865-870.
8. WAUGH, W. A. & C. G. KING. 1932. J. Biol. Chem. **97:** 325-331.
9. LEE, W., P. HAMERNIJIK, M. HUTCHINSON, V. RAIYS & R. R. LABA. 1982. Clin. Chem. **28:** 2165-2169.
10. BLONDIN, J., J. SADOWSKI, V. J. BARAGI, E. SCHWARTZ & A. TAYLOR. 1986. Free Rad. Biol. Med. **2:** 275-281.

Effect of Acute Ascorbic Acid Deficiency on the Plasma Lipids and Postheparin Lipolytic Activity in Guinea Pigs

VED P. S. CHAUHAN,[a] ABHA CHAUHAN,[a] AND A. K. SARKAR

Department of Biochemistry
Postgraduate Institute of Medical Education and Research
Chandigarh 160012, India

Various studies[1-3] have shown occurrence of hypertriglyceridemia and hypercholesterolemia in animals with vitamin C deficiency. On the other hand, supplementation of megadoses of vitamin C can lower the serum cholesterol level. It has therefore been suggested that megadoses may be useful in reducing the risk of heart diseases. In the present study, we have shown that acute vitamin C deficiency can cause a reduction in the postheparin-lipolytic activity, which may have a role in the hyperlipidemia in vitamin C deficiency.

Acute scurvy in guinea pigs was produced as described earlier.[4] Lipids[5] and postheparin lipolytic activity[6] were estimated by standard methods. Triton WR 1339 (5%) was injected intraperitoneally at a dose of 500 mg/kg body weight. Lipids were analyzed in blood samples collected before and after 8 h of Triton injection to animals.

TABLE 1 shows increased level of plasma free fatty acids, free and esterified

TABLE 1. Plasma Lipids in Various Groups of Guinea Pigs (mg/100 ml Plasma)[a]

Lipids	Control	Scorbutic	Pairfed Control	Excess Vitamin Fed
Total lipid	85.0 ± 2.4	161.0 ± 9.0[b]	82.0 ± 2.3	93.0 ± 7.93
Triglycerides	14.1 ± 0.5	25.0 ± 1.3[b]	11.8 ± 1.2	16.7 ± 2.22
Total cholesterol	27.5 ± 1.8	46.6 ± 2.7[b]	25.2 ± 0.9	23.9 ± 1.43
Free cholesterol	9.2 ± 0.7	14.7 ± 0.7[b]	7.5 ± 0.5	6.6 ± 0.69
Esterified cholesterol	18.1 ± 0.5	31.9 ± 2.0[b]	17.7 ± 0.5	18.7 ± 0.50
Free fatty acid	14.1 ± 1.2	29.7 ± 2.1[b]	14.7 ± 1.7	13.4 ± 2.60
Phospholipids	30.8 ± 1.8	54.7 ± 3.0[b]	30.3 ± 1.7	32.3 ± 1.99

[a] Values are mean ± SE of four pooled samples. Each pool consisted of two animals.
[b] $p < 0.01$ as compared to pairfed control and control.

[a] Present Address: Department of Neurochemistry, New York State Institute of Developmental Disabilities, 1050 Forest Hill Road, Staten Island, New York, N.Y. 10314.

TABLE 2. Postheparin Lipolytic Activity and Post-Triton Plasma Lipids in Various Groups

Groups	Postheparin Lipolytic Activity (μmol FFA Released/100 ml Plasma/h)			Post-Triton Lipids (mg/100 ml Plasma)		
	Total	Protamine Resistant	Protamine Sensitive	Cholesterol	Phospholipids	Triglycerides
Control	36.9 ± 1.0	3.3 ± 0.5	33.6 ± 0.7	50 ± 3	100 ± 3	75 ± 4
Scorbutic	22.9 ± 2.8[a,b]	8.7 ± 1.2[b]	14.2 ± 1.7[a,b]	110 ± 5[a,b]	160 ± 4[a,b]	149 ± 5[a,b]
Pairfed control	37.2 ± 1.4	6.4 ± 0.6	30.4 ± 1.8	60 ± 5	105 ± 4	70 ± 4
Excess vitamin fed	40.4 ± 1.8	4.1 ± 1.1	36.3 ± 1.5	48 ± 4	98 ± 4	73 ± 3

Values are mean ± SE of four observations; FFA, free fatty acid.
[a] $p < 0.05$ as compared to pairfed control.
[b] $p < 0.05$ as compared to control.

cholesterol, triglycerides, and phospholipids in acute scorbutic guinea pigs. This is consistent with the study of Fuginami *et al.*[7] that has reported an increase in plasma triglycerides and fatty acids of scorbutic guinea pigs fed atherogenic diet. Our results on [1-^{14}C]acetate incorporation into the lipids shows a decrease in the specific activity of cholesterol and phospholipids and an increase in the specific activity of triglycerides (data not shown). The decreased acetate incorporation could be due to the increased acetate pool as observed by Banerjee and Ghosh,[8] which then can have dilution effect on the isotope or it could be due to the decreased synthesis.

Since the plasma lipids represent the overall metabolic effect of all tissues, this study has examined the lipolytic effect and secretion of the lipids from liver into the plasma. Triton WR 1339 blocks the uptake of lipids by extrahepatic tissues. Significantly higher phospholipid, cholesterol, and triglyceride in the post-Triton plasma of scorbutic guinea pigs (TABLE 2) suggests a role of increased secretion of lipids in the observed hyperlipidemia. Postheparin lipolytic activity and lipids were similar in the plasma of animals with and without megadoses of vitamin C. Decreased postheparin lipolytic activity (protamine sensitive) indicates that the removal of plasma triglycerides by extrahepatic tissues is also impaired (TABLE 2). A decrease in the triglyceride removal also has been revealed after [^{3}H]glycerol administration in scorbutic guinea pigs.[9]

The data shows that acute avitaminosis C may lead to hyperlipidemia due to increase in the synthesis, decreased catabolism, and increased secretion of lipid into the plasma. Megadoses of ascorbic acid supplementation for 3 weeks does not seem to be a sufficient period for its hypolipidemic effect in normal animals.

REFERENCES

1. GINTER, E. *et al.* 1973. Lipids **8:** 135.
2. FYFE, T. *et al.* 1968. J. Atheroscler. Res. **8:** 591.
3. GINTER, E. *et al.* 1976. Fed. Chem. **1:** 23.
4. MAHMOOD, A. *et al.* 1979. Experientia **35:** 1059.
5. CHAUHAN, V. P. S. *et al.* 1977. Experientia **33:** 22.
6. FIELDING, C. J. 1968. Biochim. Biophys. Acta **159:** 94.
7. FUJINAMI, T. *et al.* 1971. Jpn. Circ. J. **35:** 1559.
8. BANERJEE, S. & P. K. GHOSH. 1960. Am. J. Physiol. **199:** 1064.
9. BOBEK, P. *et al.* 1983. Biomed. Biochim. Acta **42:** 413.

Vitamin C Levels in the Tissues of Cigarette-smoked Guinea Pigs and Rats

CHING K. CHOW, GERRY R. AIRRIESS, LI-CHUAN CHEN, AND CHARUS CHANGCHIT

Department of Nutrition and Food Science
and
Graduate Center for Toxicology
University of Kentucky
Lexington, Kentucky 40506

INTRODUCTION

Cigarette smoking has been implicated to be an important contributing factor to the causation of respiratory diseases and other disorders. However, the mechanism by which smoking contributes to the development of various diseases is not yet clear. The nutritional status of an individual may influence cellular susceptibility to the effects of chemicals, drugs, and environmental agents. Cigarette smokers have lower levels of plasma and leukocyte vitamin C.[1,2] Exposure of rats to cigarette smoke for 3-7 days causes a significant alteration in the plasma and lung levels of vitamin C.[3,4] In the present study, both guinea pigs and rats were employed to determine the long-term smoking effects on the tissue status of vitamin C.

MATERIALS AND METHODS

Male Hartley guinea pigs and Sprague-Dawley rats, 45-50 days old, maintained on commercial diets were exposed to either mainstream, sidestream, or sham smoke, or served as room controls for 16-24 weeks. Smoke was generated by a peristaltic pump machine,[5] using the Kentucky reference cigarette, 2R1. Guinea pigs were smoked twice daily, two cigarettes per session at 10 puffs per cigarette for a total of 40 puffs/day. Rats were exposed to 10 puffs per session daily. Smoking status was monitored by measuring the concentration of carboxyhemoglobin and intake of total particulate matter.[5] At the end of each experiment, portions of lung, liver, kidney, and blood plasma from each animal were processed for the measurement of vitamin C.[6]

TABLE 1. Vitamin C Levels in the Lung, Kidney, Liver, and Plasma of Cigarette-smoked Guinea Pigs

Exposure Time	Tissue	Mainstream	Sidestream	Sham	Room Control
17 weeks	lung[a]	174 ± 27[b,*]	119 ± 30	136 ± 4	124 ± 40
	kidney[a]	181 ± 26	194 ± 19	174 ± 45	195 ± 25
	liver[a]	297 ± 55	323 ± 28	373 ± 53	384 ± 7
	plasma[c]	1.04 ± 0.13	1.14 ± 0.07	1.05 ± 0.17	1.23 ± 0.10
20 weeks	lung[a]	129 ± 16	104 ± 16	146 ± 50	130 ± 43
	kidney[a]	159 ± 14	170 ± 29	169 ± 39	158 ± 10
	liver[a]	266 ± 37	305 ± 74	282 ± 22	265 ± 7
	plasma[c]	1.51 ± 0.10	1.39 ± 0.20	1.40 ± 0.18	1.39 ± 0.10

[a] μg L-ascorbic acid/g tissue.
[b] Mean ± standard deviation. Four animals were in each group except two in sham and room control groups of 17-week exposure.
[c] mg L-ascorbic acid/dl.
* No significant difference ($p > 0.05$) resulted from smoking treatment in all tissues as determined by analysis of variance.

RESULTS AND DISCUSSION

The vitamin C levels in lung, kidney, liver, and plasma of guinea pigs are shown in TABLE 1. Relative to the sham or room control group, exposure of guinea pigs to mainstream or sidestream smoke for 17 or 20 weeks did not cause significant changes in plasma and tissues studied. Similarly, exposure of rats to mainstream or sidestream smoke for 16 or 24 weeks did not significantly alter the levels of vitamin C in plasma, lung, liver, and kidney (TABLE 2).

TABLE 2. Vitamin C Levels in the Lung, Kidney, Liver and Plasma of Cigarette-smoked Rats

Exposure Time	Tissue	Mainstream	Sidestream	Sham	Room Control
16 weeks	lung[a]	124 ± 11[b,*]	129 ± 8	133 ± 12	126 ± 7
	kidney[a]	80 ± 8	90 ± 4	83 ± 6	80 ± 5
	liver[a]	122 ± 14	145 ± 3	141 ± 8	142 ± 8
	plasma[c]	1.03 ± 0.12	1.24 ± 0.13	1.21 ± 0.09	1.19 ± 0.07
24 weeks	lung[a]	128 ± 9	139 ± 12	128 ± 11	120 ± 8
	kidney[a]	93 ± 9	94 ± 10	92 ± 13	98 ± 11
	liver[a]	117 ± 13	128 ± 16	131 ± 13	130 ± 9
	plasma[c]	1.20 ± 0.12	1.25 ± 0.11	1.29 ± 0.12	1.24 ± 0.08

[a] μg L-ascorbic acid/g tissue.
[b] Mean ± standard deviation. Eight animals in each group.
[c] mg L-ascorbic acid/dl.
* No significant difference ($p > 0.05$) resulted from smoking treatment in all tissues as determined by analysis of variance.

In addition to a metabolic adaptation following the long-term smoking exposure, the lack of effect on the plasma and lung levels of vitamin C in rats observed in this research as opposed to those of the short-term animal studies[3,4] can be partly attributable to the ability of rats to synthesize vitamin C. Since guinea pigs cannot synthesize vitamin C, as in the case of humans, the abundant source of vitamin C (1000 ppm) in the experimental diet employed is probably the major factor responsible for the lack of smoking effect on vitamin C status in guinea pigs observed. Thus it is possible that the lower levels of vitamin C observed in plasma or leukocytes of smokers[1,2] may not be due to a single factor but rather to a combination of the following factors: (1) direct oxidation of vitamin C by oxidants or free radicals in smoke,[7] (2) increased vitamin C excretion attributable to smoking,[1,2] and (3) inadequate dietary vitamin C intake of smokers.[8] The results obtained also suggest that adequate dietary vitamin C may overcome the adverse effect of cigarette smoking on the nutritional status of vitamin C.

REFERENCES

1. PELLETIER, O. 1970. Am. J. Clin. Nutr. **23:** 520-528.
2. PELLETIER, O. 1977. Int. J. Vitam. Nutr. Res. **16:** 147-169.
3. CHOW, C. K., L. H. CHEN & R. R. THACKER. 1984. Environ. Res. **34:** 8-17.
4. CHEN, L. H. & C. K. CHOW. 1980. Nutr. Rep. Int. **22:** 301-309.
5. GRIFFITH, R. B. & S. STANDAFER. 1985. Toxicology **35:** 13-24.
6. OMAYE, S. T., J. D. TURNBULL & H. E. SAUBERLICH. 1979. Methods Enzymol. **62:** 3-11.
7. PRYOR, W. A., K. I. TERAUCHI & W. H. DAVIS, JR. 1976. Environ. Health Persp. **16:** 161-175.
8. FEHILY, A. M., K. M. PHILLOPS & J. W. G. YARNELL. 1980. Am. J. Clin. Nutr. **40:** 827-833.

Some Effects of Vitamin C May Be Indirect, Since It Affects the Blood Levels of Cortisol and Thyroid Hormones[a]

E. DEGKWITZ

Biochemical Institute
University of Giessen
D-63 Giessen, Federal Republic of Germany

The results presented here were obtained from guinea pigs adapted to diets containing 5-680 mg% vitamin C for 6-8 weeks, with final body weights of 600-750 g. The margin for survival seems to be 5 mg%. The animals do so without signs of scurvy. They show equal contents of cytochrome P-450 in the adrenal glands like animals adapted to 10-680 mg%; the content is even raised after cessation of supply. But they show reduced contents in hepatic cytochrome P-450 and accordingly prolonged sleeping times in response to evipan. In animals supplied by 10-90 mg% there are transient decreases in hepatic cytochrome P-450 during the adapting phase. These different modifications, which in part conform to findings in unadapted guinea pigs submitted to diet without vitamin C,[1-2] seem incompatible with a direct influence of vitamin C. They may be explained by alterations in the hormonal regimen and differently sensitive responses of individual organs during adaptation to low vitamin supply. Serum cortisol levels increase with a decline of the vitamin C supply below 90 mg% (FIG. 1), accompanied by a reduced interval between the daily maximum and minimum levels. Since connective tissue is a target organ for cortisol, the symptoms of scurvy, manifest 2-3 weeks after cessation of vitamin C in guinea pigs,[1] a period of time far less than half-life of collagen, might also be initiated by an increased catabolic state due to raised cortisol levels. The portion of vitamin C excreted into the urine is increasingly reduced in guinea pigs supplied with vitamin C beyond 90 mg%. The finding indicates that high vitamin C supply induces an increased rate of its catabolism. This symptom becomes effective if guinea pigs, adapted to high dietary contents of vitamin C, are submitted to a reduction of the supply. Their organ levels fall off rapidly for about 2 weeks and start to increase thereafter, arriving at a new steady state at latest after 4-5 more weeks.[3] The key reaction seems to be the transport of vitamin C into the hepatic mitochondria, since isolated hepatic mitochondria of guinea pigs adapted to 680 mg% accumulate vitamin C more rapidly than mitochondria of animals adapted to 90 mg%. The total amounts of hepatic (not muscular) mitochondrial cytochromes (a, a_3, b, c, and c_1) are increased in guinea pigs adapted to 680 mg%, compared to

[a] The support of these investigations by the Deutsche Forschungsgemeinschaft is gratefully acknowledged.

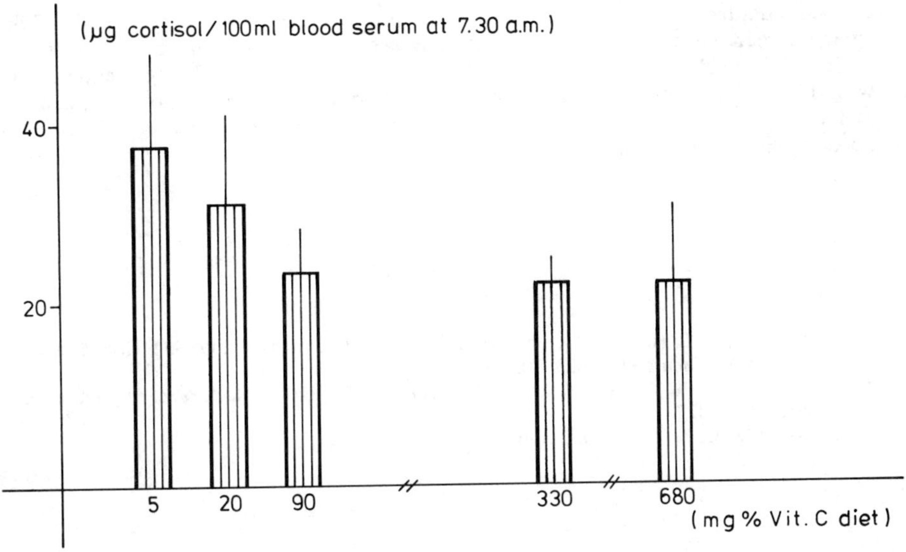

FIGURE 1. Morning serum levels of cortisol in guinea pigs adapted to constant contents of vitamin C in the diet for 6-8 weeks.

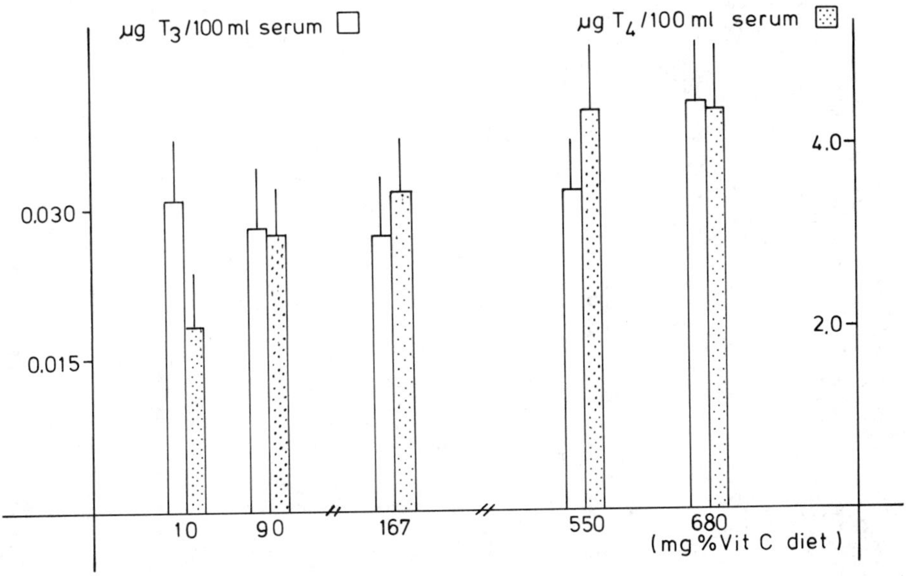

FIGURE 2. Serum levels of thyroid hormones in guinea pigs adapted to constant contents of vitamin C in the diet for 6-8 weeks.

animals adapted to 176 or less mg%. This improvement may be the result of an enlarged regimen of thyroid hormones. The serum levels of thyroid hormones, especially T_4, are about equally high in guinea pigs adapted to 550-680 mg% and decline with diminished supply (FIG. 2). Since thyroid hormones are known to raise the basal metabolic rate, their increase may be the mechanism for the claimed benefits of high doses of vitamin C in man.

REFERENCES

1. DEGKWITZ, E., L. HÖCHLI-KAUFMANN, D. LUFT & HJ. STAUDINGER. 1971. Hoppe-Seyler's Z. Physiol. Chem. **353:** 1023-1033.
2. DEGKWITZ, E., S. WALSCH & M. DUBBERSTEIN. 1974. Hoppe Seyler's Z. Physiol. Chem. **355:** 1420-1422.
3. DEGKWITZ, E. 1985. Z. Ernährungswiss. **24:** 219-230.

Chronic Vitamin C Deficiency Lowers Fractional Catabolic Rate of Low-Density Lipoproteins in Guinea Pigs

EMIL GINTER AND MÁRIA JURČOVIČOVÁ

Institute of Human Nutrition Research
812 30 Bratislava, Czechoslovakia

Vitamin C is necessary for the transformation of cholesterol to bile acids, for it affects the rate-limiting reaction of cholesterol catabolism, the microsomal 7-α-hydroxylation of cholesterol in the liver.[1–3] In guinea pigs with chronic marginal vitamin C deficiency, this reaction becomes slowed down, thus bringing about an accumulation of cholesterol in the liver and an increase of plasma cholesterol in the fraction of low-density lipoproteins (LDL). The concentration of LDL is controlled in man and guinea pig by hepatic LDL receptors that are subject to end-product feedback regulation by the level of cholesterol in liver cells.[4] LDL receptors are currently measured *in vivo* by determining the rate of disappearance of ^{125}I-labeled LDL from the circulation. This

TABLE 1. The Influence of Chronic Vitamin C Deficiency on Cholesterol Levels and Turnover of Plasma Low-Density Lipoproteins (LDL) in Guinea Pigs[a]

Parameter	Control Group	Vitamin-C-Deficient Group	Significance
Number of animals	6	5	
Final body weight (g)	514 ± 47	549 ± 41	ns
Vitamin C in liver (μmol/kg)	397 ± 43	86 ± 10	$p < 0.001$
Total Cholesterol			
In blood serum (mmol/l)	3.1 ± 0.4	4.7 ± 0.5	$p < 0.05$
In liver (mmol/kg)	9.7 ± 1.1	14.8 ± 2.2	$p = 0.05$
LDL Turnover			
$t_{1/2}\alpha$ (h)	0.73 ± 0.08	1.65 ± 0.45	ns ($p < 0.1$)
$t_{1/2}\beta$ (h)	11.2 ± 0.6	14.0 ± 0.5	$p < 0.01$
FCR (h^{-1})	0.094 ± 0.004	0.077 ± 0.003	$p < 0.02$

[a] Means ± SEM; ns = not significant.

technique was used to elucidate the effect of vitamin C deficiency on plasma LDL turnover.

Male short-hair guinea pigs weighing 300-350 g were randomly divided into two groups and fed ad libitum vitamin-C-free diet.[5] The vitamin-C-deficient group was given this diet for 2 weeks and then was supplemented with an oral maintenance dose of ascorbic acid (0.5 mg per animal per day) with the aim of maintaining normal growth and preventing symptoms of acute scurvy. The control group was fed the same diet with addition of 0.5% of ascorbic acid. The food intake and body weight curves were identical in both groups. After 16 weeks LDL was isolated from the plasma of male guinea pigs fed a standard diet by ultracentrifugation and iodinated by the chloramine-T procedure.[6] [^{125}I]LDL were injected intracardially into the control and deficient animals and plasma samples were obtained from retroorbital venous plexus at intervals 5 min, 1.5, 3, 4.5, 8, 12, 16, and 24 h after [^{125}I]LDL injection. The die-away curves of plasma [^{125}I]LDL were analyzed in terms of a two-pool model and fractional catabolic rate (FCR) of LDL was calculated.[7]

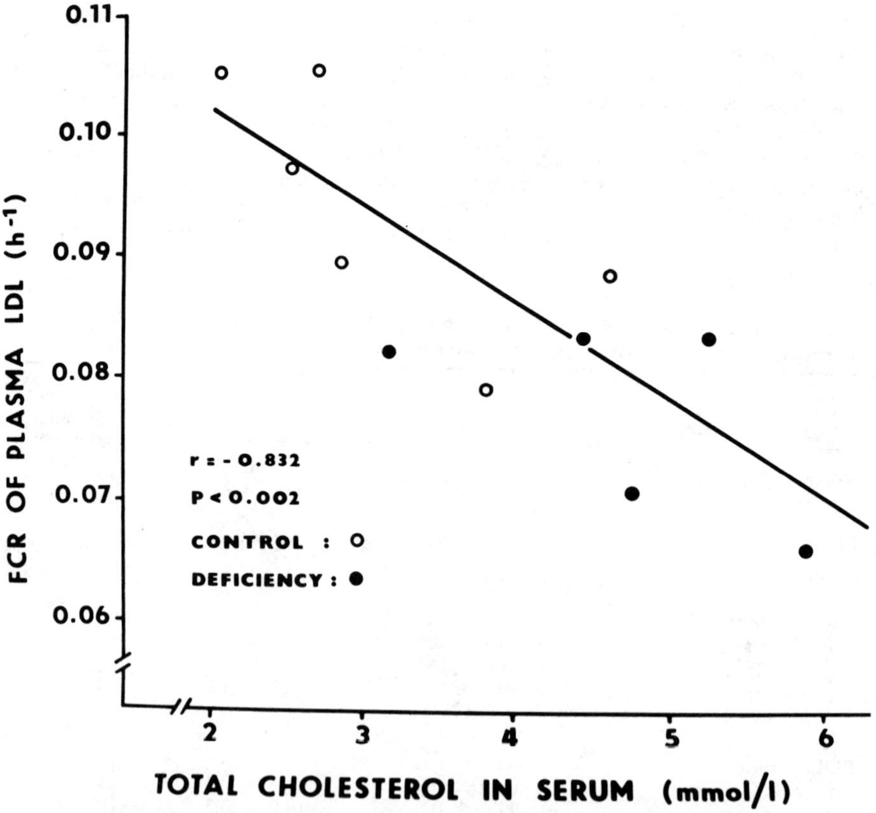

FIGURE 1. Indirect correlation between the level of total cholesterol in blood serum and the value of fractional catabolic rate (FCR) of plasma low-density lipoproteins (LDL) in control and vitamin-C-deficient guinea pigs.

In the vitamin-C-deficient group significantly lower levels of liver ascorbate and accumulation of total cholesterol in the blood serum and liver were found (TABLE 1). In the deficient group the course of die-away curves and the turnover of [^{125}I]LDL were changed: there was (1) prolonged half-life of the linear segment of die-away curve, and (2) lowered FCR of plasma LDL (TABLE 1). Indirect correlation between serum cholesterol concentration and FCR of LDL was found; control animals with low cholesterolemia had significantly higher FCR than vitamin-C-deficient animals (FIG. 1). Direct correlation between the hepatal vitamin C concentration and FCR was observed ($r = 0.717$, $p < 0.02$): animals with higher vitamin C levels had higher FCR of LDL than animals with low vitamin C status.

CONCLUSION

The probable string of events during the development of chronic vitamin C deficiency is: (1) decrease of vitamin C concentration in hepatocytes, (2) lowered activity of cholesterol 7-α-hydroxylase in the liver, (3) lowered transformation of cholesterol into bile acids, (4) accumulation of cholesterol in the liver, (5) decreased production of LDL receptors on the surface of liver cells due to feedback, and (6) prolonged half-life and lowered turnover (FCR) of plasma LDL.

REFERENCES

1. GINTER, E. 1975. Ann. N. Y. Acad. Sci. **258:** 410-421.
2. BJÖRKHEM, I. & A. KALLNER. 1976. J. Lipid Res. **17:** 360-365.
3. HOLLOWAY, D. E. & J. M. RIVERS. 1981. J. Nutr. **111:** 412-424.
4. GOLDSTEIN, J. L. & M. S. BROWN. 1984. J. Lipid Res. **25:** 1450-1461.
5. GINTER, E., A. KOSINOVA, A. HUDECOVA & U. MLYNARCIKOVA. 1984. J. Nutr. **114:** 485-492.
6. WEECH, P. K., F. MCTAGGARE & G. L. MILLS. 1978. Biochem. J. **169:** 687-695.
7. MATTHEWS, C. M. E. 1957. Phys. Med. Biol. **2:** 36-53.

Effect of Ascorbic and Dehydroascorbic Acids on Lipid Composition of $C_3H/10T\frac{1}{2}$ Cells[a]

LUMINITA L. V. IBRIC[b] AND ALEX SEVANIAN[c]

[b]*Childrens Hospital of Los Angeles*
Los Angeles, California 90027

[c]*Institute for Toxicology*
School of Pharmacy
University of Southern California
Los Angeles, California 90033

INTRODUCTION

Chemical transformation induced by methylcholanthrene (MCA) can be prevented by daily treatments with ascorbic acid (AA). A gradual accumulation of lipophilic bodies has been previously observed,[1] but the role of lipid synthesis was not considered in the proposed scheme for the inhibition. Our studies on the effect of other derivatives of AA on transformation, collagen synthesis, and on redox potentials in the cells were intended to elucidate the mechanism for the inhibition of transformation (Ibric *et al.*, Cancer Res., in preparation). The inhibition of 3T3 fibroblast conversion to adipocytes by tumor promoters has been described.[2] It has been shown previously that the lipid content can vary in tumors with different growth rates.[3,4] The malignant tumors or chemically induced skin carcinomas exhibit increased content in phospholipids (PL) and cholesterol.[5–7] Dietary antioxidants (*e.g.*, AA) have been shown to play an important role in modulating the biological aging processes[8,9] and lipid peroxidation.[3,4,10,11] A modulation of the lipid composition by AA may also produce changes in phenotype, such as in membrane fluidity, and affect the expression of membrane-dependent functions. Our present studies focused on the effects of AA and dehydroascorbic acid (DHA) on lipid synthesis in confluent $C_3H/10T\frac{1}{2}$ cells.

[a]This work was supported by the National Cancer Institute (grant no. 09320) and the National Institutes of Health (grant no. ES03466).

EXPERIMENTAL

$C_3H/10T\frac{1}{2}$ cells were plated at 6000 cells/75 cm^2 flasks in Basal Eagle's medium supplemented with 10% heat-inactivated fetal calf serum. Daily treatments with 5 μg/ml AA or 50 μg/ml DHA began after 48 h in culture, as previously described,[1] for prevention of MCA transformation. Media was changed twice a week. The lipid analyses were performed on cells and in media after extraction and resolution of lipid classes by thin-layer and gas chromatography as described previously.[13]

RESULTS AND DISCUSSION

Confluent monolayers of $C_3H/10T\frac{1}{2}$ cells, and media from cultures untreated or treated daily with AA or DHA, were analyzed for lipid composition after 7, 14, and 21 days. Total lipid content in cells increased with time while in confluency. Both AA and DHA reduced the accumulation of the total lipids over 21 days (TABLE 1A). The triglycerides (TG) increased twofold in the AA- or DHA-treated cells as compared to controls. The free fatty acids (FFA) also increased in treated cells. Moreover, after the 7 days of treatment, the neutral lipids were largely released into media (TABLE 1B). PL increased over the 21-day period in control cells by approximately 200-fold; however, AA and DHA reduced this increase by 30% (TABLE 1A). The cholesterol content also increased in control cells, but this increase was not significantly affected by AA or DHA (a slight decrease was encountered after 14 days, TABLE 1A). These effects would be expected to normalize changes in fluidity due to increased proportions of unsaturated fatty acids. In comparison to $10T\frac{1}{2}$ cells, MCA clone 15 (MCA Cl 15) had a highly modified pattern of lipid composition. After 7 days in culture, MCA Cl 15 had very high lipid content (108.43 ± 6.31 $\mu g/10^6$ cells). The lipids consisted mainly of PL and cholesterol (81.65 ± 0.41 and 21.74 ± 4.40 $\mu g/10^6$ cells, respectively, TABLE 1A). Our findings are in agreement with previous reports.[5–7] The PL composition shown in TABLE 2 indicates that the amounts of all PL species were elevated in all study groups over the 21-day period. By 21 days of treatment AA and DHA depleted the sphingomyelin (SPH), phosphatidylcholine (PC) and lysophosphatidylcholine (LPC) content, but had a transitory effect on phosphatidylinositol + serine (PI + PS) levels. DHA elevated the content of phosphatidylethanolamine (PE), while AA had no effect. In comparison, MCA Cl 15 cells contain increased amounts of SPH and PE, and low levels of PC and LPC. It has been previously reported[4,6,13] that hepatoma cells contain high levels of SPH and cholesterol in comparison to normal liver cells. FIGURE 1 shows the ratios of unsaturated to saturated fatty acids (US/S) in PL, TG, and FFA from cells (A-C) and in media (D-F). Unsaturation increased significantly in PL of treated cells at 7 days, while TG and FFA were affected only by AA or DHA, respectively (B, C). However, there was a large release of these neutral lipids during this period (E, F). By 21 days, treatment with AA and DHA produced an elevation in the unsaturation index of these cells (C). An inverse correlation between US/S and cholesterol/PL ratios was apparent (TABLE 2). MCA Cl 15 cells also possess a high cholesterol content and a high degree of saturation. In conclusion, our findings show that changes in lipid composition take place following carcinogen-induced transformation, in agreement with previous re-

TABLE 1A. The Lipid Composition in 10T½ and MCA Clone 15 Cells

Treatment	Total Lipids	Triglycerides	Free Fatty Acids	Phospholipids	Cholesterol
Control					
7 Days	2.72 ± 0.38	0.31 ± 0.01	0.48 ± 0.05	0.48 ± 0.23	1.45 ± 0.10
14 Days	54.79	0.86	14.14	30.23	9.66
21 Days	183.55 ± 7.55	12.79 ± 2.42	13.67 ± 1.45	148.73 ± 5.80	8.37 ± 0.77
Ascorbic Acid					
7 Days	4.18 ± 0.72	0.28 ± 0.04	0.56 ± 0.24	0.68 ± 0.36	2.67 ± 0.08
14 Days	83.35 ± 37.30	20.68 ± 14.23	29.68 ± 11.91	26.30 ± 10.94	6.69 ± 0.22
21 Days	141.03 ± 2.69	23.75 ± 1.76	16.39 ± 0.93	89.20	11.69
Dehydroascorbic Acid					
7 Days	13.16 ± 3.89	0.77 ± 0.06	7.15 ± 0.84	2.40 ± 2.27	2.84 ± 0.26
14 Days	90.80 ± 5.20	2.32 ± 0.02	16.62 ± 4.69	64.49 ± 0.21	7.51 ± 0.44
21 Days	155.86 ± 24.56	22.59 ± 7.38	23.81 ± 5.79	100.78 ± 10.61	8.68 ± 0.78
MCA Cl 15					
7 Days	108.43 ± 6.31	1.57 ± 0.68	3.47 ± 0.82	81.65 ± 0.41	21.74 ± 4.40

Data represent the mean ± SEM of two data points experiment. Concentrations are expressed as $\mu g/10^6$ cells. Daily treatments with 5 $\mu g/ml$ of AA or 50 $\mu g/ml$ of DHA were applied only to the 10T½ cells.

TABLE 1B. The Lipid Composition in Media from 10T½ and MCA Clone 15 Cells

Treatment	Total Lipids	Triglycerides	Free Fatty Acids	Phospholipids	Cholesterol
Control					
7 Days	15.15 ± 4.67	0.64 ± 0.46	5.40 ± 2.55	6.36 ± 1.19	2.75 ± 0.47
14 Days	43.17 ± 14.05	nd	16.67 ± 7.75	23.67 ± 3.92	2.83 ± 2.38
21 Days	25.76 ± 3.96	2.21 ± 1.11	11.62 ± 0.89	0.98 ± 0.01	10.95 ± 1.95
Ascorbic Acid					
7 Days	119.57 ± 5.42	7.66	93.05 ± 0.12	14.81 ± 3.99	4.11 ± 1.37
14 Days	33.97 ± 11.02	2.33 ± 1.23	16.67 ± 7.75	14.62 ± 1.82	0.35 ± 0.21
21 Days	21.91 ± 3.32	1.48 ± 1.21	10.08 ± 0.69	1.88 ± 0.09	8.47 ± 1.33
Dehydroascorbic Acid					
7 Days	59.84 ± 7.55	4.72	38.43 ± 6.05	13.59 ± 1.11	3.10 ± 0.39
14 Days	54.86 ± 16.83	nd	20.64 ± 12.80	27.84 ± 3.43	6.38 ± 0.60
21 Days	29.51 ± 1.42	3.68 ± 0.70	14.39	5.28 ± 0.35	6.16 ± 0.37
MCA Cl 15					
7 Days	47.71 ± 12.09	2.29 ± 0.70	18.97 ± 2.42	23.43 ± 8.61	3.02 ± 0.36

The data represent the mean ± SEM of two data points experiment. The concentrations are expressed as $\mu g/10^6$ cells.

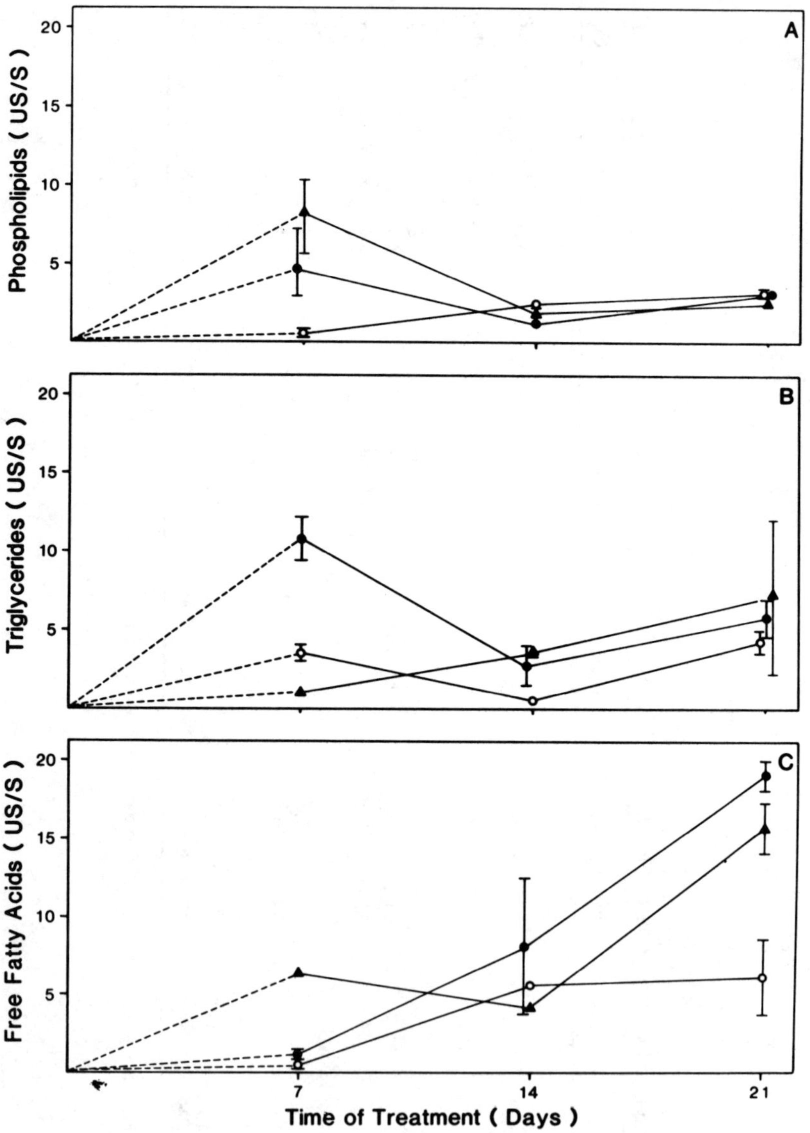
A
Phospholipids (US/S)
20
15
10
5
B
Triglycerides (US/S)
20
15
10
5
C
Free Fatty Acids (US/S)
20
15
10
5
7
14
21
Time of Treatment (Days)

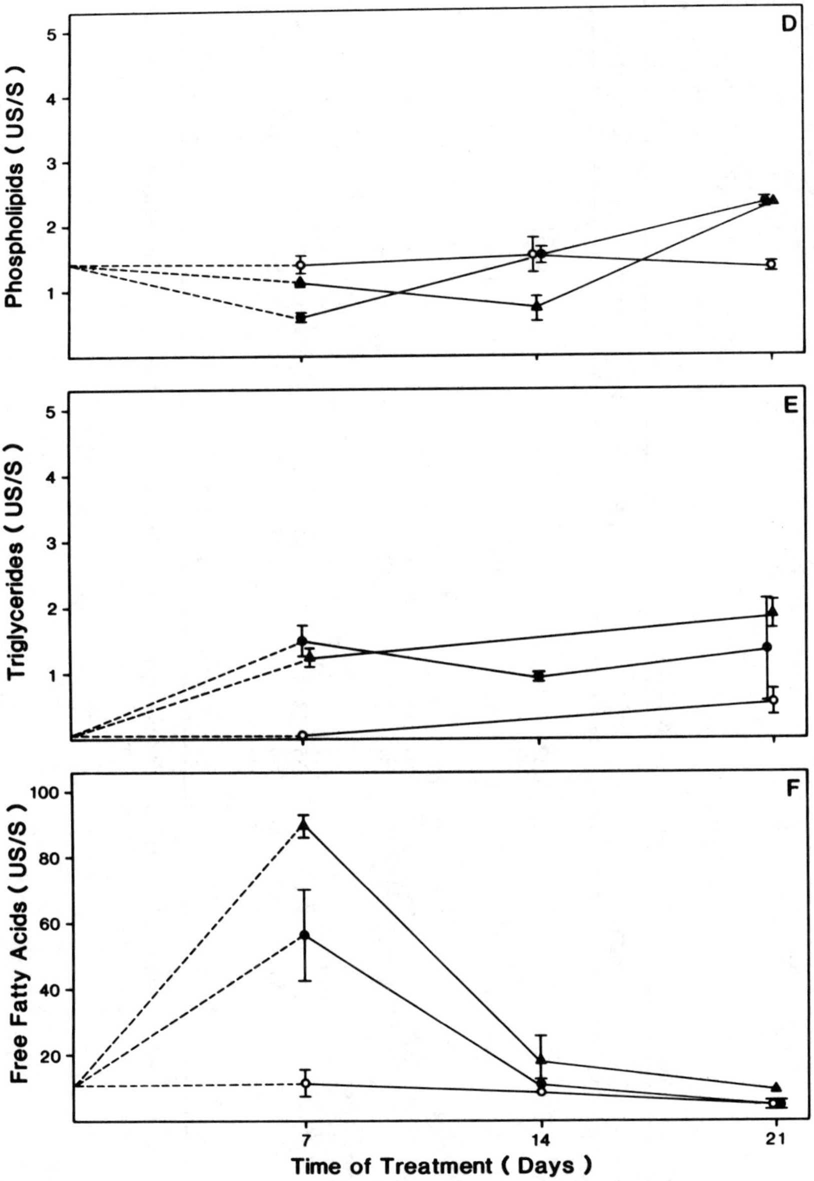

FIGURE 1. The unsaturated/saturated fatty acid ratios of phospholipids, triglycerides, and free fatty acids in $C_3H/10T\frac{1}{2}$ cells (**A-C**) and in media (**D-F**). The US/S ratios for controls (○), AA (●), and DHA (▲) represent the mean and standard errors per two data points.

TABLE 2. The Phospholipid Class Composition of 10T½ and MCA Cl 15 Cells

Treatment	SPH	PC	PE	LPC	PI + PS
Control					
7 Days	0.12	0.19 ± 0.002	0	0.03 ± 0.009	0.08 ± 0.006
14 Days	8.87	5.93	1.52	9.10	0.10
21 Days	13.28 ± 1.72	22.55 ± 0.92	17.61 ± 0.31	48.96 ± 12.97	15.71 ± 2.53
Ascorbic Acid					
7 Days	0.33	0.06 ± 0.007	0.06 ± 0.02	0.24 ± 0.10	0.14 ± 0.13
14 Days	4.82 ± 0.91	3.74 ± 1.80	1.53 ± 0.17	7.90 ± 1.04	5.06 ± 0.26
21 Days	8.57 ± 0.20	7.82 ± 2.04	16.36 ± 1.78	30.60 ± 9.81	10.48 ± 2.68
Dehydroascorbic Acid					
7 Days	0.11	0.19	0.36	0.58 ± 0.003	0.17
14 Days	13.22 ± 3.08	6.85 ± 2.14	9.63 ± 6.38	4.34 ± 0.35	0.56 ± 0.26
21 Days	8.81 ± 5.47	15.54	26.95 ± 2.94	14.22 ± 11.94	18.99 ± 2.35
MCA Cl 15					
7 Days	17.52 ± 6.74	6.26 ± 0.38	19.25 ± 1.26	5.22	14.28 ± 1.28

Concentrations are expressed as $\mu g/10^6$ cells. Abbreviations: SPH, sphingomyelin; PC, phosphatidylcholine; PE, phosphatidylethanolamine; LPC, lysophosphatidylcholine; PI + PS, phosphatidylinositol + phosphatidylserine.

ports, and that AA and DHA may prevent many of these changes in cells during the attendant inhibition of chemical transformation.

ACKNOWLEDGMENTS

The authors wish to thank Ms. Laurie McLeod and Mr. Dan A. Potter for their technical support.

REFERENCES

1. BENEDICT, W. F., W. L. WHEATLY & P. A. JONES. 1980. Cancer Res. **40:** 2796-2801.
2. DIAMOND, L., T. G. O'BRIEN & G. ROVERA. 1977. Nature **269:** 247-248.
3. BARTOLI, G. M. & T. GALEOTTI. 1979. Biochim. Biophys. Acta **574:** 537-541.
4. BARTOLI, G. M., S. BARTOLI, T. GALEOTTI & E. BERTOLI. 1980. Biochim. Biophys. Acta **620:** 205-211.
5. HAVEN, F. L. & W. R. BLOOR. 1956. Adv. Cancer Res. **4:** 237-314.
6. BORRELLO, S., G. MINOTTI, G. PALOMBINI, A. GRATTAGLIANO & T. GALLEOTTI. 1985. Arch. Biochim. Biophys. **238**(2): 588-595.
7. BULLOCK, W. E. & W. CRAMER. 1914. Proc. R. Soc. Lond. Ser. B. **87:** 236-239.
8. SUN, A. Y. & G. Y. SUN. 1982. *In* Nutritional Approaches to Aging Research. G. B. Moment, Ed.: 135-136. CRC Press. Boca Raton, Fla.
9. AMES, B. N., R. CATHCART, E. SCHWIERS & P. HOCHSTEIN. 1981. Proc. Natl. Acad. Sci. **78:** 6858-6862.
10. SEVANIAN, A. & P. HOCHSTEIN. 1985. Ann. Rev. Nutr. **5:** 365-390.
11. KING, M. M. & P. MCCAY. 1981. Cancer Res. **41**(Suppl.): 2485-2490.
12. GOULD, R. J. & B. H. GINSBERG. 1984. *In* Membrane, Detergents, and Receptor Solubilization. Alan R. Liss. New York. pp. 65-83.
13. WOOD, R., G. C. UPRETI & R. J. ANTUENO. 1986. Lipids **21**(4):292-300.
14. CRISTENSEN LOU, H. O., J. CLAUSE & F. BIERRING. 1965. J. Neurochem. **12:** 619-627.

Vitamin C and Sickle Cell Disease[a]

SUSHIL K. JAIN AND DARRYL M. WILLIAMS

Departments of Pediatrics and Medicine
Louisiana State University School of Medicine
Shreveport, Louisiana 71130

The role of oxidant damage to the red cell membrane in sickle cell disease has been of interest in recent years. The observation that sickle cells are more vulnerable to peroxidant threat than normal cells was first reported by Stocks *et al.*[1] Natta and Machlin[2] have reported low plasma vitamin E in sickle cell patients. Subsequent studies of Chiu and Lubin[3] have shown that red cell vitamin E is also low in sickle cell patients in comparison to normals. Chiu and Lubin[3] have also shown that the increased vulnerability of sickle cells to peroxidant threat is blocked *in vitro* when the cells are preincubated with vitamin E.

Vitamin C is another antioxidant that is known to participate in scavenging oxidative reactions along with vitamin E. This study reports investigations on vitamin C and sickle cell disease. Details of this study are given in refs. 4 and 5.

FIGURE 1 illustrates that plasma vitamin C levels are significantly lower in sickle cell disease patients in comparison to control subjects. Previous studies[6] have shown that sickle cells undergo significant membrane lipid peroxidation *in vivo*[6,7] and have externalization of phosphatidylserine (PS) in their membrane bilayer.[8] When peroxidation lipid damage was induced in red cells in rats by the administration of phenylhydrazine, an oxidant drug, it externalized PS in red cells. Data in TABLE 1 show that membrane lipid peroxidation as well as externalization of PS was blocked in red cells of rats supplemented with vitamin C. This suggests that in rats vitamin C can scavenge peroxidation reactions and prevent the associated red cell membrane alterations. Thus, it seems that low plasma vitamin C may be a factor in the increased vulnerability of sickle cells to oxidant damage *in vivo.*

Decreased vitamin C levels may be a consequence of increased utilization of the vitamin in the detoxification of oxygen radicals that are generated by sickle hemoglobin.[9] The resultant low vitamin C levels could then impair the oxygen radical detoxification mechanisms, which can then cause peroxidation of membrane lipids in sickle cells. It is known that vitamin C can protect the sickle erythrocytes against oxidant damage *in vitro* induced by exposure to H_2O_2[4] or acetylphenylhydrazine.[10] It would be interesting to examine the effects of vitamin C supplementation on disease severity in sickle cell patients.

[a] This study was supported by grant no. HL 30247 from the NIH and by Hoffmann-La Roche, Inc.

TABLE 1. Phospholipid Hydrolysis by Phospholipase A_2 and Russell's Viper Venom Clotting Time of Erythrocytes from Rats Administered with Phenylhydrazine and/or Vitamin C

	Malonyldialdehyde (nmol/ml PCV)	Phospholipid Hydrolysis (%)			RVV Clotting Time (% of Control)
		PC	PE	PS	
Control	2.2 ± 0.3	61.1 ± 1.7	11.9 ± 1.0	0	100 ± 1
Vitamin C (10 mg/kg BW)	2.3 ± 0.5	60.9 ± 0.6	12.1 ± 0.6	0	102 ± 2
Phenylhydrazine (20 mg/kg BW)	4.2 ± 0.3*	52.1 ± 1.9*	28.9 ± 1.3*	8.8 ± 0.8*	90 ± 1*
Vitamin C + phenylhydrazine	3.2 ± 0.4†	57.2 ± 1.3†	18.7 ± 1.0†	3.9 ± 0.6†	95 ± 2†

Values are mean ± SE of five observations. Differences in values marked * in comparison to controls and between * and † in the same column are statistically significant ($p < 0.05$). Rats were administered intramuscularly with phenylhydrazine or vitamin C for 2 days; RVV: Russell's viper venom; PCV: packed cell volume; PC: phosphatidylcholine; PE: phosphatidylethanolamine; PS: phosphatidylserine. (Data taken from ref. 5.)

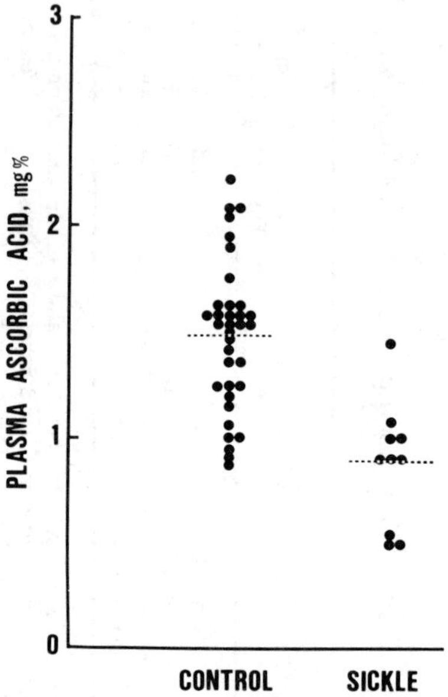

FIGURE 1. Plasma ascorbic acid levels of sickle cell patients and control subjects. Values are from 10 sickle cell patients and 35 control subjects. Differences in the two groups are statistically significant ($p < 0.001$). (Data taken from ref. 4.)

REFERENCES

1. STOCKS, J., J. KEMP & T. L. DORMANDY. 1971. Lancet **1:** 266-269.
2. NATTA, C. & L. MACHLIN. 1979. Am. J. Clin. Nutr. **32:** 1359-1362.
3. CHIU, D. & B. LUBIN. 1979. J. Lab. Clin. Med. **94:** 542-548.
4. JAIN, S. K. & D. M. WILLIAMS. 1985. Clin. Chim. Acta **149:** 257-261.
5. JAIN, S. K. 1985. J. Clin. Invest. **76:** 281-286.
6. DAS, S. K. & R. C. NAIR. 1980. Brit. J. Haematol. **44:** 87-92.
7. JAIN, S. K. & S. B. SHOHET. 1984. Blood **63:** 362-367.
8. LUBIN, B. *et al.* 1981. J. Clin. Invest. **67:** 1643-1649.
9. HEBBEL, R. P. *et al.* 1982. J. Clin. Invest. **70:** 1253-1259.
10. LACHANT, N. A. & K. R. TANAKA. 1986. Am. J. Med. Sci. **292:** 3-10.

Bimodal Effects of Megadose Vitamin C on Adrenal Steroid Production in Man

An *in Vivo* Study[a]

SURAT KOMINDR,[b] GEORGE E. NICHOALDS,[c] AND ABBAS E. KITABCHI[b]

[b]*Division of Endocrinology and Metabolism*
Department of Medicine

[c]*Clinical Nutrition Laboratory*
Department of Obstetrics and Gynecology
and
Clinical Research Center
The University of Tennessee, Memphis
Memphis, Tennessee 38163

Much evidence supports the concept that vitamin C has an inhibitory effect on steroid production *in vitro* in experimental animals (for reviews, see refs. 1 and 2). However, except for one study on the role of vitamin C in adrenal steroids in children,[3] the role of this vitamin in adrenal steroid production in humans is lacking. Therefore, the purposes of the present study were to investigate whether consumption of megadose vitamin C for a short period affects the adrenal glucocorticoid and androgen levels of adult men and to locate the possible site(s) of blockage of the steroidogenesis pathway.

The subjects were 13 healthy, lean male volunteers (25-35 years of age) on no medicine for at least 4 weeks prior to the study. After giving written consent, the subjects were admitted to the Clinical Research Center for 10 days. After 3 days of baseline hormonal studies (control period), each subject received two 500 mg vitamin C tablets q 6 h until the end of the study (days 3 to 10). Hormonal studies consisted of the following measurements:

1. The diurnal variations of plasma cortisol at 0400, 0600, 0800, 1200, 1600, 2000, and 2400 hours and plasma dehydroepiandrosterone (DHEA) during the morning cortisol peaks on days 1 and 8.

[a]This work was supported in part by Training Grant no. AM07088 and General Clinical Research Center Grant no. RR00211 from the U.S. Public Health Service and by the Department of Obstetrics and Gynecology, the University of Tennessee, Memphis, Tenn.

2. Plasma cortisol, DHEA, 11-deoxycortisol, and 17-α-hydroxyprogesterone (17[OH]P) responses to intravenous physiologic (small) bolus of β1-24 ACTH (150 ng/1.5 M^2) on days 2 and 9.
3. Plasma cortisol and DHEA responses to intravenous maximal effective (large) bolus (500 ng/1.5 M^2) of synthetic β1-24 ACTH (Cortrosyn) on days 3 and 10 after ingestion of 1 mg dexamethasone orally at 11 P.M. the night before. Blood was drawn at 0, 7.5, and 15 minutes.

All the subjects tolerated the vitamin C dose (4 g/day) well. Serum ascorbate levels were 2.5 times higher during vitamin C ingestion than in the control period (1.82 ± 0.18 vs. 4.42 ± 0.29 mg/dl, $p < 0.001$).

Although ingestion of vitamin C did not alter the normal pattern of diurnal variation of plasma cortisol levels, it significantly lowered mean plasma cortisol levels at 0400, 0800, 1600, and 2000 hours ($p < 0.05$). Moreover, a significant decrease in the overall 24-hour plasma cortisol curve during vitamin C ingestion compared to the control period was observed ($p < 0.02$). TABLE 1 demonstrates that there was a reduction in the maximum daily cortisol levels during vitamin C ingestion ($p < 0.05$) with a reciprocal increase of concomitant DHEA levels in association with vitamin C ingestion ($p < 0.001$).

Plasma cortisol, DHEA, 11-deoxycortisol, and 17(OH)P responses to small ACTH boluses are depicted in FIGURE 1. The increment of plasma cortisol in response to 150 ng/1.5 M^2 ACTH bolus was blunted (10.4 ± 1.3 vs. 8.3 ± 0.9 μg/dl, $p <$ 0.005, at 15 min). There was no significant difference between the DHEA increments during both periods. Although there was no difference in the increment of 11-deoxycortisol during control and vitamin C, a greater than twofold increase in 17(OH)P level was noted at 7.5 minutes after vitamin C supplementation as compared to the control period. This pattern of response was noted in 6 of the 7 subjects studied.

TABLE 1. Individual Maximum Daily Plasma Cortisol Levels and Concomitant Plasma DHEA Levels before and during Vitamin C Ingestion

	Cortisol (μg/dl)		DHEA (ng/dl)	
Subject	Control	Vitamin C	Control	Vitamin C
1	23.2	18.1	1140	1450
2	21.7	17.9	220	540
3	26.0	11.1	470	530
4	18.3	13.8	690	1150
5	11.8	11.6	70	170
6	22.2	11.5	820	1130
7	16.2	20.0	250	190
8	12.4	10.4	390	1170
9	18.7	16.2	150	320
10	14.6	22.1	310	390
11	16.6	17.3	690	1160
12	13.6	11.7	150	870
13	15.9	11.1	190	740
Mean	19.6	14.8	426	755
SEM	1.4	1.1	85	114
p value	<0.05		<0.001	

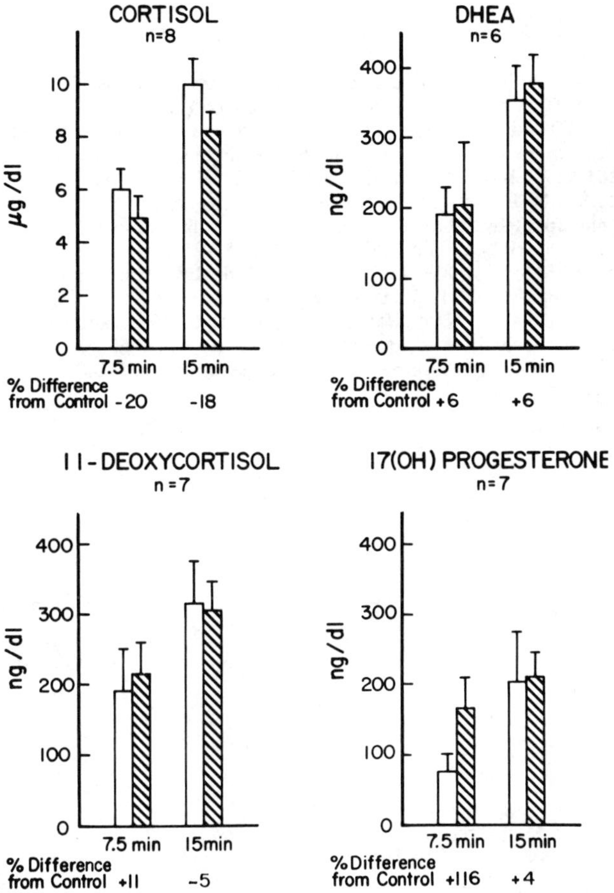

FIGURE 1. The responses of plasma cortisol (μg/dl), 11-deoxycortisol (ng/dl), 17(OH)P (ng/dl), and DHEA (ng/dl) to small-dose ACTH bolus (150 ng/1.5 M^2) during the control (*open bars*) and vitamin C (*hatched bars*) periods. The percent difference between the two periods is expressed underneath each graph.

When a higher dose of ACTH bolus (500 ng/1.5 M^2) was given, there was no difference in the cortisol or DHEA increment between both periods (data not shown).

Our studies indicate that megadose vitamin C decreases the sensitivity but not the capacity of adrenal gland to respond to ACTH. The present findings are in support of our earlier work *in vitro,* in that vitamin C in beef adrenal inhibited 21-hydroxylase[4] and shifted the ACTH dose-response curve of corticosterone in rat adrenal to the right.[5] It is tempting to postulate that with vitamin C ingestion there may be a partial blockage of 21-hydroxylase so that more substrate becomes available for the androgenic pathway. This blockage is incomplete since a higher ACTH dose reverses the abnormal cortisol response to the low dose. Additional studies are required to determine the long-term effect of ascorbic acid ingestion on the hypothalamic-pituitary-adrenal axis in young adults and older age groups.

REFERENCES

1. Kitabchi, A. E. 1967. Ascorbic acid in steroidogenesis. Nature **215:** 1385-1386.
2. Kitabchi, A. E., A. Nathans, J. A. Barker, L. Kitchell & B. Watson. 1978. Adenylate cyclase (AC) activity in isolated adrenal cells of vitamin E-deficient (−E) and -sufficient (+E) rats. *In* Tocopherol, Oxygen and Biomembranes, Vol. 1. C. de Duve & O. Hayaishi, Eds.: 201-219. Elsevier/North Holland Biomedical Press. Amsterdam.
3. Liakakos, D., N. L. Doulas, D. Ikkos, C. Anoussakis, P. Vlachos & G. Jouramani. 1975. Inhibitory effect of ascorbic acid (vitamin C) on cortisol secretion following adrenal stimulation in children. Clin. Chim. Acta **65:** 251-255.
4. Kitabchi, A. E. 1967. Inhibition of steroid C-21 hydroxylase by ascorbate: Alterations of microsomal lipids in beef adrenal cortex. Steroids **10:** 567-576.
5. Kitabchi, A. E., A. H. Nathans & L. C. Kitchell. 1972. Adrenal gland in vitamin E deficiency. III. Inhibition of adrenocorticotropic hormone-induced steroidogenesis in isolated adrenal cells by ascorbic acid. J. Biol. Chem. **248:** 835-840.

Megadoses of Vitamin C

A Pharmacokinetic Evaluation

S. MELETHIL,[a] M. B. SUBRAHMANYAM,[b]
C. J. CHANG,[a] AND W. D. MASON [a]

[a]*Schools of Pharmacy and Medicine*
University of Missouri
Kansas City, Missouri 64108

[b]*Department of Electrical and Computer Engineering*
University of Missouri
Columbia, Missouri 65211

A recent report from our laboratory involving four human subjects showed that steady-state plasma vitamin C concentrations following oral doses of 0.5, 1, and 2 g were essentially equal.[1] A pharmacokinetic model where both gastrointestinal absorption and renal tubular reabsorption of the vitamin are saturable was found to describe the data adequately. The model in equation form is:

$$dCp/dt = V_{\max 1} \cdot A_{gi} / Vd(K_{m1} + A_{gi}) - k_1 \cdot Cp + V_{\max 2},$$

where dCp/dt is the rate of change of plasma vitamin concentration, $V_{max\,1}$ and K_{m1} are the maximum capacity and Michaelis constant for the absorption process, A_{gi} is the amount of vitamin in the gastrointestinal tract (site of absorption), Vd is the volume of distribution, k_1 is a first-order rate constant representing the glomerular filtration process, and $V_{max\,2}$ is the maximum capacity for tubular reabsorption.

Parameters describing the equation were estimated from experimental data. Absorption of vitamin C was significantly lower in subject 2 and was excluded from parameter estimation. Mean values derived from data obtained from the remaining three subjects were: $V_{max} = 220$ mg/h, $K_{m1} = 123$ mg, $k_1 = 0.2\ h^{-1}$, and $V_{max2} =$ 2.6 mg/(l · h). By using these mean values and assuming an absorption time of 3 h and a V_d of 30 liters, plasma vitamin C concentration-time profiles were computer generated as a function of dose. Based on reported consumption,[2] a dose range 0.5 to 20 g was selected and each dose was assumed to be ingested once daily.

Simulations showed that steady state is approached within 24-48 hours (1-2 doses). As shown in TABLE 1 and FIGURE 1, simulated data for the 0.5, 1, and 2 g are in good agreement with experimental values. Peak plasma concentrations obtained at steady state, *Cp*(max), are similar and practically independent of dose. Model simulations are in good agreement with data published by Yew,[2] who found steady-state minimum concentrations of 15.4 ± 0.16 mg/l ($n = 11$) and 19.5 ± 0.20 ($n = 6$) in subjects ingesting 1 to 3 g and 8 to 12 g, respectively, and Hornig *et al.*[3] who reported urinary recoveries of 75.44, and 20.9% of dose following 1, 2, and 5g daily

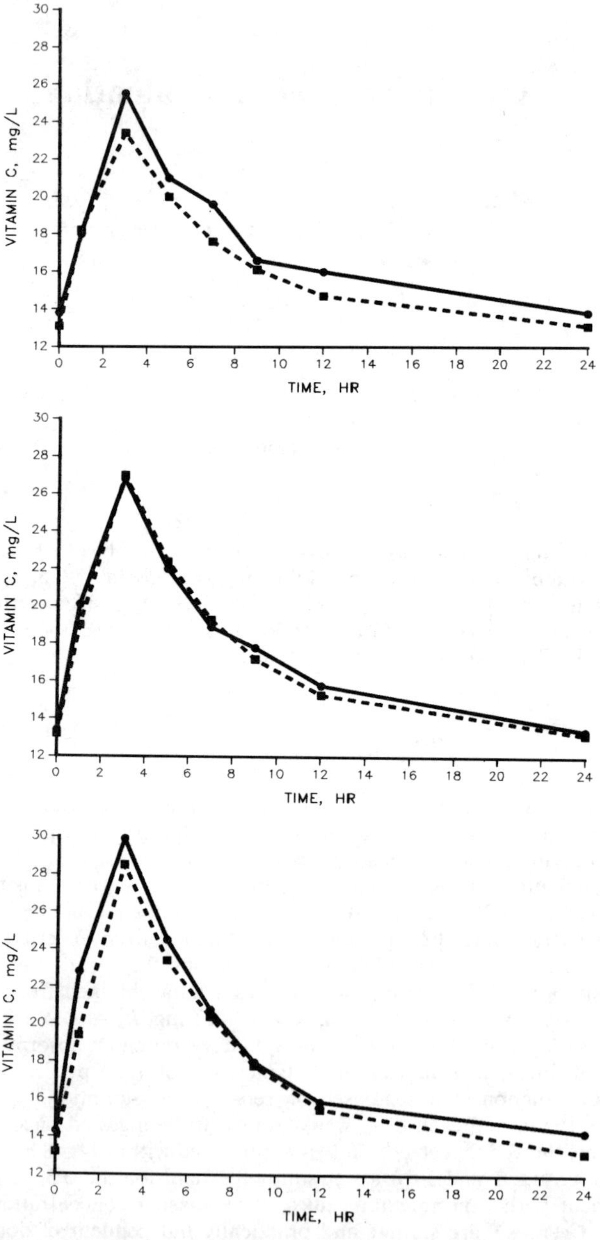

FIGURE 1. Steady-state plasma vitamin C concentration as a function of time over a dosing interval. *Top:* 0.5 g daily; *Middle:* 1.0 g daily; *Bottom:* 2.0 g daily; ● = experimental; ■ = simulated.

TABLE 1. Effect of Dose on Vitamin C Disposition in Humans

Daily Dose (g)	F		Cp (max), mg/l		$(AUC)_0^{24}$, mg · h/l	
	E	S	E	S	E	S
0.5	85.2 ± 11.8[a]	85.1	25.5 ± 5.87	23.4	410.9 ± 98.54	382.4
1	56.0 ± 14.5	55.9	26.8 ± 3.68	27.0	416.0 ± 62.13	405.1
2	41.4 ± 6.76	30.7	29.9 ± 6.92	28.5	441.8 ± 72.78	414.2
5	nd	12.8	nd	29.2	nd	419.0
10	nd	6.52	nd	29.4	nd	420.4
20	nd	3.28	nd	29.6	nd	421.1

AUC = area under the curve; E = experimental; F = percent of dose absorbed; nd = not determined; S = simulated.
[a] $n = 3$.

doses, respectively, in 1 subject. It is interesting to note that at daily doses of 1 g and higher the actual amount absorbed as predicted by the model is independent of dose, with the maximum amount absorbed being in the 500-600 mg range (a previous study[3] reported a value of 1160 mg for maximum daily absorption). This is one reason why plasma concentrations are essentially independent of dose (FIG. 1). A second reason is the saturation of the renal tubular reabsorptive process, which causes more rapid urinary excretion of the vitamin as its plasma concentration is increased.[1] More detailed studies aimed at establishing dose-concentration relationships are needed to confirm this model. The results from this report question the systemic benefits to be derived from the popular practice of ingesting megadoses of the vitamin.

REFERENCES

1. MELETHIL, S., W. D. MASON & C. J. CHANG. 1986. Int. J. Pharmaceut. **31:** 83-89.
2. YEW, M. S. 1984. Nutr. Rep. Intl. **30:** 597-601.
3. HORNIG, D., J.-P. VUILLEUMIER & D. HARTMANN. 1980. Int. J. Vitam. Nutr. Res. **50:** 309-314.

Vitamin C Increases Hexose Monophosphate Pathway Activity but Not Chemiluminescence in Neutrophils from Chronic Hemodialysis Patients[a]

JOSEPH NICOTRA, PAULETTE NOVEMBRE, MARK A. NEEDLE, AND VINCENT A. DEBARI[b]

The Renal Laboratory and Service
Department of Medicine
St. Joseph's Hospital and Medical Center
University of Medicine and Dentistry of New Jersey
Paterson, New Jersey 07503

Although much has been learned in recent years about the immunological defects of chronic hemodialysis patients (CHD), infection continues as a cause of significant morbidity and mortality.[1] Because several cellular responses associated with polymorphonuclear neutrophil (PMN) phagocytosis are now known to be decreased in CHD,[2,3] their elucidation is difficult. It has been proposed that increased levels of regulatory factors such as cyclic nucleotides[4] and decreased levels of nutritional factors, especially vitamins, may play important roles in defective PMN function.[2] It has been shown[5] that the PMN of CHD contains significantly decreased levels of several water-soluble vitamins, the greatest being that of vitamin C (ascorbate). Although the exact role of ascorbate in contributing to a bactericidal response is not entirely understood, it has been demonstrated to be a dependent factor in increasing hexose monophosphate pathway (HMP) in human PMN[6] and may enhance direct bactericidal mechanisms as well.[7] During phagocytosis, the PMN undergoes an increase in oxidative metabolism including activation of HMP and production of "active" oxygen species, which include a number of oxygen-containing free radicals and hydrogen peroxide. These substances participate in a variety of reactions that ultimately lead to the destruction of invading organisms.[8] The use of luminol as a photon-emitting redox reactant has become a commonly applied technique (chemiluminescence, CL) to study active oxygen generation; the HMP activity can be assessed by the measurement of CO_2 generated from the C-1 position of glucose. In this work we report the *in vitro* effect of ascorbate on these two metabolic events in the PMN of CHD.

[a] This work was supported, in part, by a grant-in-aid from the New Jersey Lions District 16A Charitable Foundation.

[b] To whom all correspondence should be addressed at the Renal Lab., St. Joseph's Hospital and Medical Center, 703 Main St., Paterson, N.J.

MATERIALS AND METHODS

Patients and PMN Harvesting

Five CHD ranging from 23 to 69 years of age and receiving no vitamin supplementation other than the 1 mg/day regimen of folate routinely given to all patients in the hemodialysis unit were chosen for this study. After informed consent was given, blood was withdrawn from CHD immediately before treatment. PMN were obtained by dextran sedimentation of the whole blood, followed by NH_4Cl lysis and layering over fetal bovine serum as previously described.[5]

Measurement of HMP Activity

HMP activity was assessed by the production of $^{14}CO_2$ from [1-^{14}C]glucose using a radiometric Warburg apparatus as described in earlier publications.[9] Polystyrene microspheres (0.81 μm diameter, Difco Laboratories, Detroit, Mich.) were used as the particulate stimulus. Incubation mixtures contained a total volume of 125 μl, of which 25 μl was the cell suspension (10^6 PMN) in Hanks-buffered minimum essential medium (MEM) containing 0.1% gelatin and ascorbate to the indicated concentration, 50 μl was the polystyrene latex (1.5% wt/vol), and 50 μl was the glucose solution (0.1 mg/ml). Ascorbate was added to the cell suspension either 30 min prior to the initial measurements (preincubation) or during the period in which the measurements were made (coincubation). Reaction vials were flushed with an oxygen (95%) and carbon dioxide (5%) gas mixture in all but preliminary experiments. Incubations were carried out at 37°C.

CL Measurements

A photon-counting luminometer (Model 2001, Analytical Luminescence Lab., San Diego, Calif.) was coupled to a recorder and photon counts continuously recorded upon the addition of luminol to the incubation mixture as previously described.[10] Cells (10^7 PMN) were suspended in 0.5 ml Hepes-buffered saline (HBS) either with or without ascorbate. Latex particles were added as a 1.5% wt/vol suspension in 0.5 ml of HBS. All CL experiments were conducted in an atmosphere of 95% O_2, 5% CO_2. The sample compartment thermostat was maintained at 37°C.

RESULTS AND DISCUSSION

Preliminary HMP experiments comparing incubations in room air with those in 95% O_2 demonstrated that no difference was seen in the presence or absence of

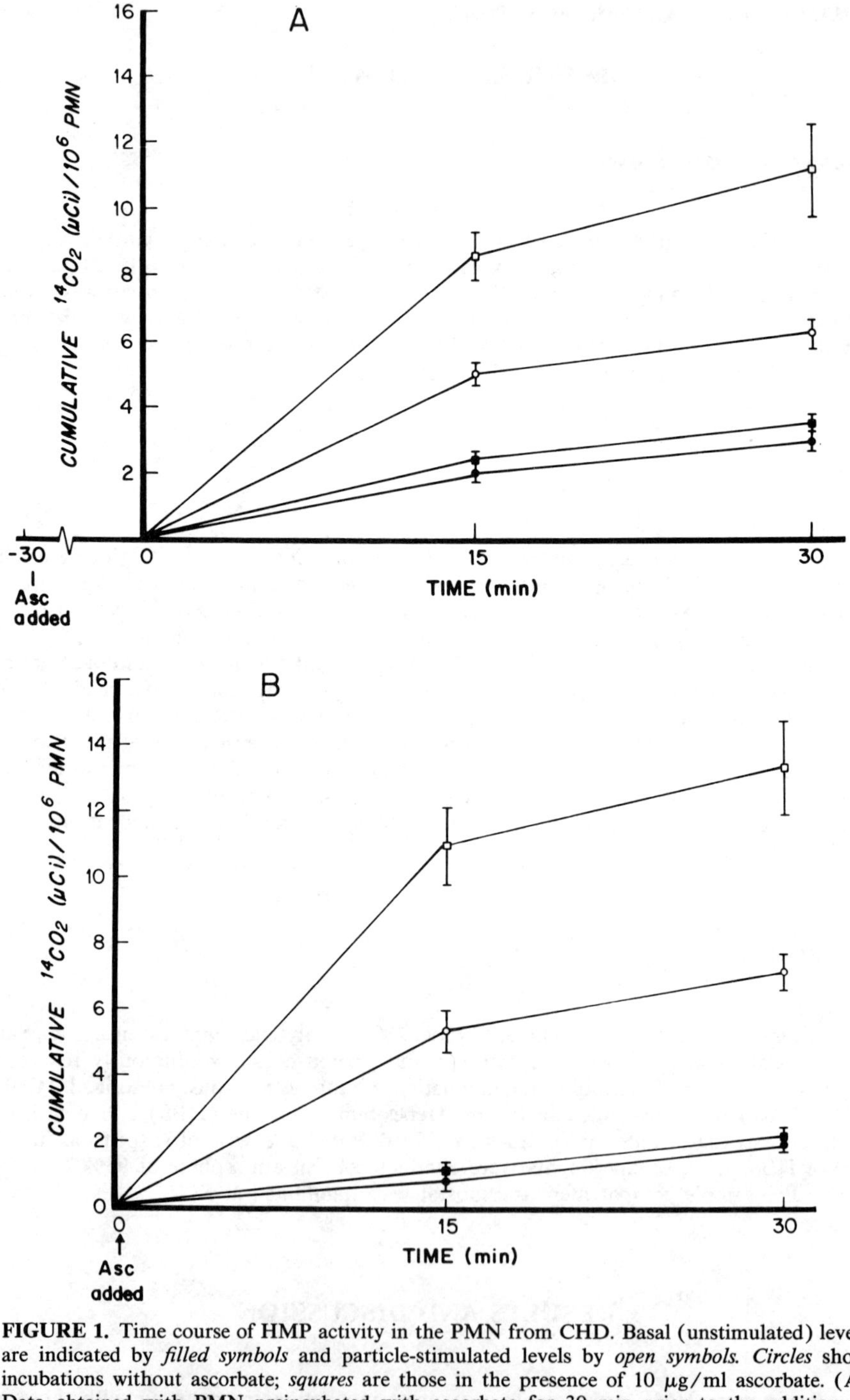

FIGURE 1. Time course of HMP activity in the PMN from CHD. Basal (unstimulated) levels are indicated by *filled symbols* and particle-stimulated levels by *open symbols. Circles* show incubations without ascorbate; *squares* are those in the presence of 10 μg/ml ascorbate. **(A)** Data obtained with PMN preincubated with ascorbate for 30 min prior to the addition of particles or buffer. Data in **B** were obtained by adding ascorbate simultaneously with the buffer or latex suspension.

ascorbate in room air incubated PMN (−ascorbate 6.6 ± 2.8 μCi $^{14}CO_2/10^6$ PMN; + ascorbate 5.9 ± 3.4 μCi/10^6 PMN), whereas in 95% O_2, 7.1 ± 2.5 μCi/10^6 PMN were produced in the absence of ascorbate and 13.2 ± 1.5 μCi/10^6 PMN were generated in the presence of ascorbate ($n = 3$, data are means ± 1 SEM, 10 μg ascorbate/ml incubation mixture). This confirms the observation that ascorbate must be oxidized to dehydroascorbate in order for it to penetrate the PMN.[6] Cells were both preincubated (30 min @ 37°C) and coincubated with ascorbate (10 μg/ml ascorbate) in a 95% O_2 atmosphere (FIG. 1). FIGURE 1A, showing the preincubation results, clearly demonstrates a large and statistically significant ($p < 0.05$) difference between the particle-stimulated PMN in the presence and absence of ascorbate. An even larger increase (approximately twofold) is seen with the coincubated PMN during

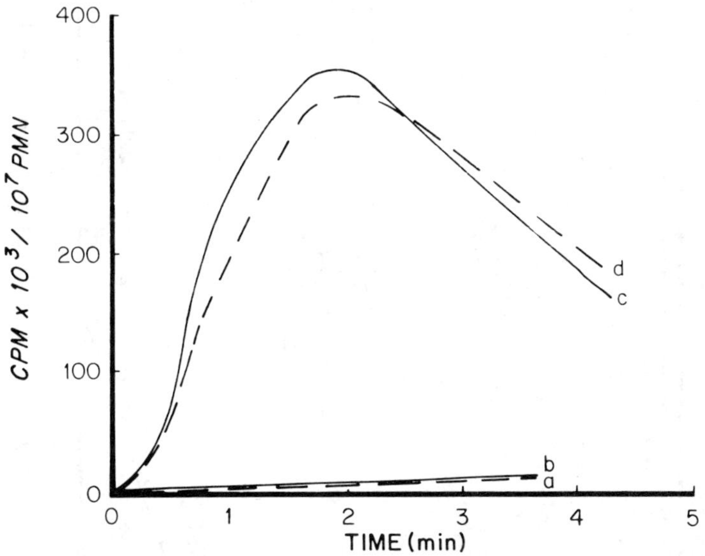

FIGURE 2. Recording of CL generation from 10^7 PMN from a single patient. The two lower curves were produced by unstimulated cells: *a,* basal (buffer alone) and *b,* buffer + ascorbate (10 μg/ml); upper curves are data from latex-stimulated cells with: *c,* no ascorbate and *d,* 10 μg/ml ascorbate.

particle stimulation. Small, insignificant ($p > 0.1$) increases in basal HMP activity were observed in both experiments upon the addition of ascorbate. These data, when compared to other HMP activity measurements from our laboratory,[11] albeit utilizing a slightly different methodology, indicated that ascorbate increases the HMP activity of PMN from CHD to roughly the same level observed in healthy control subjects.

The CL response of PMN from CHD was studied by using the same basic protocol in coincubation experiments. The results of a rather typical experiment are shown in FIGURE 2. The low basal CL output is not appreciably changed when ascorbate is added to the incubation medium. The much higher particle-stimulated response does not appear to be much affected by ascorbate. In fact, when all five patients were studied, no significant difference ($p > 0.2$) was observed between peak heights (max-

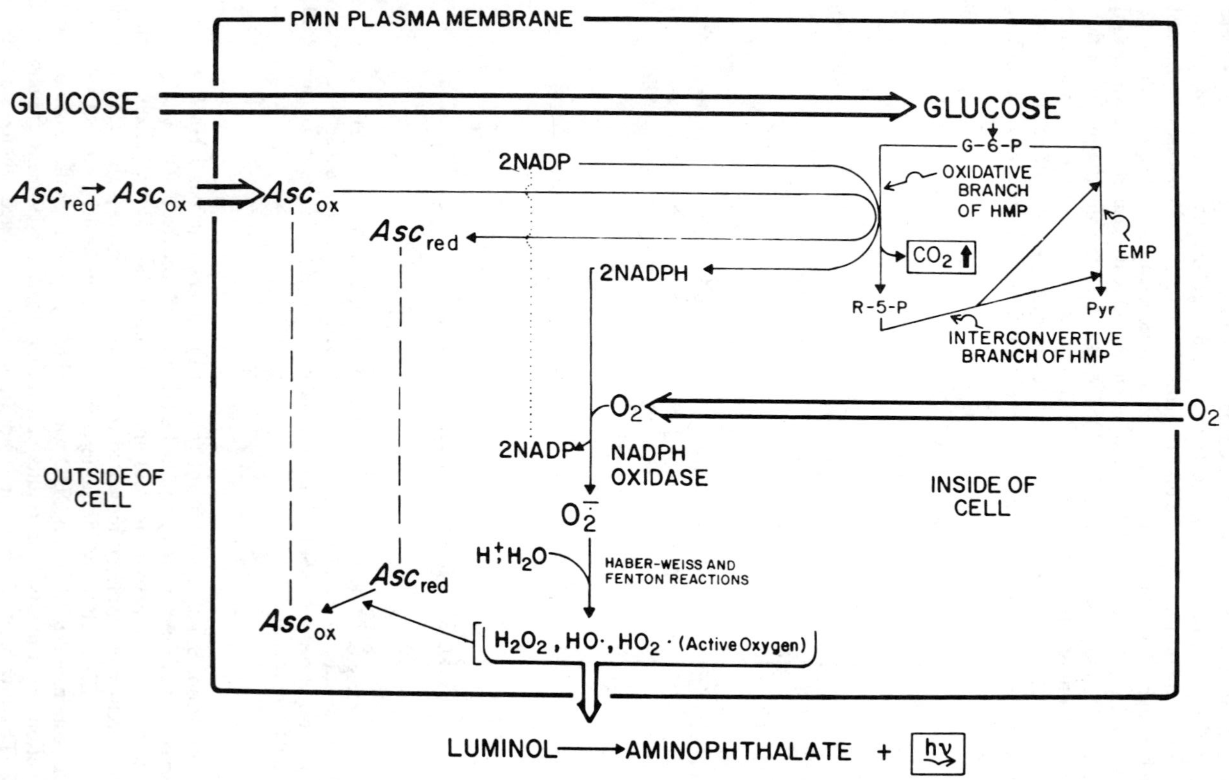

FIGURE 3. Proposed scheme of ascorbate utilization in PMN. Glucose, during phagocytosis, is primarily utilized by the HMP rather than the Embden-Meyerhoff pathway (EMP). During the initial sequence of reactions in the HMP (the oxidative branch), Asc_{ox}, which has traversed the plasma membrane, may act to increase the total amount of redox cosubstrate and "drive" the HMP to a greater degree, thus producing the observed increase (⬆) in CO_2. This increase is not associated with concomitant increase in CL, presumably due to the scavenging of oxidants by Asc_{red}.

imum cpm) obtained with ascorbate (299 ± 48 cpm/10^7 PMN) and those obtained in its absence (313 ± 59 cpm/10^7 PMN). A possible explanation for these observations is depicted in the metabolic scheme shown in FIGURE 3. The overall effect of these processes is a frustrated increase in intracellular phagocytic metabolism wherein ascorbate participates in a futile redox cycle, increasing HMP activity (dehydroascorbate to ascorbate) but scavenging active oxygen species (ascorbate to dehydroascorbate), thus decreasing luminol-dependent CL. In light of these experiments, the possibility that ascorbate may correct the depressed bactericidal activity associated with the PMN from CHD seems unlikely.

REFERENCES

1. MATTERN, W. D., L. J. HAK, R. W. LAMANNA, K. M. TEASLEY & M. S. LAFFELL. 1982. Malnutrition, altered immune function, and the risk of infection in maintenance hemodialysis patients. Am. J. Kidney Dis. **1:** 206-216.
2. DEBARI, V. A. & M. A. NEEDLE. 1984. Polymorphonuclear leukocyte defects in chronic hemodialysis patients. Clin. Nephrol. **22:** 51-52.
3. RITCHEY, E. E., J. D. WALLIN & S. V. SHAH. 1981. Chemiluminescence and superoxide anion production by leukocytes from chronic hemodialysis patients. Kidney Internat. **19:** 349-358.
4. CHARPENTIER, B., P. LANG, B. MARTIN, J. NOURY, D. MATHIEU & D. FRIES. 1983. Depressed polymorphonuclear leukocyte functions associated with normal cytotoxic functions of T and natural killer cells during chronic hemodialysis. Clin. Nephrol. **19:** 288-294.
5. DEBARI, V. A., O. FRANK, H. BAKER & M. A. NEEDLE. 1984. Water soluble vitamins in granulocytes, erythrocytes and plasma obtained from chronic hemodialysis patients. Am. J. Clin. Nutr. **39:** 410-415.
6. DECHATELET, L. R., M. R. COOPER & C. E. MCCALL. 1972. Stimulation of the hexose monophosphate shunt in human neutrophils by ascorbic acid: Mechanism of action. Antimicrob. Agents Chemother. **1:** 12-16.
7. MILLER, T. E. 1969. Killing lysis of gram-negative bacteria through the synergistic affect of hydrogen peroxide, ascorbic acid and lysozyme. J. Bacteriol. **98:** 949-955.
8. BABIOR, B. M. 1984. The respiratory burst of phagocytes. J. Clin. Invest. **73:** 599-601.
9. DEBARI, V. A., N. A. NOVAK & A. BENNUN. 1984. Cyclic nucleotide metabolism in the human erythrocyte. Clin. Physiol. Biochem. **2:** 227-238.
10. NICOTRA, J., A. J. ORSINI & V. A. DEBARI. 1985. Chemiluminescence response of the human polymorphonuclear neutrophil to lipopolysaccharides. Cell Biophys. **7:** 283-291.
11. DEBARI, V. A., M. A. FINGERHUT, L. B. KEIL & M. A. NEEDLE. 1980. Progressive peripheral phagocyte deficiency in chronic hemodialysis patients. *In* Internal Medicine. L. Condorelli, U. Teodori, A. Beretta-Anguissola & M. Sangiorgi, Eds.: 773-777. Excerpta Medica. Amsterdam.

Effect of Cigarette Smoking on the Vitamin C Status of Pregnant Women and Their Offspring

E. P. NORKUS,[a] H. HSU,[b] AND M. R. CEHELSKY[b]

[a]*Department of Biomedical Research*
[b]*Department of Obstetrics and Gynecology*
Our Lady of Mercy Medical Center
Bronx, New York 10466

INTRODUCTION AND METHODS

Cigarette smokers have significantly lower serum/plasma vitamin C (ascorbic acid) levels than nonsmokers.[1-3] Among these individuals, impaired vitamin C status persists regardless of the socioeconomic group that is studied.[4-5] Most often, decreased food intake with an associated decreased vitamin C intake results in a decreased vitamin C status. However, among cigarette smokers, experimental studies suggest that the decrease in vitamin C status is caused by a decreased ability to absorb ascorbic acid[6] and/or an increased total body turnover for the vitamin.[7] These reports suggest to us that pregnant women who smoke cigarettes regularly, and their offspring, may be at risk for low vitamin C status as a result of a mechanism other than decreased vitamin C intake.

The method of Bessey *et al.*[8] was used to determine the ascorbic acid levels in maternal plasma, cord blood plasma, and placental tissue from women who either smoked cigarettes or were nonsmokers during pregnancy. All samples were obtained at delivery. All samples were deproteinized and acidified within two hours of collection and were either assayed on the same day or stored at −85°C for assay within a 2-week period.

RESULTS AND DISCUSSION

Twenty-six women (14 smokers and 12 nonsmokers) and their offspring were examined. TABLE 1 demonstrates that both smoker and nonsmoker groups had similar daily vitamin C intakes. Thus, the significantly lower vitamin C levels ($p < 0.01$) found in maternal plasma, cord blood plasma, and placental tissue in the smoker group (TABLE 2) are not the result of a decreased daily intake of the vitamin. These findings support the conclusions of previous studies[6,7] that suggest cigarette smoking decreases vitamin C status by altering the absorption or metabolism of this nutrient.

TABLE 1. Comparison of Age and Daily Ascorbic Acid Intake between Cigarette Smokers and Nonsmokers during Pregnancy

Group	n	Age Years	Ascorbic Acid Intake (mg per Day)
Smokers[a]	14	29.8 ± 7.0	76 ± 25 (30-120 mg)
Nonsmokers	12	28.5 ± 6.4	87 ± 26 (30-120 mg)

Means ± SD are given and the ranges of ascorbic acid intake are in parentheses; n equals the number of subjects in each group.

[a] Women who smoke more than five cigarettes per day.

TABLE 2. Comparison of Maternal Plasma, Cord Blood Plasma, and Placental Tissue Ascorbic Acid Levels between Cigarette Smokers and Nonsmokers

		Ascorbic Acid Concentration (mg%)		
Group	n	Maternal Plasma	Cord Plasma	Placental Tissue
Smokers[a]	14	0.40 ± 0.32[b]	0.61 ± 0.56[b]	10.1 ± 4.8[b]
Nonsmokers	12	0.89 ± 0.27	1.68 ± 0.58	20.9 ± 4.9

The means ± SD are given; n equals the number of subjects in each group.

[a] Women who smoke more than 5 cigarettes per day.

[b] Mean ± SD values differ from the mean ± SD values from nonsmokers, $p < 0.01$.

In conclusion, our data suggest that moderate-to-heavy cigarette use (five or more cigarettes per day) during pregnancy places both mother and newborn at risk for low vitamin C status and its consequences.

REFERENCES

1. McLean, H. E., P. M. Dodds, M. N. Abernathy, A. W. Stewart & D. W. Blaven. 1976. N. Z. Med. J. **83:** 226-229.
2. Pelletier, O. 1975. Ann. N.Y. Acad. Sci. **258:** 156-168.
3. Chow, C. K., R. R. Thacker, C. Changchit, R. B. Bridges, S. R. Rehm, J. Humble & J. Turbek. 1986. J. Am. Coll. Nutr. **5:** 305-312.
4. Hoefel, O. S. 1977. Int. J. Vitam. Nutr. Res. (Suppl. 16): 127-137.
5. Schorah, C. J. 1981. Vitamin Status in Population Groups. *In* Vitamin C (Ascorbic Acid). N.J. Counsell & D. H. Hornig, Eds.: 23-48. Applied Science Publ. Englewood, N.J.
6. Pelletier, O. 1977. Int. J. Vitam. Nutr. Res. (Suppl. 16): 147-170.
7. Kallner, A., D. Hartmann & D. Hornig. 1981. Am. J. Clin. Nutr. **34:** 1347-1355.
8. Bessey, O. A., O. H. Lowry & M. J. Brock. 1947. J. Biol. Chem. **168:** 197-203.

Activation of Serum Complement Generates Inhibitor of Ascorbate Transport

HARISH PADH[a] AND JOSEPH J. ALEO

Department of Pathology
Temple University Health Sciences Center
Philadelphia, Pennsylvania 19140

Ascorbic acid (vitamin C) is transported into 3T6 fibroblasts, which synthesize and secrete collagen, by an energy-dependent saturable active process with a K_m of 112 μM and V_{max} of 158 pmol/min/mg protein. The transport is dependent on the extracellular Na^+ concentration, which reduces K_m.[1] While studying the ascorbate transport we observed that serum contained a heat-labile factor which, when interacted with bacterial endotoxin, inhibited ascorbate transport.[2]

There is a possibility that the endotoxin activated the complement system in the unheated serum and the activation of complement-generated factor that was observed to be inhibitory for ascorbic acid transport. The basis for this suggestion is the endotoxin's ability to activate complement pathways and the similar heat lability of complement and the inhibitory factor (56°C for 30 minutes). Therefore, we have examined if it is the activation of serum complement that generated the "factor" inhibitory for ascorbate transport. This was done by activation of complement by other activators like zymosan and immunocomplexes (*e.g.,* albumin and antibodies to albumin). This was further investigated by using serum devoid of component 3 of complement (C3). In addition, the effect of inhibitory factor on the kinetic parameters was also examined.

Since our initial studies were done with bovine serum, we examined whether sera from different species also contained the inhibitory factor for ascorbate transport. As shown in TABLE 1, sera from different species showed endotoxin-dependent inhibition of ascorbate transport. Guinea pig serum exhibited the maximum inhibition and was therefore used for further studies. The next thing we examined was to see whether activation of serum complement by immunocomplexes leads to inhibition of ascorbate transport. As shown in FIGURE 1, immunocomplexes in the presence of guinea pig serum inhibited ascorbate transport by 3T6 fibroblasts. Immunocomplexes in the absence of serum or in the presence of heat-inactivated serum were ineffective. Then we examined the effect of endotoxin in guinea pig serum devoid of C3 of complement. As shown in FIGURE 2, guinea pig serum devoid of C3 did not show any inhibition of ascorbate transport by endotoxin, unless reconstituted with purified C3. These results clearly demonstrate that activation of complement by endotoxin and immunocomplexes generates a factor that inhibits ascorbate transport. It is not known at which stage of complement activation the inhibitor is generated. However, from the

[a] Present address: Department of Biochemistry, The University of Chicago, Chicago, Ill. 60637.

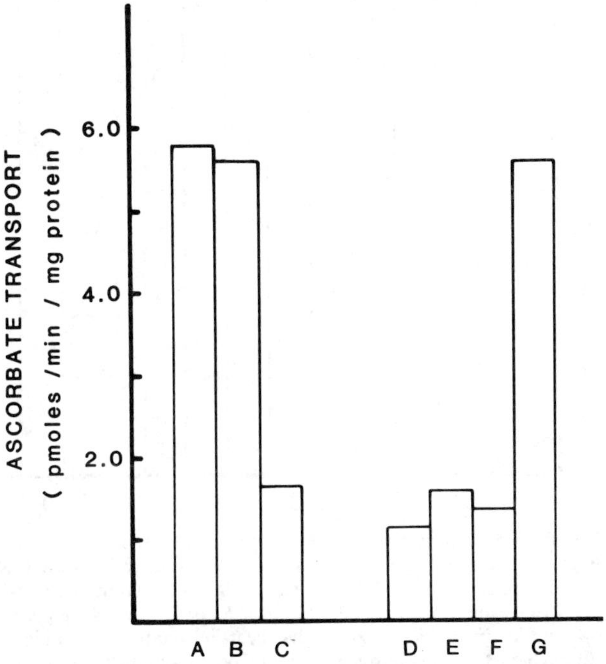

FIGURE 1. Effect of immunocomplexes on the ascorbate transport. 3T6 fibroblast cells were obtained from American Type Culture Collection (ATCC #CCL 96, Rockville, Md.) and grown in Dulbecco's minimal essential medium as described earlier.[2] When cells reached near confluency, the ascorbate transport was carried out in a total volume of 2.5 ml of balanced salt solution (BSS; 150 mM NaCl, 4.2 mM KCl, 1.0 mM NaH_2PO_4, 11.2 mM D-glucose, 0.7 mM $MgCl_2$, 2.0 mM $CaCl_2$, and 10 mM Hepes, pH 7.2) containing 1 mM of thiourea to prevent oxidation of ascorbic acid. Cells were incubated with respective additions as indicated below for 15 minutes before starting ascorbate transport by addition of L-[1-^{14}C]ascorbic acid. The concentration of radiolabeled ascorbate was 2.75 μM. Other details of the transport procedure have been published earlier.[2] All antigens and antisera were obtained from Cooper Biomedicals, Malvern, Pa. Guinea pig serum was purchased from Miles Laboratories. Before making immunocomplexes, antisera were heat-inactivated at 56°C for 30 minutes. Usually 10 μl aliquots of antigen solution (10 mg/ml) were added to 2 ml of ice-cold, heat-inactivated respective antiserum until it turned cloudy. It took between 50 to 100 μl of antigen solution to make visible immunocomplexes in 2 ml of antiserum. After keeping it on ice for 15 minutes, it was added to flasks as shown. When indicated, guinea pig serum was added at 20% concentration and endotoxin at 200 μg/ml. *A,* Balanced salt solution (BSS); *B,* BSS + guinea pig serum; *C,* BSS + guinea pig serum + endotoxin; *D,* BSS + guinea pig serum + immunocomplex-1 (human albumin + goat antiserum against human albumin), 250 μl; *E,* BSS + guinea pig serum + immunocomplex-2 (human IgG + goat antiserum against human IgG), 250 μl; *F,* BSS + guinea pig serum + immunocomplex-3 (guinea pig IgG + rabbit antiserum against guinea pig IgG), 250 μl; *G,* BSS + immunocomplex-1, -2 or -3, 250 μl.

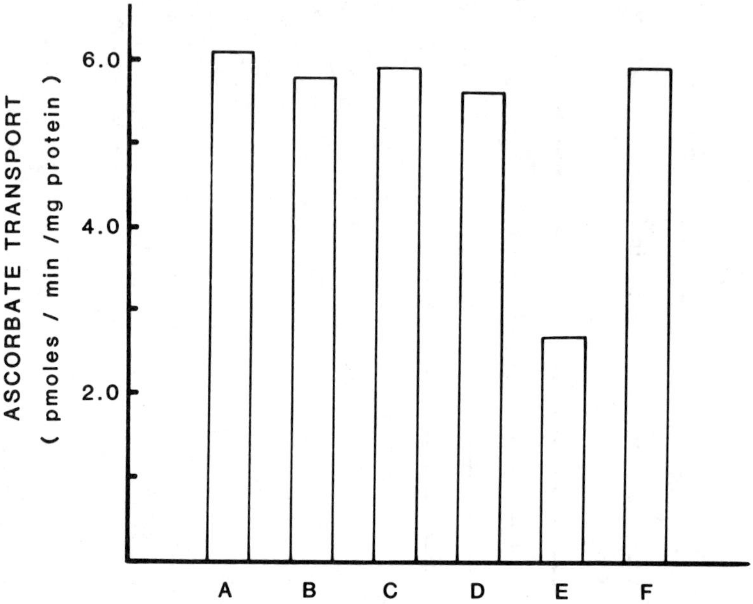

FIGURE 2. Effect of C3-deficient serum on the endotoxin-induced inhibition of ascorbate transport. The details of the transport procedure were same as given in FIGURE 1. *A,* BSS; *B,* BSS + guinea pig serum (minus C3); *C,* BSS + guinea pig serum (minus C3) + endotoxin; *D,* BSS + guinea pig serum (minus C3) + C3 (0.2 mg); *E,* BSS + guinea pig serum (minus C3) + C3 (0.2 mg) + endotoxin; *F,* like *E* except that heat-inactivated guinea pig serum was used. Guinea pig serum devoid of C3 was purchased from Cooper Biomedicals, Malvern, Pa. Guinea pig C3 was purchased from Diamedix Corporation, Miami, Fla.

TABLE 1. Inhibition of Ascorbate Transport by Endotoxin: Effect of Sera from Different Species

	Ascorbate Transport (Percentage of Control)
Bovine serum (GIBCO)	100
Bovine serum + endotoxin	64
Guinea pig serum (Miles)	100
Guinea pig serum + endotoxin	37
Rabbit serum (Cappel)	100
Rabbit serum + endotoxin	69
Mouse serum (Cappel)	100
Mouse serum + endotoxin	78

When near confluency cells were incubated for 4 hours in a fresh medium containing 20% of the indicated serum ± endotoxin (200 μg/ml). The ascorbic acid transport was then studied by addition of radioactive ascorbate (2.75 μM) and incubated for 1 hour. Other experimental details have been published earlier.[2] The ascorbate transport in the controls was 6.0 ± 0.25 pmol/min/mg protein.

data in FIGURE 2, it is suggested that the inhibitor is produced at or beyond the C3 step of complement activation. The details of the mechanism of inhibition are not known. What is known is that the inhibitor increases the K_m without affecting the V_{max} of the process (data not presented), indicating that the inhibitor decreases the affinity of transporter for ascorbate without affecting the translocation process. Since the characteristics of ascorbate transport in 3T6 cells are remarkably similar to other cell types studied,[3–5] it is possible that the inhibitor generated during activation of serum complement also inhibits ascorbate transport in other tissues. Finally, the cells during the study remained firmly attached to the flasks and alive, as judged by Trypan Blue Exclusion.

There are several implications of these findings. First, it demonstrates the presence of serum complement factor, which inhibits ascorbic acid transport when activated. Secondly, it is fascinating to realize that activation of complement leads to inhibition of ascorbate transport, which could lower tissue supply and level of vitamin C. It opens up the question of adequacy of vitamin C supply to tissues under the conditions that activate serum complement, such as infections and autoimmune diseases. Among all nutrients, the determination of proper nutritional requirement for vitamin C has sparked the most intense debate. While the officially recommended daily allowance for adults in the United States is 60 mg, up to several grams quantity has been recommended by some experts in the professional as well as public press. Our results suggest that what may be considered adequate intake of vitamin C in a healthy state may not be enough during infections because of the disrupted supply of the vitamin to tissues. This could be the case even when vitamin C intake and its blood level are normal.

REFERENCES

1. PADH, H. & J. J. ALEO. 1987. Biochim. Biophys. Acta In press.
2. ALEO, J. J. & H. PADH. 1985. Proc. Soc. Exp. Biol. Med. **179:** 128-131.
3. STEVENSON, N. R. 1974. Gastroenterology **67:** 952-956.
4. CHATTERJEE, I. B. 1978. World Rev. Nutr. Diet **30:** 69-87.
5. DILIBERTO, E. J., JR., G. C. HECKMAN & A. J. DANIELS. 1983. J. Biol. Chem. **258:** 12886-12894.

Ascorbic Acid Protection against Free Radicals

RICHARD C. ROSE

The Pennsylvania State University
College of Medicine
Hershey, Pennsylvania 17033

The highly reactive hydroxyl free radical (OH·) is apparently generated in all aerobic cells. The possibility is currently considered that ascorbate scavenges OH· (FIG. 1), resulting in formation of the ascorbate free radical (AFR). Pairs of AFR spontaneously disproportionate to form ascorbate and dehydroascorbic acid (DHA), the latter being a toxic molecule that is normally maintained at a low concentration in plasma. Dehydroascorbic acid spontaneously degrades to diketogulonic acid, which is devoid of antiscorbutic or other metabolic activity. A direct and energetically efficient pathway for reforming the nontoxic and useful ascorbate would be through direct reduction of AFR by one-electron transfer. The present study is an evaluation of whether mammalian tissues have the capacity to use ascorbate to protect against OH·.

We evaluated first whether free radicals that result from ionizing radiation are scavenged by ascorbate. An analysis was made with [^{14}C]ascorbate (10 μM) dissolved in physiologic buffer (pH 7.2) at 23°C. The buffer contained 1 mM thiourea to slow spontaneous degradation of [^{14}C]ascorbate. Samples were exposed to cobalt-60 irradiation (1500 rad/min) to generate OH· by radiolysis of water (FIG. 2, closed squares); control samples were not irradiated (closed circles). Aliquots taken at time 0 and after 2 and 5 minutes of radiation were assayed by HPLC for identification of the [^{14}C]label. Any OH· generated by ionizing radiation that are scavenged by [^{14}C]ascorbic acid result in formation of [^{14}C]AFR. In the absence of an enzyme system to reduce the free radical, radiation is seen to promote net formation of [^{14}C]DHA from [^{14}C]ascorbate (compare closed squares with closed circles in FIG. 2).

It was then evaluated whether a tissue enzyme exists that reduces AFR. An evaluation was made on rat or human colonic mucosa. An aliquot of tissue homogenate in buffer was centrifuged at 5,000 $\times$g and the supernatant (termed tissue extract) was exposed to radiation. Tissue extract slowed the rate of ascorbate degradation (FIG. 2, triangles). Tissue extracts heated 10 minutes to 100°C (open squares) or precipitated with PCA (open circles), were no longer active in preventing ascorbate degradation.

Samples of tissue extract were dialyzed 16 hours at 4°C in 1,000 MW tubing against 30 volumes of buffer. Aliquots of dialysate were then exposed to radiation in the presence of [^{14}C]ascorbate. Dialysate alone maintained little ascorbate in the reduced state. However, dialysate with GSH and cysteine was nearly as effective as nondialyzed samples in keeping ascorbic acid reduced. NADPH, however, was far less effective as a cofactor. Reducing equivalents in the absence of dialysate were ineffective.

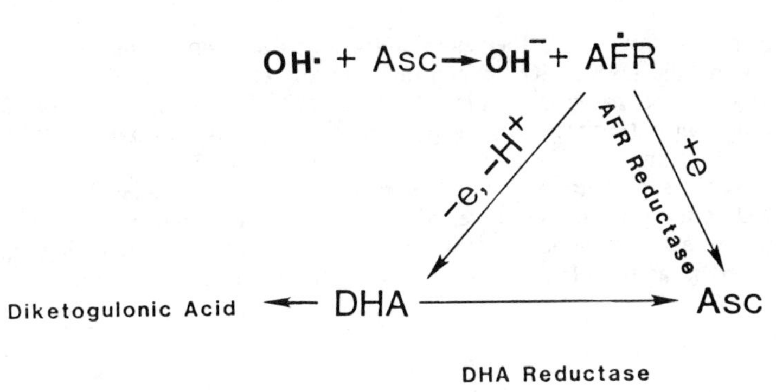

FIGURE 1. A proposed scheme by which ascorbate metabolism participates in detoxification of OH·.

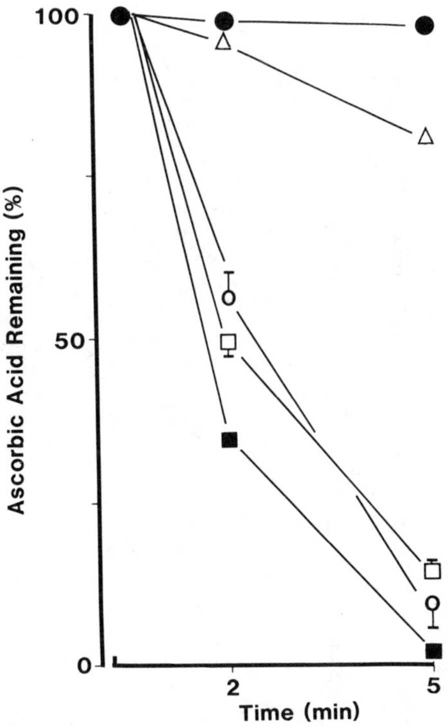

FIGURE 2. Stability of 10 μM [^{14}C]ascorbate in buffer at 23°C in the absence of radiation (●) or in buffer with radiation (■). Samples of human colonic extract under radiation protected ascorbate (control, △) unless they were previously boiled (□) or acid precipitated (○). All samples had GSH (600 μM) and cysteine (100 μM).

The tissue factor was further characterized as fractionating between 55-70% saturation with solid $(NH_4)_2SO_4$. When this fraction was redissolved in buffer, desalted by dialysis, and centrifuged 45 minutes at 100,000 × g, the apparent purification of the soluble factor (activity per mg protein, Lowry) was 3.5-fold (± 0.7) in colon from three rats and 2.5-fold in a single study on human colon.

Because the soluble tissue factor that reduces AFR has a molecular weight greater than 1000, is temperature sensitive, pH sensitive, fractionated by ammonium sulfate precipitation, and functionally dependent on the availability of reducing equivalents, it is referred to as soluble ascorbate free radical-reductase (sAFR-reductase).

Ascorbate Inactivates Tyrosine Hydroxylase from Rat Brain under Cyclic-AMP-Dependent Protein Phosphorylation Conditions[a]

ROBERT ROSKOSKI, JR. AND HARVEY WILGUS

Department of Biochemistry and Molecular Biology
LSU Medical Center
New Orleans, Louisiana 70119

Tyrosine hydroxylase activity is reversibly modulated by the actions of several protein kinases and phosphoprotein phosphatase systems. Following gel filtration to remove low molecular weight components, incubation of rat corpus striatal extracts with MgATP and the catalytic subunit of cyclic-AMP-dependent protein kinase produces a 5- to 10-fold activation of tyrosine hydroxylase. Ascorbate produces a dose-dependent inactivation of tyrosine hydroxylase when included in the phosphorylation reaction. The effective concentration that inhibits 50% (EC_{50}) for this effect is 5.9×10^{-6} M. This is much lower than the concentrations of ascorbate in the rat central nervous system of about 3 mM. The effect of ascorbate occurs only under phosphorylation conditions (MgATP plus catalytic subunit of cyclic-AMP-dependent protein kinase). In the absence of phosphorylation conditions, even 0.1 M ascorbate is without an inhibitory effect. Similar results were obtained with enzyme isolated from a transplantable rat pheochromocytoma. Ascorbate inactivation occurred at pH 6.0 or 7.2.

Because ascorbate chelates metals and tyrosine hydroxylase is most likely an iron-containing enzyme, we examined the effects of other metal-chelating agents. 1,10-Phenanthroline exhibited an EC_{50} of 1.2×10^{-5} M under phosphorylating conditions and 9.2×10^{-6} M under nonphosphorylating conditions. Inhibition under nonphosphorylating conditions clearly distinguished its effect from that of ascorbate. EDTA gave similar results by inhibiting under phosphorylating ($EC_{50} = 4.8 \times 10^{-3}$ M) and nonphosphorylating conditions ($EC_{50} = 2.1 \times 10^{-2}$ M).

In addition to ascorbate, other biological reducing agents also produced a decrease in tyrosine hydroxylase activity. Tetrahydrobiopterin, the natural reducing cosubstrate for tyrosine hydroxylase, also decreased tyrosine hydroxylase activity under phosphorylating ($EC_{50} = 6.8 \times 10^{-7}$ M) and nonphosphorylating ($EC_{50} = 4.2 \times 10^{-4}$ M) conditions. Glutathione also decreased tyrosine hydroxylase activity ($EC_{50} = 7.7 \times 10^{-4}$ M) under phosphorylating conditions.

Inactivation by ascorbate, tetrahydrobiopterin, and glutathione was apparently irreversible in nature. Activity was not regained by gel filtration, dialysis, or incubation at 30° or 0° for short or long periods of time. Tyrosine hydroxylase purified from a transplantable rat pheochromocytoma behaved similarly. With the latter preparation,

[a]This work was supported by U.S. Public Health Service grant no. NS-15994.

inactivation was not associated with proteolytic degradation as determined by sodium dodecylsulfate polyacrylamide gel electrophoresis.

The possibility exists that ascorbate or tetrahydrobiopterin reacts with oxygen to yield a reactive species that is the active agent in destroying tyrosine hydroxylase activity. High activities of catalase, peroxidase, or superoxide dismutase, however, failed to protect tyrosine hydroxylase against inactivation. The molecular mechanism of inactivation remains obscure; whether this effect has a counterpart in the physiological situation requires additional study.

Physiological Concentrations of Ascorbic Acid and Myeloperoxidase

ELLEN E. SCHAUS AND STANLEY T. OMAYE

Biochemistry Research Unit
Western Human Nutrition Research Center
U.S. Department of Agriculture, ARS
Presidio of San Francisco, California 94129

Ascorbic acid (AA) has been reported to stimulate many aspects of phagocytic cell function, but the mechanism by which this occurs is unknown. It has been proposed[1,2] that AA directly inhibits the activity of myeloperoxidase (MPO), an enzyme found in human neutrophils (PMN). Neutrophils release MPO into the extracellular environment,[3] where its products have been shown to inactivate many soluble inflammatory mediators and thus modulate immune responses.[4] Any effect AA has on the MPO reaction could therefore potentially affect many aspects of immune function and could explain at least in part the mechanism by which AA stimulates phagocytes.

In vivo, MPO catalyzes the production of $HOCl^-$ from H_2O_2 and Cl^-. Two *in vitro* techniques have commonly been used when measuring how AA affects MPO; the first uses radiolabeled I^- to replace Cl^- and measures the nonspecific iodination of protein by the reaction product. The second measures MPO activity by a spectrophotometric procedure where the reaction products oxidize a dye to form a colored compound. Since AA can also be oxidized, any measurements made when AA is present in the system must consider this factor, especially since this procedure is currently being used to study AA-MPO relationships.[5] The purpose of this study was to evaluate how physiological concentrations of AA affect MPO measurements using a spectrophotometric procedure[5] in (1) MPO extracted from 9.9×10^6 PMN, and (2) MPO released into the extracellular environment by the same amount of PMN stimulated by *N*-formyl-L-methionyl-L-leucyl-L-phenylalanine (FMLP). In addition, to evaluate whether AA competes with the dye for oxidation in the *in vitro* system, AA was allowed to oxidize and MPO measurement taken. Also, AA and isoascorbate (IAA) were compared to evaluate whether AA inhibits MPO by a coenzyme role.

Human neutrophils were isolated and MPO measured as described.[5] The onset of the reaction and MPO activity were inversely related (FIG. 1) at all physiological concentrations of AA tested, which suggests that the decreased activity as AA increases is due to competition with the dye. TABLE 1 shows the values obtained. At deficient (0.9 μg/ml) and marginally deficient (2 μg/ml) AA levels, no effect on the MPO reaction was seen, whereas normal physiological levels of 5-20 μg/ml affected both parameters in a dose-dependent manner. MPO released from stimulated PMN was about 10% of the total extracted and was completely inhibited by all levels of AA. Ascorbic acid allowed to oxidize in pH 6.0 buffer with time reversed the delay in the onset of the reaction but did not change the decreased MPO activity when compared to controls. We conclude that AA inhibits the onset of the MPO reaction by competing with the dye and this inhibition is an *in vitro* artifact. There was no difference in MPO

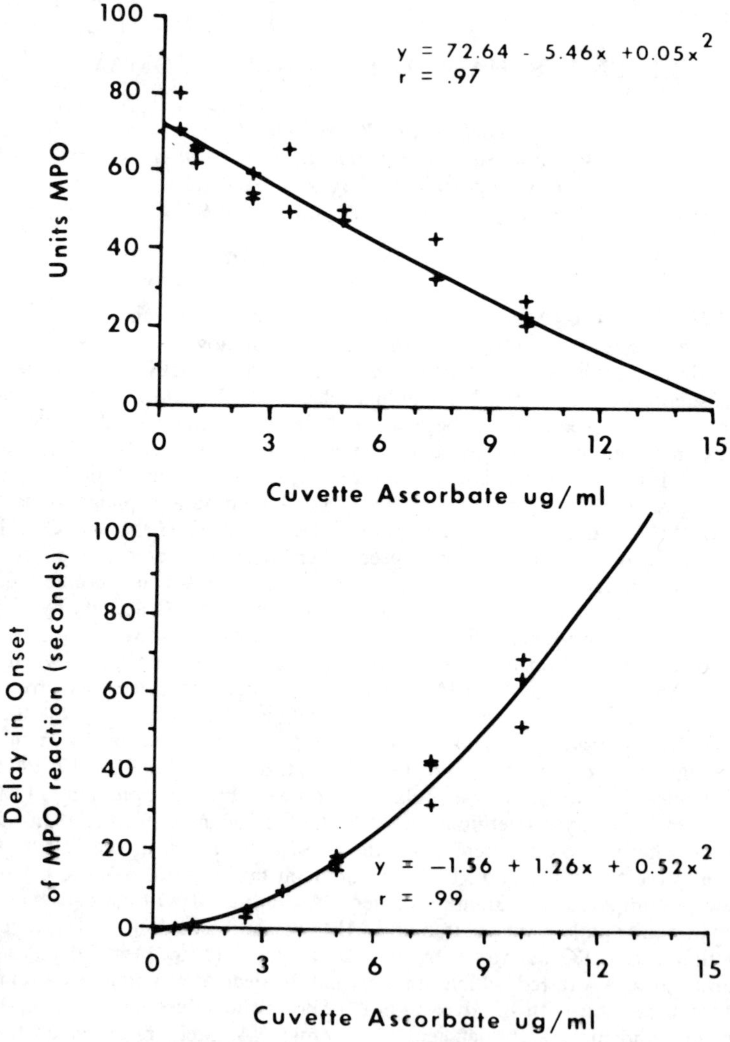

FIGURE 1. Relationship between delay in onset of MPO reaction and MPO activity when varying physiological concentrations of ascorbate.

TABLE 1. Effect of Physiological Concentrations of Ascorbate (AA) on Myeloperoxidase (MPO)[a]

Cuvette AA (μg/ml)	MPO Activity	Delay (s)
I. Effect on MPO Extracted from 9.9 × 10^6 PMN		
0	20 ± 1	0
0.9	20 ± 1	0
2	18 ± 1	0
5	16 ± 1	20 ± 3
10	8 ± 1	94 ± 18
20	nm[b]	no onset
II. Effect on MPO Released by 9.9 × 10^6 PMN Stimulated by FMLP		
0	1.8 ± 0.2	0
0.9-20	nm	nm

III. Effect of AA and Iso-AA (IAA) over Time on MPO Extracted from 9.9 × 10^7 PMN (Both AA and IAA at 10 μg/ml) Time	MPO Activity	Delay (s)
0 (−AA)	63 ± 1	0
0 (+AA)	44 ± 3	60 ± 0
0 (+IAA)	40 ± 2	50 ± 3
1 (+AA)	41 ± 1	50 ± 3
1 (+IAA)	37 ± 3	42 ± 3
2 (+AA)	44 ± 2	38 ± 5
3 (+AA)	43 ± 1	34 ± 3
4 (+AA)	40 ± 2	28 ± 0

[a] $n = 4$ for each measurement.
[b] nm = Not measurable.

when AA or IAA were used, which suggests the vitamin does not inhibit MPO via a coenzyme role. The lowered MPO activity is most likely due to a decrease in substrate (H_2O_2) availability.

REFERENCES

1. ANDERSON, R. & P. T. JONES. 1982. Clin. Exp. Immunol. **47:** 487-496.
2. THERON, A. & R. ANDERSON. 1985. Arch. Biochem. Biophys. Acta **111:** 73-79.
3. HENSON, P. M., M. H. GUISBERG & D. C.. MORRISON. 1978. Cell Surg. Rev. **5:** 407-508.
4. HENDERSON, W. R. & S. J. KLEBANOFF. 1983. J. Biol. Chem. **258:** 13522-13527.
5. WILLIAMS, R. N., C. A. PATERSON, K. E. EAKINS & P. BHATTACHERJEE. 1984. Exp. Eye Res. **39:** 261-265.

Dietary Ascorbic Acid Increases Collagen Production in Skin of Streptozotocin-induced Diabetic Rats by Normalizing Ribosomal Efficiency

MICHAEL SCHNEIR,[a] MICHAEL IMBERMAN,[a] LORNE GOLUB,[b] AND NANGAVARAM RAMAMURTHY[b]

[a]*University of Southern California School of Dentistry*
Los Angeles, California 90089

[b]*School of Dental Medicine*
State University of New York
Stony Brook, New York 11794

As previously shown, dietary ascorbic acid normalizes in skin nascent collagen hydroxylation of streptozotocin-induced diabetic rats concomitantly with a partial correction of intracellular procollagen degradation, the latter directly increasing the depressed fractional rate of collagen production attending the diabetic state.[1] As an extension of these observations, in the present study the effect of diabetes and of ascorbic acid "therapy" was quantified by use of two important measurements of collagen production: absolute rate and ribosomal efficiency. Similar measurements were made for noncollagen protein (NCP).

Male Sprague-Dawley rats ($\sim$400 g, n = 54) were equally allocated to the following experimental groups: nondiabetic controls, untreated diabetics, ascorbic-acid-treated diabetics (in drinking water, 10 mg/100 ml, daily dose $\sim$20 mg or 3.5 g/kg body weight). In addition, a group of nondiabetic controls (n = 6) were killed on day 0. On day 20, all rats (in the three experimental groups) were injected with a pool-expanding dose of [^{3}H]proline (400 mg, 1 mCi) and killed 2 h later. Additional information pertinent to animal treatment has been described elsewhere.[2] Excised entire skins were weighed and analyzed for the following: RNA (a measure of ribosome amount)[3]; TCA-insoluble hydroxyproline (a measure of collagen amount); TCA-insoluble proline (a measure of NCP proline, see legend to TABLE 1). An absolute production rate was calculated from the fractional rate of collagen and of NCP production (data not shown) multiplied by tissue collagen mass (hydroxyproline $\times$ 7.1) or by tissue NCP mass (tissue NCP proline $\times$ 25), respectively.[4] Ribosomal efficiency for collagen or for NCP production was calculated as described (legend to TABLE 2).

TABLE 1. Body Weight, Skin Weight, and Skin Constituents

Category	Body Weight (g)	Skin Weight (wt/wt, g)	Skin RNA (mg)	Collagen (mg Hyp)[a]	NCP (mg Pro)[b]
Nondiabetic Controls					
Day 0	359 ± 4[c]	47 ± 3	38 ± 5	853 ± 82[c]	495 ± 58
Day 20	489 ± 18	63 ± 4	63 ± 7	1186 ± 76	770 ± 73
Diabetics (Day 20)					
Untreated	334 ± 32[c]	28 ± 4	17 ± 4[c]	669 ± 89	279 ± 81[c]
Ascorbic acid treated	401 ± 11	38 ± 5	24 ± 4[c]	856 ± 110[c]	337 ± 127[c]

[a] TCA-insoluble hydroxyproline.
[b] [TCA-insoluble proline (mmoles) − 1.1 × TCS-insoluble Hyp (mmoles)] × 115.
[c] Values are mean ± SD, $n = 4$ (day 0 controls), $n = 18$ (all day 20 rats). Values with superscript *c* are not significantly different ($p > 0.05$) from each other, whereas all others are significantly different ($p < 0.05$) from each other.

Diabetes caused a decrease in skin wet weight and skin constituents (RNA, collagen, NCP) (TABLE 1). Ascorbic acid treatment increased tissue collagen mass (also wet weight and body weight), but not RNA or NCP mass. "Dissection" of the ascorbic acid effect into measurement of both ribosome content (TABLE 1) and ribosomal efficiency (TABLE 2) indicates a normalization of the latter, so that the decrease in absolute rate of collagen production in ascorbic-acid-treated diabetics is entirely caused by a decrease in ribosome content.

For NCP, ribosomal efficiency was unchanged by the diabetic state, a decrease in ribosome content solely accounting for the reduced NCP production (TABLE 2). Ascorbic acid treatment of diabetic rats increased ribosomal efficiency for NCP, but not to the same extent as for collagen.

In conclusion, dietary ascorbic acid normalized the depressed ribosomal efficiency for collagen production in skin of diabetic rats, contributing to a net increase of tissue collagen mass. Although ribosomal efficiency was also increased for skin NCP, no net increase in skin NCP mass was effected, indicating NCP degradation must have also been increased.

TABLE 2. Absolute Rate and Ribosomal Efficiency for Collagen and NCP Production

Category	Collagen		NCP	
	Absolute Rate[a]	Efficiency[b]	Absolute Rate[a]	Efficiency[b]
Nondiabetic controls	275 ± 48	4.4 ± 0.6[c]	1937 ± 394	31 ± 5[c]
Diabetics				
Untreated	22 ± 12	1.3 ± 0.6	523 ± 127	32 ± 4[c]
Ascorbic acid treated	103 ± 33	4.2 ± 1.2[c]	1104 ± 256	46 ± 8

[a] mg collagen (or mg NCP) per skin per day.
[b] mg collagen (or mg NCP) per mg RNA.
[c] All values are mean ± SD, $n = 18$. Values with superscript *c* are not significantly different ($p > 0.05$) from each other, whereas all others are significantly different from each other ($p < 0.05$).

REFERENCES

1. Schneir, M., L. Golub & N. Ramamurthy. 1985. Dietary ascorbic acid increases nascent collagen production in skins of diabetic rats by reversing procollagen underhydroxylation and concomitant intracellular degradation. Ann. N. Y. Acad. Sci. **460:** 500-502.
2. Schneir, M., N. Ramamurthy & L. Golub. 1985. Dietary ascorbic acid normalizes diabetes-induced underhydroxylation of nascent type I collagen molecules. Collagen Rel. Res. **5:** 415-422.
3. Brown, J. G., J. van Buren & D. J. Millward. 1983. The effect of triiodothyronine administration on protein synthesis in the diabetic rat. Biochem. J. **214:** 637-640.
4. Diegelmann, R. F. & B. Peterkofsky. 1972. Collagen biosynthesis during connective tissue development in chick embryo. Dev. Biol. **28:** 443-453.

Protective Effects of Ascorbic Acid on Murine Frostbite

CHARLES R. SPILLERT,[a] MAY B. HOLLINSHEAD,[b] AND ERIC J. LAZARO[a]

[a]*Department of Surgery*
[b]*Department of Anatomy*
University of Medicine and Dentistry of New Jersey
New Jersey Medical School
Newark, New Jersey 07103-2757

In a previous study[1] we demonstrated that ascorbic acid markedly reduces the severity of murine burns. Since frostbite is a form of low temperature thermal injury, this study was undertaken to determine whether ascorbic acid can also be protective in frostbite.

Swiss Webster mice (27 g) with initial tail lengths of 9.1 ± 2 cm were anesthetized with pentobarbital sodium (60 mg/kg) and their tails totally immersed in a freezing solution (−17°C) for 6 minutes. Thereafter, the tails were thawed at room temperature (24°C). The mice (n = 30) were assigned randomly to one of three equal groups A, B, and C. Thirty minutes after the termination of freezing either 25 mg ascorbic acid (group A), 3.3 mg ascorbic acid (group B), or an equal volume of saline (0.25 cc, group C) was administered intraperitoneally. Then supplemental ascorbic acid was dissolved in 5% dextrose and given to groups A and B for drinking water, whereas group C was given only 5% dextrose. Doses of ascorbic acid given to mice are equivalent to the doses of 4.25 g and 70 g for an average human (70 kg).

Twenty-four hours postfrostbite, the distance of vascular perfusion of the tail of each mouse was determined by administration of 10 mg patent violet intraperitoneally. Remaining tail lengths were measured at 12 days following freezing for each treatment regimen. By this time complete autoamputation of gangrenous tails had occurred.

Length of tail perfusion of dye 10 minutes after dye injection and 24 hours following freeze injury is shown in TABLE 1. Remaining tail lengths post autoamputation are found in TABLE 2.

The data obtained in this study clearly demonstrate the beneficial effect of ascorbic acid given 30 minutes after cold exposure. Since the remaining tail lengths were directly proportional to the degree of tail perfusion with dye patent violet, it is reasonable to assume that ascorbic acid accomplished this by protecting the integrity of the microcirculation. One or more of the following mechanisms may have contributed to this effect: (1) inhibition of microvascular thrombosis; (2) preservation of capillary integrity; (3) alteration of the immune system; and (4) antioxidant properties of ascorbate or its derivatives. Whether this phenomenon can be achieved in humans remains to be determined.

TABLE 1. Distance of Tail Perfused by Dye

Drug Regimen	Tail Perfusion of Dye (in cm ± SEM) (10 minutes Post Dye Injection 24-h after Freeze Injury)
Group A (25 mg ascorbic acid)	4.23 ± .26
Group B (3.3 mg ascorbic acid)	1.83 ± .12
Group C (0.25 cc saline)	1.21 ± .10
Significance (ANOVA)	
Group A vs. Group C: $p < .001$	
Group B vs. Group C: $p < .01$	
Group A vs. Group B: $p < .001$	

TABLE 2. Final Tail Length

Drug Regimen	Tail Length Post Autoamputation (cm ± SEM, 12 Days Postfrostbite)
Group A (25 mg ascorbic acid)	8.03 ± .73
Group B (3.3 mg ascorbic acid)	4.84 ± 1.05
Group C (0.25 cc saline)	1.70 ± .24
Significance (ANOVA)	
Group A vs. Group C: $p < 0.001$	
Group B vs. Group C: $p < 0.01$	
Group A vs. Group B: $p < 0.05$	

REFERENCES

1. HOLLINSHEAD, M. B., C. R. SPILLERT & E. J. LAZARO. 1985. The beneficial effects of ascorbic acid on murine burns. J. Burn Care Res. **6:** 50-54.

Protective Effect of Ascorbate on Radiation-Sensitive Thymidine Uptake by Lymphocytes

J. C. STANDEFER, DOROTHY VANDERJAGT, ROBERT E. ANDERSON, P. J. GARRY, AND LAURA POGUE

Department of Pathology
University of New Mexico School of Medicine
Albuquerque, New Mexico 87131

Reducing agents can protect certain cellular processes from injury by oxidants or free radicals: ascorbate will protect the PHA-induced proliferative response of T lymphocytes against oxidant injury by the chloride ion.[1] Since x-irradiation induces free radical formation[2] and can inhibit mitogen-stimulated lymphocyte proliferation,[3] an antioxidant such as ascorbate may prevent the effect of irradiation on lectin-stimulated thymidine uptake. We examined this hypothesis by isolating murine lymphocytes from spleen and irradiating them (50 to 200 rads) in the presence of ascorbate (5 to 50 mg/l). After 4 hours, the cells were incubated in the presence of lectin (PHA or LPS) for 72 hours and then pulsed with tritiated thymidine for 24 hours. The incorporation of tritiated thymidine in the nonirradiated, untreated cells was compared with that in cells at each radiation dose and each ascorbate concentration. Ascorbate concentrations in the media were determined by HPLC.

In agreement with other reports,[4] ascorbate enhances the response to PHA by nonirradiated cells in a dose-dependent fashion (FIG. 1). With irradiation, the response of lymphocytes to PHA and LPS stimulation decreases with increasing radiation dose. However, when sufficient ascorbate is added the radiation effect is ameliorated for PHA-stimulated cells (T cells), (FIG. 1), but not for LPS-stimulated cells (B cells, data not shown). At ascorbate concentrations of 25 mg/l or greater, the percent stimulation of thymidine uptake is higher at each radiation dose compared to the cells with no ascorbate. Indeed, in cells receiving 50 and 100 rads, thymidine uptake is enhanced compared to the nonirradiated controls. Ascorbate concentration in the extracellular media decreased approximately 40% during a 4-h postirradiation incubation period and ascorbate in the media from irradiated cells is 10-20% lower than nonirradiated controls (TABLE 1). Ascorbate may be entering the cells during this incubation, but we were unable to detect significant intracellular accumulation. Glutathione (250 μg/ml) had no protective effect against irradiation; however, D ascorbate (50 mg/l) stimulated nonirradiated cells and provided some protection against irradiation, but was not as effective as the same concentration of L-ascorbate.

This study demonstrates a dose-dependent protection of lectin-stimulated lymphocyte proliferation by L-ascorbate against the effects of low-dose irradiation. This protective effect is limited to a subpopulation of T lymphocytes, *i.e.,* cells responding

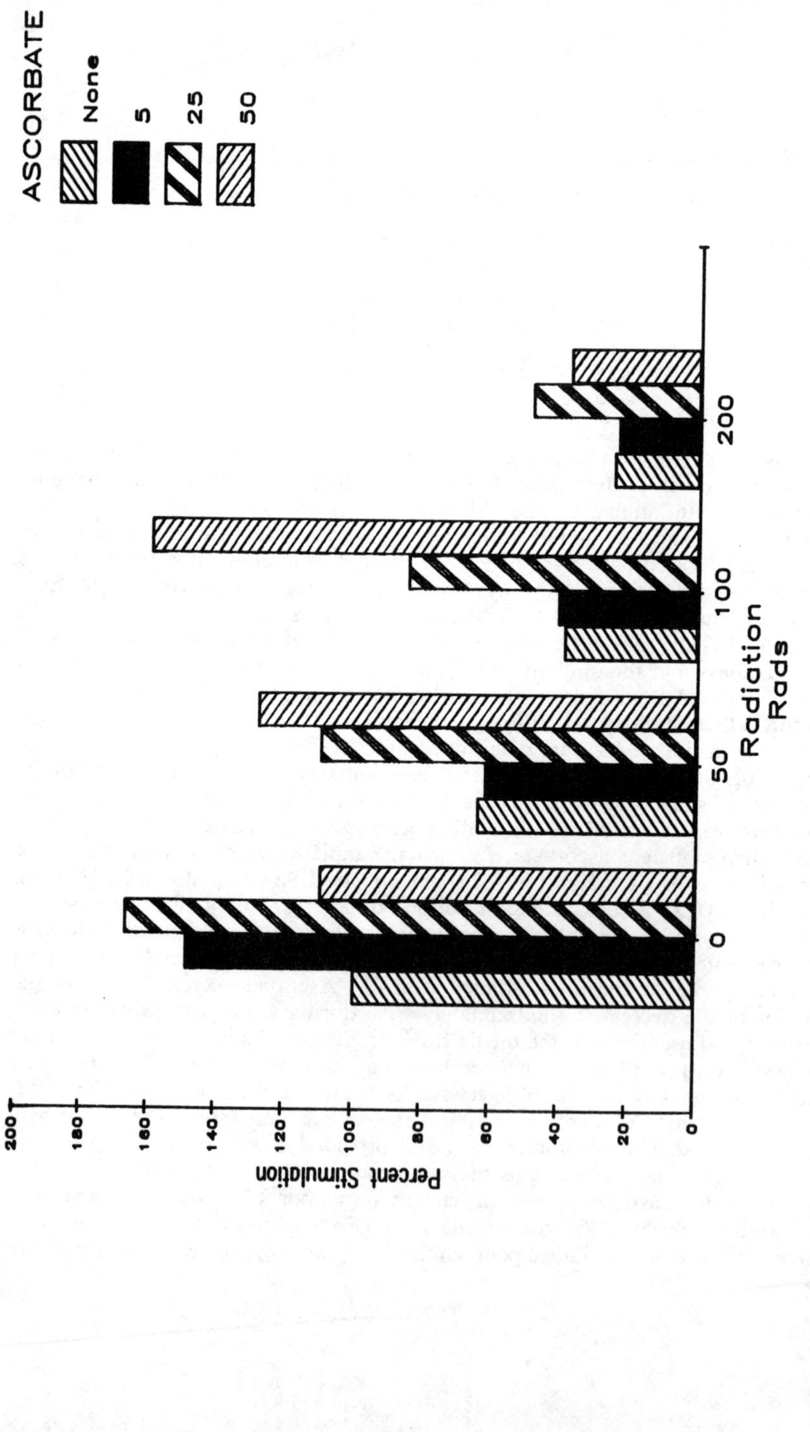

FIGURE 1. Effect of L-ascorbate on irradiated murine lymphocytes stimulated by PHA (78 mg/l). L-ascorbate (mg/l) is added prior to irradiation and lectin-stimulated thymidine uptake for each radiation dose is compared to nonirradiated cells.

TABLE 1. Ascorbate in Extracellular Media following Irradiation

Ascorbate (mg/l)	Radiation (rads)	Ascorbate (mg/l) Postirradiation			
		0 h	1 h	2 h	4 h
5	0	5	1.4	na[a]	1.2
	50	—	1.4	na	1.2
	100	—	1.4	na	1.2
	200	—	0.8	na	<0.5
25	0	23.7	20.5	19.8	16.5
	50	—	18.4	16.4	14.4
	100	—	18.9	18.1	14.4
	200	—	18.3	16.0	14.0
50	0	50.1	45.5	41.3	37.5
	50	—	41.3	38.4	38.4
	100	—	44.4	38.7	36.4
	200	—	43.3	35.8	34.1

[a] Not analyzed.

to PHA but not LPS. These results are consistent with the hypothesis that low-dose irradiation inhibits lymphocyte proliferation by initiating oxidative processes in the plasma membrane that can be obviated by specific events requiring L-ascorbate.

REFERENCES

1. Panush, R. S., J. C. Delafuente, P. Katz & J. Johnson. 1981. Int. J. Vitam. Nutr. Res. (Suppl. 20): 35-37.
2. Gee, C. A., K. J. Kittridge & R. L. Wilson. 1985. Br. J. Radiology **58:** 251-256.
3. Anderson, R. E. & J. C. Standefer. 1981. *In* The Immune System, Vol. 2. C. M. Steinberg & I. Lefkovits, Eds.: 258-264. S. Karger. Basel.
4. Munster, A. M., C. B. Loadholdt, A. G. Leary & M. A. Barnes. 1977. Surgery **81:** 692-695.

Interaction between Ascorbate and the 5-Lipoxygenase Pathway of Arachidonic Acid in Human PMNL

D. STEINHILBER,[a] U. MOSER,[c] H.-J. ROTH,[a] AND K. H. SCHMIDT[b]

[a]*Department of Pharmaceutical Chemistry*
and
[b]*Department of Surgery*
University of Tübingen
Tübingen, Federal Republic of Germany

[c]*Department of Vitamin and Nutrition Research*
F. Hoffmann-La Roche
Basel, Switzerland

Antioxidants such as α-tocopherol, flavonoids, or dihydronorguaretic acid (NDGA) are well-known inhibitors of 5-lipoxygenase in human polymorphonuclear neutrophils (PMNL).[1] L-Ascorbic acid as an antioxidant and immunoreactive compound was investigated with respect to its potential interactions with 5-lipoxygenase activity. As a parameter for 5-lipoxygenase activity, leukotriene B4 (LTB4) production was measured in resting and preincubated PMNL after stimulation by the Ca^{2+} ionophore A 23187. In resting PMNL, physiological extracellular concentrations of ascorbate did not alter Ca^{2+}-ionophore-stimulated LTB4 production (FIG. 1). However, extracellular ascorbate concentrations of 50 mg/dl (2.8 mM), corresponding to 180 mg/dl intracellular concentration, caused a 50% inhibition of LTB4 release (FIG. 1).

In PMNL preincubated with opsonized zymosan a rapid decline of the ionophore-induced LTB4 release was measured, indicating a time-dependent deactivation of 5-lipoxygenase by this phagocytic stimulus. The ionophore-induced LTB4 production decreased to 33% and 16% of control 10 and 20 min after incubation of PMNL with opsonized zymosan respectively. Opsonized zymosan incubation alone caused negligible LTB4 release by PMNL.

In PMNL preincubated with opsonized zymosan, L-ascorbic acid in a concentration range from 5 to 20 mg/dl in the extracellular medium, corresponding to 31-74 mg/dl intracellularly, significantly enhanced LTB4 production as compared to the untreated controls. (TABLE 1).

The unphysiologically high extracellular concentrations of L-ascorbic acid were used in order to minimize incubation time and thus to maintain the viability of the cells *in vitro.* Under *in vivo* conditions the corresponding intracellular levels of L-ascorbic acid are only reached by the active and stereospecific transport of L-ascorbic acid through the plasma membrane of PMNL even at very low extracellular concentrations.

TABLE 1. Effect of L-Ascorbic Acid on Ca^{2+}-Ionophore-induced LTB4 Production by PMNL after Preincubation (10 min) with Opsonized Zymosan[a]

Concentration of Ascorbic Acid	LTB4 Production by 10^7 PMNL
0 mg/dl	22.7 ± 2 ng
2 mg/dl	19.9 ± 1 ng
5 mg/dl	35.3 ± 5 ng
10 mg/dl	44.3 ± 6 ng
20 mg/dl	63.6 ± 6 ng

[a] Mean values and standard deviations from three independent experiments.

In contrast to L-ascorbic acid, the isomers D-ascorbic acid and D-isoascorbic acid (erythorbic acid) are accumulated in PMNL only in negligible amounts. Significant uptake of the isomers was only observed at extracellular concentrations higher than 5 mg/dl.

This mechanism may also account for the fact that the isomers of L-ascorbic acid, D-ascorbic acid and D-isoascorbic acid, showed similar effects on the LTB4 production

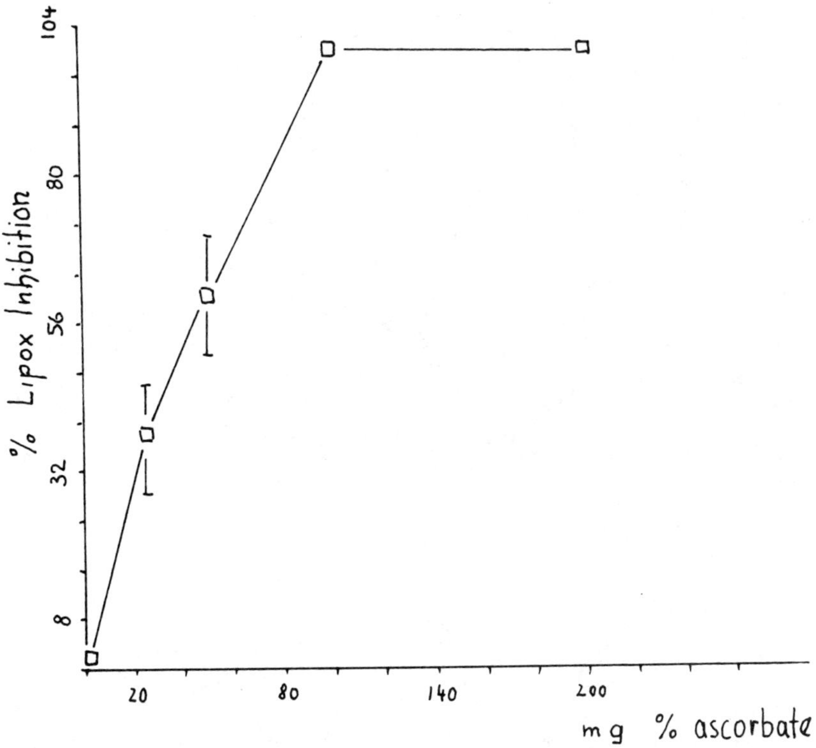

FIGURE 1. Inhibition of calcium-ionophore-induced LTB4 production by L-ascorbic acid.

under these unphysiological *in vitro* conditions, whereas they do not exhibit any immunomodulatory effect *in vivo*.

The inhibitory effect of L-ascorbic acid on the self-deactivation of 5-lipoxygenase in PMNL provides for the first time direct evidence of a molecular mechanism for the efficacy of vitamin C as an immunomodulator.[2]

REFERENCES

1. REDDANNA, P., M. K. RAO & C. C. REDDY. 1985. FEBS Lett. **193:** 39-43.
2. SCHMIDT, K. H. & U. MOSER. 1985. Int. J. Vitam. Nutr. Res. (Suppl. 27): 363-379.

Bioavailability of Synthetic Ascorbic Acid and a Citrus Extract

JOE A. VINSON AND PRATIMA BOSE

Chemistry Department
University of Scranton
Scranton, Pennsylvania 18510

INTRODUCTION

There is now considerable general interest in vitamin C supplementation due to recent popular books on the treatment of the common cold and cancer by ascorbic acid (AA). Also, certain populations such as smokers and the elderly are known to be deficient in AA[1,2] and would benefit from supplementation. Although natural and synthetic AA are chemically identical, citrus fruits contain bioflavonoids that might affect the bioavailability of AA.

A new formulation of ascorbate has recently become available, Renatured Vitamin C in Citrus Fruit Media (CF), from Grow Company, Hackensack, N.J. CF was compared with AA in animal and human studies to determine relative bioavailability.

RESULTS AND DISCUSSION

An *in vitro* accelerated aging study was conducted at 45°C in a dry, dark oven to determine the relative stability of CF vs. AA. Powdered AA decomposed 15% after 8 weeks and CF 1%. At pH 7 in solution, AA decomposed 99% and CF 65%. At pH 4, where AA is less stable, it decomposed 100% while CF only lost 82%. Thus, the CF is more stable than AA alone.

Adult male guinea pigs (2-5 per group) were fed an ascorbate-deficient diet for a period of 1 month. Then three doses of ascorbate alone or in CF were given, 8 mg, 16 mg, and 32 mg in 1 ml of Gatorade three times a week. At the end of 6 weeks, the animals were sacrificed and serum analyzed for AA. A comparison of the slopes of the dose-response lines showed that CF was three times more absorbed than AA.

A single dose of 50 mg AA alone or in CF was given orally to 10 fasting guinea pigs divided into two groups. Blood samples were taken periodically from 0 to 4 hours. The CF was 48% more absorbed than the AA by measurement of the area under the curves. The half-life of the AA elimination was 1.0 hours and the CF 1.6 hours. Thus, the CF was more absorbed than AA, took longer to reach its maximum concentration (2.0 vs. 1.5 hours, respectively) and lasted for a longer time in the blood.[3]

Most recently, a human study was done to confirm the animal results. Eight healthy nonsmoker individuals (five males and three females) with an average age of 22 years participated with informed consent. CF showed a 68% greater area under the curve than AA ($p < 0.005$) and took longer to reach its maximum concentration (4.1 vs. 2.9 hours).

CF has been shown in guinea pigs and humans to be more absorbed than AA and to remain in the body for a longer period of time. CF is thus the preferred form of ascorbate for supplementation.

REFERENCES

1. Pelletier, O. 1970. Am. J. Clin. Nutr. **23:** 520-524.
2. Burr, M. L., P. C. Elwood, D. J. Hole, R. J. Hurley & R. E. Hughes. 1974. Am. J. Clin. Nutr. **27:** 144-151.
3. Vinson, J. A. & P. Bose. 1983. Nutr. Rep. Intl. **27:** 875-880.

Ascorbic Acid Reduces and Diethyldithiocarbamate Potentiates Methamphetamine-induced Dopamine and Serotonin Depletions

GEORGE C. WAGNER AND MICHAEL J. DEVITO

Department of Psychology
and
Department of Toxicology
Rutgers University
New Brunswick, New Jersey 08903

Repeated administration of methamphetamine (MA) causes a depletion of central nervous system dopamine (DA) and serotonin (5HT) and nerve terminal degeneration.[1] Recently it has been demonstrated that MA caused the *in vivo* production of 6-hydroxydopamine[2] and 5,6-dihydroxytryptamine.[3] It has been speculated that the release of DA and 5HT combined with the inhibition of monoamine oxidase by MA leads to autoxidation of DA and 5HT to their respective neurotoxins. This theory has been evaluated indirectly by pretreating rats with high doses of ascorbic acid (AA, an antioxidant) and diethyldithiocarbamate (DDC, a superoxide dismutase inhibitor).

Methamphetamine (12.5 or 25.0 mg/kg) or saline was administered to male Sprague-Dawley rats (350-450 g) pretreated with AA (100 mg/kg), DDC (200 mg/kg), or saline. MA was delivered in four subcutaneous injections spaced 2 h apart. AA was administered i.p. 30 min before the first dose of MA and 1 h after each MA dose. DDC was delivered i.p. 2 h before the first dose of MA. Rats were allowed to survive for 2 weeks and were then sacrificed. The brains were rapidly removed and the striatal region dissected. Brain parts were stored in liquid nitrogen until assayed by HPLC with electrochemical detection.[4]

The repeated administration of MA (12.5 or 25.0 mg/kg) resulted in a significant depletion of striatal DA and 5HT compared to controls (FIGS. 1 and 2, respectively). There were no significant differences between the AA-saline, the DDC-saline, and the saline-saline DA and 5HT levels and therefore these groups were treated as one in the statistical analyses. Pretreatment with AA significantly reduced the extent of the DA and 5HT depletions caused by MA (12.5 or 25.0 mg/kg). However, it should be noted that the AA-MA treated rats did have significantly lower DA and 5HT levels than the controls. DDC significantly potentiated the DA and 5HT depletions produced by MA (12.5 mg/kg). The combination of MA (25 mg/kg) and DDC (200 mg/kg) proved to be lethal.

The neurotoxic effects following the repeated administration of MA are manifest as long-lasting depletions of DA and 5HT.[1] The recent finding that MA leads to the *in vivo* production of 6-hydroxydopamine and 5,6-dihydroxytryptamine provides insight into the mechanism of MA neurotoxicity. In the present study, it was found

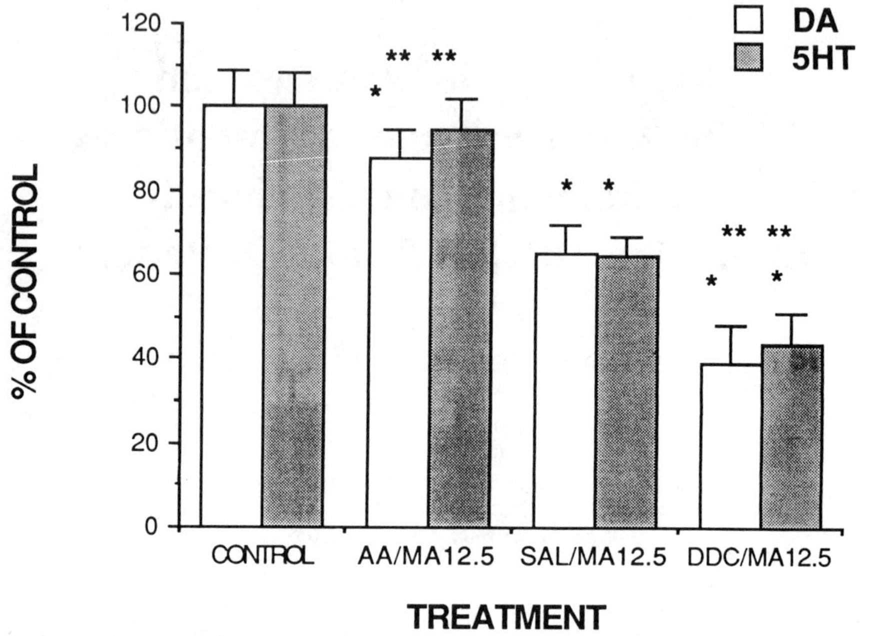

FIGURE 1. The effects of ascorbic acid pretreatment on methamphetamine-induced striatal dopamine and serotonin depletion expressed as percentage of control. Mean control levels were dopamine, 13.11 μg/g tissue; serotonin, 0.94 μg/g tissue. Bars represent 1 SEM; * = significant from controls, $p<0.05$; ** = significant from saline (SAL)/MA (12.5 mg/kg), $p<0.05$.

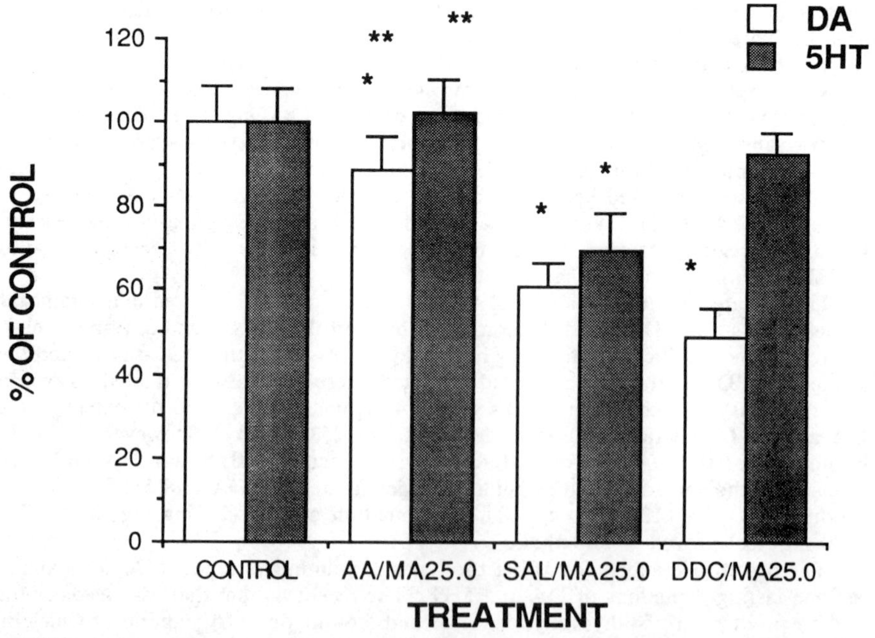

FIGURE 2. The effects of ascorbic acid pretreatment on methamphetamine-induced striatal dopamine and serotonin depletion expressed as percentage of control. Mean control levels were dopamine, 13.11 μg/g tissue; serotonin, 0.94 μg/g tissue. *Bars* represent 1 SEM; *significant from controls, $p < 0.05$; **significant from saline (SAL)/MA (25.0 mg/kg), $p < 0.05$.

that pretreatment with AA attenuated and DDC potentiated the neurotoxic actions of MA. The conclusions that AA reduced the neurotoxicity by acting as an antioxidant and DDC potentiated the neurotoxicity by inhibiting superoxide dismutase and impairing the ability of the cell to eliminate reactive intermediates such as superoxide radicals is tentative. It is possible that AA or DDC could have altered the distribution or the metabolism of MA, causing either an attenuation or a potentiation of the neurotoxicity of MA. The mechanism through which AA and DDC alter MA neurotoxicity requires further investigation.

REFERENCES

1. RICAURTE, G. A., R. W. GUILLERY, L. S. SEIDEN, C. R. SCHUSTER & R. Y. MOORE. 1980. Dopamine nerve terminal degeneration produced by high doses of methylamphetamine in the rat brain. Brain Res. **235:** 93-102.
2. SEIDEN, L. S. & C. VOSMER. 1984. Formation of 6-hydroxydopamine in caudate-nucleus of the rat brain after a large dose of methamphetamine. Pharmacol. Biochem. Behav. **21:** 29-31.
3. AXT, K. J., D. L. COMMINS & L. S. SEIDEN. 1984. Alpha-methyl-para-tyrosine prevents the methylamphetamine (MA)-induced formation of endogenous 5,6-dihydroxytryptamine in the rat hippocampus. Soc. Neurosci. Abstr. **11:** 350.2.
4. WAGNER, G. C., J. GARDENER, D. J. TSIGAS & D. B. MASTERS. 1984. Tolerance following the repeated administration of high doses of phencyclidine: No relation to central catecholamine depletion. Drug and Alcohol Dependence **13:** 224-234.

Supplementation of Vitamins C and E and Cellular Immune Function in Young and Aging Men[a]

RONALD ROSS WATSON,[b] J. BENEDICT,[b]
J. C. MAYBERRY,[b] M. J. HICKS,[c] AND
S. MORIGUCHI[b,d]

[b]*Department of Family and Community Medicine*
[c]*Department of Pathology*
University of Arizona
Tucson, Arizona 85724

Although most diets contain enough vitamin C and E to prevent deficiency, the optimum amount for enhanced longevity, immune responses, and tumor growth is poorly defined.[1] High vitamin E intake stimulated humoral immune responses and enhanced helper T cell activity and maturation of cellular immune functions in young mice.[1] In addition, vitamin C or exercise may enhance some immune functions as an additional modifier.[2]

MATERIALS AND METHODS

Two groups of subjects were used: males aged 40-55, exercised for 30 days and concomitantly supplemented with vitamins E and C, and 50 males between the ages of 21 to 30. The subjects were divided into trained and untrained groups, according to the nature and frequency of physical activity. They were randomly assigned under single blind conditions. L-ascorbic acid (1 g) and D-α-tocopherol (800 IU) were taken every day by the subjects in the young group; one half that amount was taken by the older group. Placebos were provided for the control groups.

Peripheral blood was collected in sodium heparinized vacutainers. Lymphocytes were separated by density gradient centrifugation procedures.[2] The T subset analysis with fluorescently labeled antisera on a cell sorter, lymphocyte mitogenesis, and NK cell assays were as previously described.[2] The extraction of vitamin E from plasma was based on a method described by Desai,[3] and subsequent determination of vitamin E concentration was based on methods reported by Buttriss and Diplock.[4] All data were analyzed using SPSS repeated measures MANOVA (multivariate analysis of variance), with trials as the within-subjects factor and groups as the between-subjects

[a]This work was supported by a grant from Hoffman-LaRoche, Inc.

[d]Current address: Department of Nutrition, Tokushima University, Tokushima, Japan 770.

factor. In order to analyze the pre- and post-changes within any single group, paired Tukey Post Hoc Tests were employed.

RESULTS AND DISCUSSION

The fit aging subjects exhibited a significant increase, whereas nonfit subjects given the vitamin did not exhibit an increase in numbers of T cells. The mitogenic activity of T lymphocytes after exposure to a polyclonal T lymphocyte mitogen was not significantly affected by consumption of vitamins. Stimulation units are defined as the cpm of tritiated thymidine incorporated into unstimulated lymphocytes. There was a significant effect due to consumption of the vitamins on stimulation units. Nonphysically fit subjects given vitamins exhibited an increase in their lymphocytic PHA-driven stimulation units, whereas placebo-fed subjects exhibited no significant increase in this parameter. One mechanism by which lymphocytes control the immune response

TABLE 1. T-Lymphocyte Subsets in Young Men Supplemented with Vitamins E and C[a,b]

	Supplemented	Week 0	Week 7	Week 14
		NATURAL KILLER CELLS (%)[c]		
Physically fit	−	78.0 ± 8.9	80.3 ± 5.0	80.0 ± 7.2
Physically fit	+	81.3 ± 8.9	82.5 ± 3.9	80.8 ± 3.2
Untrained	−	79.3 ± 3.1	82.5 ± 3.8	81.5 ± 2.3
Untrained	+	76.5 ± 4.2	81.3 ± 4.7	79.1 ± 4.6
		T HELPER CELLS (%)		
Physically fit	−	35.7 ± 5.5	39.7 ± 8.0	43.7 ± 11.1
Physically fit	+	41.5 ± 8.1	42.9 ± 5.1	42.5 ± 7.6
Untrained	−	38.5 ± 8.6	40.7 ± 4.1	49.7 ± 6.2
Untrained	+	38.8 ± 6.7	39.8 ± 5.6	41.6 ± 5.4
		T SUPPRESSOR-CYTOTOXIC CELLS (%)		
Physically fit	−	17.0 ± 3.0	20.7 ± 5.1	20.7 ± 6.0
Physically fit	+	20.9 ± 6.6	18.5 ± 5.3	15.4 ± 5.5
Untrained	−	21.8 ± 7.8	22.7 ± 4.8	26.7 ± 9.6
Untrained	+	17.5 ± 5.3	19.5 ± 6.6	20.9 ± 6.9

[a] Mean ± SD with a mean of seven individuals per group.

[b] Vitamin E levels in serum of treated groups increased 97% in supplemented, physically fit group; untrained group increased 76% after 7 weeks. Placebo controls increased 2% and 17% for fit and untrained, respectively.

[c] T helper cells were identified with OKT4 and T suppressor cells were identified with OKT8 antisera. No significant change in T helper : T suppressor ratio was found.

TABLE 2. T Cell Mitogenesis and NK Functions of Peripheral Blood Lymphocytes from Young Men Supplemented with Vitamins E and C[a,b]

	Supplemented	Week 0	Week 7	Week 14
			NK FUNCTION[c]	
Physically fit	−	44.7 ± 18.9	72.7 ± 20.7	83.3 ± 15.6
Physically fit	+	46.7 ± 10.1	58.3 ± 14.2	86.7 ± 11.7
Untrained	−	17.3 ± 13.6	65.6 ± 18.0	79.4 ± 14.6
Untrained	+	25.1 ± 15.3	68.1 ± 24.4	80.3 ± 13.1
			STIMULATION UNITS	
Physically fit	−	89 ± 99	143 ± 87	31 ± 31
Physically fit	+	168 ± 91	128 ± 93	31 ± 21
Untrained	−	238 ± 191	62 ± 93	48 ± 47
Untrained	+	267 ± 28	45 ± 68	41 ± 30

[a] Mean ± SD with a mean number of 11 individuals per group.

[b] Compliance based upon pill count was 90.8 ± 7.5% and 92.0 ± 8.3% in placebo and supplemented, respectively.

[c] NK functions measure percentage of K562 tumor cells killed by *in vitro* incubation with lympohocytes.

is by production of prostaglandins (PG) that stimulate suppressor lymphocytes. We found that lymphocytes of aging subjects given the vitamin supplementation demonstrated a significant increase in PGE_1 production after exposure to PHA, while subjects given placebo showed no alteration in this parameter. These changes were difficult to confirm in the young subjects, which may be due to smaller sample size, aging, or shorter exposure to supplementation in the aging group.

Vitamins C and E had no significant effect on the production of γ-interferon by nonmitogen-stimulated cells (unstimulated) from fit or unfit young men. The mean titer after days of supplementation was 4.8 ± 7.24 for the supplemented group and 4.0 ± 8.02 for controls. The placebo and treatment groups were not significantly different in the responses of Con-A-stimulated cells.

In the young subjects the numbers of T suppressor and T helper cells did not change significantly with vitamin intakes (TABLE 1). There was no significant change with supplementation or fitness state in T helper : T suppressor ratio. In addition, T cell mitogenesis and NK functions were not significantly altered, although serum vitamin E increased significantly in the supplemented subjects (TABLE 2). Based upon our work, we believe that enhanced disease and tumor resistance in mice is due to altered cellular and secretory immune functions.[5] We found, upon supplementation of high dietary vitamin E in young mice, elevated numbers of cytotoxic lymphocytes that could be important in anticancer defenses, perhaps due to more rapid maturation of T cell functions.[5,6] This hypothesis is supported by Corwin and Shloss,[7] who showed high dietary vitamin E increased the PHA : Con A response ratio, suggesting an effect of the vitamin on maturation of T cells.

Our human experiments in young men showed little effect of the combined regimen of vitamin C and E on the lymphocyte functions measured, except in aging men with somewhat reduced immune functions. This was not found here in men fed vitamin E and C. However, this may be due to vitamin E accelerating maturation of only certain T cell functions.[5,6]

ACKNOWLEDGMENTS

The assistance of Dr. T. Petro and Dr. H. Ismail (deceased) is appreciated.

REFERENCES

1. WATSON, R. R. 1984. Nutrition, Disease Resistance, and Immune Function. Marcel Dekker. New York. pp. 299-312.
2. WATSON, R. R., J. C. JACKSON, L. WERNER & J. WILMORE. 1986. Med. Sci. Sports Exer. **18:** 95.
3. DESAI, I. 1984. Vitamin E analysis methods for animal tissues. Methods in Enzymology, Vol. 105. Academic Press. New York.
4. BUTTRISS, J. & A. DIPLOCK. 1984. High-performance liquid chromatography methods for vitamin E in tissues. Methods in Enzymology, Vol. 105. Academic Press. New York.
5. LIM, T. S., N. PUTT, D. SAFRANSKI, C. CHUNG & R. R. WATSON. 1981. Immunology **44:** 289.
6. WATSON, R. R., C. CHUNG & T. M. PETRO. 1982. Ann. N. Y. Acad. Sci. **393:** 205.
7. CORWIN, L. M. & J. SHLOSS. 1980. J. Nutr. **110:** 916.

Concluding Remarks

JOHN J. BURNS

The Rockefeller University
New York, New York 10021

This conference has added much new information to our knowledge of ascorbic acid. Newly discovered mechanisms for its action and possible roles in human health have been pointed out. One of the important areas is the neurochemical function of ascorbic acid. Involvement of ascorbic acid in hormone and neurotransmitter synthesis is no longer speculation; several investigators presented evidence that ascorbic acid acts as a cofactor in two important systems, dopamine β-hydroxylase and peptidylglycine α-amidating monooxygenase. It will be important to understand the physiological significance of this neurohumoral action of ascorbic acid. Consequences of this action may not be readily evident in ascorbic-acid-deficient animals, since the particular tissues involved, especially in the central nervous system, have the ability to conserve their ascorbic acid levels.

It has been thought for many years that ascorbic acid has a role in collagen formation. Papers presented at the conference proposed intriguing new mechanisms for the involvement of ascorbic acid as a neuronal factor in the synthesis of procollagen and collagen. These observations should stimulate interesting investigations on this action of ascorbic acid.

An important topic considered at the conference was the antioxidant role played by ascorbic acid in neutralizing extracellular reactive oxidants generated during phagocytic activity of leukocytes and in protecting cellular membranes from free radical damage in general. In this capacity, ascorbic acid could act as an antioxidant itself or could act in conjunction with vitamin E. This may explain earlier observations in deficient guinea pigs that show an interaction of vitamin C and vitamin E.

A number of epidemiological studies to evaluate the role of ascorbic acid in human health were reported. These controlled studies included measurements of ascorbic acid blood levels to determine nutritional status of the individuals. Among the groups studied were the elderly, individuals with cervical dysplasia, and individuals with increased risk of cancer and cardiovascular disease. Although the results in most instances suggest a role for the vitamin, clinical intervention trials are required to give definitive results.

Other studies showed the effect of ascorbic acid on airway constriction in asthmatics and in individuals exposed to air pollution. A role for ascorbic acid in male fertility and in maintaining periodontal health was reported. Information was presented that could justify additional clinical studies on the relationship of ascorbic acid intake to the occurrence of cataracts.

Epidemiological studies demonstrate the effect of smoking in reducing ascorbic acid levels in blood. Other studies show the effect of ascorbic acid in neutralizing phagocytic-derived extracellular oxidants in smokers. These observations point to the need for further research to evaluate the possibility that ascorbic acid intake can mitigate the harmful consequences of smoking.

Information was presented on the ability of ascorbic acid to inhibit nitrosamine formation. Of particular interest are observations indicating that ascorbic acid blocks the formation of nitrosamines by macrophages. Such nitrosamines could have deleterious effects. A marked influence of ascorbic acid on adverse effects of alcohol consumption was reported. This should encourage further research to determine the relevance of ascorbic acid intake to some of the consequences of alcoholism.

The many actions of ascorbic acid must be considered not only in understanding its role as a vitamin to prevent scurvy, but also in accounting for its action on a number of diverse biological systems. Any explanation for the overall actions of ascorbic acid must take into consideration that it is a vitamin in only a few animal species such as man, monkey, and guinea pig. These species lack the enzyme L-glulonolactone oxidase, which is required for the synthesis of ascorbic acid.

Other animal species, such as the rat, have the capability of synthesizing ascorbic acid in quantities that seem quite large when compared to the amount usually consumed by humans in their diet. Turnover rate studies that we carried out some 30 years ago indicated that rats synthesize about 26 mg per kg per day as compared to about 1 mg per kg per day available in the usual human diet. On this basis, a 70-kg human would have to consume daily almost 2 g of ascorbic acid to obtain the amount the rat derives from its own synthesis. Furthermore, a number of factors, especially the administration of drugs and a wide variety of chemicals, increase the synthesis of ascorbic acid up to eightfold in the rat.

It is of interest to speculate why the rats maintain such a high level of ascorbic acid in their body when one compares this with the very small amount available in the diet of humans. Do the high levels of ascorbic acid in animals capable of their own synthesis suggest the vitamin has, in addition to scurvy prevention, other as yet undiscovered functions? A number of studies suggest that in order to ensure optimal health, ascorbic acid intake in humans should be substantially higher than that which is now recommended. This subject has been a matter of controversy and certainly requires more research. The many actions of ascorbic acid, indicated by the results highlighted in this conference, should be a spur to the research needed to answer this important question.

Much progress has been made in our knowledge of ascorbic acid since the New York Academy of Sciences sponsored the First International Conference on Vitamin C in 1960 and the Second International Conference in 1974. I had the honor of cochairing these conferences with C. G. King. Since he was not able to attend this conference, I would like to recall the many contributions that he made to our knowledge of vitamin C, particularly his important role in the discovery of the vitamin and the research carried out in his laboratory over many years that led to the understanding of its biosynthesis and metabolism. There is no question that C. G. King deserves recognition by the scientific community for these many contributions to our overall knowledge of ascorbic acid.

Index of Contributors